Lecture Notes in Artificial Intelligence   6401

Edited by R. Goebel, J. Siekmann, and W. Wahlster

Subseries of Lecture Notes in Computer Science

W0036831

Jian Yu   Salvatore Greco   Pawan Lingras
Guoyin Wang   Andrzej Skowron (Eds.)

Rough Set and Knowledge Technology

5th International Conference, RSKT 2010
Beijing, China, October 15-17, 2010
Proceedings

 Springer

Volume Editors

Jian Yu
Beijing Jiaotong University, Beijing 100044, China
E-mail: jianyu@bjtu.edu.cn

Salvatore Greco
University of Catania, Corso Italia, 55, 95129 Catania, Italy
E-mail: salgreco@unict.it

Pawan Lingras
Saint Mary's University, Halifax, Nova Scotia, B3H 3C3, Canada
E-mail: pawan@cs.smu.ca

Guoyin Wang
Chongqing University of Posts and Telecommunications, Chongqing 400065, China
E-mail: wanggy@cqupt.edu.cn

Andrzej Skowron
Warsaw University, Banacha 2, 02-097 Warsaw, Poland
E-mail: skowron@mimuw.edu.pl

Library of Congress Control Number: 2010935669

CR Subject Classification (1998): I.2, H.2.4, H.3, F.4.1, F.1, I.5, H.4

LNCS Sublibrary: SL 7 – Artificial Intelligence

ISSN	0302-9743
ISBN-10	3-642-16247-9 Springer Berlin Heidelberg New York
ISBN-13	978-3-642-16247-3 Springer Berlin Heidelberg New York

springer.com

© Springer-Verlag Berlin Heidelberg 2010
Printed in Germany

Typesetting: Camera-ready by author, data conversion by Scientific Publishing Services, Chennai, India
Printed on acid-free paper   06/3180

Preface

The International Conference on Rough Set and Knowledge Technology (RSKT) has been held every year since 2006. RSKT serves as a major forum that brings researchers and industry practitioners together to discuss and deliberate on fundamental issues of knowledge processing and management and knowledge-intensive practical solutions in the current knowledge age. Experts from around the world meet to present state-of-the-art scientific results, to nurture academic and industrial interaction, and to promote collaborative research in rough sets and knowledge technology. The first RSKT was held in Chongqing, China, followed by RSKT 2007 in Toronto, Canada, RSKT 2008 in Chengdu, China and RSKT 2009 in Gold Coast, Australia. RSKT 2010, the 5th in the series, was held in Beijing, China, October 15–17, 2010.

This volume contains 98 papers selected for presentation at RSKT 2010. Following the success of the previous conferences, RSKT 2010 continued the tradition of a very rigorous reviewing process. Every submission was reviewed by at least two reviewers. Moreover, RSKT 2010 invited several area chairs to supervise the review process of every submission. Most submissions were reviewed by three experts. The Program Committee members were deeply involved in a highly engaging selection process with discussions among reviewers and area chairs. When necessary, additional expert reviews were sought. As a result, only top-quality papers were chosen for presentation at the conference, including 49 regular papers (acceptance rate of 28%) and 25 short papers (acceptance rate of 14.3%). We would like to thank all the authors for contributing their best papers. Without their support, this conference would not have been possible.

The RSKT program was further enriched by six keynote speeches. We are grateful to our keynote speakers, Bo Zhang, Ian H. Witten, Roman Slowinski, Deyi Li, Jianchang Mao, and Sankar K. Pal, for their visionary talks on rough sets and knowledge technology. The RSKT 2010 program included four special sessions with 24 papers: Data Mining in Cloud Computing, Decision-Theoretic Rough Set (DTRS) Model, Quotient Space Theory and Application, and Cloud Model and Application.

RSKT 2010 would not have been successful without the support of many people and organizations. We wish to thank the members of the Steering Committee for their invaluable suggestions and support throughout the organizational process. We are indebted to the area chairs, Program Committee members, and external reviewers for their effort and engagement in providing a rich and rigorous scientific program for RSKT 2010. We express our gratitude to our Special Session Chairs (Zhongzhi Shi, Yong Yang, Fan Yang, Guisheng Chen, Jingtao Yao, Tianrui Li, Xiaoping Yang, Yanping Zhang) for selecting and coordinating the exciting sessions. We are also grateful to the Local Arrangement Chairs

Liping Jing and Zhen Han as well as the Local Organizing Committee, whose great effort ensured the success of the conference.

We greatly appreciate the cooperation, support, and sponsorship of various institutions, companies, and organizers, including Beijing Jiaotong University, China, National Natural Science Foundation of China (NSFC), International Rough Set Society (IRSS), and the Rough Sets and Soft Computation Society of the Chinese Association for Artificial Intelligence (CRSSC).

We are thankful to Alfred Hofmann and the excellent LNCS team at Springer for their support and cooperation in publishing the proceedings as a volume of the *Lecture Notes in Computer Science*.

October 2010

Jian Yu
Salvatore Greco
Pawan Lingras
Guoyin Wang
Andrzej Skowron

Organization

Organizing Committee

Conference Chairs	Bin Ning (China)
	Sankar K. Pal (India)
	Zhi-Hua Zhou (China)
Program Chairs	Jian Yu (China)
	Salvatore Greco (Italy)
	Pawan Lingras (Canada)
Organizing Chairs	Liping Jing (China)
	Zhen Han (China)
Special Session Chairs	Hung Son Nguyen (Poland)
Publicity Chairs	Jerzy W. Grzymala-Busse (USA)
	Sushmita Mitra (India)
Steering Committee Chairs	Andrzej Skowron (Poland)
	Guoyin Wang (China)
	Yiyu Yao (Canada)

Program Committee

Esma Aimeur (Canada)
Kankana Chakrabarty (Australia)
Cornelis Chris (Belgium)
Davide Ciucci (Italy)
Jianhua Dai (China)
Martine De Cock (Belgium)
Jitender Deogun (USA)
Patrick Doherty (Sweden)
Yang Gao (China)
Jerzy Grzymala-Busse (USA)
Zhimin Gu (China)
Jianchao Han (USA)
Aboul E. Hassanien (Egypt)
Joseph P. Herbert (Canada)
Tzung-Pei Hong (Taiwan)
Xiaohua Tony Hu (USA)
Masahiro Inuiguchi (Japan)
Ryszard Janicki (Canada)
Richard Jensen (UK)
Chaozhe Jiang (China)
Etienne Kerre (Belgium)

Taghi M. Khoshgoftaar (USA)
Tai-hoon Kim (Korea)
Jan Komorowski (Sweden)
Raymond Y. K. Lau (Hong Kong)
Els Lefever (Belgium)
Yee Leung (Hong Kong)
Guohe Li (China)
Guozheng Li (China)
Zou Li (China)
Jiye Liang (China)
Tsau Young Lin (USA)
Jie Lu (Australia)
Victor Marek (USA)
Nicolas Marin (Spain)
German Hurtado Martin (Belgium)
Benedetto Matarazzo (Italy)
Rene Mayorga (Canada)
Ernestina Menasalvas-Ruiz (Spain)
Jusheng Mi (China)
Duoqian Miao (China)
Wojtek Michalowski (Canada)

Sadaaki Miyamoto (Japan)
Hongwei Mo (China)
Tetsuya Murai (Japan)
Michinori Nakata (Japan)
Krzysztof Pancerz (Poland)
Witold Pedrycz (Canada)
Georg Peters (Germany)
J F Peters (Canada)
Mieczysw A.K Potek (Poland)
Keyun Qin (China)
Anna M. Radzikowska (Poland)
Sheela Ramanna (Canada)
Kenneth Revett (UK)
Leszek Rutkowski (Poland)
Henryk Rybinski (Poland)
Hiroshi Sakai (Japan)
B. Uma Shankar (India)
Wladyslaw Skarbek (Poland)
Andrzej Skowron (Poland)
Dominik Slezak (Canada)
Nguyen Hung Son (Poland)
Jaroslaw Stepaniuk (Poland)
Zbigniew Suraj (Poland)
Piotr Synak (Poland)
Andrzej Szalas (Sweden)
Li-Shiang Tsay (USA)

I. Burhan Turksen (Canada)
Dimiter Vakarelov (Bulgaria)
Anita Wasilewska (USA)
Peng (Paul) Wen (Australia)
Alicja Wieczorkowska (Poland)
Marcin Wolski (Poland)
S. K. Michael Wong (USA)
Dan Wu (Canada)
Weizhi Wu (China)
Zhaocong Wu (China)
Wei Xiang (Australia)
Jiucheng Xu (China)
Ron Yager (USA)
Jie Yang (China)
Simon X. Yang (Canada)
Dongyi Ye (China)
Bonikowski Zbigniew (Poland)
Justin Zhan (USA)
Songmao Zhang (China)
Yanqing Zhang (USA)
Yan Zhao (Canada)
Ning Zhong (Japan)
William Zhu (China)
Yan Zhu (China)
Wojciech Ziarko (Canada)

Sponsoring Institutions

Beijing Jiaotong University
National Natural Science Foundation of China
International Rough Set Society
Rough Sets and Soft Computation Society of the Chinese Association for
    Artificial Intelligence

Table of Contents

Fuzzy Sets

Knowledge Technology

Intelligent Information Processing

Health Informatics and Biometrics Authentication

Neural Networks

Complex Networks

Granular Computing

Metaheuristic

Special Session: Cloud Model and Its Application

Special Session: Data Mining in Cloud Computing

Special Session: Decision-Theoretic Rough Set Model

Special Session: Quotient Space Theory Research and Application

Comparative Study on Mathematical Foundations of Type-2 Fuzzy Set, Rough Set and Cloud Model

Deyi Li

Tsinghua University, China

Abstract. Mathematical representation of a concept with uncertainty is one of foundations of Artificial Intelligence. The type-2 fuzzy set introduced by Mendel studies fuzziness of the membership grade of a concept. Rough set proposed by Pawlak defines an uncertain concept through two crisp sets. Cloud model, based on probability measure space, automatically produces random membership grades of a concept through a cloud generator. The three methods all concentrate on the essentials of uncertainty and have been applied in many fields for more than ten years. However, their mathematical foundations are quite different. The detailed comparative study on the three methods will discover the relationship in the betweens, and provide a fundamental contribution to Artificial Intelligence with uncertainty.

J. Yu et al. (Eds.): RSKT 2010, LNAI 6401, p. 1, 2010.
© Springer-Verlag Berlin Heidelberg 2010

Scientific Challenges in Contextual Advertising

Jianchang Mao

Advertising Sciences, Yahoo! Labs

Abstract. Online advertising has been fueling the rapid growth of the Web that offers a plethora of free web services, ranging from search, email, news, sports, finance, and video, to various social network services. Such free services have accelerated the shift in people's media time spend from offline to online. As a result, advertisers are spending more and more advertising budget online. This phenomenon is a powerful ecosystem play of users, publishers, advertisers, and ad networks. The rapid growth of online advertising has created enormous opportunities as well as technical challenges that demand computational intelligence. Computational Advertising has emerged as a new interdisciplinary field that studies the dynamics of the advertising ecosystem to solve challenging problems that rise in online advertising.

In this talk, I will provide a brief introduction to various forms of online advertising, including search advertising, contextual advertising, guaranteed and non-guaranteed display advertising. Then I will focus on the problem of contextual advertising, which is to find the best matching ads from a large ad inventory to a user in a given context (e.g., page view) to optimize the utilities of the participants in the ecosystem under certain business constraints (blocking, targeting, etc). I will present a problem formulation and describe scientific challenges in several key technical areas involved in solving this problem, including user understanding, semantic analysis of page content, user response prediction, online learning, ranking, and yield optimization.

J. Yu et al. (Eds.): RSKT 2010, LNAI 6401, p. 2, 2010.

F-granulation, Generalized Rough Entropy and Pattern Recognition

Sankar K. Pal

Indian Statistical Institute

Abstract. The role of rough sets in uncertainty handling and granular computing is described. The significance of its integration with other soft computing tools and the relevance of rough-fuzzy computing, as a stronger paradigm for uncertainty handling, are explained. Different applications of rough granules and certain important issues in their implementations are stated. Three tasks such as class-depedendent rough-fuzzy granulation for classification, rough-fuzzy clustering and defining generalized rough sets for image ambiguity measures and anlysis are then addressed in this regard, explaining the nature and characteristics of granules used therein.

Merits of class dependentgranulation together with neighborhood rough sets for feature selection are demonstreted in terms of different classification indices. Significance of a new measure, called "dispersion" of classification performance, which focuses on confused classes for higher level analysis, is explained in this regard. Superiority of rough-fuzzy clustering is illustrated for determining bio-bases (c-medoids) in encoding protein sequence for analysis. Generalized rough sets using the concept of fuzziness in granules and sets are defined both for equivalence and tolerance relations. These are followed by the definitions of entropy and different image ambiguities. Image ambiguity measures, which take into account both the fuzziness in boundary regions, and the rough resemblance among nearby gray levels and nearby pixels, have been found to be useful for various image analysis operations

The talk concludes with stating the future directions of research and challenges.

J. Yu et al. (Eds.): RSKT 2010, LNAI 6401, p. 3, 2010.
© Springer-Verlag Berlin Heidelberg 2010

Knowledge Discovery about Preferences Using the Dominance-Based Rough Set Approach

Roman Slowinski

Laboratory of Intelligent Decision Support Systems, Institute of Computing Science,
Poznan University of Technology, Poland

Abstract. The aim of scientific decision aiding is to give the decision maker a recommendation concerning a set of objects (also called alternatives, solutions, acts, actions, . . .) evaluated from multiple points of view considered relevant for the problem at hand and called attributes (also called features, variables, criteria, . . .). On the other hand, a rational decision maker acts with respect to his/her value system so as to make the best decision. Confrontation of the value system of the decision maker with characteristics of the objects leads to expression of preferences of the decision maker on the set of objects. In order to recommend the most-preferred decisions with respect to classification, choice or ranking, one must identify decision preferences. In this presentation, we review multi-attribute preference models, and we focus on preference discovery from data describing some past decisions of the decision maker. The considered preference model has the form of a set of *if..., then...* decision rules induced from the data. In case of multi-attribute classification the syntax of rules is: **if** *performance of object a is better (or worse) than given values of some attributes*, **then** *a belongs to at least (at most) given class*, and in case of multi-attribute choice or ranking: **if** *object a is preferred to object b in at least (at most) given degrees with respect to some attributes*, **then** *a is preferred to b in at least (at most) given degree*. To structure the data prior to induction of such rules, we use the Dominance-based Rough Set Approach (DRSA). DRSA is a methodology for reasoning about ordinal data, which extends the classical rough set approach by handling background knowledge about ordinal evaluations of objects and about monotonic relationships between these evaluations. We present DRSA to preference discovery in case of multi-attribute classification, choice and ranking, in case of single and multiple decision makers, and in case of decision under uncertainty and time preference. The presentation is mainly based on publications [1,2,3].

References

1. Greco, S., Matarazzo, B., Slowinski, R.: Dominance-based rough set approach to decision involving multiple decision makers. In: Greco, S., Hata, Y., Hirano, S., Inuiguchi, M., Miyamoto, S., Nguyen, H.S., Słowiński, R. (eds.) RSCTC 2006. LNCS (LNAI), vol. 4259, pp. 306–317. Springer, Heidelberg (2006)

J. Yu et al. (Eds.): RSKT 2010, LNAI 6401, pp. 4–5, 2010.

2. Greco, S., Matarazzo, B., Slowinski, R.: Dominance-based rough set approach to decision under uncertainty and time preference. Annals of Operations Research 176, 41–75 (2010)
3. Slowinski, R., Greco, S., Matarazzo, B.: Rough Sets in Decision Making. In: Meyers, R.A. (ed.) Encyclopedia of Complexity and Systems Science, pp. 7753–7786. Springer, New York (2009)

Wikipedia and How to Use It for Semantic Document Representation

Ian H. Witten

University of Waikato, New Zealand

Abstract. Wikipedia is a goldmine of information; not just for its many readers, but also for the growing community of researchers who recognize it as a resource of exceptional scale and utility. It represents a vast investment of manual effort and judgment: a huge, constantly evolving tapestry of concepts and relations that is being applied to a host of tasks. This talk focuses on the process of "wikification"; that is, automatically and judiciously augmenting a plain-text document with pertinent hyperlinks to Wikipedia articlesas though the document were itself a Wikipedia article. I first describe how Wikipedia can be used to determine semantic relatedness between concepts. Then I explain how to wikify documents by exploiting Wikipedia's internal hyperlinks for relational information and their anchor texts as lexical information. Data mining techniques are used throughout to optimize the models involved.

I will discuss applications to knowledge-based information retrieval, topic indexing, document tagging, and document clustering. Some of these perform at human levels. For example, on CiteULike data, automatically extracted tags are competitive with tag sets assigned by the best human taggers, according to a measure of consistency with other human taggers. All this work uses English, but involves no syntactic parsing, so the techniques are language independent.

J. Yu et al. (Eds.): RSKT 2010, LNAI 6401, p. 6, 2010.

Granular Computing and Computational Complexity

Bo Zhang

Computer Science and Technology Department of Tsinghua University, China

Abstract. Granular computing is to imitate humans multi-granular computing strategy to problem solving in order to endow computers with the same capability. Its final goal is to reduce the computational complexity. To the end, based on the simplicity principle the problem at hand should be represented as simpler as possible. From structural information theory, its known that if a problem is represented at different granularities, the hierarchical description of the problem will be a simpler one. The simpler the representation the lower the computational complexity of problem solving should be. We presented a quotient space theory to multi-granular computing. Based on the theory a problem represented by quotient spaces will have a hierarchical structure. Therefore, the quotient space based multi-granular computing can reduce the computational complexity in problem solving. In the talk, well discuss how the hierarchical representation can reduce the computational complexity in problem solving by using some examples.

J. Yu et al. (Eds.): RSKT 2010, LNAI 6401, p. 7, 2010.

Some Comparative Analyses of Data in the RSDS System*

Zbigniew Suraj and Piotr Grochowalski

Chair of Computer Science
University of Rzeszów, Poland
{zsuraj,piotrg}@univ.rzeszow.pl

Abstract. The paper concerns the analysis of the data included in the *Rough Set Database System* (called in short the RSDS system). The data is analyzed by means of statistical and graph-theoretical methods. The results of the analysis show interesting relations among the authors of publications whose descriptions are included in the system. The results may be useful for determining the structure of research groups related to the rough sets, the scientific interests of the members of those groups, mutual collaboration between groups, as well as for identifying the trends appearing in research.

Keywords: rough sets, collaboration graph, patterns of collaboration, social networks.

1 Introduction

Rough set theory was proposed by prof. Zdzisław Pawlak [5] in the early 1980's. A rapid growth of interest in the rough set theory and its applications can be seen in the number of international workshops, conferences and seminars that are either directly dedicated to rough sets or include the subject-matter in their programs. A large number of high quality papers on various aspects of rough sets and their applications have been published in period of 1981 - 2010 as a result of this attention. In 2002 the first version of the RSDS system appeared. The system was created in order to spread information about publications, software and people related to the theory of rough sets and its applications [6,7,8].

The paper is devoted to graph-statistical analysis of the data included in the RSDS system. The aim of the analysis is to get information concerning progress in the development of the theory of rough sets and its application within the space of years since its beginning until the present, i.e., 1981 - 2010. Analyzing the data included in the system there have been gained a lot of interesting information concerning hidden patterns for publications as well as patterns for collaboration between the authors of publications. The results may be useful for determining the structure of research groups related to the rough sets, the

* This work has been partially supported by the grant N N516 368334 from Ministry of Science and Higher Education of the Republic of Poland.

J. Yu et al. (Eds.): RSKT 2010, LNAI 6401, pp. 8–15, 2010.

scientific interests of the members of those groups, mutual collaboration between groups, as well as for identifying the trends appearing in research [1,2,3,4].

In order to detect the links between the authors of particular publication as well as the changes in them there are used the so called collaboration graph and standard statistical methods. The nodes of the collaboration graph represent the authors of publications described in the RSDS system, and the edges link two nodes of this graph when the authors ascribed to those nodes wrote a common publication. The analysis of the collaboration graph consists in describing the structure of the graph and its basic properties, as well as comparing estimated quantities with the results of a similar analysis described in the papers [7,8]. In the result of the conducted data analysis there have been discovered and interpreted processes and phenomena observed in the system. The data has been analyzed by means of the own software prepared particularly for this aim.

The structure of the paper is as follows. In Section 2 general characteristics of the RSDS system and the results of the statistical data analysis are presented. Section 3 is devoted to the description of a collaborative graph and the methods of its analysis. In Section 4 the problem of data changes in time is discussed. Section 5 concludes the paper.

2   Some Basic Analyses

The RSDS system has been created in order to catalogue and facilitate the access to information about publications concerning the rough sets and related fields. It is available at: http://rsds.univ.rzeszow.pl. Detailed information about the system can be found in papers [6,7,8]. Nowadays, there are 4098 descriptions of publications available in the system, which have been developed by 2453 authors. As compared to the year 2007, when creating thorough statistics of the content of the system was initiated, the data shows a significant increase both in the number of publications and the number of authors. In 2007 the number of publications registered in the RSDS system was 3079, and the number of their authors totaled 1568. All the publications included in the system were divided into 12 generic groups (i.e., articles, books, proceedings, etc.) according to the BibTeX standard. Table 1 shows both the division of publications and the numerical statistics concerning the years 2007 and 2010.

The above statistics depict the significant increase in the number of publications from 2007 to 2010 observed in the system.

What follows from the presented statistics is that people working on rough sets eagerly present their works by taking part in scientific conferences, what results in a significant growth in the number of publications in the conference materials (inproceedings). Analyzing the results in a given table one can notice a visible growth in the number of publications from many categories, which may be due to a great interest in the theory of rough sets. This thesis is also proved by the breakdown concerning the number of publications in the ten-year-periods specified as follows: 1981 - 1990: 121 publications, 1991 - 2000: 1492 publications, and 2001 - 2010: 2420 publications.

Table 1. The division of publications considering their characteristics

Type of publication	Year 2007	Year 2010
article	813	975
book	125	182
inbook	17	19
incollection	282	401
inproceedings	1598	2268
manual	2	2
masterthesis	12	12
phdthesis	14	16
proceedings	67	73
techreport	148	148
unpublished	1	2
Total	3079	4098

Remark: The amount of all publications with their division into years comes to 4033. This number differs from the general number of publications given in Table 1 (see column: Year 2010). This difference is due to the fact that the year of publishing is not always known for all publications described in the RSDS system.

The authors of publications included in the system come from many countries, i.e., from Poland - 99, Japan - 28, Canada - 24, USA - 16, Italy - 6, China - 5, Sweden - 5, Spain - 4, France - 3, India - 3, Russia - 2, and Egypt, Finland, Ireland, Netherlands, Norway, Romania, Taiwan, United Kingdom - 1. In this statistics not all of the authors of publications included in the system are taken into consideration, because in many cases it is difficult to determine which country a particular author comes from.

Analyzing the data available in the RSDS system one can notice a significant interest in the resources of the system. The greatest interest in the system until 2010 (from the moment of activating the system) has been observed in Poland - 2558 references, the next positions in this category are taken by Japan - 133 references, Italy - 111 references, China - 95 references, Canada - 86 references, Germany - 62 references, etc. The ranking is very similar to the ranking of the authors of publications presented above.

On the basis of the collected data there was conducted an advanced statistical analysis allowing for detecting a lot of interesting correlations between authors, publications and the way of creating publications. In Table 2 there are presented the results of the data analysis, which take into consideration incremental time periods in a form of five-year-periods, i.e., periods: 1981 - 1985, 1981 - 1990, 1981 - 1995, 1981 - 2000, 1981 - 2005, 1981 - 2010. As a unit for measuring the increase in time the researchers adopted a five-year-period, as in such a period of time one can observe some changes in time as permanent and not temporary.

In Table 2 one can notice that the number of publications as well as the number of authors has the tendency to grow. Unfortunately, this upturn does not translate

Table 2. The results of the data analysis taking into consideration increasing time period

Year of completion	1981-85	1981-90	1981-95	1981-00	1981-05	1981-10
Number of publications	35	121	610	1613	3124	4033
Number of authors	12	47	238	671	1603	2414
The average: publications/author	4	3.87	4.18	6.64	5.67	5.03
Standard deviation publication/author	4.93	5.25	8.64	63.02	78.95	82.74
The average: authors/publications	0.37	0.5	0.63	0.78	0.93	1.03
Standard deviation authors/publication	0.76	0.83	0.97	1.09	1.19	1.24
Percentage share of publications with n co-authors						
Absence of authors (a publication under edition only)	2.86%	2.48%	5.41%	5.58%	6.59%	5.98%
$n = 0$	68.57%	58.68%	47.38%	41.6%	35.08%	31.94%
$n = 1$	20%	28.93%	33.28%	32.42%	31.47%	32.04%
$n = 2$	5.71%	6.61%	8.52%	12.83%	17%	18.87%
$n > 2$	2.86%	3.31%	5.41%	7.56%	9.86%	11.18%
Number of authors sharing common publications	12	36	195	585	1475	2274
Their percentage share	100%	76.6%	81.93%	87.18%	92.01%	94.2%
The average: co-authors/author	2.17	2.13	2.38	2.8	3.12	3.33
The average: co-authors/author sharing a common publication	2.17	2.78	2.9	3.22	3.39	3.54

into the growth of the average number of publications per author, i.e., from the point of view of the average publication/author follows that the best period was 1981 - 2000. After this period one can notice a decreasing tendency of the value of this index, what may indicate either the decreasing number of publications or an increasing number of authors of a given publication. However, the standard deviation of a publication towards an author shows a great upturn which means a considerable deviation from a fixed average value for a given time period.

What may be another index, is the average number of authors falling to one publication taking into consideration the standard deviation. The value of this index shows that in the first period of the development of the rough set theory one author usually fell to one publication, whereas as the time has been passing works by individual authors have been supplanted by collaborative works, especially the ones written by two authors. This observation is also proved by the number of authors falling to a particular publication presented in percentages. Another observation is that a considerable number of authors (as many as 90%) write publications with co-authors; what follows from the data analysis is that on average 2 - 3 co-authors fall to one author.

3   Analysis of Collaboration Graph

On the basis of the data about authors and their publications a collaboration graph C was defined in which the nodes stand for the authors of publications.

The edges between the nodes mean that the authors represented by particular nodes connected with one another, are the authors of a common publication, hence they are its co-authors. In order to avoid distortions in the analysis there have been made a correction of some anomalies which have appeared in the available data, i.e., works with the description of authors with the abbreviation "et al." and the works described as "edited by". The abbreviation "et al." is used to describe collaborative works written by many authors. On the basis of such a description it is impossible to determine unambiguously all the co-authors of a given publication. A work with a description "edited by" does not have an author; it has an editor or editors who are not included in a created collaboration graph.

A defined collaboration graph C reflects relations between the authors which appear during creating common publications. Taking into consideration the whole period in which publications have been created, i.e., 1981 - 2010 it turns out that for such a time period graph C contains 2414 nodes and 4022 edges. The average degree of a node in this graph is 3,33, which means that each author writing a publication cooperates with 3 other authors. A developed graph C contains also isolated nodes, that is nodes not connected to any other node. This result can be interpreted in such a way, that in the system there exist publications with one author, i.e., their authors do not cooperate in this field; there are 140 such authors in the system. In further considerations the nodes of this type in graph C were eliminated. After deleting such nodes a collaboration graph C contains 2274 nodes, and the average degree of each node is 3,54.

While comparing the current results of graph C analysis with the results of the analysis of the collaboration graph constructed on the basis of the data from 2007, one can notice both the increase in the number of nodes in graph C from 1568 to 2274 and the increase in the number of edges in the graph from 2466 to 4022, which means the extension of collaboration between authors assigned to those nodes.

A further analysis of the number of authors with the reference to the number of publications (such correlations can also be derived from a collaboration graph) shows that 64,8% of authors have written 1 publication, 14,2% - 2 publications, 6,2% - 3 publications. The first result of this analysis could mean that the majority of authors of publications is still not directed in their research and probably they are still looking for a subject which they would like to cover. There are only 10 people among all the authors represented in the system who wrote more than 100 publications (which constitutes 0,4% of the total number of authors).

A detailed breakdown of the results of these statistics is presented in Table 3.

In a collaboration graph there can be distinguished connected components, i.e., sub-graphs in which there exists a link connecting any single node with another node. A defined collaboration graph C has one maximal connected component including 1034 nodes. Apart from the maximal connected component this graph contains also 348 further components. Those components can be interpreted as research groups gathered around one person, who is a leader of this group. Moreover, from those components one can find out about interests of

Table 3. The percentage share of authors with a given number of publications

Number of publications	Percentage of authors
1	64.8%
2	14.2%
3	6.2%
4	3.1%
5	2.1%
6-10	3.6%
11-20	2.8%
21-50	1.2%
51-100	0.5%
101-200	0.3%
> 200	0.1%

a given research group. Those interests are expressed through the subjects of publications written by members of a given group.

Analyzing the maximal connected component we can define the average distance between nodes. In case of the analyzed graph C the average distance is 5,15, and the standard deviation equals 1,68. This distance defines the authors closely cooperating with each other. For the data from 2007 the average distance between nodes of the maximal connected component of a collaboration graph was 4,54, with a deviation 1,41. The present results of the analysis of a maximal connected component of a collaboration graph show that the authors of publications collaborate in bigger and bigger groups in order to write common publications. A diameter of a maximal connected component, i.e., the maximal distance between two nodes of the maximal connected component amounts to 13, and a radius of this component, i.e., the maximal distance from one node to any other node equals 7. Those results characterize a research group defined by means of a maximal connected component in a following way:

- taking into consideration the radius of the component, and then describing the sphere on each node of the component with the radius equal to the radius of the component, we can determine the leaders of groups,
- on the other hand, if we take into consideration a diameter of the component and also on each node we describe a sphere whose radius equals the diameter, we will describe the so called satellites of a group, i.e., authors the most loosely related to a given group.

The way of conducting the above analyses was described in the papers [7,8].

4   Analysis of Data Changes in Time

In order to reflect changes happening in a described process of creating various publications related to the rough set theory and their applications, the whole time period (1981 - 2010) was divided and defined in the system by means of five-year-periods, i.e., 1981 - 1985, 1986 - 1990, etc. For each such period there was conducted another analysis of the possessed data, according to the indices provided in Table 2. The results of this analysis are presented in Table 4.

Table 4. The results of the data analysis for each defined time period

In years	1981-85	1986-90	1991-95	1996-00	2001-05	2006-10
Number of publications	35	86	489	1003	1511	909
Number of authors	12	45	219	533	1127	1066
The average: publications/author	4	2.98	3.71	5.35	4.12	2.01
Standard deviation publication/author	4.93	3.39	7.01	43.52	45.4	3.53
The average: authors/publication	0.37	0.56	0.66	0.87	1.08	1.36
Standard deviation authors/publication	0.76	0.84	1	1.15	1.28	1.36
Percentage share of publications with n coauthors						
Absence of authors (a publication under edition only)	2.86%	2.33%	6.13%	5.86%	7.68%	3.85%
$n = 0$	68.57%	54.65%	44.58%	38.09%	28.13	21.12%
$n = 1$	20%	32.56%	34.36%	31.9%	30.44%	33.99%
$n = 2$	5.71%	6.98%	9%	15.45%	21.44%	25.3%
$n > 2$	2.86%	3.49%	5.93%	8.87%	12.31%	15.73%
Number of authors sharing common publications	12	33	180	475	1051	1020
Their percentage share	100%	73.33%	82.19%	89.12%	93.26%	95.68%
The average: coauthors/author	2.17	1.96	2.27	2.78	3.04	3.14
The average: coauthors/author sharing a common publication	2.17	2.67	2.77	3.12	3.26	3.28

Analyzing the defined periods according to the number of written publications one can notice that the most favorable period is between 2001 - 2005, as there were written 1511 publications then. Taking into consideration the average number of publications falling to one author the most favorable seems to be the period between 1996 - 2000. Unfortunately, in that period there was a huge discrepancy (i.e., a huge standard deviation) in the number of publications. Taking into account the collaboration between authors the most successful are the years 2006 - 2010. These results show the appearance of a new trend related to a domination of a number of common works over a number of individual works. This phenomenon is characteristic in the development of many branches of science, when their development reaches such a level, that a group research is more significant than an individual research. Given statistics prove the appearance of this phenomenon in the development of a rough set theory. Only about 4,5% of all authors create their works individually.

5   Final Remarks

From the conducted data analysis included in the RSDS system results that a rough set theory and its applications still develop dynamically, however the rate of this development throughout a couple of years recently has weakened a bit as compared to the preceding period. It seems that a more thorough analysis of a gathered data will allow for deriving even more subtle conclusions. Even now one can notice that in a considered rough set community there are certain authors who are distinguished from the point of view of the number of publications and their co-authors, and hence can be treated as the centers of some smaller communities around which another authors are gathered. The reality shows that they are significant people who have an established scientific position and hence they constitute a pattern for young scientists. Those people gather around themselves significant groups of people, the so called research groups, what comes out while analyzing a collaboration graph. This type of problems will be discussed in our further work.

Moreover, one can observe in the analysis of the statistics of the references to the RSDS system that the observed interest in the system stays on "a very stable level", i.e., there are 50 - 130 visits to the system each month, which can be interpreted as a need for its further development in order to spread its resources and facilitate making new scientific contacts.

Acknowledgments. We are grateful to Professor Andrzej Skowron for stimulating discussions and interesting suggestions about the RSDS system. We also wish to thank our colleagues for their valuable support, inspiring ideas and fruitful cooperation.

References

1. Aiello, W., Chung, F., Lu, L.: A random graph model for power law graphs. Experimental Mathematics 10, 53–66 (2001)
2. Barabasi, A.: Linked: The New Science of Networks. Perseus, New York (2002)
3. Buchanan, M.: Nexus: Small Worlds and the Groundbreaking Science of Networks. W.W. Norton, New York (2002)
4. Grossman, J.W.: Patterns of Collaboration in Mathematical Research. SIAM New 35(9) (2002)
5. Pawlak, Z.: Rough Sets. Int. J. of Comp. and Inf. Sci. 11, 341–356 (1983)
6. Suraj, Z., Grochowalski, P.: The Rough Set Database System: An Overview. In: Peters, J.F., Skowron, A. (eds.) Transactions on Rough Sets III. LNCS, vol. 3400, pp. 190–201. Springer, Heidelberg (2005)
7. Suraj, Z., Grochowalski, P.: Functional Extension of the RSDS System. In: Greco, S., Hata, Y., Hirano, S., Inuiguchi, M., Miyamoto, S., Nguyen, H.S., Słowiński, R. (eds.) RSCTC 2006. LNCS (LNAI), vol. 4259, pp. 786–795. Springer, Heidelberg (2006)
8. Suraj, Z., Grochowalski, P.: Patterns of Collaborations in Rough Set Research. In: Proc. of the Int. Symp. on Fuzzy and Rough Sets, ISFUROS 2006, Santa Clara, Cuba (2006)

Rough Temporal Vague Sets in Pawlak Approximation Space

Yonghong Shen

School of Mathematics and Statistics, Tianshui Normal University,
Tianshui 741001, P.R. China
shenyonghong2008@hotmail.com

Abstract. The combination of temporal vague set theory and rough set theory is developed in this paper. The lower and upper approximation operators of a temporal vague set are constructed, which is partitioned by an indiscernibility relation in Pawlak approximation space, and the concept of rough temporal vague sets is proposed as a generalization of rough vague sets. Further properties associated with the lower and upper approximations of temporal vague sets are studied. Finally, the roughness measure of a temporal vague set is defined as an extension of the parameterized roughness measure of a vague set. Meantime, some properties of roughness measure are established.

Keywords: Temporal vague sets, Pawlak approximation space, rough temporal vague sets, temporal $\alpha\beta$-level sets, roughness measure.

1 Introduction

Vague set theory was first introduced by Gau and Buehrer in 1993 [1], which was regarded as a promotion and extension for the theory of fuzzy sets. A vague set A is characterized by a truth-membership function t_A and a false-membership function f_A. For any $x \in U$ (U is a universe), $t_A(x)$ is a lower bound on the grade of membership of x derived from the evidence for x, $f_A(x)$ is a lower bound on the negation of x derived from the evidence against x, and $t_A(x) + f_A(x) \leq 1$. In fact, every decision-maker hesitates more or less in every evaluation activity. For instance, in order to judge whether a patient has cancer or not, a doctor (the decision-maker) will hesitate because of the fact that a fraction of evaluation he thinks in favor of truth, another fraction in favor of falseness and rest part remains undecided to him. Therefore, vague sets can realistically reflect the actual problem. Whereas, sometimes we will meet the truth-membership and false-membership are certain variable values associated with the time in real life. To solve this uncertain and vague problems, the concept of the temporal vague sets is introduced. Rough set theory was proposed by Pawlak in 1982 [2], which is an extension of set theory for the research of intelligent systems characterized by insufficient and incomplete information. The rough set theory approximates any subset of objects of the universe by the lower and upper approximations. It focuses on the ambiguity caused by the limited discernibility of objects in the

J. Yu et al. (Eds.): RSKT 2010, LNAI 6401, pp. 16–24, 2010.

universe of discourse. And the ambiguity was characterized by the roughness measure introduced by Pawlak [3]. The combination of fuzzy set theory and rough set theory has been done by many researchers in recent years [4-10]. And many new mathematical methods are generated for dealing with the uncertain and imprecise information, such as the fuzzy rough sets and rough fuzzy sets, etc. Additionally, many measure methods are proposed and investigated by different authors in order to characterize the uncertainty and ambiguity of different sets [11-17]. In this paper, we mainly focus on the construction of the lower and upper approximation operators of a temporal vague set based on the works of Wang et al. [11] and Al-Rababah and Biswas [18]. Meantime, some new notions are presented and the corresponding properties associated with the approximation operators and roughness measure are examined.

2   Temporal Vague Sets(TVS)

In the theory of vague sets, the truth(false)-membership function is not related to the time. However, the decision-making results are possible to be mutative along with the time in many practical situations. To deal with this uncertainty and imprecision decision questions, the concept of the temporal truth(false)-membership function is introduced by adding a time parameter in classical membership function.

Definition 1. *Let U be the universe of discourse, T be a non-empty time set. The triple $A(T) = \{< x, t_A(x,t), f_A(x,t) >: (x,t) \in U \times T\}$ be called temporal vague sets(or in short TVS) on $U \times T$, where $t_A(x,t) \subseteq [0,1]$ and $f_A(x,t) \subseteq [0,1]$ denote the membership and non-membership of the element $x \in U$ at $t \in T$, respectively. For all $(x,t) \in U \times T$, $t_A(x,t) + f_A(x,t) \leq 1$.*

Definition 2. *Let $A(T_1) = \{< x, t_A(x,t), f_A(x,t) >: (x,t) \in U \times T_1\}$ and $B(T_2) = \{< x, t_B(x,t), f_B(x,t) >: (x,t) \in U \times T_2\}$ be two temporal vague sets of the universe of discourse U, then*
(a) equality: $A(T_1) = B(T_2)$ iff $T_1 = T_2, \forall (x,t) \in U \times (T_1 \cup T_2)$,
        $t_A(x,t) = t_B(x,t)$ and $f_A(x,t) = f_B(x,t)$;
(b) inclusion: $A(T_1) \subseteq B(T_2)$ iff $T_1 = T_2, \forall (x,t) \in U \times (T_1 \cup T_2)$,
        $t_A(x,t) \leq t_B(x,t)$ and $f_A(x,t) \geq f_B(x,t)$;
(c) intersection: $C(T) = A(T_1) \wedge B(T_2)$ iff $T = T_1 \cup T_2,\ \forall (x,t) \in U \times (T_1 \cup T_2)$,
        $t_C(x,t) = \bar{t}_A(x,t) \wedge \bar{t}_B(x,t)$ and $f_C(x,t) = \bar{f}_A(x,t) \vee \bar{f}_B(x,t)$;
In fact, it is easy to see that $C(T) = A(T_1) \wedge B(T_2) = A \wedge B(T_1 \cup T_2)$.

(d) union: $D(T) = A(T_1) \vee B(T_2)$ iff $T = T_1 \cup T_2,\ \forall (x,t) \in U \times (T_1 \cup T_2)$,
        $t_C(x,t) = \bar{t}_A(x,t) \vee \bar{t}_B(x,t)$ and $f_C(x,t) = \bar{f}_A(x,t) \wedge \bar{f}_B(x,t)$;
Analogously, we also have $D(T) = A(T_1) \vee B(T_2) = A \vee B(T_1 \cup T_2)$.

(e) complement: $A^c(T_1) = \{< x, t_{A^c}(x,t), f_{A^c}(x,t) >: (x,t) \in U \times T_1\}$
        $= \{< x, f_A(x,t), t_A(x,t) >: (x,t) \in U \times T_1\}$.
where

$$\bar{t}_A(x,t) = \begin{cases} t_A(x,t), & t \in T_1 \\ 0, & t \in T_2 - T_1 \end{cases} \quad , \quad \bar{f}_A(x,t) = \begin{cases} f_A(x,t), & t \in T_1 \\ 1, & t \in T_2 - T_1 \end{cases} \quad ,$$

$$\bar{t}_B(x,t) = \begin{cases} t_B(x,t), & t \in T_2 \\ 0, & t \in T_1 - T_2 \end{cases} \quad , \quad \bar{f}_B(x,t) = \begin{cases} f_B(x,t), & t \in T_2 \\ 1, & t \in T_1 - T_2 \end{cases} \quad .$$

3   Rough Approximation of a Temporal Vague Set

The rough approximations of a temporal vague set are constructed. Meanwhile, some related properties are examined in this section.

3.1   Rough Temporal Vague Sets(RTVS)

Definition 3. *Let U be the universe of discourse, T be a non-empty time set. The triple $A(T) = \{< x, t_A(x,t), f_A(x,t) >: (x,t) \in U \times T\}$ be a temporal vague set on $U \times T$. The* lower approximation $\underline{A}(T)$ *and the* upper approximation $\overline{A}(T)$ *of the temporal vague set $A(T)$ in Pawlak approximation space (U, R) are defined, respectively, by*

$$\underline{A}(T) = \{< x, t_{\underline{A}}(x,t), f_{\underline{A}}(x,t) >: (x,t) \in U \times T\}$$
$$\overline{A}(T) = \{< x, t_{\overline{A}}(x,t), f_{\overline{A}}(x,t) >: (x,t) \in U \times T\}$$

where $\forall (x,t) \in U \times T$,

$$t_{\underline{A}}(x,t) = inf\{t_A(y,t) : y \in [x]_R, (x,t) \in U \times T\},$$
$$f_{\underline{A}}(x,t) = sup\{f_A(y,t) : y \in [x]_R, (x,t) \in U \times T\},$$
$$t_{\overline{A}}(x,t) = sup\{t_A(y,t) : y \in [x]_R, (x,t) \in U \times T\},$$
$$f_{\overline{A}}(x,t) = inf\{f_A(y,t) : y \in [x]_R, (x,t) \in U \times T\}.$$

Here, $[x]_R$ denotes the equivalence class which belongs to x.

For any $x \in U$, $t_A(x,t)(t_{\overline{A}}(x,t))$ may be viewed as the evidence in favor of the fact that x definitely(possibly) belongs to the temporal vague set $A(T)$ at $t \in T$. On the contrary, $f_{\underline{A}}(x,t)(f_{\overline{A}}(x,t))$ represent the evidence against the fact that x definitely(possibly) belongs to the temporal vague set $A(T)$ at $t \in T$.

Definition 4. *Let $A(T)$ be a temporal vague set of the universe of discourse U, If $\underline{A}(T) = \overline{A}(T)$, the temporal vague set $A(T)$ is called a* definable temporal vague set *in (U, R). Otherwise, it is called a rough temporal vague set in (U, R).*

Example. Let $U = \{x_i : i = 1, 2, \cdots, 8\}$ be a universe, the time set $T = [0, 2]$, there are four equivalence classes, $\{x_1, x_4\}, \{x_2, x_3, x_6\}, \{x_5\}, \{x_7, x_8\}$, which is partitioned by an equivalence relation R. Now, we consider the following two temporal vague sets

$$A(T) = < [-\tfrac{1}{4}(t-3), \tfrac{1}{5}(t+1)]/x_1 + [-\tfrac{1}{10}(2t-5), \tfrac{1}{10}(2t+1)]/x_4$$
$$+ [\tfrac{1}{8}(t+4), -\tfrac{1}{10}(t-4)]/x_5 + [\tfrac{3}{20}(t+2), -\tfrac{1}{10}(2t-5)]/x_7 : t \in [0, 2] >,$$
$$B(T) = < [-\tfrac{1}{4}(t-3), \tfrac{1}{5}(t+1)]/x_1 + [-\tfrac{1}{4}(t-3), \tfrac{1}{5}(t+1)]/x_4 +$$
$$[\tfrac{1}{8}(t+4), -\tfrac{1}{10}(t-4)]/x_5 : t \in [0, 2] >.$$

According to Definition 3, we can obtain

$$\underline{A}(T) =< [-\tfrac{1}{10}(2t-5), \tfrac{1}{10}(2t+1)]/x_1 + [-\tfrac{1}{10}(2t-5), \tfrac{1}{10}(2t+1)]/x_4$$
$$+[\tfrac{1}{8}(t+4), -\tfrac{1}{10}(t-4)]/x_5 : t \in [0,2] >,$$
$$\overline{A}(T) =< [-\tfrac{1}{4}(t-3), \tfrac{1}{5}(t+1)]/x_1 + [-\tfrac{1}{4}(t-3), \tfrac{1}{5}(t+1)]/x_4 + [\tfrac{1}{8}(t+4), -\tfrac{1}{10}(t-$$
$$4)]/x_5 + [\tfrac{3}{20}(t+2), -\tfrac{1}{10}(2t-5)]/x_7 + [\tfrac{3}{20}(t+2), -\tfrac{1}{10}(2t-5)]/x_8 : t \in [0,2] >,$$
$$\underline{B}(T) =< [-\tfrac{1}{4}(t-3), \tfrac{1}{5}(t+1)]/x_1 + [-\tfrac{1}{4}(t-3), \tfrac{1}{5}(t+1)]/x_4$$
$$+[\tfrac{1}{8}(t+4), -\tfrac{1}{10}(t-4)]/x_5 : t \in [0,2] >,$$
$$\overline{B}(T) =< [-\tfrac{1}{4}(t-3), \tfrac{1}{5}(t+1)]/x_1 + [-\tfrac{1}{4}(t-3), \tfrac{1}{5}(t+1)]/x_4$$
$$+[\tfrac{1}{8}(t+4), -\tfrac{1}{10}(t-4)]/x_5 : t \in [0,2] >.$$

Obviously, the results show that $A(T)$ is a rough temporal vague set and $B(T)$ is a definable temporal vague set in Pawlak approximation space (U, R).

3.2 The Properties of RTVS

Proposition 1. *Let $A(T_1), B(T_2)$ be two temporal vague sets of the discourse U with respect to the time sets T_1 and T_2, then the following conclusions hold*
(P1) $\underline{A}(T_1) \subseteq A(T_1) \subseteq \overline{A}(T_1)$;
(P2) $A(T_1) \subseteq B(T_2) \Rightarrow \underline{A}(T_1) \subseteq \underline{B}(T_2)$ and $\overline{A}(T_1) \subseteq \overline{B}(T_2)$;
(P3) $\overline{A \vee B}(T_1 \cup T_2) = \overline{A} \vee \overline{B}(T_1 \cup T_2)$, $\underline{A \wedge B}(T_1 \cup T_2) = \underline{A} \wedge \underline{B}(T_1 \cup T_2)$;
(P4) $\overline{A \wedge B}(T_1 \cup T_2) \subseteq \overline{A} \wedge \overline{B}(T_1 \cup T_2)$, $\underline{A \vee B}(T_1 \cup T_2) \supseteq \underline{A} \vee \underline{B}(T_1 \cup T_2)$;
(P5) $\underline{A^c}(T_1) = \overline{A}^c(T_1)$, $\overline{A^c}(T_1) = \underline{A}^c(T_1)$.

Proof. (P1) For $\forall (x,t) \in U \times T_1$, since
$$inf\{t_A(y,t) : y \in [x]_R\} \le t_A(x,t) \le sup\{t_A(y,t) : y \in [x]_R\},$$
$$inf\{f_A(y,t) : y \in [x]_R\} \le f_A(x,t) \le sup\{f_A(y,t) : y \in [x]_R\}.$$
By the Definition 3, we have
$$t_{\underline{A}}(x,t) \le t_A(x,t) \le t_{\overline{A}}(x,t) \quad \text{and} \quad f_{\overline{A}}(x,t) \le f_A(x,t) \le f_{\underline{A}}(x,t).$$
So we obtain $\underline{A}(T_1) \subseteq A(T_1) \subseteq \overline{A}(T_1)$.
(P2) According to Definition 2, if $A(T_1) \subseteq B(T_2)$, then $T_1 = T_2$, and for $\forall (x,t) \in U \times T_1$, we have $t_A(x,t) \le t_B(x,t)$ and $f_A(x,t) \ge f_B(x,t)$.
Therefore, we easily know
$$t_{\underline{A}}(x,t) = inf\{t_A(y,t) : y \in [x]_R\} \le inf\{t_B(y,t) : y \in [x]_R\} = t_{\underline{B}}(x,t)$$
$$f_{\underline{A}}(x,t) = sup\{f_A(y,t) : y \in [x]_R\} \ge sup\{f_B(y,t) : y \in [x]_R\} = f_{\underline{B}}(x,t)$$
Similarly, we can obtain the following results
$$t_{\overline{A}}(x,t) = sup\{t_A(y,t) : y \in [x]_R\} \le sup\{t_B(y,t) : y \in [x]_R\} = t_{\overline{B}}(x,t)$$
$$f_{\overline{A}}(x,t) = inf\{f_A(y,t) : y \in [x]_R\} \ge inf\{f_B(y,t) : y \in [x]_R\} = f_{\overline{B}}(x,t)$$
So we can get $\underline{A}(T_1) \subseteq \underline{B}(T_2)$ and $\overline{A}(T_1) \subseteq \overline{B}(T_2)$.
(P3) For $\forall (x,t) \in U \times (T_1 \cup T_2)$,
$$\overline{A \vee B}(T_1 \cup T_2) = \{< x, t_{\overline{A \vee B}}(x,t), f_{\overline{A \vee B}}(x,t) >\}$$
$$= \{< x, sup\{t_{A \vee B}(y,t) : y \in [x]_R\}, inf\{f_{A \vee B}(y,t) : y \in [x]_R\} >\}$$
$$= \{< x, sup\{\bar{t}_A(y,t) \vee \bar{t}_B(y,t) : y \in [x]_R\}, inf\{\bar{f}_A(y,t) \wedge \bar{f}_B(y,t) : y \in [x]_R\} >\}$$
$$= \{< x, sup\{\bar{t}_A(y,t) : y \in [x]_R\} \vee sup\{\bar{t}_B(y,t) : y \in [x]_R\},$$
$$inf\{\bar{f}_A(y,t) : y \in [x]_R\} \wedge inf\{\bar{f}_B(y,t) : y \in [x]_R\} >\}$$
$$= \{< x, t_{\overline{A}}(x,t) \vee t_{\overline{B}}(x,t), f_{\overline{A}}(x,t) \wedge f_{\overline{B}}(x,t) >\}$$
$$= \{< x, t_{\overline{A \vee B}}(x,t), f_{\overline{A \vee B}}(x,t) >\} = \overline{A} \vee \overline{B}(T_1 \cup T_2).$$
Hence, $\overline{A \vee B}(T_1 \cup T_2) = \overline{A} \vee \overline{B}(T_1 \cup T_2)$. Similarly, we can get $\underline{A \wedge B}(T_1 \cup T_2) = \underline{A} \wedge \underline{B}(T_1 \cup T_2)$ by the same schemer above.

(P4) For $\forall\,(x,t) \in U \times (T_1 \cup T_2)$,

$\underline{A \vee B}(T_1 \cup T_2) = \{< x, t_{\underline{A \vee B}}(x,t), f_{\underline{A \vee B}}(x,t) >\}$

$= \{< x, \bar{t}_{\underline{A}}(x,t) \vee \bar{t}_{\underline{B}}(x,t), \bar{f}_{\underline{A}}(x,t) \wedge \bar{f}_{\underline{B}}(x,t) >\}$

$= \{< x, inf\{\bar{t}_{\underline{A}}(y,t) : y \in [x]_R\} \vee inf\{\bar{t}_{\underline{B}}(y,t) : y \in [x]_R\},$

$\quad sup\{\bar{f}_{\underline{A}}(y,t) : y \in [x]_R\} \wedge sup\{\bar{f}_{\underline{B}}(y,t) : y \in [x]_R\} >\}$

$\subseteq \{< x, inf\{\bar{t}_{\underline{A}}(y,t) \vee \bar{t}_{\underline{B}}(y,t) : y \in [x]_R\}, sup\{\bar{f}_{\underline{A}}(y,t) \wedge \bar{f}_{\underline{B}}(y,t) : y \in [x]_R\} >\}$

$= \{< x, inf\{t_{\underline{A \vee B}}(y,t) : y \in [x]_R\}, sup\{f_{\underline{A \vee B}}(y,t) : y \in [x]_R\} >\}$

$= \{< x, t_{\underline{A \vee B}}(x,t), f_{\underline{A \vee B}}(x,t) >\} = \underline{A \vee B}(T_1 \cup T_2).$

In addition, the relation $\overline{A \wedge B}(T_1 \cup T_2) \subseteq \overline{A} \wedge \overline{B}(T_1 \cup T_2)$ can be derived similarly.

(P5) For $\forall\,(x,t) \in U \times T_1$, since $A^c(T_1) = \{< x, f_A(x,t), t_A(x,t) >\}$, we have

$\overline{A}^c(T_1) = \{< x, f_{\overline{A}}(x,t), t_{\overline{A}}(x,t) >\}$

$\quad = \{< x, inf\{f_A(x,t) : y \in [x]_R, sup\{t_A(x,t) : y \in [x]_R\} >\}$

$\quad = \{< x, inf\{t_{A^c}(x,t) : y \in [x]_R, sup\{f_{A^c}(x,t) : y \in [x]_R\} >\}$

$\quad = \{< x, t_{A^c}(x,t), f_{A^c}(x,t) >\} = \underline{A^c}(T_1).$

Thus, we have $\underline{A^c}(T_1) = \overline{A}^c(T_1)$. Similarly, the proof of the second conclusion $\overline{A^c}(T_1) = \underline{A}^c(T_1)$ is the same as the previous one.

From the above conclusions, we can see that the rough approximations of the temporal vague sets have the same properties as the classical rough sets.

In addition, if the temporal vague set $A(T_1)$ or $B(T_2)$ satisfies certain condition, then the property (P4) of the above proposition will become equality.

Proposition 2. *Let $A(T_1), B(T_2)$ be two temporal vague sets of the discourse U, if either $A(T_1)$ or $B(T_2)$ is a definable temporal vague set, then*

$\overline{A \wedge B}(T_1 \cup T_2) = \overline{A} \wedge \overline{B}(T_1 \cup T_2) \quad and \quad \underline{A \vee B}(T_1 \cup T_2) = \underline{A} \vee \underline{B}(T_1 \cup T_2).$

Proof. Without loss of generality, we assume that $A(T_1)$ is a definable temporal vague set. According to Definitions 3 and 4, for $\forall\,(x,t) \in U \times T_1$, $\{t_A(y,t) : y \in [x]_R\}$ and $\{f_A(y,t) : y \in [x]_R\}$ are constants, written as a, b, respectively. Obviously, $0 \le a, b \le 1$, $0 \le a + b \le 1$. On the other hand, by the Definition 2, we can easily know that

$$\bar{t}_A(x,t) = \begin{cases} a, & t \in T_1 \\ 0, & t \in T_2 - T_1 \end{cases} \quad , \quad \bar{f}_A(x,t) = \begin{cases} b, & t \in T_1 \\ 1, & t \in T_2 - T_1 \end{cases}$$

Since the $\bar{t}_A(x,t)$ and $\bar{f}_A(x,t)$ are piecewise constant functions with respect to the time parameter t, so we divide the proof into two parts.

(i) For $\forall\,x \in U$, if $x \in T_1$, we have

$\underline{A \vee B}(T_1 \cup T_2) = \{< x, t_{\underline{A \vee B}}(x,t), f_{\underline{A \vee B}}(x,t) >\}$

$\quad = \{< x, \bar{t}_{\underline{A}}(x,t) \vee \bar{t}_{\underline{B}}(x,t), \bar{f}_{\underline{A}}(x,t) \wedge \bar{f}_{\underline{B}}(x,t) >\}$

$\quad = \{< x, inf\{\bar{t}_{\underline{A}}(y,t) : y \in [x]_R\} \vee inf\{\bar{t}_{\underline{B}}(y,t) : y \in [x]_R\},$

$\quad\quad sup\{\bar{f}_{\underline{A}}(y,t) : y \in [x]_R\} \wedge sup\{\bar{f}_{\underline{B}}(y,t) : y \in [x]_R\} >\}$

$\quad \overset{t \in T_1}{=} \{< x, inf\{a : y \in [x]_R\} \vee inf\{\bar{t}_{\underline{B}}(y,t) : y \in [x]_R\},$

$\quad\quad sup\{b : y \in [x]_R\} \wedge sup\{\bar{f}_{\underline{B}}(y,t) : y \in [x]_R\} >\}$

$\quad = \{< x, inf\{a \vee \bar{t}_{\underline{B}}(x,t) : y \in [x]_R\}, sup\{b \wedge \bar{f}_{\underline{B}}(x,t) : y \in [x]_R\} >\}$

$\quad = \{< x, inf\{t_{\underline{A \vee B}}(x,t) : y \in [x]_R\}, sup\{f_{\underline{A \vee B}}(x,t) : y \in [x]_R\} >\}$

$$= \{< x, t_{\underline{A \vee B}}(x,t), f_{\underline{A \vee B}}(x,t) >\} = \underline{A \vee B}(T_1 \cup T_2).$$

(ii) For $\forall\, x \in U$, if $x \in T_2 - T_1$, the same as the previous proof, we can also get the same conclusion.

Thus, for $\forall\, (x,t) \in U \times (T_1 \cup T_2)$, we have $\underline{A \vee B}(T_1 \cup T_2) = \underline{A} \vee \underline{B}(T_1 \cup T_2)$. Similarly, we can obtain the other relation.

Let P, R be two equivalence relations on the universe U. For any $x \in U$, if $[x]_P \subseteq [x]_R$, we call that the relation P is thinner than R, or R is coarser than P alternatively. It is denoted by $P \prec R$.

Proposition 3. *Let $\underline{A}(T)$ be a temporal vague set on $U \times T$. If $P \prec R$, then $\underline{A}(T) \subseteq \underline{A}_P(T)$ and $\overline{A}_P(T) \subseteq \overline{A}(T)$.*

Proof. For $\forall\, (x,t) \in U \times (T_1 \cup T_2)$, since $P \prec R \Leftrightarrow [x]_P \subseteq [x]_R$, so we have
$$inf\{t_A(y,t) : y \in [x]_R\} \leq inf\{t_A(y,t) : y \in [x]_P\}$$
$$sup\{f_A(y,t) : y \in [x]_R\} \geq sup\{f_A(y,t) : y \in [x]_P\}$$
$$sup\{t_A(y,t) : y \in [x]_R\} \geq sup\{t_A(y,t) : y \in [x]_P\}$$
$$inf\{f_A(y,t) : y \in [x]_R\} \leq inf\{f_A(y,t) : y \in [x]_P\}$$
Hence, we easily get $\underline{A}(T) \subseteq \underline{A}_P(T)$ and $\overline{A}_P(T) \subseteq \overline{A}(T)$.

4   Roughness Measure of a Temporal Vague Set

The roughness measure of a temporal vague set is presented based on the one of a vague set.

Definition 5. *Let $A(T)$ be a temporal vague set on $U \times T$. For $\forall\, t \in T$, the temporal $\alpha\beta$-level sets of $\underline{A}(T)$ and $\overline{A}(T)$ at time t, denoted by $\underline{A}_{\alpha\beta}(t)$ and $\overline{A}_{\alpha\beta}(t)$, are defined, respectively, as follows*
$$\underline{A}_{\alpha\beta}(t) = \{x \in U : t_{\underline{A}}(x,t) \geq \alpha, f_{\underline{A}}(x,t) \leq \beta\}$$
$$\overline{A}_{\alpha\beta}(t) = \{x \in U : t_{\overline{A}}(x,t) \geq \alpha, f_{\overline{A}}(x,t) \leq \beta\}$$
where $0 < \alpha, \beta \leq 1, \alpha + \beta \leq 1$.

The roughness measure of the temporal vague set $A(T)$ can be obtained using the temporal $\alpha\beta$-level sets of $\underline{A}(T)$ and $\overline{A}(T)$.

Definition 6. *Let $A(T)$ be a temporal vague set on $U \times T$. For $\forall\, t \in T$, a roughness measure $\rho_A^{\alpha\beta}(t)$ of the temporal vague set $A(T)$ with respect to the parameters α, β at t, in Pawlak approximation space (U, R), is defined as*
$$\rho_A^{\alpha\beta}(t) = 1 - \frac{|\underline{A}_{\alpha\beta}(t)|}{|\overline{A}_{\alpha\beta}(t)|} \quad (t \in T)$$
where the notation $|\cdot|$ denotes the cardinality of the set.

Obviously, for every $t \in T$, $0 \leq \rho_A^{\alpha\beta}(t) \leq 1$. Especially, $\rho_A^{\alpha\beta}(t) = 0$ when $|\underline{A}_{\alpha\beta}(t)| = 0$.

For the roughness measure of a temporal vague set, we have the following properties.

Proposition 4. *If $A(T)$ be a definable temporal vague set, then $\rho_A^{\alpha\beta}(t) \equiv 0$.*

Proposition 5. *Let $A(T_1)$ and $B(T_2)$ be two temporal vague sets, if $A(T_1) = B(T_2)$, then $\rho_A^{\alpha\beta}(t) \equiv \rho_B^{\alpha\beta}(t)$.*

Proposition 6. *Let $A(T_1)$ and $B(T_2)$ be two temporal vague sets, if $A(T_1) \subseteq B(T_2)$, then*
(a) if $\overline{A}_{\alpha\beta}(t) \equiv \overline{B}_{\alpha\beta}(t)$, then $\rho_A^{\alpha\beta}(t) \geq \rho_B^{\alpha\beta}(t)$;
(b) if $\underline{A}_{\alpha\beta}(t) \equiv \underline{B}_{\alpha\beta}(t)$, then $\rho_A^{\alpha\beta}(t) \leq \rho_B^{\alpha\beta}(t)$.

Proof. It can be easily proved by the Proposition 1 and Definitions 2 and 6.

Proposition 7. *Let $A(T)$, $B(T)$ be two temporal vague sets, for $\forall\, t \in T$, we have*
$$\rho_{A \vee B}^{\alpha\beta}(t)|\overline{A}_{\alpha\beta}(t) \vee \overline{B}_{\alpha\beta}(t)| \leq \rho_A^{\alpha\beta}(t)|\overline{A}_{\alpha\beta}(t)| + \rho_B^{\alpha\beta}(t)|\overline{B}_{\alpha\beta}(t)|$$
$$-\rho_{A \wedge B}^{\alpha\beta}(t)|\overline{A}_{\alpha\beta}(t) \wedge \overline{B}_{\alpha\beta}(t)|.$$

Proof. The proof is the same as the Proposition 4.5 in [11], so we omit it here.

Corollary 1. *If either $A(T)$ or $B(T)$ is a definable temporal vague set, for $\forall\, t \in T$, then we have*
$$\rho_{A \vee B}^{\alpha\beta}(t)|\overline{A}_{\alpha\beta}(t) \vee \overline{B}_{\alpha\beta}(t)| = \rho_A^{\alpha\beta}(t)|\overline{A}_{\alpha\beta}(t)| + \rho_B^{\alpha\beta}(t)|\overline{B}_{\alpha\beta}(t)|$$
$$-\rho_{A \wedge B}^{\alpha\beta}(t)|\overline{A}_{\alpha\beta}(t) \wedge \overline{B}_{\alpha\beta}(t)|.$$

Proposition 8. *Let $A(T)$ be a temporal vague set on $U \times T$, for $\forall\, t \in T$, if $\alpha_1 \geq \alpha_2$, then*
(a) if $\overline{A}_{\alpha_1\beta}(t) \supseteq \overline{A}_{\alpha_2\beta}(t)$, then $\rho^{\alpha_1\beta}(t) \geq \rho^{\alpha_2\beta}(t)$;
(b) if $\underline{A}_{\alpha_1\beta}(t) \supseteq \underline{A}_{\alpha_2\beta}(t)$, then $\rho^{\alpha_1\beta}(t) \leq \rho^{\alpha_2\beta}(t)$.

Proof. For $\forall\, t \in T$, if $\alpha_1 \geq \alpha_2$, then we have $\underline{A}_{\alpha_1\beta}(t) \subseteq \underline{A}_{\alpha_2\beta}(t)$ and $\overline{A}_{\alpha_1\beta}(t) \subseteq \overline{A}_{\alpha_2\beta}(t)$. According to Definition 6, it can easily be verified that $\rho^{\alpha_1\beta}(t) \geq \rho^{\alpha_2\beta}(t)$ when $\overline{A}_{\alpha_1\beta}(t) \supseteq \overline{A}_{\alpha_2\beta}(t)$ and $\rho^{\alpha_1\beta}(t) \leq \rho^{\alpha_2\beta}(t)$ when $\underline{A}_{\alpha_1\beta}(t) \supseteq \underline{A}_{\alpha_2\beta}(t)$.

Proposition 9. *Let $A(T)$ be a temporal vague set on $U \times T$, for $\forall\, t \in T$, if $\beta_1 \geq \beta_2$, then*
(a) if $\overline{A}_{\alpha\beta_1}(t) \subseteq \overline{A}_{\alpha\beta_2}(t)$, then $\rho^{\alpha\beta_1}(t) \leq \rho^{\alpha\beta_2}(t)$;
(b) if $\underline{A}_{\alpha\beta_1}(t) \subseteq \underline{A}_{\alpha\beta_2}(t)$, then $\rho^{\alpha\beta_1}(t) \geq \rho^{\alpha\beta_2}(t)$.

Proof. The proof is the same as the Proposition 8.

For convenience, we introduce several notations as follows.
$$\underline{A}_\alpha^T(t) = \{x \in U : t_{\underline{A}}(x,t) \geq \alpha\}; \quad \underline{A}_\beta^F(t) = \{x \in U : f_{\underline{A}}(x,t) \leq \beta\};$$
$$\overline{A}_\alpha^T(t) = \{x \in U : t_{\overline{A}}(x,t) \geq \alpha\}; \quad \overline{A}_\beta^F(t) = \{x \in U : f_{\overline{A}}(x,t) \leq \beta\}.$$

Proposition 10. *Let $A(T)$ be a temporal vague set on $U \times T$, for $\forall\, t \in T$, if $\alpha_1 \geq \alpha_2, \beta_1 \geq \beta_2$, then*
(a) if $\forall\, t \in T$, $\underline{A}_{\alpha_1}^T(t) \supseteq \underline{A}_{\alpha_2}^T(t)$, $\overline{A}_{\alpha_1\beta_1}(t) \equiv \overline{A}_{\alpha_2\beta_2}(t)$, then $\rho_A^{\alpha_1\beta_1}(t) \leq \rho_A^{\alpha_2\beta_2}(t)$;
(b) if $\forall\, t \in T$, $\underline{A}_{\beta_1}^F(t) \subseteq \underline{A}_{\beta_2}^F(t)$, $\overline{A}_{\alpha_1\beta_1}(t) \equiv \overline{A}_{\alpha_2\beta_2}(t)$, then $\rho_A^{\alpha_1\beta_1}(t) \geq \rho_A^{\alpha_2\beta_2}(t)$;
(c) if $\forall\, t \in T$, $\overline{A}_{\alpha_1}^T(t) \supseteq \overline{A}_{\alpha_2}^T(t)$, $\underline{A}_{\alpha_1\beta_1}(t) \equiv \underline{A}_{\alpha_2\beta_2}(t)$, then $\rho_A^{\alpha_1\beta_1}(t) \geq \rho_A^{\alpha_2\beta_2}(t)$;
(d) if $\forall\, t \in T$, $\overline{A}_{\beta_1}^F(t) \subseteq \overline{A}_{\beta_2}^F(t)$, $\underline{A}_{\alpha_1\beta_1}(t) \equiv \underline{A}_{\alpha_2\beta_2}(t)$, then $\rho_A^{\alpha_1\beta_1}(t) \leq \rho_A^{\alpha_2\beta_2}(t)$.

Proof. For $t \in T$, if $\alpha_1 \geq \alpha_2$, we then obtain $\underline{A}_{\alpha_1}^T(t) \subseteq \underline{A}_{\alpha_2}^T(t)$, $\overline{A}_{\alpha_1}^T(t) \subseteq \overline{A}_{\alpha_2}^T(t)$. Similarly, for $t \in T$, if $\beta_1 \geq \beta_2$, we have $\underline{A}_{\beta_1}^F(t) \supseteq \underline{A}_{\beta_2}^F(t)$, $\overline{A}_{\beta_1}^F(t) \supseteq \overline{A}_{\beta_2}^F(t)$. According to Definition 6 and the above conditions, the conclusions hold.

Proposition 11. *Let $A(T)$ be a temporal vague set on $U \times T$, for $\forall\, t \in T$, if $\alpha_1 \geq \alpha_2, \beta_1 \leq \beta_2$, then*
(a) if $\forall\, t \in T$, $\overline{A}_{\alpha_1\beta_1}(t) \supseteq \overline{A}_{\alpha_2\beta_2}(t)$, then $\rho_A^{\alpha_1\beta_1}(t) \geq \rho_A^{\alpha_2\beta_2}(t)$;
(b) if $\forall\, t \in T$, $\underline{A}_{\alpha_1\beta_1}(t) \supseteq \underline{A}_{\alpha_2\beta_2}(t)$, then $\rho_A^{\alpha_1\beta_1}(t) \leq \rho_A^{\alpha_2\beta_2}(t)$.

Proof. If $\alpha_1 \geq \alpha_2$ and $\beta_1 \leq \beta_2$, then we have $\underline{A}_{\alpha_1\beta_1}(t) \subseteq \underline{A}_{\alpha_2\beta_2}(t)$ and $\overline{A}_{\alpha_1\beta_1}(t) \subseteq \overline{A}_{\alpha_2\beta_2}(t)$. It is obvious that the conclusions hold.

Proposition 12. *Let $A(T)$ be a temporal vague set on $U \times T$, for $\forall\, t \in T$, if $\alpha_1 \leq \alpha_2, \beta_1 \geq \beta_2$, then*
(a) if $\forall\, t \in T$, $\overline{A}_{\alpha_1\beta_1}(t) \subseteq \overline{A}_{\alpha_2\beta_2}(t)$, then $\rho_A^{\alpha_1\beta_1}(t) \leq \rho_A^{\alpha_2\beta_2}(t)$;
(b) if $\forall\, t \in T$, $\underline{A}_{\alpha_1\beta_1}(t) \subseteq \underline{A}_{\alpha_2\beta_2}(t)$, then $\rho_A^{\alpha_1\beta_1}(t) \geq \rho_A^{\alpha_2\beta_2}(t)$.

Proof. The proof is the same as the Proposition 11.

Proposition 13. *Let $A(T)$ be a temporal vague set on $U \times T$, for $\forall\, t_1, t_2 \in T$ and $t_1 < t_2$, and if $t_A(x, t_1) \leq t_A(x, t_2), f_A(x, t_1) \geq f_A(x, t_2)$, then*
(a) if $\underline{A}_{\alpha\beta}(t_1) \supseteq \underline{A}_{\alpha\beta}(t_2)$, then $\rho_A^{\alpha\beta}(t_1) \leq \rho_A^{\alpha\beta}(t_2)$;
(b) if $\overline{A}_{\alpha\beta}(t_1) \supseteq \overline{A}_{\alpha\beta}(t_2)$, then $\rho_A^{\alpha\beta}(t_1) \geq \rho_A^{\alpha\beta}(t_2)$.

Proof. Since $t_1 < t_2(\in T)$, we can easily obtain $\underline{A}_{\alpha\beta}(t_1) \subseteq \underline{A}_{\alpha\beta}(t_2)$ and $\overline{A}_{\alpha\beta}(t_1) \subseteq \overline{A}_{\alpha\beta}(t_2)$. According to Definition 6, the above conclusions hold.

Analogously, we can obtain another proposition

Proposition 14. *Let $A(T)$ be a temporal vague set on $U \times T$, for $\forall\, t_1, t_2 \in T$ and $t_1 < t_2$, and if $t_A(x, t_1) \geq t_A(x, t_2), f_A(x, t_1) \leq f_A(x, t_2)$, then*
(a) if $\underline{A}_{\alpha\beta}(t_1) \subseteq \underline{A}_{\alpha\beta}(t_2)$, then $\rho_A^{\alpha\beta}(t_1) \geq \rho_A^{\alpha\beta}(t_2)$;
(b) if $\overline{A}_{\alpha\beta}(t_1) \subseteq \overline{A}_{\alpha\beta}(t_2)$, then $\rho_A^{\alpha\beta}(t_1) \leq \rho_A^{\alpha\beta}(t_2)$.

It is evident that the roughness measure is monotonic with respect to the time t and the parameters α, β under certain conditions.

Let P be another equivalence relation of the discourse U, we write the temporal $\alpha\beta$-level sets of $\underline{A}(T), \overline{A}(T)$ and the roughness measure as $\underline{A}_{\alpha\beta}^P(t), \overline{A}_{\alpha\beta}^P(t), \rho_{A_P}^{\alpha\beta}(t)$ at t, respectively.

Proposition 15. *For $\forall\, t \in T$, if $P \prec R$, then $\rho_{A_P}^{\alpha\beta}(t) \leq \rho_A^{\alpha\beta}(t)$.*

Proof. If $P \prec R$, by the Proposition 3, it is easy to show that $\underline{A}(T) \subseteq \underline{A}_P(T)$ and $\overline{A}(T) \subseteq \overline{A}_P(T)$. For $\forall\, t \in T$, let us suppose that $x \in \underline{A}_{\alpha\beta}(t)$, then $t_{\underline{A}}(x, t) \geq \alpha$ and $f_{\underline{A}}(x, t) \leq \beta$. Since $t_{\underline{A}_P}(x, t) \geq t_{\underline{A}}(x, t) \geq \alpha$ and $f_{\underline{A}_P}(x, t) \leq f_{\underline{A}}(x, t) \leq \beta$, namely, $x \in \underline{A}_{\alpha\beta}^P(t)$. Consequently, we have $\underline{A}_{\alpha\beta}(t) \subseteq \underline{A}_{\alpha\beta}^P(t)$. Similarly, we can get $\overline{A}_{\alpha\beta}(t) \supseteq \overline{A}_{\alpha\beta}^P(t)$. So we have $\rho_{A_P}^{\alpha\beta}(t) \leq \rho_A^{\alpha\beta}(t)$.

Obviously, the proposition shows that the roughness of temporal vague sets decreases when the equivalence relation becomes thinner in the universe.

5 Conclusions

The combination of rough sets and temporal vague sets is established based on the previous research results. The obtained results are viewed as a generalization of RVS. The concepts of RTVS combines two different theories: rough set theory and vague set theory. Consequently, it is obvious that RTVS can be utilized for handling the vague and incomplete information associated with time.

Acknowledgments. This work is supported by Scientific Research Foundation of Tianshui Normal University (No.TSA0940).

References

1. Gau, W.L., Buehrer, D.J.: Vague sets. IEEE Transactions on Systems, Man and Cybernetics 23, 610–614 (1993)
2. Pawlak, Z.: Rough sets. International Journal of Computer and Information Sciences 11, 341–356 (1982)
3. Pawlak, Z.: Rough sets: Theoretical aspects of reasoning about data. Kluwer academic publishers, Dordrecht (1991)
4. Pawlak, Z.: Rough sets and fuzzy sets. Fuzzy Sets and Systems 17, 99–102 (1985)
5. Dubois, D., Prade, H.: Rough fuzzy sets and fuzzy rough sets. International Journal of General Systems 17, 191–209 (1990)
6. Dubois, D., Prade, H.: Two fold fuzzy sets and rough sets-some issues in knowledge representation. Fuzzy Sets and Systems 23, 3–18 (1987)
7. Wygralak, M.: Rough sets and fuzzy sets-some remarks on interrelations. Fuzzy Sets and Systems 29, 241–243 (1989)
8. Yao, Y.Y.: A comparative study of fuzzy sets and rough sets. Information Sciences 109, 227–242 (1998)
9. Yao, Y.Y.: Semantics of fuzzy sets in rough set theory. LNCS Transactions on Rough Sets 2, 310–331 (2004)
10. Yao, Y.Y.: Combination of rough and fuzzy sets based on α-level sets. In: Lin, T.Y., Cercone, N. (eds.) Rough Sets and Data Mining: Analysis for Imprecise Data, pp. 301–321. Kluwer Academic Publishers, Boston (1997)
11. Wang, J., Liu, S.Y., Zhang, J.: Roughness of a vague set. International Journal of Computational Cognition 3(3), 83–87 (2005)
12. Banerjee, M., Sankar, K.P.: Roughness of a fuzzy set. Information Sciences 93, 235–246 (1996)
13. Eswarlal, T.: Roughness of a Boolean vague set. International Journal of Computational Cognition 6(1), 8–11 (2008)
14. Gu, S.M., Gao, J., Tian, X.Q.: A fuzzy measure based on variable precision rough sets. In: Cao, B.Y. (ed.) Fuzzy Information and Engineering (ICFIE). ASC, vol. 40, pp. 798–807 (2007)
15. Huynh, V.N., Nakamori, Y.: A roughness measure for fuzzy sets. Information Sciences 173, 255–275 (2005)
16. Zhang, X.Y., Xu, W.H.: A Novel Approach to Roughness Measure in Fuzzy Rough Sets. Fuzzy Information and Engineering (ICFIE) ASC 40, 775–780 (2007)
17. Pattaraintakorn, P., Naruedomkul, K., Palasit, K.: A note on the roughness measure of fuzzy sets. Applied Mathematics Letters 22, 1170–1173 (2009)
18. Al-Rababah, A., Biswas, R.: Rough vague sets in an approximation space. International Journal of Computational Cognition 6(4), 60–63 (2008)

Poset Approaches to Covering-Based Rough Sets

Shiping Wang[1], William Zhu[2], and Peiyong Zhu[1]

[1] School of Mathematical Sciences,
University of Electronic Science and Technology of China, Chengdu, China
[2] School of Computer Science and Engineering,
University of Electronic Science and Technology of China, Chengdu, China

Abstract. Rough set theory is a useful and effective tool to cope with granularity and vagueness in information system and has been used in many fields. However, it is hard to get the reduct of a covering in rough sets. This paper attempts to get the reduct of a covering at a high speed in theory. It defines upset and downset based on a poset in a covering, studies the relationship between reducible element and downset, and obtains some good results such as sufficient and necessary condition about the reducible element in a covering.

1 Introduction

Rough set theory is a very useful and meaningful tool for dealing with granularity and vagueness in data analysis. Rough set theory has already been applied to various fields such as process control, economics, medical diagnosis, biochemistry, environmental science, biology, chemistry psychology, conflict analysis and so on. Covering-based rough set theory which is an extension to classical rough set theory has been investigated by many scholars. This paper also studies covering-based rough sets. In order to integrate rough set theory with practice, it is meaningful to combine rough sets with other sciences, such as with information security, with social computing, and with computing in words. Some significant results have been achieved, such as Lin's model and the Great Wall security's model.

From the statement above, to study rough sets is of great importance and significance. Rough set theory has been investigated from different angles and with various approaches, from the standpoint of binary relation [1,2,3], from the standpoint of axiomatization [7,8], with topological approaches [4], with algebraic methods [5,6].

In this paper, inspired by some mathematical concepts, we define posets in a covering. Naturally upsets and downsets based on posets are defined accurately. In this case, we can get some useful and interesting conclusions. For example, a reducible element in a covering is well-described with downset.

The remaining part of this paper is arranged as follows: In Section 2, we present some fundamental concepts of the covering-based rough sets and definitions to be used. Section 3 is the focus of this paper. We discuss some concepts

J. Yu et al. (Eds.): RSKT 2010, LNAI 6401, pp. 25–29, 2010.

such as reducible elements in a covering. We also get some results in covering-based rough sets from the standpoint of poset. This paper is concluded in section 4 with remarks for future works.

2 Basic Definitions

First we list some definitions about covering-based rough sets to be used in this paper.

Definition 1. *(Covering)* $\mathbf{C}$ *is called a covering of a domain* U *if* $\bigcup \mathbf{C} = U$.

Definition 2. *(A reducible element) Let* $\mathbf{C}$ *be a covering of a domain* U *and* $K \in \mathbf{C}$. *If* K *is a union of some sets in* $\mathbf{C} - \{K\}$, *we say* K *is reducible in* $\mathbf{C}$, *otherwise* K *is irreducible.*

Example 1. Let $U = \{a, b, c, d\}$, $K_1 = \{a, b\}$, $K_2 = \{a, b, c\}$, $K_3 = \{c\}$, $K_4 = \{a, b, c, d\}$, $\mathbf{C} = \{K_1, K_2, K_3, K_4\}$. K_2 is a reducible element in the covering $\mathbf{C}$ since $K_2 = K_1 \cup K_3$. K_4 is not a reducible element in the covering $\mathbf{C}$.

Definition 3. *(Minimal description [10]) Let* $\mathbf{C}$ *be a covering of a domain* U. $x \in U$,

$$Md(x) = \{K \in \mathbf{C} \mid x \in K \wedge (\forall A \in \mathbf{C} \wedge x \in A \wedge A \subseteq K \Rightarrow K = A)\}$$

is called the minimal description of x.

In order to describe an object or a system, we need only the essential characteristics related to it. That is the purpose of the minimal description.

Definition 4. *(Poset) A poset is an ordered pair* $(P, \prec)$ *including a finite set* P *and a relation* $\prec$ *on* P, *and satisfying the following three conditions:* $\forall x, y, z \in P$,
(1) $x \prec x$;
(2) $x \prec y$ *and* $y \prec x \Rightarrow x = y$;
(3) $x \prec y$ *and* $y \prec z \Rightarrow x \prec y$.

Definition 5. *(upset and downset [9]) For a poset* $(P, \prec)$, *we define*

$$\downarrow A = \{x \in P \mid \exists a \in A, x \prec a\}$$
$$\uparrow A = \{x \in P \mid \exists a \in A, a \prec x\}.$$

$\downarrow A$ *is called a downset of* P; $\uparrow A$ *is called an upset of* P. A *is downdefinable if* $\downarrow A = A$; A *is updefinable if* $\uparrow A = A$.

3 Main Results

Covering-based rough set theory involves several basic concepts such as minimal description, reducible element. The work of this section is to depict minimal description and reducible element with downset.

Proposition 1. *If $\mathbf{C}$ is a covering of a domain U, then $(\mathbf{C}, \subseteq)$ is a poset.*

The main task of the later part is to study the reduct of a covering with upsests and downsets defined on posets.

Let $\mathbf{C}$ is a covering of a domain U. We denote the following marks:

$$S_1 = \{C_i \in C | \downarrow \{C_i\} = \{C_i\}\},$$
$$S_2 = \{C_i \in C | \uparrow \{C_i\} = \{C_i\}\},$$
$$S = \uparrow (S_1 - S_2).$$

Example 2. If $U = \{x_1, \cdots, x_{10}\}$,
$K_1 = \{x_1, x_2\}$, $K_2 = \{x_1, x_2, x_3\}$, $K_3 = \{x_3, x_4\}$,
$K_4 = \{x_4, x_5\}$, $K_5 = \{x_4, x_5, x_6\}$, $K_6 = \{x_7, x_8, x_9, x_{10}\}$
$K_7 = \{x_9, x_{10}\}$, $\mathbf{C} = \{K_1, \cdots, K_7\}$,
then
$S_1 = \{K_i \in C | \downarrow \{K_i\} = \{K_i\}\} = \{K_1, K_3, K_4, K_7\}$,
$S_2 = \{K_i \in C | \uparrow \{K_i = \{K_i\}\} = \{K_2, K_3, K_5, K_6\}\}$,
$S = \uparrow (S_1 - S_2) = \uparrow \{K_1, K_4, K_7\} = \{K_1, K_2, K_4, K_5, k_6, K_7\}$.

In order to show the above definitions being interesting, we firstly depict one of core concepts in covering-based rough sets.

Proposition 2. *Let $\mathbf{C}$ be a covering of a domain U. $\forall x \in U$,*

$$Md(x) = \{K \in S_1 | x \in K\}.$$

Proof. $Md(x) = \{K \in \mathbf{C} \mid x \in K \wedge (\forall A \in \mathbf{C} \wedge x \in A \wedge A \subseteq K \Rightarrow K = A)\} = \{K \in \mathbf{C} | x \in K \wedge (\downarrow \{K\} = \{K\})\} = \{K \in S_1 | x \in K\}$.

Proposition 3. *Let $\mathbf{C}$ be a covering of a domain U and $K \in \mathbf{C}$. K is reducible in $\mathbf{C}$ if and only if $K = \cup(\downarrow \{K\} - \{K\})$.*

Proof. On the one hand, because K is reducible, there exists $\{C_1, \cdots, C_m\}$, such that $K = \cup_{i=1}^{m} C_i$. So $\forall x \in K$, there exists $i_0 \in \{1, \cdots, m\}$, such that $x \in C_{i_0}$. Also because $C_{i_0} \subseteq K$ and $C_{i_0} \neq K$, then $C_{i_0} \in \downarrow \{K\} - \{K\}$. So $x \in C_{i_0} \subseteq \cup(\downarrow \{K\} - \{K\})$. Then we can get $K = \cup_{x \in K} \{x\} \subseteq \cup(\downarrow \{K\} - \{K\})$. Conversely, it is obvious that $\cup(\downarrow K - \{K\}) \subseteq K$. In a word, $K = \cup(\downarrow \{K\} - \{K\})$.
On the other hand, if $K = \cup(\downarrow \{K\} - \{K\})$, then K is reducible since $\cup(\downarrow \{K\} - \{K\}) \subseteq \mathbf{C}$.

The above proposition is very important, because it shows accurately the relationship between reducible element and downset.

In a information system of vast amounts of data, the following conclusions are of great importance.

Proposition 4. *Let $\mathbf{C}$ be a covering of a domain U and $K \in \mathbf{C}$. If K is reducible in $\mathbf{C}$, then $K \in S$.*

Proof. If K is reducible in $\mathbf{C}$, then K can be expressed as a union of some sets in $\mathbf{C} - \{K\}$. There exists $C_1, C_2, \cdots, C_m$ such that $K = \cup_{i=1}^{m} \{C_i\}$. It is easy to see that $C_i \subseteq K$ for each i, then $C_i \notin S_2, i = 1, \cdots, m$. For a particular i, we can discuss on the following two situations. If $C_i \in S_1$, then $C_i \in S_1 - S_2$, so $K \in S = \uparrow (S_1 - S_2)$. If $C_i \notin S_1$, then there exists $C_0 \in \mathbf{C}, C_0 \subseteq C_i$ *and* $C_0 \in S_1 - S_2$. Obviously $C_0 \subseteq K$, so $K \in S = \uparrow (S_1 - S_2)$. To sum up, $K \in S$.

Proposition 5. *Let $\mathbf{C}$ be a covering of a domain U and $K \in \mathbf{C}$. K is reducible in $\mathbf{C}$ if and only if K is reducible in S.*

Proof. If K is reducible in $\mathbf{C}$, then there exists $\{C_1, \cdots, C_m\} \subseteq \mathbf{C}$, such that $K = \cup_{i=1}^{m} \{C_i\}$. The following work is to prove $\{K, C_1, \cdots, C_m\} \subseteq S$. According to the above proposition, $K \in S$, then we only need to prove $\{C_1, \cdots, C_m\} \subseteq S$. In fact, it is easy to get $C_i \subseteq K$ for every i, then obviously $C_i \notin S_2$. For a particular i, in the one circumstance, if $C_i \in S_1$, then $C_i \in S_1 - S_2$, and then $C_i \in \uparrow (S_1 - S_2) = S$. In the other circumstance, if $C_i \notin S_1$, then $\exists C_0 \in \mathbf{C}, C_0 \subseteq C_i$ *and* $C_0 \in S_1 - S_2$, and then $C_i \in \uparrow (S_1 - S_2) = S$. To summarize, $C_i \in S$, $\{C_1, \cdots, C_m\} \subseteq S$.

On the other hand, suppose K is reducible in S, then K is reducible in $\mathbf{C}$ since $S \subseteq \mathbf{C}$.

The two above propositions are meaningful, because getting $Reduct(\mathbf{C})$ is diverted to getting its subset $Reduct(S)$.

Remark 1. If $\mathbf{C}$ be a covering of a domain U, then $(\mathbf{C}, \subseteq)$ is a poset, S a covering of $\cup S$, and $(S, \subseteq)$ a poset. So we can define similarly upsets and downsets on S.

Corollary 1. *Let $\mathbf{C}$ be a covering of a domain U. If K is reducible in $\mathbf{C}$, then there exists $\{A_1, A_2, \cdots, A_m\} \subseteq S, m \geq 2$, such that $K \in \bigcap_{i=1}^{m} \uparrow \{A_i\}$.*

Corollary 2. *Let $\mathbf{C}$ be a covering of a domain U. If $S_1 = S_2$, then $Reduct(\mathbf{C}) = S_1$.*

Corollary 3. *Let $\mathbf{C}$ be a covering of a domain U. If $Reduct(S)$ is the reduct of S, then $Reduct(\mathbf{C}) = \mathbf{C} - (S - Reduct(S))$*

Example 3. If $U = \{x_1, x_2, \cdots, x_{20}\}$,
$K_1 = \{x_1, x_2, \cdots, x_5\}$, $K_2 = \{x_6, x_7\}$, $K_3 = \{x_8, x_9, x_{10}\}$,
$K_4 = \{x_{11}, x_{12}, x_{13}, x_{14}\}$, $K_5 = \{x_{15}, x_{16}\}$, $K_6 = \{x_{16}, \cdots, x_{20}\}$,
$K_7 = \{x_1, x_2, \cdots, x_7\}$, $K_8 = \{x_{15}, \cdots, x_{20}\}$,
$\mathbf{C} = \{K_1, \cdots, K_8\}$,
then
$S_1 = \{C_i \in \mathbf{C} | \downarrow \{C_i\} = \{C_i\}\} = \{K_1, K_2, K_3, K_4, K_5, K_6\}$,
$S_2 = \{C_i \in \mathbf{C} | \uparrow \{C_i\} = \{C_i\}\} = \{K_3, K_4, K_7, K_8\}$,
$S_1 - S_2 = \{K_1, K_2, K_5, K_6\}$,
$S = \uparrow (S_1 - S_2) = \{K_1, K_2, K_5, K_6, K_7, K_8\}$,
$\uparrow \{K_1\} = \{K_1, K_7\}$, $\uparrow \{K_2\} = \{K_2, K_7\}$,
$\uparrow \{K_5\} = \{K_5, K_8\}$, $\uparrow \{K_6\} = \{K_6, K_8\}$,

$K_7 \in \uparrow \{K_1\} \cap \uparrow \{K_2\}$,
$K_8 \in \uparrow \{K_5\} \cap \uparrow \{K_6\}$.
So the possible reducible elements are K_7, K_8. Furthermore, K_7, K_8 are reducible since $K_7 = K_1 \cup K_2$, $K_8 = K_5 \cup K_6$.

4   Conclusions

This paper studies some core concepts such as reducible elements in covering-based rough sets. With those upsets and downsets based on a poset, we get some important and interesting conclusions which can be seen a new way to describe reducible elements. To investigate further properties and other concepts in covering-based rough sets is our future works.

Acknowledgments

This work is in part supported by National Nature Science Foundation of China under grant No.60873077/F020107 and the Public Fund from the Key Laboratory of Complex System and Intelligence Science, Institute of Automation, Chinese Academy of Science under grant No.20070105.

References

1. Zhu, W.: Generalized rough sets based on relations. Information Sciences 177, 4997–5011 (2007)
2. Zhu, W.: Relationship between generalized rough sets based on binary relation and covering. Information Sciences 179, 210–225 (2009)
3. Zhu, W.: Relationship among basic concepts in covering-based rough sets. Information Sciences 17, 2478–2486 (2009)
4. Zhu, W.: Topological approaches to covering rough sets. Information Sciences 177, 1499–1508 (2007)
5. Pawlak, Z., Skowron, A.: Rough sets and boolean reasoning. Information Sciences 177, 41–73 (2007)
6. Yao, Y.Y.: Constructive and algebraic methods of theory of rough sets. Information Sciences 109, 21–47 (1998)
7. Zhu, W., Wang, F.Y.: On three types of covering rough sets. IEEE Transactions on Knowledge and Data Engineering 19, 1131–1144 (2007)
8. Zhu, W., Wang, F.Y.: Reduction and axiomization of covering generalized rough sets. Information Sciences 152, 217–230 (2003)
9. Lu, S.: The measure on the continuous poset. Journal of Sichuan University 41, 1137–1142 (2004)
10. Bonikowski, Z., Bryniarski, E., Wybraniec-Skardowska, U.: Extensions and intentions in the rough set theory. Information Sciences 107, 149–167 (1998)

1-vs-Others Rough Decision Forest

Jinmao Wei[1], Shuqin Wang[2], and Guoying Wang

[1] College of Information Technical Science, Nankai University,
Tianjin 300071, China
weijm@nankai.edu.cn
[2] College of Computer and Information Engineering, Tianjin Normal University,
Tianjin 300387, China
wangsq562@nenu.edu.cn

Abstract. Bootstrap, boosting and subspace are popular techniques for inducing decision forests. In all the techniques, each single decision tree is induced in the same way as that for inducing a decision tree on the whole data, in which all possible classes are dealt with together. In such induced trees, some minority classes may be covered up by others when some branches grow or are pruned. For a multi-class problem, this paper proposes to induce individually the 1-vs-others rough decision trees for all classes, and finally construct a rough decision forest, intending to reduce the possible side effects of imbalanced class distribution. Since all training samples are reused to construct the rough decision trees for all classes, the method also tends to have the merits of bootstrap, boosting and subspace. Experimental results and comparisons on some hard gene expression data show the attractiveness of the method.

Keywords: Classification, Decision tree, Decision forest, Rough set.

1 Introduction

Decision tree technique has been widely studied and exploited in classifications for its simplicity and effectiveness. Many methods have been introduced for constructing decision trees[1], [2], [3], [4], [5]. For remedying the side effects caused by the inductive bias of single decision trees and for inducing decision trees from large scale of data, etc., many researchers have investigated to construct decision forests for classification. In [6], [7], the author proposed to form multiple versions of trees by making bootstrap replicates of the learning set for improving decision making. In [8], the authors also suggested to use two voting methods, i.e. bootstrap and randomization, for generating multiple hypotheses to reduce variance of individual decision trees. In [9], the authors investigated to improve the predictive power of classifiers by boosting. And in [10], the author exploited both bootstrap and boosting for enhancing the predictive power of C4.5. Both bootstrap and boosting techniques have been introduced for remedying the side effects of changes of sample distribution. In the two techniques, training samples are reused by resampling randomly or emphasizing some samples. In another line, the author proposed to pseudo-randomly select subsets of components of

J. Yu et al. (Eds.): RSKT 2010, LNAI 6401, pp. 30–37, 2010.

the feature vector to construct subtrees[11]. In this technique, single decision trees are induced from different projected spaces which are expanded by different subsets of condition attributes. In all the above techniques, the process of inducing the single decision trees of the constructing decision forests is the same as the process of inducing decision trees on the whole training data sets, in which all possible classes are dealt with together with each other. As one may agree, the number of samples of one class may be considerably larger than that of some other classes in a real data. When inducing a single decision tree with all classes dealt with together, the majority classes may cover up those minority classes in some branches or in the process of pruning. In light of this observation, this paper proposes to construct rough decision forests in a 1-vs-others way. By individually dealing with different classes and constructing 1-vs-others rough decision tree for each class, the induced rough decision forest may reduce the possible bias caused by imbalanced class distribution or minority classes.

2 1-vs-Others Rough Decision Forest

In this section, we will discuss how to construct 1-vs-others rough decision forests based on the rough set theory. Rough set theory was introduced by Polish mathematician Pawlak in 1982[12]. It has been taken as a new mathematic tool to deal with vagueness and uncertainty after the fuzzy set theory. The theory has been widely studied and employed in the fields of machine learning, pattern recognition, data mining, etc. In [13], [14], the authors proposed to induce decision trees based on the rough set theory. The criterion for forming the set of attributes for splitting is based on sizes of explicit regions. If the size of the explicit region of a condition attribute is larger than that of the other attributes, this attribute will be chosen for splitting. In [15], [16], [17], etc, the authors discussed the construction of another type of rough set(or reduct) based decision tree(RDT) based on the reduct of condition attribute set. Compared with such a decision tree, the process of inducing a rough decision tree using the criterion of maximal explicit region is a wrapper method. For inducing 1-vs-others rough decision forests for multi-class problems, we firstly review the basic definition of explicit region for constructing single decision trees. Given a knowledge representation system $S = (U, C \bigcup D, V, \rho)$, the explicit region was defined as[14]:

Definition:

Let $A \subseteq C$, $B \subseteq D$, $A^* = \{X_1, X_2, \ldots, X_n\}$, $B^* = \{Y_1, Y_2, \ldots, Y_m\}$ denote the partitions of U induced by equivalence relation $\widetilde{A}$ and $\widetilde{B}$ respectively, where equivalence relation $\widetilde{A}$ and $\widetilde{B}$ are induced from A and B. The explicit region is defined as $Exp_A(B^*) = \bigcup_{Y_i \in B^*} \underline{A}(Y_i)$, where $\underline{A}(Y_i)$ is the lower approximation of Y_i with respect to $\widetilde{A}$.

As other decision tree techniques, rough decision trees induced according to the criterion of maximal explicit region also deal with all classes of a multi-class problem together with each other. One can also employ bootstrap, boosting and subspace techniques to construct a rough decision forest. Yet, as mentioned above, for a real problem, things can be more complicated. As is well known to any practitioners, the samples within a class label(a leaf node) of a decision tree are often impure. The class label is determined according to the criterion of majority. The minority samples with the same class label within different leaf nodes may be covered up by others, which may consequently affect the performance of the induced tree in some way. For removing such possible side effects, we propose to construct single rough decision trees in the 1-vs-others way as shown below.

Given a k-class problem, the set of all possible classes is $\{d_1, d_2, \ldots, d_k\}$. We firstly convert the problem into k 1-vs-others sub-problems. Each converted sub-problem is denoted as S_i, $D_{S_i} = \{d_i, \bigcup_{k \neq i} d_k\}$, $D_{S_i} = \{d_i, \overline{d_i}\}$ for simplicity. For each sub-problem S_i, the explicit region of each of all candidate condition attributes, $a \in C$, can then be computed as $Exp_a(D_{S_i}^*) = \bigcup_{Y_i \in D_{S_i}^*} \underline{a}(Y_i)$. If an attribute a has the biggest explicit region, it is sifted to split the data sets. We now have constructed one 'class d_i leaf node'. The data set is subsequently changed to be $U - \underline{a}(d_i)$. And C is changed to be $C - \{a\}$. Here, $\underline{a}(d_i)$ simply denotes the lower approximation of equivalence class $\{s|_{s \in U} \wedge \rho(s,d) = d_i\}$ with respect to attribute a. If no sample of class d_i exists in the new data set, the tree stops growing. Otherwise, the tree continues growing based on the new data set. Consequently, a 1-vs-others rough decision tree for class d_i can be induced according to the criterion of maximal explicit region. For each other class, we evaluate similarly to construct the corresponding 1-vs-others tree, and finally construct the rough decision forest.

The induced subtrees are denoted as $\{T_1, T_2, \ldots, T_k\}$. T_i can be called 'the d_i-vs-other classes tree', which has one 'other classes node' and some 'class d_i node'. An unseen sample is tested on all the k subtrees and predicted to be class d_i if it reaches one of the class d_i nodes of subtree T_i. The constructed subtree T_i and the constructed decision forest may look like that shown in Fig. 1.

Since each class d_i corresponds to a d_i-vs-others subtree induced based upon the whole original data set, which is used exclusively for predicting class d_i, the possible side effects, such as some majority classes may cover up some minority classes, tend to be reduced. In addition, in the process of inducing the rough decision forest, all samples are used k times. From this point of view, such process tends to have the merits of bootstrap and boosting techniques, in which

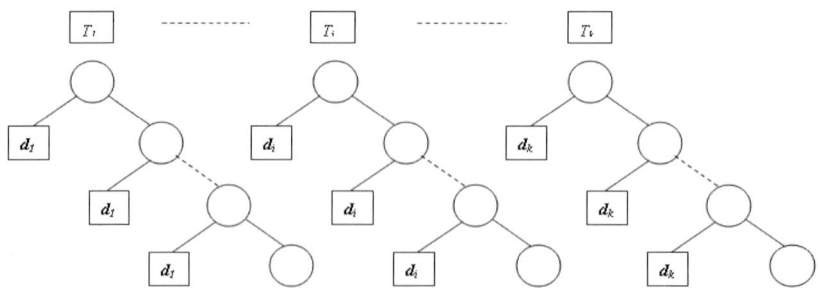

Fig. 1. 1-vs-others rough decision forest

some samples are reused for more than one time. Furthermore, when inducing each 1-vs-others subtree, all possible condition attributes are evaluated. The sifted splitting attributes in different subtrees for different classes may also be different from that of each other. In this sense, the process of inducing the 1-vs-others rough decision forest also has the merit of the subspace technique, in which data set is projected to different condition attribute subset spaces. Hence rough decision forest technique may be employed for achieving high classification performance.

It should be mentioned, the constructed decision forest is different from another type of 1-vs-others decision tree. Such a tree is also a binary tree, which has all kinds of class nodes. As any other kinds of decision trees, all classes are predicted by the single tree. In comparison, a 1-vs-others rough decision tree corresponds to only one class. It in fact can be taken as a one-class classifier. Consequently, for an unseen sample, only one class subtree of all the 1-vs-others subtrees outputs prediction. This is obviously different from all kinds of decision forests in which classification are made by voting. Since the number of a 1-vs-others rough decision tree is equivalent to the number of classes, the scale of the rough decision forest for a given problem is hence deterministic. This implies, such 1-vs-others rough decision forest technique is merely proper for multi-class problems.

3   Experimental Comparison

For demonstrating the feasibility of the proposed method, we conducted some experiments on some biomedicine data sets, which are hard for classification. We chose four gene expression data sets about cancer classification from [18]. The detailed descriptions of the data sets are listed in Table 1. All the original data sets and their descriptions can be found in [19], [20], [21], [22].

Table 1. Gene expression Data sets

Data sets	TrN	TeN	CN	DN
SRBCT	62	24	2308	5
GCM	143	45	16063	15
Cancers	99	73	12533	11
Breast	53	29	9216	5

In the table, 'TrN', 'TeN', 'CN' and 'DN' indicate the number of training samples, test samples, condition attributes(genes) and decision attributes(cancer types) respectively. Generally speaking, all the four problems tend to be hard for classification, for the number of training samples of each problem is relatively smaller than the number of condition attributes. We tried to construct 1-vs-others rough decision forests for the four problems. Firstly before decision tree construction, we discretized each gene attribute with DiscretVPRS, which was introduced as a new discretization method based on the rough set theory[23]. After inducing the decision forests we also employed ESP pruning method, which is a new pruning method taking into account both accuracy and structural complexity of decision trees[24], to prune all the induced 1-vs-others decision trees. All the classification results before and after pruning of the rough decision forests obtained by the proposed method are shown in Table 2. For comparison, we also list the results obtained by single rough decision trees constructed by the method proposed in [14]. In the table, 'acc1' and 'acc2' indicate the classification accu-

Table 2. Classification accuracies before and after pruning

Data sets	acc1	acc2	acc3	acc4
SRBCT	87.1	87.1	91.9	100
GCM	75.6	44.6	36	100
Cancers	78.1	21	64.4	95.9
Breast	41.3	35	69	93.1

racy of single rough decision tree before and after pruning. 'acc3' and 'acc4' indicate the classification accuracy of 1-vs-others rough decision forest before and after pruning. For comparison, the classification results obtained by some popular classification models are listed in Table 3.

Table 3. Classification accuracies of other classification methods

Data sets	C4.5	NB	k-NN	SVM	Randomforest
SRBCT	75	60	30	100	64
GCM	52.17	52.17	34.78	65.22	34.78
Cancers	68.92	79.73	64.86	83.78	39.19
Breast	73.33	66.67	63.33	83.33	63.33

From the results shown in Table 3 we can clearly see, all the four problems are indeed hard for classification. Except for SVM, the accuracies of all the other four classification methods for the four problems are under 80%. In comparison, the 1-vs-others rough decision forests after pruning obtained relatively higher accuracies for all the four problems. One may notice in Table 2, the accuracies before pruning are higher than the accuracies after pruning for the single rough decision trees. Such results are negative to what one may have expected by exploiting pruning methods. That is, in general cases, one may expect to achieve high classification performance by pruning the constructed decision trees or decision forests. Many research works on pruning decision trees have shown that decision trees after pruning tend to achieve high classification accuracies. Pruning methods are usually employed for overwhelming the problem of overfitting in the training stage and enhancing the generalization ability of classifiers for predicting unseen data. With the results reported in this paper we don't intend to announce that the conclusions from the many concrete experimental works had reached the wrong end. In stead, we guess such results may be caused by the data sets themselves. As we have seen, the number of training samples of each of the four problems is relatively small. Pruning may neglect minority classes for reducing the scale of decision trees. Furthermore, small training data sets also mean even small training data for each class. Such small training data for each class may affect the construction of decision trees. In addition, if the class distribution is imbalanced, the majority of some classes may probably affect some other minority classes when evaluating attributes and inducing single decision trees. As we have discussed above, 1-vs-others rough decision forest technique deals with different classes individually. The side effects of minority class and imbalanced class distribution tend to be reduced. In addition, 1-vs-others rough decision forests reuse the whole data sets k times for training classifiers. The side effects of small size samples also tend to be reduced.

Akin to [18], for considering biologically meaningful rules, we also report the number of condition attributes(genes) sifted for constructing the rough decision forests. The numbers of genes sifted for each of all classes by the rough decision forests after pruning for the four problems are shown in Table 4. The numbers of genes sifted by other classification methods are shown in Table 5.

From Table 4 one can see, the number of genes for each pruned rough decision tree is different from that of the others. Yet, for most classes of each problem, the number of genes for one class does not change more violently than that for the other classes. For the case of one gene for some classes, the constructed decision trees only sifted one gene for classifying the samples of that classes. For the case of more than one gene, it means more genes were sifted. It should be noticed, all the sifted genes for one class will not be simultaneously evaluated for predicting unseen data of this class. For example, if the gene expression value is within the range for reaching the leaf node under the root node, only the gene at the root node will be used. Otherwise, more other genes will be used for

Table 4. Number of genes sifted by rough decision forests

Data sets	c1	c2	c3	c4	c5	c6	c7	c8	c9	c10	c11	c12	c13	c14	c15
SRBCT	9	8	8	9	1										
GCM	4	5	1	5	5	5	4	1	1	5	1	5	5	5	4
Cancers	4	5	5	5	5	5	5	5	4	5	9				
Breast	20	1	5	7	7										

Table 5. Number of genes sifted by other classification methods

Data sets	C4.5	NB	k-NN	SVM	Randomforest
SRBCT	3	All	All	All	120
GCM	18	All	All	All	140
Cancers	10	All	All	All	140
Breast	4	All	All	All	140

predicting. From Table 5 we can see, 'NB', 'k-NN' and 'SVM' used all genes. Though 'Randomforest' did not use all genes, it in fact randomly sifted genes. Hence, from the biological point of view, all the four methods are not suitable for sifting functional genes and for generating biologically meaningful rules for classification. Though the number of genes sifted by 'C4.5' is smaller than that sifted by the 1-vs-others rough decision forests, the accuracies are relatively lower. All the results show the proposed method is feasible for handling such problems.

From Table 4 we can see, for most classes of the problems, several genes were involved. That is to say, for those classes, the involved genes may be functional for those types of cancers. Yet, what biological and medical information for classifying cancers were informed by the genes sifted in the process of inducing rough decision forests may need further biological efforts, which is beyond the topic of this work.

4   Conclusions

In this paper, the technique of 1-vs-others rough decision forest is introduced for classification. A rough decision forest can be induced based on the 1-vs-others rough decision trees, which are constructed individually for each of all classes based on the criterion of maximal explicit region. With their class decision trees induced individually, minority classes may not be covered up by other classes or affect the construction of decision trees. Since all training samples are reused for inducing the 1-vs-others decision trees for all classes, the rough decision forest technique also tends to own the merits of bootstrap and boosting as well as the merit of subspace technique. Experimental results on some gene expression data demonstrate the attractiveness of the proposed method. Further efforts can be devoted to investigating how minority classes may affect in the process of inducing single decision trees when all classes are dealt with together.

Acknowledgments

This work was supported by the Science Foundation of Tianjin under grant 10JCZDJC15800, and the National 863 High Technology Research and Development Program of China under grant 2009AA01Z152.

References

1. Breiman, L., Friedman, J.H., Olshen, R.A., Stone, C.J.: Classification and Regression Trees. Technical report. Wadsworth International, Monterey, CA (1984)
2. Quinlan, J.R.: Introduction of Decision Trees. Machine Learning 3, 81–106 (1986)
3. Quinlan, J.R.: C4.5: Programs for Machine Learning. Morgan Kaufmann, San Francisco (1993)
4. Quinlan, J.R., Rivest, R.: Inferring Decision Trees Using the Minimum Description Length Principle. Information and Computation 80, 227–248 (1989)

5. Fayyad, U.M., Irani, K.B.: On the Handling of Continuous-valued Attributes in Decision Tree Generation. Machine Learning 8, 87–102 (1992)
6. Breiman, L.: Bagging Predictors. Machine Learning 24, 123–140 (1996)
7. Breiman, L.: Random Forests. Machine Learning 45, 5–32 (2001)
8. Dietterich, T.G., Kong, E.B.: Machine Learning Bias, Statistical Bias and Statistical Variance of Decision Tree Algorithms. Technical Report. Oregon State University (1995)
9. Freund, Y., Schapire, R.E.: A Decision-theoretic Generalization of On-line Learning and an Application to Boosting. Journal of Computer and System Sciences 55, 119–139 (1997)
10. Quinlan, J.R.: Bagging, Boosting, and C4.5. In: Proceedings, Fourteenth National Conference on Artificial Intelligence (1996)
11. Ho, T.K.: The Random Subspace Method for Constructing Decision Forests. IEEE Transactions on Pattern Analysis and Machine Intelligence 20, 832–844 (1998)
12. Pawlak, Z.: Rough sets. International Journal of Computer and Information Science 11, 341–356 (1982)
13. Wei, J.M.: Rough Set Based Approach to Selection of Node. International Journal of Computational Cognition 1, 25–40 (2003)
14. Wei, J.M., Wang, S.Q., et al.: Rough Set Based Approach for Inducing Decision Trees. Knowledge-Based Systems 20, 695–702 (2007)
15. Yellasiri, R., Rao, C.R., Reddy, V.: Decision Tree Induction Using Rough Set Theory: Comparative Study. Journal of Theoretical and Applied Information Technology 3, 110–114 (2005)
16. Minz, S., Jain, R.: Rough Set Based Decision Tree Model for Classification. In: Kambayashi, Y., Mohania, M., Wöß, W. (eds.) DaWaK 2003. LNCS, vol. 2737, pp. 172–181. Springer, Heidelberg (2003)
17. Minz, S., Jain, R.: Refining Decision Tree Classifiers Using Rough Set Tools. International Journal of Hybrid Intelligent Systems 2, 133–148 (2005)
18. Tan, A.C., Naiman, D.Q., Xu, L., Winclow, R.L., Geman, D.: Simple Decision Rules for Classifying Human Cancers from Gene Expression Profiles. Bioinformatics 21, 3896–3904 (2005)
19. Khan, J., et al.: Classification and Diagnostic Prediction of Cancers Using Gene Expression Profiling and Artificial Neural Networks. Nat. Med. 7, 673–679 (2001)
20. Ramaswamy, S., et al.: Multiclass Cancer Diagnosis Using Tumor Gene Expression Signatures. Proc. Natl. Acad. Sci. USA 98, 15149–15154 (2001)
21. Su, A.I., et al.: Molecular Classification of Human Carcinomas by Use of Gene Expression Signatures. Cancer Res. 61, 7388–7393 (2001)
22. Perou, C.M., et al.: Molecular Portraits of Human Breast Tumours. Nature 406, 747–752 (2000)
23. Wei, J.M., Wang, G.Y., Kong, X.M., Li, S.J., Wang, S.Q.: A New Method for Discretization of Continuous Attributes Based on VPRS. LNCS, pp. 183–190. Springer, Heidelberg (2006)
24. Wei, J.M., Wang, S.Q., Yu, G., Gu, L., Wang, G.Y., Yuan, X.J.: A Novel Method for Pruning Decision Trees. In: ICMLC 2009, pp. 339–343 (2009)

Knowledge Reduction in Random Incomplete Information Systems via Evidence Theory

Wei-Zhi Wu

School of Mathematics, Physics and Information Science,
Zhejiang Ocean University, Zhoushan, Zhejiang, 316004, P.R. China
wuwz@zjou.edu.cn

Abstract. Knowledge reduction is one of the main problems in the study of rough set theory. This paper deals with knowledge reduction in random incomplete information systems based on Dempster-Shafer theory of evidence. The concepts of random belief reducts and random plausibility reducts in random incomplete information systems are introduced. The relationships among the random belief reduct, the random plausibility reduct, and the classical reduct are examined. It is proved that, in a random incomplete information system, an attribute set is a random belief reduct if and only if it is a classical reduct, and a random plausibility consistent set must be a consistent set.

Keywords: Belief functions, incomplete information systems, knowledge reduction, random incomplete information systems, rough sets.

1 Introduction

Imprecision and uncertainty are two important aspects of incompleteness of information. One theory for the study of insufficient and incomplete information in intelligent systems is rough set theory [9]. The basic structure of rough set theory is an approximation space consisting of a universe of discourse and a binary relation imposed on it. Based on the approximation space, the primitive notion of a pair of lower and upper approximations can be induced. Using the concepts of lower and upper approximations, knowledge hidden in information systems may be unravelled and expressed in the form of decision rules. Another important method used to deal with uncertainty in information systems is the Dempster-Shafer theory of evidence [10]. The basic representational structure in this theory is a belief structure, which consists of a family of subsets, called focal elements, with associated individual positive weights summing to one. The fundamental numeric measures derived from the belief structure are a dual pair of belief and plausibility functions. It has been demonstrated that various belief structures are associated with various approximation spaces such that the different dual pairs of lower and upper approximation operators induced by approximation spaces may be used to interpret the corresponding dual pairs of belief and plausibility functions induced by belief structures [11,14,15,18].

J. Yu et al. (Eds.): RSKT 2010, LNAI 6401, pp. 38–45, 2010.

The basic idea of rough set data analysis is to unravel an optimal set of decision rules from an information system (basically an attribute-value table) via an objective knowledge induction process which determines the necessary and sufficient attributes constituting the rules for decision making. Knowledge reduction is thus an outstanding contribution made by rough set research to data analysis. In recent years, many authors proposed various concepts of reducts in classical information systems in rough set research, each of which aimed at some basic requirements [1,5,7,8,12,17,20,21]. These types of reducts are based on the classical Pawlak rough-set data analysis which uses equivalence relations in complete information systems. Pawlak's rough set model may be generalized to nonequivalence relations which may be used in reasoning and knowledge acquisition in incomplete information systems [2,3,4,6].

One can see that the lower and upper approximations of a set characterize the non-numeric aspect of the set expressed by the available information, and the belief and plausibility measures of the same set reflect numeric aspects of uncertain knowledge. Thus, the Dempster-Shafer theory of evidence may be used to analyze knowledge acquisition in information systems. Zhang et al. [20] are the first who proposed the concepts of belief reduct and plausibility reduct in complete information systems. Wu et al. [17] discussed knowledge reduction in random complete decision systems via the Dempster-Shafer theory of evidence. Lingras and Yao [6] employed two different generalizations of rough set models to generate plausibility rules with incomplete databases instead of probabilistic rules generated by a Pawlak's rough set model with complete decision tables. Wu and Mi [16] studied knowledge reduction in incomplete information systems within evidence theory. Wu [13] further used belief and plausibility functions to discuss attribute reducts in incomplete decision systems. Since the available incomplete database may be obtained by a randomization method, in this paper we discuss attribute reduction in random incomplete systems within the Dempster-Shafer theory of evidence.

2   Random Incomplete Information Systems and Rough Set Approximations

The notion of information systems provides a convenient tool for the representation of objects in terms of their attribute values.

An information system is a pair (U, AT), where $U = \{x_1, x_2, \ldots, x_n\}$ is a nonempty, finite set of objects called the universe and $AT = \{a_1, a_2, \ldots, a_m\}$ is a non-empty, finite set of attributes, such that $a : U \to V_a$ for any $a \in AT$, where V_a is called the domain of a. When the precise values of some of the attributes in an information system are not known, i.e., missing or known partially, then such a system is called an incomplete information system (IIS) and is still denoted without confusion by $S = (U, AT)$. Such a situation can be described by a set-based information system [4] in which the attribute value function a is defined as a mapping from U to the power set $\mathcal{P}(V_a)$ of V_a where there is an uncertainty on what values an attribute should take but the set of acceptable values can be clearly specified. For example, the missing values $a(x)$ can be represented by the set of all possible values for the attribute, i.e., $a(x) = V_a$; and if $a(x)$ is known partially, for instance, if we know that $a(x)$ is not $b, c \in V_a$ (for example, "the color was red or yellow but not black or white"), then the value $a(x)$ is specified as $V_a - \{b, c\}$.

If P is a normal probability measure on U, that is, $P(x) := P(\{x\}) > 0$ for all $x \in U$ and $\sum_{x \in U} P(x) = 1$, then the triple (U, P, AT) is called a random IIS. It should be noted that an incomplete information system may be treated as a random IIS with a special probability $P(x) = 1/|U|$ for all $x \in U$.

Example 1. Table 1 depicts a random IIS (U, P, AT) with missing values containing information about cars in which (U, AT) is a modification in [3]. The associated random set-based information system is given as Table 2. From Table 1 we have: $U = \{x_1, x_2, x_3, x_4, x_5, x_6\}$, $AT = \{P, M, S, X\}$, where P, M, S, X stand for Price, Mileage, Size, Max-Speed respectively, and the probability P on U is defined as follows (see the second column of the Table 1): $P(x_1) = 3/24$, $P(x_2) = \pi/24$, $P(x_3) = (5 - \pi)/24$, $P(x_4) = 4/24$, $P(x_5) = 7/24$, and $P(x_6) = 5/24$. The attribute domains are as follows: $V_X = \{\text{High, Low}\}$, $V_M = \{\text{High, Low}\}$, $V_S = \{\text{Full, Compact}\}$, $V_X = \{\text{High, Low}\}$.

For a random IIS (U, P, AT) which is represented as a set-based information system, each nonempty subset $A \subseteq AT$ determines a similarity relation:

$$R_A = \{(x, y) \in U \times U : a(x) \cap a(y) \neq \emptyset, \forall a \in A\}.$$

We denote $S_A(x) = \{y \in U : (x, y) \in R_A\}$, $S_A(x)$ is called the similarity class of x w.r.t. A in S, the family of all similarity classes w.r.t. A is denoted by U/R_A, i.e., $U/R_A = \{S_A(x) : x \in U\}$.

Table 1. An exemplary random incomplete information system

Car	P	Price	Mileage	Size	Max-Speed
x_1	3/24	High	Low	Full	Low
x_2	$\pi/24$	Low	*	Full	Low
x_3	$(5 - \pi)/24$	*	*	Compact	Low
x_4	4/24	High	*	Full	High
x_5	7/24	*	*	Full	High
x_6	5/24	Low	High	Full	*

Table 2. A random set-based information system

Car	P	Price	Mileage	Size	Max-Speed
x_1	3/24	{High}	{Low}	{Full}	{Low}
x_2	$\pi/24$	{Low}	{Low, High}	{Full}	{Low}
x_3	$(5 - \pi)/24$	{Low, High}	{Low, High}	{Compact}	{Low}
x_4	4/24	{High}	{Low, High}	{Full}	{High}
x_5	7/24	{Low, High}	{Low, High}	{Full}	{High}
x_6	5/24	{Low}	{High}	{Full}	{Low, High}

Property 1. Let (U, P, AT) be a random IIS, and $B \subseteq A \subseteq AT$, then $S_A(x) \subseteq S_B(x)$ for all $x \in U$.

Example 2. In Example 1, the similarity classes determined by AT are:
$S_{AT}(x_1) = \{x_1\}$, $S_{AT}(x_2) = \{x_2, x_6\}$, $S_{AT}(x_3) = \{x_3\}$,
$S_{AT}(x_4) = \{x_4, x_5\}$, $S_{AT}(x_5) = \{x_4, x_5, x_6\}$, $S_{AT}(x_6) = \{x_2, x_5, x_6\}$.

For the random IIS (U, P, AT), $A \subseteq AT$, and $X \subseteq U$, one can characterize X by a pair of lower and upper approximations w.r.t. A:

$$\underline{A}(X) = \{x \in U : S_A(x) \subseteq X\}, \quad \overline{A}(X) = \{x \in U : S_A(x) \cap X \neq \emptyset\}.$$

Since a similarity relation is reflexive and symmetric, the approximations have the following properties [19]:

Property 2. Let (U, P, AT) be a random IIS, and $A, B \subseteq AT$, then: $\forall X, Y \in \mathcal{P}(U)$,
   (1) $\underline{A}(X) \subseteq X \subseteq \overline{A}(X)$,
   (2) $\underline{A}(\sim X) =\sim \overline{A}(X)$, where $\sim X$ is the complement of the set X in U,
   (3) $\underline{A}(U) = \overline{A}(U) = U$, $\underline{A}(\emptyset) = \overline{A}(\emptyset) = \emptyset$,
   (4) $\underline{A}(X \cap Y) = \underline{A}(X) \cap \underline{A}(Y)$, $\overline{A}(X \cup Y) = \overline{A}(X) \cup \overline{A}(Y)$,
   (5) $X \subseteq Y \Longrightarrow \underline{A}(X) \subseteq \underline{A}(Y)$, $\overline{A}(X) \subseteq \overline{A}(Y)$,
   (6) $A \subseteq B \subseteq AT \Longrightarrow \underline{A}(X) \subseteq \underline{B}(X)$, $\overline{B}(X) \subseteq \overline{A}(X)$.

Example 3. In Example 1, if we set $X = \{x_2, x_5, x_6\}$, then we can obtain that $\underline{AT}(X) = \{x_2, x_6\}$, and $\overline{AT}(X) = \{x_2, x_4, x_5, x_6\}$.

3   Belief and Plausibility Functions

The basic representational structure in the Dempster-Shafer theory of evidence is a belief structure [10].

Definition 1. *Let U be a non-empty finite set, a set function $m : \mathcal{P}(U) \to [0, 1]$ is referred to as a basic probability assignment, if it satisfies axioms:*

(M1) $m(\emptyset) = 0$, (M2) $\sum\limits_{X \subseteq U} m(X) = 1$.

A set $X \in \mathcal{P}(U)$ with nonzero basic probability assignment is referred to as a focal element. We denote by $\mathcal{M}$ the family of all focal elements of m. The pair $(\mathcal{M}, m)$ is called a belief structure. Associated with each belief structure, a pair of belief and plausibility functions can be defined [10].

Definition 2. *Let $(\mathcal{M}, m)$ be a belief structure. A set function* Bel $: \mathcal{P}(U) \to [0, 1]$ *is called a belief function on U if*

$$\mathrm{Bel}(X) = \sum_{\{M : M \subseteq X\}} m(M) \quad \forall X \in \mathcal{P}(U).$$

A set function Pl $: \mathcal{P}(U) \to [0, 1]$ *is called a plausibility function on U if*

$$\mathrm{Pl}(X) = \sum_{\{M : M \cap X \neq \emptyset\}} m(M) \quad \forall X \in \mathcal{P}(U).$$

Belief and plausibility functions based on the same belief structure are connected by the dual property $\mathrm{Pl}(X) = 1 - \mathrm{Bel}(\sim X)$, and moreover, $\mathrm{Bel}(X) \leq \mathrm{Pl}(X)$ for all $X \in \mathcal{P}(U)$.

Notice that a similarity relation is reflexive, then, according to [14,18], we can conclude the following theorem, which shows that the pair of lower and upper approximation operators w.r.t. an attribute set in a random IIS generates a pair of belief and plausibility functions.

Theorem 1. *Let (U, P, AT) be a random IIS, $A \subseteq AT$, for any $X \subseteq U$, denote*

$$\mathrm{Bel}_A(X) = \mathrm{P}(\underline{A}(X)), \qquad \mathrm{Pl}_A(X) = \mathrm{P}(\overline{A}(X)). \tag{1}$$

Then Bel_A and Pl_A are a dual pair of belief and plausibility functions on U, and the corresponding basic probability assignment is

$$m_A(Y) = \mathrm{P}(j_A(Y)), \quad Y \in \mathcal{P}(U),$$

where $j_A(Y) = \{u \in U : S_A(u) = Y\}$.

Combining Theorem 1 and Property 2 we obtain following

Lemma 1. *Let (U, P, AT) be a random IIS, $B \subseteq A \subseteq AT$, then for any $X \subseteq U$,*

$$\mathrm{Bel}_B(X) \leq \mathrm{Bel}_A(X) \leq \mathrm{P}(X) \leq \mathrm{Pl}_A(X) \leq \mathrm{Pl}_B(X).$$

4   Attribute Reducts in Random Incomplete Information Systems

In this section, we introduce the concepts of random belief reducts and random plausibility reducts in a random IIS and compare them with the existing classical reduct.

Definition 3. *Let $S = (U, \mathrm{P}, AT)$ be a random IIS, then*

(1) an attribute subset $A \subseteq AT$ is referred to as a consistent set of S if $R_A = R_{AT}$. If $B \subseteq AT$ is a consistent set of S and no proper subset of B is a consistent set of S, then B is clled a classical reduct of S.

(2) an attribute subset $A \subseteq AT$ is referred to as a random belief consistent set of S if $\mathrm{Bel}_A(X) = \mathrm{Bel}_{AT}(X)$ for all $X \in U/R_{AT}$. If $B \subseteq AT$ is a random belief consistent set of S and no proper subset of B is a random belief consistent set of S, then B is called a random belief reduct of S.

(3) an attribute subset $A \subseteq AT$ is referred to as a random plausibility consistent set of S if $\mathrm{Pl}_A(X) = \mathrm{Pl}_{AT}(X)$ for all $X \in U/R_{AT}$. If $B \subseteq AT$ is a random plausibility consistent set of S and no proper subset of B is a random plausibility consistent set of S, then B is called a random plausibility reduct of S.

Theorem 2. *Let $S = (U, \mathrm{P}, AT)$ be a random IIS and $A \subseteq AT$. Then*

(1) A is a consistent set of S iff A is a random belief consistent set of S.

(2) A is a classical reduct of S iff A is a random belief reduct of S.

Proof. (1) Assume that A is a consistent set of S. For any $C \in U/R_{AT}$, since $S_A(x) = S_{AT}(x)$ for all $x \in U$, we have

$$S_A(x) \subseteq C \iff S_{AT}(x) \subseteq C.$$

Then, by the definition of lower approximation, we have

$$x \in \underline{A}(C) \iff x \in \underline{AT}(C), \quad x \in U.$$

Hence $\underline{A}(C) = \underline{AT}(C)$ for all $C \in U/R_{AT}$. By Eq. (1), it follows that $\mathrm{Bel}_A(C) = \mathrm{Bel}_{AT}(C)$ for all $C \in U/R_{AT}$. Thus A is a random belief consistent set of S.

Conversely, if A is a belief consistent set of S, that is,

$$\mathrm{Bel}_A(S_{AT}(x)) = \mathrm{Bel}_{AT}(S_{AT}(x)) \quad \forall x \in U.$$

Then

$$\mathrm{P}(\underline{A}(S_{AT}(x))) = \mathrm{P}(\underline{AT}(S_{AT}(x))) \quad \forall x \in U.$$

Notice that P is a normal probability measure on U, by Lemma 1 and Property 2, it can be easy to verify that $\underline{A}(S_{AT}(x)) = \underline{AT}(S_{AT}(x))$ for all $x \in U$. Hence, by the definition of lower approximation, we have

$$\{y \in U : S_A(y) \subseteq S_{AT}(x)\} = \{y \in U : S_{AT}(y) \subseteq S_{AT}(x)\} \quad \forall x \in U.$$

That is,

$$S_A(y) \subseteq S_{AT}(x) \iff S_{AT}(y) \subseteq S_{AT}(x) \quad \forall x, y \in U. \tag{2}$$

Let $y = x$, clearly, $S_{AT}(y) = S_{AT}(x) \subseteq S_{AT}(x)$. Then, by Eq. (2), we have $S_A(x) \subseteq S_{AT}(x)$ for all $x \in U$. Therefore, by Property 1, we conclude that $S_A(x) = S_{AT}(x)$ for all $x \in U$. Thus, A is a consistent set of S.

(2) It follows immediately from (1).

Denote

$$U/R_{AT} = \{C_1, C_2, \ldots, C_t\}, \quad M = \sum_{i=1}^{t} \mathrm{Bel}_{AT}(C_i).$$

In terms of Theorem 2, we can conclude following

Theorem 3. *Let $S = (U, \mathrm{P}, AT)$ be a random IIS and $A \subseteq AT$. Then*

(1) *A is a consistent set of S iff $\sum_{i=1}^{t} \mathrm{Bel}_A(C_i) = M$.*

(2) *A is a classical reduct of S iff $\sum_{i=1}^{t} \mathrm{Bel}_A(C_i) = M$, and for any nonempty proper subset $B \subset A$, $\sum_{i=1}^{t} \mathrm{Bel}_B(C_i) < M$.*

Example 4. In Example 1, it can be calculated that

$$\sum_{i=1}^{6} \mathrm{Bel}_{\{P,S,X\}}(C_i) = \sum_{i=1}^{6} \mathrm{P}(\{\underline{P, S, X}\}(C_i)) = \sum_{i=1}^{6} \mathrm{Bel}_{\{AT\}}(C_i) = (28 + \pi)/24.$$

On the other hand, it can be verified that

$$\sum_{i=1}^{6} \mathrm{Bel}_{\{P,S\}}(C_i) = \mathrm{P}(\{x_2, x_3, x_6\}) = 10/24,$$

$$\sum_{i=1}^{6} \mathrm{Bel}_{\{P,X\}}(C_i) = 2\mathrm{P}(\{x_4\}) = 8/24,$$

$$\sum_{i=1}^{6} \mathrm{Bel}_{\{S,X\}}(C_i) = \mathrm{P}(\{x_3, x_4, x_5\}) = (16 - \pi)/24.$$

Thus, by Theorem 3, we see that $\{P, S, X\}$ is the unique random belief reduct of S. It can also be calculated by the discernibility matrix method [2] that the system has the unique classical reduct $\{P, S, X\}$.

Theorem 4. *Let* $S = (U, \mathrm{P}, AT)$ *be a random IIS and* $A \subseteq AT$. *If* A *is a consistent set of* S, *then* A *is a random plausibility consistent set of* S.

Proof. It is similar to the proof of the analogous result in incomplete information systems as shown in [16].

5   Conclusion

In real-life world, the available incomplete database may be obtained by a randomization method, such an incomplete information system is called a random incomplete information system. In this paper, we have introduced the notions of random belief reduct and random plausibility reduct in a random incomplete information system. We have examined that an attribute set in a random incomplete information system is a classical reduct if and only if it is a random belief reduct. In [16], Wu and Mi studied knowledge reduction in incomplete information systems within evidence theory. This paper may be viewed as an extension of [16] from an incomplete information system to an arbitrary random incomplete information system. We will investigate knowledge reduction and knowledge acquisition in random incomplete decision systems in our further study.

Acknowledgement

This work was supported by grants from the National Natural Science Foundation of China (Nos. 60673096 and 60773174) and the Natural Science Foundation of Zhejiang Province in China (No. Y107262).

References

1. Beynon, M.: Reducts within the variable precision rough sets model: A further investigation. European Journal of Operational Research 134, 592–605 (2001)
2. Kryszkiewicz, M.: Rough set approach to incomplete information systems. Information Sciences 112, 39–49 (1998)
3. Kryszkiewicz, M.: Rules in incomplete information systems. Information Sciences 113, 271–292 (1999)

4. Leung, Y., Wu, W.-Z., Zhang, W.-X.: Knowledge acquisition in incomplete information systems: A rough set approach. European Journal of Operational Research 168, 164–180 (2006)
5. Li, D.Y., Zhang, B., Leung, Y.: On knowledge reduction in inconsistent decision information systems. International Journal of Uncertainty, Fuzziness and Knowledge-Based Systems 12, 651–672 (2004)
6. Lingras, P.J., Yao, Y.Y.: Data mining using extensions of the rough set model. Journal of the American Society for Information Science 49, 415–422 (1998)
7. Mi, J.-S., Wu, W.-Z., Zhang, W.-X.: Approaches to knowledge reductions based on variable precision rough sets model. Information Sciences 159, 255–272 (2004)
8. Nguyen, H.S., Slezak, D.: Approximation reducts and association rules correspondence and complexity results. In: Zhong, N., Skowron, A., Ohsuga, S. (eds.) RSFDGrC 1999. LNCS (LNAI), vol. 1711, pp. 137–145. Springer, Heidelberg (1999)
9. Pawlak, Z.: Rough Sets: Theoretical Aspects of Reasoning about Data. Kluwer Academic Publishers, Boston (1991)
10. Shafer, G.: A Mathematical Theory of Evidence. Princeton University Press, Princeton (1976)
11. Skowron, A.: The rough sets theory and evidence theory. Fundamenta Informaticae 13, 245–262 (1990)
12. Slezak, D.: Searching for dynamic reducts in inconsistent decision tables. In: Proceedings of IPMU 1998, Paris, France, vol. 2, pp. 1362–1369 (1998)
13. Wu, W.-Z.: Attribute reduction based on evidence theory in incomplete decision systems. Information Sciences 178, 1355–1371 (2008)
14. Wu, W.-Z., Leung, Y., Zhang, W.-X.: Connections between rough set theory and Dempster-Shafer theory of evidence. International Journal of General Systems 31, 405–430 (2002)
15. Wu, W.-Z., Leung, Y., Mi, J.-S.: On generalized fuzzy belief functions in infinite spaces. IEEE Transactions on Fuzzy Systems 17, 385–397 (2009)
16. Wu, W.-Z., Mi, J.-S.: Knowledge reduction in incomplete information systems based on Dempster-Shafer theory of evidence. In: Wang, G.-Y., Peters, J.F., Skowron, A., Yao, Y. (eds.) RSKT 2006. LNCS (LNAI), vol. 4062, pp. 254–261. Springer, Heidelberg (2006)
17. Wu, W.-Z., Zhang, M., Li, H.-Z., Mi, J.-S.: Knowledge reduction in random information systems via Dempster-Shafer theory of evidence. Information Sciences 174, 143–164 (2005)
18. Yao, Y.Y.: Interpretations of belief functions in the theory of rough sets. Information Sciences 104, 81–106 (1998)
19. Yao, Y.Y.: Generalized rough set models. In: Polkowski, L., Skowron, A. (eds.) Rough Sets in Knowledge Discovery: 1. Methodology and Applications, pp. 286–318. Physica, Heidelberg (1998)
20. Zhang, M., Xu, L.D., Zhang, W.-X., Li, H.-Z.: A rough set approach to knowledge reduction based on inclusion degree and evidence reasoning theory. Expert Systems 20, 298–304 (2003)
21. Zhang, W.-X., Mi, J.-S., Wu, W.-Z.: Approaches to knowledge reductions in inconsistent systems. International Journal of Intelligent Systems 18, 989–1000 (2003)

Knowledge Reduction Based on Granular Computing from Decision Information Systems

Lin Sun, Jiucheng Xu, and Shuangqun Li

College of Computer and Information Technology
Henan Normal University, Xinxiang Henan 453007, China
slinok@126.com, xjch3701@sina.com, lisq001@126.com

Abstract. Efficient knowledge reduction in large inconsistent decision information systems is a challenging problem. Moreover, existing approaches have still their own limitations. To address these problems, in this article, by applying the technique of granular computing, provided some rigorous and detailed proofs, and discussed the relationship between granular reduct introduced and knowledge reduction based on positive region related to simplicity decision information systems. By using radix sorting and hash methods, the object granules as basic processing elements were employed to investigate knowledge reduction. The proposed method can be applied to both consistent and inconsistent decision information systems.

Keywords: Granular computing, Rough set theory, Knowledge reduction, Decision information systems, Granular.

1 Introduction

Granular computing (GrC) is thus a basic issue in knowledge representation and data mining. The root of GrC comes from the concept of information granularity presented by Zadeh in the context of fuzzy set theory [1]. In coming years, GrC, rough set theory and NL-Computation are likely to become a part of the mainstream of computation and machine intelligence. Rough set models enable us to precisely define and analyze many notions of GrC [2-4]. Recently, many models and methods of GrC have been proposed and studied. For example, Yao [5] proposed a partition model of GrC. Lin [6] presented a fast association rule algorithm based on GrC. But in his work, generating different levels of association rules were not considered. Furthermore, how to store bit maps was not very clear. Up to now, many types of attribute reduction have been proposed in the analysis of information systems and decision tables, each of them aimed at some basic requirements. Various approaches have also been developed to perform knowledge reduction and obtain optimal true, certain, and possible decision rules in decision systems. For example, a maximum distribution reduct [7] preserved all maximum decision rules, but the degree of confidence in each uncertain decision rule might not be equal to the original one. Moreover, many reduction algorithms have some limitations. Lately, Hu [8] have proposed methods based on positive

J. Yu et al. (Eds.): RSKT 2010, LNAI 6401, pp. 46–53, 2010.

regions, the δ neighborhood rough set model and the k-nearest-neighbor rough set model. Both have the advantage of being able to deal with mixed attributes. However, these methods are still inefficient, and thus unsuitable for the reduction of voluminous data. Therefore, proposing an efficient and effective approach to knowledge reduction for both consistent and inconsistent decision information system is very desirable.

2   Preliminaries

The quintuple (U, A, F, D, G) is called a decision information system (DIS), where (U, A, F) is an information system (IS); A is a condition attribute set; D is a decision attribute set with $A \cap D = \emptyset$; $G = \{g_d | \forall d \in D\}$, where $g_d : U \rightarrow V_d$, for any $d \in D$; and V_d is the domain of the decision attribute d. Each non-empty subset $P \subseteq A \cup D$ determines a binary discernibility relation, i.e. $R_P = \{(x, y) \in U \times U | f(x, a) = f(y, a), \forall a \in P\}$. R_P determines a partition on U, that is U/R_P, and any element $[x]_P = \{y | f(x, a) = f(y, a), \forall a \in P\}$ in U/R_P is called an equivalent class.

Now we define a partial order on all partition sets of U. If P and Q are two equivalence relations on U, $\forall P_i \in U/R_P$, and $\exists Q_j \in U/R_Q$, so that $P_i \subseteq Q_j$, then we call the partition U/R_Q coarser than U/R_P, denoted by $P \preceq Q$.

Theorem 1. Let (U, A, F, D, G) be a DIS. If $Q \subseteq P$, for any $P, Q \subseteq A \cup D$, then we have $P \preceq Q$.

Proof. It is straightforward.

Theorem 2. Let (U, A, F, D, G) be a DIS. If there exists $R_A \subseteq R_D$, then (U, A, F, D, G) is referred to as a consistent DIS. Otherwise, (U, A, F, D, G) is referred to as an inconsistent one.

Proof. It is straightforward.

In a DIS, then, the positive region of the partition U/R_D with respect to P, $POS_P(D) = \cup\{\underline{P}Y | Y \in U/R_D\}$. For any $P \subseteq A$, to make $r \in P$, and r in P is unnecessary for D, if $POS_P(D) = POS_{P-\{r\}}(D)$. Otherwise r is necessary. Then, P is independent relative to D, if every element in P is necessary for D.

Theorem 3. Let (U, A, F, D, G) be a DIS, then, the positive region of the partition U/R_D with respect to A, denoted by $POS_A(D)$, that is,

$$POS_A(D) = \cup \{X | X \in U/R_A \wedge \forall x, y \in X \Rightarrow g_d(x) = g_d(y), \forall d \in D\}. \quad (1)$$

Proof. It is straightforward.

Theorem 4. If (U, A, F, D, G) is a consistent DIS, we have $POS_A(D) = U$.

Proof. It is straightforward.

Theorem 5. Let (U, A, F, D, G) be a DIS, for any $P \subseteq A$, and we then have $POS_P(D) \subseteq POS_A(D)$.

Proof. It is straightforward.

Definition 1. Let (U, A, F, D, G) be a DIS, $U/R_A = \{[U'_1]_A, [U'_2]_A, \cdots, [U'_n]_A\}$, where $U = \{u_1, u_2, \cdots, u_m\}$, $n \le m$, and $U'_i \in U$, then $U' = \{U'_1 \cup U'_2 \cup \cdots \cup U'_n\}$, $F' : U' \times (A \cup D) \to V'$ is called a new information function. It is said that the 6-tuple (U', A, F', D, G, V') is a simplicity decision information system (SDIS).

Proposition 1. Let (U', A, F', D, G, V') be a SDIS, then the positive region of the partition U/R_D with respect to A, i.e., $POS_A(D) = \{[U'_{i_1}]_A \cup [U'_{i_2}]_A \cup \cdots \cup [U'_{i_t}]_A\}$, where $U'_{i_s} \in U$, and $|[U'_{i_s}]_A/R_D| = 1$, $s = 1, 2, \cdots, t$. Thus, there exists $U'_{POS} = \{U'_{i_1}, U'_{i_2}, \cdots, U'_{i_t}\}$, and then we have $U'_{NEG} = U' - U'_{POS}$.

Theorem 6. Let (U', A, F', D, G, V') be a SDIS, then one has that $U'_{POS} \subseteq POS_A(D)$, where $POS_A(D)$ is the positive region of DIS.

Proof. It is straightforward.

Definition 2. Let (U', A, F', D, G, V') be a SDIS, for any $P \subseteq A$, then, the positive region of the partition U'/R_D with respect to P is defined as

$$POS'_P(D) = \cup \{X | X \in U'/R_P, X \subseteq U'_{POS} \wedge |X/R_D| = 1\}. \qquad (2)$$

3   The Proposed Approach

3.1   GrC Representation in SDIS

For an individual $x \in U$ on IS, if it satisfies an atomic formula φ, then we write $x \models \varphi$. The satisfiablity of an atomic formula by individuals of U is viewed as the knowledge describable. Thus, if φ is a formula, the set $m(\varphi) = \{x \in U | x \models \varphi\}$ is called the meaning of the formula φ. The meaning of a formula φ is indeed the set of all objects having the properties expressed by the formula φ. As a result, a concept can be expressed by a pair $(\varphi, m(\varphi))$, where φ is the intension of a concept, while $m(\varphi)$ is the extension of a concept. Thus, a connection between formulas and subsets of U is established. Similarly, a connection between logic connectives and set-theoretic operations can be stated [2]. Thus, a definable (information) granule is represented by a pair $Gr = (\varphi, m(\varphi))$, where φ refers to the intension of (information) granule Gr, and $m(\varphi)$ represents the extension of (information) granule Gr. Thus, a set composed of all elementary granules (Gr), separated from a SDIS, is regarded as a granular space, simply denoted by GrS. Meanwhile, there also exists a map between the granules of GrS and U', denoted by $gs : GrS \to U'$, such that $gs(Gr) = m(\varphi)$, for any $Gr \in GrS$.

3.2   Granular Computing Based Knowledge Reduction in SDIS

Definition 3. Let (U', A, F', D, G, V') be a SDIS, and GrS be its granular space. The objects set, separated from the positive region $POS_A(D)$ in a SDIS, is regarded as a granular space of positive region, denoted by $GrSP$; and one separated from the negative region $NEG_A(D)$ in a SDIS, is regarded as a granular space of negative region, denoted by $GrSN$.

Definition 4. Let (U', A, F', D, G, V') be a SDIS, and GrS be its granular space, $P \subseteq A$, then, for any $Gr_i \in GrSP$, if there exist $Gr_j \in GrSP$, and $Gr_i \neq Gr_j$ such that the intension of Gr_i and Gr_j has the equal sets of attribute values, however, they have different sets of decision values, then, the granule Gr_i is called the conflict granule of $GrSP$ with respect to P. On the other hand, there exist $Gr_j \in GrSN$, and $Gr_i \neq Gr_j$ such that the intension of Gr_i and Gr_j has the equal sets of attribute values, then, the granule Gr_i is called the conflict granule of $GrSP$ with respect to P. Otherwise, Gr_i is called the non-conflict granule of $GrSP$ with respect to P.

Theorem 7. Let (U', A, F', D, G, V') be a SDIS, and GrS be its granular space. Suppose $U'_{GrSP} = \cup \{gs(Gr)|Gr \in GrSP\}$, and $U'_{GrSN} = \cup \{gs(Gr)|Gr \in GrSN\}$, then there must exist $U'_{GrSP} = U'_{POS}$, and $U'_{GrSN} = U'_{NEG}$.

Proof. It can be achieved by Proposition 1 and Definition 3.

Definition 5. Let (U', A, F', D, G, V') be a SDIS, and GrS be its granular space. Suppose $U'_{GrSP} = gs(GrS)$ or $U'_{GrSN} = \emptyset$, then (U', A, F', D, G, V') is referred to as a consistent SDIS. Otherwise, (U', A, F', D, G, V') is referred to as an inconsistent one.

Theorem 8. Let (U', A, F', D, G, V') be a SDIS, and GrS be its granular space. Then, (U', A, F', D, G, V') is referred to as a consistent SDIS if and only if for any $Gr \in GrS$, there must exist $x \in gs(Gr) \wedge y \in gs(Gr) \Rightarrow f_D(x) = f_D(y)$, for any $x, y \in U'$.

Proof. It can be achieved by Theorem 7 and Definition 5.

Definition 6. Let (U', A, F', D, G, V') be a SDIS, and GrS be its granular space. For any $P \subseteq A$, then, the positive region of the partition U'/R_D with respect to P in GrS, denoted by $POS'_P(D)$, is defined as

$$POS'_P(D) = \cup \{X|X \in gs(GrS)/R_P \wedge X \subseteq U'_{GrSP}\}. \tag{3}$$

Theorem 9. Let (U', A, F', D, G, V') be a SDIS, GrS be its granular space, and $P \subseteq A$. If there exist $POS'_P(D) = U'_{GrSP}$, and $POS'_P(D) \neq POS'_{P-p}(D)$, for any $p \in P$, then it follows that P is a more precise relative reduct of A with respect to D.

Proof. For any $p \in P$, there exists $P - \{p\} \subseteq P$, by Theorem 5, then, we have that $POS'_{P-\{p\}}(D) \subseteq POS'_P(D)$. Since $POS'_P(D) \neq POS'_{P-\{p\}}(D)$, there must exist some $u'_l \in POS'_P(D) = U'_{GrSP}$ such that $u'_l \in POS'_{P-\{p\}}(D)$ doesn't hold. Then, one must have that $[u'_l]_{P-\{p\}} = X_i \in gs(GrS)/R_{P-\{p\}}$. From the definition of the positive region in a SDIS, it follows that $X_i \subset U'_{GrSP}$

doesn't hold. Thus, we know that there must exist some $u'_k \in X_i$ such that $u'_k \in gs(Gr_t) \in U'_{GrSN} \neq \emptyset$, where $Gr_t \in GrS$. From the definition of the inconsistent SDIS, we have that $x \in gs(Gr_t) \wedge y \in gs(Gr_t)) \Rightarrow f_D(x) \neq f_D(y)$, for any $x, y \in U'$. Hence, one can obtain that $[u'_k]_A \subseteq [gs(Gr_t)]_A \subseteq U'_{GrSN}$, i.e. $[u'_k]_A \subset U'_{GrSP}$ doesn't hold. Hence, it follows from Proposition 1 and Theorem 7 that we can't get $[u'_k]_A \subset POS_A(D)$. Since $P - \{p\} \subseteq P \subseteq A$, there must exist $POS_{P-\{p\}}(D) \subseteq POS_P(D) \subseteq POS_A(D)$. Thus, we have that $[u'_k]_A \subset POS_{P-\{p\}}(D)$ is not. Hence, we obtain $[u'_k]_{P-\{p\}} = [u'_l]_{P-\{p\}}$, when some $u'_k \in X_i = [u'_l]_{P-\{p\}}$. For any $P \subseteq A$, we have that $P - \{p\} \subseteq A$, then it is obvious that $[u'_k]_A \subseteq [u'_l]_{P-\{p\}}$. Clearly, there doesn't exist $[u'_l]_{P-\{p\}} \subset POS_{P-\{p\}}(D)$ such that $[u'_l]_{P-\{p\}} \subset POS_A(D)$ is also not. In this situation, we then easily obtain that $u'_l \in POS_{P-\{p\}}(D)$ and $u'_l \in POS_A(D)$ don't hold. Since $POS'_P(D) = U'_{GrSP}$, one has that $POS_P(D) = POS_A(D)$. Thus, we have that $u'_l \in POS_P(D)$ is not. So, one can obtain that $POS_{P-\{p\}}(D) \neq POS_P(D)$. That is, every element p in P is necessary for D. Therefore, it is easy to see that $P \subseteq A$ is independent with respect to D. Thus, it follows that P is a more precisely relative reduct of A with respect to D.

Proposition 2. Let (U', A, F', D, G, V') be a SDIS, and GrS be its granular space. For any $P \subseteq A$, if there does not exist any conflict granule with respect to P in $GrSP$, then P is a more precise relative reduct of A with respect to D. *Proof.* It can be achieved from the definition of positive region and Theorem 9.

Proposition 3. Let (U', A, F', D, G, V') be a SDIS, and GrS be its granular space. If any $P \subseteq A$, and $p \in A - P$, we then obtain a granular partition of U', for any $Gr \in GrS$, denoted by

$$gs(Gr)/R_{P-\{p\}} = \cup \{X/R_{\{p\}} | X \in gs(Gr)/R_P\}. \tag{4}$$

Proof. It can be achieved under Property 2 in [11].

In a DIS, if $P \subseteq A$, $U/R_P = \{P_1, P_2, \cdots, P_n\}$, and $P_i \subseteq POS_P(D)$, for any equivalence class $P_i \in U/R_P$, then one has that $U = U - P_i$, and $POS_A(D) = POS_A(D) - P_i$. Thus, it will help to reduce the quantity of computation, time complexity, and space complexity of search. In this case, if $P_i \subseteq POS_P(D)$ doesn't hold, we then use $U/R_{P\cup\{r\}} = \cup\{X/R_{\{r\}} | X \in U/R_P\}$ to form a new partition of U again, and if $P_i \subseteq U - POS_A(D)$, then P_i is deleted from U, because it contributes nothing to computing the positive region of DIS. Hence, we have the conclusion that it only considers the information, which can distinguish one class from the remaining classes, and then omits that of distinguishing one class from another class. To address these problems, we define granular reduct in a SIDS. In the following, on the basis of the idea analyzed in detail, we investigate some definition and judgment methods of granular reduct in a SIDS.

Definition 7. Let (U', A, F', D, G, V') be a SDIS, and GrS be its granular space. If any $P \subseteq A$, then, the significance measure of p in P, for any $p \in A - P$, denoted by $SIG_P(p)$, is defined as

$$SIG_P(p) = \frac{|GrS_{P\cup\{p\}} - GrS_P|}{|U'|}, \tag{5}$$

where $GrS_P = \{\cup\{X|X \in gs(GrS)/R_P \wedge X \subseteq U'_{GrSP}\}\} \cup \{\cup\{X|X \in gs(GrS)/R_P \wedge X \subseteq U'_{GrSN}\}\}$. Noticing that if $\{p\} = \emptyset$, then we have that $SIG_P(\emptyset) = 0$.

Theorem 10. Let (U', A, F', D, G, V') be a SDIS, and GrS be its granular space. If $P \subseteq A$, $GrS_P = gs(GrS)$, and $GrS_{P-\{p\}} \subset gs(GrS)$, for any $p \in P$, then, it follows that P is a more precise relative reduct of A with respect to D.

Proof. For any $P \subseteq A$, by Theorem 5, then there exists $POS'_P(D) \subseteq POS'_A(D)$. Thus, since $P - \{p\} \subseteq P$, we have that $POS'_{P-\{p\}}(D) \subseteq POS'_P(D)$. We assume that $gs(Gr_i) \in POS'_A(D)$, for any $Gr_i \in GrS$. Since $P \subseteq A$, then, one has that $[gs(Gr_i)]_A \subseteq [gs(Gr_i)]_P$. On the other hand, for any $GrS_P = gs(GrS)$, by Definition 7, it is obvious that $[gs(Gr_i)]_P \subseteq U'_{GrSP}$, or $[gs(Gr_i)]_P \subseteq U'_{GrSN}$. However, if $[gs(Gr_i)]_P \subseteq U'_{GrSN}$, then , there must exist $[gs(Gr_i)]_A \subseteq U'_{GrSN}$ such that $[gs(Gr_i)]_A \subset U'_{GrSP}$ doesn't hold, i.e. $gs(Gr_i) \in POS'_A(D)$ is not. Obviously, this yields a contradiction. Therefore, we have that $[gs(Gr_i)]_P \subseteq U'_{GrSP}$, by Definition 6, then, we can find $gs(Gr_i) \in POS'_P(D)$ such that $POS'_A(D) \subseteq POS'_P(D)$. Hence, we have that $POS'_P(D) = POS'_A(D)$.

For any $p \in P$, we have that $GrS_{P-\{p\}} \subset gs(GrS)$, then, there must exist $Gr_l \in GrS$ such that $gs(Gr_l) \in GrS_{P-\{p\}}$ doesn't hold. It is clear that we have that $gs(Gr_l) \in \{\cup\{X|X \in gs(GrS)/R_{P-\{p\}} \wedge X \subseteq U'_{GrSP}\}\}$ and $gs(Gr_l) \in \{\cup\{X|X \in gs(GrS)/R_{P-\{p\}} \wedge X \subseteq U'_{GrSN}\}\}$ don't hold. Therefore, there must exist $Gr_j \in GrS$ such that $gs(Gr_l) \in [gs(Gr_j)]_{P-\{p\}}$. Thus, since $[gs(Gr_j)]_{P-\{p\}} \subseteq U'_{GrSP}$ is not, by Definition 6, it easily follows that $[gs(Gr_j)]_{P-\{p\}} \subset POS'_{P-\{p\}}(D)$ is not. In this case, it is easy to know that $gs(Gr_l) \in [gs(Gr_j)]_{P-\{p\}}$, and $[gs(Gr_j)]_{P-\{p\}} \subset U'_{GrSN}$ is not, and then there must exist some $Gr_k \in GrS$ such that $gs(Gr_k) \in [gs(Gr_j)]_{P-\{p\}}$, and $gs(Gr_k) \subseteq U'_{GrSP}$, by Definition 5, it follows that $gs(Gr_k)$ is the objects of positive region in a SDIS, i.e., $gs(Gr_k) \in POS_A(D)$. Hence, we have that $[gs(Gr_k)]_A \subseteq U'_{GrSP}$. Thus, it is clear that we have $gs(Gr_k) \in POS'_A(D)$. Since $[gs(Gr_k)]_{P-\{p\}} = [gs(Gr_j)]_{P-\{p\}}$, one can find that $gs(Gr_k) \in POS'_{P-\{p\}}(D)$ is not, and then we have that $POS'_A(D) \neq POS'_{P-\{p\}}(D)$, i.e., $POS'_P(D) \neq POS'_{P-\{p\}}(D)$. From the necessary attribute definition of positive region in a SDIS, it is clear that any $p \in P$ is independent relative to D. Therefore, it follows that P is a more precise algebraic reduct of A with respect to D.

Thus, to reduce the quantity of computation, and space complexity of search, GrS is separated into $GrSP$ and $GrSN$. From the algebraic point of view in [3, 4], this idea can compensate for these current disadvantages of the classical reduction algorithms. Moreover, $SIG_P(p)$ is called a heuristic function from the algebraic point of view in a SDIS, and then we assume that $P = \emptyset$, adding attributes bottom-up on P. Thus, if a new corresponding granule $Gr = (\varphi, m(\varphi))$, separated from a SIDS, is a non-conflict granule with respect to A, and contained by $GrSP$, or a conflict granule with respect to A, is contained by $GrSN$. Then, we have $SIG_P(p) \neq 0$. That is to say, when we continually add any attribute p to P given, if the radix conflict granule of $GrSP$ with respect to $P \cup \{p\}$ is not changed, then $SIG_P(p) \neq 0$. Hence, $SIG_P(p)$ thus describes a decrease in the dependency degree of P. However, in a SDIS, to improve efficiency, we must employ some effective measures for computing partitions, positive region, and separating a SDIS to acquire its granular space GrS. Thus, through computing partitions, based on radix sorting in [9], costing the time complexity $O(|A||U|)$, and positive region, based on twice-hash in [10], costing the time complexity $O(|U|)$, we make it easy to construct the efficient method of computing partitions and positive region based on radix sorting and hash, so that the worst time complexity can be cut down to $O(|A|^2|U/A|)$.

4   Experimental Results

The experiments on PC (AMD Dual Core 2.71GHz, 2GB RAM, WINXP) under JDK1.4.2, are performed on three data sets from UCI Machine Learning Repository, then we choose two algorithms such as the Algorithm A in [12], Algorithm 4 in [11], compared with the proposed method, denoted by Algorithm_a, Algorithm_b, Algorithm_c respectively. Thus we obtain the results of reduct comparison in Table 1, where m and n are the numbers of primal condition attributes and after reduction respectively.

Table 1. Comparison of Reduct Results

Data Sets	number of objects	m	Algorithm_a		Algorithm_b		Algorithm_c	
			n	time (s)	n	time (s)	n	time (s)
Iris	150	4	3	0.09	3	0.05	3	0.03
Zoo	101	17	11	0.35	10	0.11	10	0.05
Mushroom	8142	22	5	470.5	5	5.25	4	4.75

5   Conclusion

Knowledge reduction is winning more and more attention from researchers for its own advantages. Although a few algorithms for dealing with DIS have been proposed, their complexities are always no less than $O(|A|^2|U|^2)$. In this article, we presented a SDIS, laying the theoretical foundation for our method. Thus, by applying the GrC technique, we defined a granule as a set of entities, had the same properties in relational databases, so that a granule could be considered as an equivalent class of attribute values, and then we investigated the components of granule, GrS, $GrSP$, and $GrSN$ in a SDIS, and provided rigorous and detailed proofs to discuss the relationship between granular reduct and knowledge reduction based on the positive region related to SDIS. Using radix sorting and hash techniques, we employed the object granules as basic processing elements to investigate knowledge reduction, based on GrC. This method was suitable and complete for both consistent and inconsistent SDIS. In sum, the proposed theory of GrC is effective and feasible and with high applicability, and the knowledge reduction method is an efficient means of attribute reduction in both consistent and inconsistent DIS, especially large ones.

Acknowledgment. This paper was supported by the National Natural Science Foundation of China under Grant (No. 60873104) and New Century Excellence Genius Support Plan of Henan Province of China (No. 2006HANCET-19), and Natural Science Foundation of Educational Department of Henan Province of China (No. 2008B520019).

References

1. Bargiela, A., Pedrycz, W.: Granular Computing: An Introduction. Kluwer Academic Publishers, Dordrecht (2002)
2. Miao, D.Q., Wang, G.Y., Liu, Q., Lin, T.Y., Yao, Y.Y.: Granular Computing: Past, Present, and the Future Perspectives. Academic Press, Beijing (2007)
3. Xu, J.C., Sun, L.: New Reduction Algorithm Based on Decision Power of Decision Table. In: Wang, G., Li, T., Grzymala-Busse, J.W., Miao, D., Skowron, A., Yao, Y. (eds.) RSKT 2008. LNCS (LNAI), vol. 5009, pp. 180–188. Springer, Heidelberg (2008)
4. Xu, J.C., Sun, L.: Research of Knowledge Reduction Based on New Conditional Entropy. In: Wen, P., Li, Y., Polkowski, L., Yao, Y., Tsumoto, S., Wang, G. (eds.) RSKT 2009. LNCS, vol. 5589, pp. 144–151. Springer, Heidelberg (2009)
5. Yao, Y.Y.: A Partition Model of Granular Computing. LNCS Transactions on Rough Sets 1, 232–253 (2004)
6. Lin, T.Y., Louie, E.: Finding Association Rules by Granular Computing: Fast Algorithms for Finding Association Rules. In: Proceedings of the 12th International Conference on Data Mining, Rough Sets and Granular Computing, Berlin, German, pp. 23–42 (2002)
7. Kryszkiewicz, M.: Comparative Study of Alternative Types of Knowledge Reduction in Insistent Systems. International Journal of Intelligent Systems 16, 105–120 (2001)
8. Hu, Q.H., Yu, D.R., Xie, Z.X.: Neighborhood Classifiers. Expert Systems with Applications 34, 866–876 (2008)
9. Xu, Z.Y., Liu, Z.P., et al.: A Quick Attribute Reduction Algorithm with Complexity of $Max(O(|C||U|),O(|C|^2|U/C|))$. Journal of Computers 29(3), 391–399 (2006)
10. Liu, Y., Xiong, R., Chu, J.: Quick Attribute Reduction Algorithm with Hash. Chinese Journal of Computers 32(8), 1493–1499 (2009)
11. Liu, S.H., Sheng, Q.J., Wu, B., et al.: Research on Efficient Algorithms for Rough Set Methods. Chinese Journal of Computers 26(5), 524–529 (2003)
12. Guan, J.W., Bell, D.A.: Rough Computational Methods for Information Systems. International Journal of Artificial Intelligences 105, 77–103 (1998)

Pattern Classification Using
Class-Dependent Rough-Fuzzy Granular Space

Sankar K. Pal[1], Saroj K. Meher[1,*], and Soumitra Dutta[2]

[1] Center for Soft Computing Research,
Indian Statistical Institute, Kolkata 700108, India
[2] INSEAD, Blvd de Constance, Fontainebleau 77305, France

Abstract. The article describes a new rough-fuzzy model for pattern classification. Here, class-dependent granules are formulated in fuzzy environment that preserve better class discriminatory information. Neighborhood
rough sets (NRS) are used in the selection of a subset of granulated features that explore the local/contextual information from neighbor granules. The model thus explores mutually the advantages of class-dependent
fuzzy granulation and NRS that is useful in pattern classification with overlapping classes. The superiority of the proposed model to other similar
methods is demonstrated with both completely and partially labeled data
sets using various performance measures. The proposed model learns well
even with a lower percentage of training set that makes the system fast.

Keywords: Fuzzy information granulation, rough neighborhood sets,
rough-fuzzy granular computing, pattern recognition, soft computing.

1   Introduction

Granular computing (GrC) refers to that where computation and operations
are performed on information granules (clumps of similar objects or points).
Many researchers [1] have used GrC models to build efficient computational
algorithms that can handle huge amount of data, information and knowledge.
However, the main task to be focussed in GrC is to construct and describe
information granules, a process called information granulation [2,3,4] on which
GrC is oriented. Although, modes of information granulation with crisp granules
play important roles in a wide variety of applications, they have a major blind
spot [2]. More particularly they fail to reflect most of the processes of human
reasoning and concept formation, where the granulation is more appropriately
fuzzy rather than crisp.

The process of fuzzy granulation involves the basic idea of generating a family
of fuzzy granules from numerical features and transform them into fuzzy linguistic variables. Fuzzy information granulation has come up with an important
concept in fuzzy set theory, rough set theory and the combination of both in
recent years [4,5]. In general, the process of fuzzy granulation can be broadly

---

* Corresponding author, *Email*: saroj.meher@gmail.com, Tel.: (+91) (33) 2575-3447.

J. Yu et al. (Eds.): RSKT 2010, LNAI 6401, pp. 54–61, 2010.

categorized as class-dependent (CD) and class-independent (CI). With CI granulation each feature is described with three fuzzy membership functions representing three linguistic values over the whole space as is done in [6]. However, the process of granulation does not take care of the class belonging information of features to different classes that lead to a degradation of performance in pattern classification for data sets with highly overlapping classes. On the other hand, in CD granulation feature-wise individual class information is restored by the generated fuzzy granules.

Rough set theory, as proposed by Pawlak [4] (henceforth it will be abbreviated as PaRS), has been proven to be an effective tool for feature selection, knowledge discovery and rule extraction from categorical data [4]. PaRS can also be used to deal with both vagueness and uncertainty in data sets and to perform granular computation. However for the numerical data, PaRS theory can be used with the discretisation of data that results in the loss of information. To deal with this, neighborhood rough set (NRS) [7,8] is found to be suitable that can deal with both numerical and categorical data sets without discretisation. Further, NRS facilitates to gather the possible local information through neighbor granules that is useful for a better discrimination of patterns, particularly in class overlapping environment.

In this article, we describe a rough-fuzzy granular space based on a synergistic integration of the merits of both CD fuzzy granulation and the theory of NRS. The resulting output can be used as an input to any classifier. To demonstrate the effectiveness of the proposed rough-fuzzy granular space based model, we have used here different classifiers, such as k-nearest neighbor (k-NN) ($k = 1$, 2 and 3) classifier, maximum likelihood (ML) classifier and multi-layered perceptron (MLP). However, other classifiers may also be used. We have demonstrated the potential of the proposed model with two completely labeled data sets, and two partially labeled multispectral remote sensing images using various performance measures. For multispectral images we have used the spectral (band) values as features. The experimental results show that the proposed model provides improved classification accuracy in terms of different quantitative measures, even with a smaller training set.

2   Proposed Model for Pattern Classification

The model has three steps of operation as illustrated in Fig. 1, namely, class-dependent (CD) fuzzy granule generation, rough set based feature selection using reducts, and classification based on the selected features. These are described below:

2.1   Class-Dependent Granule Generation

For CD fuzzy granulation, each feature of the input pattern vector is described in terms of its fuzzy membership values corresponding to $\mathbf{C}$ (= total number of classes) linguistic fuzzy sets using π-type MF [9]. Thus, an n-dimensional

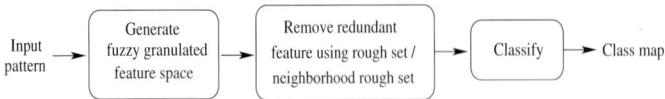

Fig. 1. Schematic flow diagram of the proposed model for pattern classification

pattern vector, $\mathbf{F} = [F_1, F_2, ..., F_n]$, is expressed as $(n \times \mathbf{C})$-dimensional vector and is given by

$$\mathbf{F} = [\mu_1^1(F_1), \mu_2^1(F_1), ..., \mu_c^1(F_1), ..., \mu_{\mathbf{C}}^1(F_1); \mu_1^2(F_2), \mu_2^2(F_2), ..., \mu_c^2(F_2), ..., \mu_{\mathbf{C}}^2(F_2);$$
$$\mu_1^n(F_n), \mu_2^n(F_n), ..., \mu_c^n(F_n), ..., \mu_{\mathbf{C}}^n(F_n)], \ (c = 1, 2, ..., \mathbf{C})$$

where $\mu_1^n(F_n), \mu_2^n(F_n), ..., \mu_c^n(F_n), ..., \mu_{\mathbf{C}}^n(F_n)$ signify the membership values of F_n to $\mathbf{C}$ number of fuzzy sets along the nth feature axis and $\mu(F_n) \in [0,1]$. It implies that each feature F_n is characterizing $\mathbf{C}$ number of fuzzy granules with fuzzy boundary among them and thus comprising $\mathbf{C}^n$ fuzzy granules in an n-dimensional feature space. The fuzzy granules thus generated explores the degree of belonging of a pattern into different classes based on individual features. This is potentially useful for improved classification for the data sets with overlapping classes.

For a better visualization of the granules generated by the proposed method, we have converted fuzzy membership values to the patterns to binary ones, i.e., fuzzy membership functions (MFs) to binary functions using α-cut. This is demonstrated in Fig. 2, where 0.5-cut is used to obtain $4^2 = 16$ crisp granules for four overlapping classes in two-dimensional feature space. For example, the region (granule no 6) in Fig. 2 indicates a crisp granule that is characterized by $\mu_{C_2}^1$ and $\mu_{C_3}^2$.

2.2   Feature Selection

In the granulation process, each feature value is represented with more than one membership value and thus the feature dimension increases. The increased

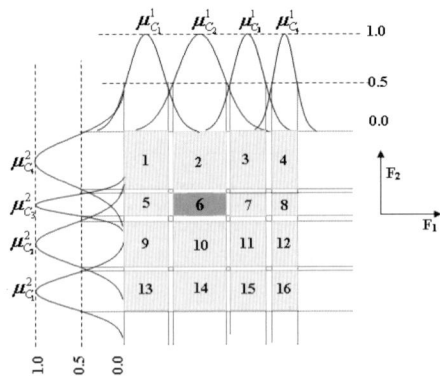

Fig. 2. Generation of class-dependent fuzzy granules

dimension brings great difficulty in solving many tasks of pattern recognition. To reduce the feature dimension, we have used neighborhood rough set [7,8] (NRS) based feature selection method in the second step of the proposed model (Fig. 1). The details of NRS theory may be referred to [7,8].

After the features are selected, we use a classifier as in the third step of Fig. 1 to classify the input pattern based on the selected features.

3   Performance Measurement Indexes

To examine the practical applicability of the proposed model various performance measures are used. For completely labeled data sets, we use measures like percentage of overall classification accuracy (PA), Kappa coefficient (KC) [10] and total computation time T_c (sum of training and testing times). For partially labeled data sets, quantitative indexes such as β [11] and Davies-Bouldin (DB) [12], are used.

4   Results and Discussion

In the present investigation we have compared the performance of the proposed model with different combinations of fuzzy granulation and rough feature selection methods. Five different combinations of classification models using granular feature space and feature selection methods those are considered for performance comparison are mentioned below. Patterns with its original feature representation are fed as input to these models.

- Model 1 : k-nearest neighbor (k-NN with k=1) classifier,
- Model 2 : CI fuzzy granulation + PaRS based feature selection + k-NN (with k=1) classifier,
- Model 3 : CI fuzzy granulation + NRS based feature selection + k-NN (with k=1) classifier,
- Model 4 : CD fuzzy granulation + PaRS based feature selection + k-NN (with k=1) classifier,
- Model 5 : CD fuzzy granulation + NRS based feature selection + k-NN (with k=1) classifier.

Two completely labeled data sets, namely, VOWEL [13] and SATIMAGE [14], and two partially labeled multispectral remote sensing images such as IRS-1A [15] and SPOT [15], are used for the present study. However, the experimental results are provided only for VOWEL data and IRS-1A image because similar trend of comparative performance is observed for the remaining data sets.

4.1   Classification of Completely Labeled Data Sets

Selection of training and test samples for all classes in case of completely labeled data sets have been made by splitting the data sets as two parts.We repeat each of these splitting sets for ten times and the final results are then averaged over them.

Table 1. Performance comparison of models using k-NN classifier (k=1) with VOWEL data set ($\mathbf{p}$=2, Φ=0.45)

Model	10% of training set			20% of training set			50% of training set		
	PA	KC	T_c	PA	KC	T_c	PA	KC	T_c
1	73.240	0.7165	0.146	75.150	0.7212	0.124	77.560	0.7256	0.101
2	76.010	0.7502	0.236	77.010	0.7704	0.223	79.030	0.7721	0.214
3	77.870	0.7621	0.263	78.810	0.7853	0.252	80.790	0.7901	0.244
4	81.030	0.8008	0.365	81.370	0.8165	0.351	82.110	0.8202	0.345
5	83.750	0.8102	0.381	83.960	0.8253	0.378	84.770	0.8301	0.354

VOWEL Data. VOWEL data [13] is a set of vowel sounds in consonant-vowel-consonant context. It has three features and six highly overlapping classes.

The classification results for this data set with five different models using k-NN classifier (k=1), are depicted in Table 1 for three different percentages (10%, 20% and 50%) of training sets. In the present experiment, we have compared the performance of models with respect to three criteria, namely, (i) granulated and non-granulated feature space, (ii) CD and CI fuzzy granulation, and (iii) PaRS and NRS based feature selection. Performance comparison with the NRS method of feature selection depends on two key parameters such as shape and size of neighborhood [8]. The shape and size are determined by distance function ($\mathbf{p}$-norm) and threshold Φ [8], respectively. In the present study we analyzed the performance of model 5 for the variation of both these parameters. It is observed that the highest accuracy is obtained for $\Phi = 0.45$ with Euclidean distance ($\mathbf{p}$=2), which are used for presenting results for all the data sets.

In a comparative analysis, it is observed from Table 1 that for all percentages of training sets the classifiers' performance measured with PA values is better for models using granulated feature space. Further, the PA values for models 4 and 5 (CD models) are superior to models 2 and 3 (CI models), respectively. This indicates the efficacy of CD granulation compared to CI granulation.

In a performance comparison of models with NRS and PaRS, it is observed from Table 1 that the PA values for models 5 and 3 are higher compared to models 4 and 2, respectively. This justifies the advantages of NRS based feature selection. All together the proposed model (model 5) provided the best performance for all percentages of the training sets as observed in Table 1. All the critically assessed improved performance of the proposed model obtained with the PA is supported by KC measure (Table 1) and is true for all three percentages of training sets.

Table 1 also reveals that the accuracy obtained with the model 5 for minimum percentage of training set is higher compared to other models at 50% training set. This is particularly important when there is a scarcity of training set (e.g., land covers classification of remote sensing images).

A comparative analysis in terms of total computational time T_c, as required by different models, is depicted in Table 1. All the simulations are done in MATLAB environment on a Pentium-IV machine with 3.19 GHz processor speed. It is seen for all the cases that the T_c values for model 5, as expected, are higher compared to others with the cost of improved performance.

Table 2. Classification accuracies (PA) of models with different classifiers at 20% training set for VOWEL data set (**p**=2, Φ=0.45)

Model	k-NN ($k = 3$)	k-NN ($k = 5$)	ML	MLP
1	74.20	73.63	75.21	76.33
2	76.11	76.34	77.34	78.06
3	77.78	77.89	78.34	79.87
4	81.03	80.86	82.25	83.75
5	83.91	84.01	83.87	84.88

So far we have described the effectiveness of the proposed rough-fuzzy granulation and feature selection model using k-NN ($k = 1$) classifier. We also described the effectiveness of the same model using some other classifiers, e.g., k-NN ($k = 3$ and 5), maximum likelihood (ML) classifier and multi-layered perceptron (MLP). The comparative results of all models with these classifiers are depicted in Table 2 for training set of 20%, as an example. The superiority of model 5 to others for different sets of classifiers is evident. Also similar improvement in performance of the models (using different classifiers) with granulated over non-granulated, CD granulation over CI granulation and NRS based feature selection over PaRS, is observed as in the case of k-NN ($k =1$) classifier.

Table 3. Performance comparison of models using k-NN classifier (k=1) with partially labeled data sets (**p**=2, Φ=0.45)

Model	β value		DB value	
	IRS-1A	SPOT	IRS-1A	SPOT
Training samples	9.4212	9.3343	0.5571	1.4893
1	6.8602	6.8745	0.9546	3.5146
2	7.1343	7.2301	0.9126	3.3413
3	7.3559	7.3407	0.8731	3.2078
4	8.1372	8.2166	0.7790	2.8897
5	8.4162	8.4715	0.7345	2.7338

4.2   Classification of Partially Labeled Data Sets

Here the classifiers are initially trained with labeled data of different land cover types of remote sensing image and then the said trained classifiers are applied on the unlabeled image data to partition into regions.

IRS-1A Image. IRS-1A image (size 512 x 512 pixels) taken from the satellite covers an area in the near infrared band having six major land cover classes: pure water (PW), turbid water (TW), concrete area (CA), habitation (HAB), vegetation (VEG) and open spaces (OS). Since the images are of poor illumination, we have presented the enhanced images in Figs. 3a for the convenience of visualizing the content of the image. However, the algorithms are implemented on the original (poorly illuminated) image.

IRS-1A image is classified with five different models using k-NN classifier (k=1), and the performance comparison in terms of β value and DB value, is

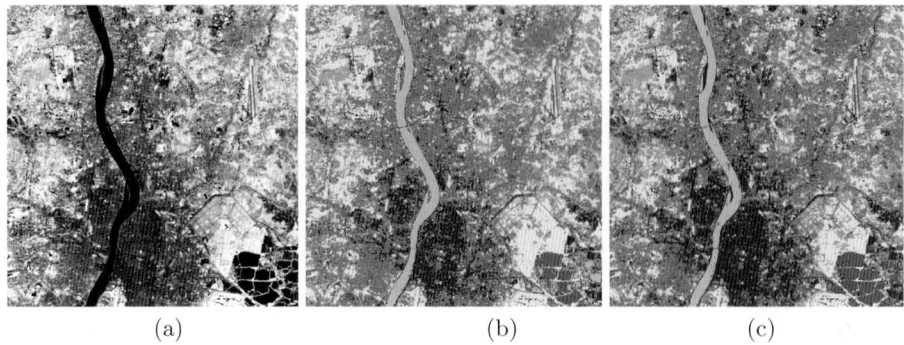

| (a) | (b) | (c) |

Fig. 3. (a) Original IRS-1A (band-4) enhanced image, and Classified IRS-1A images with (b) model 1 and (c) model 5 (proposed model)

shown in Table 3. As expected, the β value is the highest and DB value is the lowest for the training set (Table 3). It is also seen that the proposed model yields superior results in terms of both the indexes, compared to other four models. As a whole the gradation of performance of five models can be established with the following β relation: $\beta_{training} > \beta_{proposed} > \beta_{model4} > \beta_{model3} > \beta_{model2} > \beta_{model1}$. Similar gradation of performance is also observed with DB values, which further supports the superiority of the proposed model.

The significance of the proposed granular computing based model (model 5) is also demonstrated visually from Figs. 3b and 3c that shows the classified images corresponding to models 1 and 5. It is observed from Fig. 3c that different regions are properly identified and well structured compared to Fig. 3b.

5   Conclusion

In the present article, we described a rough-fuzzy model for pattern classification, where a class-dependent (CD) fuzzy granulation of input feature space was formulated and neighborhood rough sets (NRS) were used in the selection of most relevant granulated features. The model thus provides a synergistic integration of the merits of CD fuzzy granulation and the theory of NRS. Extensive experimental results with various types of completely and partially labeled data sets, the superiority of CD granulation to CI granulation and NRS based feature selection to Pawlak's rough set, is justified. It is observed that the effect of CD granulation is substantial compared to the rough feature selection methods in improving classification performance. The combined effect is further encouraging for the data sets with highly overlapping classes. The computational complexity of the proposed model is little high, but its learning ability with small percentage of training samples will make it practicably applicable to problems with a large number of overlapping classes. The model can also be useful to the data sets with large dimension because rough granular space based feature selection not only retains the representational power of the data, but also maintains its minimum redundancy.

Acknowledgments. The work was done while Prof. Pal held J. C. Bose Fellowship of the Govt. of India.

References

1. Bargiela, A., Pedrycz, W.: Granular Computing: An Introduction. Kluwer Academic Publishers, Boston (2003)
2. Zadeh, L.A.: Fuzzy sets and information granularity. In: Gupta, M., Ragade, R., Yager, R. (eds.) Advances in Fuzzy Set Theory and Applications, pp. 3–18. North-Holland Publishing Co., Amsterdam (1979)
3. Zadeh, L.A.: Toward a theory of fuzzy information granulation and its centrality in human reasoning and fuzzy logic. Fuzzy Sets and Systems 90, 111–127 (1997)
4. Pawlak, Z.: Rough sets. International Journal of Computer and Information Science 11, 341–356 (1982)
5. Pal, S.K., Skowron, A. (eds.): Rough-Fuzzy Hybridization: A New Trend in Decision Making. Springer, Singapore (1999)
6. Pal, S.K., Mitra, S.: Multilayer perceptron, fuzzy sets, and classification. IEEE Trans. Neural Networks 3, 683–697 (1992)
7. Lin, T.Y.: Granulation and nearest neighborhoods: rough set approach. In: Pedrycz, W. (ed.) Granular computing: an emerging paradigm, pp. 125–142. Physica-Verlag, Heidelberg (2001)
8. Hu, Q., Yu, D., Liu, J., Wu, C.: Neighborhood rough set based heterogeneous feature subset selection. Inf. Sciences 178, 3577–3594 (2008)
9. Zadeh, L.A.: Fuzzy sets. Inf. Control 8, 338–353 (1965)
10. Cohen, J.: A coefficient of aggrement for nominal scale. Education and psychological measurement 20, 37–46 (1960)
11. Pal, S.K., Ghosh, A., Uma Shankar, B.: Segmentation of remotely sensed images with fuzzy thresholding, and quatitative evaluation. Int. J. of Remote Sensing 21, 2269–2300 (2000)
12. Davies, D.L., Bouldin, D.W.: A cluster separation measure. IEEE Trans. Pattern Anal. Machine Intell. 1, 224–227 (1979)
13. Pal, S.K., Majumder, D.D.: Fuzzy sets and decision making approaches in vowel and speaker recognition. IEEE Trans. Syst., Man Cybern. 7, 625–629 (1977)
14. Elena database,
    ftp://ftp.dice.ucl.ac.be/pub/neural-nets/ELENA/databases/REAL/
15. NRSA, IRS data users hand book, Technical report, Document No. IRS/NRSA/NDC/HB-02/89 (1989)

Generate (F, ϵ)-Dynamic Reduct Using Cascading Hashes

Pai-Chou Wang

Department of Information Management, Southern Taiwan University,
No.1, Nantai St., Yung-Kang City, Tainan County, 710 Taiwan
pwang@mail.stut.edu.tw

Abstract. Dynamic reducts with large stability coefficients are good candidates for decision rules generation but it is time consuming to generate them. This paper presents an algorithm *dReducts* using a cascading hash function to generate (**F**, ϵ)-dynamic reducts. With the cascading hash function, an **F**-dynamic reduct can be generated in $O(m^2 n)$ time with $O(mn)$ space where m and n are total number of attributes and total number of instances of the table. Empirical results of generating (**F**, ϵ)-dynamic reducts using five of ten most popular UCI datasets are presented and they are compared to the Rough Set Exploration System (*RSES*).

Keywords: dynamic reducts, rough sets, cascading hash function.

1   Introduction

Rough set theory was developed by Zdzislaw Pawlak [8] in the early 1980's. It is a powerful tool to analyze the imprecise and ambiguous data and it has been successfully applied to many areas including information sciences, medical, finance, business, and engineering, etc [4]. Reducts in rough set serve as using minimal number of attributes to preserve the same level of determinism of the table and researchers can apply reducts to represent knowledge of the table without superfluous attributes. Reducts can be applied to generate rules but rules calculated by all reducts from a given decision table are not appreciated because unseen cases are full of real world applications. In order to be more appropriate to classify unseen cases, the most stable reducts in the process of random sampling of original decision table are suggested to generate rules and they are called dynamic reducts [1]. Empirical tests [1, 6] showed dynamic reducts with large stability coefficients are good candidates for decision rules generation but it is time consuming to generate these reducts. In order to accelerate the generation of (**F**, ϵ)-dynamic reducts, this paper proposes the reducts generation with the cascading hash function.

In generating dynamic reducts, Bazan [1] proposed two methods. One is generating all reducts from all subtables and finds reducts which satisfies the definition of dynamic reduct, the other is to generate all reducts from the original table and finds correct ones through all subatbles. Bazan[1] showed the second

J. Yu et al. (Eds.): RSKT 2010, LNAI 6401, pp. 62–69, 2010.

method can save computation time if we can find all reducts from the original table. In this paper, we adapt the second method with the cascading hash function. Initially, original table is transformed to assure the cascading hash function takes n spaces in the worst case. Then, a set of subtables are randomly selected from the transformed table and they are converted to consistent ones including the original transformed table. In term of the number of subtables, Bazan [1] showed the minimum number of subtables is no less than 268.69 in order to estimate the stability coefficient with 0.9 confidence coefficient. Theoretically, the number 268.69 can be treated as a constant if we compare it to the total number of instances of the original table. The total size of all subtables will be treated as linear order of the original table. Finally, certain reducts of the original transformed consistent table are generated using the cascading hash function and they are verified through all subtables under the ϵ constraint of (**F**, ϵ)-dynamic reducts. Operations of generating core of (**F**, ϵ)-dynamic reducts are also presented

Five of ten most popular datasets from UCI machine learning repository are used to generate (**F**, ϵ)-dynamic reducts with $\epsilon=1$. Results are compared to the Rough Set Exploration System 2.2.2 (*RSES*) [2, 11] with default dynamic reducts parameters in *RSES*. Empirical results show the generation of (**F**, ϵ)-dynamic reducts using the cascading hash function are faster than *RSES* ranging from 6 to over 1000 times. The rest of paper is organized as follow: Section 2 reviews the basic preliminaries of rough set and dynamic reducts. Section 3 presents the cascading hash function and the table transformation. Computation of core and (**F**, ϵ)-dynamic reducts using the cascading hash function are presented in section 4 and empirical results compared to *RSES* are also shown.

2 Rough Set Preliminaries and Dynamic Reducts

Let $T = (U, C, D)$ be a decision table where $U = \{x_1, ..., x_n\}$ is a nonempty finite set of instances, $C = \{c_1, ..., c_m\}$ is a nonempty finite set of attributes called condition and D is the decision attribute. The decision D is considered as a singleton set, $D \cap C = \emptyset$, and all values in D are denoted as $V_D = \{d_1, ..., d_r\}$. $IND(Q)$ of table T represents indiscernible categories of table T using a subset of condition attribute Q and it can be denoted as $IND(Q) = \{(x, y) \in U^2 : \forall a \in Q, f(x, a) = f(y, a)\}$. Function $f : U \times A \longrightarrow V_D$ is an information function which returns the attribute values in instances for $A \subseteq C$ or $A = D$. Instances indiscernible with regard to the attribute set Q are denoted as $I_Q(x)$ where $I_Q(x) = \{y \in U : (x, y) \in IND(Q)\}$. A generalized decision function is defined as $\partial_Q : U \longrightarrow g(V_D)$ to return the decision classes of instances based on a subset of condition attributes $Q \subseteq C$ and it is denoted as $\partial_Q(x) = \{f(y, D) : y \in I_Q(x)\}$. The Q-*lower* and Q-*upper* approximation of $X_i = \{x \in U : f(x, D) = d_i\}$ are defined as $\underline{Q}X_i = \{x \in U : \partial_Q(x) = \{d_i\}\}$ and $\overline{Q}X_i = \{x \in U : i \in \partial_Q(x)\}$. In rough set, region contains instances which are certainly classified using an attribute set $Q \subseteq C$ is called Q-*positive* region, $POS_Q(D)$, of decision table T and $POS_Q(D)$ is the Q-*lower* approximation of each instance of table T. Reduct and core are special attribute set in rough set. Reduct uses a minimal number of attributes to preserve the same level of determinism of the table and the removal of any attribute in a reduct will fail to maintain the same level of determinism of the table. Core exists in all reducts and it is indispensible.

In order to be more appropriate to classify unseen cases, two types of dynamic reducts, $(\mathbf{F}, \epsilon)$ and $(\mathbf{F}, \epsilon)$-generalized, [1] are proposed. In this paper, we focus on generating $(\mathbf{F}, \epsilon)$-dynamic reducts. Let $P(C)$ denote all subtables of T. For $\mathbf{F} \subseteq P(C)$ and $\epsilon \in [0, 1]$, $(\mathbf{F}, \epsilon)$-dynamic reduct is defined as

$$DR_\epsilon(T, \mathbf{F}) = \{R \in RED(T, D) : \frac{card(\{B \in \mathbf{F} : R \in RED(B, D)\})}{card(\mathbf{F})} \geq 1 - \epsilon\}$$

(1)

$DR_\epsilon(T, \mathbf{F})$ is called $(\mathbf{F}, \epsilon)$-dynamic reduct. $\frac{card(\{B \in \mathbf{F}: R \in RED(B,D)\})}{card(\mathbf{F})}$ represents the stability coefficient of a certain reduct R of table T and it is the proportion of all subtables where R is also a certain reduct of these subtables. When $\epsilon = 0$, $(\mathbf{F}, \epsilon)$-dynamic reduct is called $\mathbf{F}$-dynamic reduct. Dynamic reducts with large stability coefficients are good candidates for decision rules generation [1, 6] but it is time consuming to generate them. To accelerate the computation of $(\mathbf{F}, \epsilon)$-dynamic reducts, reduct generation using a cascading hash function is proposed.

Table 1. 5 instances decision table

U	c_1	c_2	c_3	D
x_1	1	3	2	1
x_2	2	5	1	2
x_3	2	5	0	3
x_4	2	1	0	3
x_5	1	3	2	4

Table 2. Intermediate Results of inconsistent categories using the cascading hash function for Table 1

$\{c_1\}$	$\{x_2, x_3, x_4\}, \{x_1, x_5\}$
$\{c_1, c_2\}$	$\{x_3, x_4\}, \{x_1, x_5\}$
$\{c_1, c_2, c_3\}$	$\{x_1, x_5\}$

Table 3. Consistent transformed table of Table 1

U	c_1	c_2	c_3	D
x_1	1	1	1	5
x_2	2	2	2	2
x_3	2	2	3	3
x_4	2	3	3	3
x_5	1	1	1	5

3   Cascading Hash Function

To compute $(\mathbf{F}, \epsilon)$-dynamic reducts, we need an efficient way to discern instances for a subset of attributes in the table and we propose the cascading hash function, *cHash*. In this section, we first introduce the concept of cascading hash function. Then, a transformation of the table is shown to assure the hash space takes n in the worst case. Finally, the transformed table is converted to consistent one to cover the $(\mathbf{F}, \epsilon)$-dynamic reducts generation for consistent and inconsistent original table.

For any $A \subseteq C$ in table T, the cascading hash function, $cHash(A, T)$, returns inconsistent categories for attribute set A. Instead of utilizing all attributes at once to compute the hash value, $cHash$ takes one attribute at a time. At the beginning, $cHash$ uses the first attribute as the input and indiscernible instances can be found in O(n) time using simple hash function. Next, each inconsistent category from the first attribute is cascading to the next attribute and new indiscernible categories are generated. These operations are performed iteratively until all attributes are visited. Table 2 shows intermediate results of inconsistent categories for $A = \{c_1, c_2, c_3\}$ in Table 1. The first row of Table 2 shows two inconsistent categories exist if we only use the attribute c_1 of Table 1. Inconsistencies from attribute c_1 are cascading to the next attribute and new inconsistencies are generated in the second row of the table 2 for attribute set $\{c_1, c_2\}$. With m attributes and n instances in the table, the time of performing cascading hash function for any given set of attributes of the table takes $O(mn)$.

After we introduce the concept of cascading hash function, user may wonder the size of hash space for arbitrary attribute value. In order to assure the hash space takes n in the worst case, a transformation of each condition attribute of the original table is made. For each attribute, instances with the same value are transformed to their order of occurrence of that value when we browse instances from the top to the bottom. Condition attributes in Table 3 show the transformation results of Table 1 and attribute c_2 is taken as an example. Attribute c_2 of Table 1 contains three unique values $\{3, 5, 1\}$ and the order of occurrence of each value is 1, 2, and 3 when we browse instances from the top to the bottom. Original value of the attribute is transformed to its order of occurrence and they are shown in the attribute c_2 of Table 3. By adapting the transformed table, the worst case of hash space using cascading hash function occurs when values in an attribute are totally different and this takes n spaces.

In order to cover the (**F**, ϵ)-dynamic reducts generation for consistent and inconsistent original table, the transformed original table must be converted to a consistent table by mapping decision class of all inconsistent instances to a new decision class. Wang [12] proved the correctness of generating certain reducts using the consistent transformed table and this is also applicable to the generation of (**F**, ϵ)-dynamic reducts because (**F**, ϵ)-dynamic reducts are certain reducts of the original table and all subtables with ϵ constraint. Example of mapping the decision classes of inconsistent instances $\{x_1, x_5\}$ of Table 1 to a new decision class 5 is shown in Table 3 and the conversion takes $O(n)$ time by mapping the inconsistent categories from cHash to a new decision class.

4   Generate (F, ϵ)-Dynamic Reducts Using Cascading Hash Function

After we present the concept of cascading hash function, steps of generating (**F**, ϵ)-dynamic reducts are shown. At the beginning, each attribute value of the original table is transformed using the table transformation in section 3. Then, a family of subtables is randomly selected from the transformed table and all tables are converted into consistent. After all tables are prepared, core of (**F**, ϵ)-dynamic reducts is generated before the reduct generation. Core is a special set of attributes and it exists in all reducts. Efficient core generation can accelerate the computation of reducts and core of (**F**, ϵ)-dynamic reducts is defined as follow.

Proposition 1. $T = (U, C, D)$ is the consistent transformed decision table and T' is the set of all consistent transformed subtables. An attribute $a \in C$ and $a \in Core(C, DR_\epsilon)$ if $T = (U, C - \{a\}, D)$ is inconsistent or $\frac{Const(C-\{a\},T')}{|T'|} < 1 - \epsilon$.

$Core(C, DR_\epsilon)$ represents core of $(\mathbf{F}, \epsilon)$-dynamic reducts. $Const(C - \{a\}, T')$ is the number of consistent subtables in T' after we remove attribute a and $|T'|$ is the total number of subtables in T'. The truth of proposition 1 is trivial because the removal of attribute a fails to preserve the positive region of table T and it is impossible to find any subset of attributes from $C - \{a\}$ to meet the definition of $(\mathbf{F}, \epsilon)$-dynamic reduct in section 2 when $\frac{Const(C-\{a\},T')}{|T'|} < 1 - \epsilon$ after we remove attribute a.

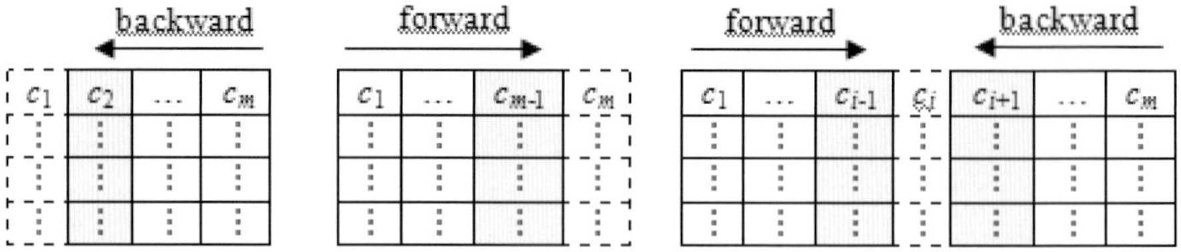

Fig. 1. Core attributes verification for attribute c_1, c_m, and c_i.

Computation of core of $(\mathbf{F}, \epsilon)$-dynamic reducts is divided into three parts which are illustrated in Fig. 1. In Fig. 1, two cascading hash tables are generated for all tables. A forward cascading hash table using attributes from c_1 to c_m is generated and the c_i^{th} column of the forward cascading hash table contains inconsistent information for attributes from c_1 to c_i. Another cascading hash table is called backward which uses attributes from c_m to c_1 and the c_i^{th} column of the backward cascading hash table contains inconsistent information for attributes from c_m to c_i. The first part of Fig. 1 shows the core verification for the first attribute c_1 and attribute c_2 of the backward cascading hash tables is used. Attribute c_2 of the backward cascading hash table contains inconsistent information from all attributes except attribute c_1. If inconsistencies exist in attribute c_2 of the backward cascading hash table or $\frac{Const(C-\{a\},T')}{|T'|}$ of attribute c_2 in the backward cascading hash tables of T' is less than 1-ϵ, attribute c_1 is a core attribute. Similar operations are applied to verify the last attribute c_m using attribute c_{m-1} of the forward cascading hash tables and this is shown in the second part of Fig. 1. To verify attribute c_i which is not the first or the last attribute, attribute c_{i-1} of the forward cascading hash table and attribute c_{i+1} of the backward cascading hash table are used to compute new inconsistent information after we remove attribute c_i and this is shown in the third part of Fig. 1. Section 3 shows the creation of forward and backward cascading hash tables takes O(mn) time with O(mn) space and the verification of each attribute takes O(n) time. The overall of core generation takes O(mn) time.

After we generate the core, certain reducts of original table are generated by merging attributes into core until they reach the definition of $(\mathbf{F}, \epsilon)$-dynamic reducts. To speed up the search of certain reducts, the current rules size, CRS, is adapted. CRS was proposed by Wang [12] and $CRS(P,T)$ is defined as the summation of the number of consistent instances and inconsistent categories using the attribute set $P \subseteq C$ in table T. With CRS, new certain reduct definition is shown.

Proposition 2. $T = (U, C, D)$ is the consistent transformed decision table, T' is the set of all consistent transformed subtables, and $P \subseteq C$. P is a certain reduct of T if $CRS(P,T)=CRS(C,T)$ and $\forall Q \subset P, CRS(Q,T) <CRS(P,T)$.

The truth of proposition 2 is trivial. When table T is consistent, $CRS(C,T)$ is equal to the table size because the positive region of table T contains all instances. For $P \subseteq C$, $CRS(P,T) =CRS(C,T)$ and $\forall Q \subset P, CRS(Q,T) <CRS(P,T)$, this shows attribute set P preserves original positive region of table T and none of its proper subset can. This proves attribute set P is a certain reduct of table T.

After we redefine the certain reduct using CRS, propositions 3 is presented to accelerate the search of candidate attributes of certain reducts.

Proposition 3. $T = (U, C, D)$ is the consistent transformed decision table, T' is the set of all consistent transformed subtables, and $P \subseteq C$. [12]

(a) Let $P \subseteq C$. An attribute $a \in \{C - P\}$ cannot be a candidate of certain reducts starting with P when $CRS(P \cup \{a\}, T) =CRS(P,T)$.
(b) Let $P \subseteq C$. Attribute $a \in \{C - P\}$ satisfying proposition 3.a cannot be a candidate of certain reducts starting with P when $\exists q \in P$ such that $CRS(\{P - \{q\}\} \cup \{a\}, T) =CRS(P \cup \{a\}, T)$.
(c) Let $P \subseteq C$. If S is a set of candidate attributes derived from propositions 3.a and 3.b for certain reducts starting with P and we merge an attribute $a \in S$ into P. New candidate attributes for certain reducts starting with $P \cup \{a\}$ can be found in $S - \{a\}$.

Wang [12] proposed proposition 3 and a certain reduct of table T is generated in $O(m^2 n log n)$ by the sorting mechanism. With the cascading hash function, the verification of proposition 3.a takes $O(n)$ time using the forward cascading hash table. Operations of verifying proposition 3.b take $O(mn)$ time by using forward and backward cascading hash tables like the core generation. Next, function $dReducts$ is presented to compute (F, ϵ)-dynamic reducts using cascading hash function.

```
Function dReducts(ϵ)
    Original table T is transformed and a family of subtables
    , T', of table T is randomly selected;
    Convert all tables into consistent ones;
    Generate Core of (F,ϵ)-dynamic reducts;
    IF Core satisfies proposition 2 and Const(C−{a},T')/|T'| ≥ 1 − ϵ THEN
        Core is the only (F,ϵ)-dynamic reduct and exit from the
        algorithm;
    ELSE
```

```
initAttrs = Core; Set candAttrs = C − Core;
    Find candAttrs for initAttrs using proposition 3.
    WHILE (candAttrs is not empty) DO
  IF (initList∪first attribute r of candAttrs) is a certain
  reduct of table T THEN
      initList∪{r} is a (F,ε)-dynamic reduct if Const(C−{a},T′)/|T′| ≥ 1 − ε;
  ELSE
      Recursively generate (F,ε)-dynamic reducts for initList∪{r}
      using proposition 3 from attributes candAttrs − {r} when
      attribute sets reach certain reducts of table T and they
      meet the definition of (F,ε)-dynamic reduct;
  END
      Remove attribute r from candAttrs;
    END
  END
END
```

Function $dReducts$ computes all certain reducts of table T and $(\mathbf{F}, \epsilon)$-dynamic reducts are generated by verifying certain reducts through all subtables under the ϵ constraints. The computation of a certain reduct of table T using proposition 3 with the cascading hash function takes $O(m^2 n)$ because we have to perform proposition 3.a and 3.b m times in the worst case. The verification of each certain reduct through all subtables takes $O(mn)$ time by applying similar operations of core generation. If user wants to compute $\mathbf{F}$-dynamic reducts, all tables can be merged and a new attribute is introduced to differentiate each table. An $\mathbf{F}$-dynamic reduct can be generated in $O(m^2 n)$ time using proposition 3 with cascading hash function. Function $dReducts$ is implemented in an uncompiled Matlab codes and five of ten most popular datasets with categorical or integer data from UCI machine learning repository are selected to generate $(\mathbf{F}, \epsilon)$-dynamic reducts with $\epsilon=1$. Missing data from Adult dataset are removed. Each dataset is executed three times and the average execution time is taken as the result. Results are compared to the Rough Set Exploration System 2.2.2 ($RSES$) [2, 11] with default dynamic reducts parameters in RSES. A 32-bit Windows 2008 based PC with a Quad Core 2.4GHz and 8G main memory is used for all tests. Table 4 shows the results of generating of $(\mathbf{F}, \epsilon)$-dynamic reducts and function $dReducts$ is faster than RSES ranging from 6 to over 1000 times.

Table 4. Execution results using five most popular UCI datasets

Datasets	Instances	Attributes	No. of reducts	$dReducts$ (seconds)	$RSES$ (seconds)
Iris	150	5	4	0.25	1.56
SPECT heart	267	23	3	0.9	9.01
Car evaluation	1728	7	1	0.84	80.58
Pokerhand(train)	25010	11	8	81.14	24950.33
Adult(train)	30162	15	2	30.69	31082.67

5   Conclusion

Dynamic reducts with large stability coefficients are good candidates for decision rules generation but it is time consuming to generate them. This paper presents an algorithm using the cascading hash function to compute core and (**F**, ϵ)-dynamic reducts. Operations of generating core takes $O(mn)$ time and the generation of an **F**-dynamic reduct may take $O(m^2n)$ time. With the algorithm in this paper, users can be more efficient to apply dynamic reducts to applications with larger datasets.

References

1. Bazan, J.G.: A comparison of dynamic and non-dynamic rough set methods for extracting laws from decision table. In: Polkowski, L., Skowron, A. (eds.) Rough Sets in Knowledge Discovery, pp. 321–365. Physica-Verlag, Heidelberg (1998)
2. Bazan, J.G., Szczuka, M.S., Wróblewski, J.: A new version of rough set exploration system. In: Alpigini, J.J., Peters, J.F., Skowron, A., Zhong, N. (eds.) RSCTC 2002. LNCS (LNAI), vol. 2475, pp. 397–404. Springer, Heidelberg (2002)
3. Hu, X., Lin, T.Y., Han, J.C.: A new rough sets model based on database systems. Fundamenta Informaticae 59, 135–152 (2004)
4. Komorowski, J., Pawlak, Z., Polkowski, L., Skowron, A.: Rough sets: a tutorial. In: Pa, S.K., Skowron, A. (eds.) Rough Fuzzy Hybridization: A New Trend in Decision-Making, pp. 3–98. Springer, Telos (1999)
5. Kumar, A.: New techniques for data reduction in database systems for knowledge discovery applications. Journal of Intelligent Information Systems 10(1), 31–48 (1998)
6. Miao, D., Hou, L.: A comparison of rough set methods and representative inductive learning algorithms. Fundamenta Informaticae 59, 203–219 (2004)
7. Nguyen, S.H., Nguyen, H.S.: Some efficient algorithms for rough set methods. In: Proc. of Information Processing and Management of Uncertainty in Knowledge-Based Systems, pp. 1451–1456 (1996)
8. Pawlak, Z.: Rough sets. International Journal of Computer and Information Sciences 11, 341–356 (1982)
9. Pawlak, Z.: Rough Sets: Theoretical Aspects of Reasoning about Data. Kluwer, Netherlands (1991)
10. Skowron, A., Rauszer, C.: The discernibility matrices and functions in information Systems. Fundamenta Informaticae 15(2), 331–362 (1991)
11. Skowron, A., Bazan, J., Szczuka, M.S., Wróblewski, J.: Rough Set Exploration System (version 2.2.2), http://logic.mimuw.edu.pl/~rses/
12. Wang, P.C.: Highly scalable rough set reducts generation. Journal of Information Science and Engineering 23, 1281–1298 (2007)

Incorporating Great Deluge with Kempe Chain Neighbourhood Structure for the Enrolment-Based Course Timetabling Problem

Salwani Abdullah[1], Khalid Shaker[1], Barry McCollum[2], and Paul McMullan[2]

[1] Data Mining and Optimization Research Group (DMO),
Center for Artificial Intelligence Technology,
Universiti Kebangsaan Malaysia, 43600 Bangi, Selangor, Malaysia
{salwani,khalid}@ftsm.ukm.my
[2] Department of Computer Science, Queen's University Belfast,
Belfast BT7 1NN United Kingdom
{b.mccollum,p.mcmullan}@qub.ac.uk

Abstract. In general, course timetabling refers to assignment processes that assign events (courses) to a given rooms and timeslots subject to a list of hard and soft constraints. It is a challenging task for the educational institutions. In this study we employed a great deluge algorithm with kempe chain neighbourhood structure as an improvement algorithm. The Round Robin (RR) algorithm is used to control the selection of neighbourhood structures within the great deluge algorithm. The performance of our approach is tested over eleven benchmark datasets (representing one large, five medium and five small problems). Experimental results show that our approach is able to generate competitive results when compared with previous available approaches. Possible extensions upon this simple approach are also discussed.

Keywords: Great Deluge, Kempe Chain, Round Robin, Course Timetabling.

1 Introduction

Course timetabling problems have long attracted the attention of the Operational Research and Artificial Intelligence communities. In addition, variations of the problem have been the subject of two competitions via the website at http://www.metaheuristics.org. A wide variety of approaches for constructing course timetables have been described and discussed in the literature. Graph colouring heuristics became the earliest approaches in solving timetabling problems. Later, metaheuristic methods have shown a success in solving this problem. These methods can be categorized into population-based and local search-based methods [12]. Population-based methods applied to course timetabling problem such as Genetic algorithm [9]; Ant algorithm [11]; and Harmony Search [5]. The local search-based methods employed on the same problem are dual simulated annealing [1]; variable neighbourhood search [2]; graph-based hyper heuristic

J. Yu et al. (Eds.): RSKT 2010, LNAI 6401, pp. 70–77, 2010.

[6]; and non-linear great deluge [8]; extended great deluge [10]. The hybridization method with an aim to take the best idea from one approach and to incorporate it with another good (or better) idea from another (or more) approach(es) have also shown some success. For example hybridization iterated local search with mutation operator [3]; electromagnetic-like mechanism with great deluge [14].

2 Problem Description

The Enrolment-based Course Timetabling Problem considered in this work was initially defined by the Metaheuristics Network[1]. This problem was discussed and an assignment of lecture events to timeslots and rooms according to a variety of hard and soft constraints. The problem description that is employed in this paper is adapted from the description presented in [11]. This problem includes four hard constraints and three soft constraints as follows:

Hard constraint:
Event conflict i.e. no student can be assigned to more than one course at the same time (coded as H_1); *Room features* i.e. the room should satisfy the features required by the event (coded as H_2); *Room capacity* i.e. the number of students attending the event should be less than or equal to the capacity of the room (coded as H_3); *Room occupancy* i.e. no more than one event is allowed at a timeslot in each room (coded as H_4).

Soft constraints:
Event in the last timeslot i.e. a student shall not have to sit a course that is scheduled in the last timeslot of the day (coded as S_1); *Two consecutive events* i.e. a student shall not have more than 2 consecutive events (coded as S_2); *One event a day* i.e. a student shall not have to sit a single course on a day (coded as S_3).

Hard constraints act an inviolable requirement. A timetable which meets the hard constraints is cognized as a *feasible* solution.

The problem has: A set of N courses, $e = \{e_1, \ldots, e_N\}$; 45 timeslots; A set of R rooms; A set of F room features; A set of M students.

The main objective is to minimise the violation of the soft constraints in a *feasible* solution that later represents the quality of the obtained solution. A solution consists of an ordered list of length $|N|$ where the position corresponds to the events i.e. position i corresponds to event e_i for $i = 1,\ldots, |N|$. The values for each position are a number between 0 to 44 corresponding to the timeslot index and 0 to $|R - 1|$ corresponding to the room index. For example a timeslot vector is given as (0,17,30,,10) and a room vector is given as (4,3,0,,3) means that event e_1 is scheduled in timeslot 0 at room 4. Event e_2 is scheduled in timeslot 17 at room 3 and finally event $e_{|N|}$ is scheduled in timeslot 10 at room 3.

---

[1] http://www.metaheuristics.net

3   The Proposed Algorithm

The algorithm consists of two phases i.e. constructive heuristic to obtain a feasible initial solution; and an improvement algorithm with an aim to optimise the violation of the soft constraints while maintaining the feasibility of the solutions that works with a set of neighbourhood structures.

3.1   Neighbourhood Structures

Neighbourhood structure is one of the key factors for a local search algorithm to get good performance. In this work, we employed three different types of neighbourhood structures. Their explanations are outlined as follows:

Nbs_1: Move a random selected event to a feasible timeslot that can generate the lowest penalty cost; Nbs_2: Swap two events at random; Nbs_3: Kempe chain neighbourhood - select one event and timeslot at random, then apply the kempe chain from [13].

Kempe chain neighbourhood involves swapping a subset of the courses in two distinct timeslots. The main reason of using kempe chain neighbourhood is to avoid the stagnation state occurs when there is no improvement on the quality of the best feasible solution obtained so far during the improvement process, as the kempe chain (when it occurs) will do shaking to the solution, thus appear a new empty timeslots. This may contributes in finding better solutions later. More details on kempe chain neighbourhood structures can be found in Burke *et al.* [7] and Thompson *et al.* [13].

3.2   Phase 1: Constructive Heuristic

The initial solution is produced using a constructive heuristic called least saturation degree which starts from an empty timetable [10]. This feasible solution is obtained by adding or removing appropriate events (courses) from the schedule based on room availability (we attempt to schedule those courses with the least room availabilities earlier on in the process), without taking into account any of the soft constraints, until the hard constraints are met. If a feasible solution is found, the algorithm stops. Otherwise, the neighbourhood moves (Nbs_1 and/or Nbs_2,) are applied to attempt to move from an infeasible to feasible solution. Nbs_1 is applied for a certain number of iterations. If a feasible solution is met, then the algorithm stops. Otherwise the algorithm continues by applying a Nbs_2 neighbourhood structure for a certain number of iterations. In this work, across all instances tested, the schedules are made feasible before starting the improvement algorithm.

3.3   Phase 2: Improvement Algorithm Using a Great Deluge with Kempe Chain Neighbourhood Structure

The pseudo code for the improvement approach which is implemented in this paper is given in Fig. 1.

Initialization Phase
$Sol_{GD} = Sol;$
$Sol_{best} = Sol;$
$f(Sol_{GD}) = f(Sol);$
$f(Sol_{best}) = f(Sol);$
Set number of iteration, NumOfIteGD;
Set not_improving_ length_GD;
Set expected final solution, Sol_{opt};
Set initial level: level = $f(Sol_{GD})$;
Set $\Delta B = ((f(Sol_{GD}) - Sol_{opt})/(NumOfIteGD)$;
Set not_improving_counter = 0;
Set iterations counter, Iter = 0;

Improvement Phase
Iter = 0;
    do while (Iter)< (NumOfIteGD)
        Define a neighbourhood of Sol_{GD} based on RR algorithm to generate $TempSol_{GD}$;
        $Sol_{hill} = TempSol_{GD}$;
        $Sol_{besthill} = TempSol_{GD}$;
        Set time limit;
        do while (not time limit)
            Define a neighbourhood Nbs_2 on Sol_{hill} to obtain $TempSol_{hill}$;
            if $(f(TempSol_{hill}) < f(Sol_{besthill}))$
                $Sol_{hill} = TempSol_{hill}$;
                $Sol_{besthill} = TempSol_{hill}$;
            end if
        end do;
        if $(Sol_{besthill} < Sol_{best}$
            $Sol_{best} = Sol_{besthill}$;
            $Sol_{GD} = Sol_{besthill}$;
            not_improving_counter = 0;
            level = level - ΔB;
        else
        if $(f(Sol_{besthill}) \leq$ level)
            $Sol_{GD} = Sol_{besthill}$;
            not_improving_counter = 0;
            level = level - ΔB;
        else
            not_improving_counter++;
            if (not_improving_counter==not_improving_length_GD)
                level= level + random(1, 3);
        end if
        Iter ++;
    end do;

Fig. 1. The pseudo code of the great deluge with kempe chain neighbourhood structure

The algorithm starts with a feasible initial solution (*Sol*) which is generated by a constructive heuristic in Phase 1. In the approach presented in this paper, a set of the neighbourhood structures is applied as in section 3.1. This application of neighbourhood structures is controlled using the Round Robin Scheduling algorithm (RR), described in section 3.4. The hard constraints are never violated during the timetabling process. The hard constraints are never violated during the course of the timetabling process.

At the initialization phase, the algorithm sets initial values for all parameters used, such as the current solution (Sol_{GD}) and best solution (Sol_{best}) are set to be Sol, the number of iteration ($NumOfIteGD$) which is set to be 200,000 (as in [1] and [14]), expected final solution ($f(Sol_{opt})$), not improving counter ($not\_improving\_length\_GD$) which is set to be 20 (based on prelimenary experiments), the "level" in the great deluge ($level$), etc.

At the improvement phase, a neighbourhood structure which is defined using a round-robin algorithm is employed on Sol_{GD} to generate a candidate solution, $TempSol_{GD}$. Set ($Sol_{hill} \leftarrow TempSol_{GD}$) and ($Sol_{besthill} \leftarrow TempSol_{GD}$). The hill climbing algorithm with Nbs_2 neighbourhood structure is then applied on Sol_{hill} for a certain time (set to be 6 seconds based on preliminary experiments) with an aim to further increase the quality of the solution obtained, called $TempSol_{hill}$. This neighbourhood structure is chosen because we believe that the operation of Nbs_2 does not take longer time. The $f(TempSol_{hill})$ is compared to the $f(Sol_{hill})$. If it is better, then the current and best solutions within the hill climbing algorithm operations are updated ($Sol_{hill} \leftarrow TempSol_{hill}$; $Sol_{besthill} \leftarrow TempSol_{hill}$).

The best solution obtained from the hill climbing ($Sol_{besthill}$) later will be compared with the best solution in hand (Sol_{best}). If a better solution is found, the current solution (Sol_{GD}) and the best solution (Sol_{best}) will be updated ($Sol_{GD} \leftarrow Sol_{hill}$, $Sol_{best} \leftarrow Sol_{hill}$) and the level will be decreased ($level \leftarrow level - \Delta B$).Otherwise $f(Sol_{besthill})$ will be compared against the $level$. If the quality of $Sol_{besthill}$ is better than the $level$, the current solution, Sol_{GD} will be updated as $Sol_{besthill}$. Otherwise, not improving counter ($not\_improving\_counter$) will be increased by 1. If $not\_improving\_counter$ is equal to a constant $not\_improving\_length\_GD$, the $level$ will be increased with a certain number (based on a random generated number between 1 and 3 in this experiment) in order to allow some flexibilities in accepting a worse solution.

3.4   Round-Robin Algorithm (RR)

The RR algorithm is employed to control the selection of the neighbourhood structures, which are ordered in sequence. In this work the neighbourhood structures are ordered as Nbs_1, Nbs_2 and Nbs_3 (see subsection 3.1). A time slice or quantum is assigned for each neighbourhood structure in equal portions, in a circular order. The neighbourhood structure is dispatched in a FIFO manner at a given quantum denoted as qtime (which is set to 5 seconds). Note that in this paper, all parameters used are based on a number of preliminary experiments. The pseudo code for the RR algorithm is presented in Fig. 2.

Set quantum time, qtime;
Set a sequence of neighbourhood structures in a queue which is ordered as Nbs_i where
$i \in 1,,K$ and $K = 3$;
Set initial value to counter_qtime;
    do while (qtime not met)
        Select a neighbourhood structure Nbs_i in the queue where $i \in 1,,K$ where $K = 3$;
    A: Apply Nbs_i on current solution,Sol_{GD} to generate new solution $TempSol_{GD}$;
       if there is an improvement on the quality of the solution
          repeat label A
       else
         insert Nbs_i into the queue;
       counter_qtime = q_time;
    end do

Fig. 2. The pseudo code for RR algorithm

4   Simulation Results

The algorithm was implemented on a Pentium 4 Intel 2.33 GHz PC Machine using Matlab on a Windows XP Operating System. For each benchmark dataset the algorithm was run for 200,000 evaluations with 10 test-runs to obtain an average value. We have evaluated our results on the instances taken from Socha et al. [17] and which are available at http://iridia.ulb.ac.be/~msampels/tt.data/. Those were proposed by the Metaheuristics Network that need to schedule 100-400 courses into a timetable with 45 timeslots corresponding to 5 days of 9 hours each, whilst satisfying room features and room capacity constraints. They are divided into three categories: small, medium and large. We deal with 11 instances: 5 small, 5 medium and 1 large.

Table 1 shows the comparison of the approach in this paper with other available approaches in the literature on the data instances. This includes iterative

Table 1. Comparison of our results with other approaches in the literature

Dataset	Our method	M1	M2	M3	M4	M5	M6	M7	M8	M9
Small1	**0**	**0**	6	0	0	0	3	0	0	0
Small2	**0**	**0**	7	0	0	0	4	0	0	0
Small3	**0**	**0**	3	0	0	0	6	0	0	0
Small4	**0**	**0**	3	0	0	0	6	0	0	0
Small5	**0**	**0**	4	0	0	0	0	0	0	0
Medium1	98	242	372	317	221	**80**	140	96	93	124
Medium2	113	161	419	313	147	105	130	**96**	98	117
Medium3	**123**	265	359	357	246	139	189	135	149	190
Medium4	100	181	348	247	165	88	112	**79**	103	132
Medium5	135	151	171	292	130	88	141	87	98	**73**
Large	610	-	1068	-	529	730	876	683	680	**424**

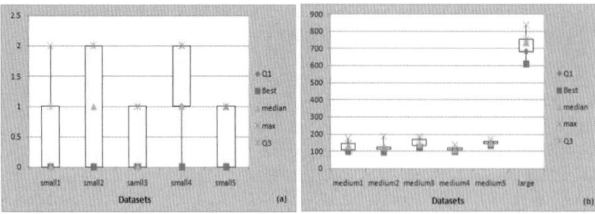

Fig. 3. (a) and (b) Box plots of the penalty costs for small, medium and large datasets

improvement algorithm by M1-Abdullah et al. [3]; M2-graph hyper heuristic by Burke et al. [6]; M3-variable neighbourhood search with tabu by Abdullah et al. [2]; M4-hybrid evolutionary approach by Abdullah et al. [4]; M5-extended great deluge by McMullan [10]; M6-non linear great deluge by Landa-Silva and Obit [8]; M7-electromagnetic-like mechanism with great deluge by Turabieh et al. [14]; M8-dual simulated annealing by Abdullah et al. [1]; and M9- harmony search by Al-Betar et al. [5]. Note that the best results are presented in bold. Our algorithm is capable to find feasible timetables for all eleven cases and obtain competitive results.

Fig. 3 (a) and (b) show the box plots of the cost when solving *small, medium* and *large* instances, respectively. The results for the *medium* datasets are less dispersed compared to *large* and *small*. We can see that the *median* is closer to the best in *small1, small3* and all *medium* datasets; however this is not the case for *small2, small4, small5* and *large* datasets. We believe that the size and the landscape of the search space is different from one problem to others even though there are form the same groups of datasets, due to the fact that the dispersion of solution points are significantly different from one to another. Thus, we believe that by combining several neighbourhood structures helps to compensate against the ineffectiveness of using each type of neighbourhood structure in isolation.

5   Conclusion and Future Work

The overall goal of this paper is to investigate a great deluge algorithm with kempe chain neighbourhood structure in the course timetabling problem. A round-robin algorithm is used to control the selection of the neighbourhood structures from the list. Preliminary comparisons indicate that our approach is able to produce both feasible and good quality timetables and competitive with other approaches in the literature applied to the same domain here. Our future work will try to intelligently select a subset of neighbourhood structures from a much larger pool of neighbourhoods and applied to curriculum-based course timetabling problems.

References

1. Abdullah, S., Shaker, K., McCollum, B., McMullan, P.: Dual sequence simulated annealing with round-robin approach for university course timetabling. In: Cowling, P., Merz, P. (eds.) EvoCOP 2010. LNCS, vol. 6022, pp. 1–10. Springer, Heidelberg (2010)
2. Abdullah, S., Burke, E.K., McCollum, B.: An investigation of variable neighbourhood search for university course timetabling. In: The 2nd Multidisciplinary International Conference on Scheduling: Theory and Applications (MISTA), pp. 413–427 (2005)
3. Abdullah, S., Burke, E.K., McCollum, B.: Using a randomised iterative improvement algorithm with composite neighbourhood structures for university course timetabling. In: Metaheuristics: Progress in complex systems optimization (Operations Research / Computer Science Interfaces Series), ch. 8. Springer, Heidelberg (2007a) ISBN:978-0-387-71919-1
4. Abdullah, S., Burke, E.K., McCollum, B.: A hybrid evolutionary approach to the university course timetabling problem. In: IEEE Congres on Evolutionary Computation, pp. 1764–1768 (2007b) ISBN: 1-4244-1340-0
5. Al-Betar, M., Khader, A., Yi Liao, I.: A Harmony Search with Multi-pitch Adjusting Rate for the University Course Timetabling. In: Geem, Z.W. (ed.) Recent Advances in Harmony Search Algorithm. SCI, vol. 270, pp. 147–161. Springer, Heidelberg (2010)
6. Burke, E.K., Meisels, A., Petrovic, S., Qu, R.: A graph-based hyper-heuristic for timetabling problems. European Journal of Operational Research 176, 177–192 (2007)
7. Burke, E., Eckersley, A., McCollum, B., Petrovic, S., Qu, R.: Hybrid variable neighbourhood approaches to university exam timetabling. Technical Report NOTTCS-TR-2006-2, University of Nottingham, School of CSiT (2006)
8. Landa-Silva, D., Obit, J.H.: Great deluge with non-linear decay rate for solving course timetabling problem. In: The fourth International IEEE conference on Intelligent Systems, Varna, Bulgaria (2008)
9. Lewis, R., Paechter, B.: New crossover operators for timetabling with evolutionary algorithms. In: Lotfi (ed.) Proceedings of the 5th International Conference on Recent Advances in Soft Computing, UK, December 16-18, pp. 189–194 (2004)
10. McMullan, P.: An extended implementation of the great deluge algorithm for course timetabling. In: Shi, Y., van Albada, G.D., Dongarra, J., Sloot, P.M.A. (eds.) ICCS 2007, Part I. LNCS, vol. 4487, pp. 538–545. Springer, Heidelberg (2007)
11. Socha, K., Knowles, J., Samples, M.: A max-min ant system for the university course timetabling problem. In: Dorigo, M., Di Caro, G.A., Sampels, M. (eds.) ANTS 2002. LNCS, vol. 2463, pp. 1–13. Springer, Heidelberg (2002)
12. Qu, R., Burke, E.K., McCollum, B., Merlot, L.T.G., Lee, S.Y.: A Survey of Search Methodologies and Automated System Development for Examination Timetabling. Journal of Scheduling 12(1), 55–89 (2009)
13. Thompson, J., Dowsland, K.: A robust simulated annealing based examination timetabling system. Computers & Operations Research 25, 637–648 (1998)
14. Turabieh, H., Abdullah, S., McCollum, B.: Electromagnetism-like Mechanism with Force Decay Rate Great Deluge for the Course Timetabling Problem. In: Wen, P., Li, Y., Polkowski, L., Yao, Y., Tsumoto, S., Wang, G. (eds.) RSKT 2009. LNCS, vol. 5589, pp. 497–504. Springer, Heidelberg (2009)

Ordered Weighted Average Based
Fuzzy Rough Sets

Chris Cornelis[1,*], Nele Verbiest[1], and Richard Jensen[2]

[1] Department of Applied Mathematics and Computer Science,
Ghent University, Gent, Belgium
{Chris.Cornelis,Nele.Verbiest}@UGent.be
[2] Department of Computer Science, Aberystwyth University, Wales, UK
rkj@aber.ac.uk

Abstract. Traditionally, membership to the fuzzy-rough lower, resp. upper approximation is determined by looking only at the worst, resp. best performing object. Consequently, when applied to data analysis problems, these approximations are sensitive to noisy and/or outlying samples. In this paper, we advocate a mitigated approach, in which membership to the lower and upper approximation is determined by means of an aggregation process using ordered weighted average operators. In comparison to the previously introduced vaguely quantified rough set model, which is based on a similar rationale, our proposal has the advantage that the approximations are monotonous w.r.t. the used fuzzy indiscernibility relation. Initial experiments involving a feature selection application confirm the potential of the OWA-based model.

Keywords: fuzzy rough sets, vaguely quantified rough sets, ordered weighted average, aggregation operators, noise tolerance, data analysis.

1 Introduction

The operations of lower and upper approximation are at the heart of rough set theory [6] and many of its applications; based on objects' indiscernibility, they determine those objects that certainly, resp. possibly belong to a given concept. More specifically, an object belongs to the lower approximation of a concept if all objects indiscernible from it belong to the concept as well, and to its upper approximation if at least one object indiscernible from it belongs to the concept.

Fuzzy rough set theory [5] extends the approximation operators by allowing the indiscernibility relation as well as the concept itself to become fuzzy. While this generalization provides for greater flexibility, the commonly used definitions of fuzzy-rough approximations (see e.g. [7]) do not address the full potential of hybridization; in particular, the process of determining the membership degrees to the approximations still depends on a single object, as dictated by the

---

* Chris Cornelis would like to thank the Research Foundation—Flanders for funding his research.

J. Yu et al. (Eds.): RSKT 2010, LNAI 6401, pp. 78–85, 2010.

quantifiers $\forall$ and $\exists$ from the crisp definitions. As a consequence of this non-compensatory approach, slight changes to the data sometimes result in drastically different approximations. This, in turn, impacts the robustness of data analysis applications based on them, such as feature selection and classification.

These considerations inspired the vaguely quantified rough set (VQRS) model [1]: by replacing the crisp quantifiers $\forall$ and $\exists$ by softer versions representing *most* and *some*, it ensures that several objects contribute to an object's membership degree to the approximations. Unfortunately, this benefit goes at the expense of certain theoretically desirable properties for (fuzzy-)rough approximations (see e.g. [2] for an overview). Most importantly, as was noted in [3], the failure of monotonicity of the VQRS lower approximation w.r.t. the fuzzy indiscernibility relation hampers feature selection based on it.

In this paper, we propose an alternative fuzzy-rough hybridization based on ordered weighted average (OWA) operators [9], in which membership degrees to the approximations are computed by an aggregation process. As such, like the VQRS approach, our proposal allows for compensation; on the other hand, compared to VQRS, the OWA-based approach has a number of important benefits: 1) it is monotonous w.r.t. the fuzzy indiscernibility relation, 2) the traditional fuzzy-rough approximations can be recovered by a particular choice of the OWA weight vectors, and 3) the VQRS rationale can be maintained by introducing vague quantifiers into the OWA model, following the proposal in [10].

The remainder of the paper is organized as follows: in Section 2, we recall necessary preliminaries about fuzzy-rough hybrizidation and OWA operators, while in Section 3 we introduce OWA-based fuzzy rough sets, give some examples and examine their monotonicity characteristics. In Section 4, we apply our model to a feature selection task, comparing it to the traditional fuzzy-rough approach as well as to the VQRS model. Finally, in Section 5, we conclude and outline future work.

2   Preliminaries

2.1   Fuzzy Rough Sets

According to Pawlak [6], the lower and upper approximation of a crisp set $A \subseteq X$ w.r.t. an equivalence relation R are defined by, for y in X,

$$y \in R{\downarrow}A \text{ iff } [y]_R \subseteq A \tag{1}$$

$$y \in R{\uparrow}A \text{ iff } [y]_R \cap A \neq \emptyset \tag{2}$$

or, equivalently,

$$y \in R{\downarrow}A \text{ iff } (\forall x \in X)((x,y) \in R \Rightarrow x \in A) \tag{3}$$

$$y \in R{\uparrow}A \text{ iff } (\exists x \in X)((x,y) \in R \wedge x \in A) \tag{4}$$

When A is a fuzzy set and R is a fuzzy relation[1] in X, equations (3) and (4) can be extended using a fuzzy implication[2] $\mathcal{I}$ and a t-norm[3] $\mathcal{T}$ to

$$(R{\downarrow}A)(y) = \inf_{x \in X} (\mathcal{I}(R(x,y), A(x))) \tag{5}$$

$$(R{\uparrow}A)(y) = \sup_{x \in X} (\mathcal{T}(R(x,y), A(x))) \tag{6}$$

for $y \in X$. [7] When A and R are both crisp, (1) and (2) are recovered.

2.2 Vaguely Quantified Rough Sets

The inf and sup operations in (5) and (6) play the same role as the $\forall$ and $\exists$ quantifiers in (1) and (2). Because of this, a change to a single object can have a large impact on the approximations. This makes fuzzy rough sets equally susceptible to noisy data —which is difficult to rule out in real-life applications— as their crisp counterparts.

To make up for this shortcoming, it was proposed in [1] to replace the universal and existential quantifier by means of vague quantifiers like *most* and *some*. Mathematically, vague quantifiers are modeled by a regularly increasing fuzzy quantifier: an increasing $[0,1] \to [0,1]$ mapping Q that satisfies the boundary conditions $Q(0) = 0$ and $Q(1) = 1$. As an example, the following parametrized formula, for $0 \le \alpha < \beta \le 1$, and x in $[0,1]$,

$$Q_{(\alpha,\beta)}(x) = \begin{cases} 0, & x \le \alpha \\ \frac{2(x-\alpha)^2}{(\beta-\alpha)^2}, & \alpha \le x \le \frac{\alpha+\beta}{2} \\ 1 - \frac{2(x-\beta)^2}{(\beta-\alpha)^2}, & \frac{\alpha+\beta}{2} \le x \le \beta \\ 1, & \beta \le x \end{cases} \tag{7}$$

generates a regularly increasing fuzzy quantifier. For instance, $Q_{(0.2,1)}$ and $Q_{(0,0.6)}$ may be used to model the vague quantifiers *most* and *some*. Once a couple (Q_l, Q_u) of fuzzy quantifiers is fixed, the Q_l-lower and Q_u-upper approximation of a fuzzy set A under a fuzzy relation R are defined by, for all y in X,

$$(R{\downarrow}_{Q_l}A)(y) = Q_l \left(\frac{|[y]_R \cap A|}{|[y]_R|} \right) \tag{8}$$

$$(R{\uparrow}_{Q_u}A)(y) = Q_u \left(\frac{|[y]_R \cap A|}{|[y]_R|} \right) \tag{9}$$

---

[1] Typically, it is assumed that R is at least a fuzzy tolerance relation, i.e., R is reflexive and symmetric.

[2] A fuzzy implication $\mathcal{I}$ is a $[0,1]^2 \to [0,1]$ mapping which is decreasing in its first argument, and increasing in its second argument, and which satisfies $\mathcal{I}(0,0) = \mathcal{I}(0,1) = \mathcal{I}(1,1) = 1$ and $\mathcal{I}(1,0) = 0$.

[3] A t-norm $\mathcal{T}$ is a commutative, associative $[0,1]^2 \to [0,1]$ mapping which is increasing in both arguments, and which satisfies $\mathcal{T}(1,x) = x$ for x in $[0,1]$.

2.3   Ordered Weighted Average Aggregation

The OWA operator [9] models an aggregation process in which a sequence A of n scalar values are ordered decreasingly and then weighted according to their ordered position by a weighting vector $W = \langle w_i \rangle$, such that $w_i \in [0,1]$ and $\Sigma_i^n w_i = 1$. In particular, if c_i represents the ith largest value in A,

$$OWA_W(A) = \sum_{i=1}^{n} w_i c_i \qquad (10)$$

The OWA's main strength is its flexibility, since it enables us to model a wide range of aggregation strategies. For example, the maximum, minimum and average can all be modelled by means of OWA operators:

1. Maximum: $W_{max} = \langle w_i \rangle$, where $w_1 = 1$, $w_i = 0$, $i \neq 1$
2. Minimum: $W_{min} = \langle w_i \rangle$, where $w_n = 1$, $w_i = 0$, $i \neq n$
3. Average: $W_{avg} = \langle w_i \rangle$, where $w_i = \frac{1}{n}$, $i = 1, \ldots, n$

The OWA operator can be analysed by several measures, among which the orness-degree and andness-degree that compute how similar its behaviour is to that of max, respectively min:

$$orness(W) = \frac{1}{n-1} \sum_{i=1}^{n} ((n-i) \cdot w_i) \qquad (11)$$

$$andness(W) = 1 - orness(W) \qquad (12)$$

Note that $orness(W_{max}) = 1$, $andness(W_{min}) = 1$ and $orness(W_{avg}) = 0.5$.

3   OWA-Based Lower and Upper Approximation

Let R be a fuzzy relation in X and A a fuzzy set in $X = \{x_1, \ldots, x_n\}$. Moreover, let $\mathcal{T}$ be a t-norm and $\mathcal{I}$ a fuzzy implication. The OWA-based lower and upper approximation of A under R with weight vectors W_l and W_u are defined as

$$(R\downarrow_{W_l} A)(y) = OWA_{W_l} \langle \mathcal{I}(R(x_i, y), A(x_i)) \rangle \qquad (13)$$

$$(R\uparrow_{W_u} A)(y) = OWA_{W_u} \langle \mathcal{T}(R(x_i, y), A(x_i)) \rangle \qquad (14)$$

In order to distinguish the behaviour of lower and upper approximation, we enforce the conditions $andness(W_l) > 0.5$ and $orness(W_u) > 0.5$. Note that the traditional lower and upper approximation are retrieved when $W_l = W^{min}$ and $W_u = W^{max}$. Below, we give two examples of constructing the OWA weight vectors in order to relax the traditional definitions.

Example 1. Let $m \leq n$. It is possible to define $W_l = \langle w_i^l \rangle$ and $W_u = \langle w_i^u \rangle$ as

$$w_{n+1-i}^l = \begin{cases} \frac{2^{m-i}}{2^m - 1} & i = 1, \ldots, m \\ 0 & i = m+1, \ldots, n \end{cases} \qquad (15)$$

$$w_i^u = \begin{cases} \frac{2^{m-i}}{2^m - 1} & i = 1, \ldots, m \\ 0 & i = m+1, \ldots, n \end{cases} \qquad (16)$$

It can be verified that $andness(W_l) > 0.5$ and $orness(W_u) > 0.5$. E.g., if $n = 8$, and $m = 3$, $W_l = \langle 0, 0, 0, 0, 0, 1/7, 2/7, 4/7 \rangle$.

Example 2. In [10], Yager proposed to simulate vague quantifiers by means of OWA operators. In particular, given regularly increasing fuzzy quantifiers Q_l and Q_u, weight vectors $W_{Q_l} = \langle w_i^l \rangle$ and $W_{Q_u} = \langle w_i^u \rangle$ can be defined as

$$w_i^l = Q_l \left(\frac{i}{n} \right) - Q_l \left(\frac{i-1}{n} \right) \tag{17}$$

$$w_i^u = Q_u \left(\frac{i}{n} \right) - Q_u \left(\frac{i-1}{n} \right) \tag{18}$$

It can be verified that if $Q_l = Q_{(\alpha,1)}$ (resp., $Q_u = Q_{(0,\beta)}$), then $andness(W_{Q_l}) > 0.5$ (resp., $orness(W_{Q_u}) > 0.5$). For instance, if $n = 8$, and $Q_l = Q_{(0.2,1)}$, $W_{Q_l} = \langle 0, 0.01, 0.09, 0.18, 0.28, 0.24, 0.14, 0.06 \rangle$.

Next, we list two important propositions which follow from the monotonicity of $\mathcal{I}$, $\mathcal{T}$, and OWA operators.

Proposition 1. *Let $A_1 \subseteq A_2$ be fuzzy sets in X, and R be a fuzzy relation in X. Then $R\downarrow_{W_l} A_1 \subseteq R\downarrow_{W_l} A_2$ and $R\uparrow_{W_u} A_1 \subseteq R\uparrow_{W_u} A_2$.*

Proposition 2. *Let $R_1 \subseteq R_2$ be fuzzy relations in X, and A be a fuzzy set in X. Then $R_1\downarrow_{W_l} A \supseteq R_2\downarrow_{W_l} A$ and $R_1\uparrow_{W_u} A \supseteq R_2\uparrow_{W_u} A$.*

Note that the second proposition does not hold for VQRS [3], which is due to the fact that R occurs both in the numerator and denominator of (8) and (9).

4   Application to Feature Selection

4.1   Fuzzy-Rough Feature Selection

The purpose of feature selection is to eliminate redundant or misleading attributes from a data set, typically with the purpose of creating faster and more accurate classifiers. For our purposes, a decision system $(X, \mathcal{A} \cup \{d\})$ consists of non-empty sets of objects $X = \{x_1, ..., x_n\}$ and conditional attributes (features) $\mathcal{A} = \{a_1, ..., a_n\}$, together with a decision attribute $d \notin \mathcal{A}$. Attributes can be either quantitative or discrete; in this paper, we assume that d is always discrete. To express the approximate equality between two objects w.r.t. a quantitative attribute a, in this paper we use the fuzzy relation R_a, defined by, for x and y in X (σ_a denotes the standard deviation of a):

$$R_a(x, y) = \max \left(\min \left(\frac{a(y) - a(x) + \sigma_a}{\sigma_a}, \frac{a(x) - a(y) + \sigma_a}{\sigma_a} \right), 0 \right) \tag{19}$$

For a discrete attribute a, $R_a(x, y) = 1$ if $a(x) = a(y)$ and $R_a(x, y) = 0$ otherwise. Given a t-norm $\mathcal{T}$, for any $B \subseteq \mathcal{A}$, the fuzzy B-indiscernibility relation can be defined by

$$R_B(x, y) = \mathcal{T}(\underbrace{R_a(x, y)}_{a \in B}) \tag{20}$$

The fuzzy B-positive region is then defined as the union of lower approximations of all decision classes w.r.t. R_B. In particular, for y in X, we have [4]

$$POS_B(y) = (R_B\downarrow[y]_{R_d})(y) \tag{21}$$

The predictive ability w.r.t. d of the attributes in B can be measured by the degree of dependency of d on B, defined as $\gamma_B = |POS_B|/|X|$.

The goal of fuzzy-rough feature selection is then to find decision reducts: subsets B of A that satisfy $\gamma_B = \gamma_A$ and cannot be further reduced, i.e., there exists no proper subset B' of B such that $\gamma_{B'} = \gamma_A$. In practice, minimality is often not necessary, and the corresponding subsets are called superreducts. In order to obtain a single superreduct of $(X, A \cup \{d\})$, in this paper we use the following hillclimbing heuristic: starting with an empty set B, we compute $\gamma_{B \cup \{a\}}$ for every attribute a and add the attribute for which this value is highest to B. This process is repeated for the remaining attributes until $\gamma_B = \gamma_A$.

4.2   VQRS- and OWA-Based Feature Selection

Since the fuzzy positive region in (21) uses the "traditional" lower approximation (5), it is sensitive to small changes in the data. For this reason, following [3], we can replace $\downarrow$ by $\downarrow_{Q_l}$, giving rise to[4]

$$POS_B^{VQRS}(y) = (R_B \downarrow_{Q_l} [y]_{R_d})(y) \tag{22}$$

Based on this, γ_B^{VQRS} may be defined analogously to γ_B. However, when using it as an evaluation measure in the hillclimbing heuristic from Section 4.1, the non-monotonicity of $\downarrow_{Q_l}$ w.r.t. R_B may result in $\gamma_{B \cup \{a\}}$ being lower than γ_B, a counterintuitive result. This problem may be mended by replacing (22) by

$$POS_B^{OWA}(y) = (R_B \downarrow_{W_l} [y]_{R_d})(y) \tag{23}$$

and defining γ_B^{OWA} analogously as before. Because of Proposition 2, applying the hillclimbing heuristic will always result in increasing evaluation values for the consecutively selected subsets, making the OWA-based feature selection intuitively sounder.

Because of the definition of $\downarrow_{W_l}$, $\gamma_B \leq \gamma_B^{OWA}$; equality holds if $W_l = W_{min}$. In general, no order relationship exists between γ_B and γ_B^{VQRS}, or between γ_B^{VQRS} and γ_B^{OWA}. On the other hand, under certain conditions it holds that subsets obtained with the hillclimbing approach using the VQRS- and OWA-based positive region, are also decision superreducts in terms of the original definition of positive region in (21), and vice versa. For VQRS, $\gamma_B = 1 \Leftrightarrow \gamma_B^{VQRS} = 1$ if $Q_l = Q_{(\alpha,1)}$ and $\mathcal{I}$ satisfies $x \leq y \Leftrightarrow \mathcal{I}(x,y) = 1$ [3]. The following proposition provides a sufficient and necessary condition for the OWA-based approach.

Proposition 3. $W_l = \langle w_i^l \rangle$ *is an OWA weight vector such that* $w_n^l > 0$ *if and only if* $\gamma_B^{OWA} = 1 \Leftrightarrow \gamma_B = 1$.

---

[4] Note that [3] also introduced an alternative definition of the VQRS-based positive region which is not considered here because of its excessive computational complexity.

4.3   Experimental Evaluation

In order to compare the performance of the traditional fuzzy-rough feature selection approach with those based on VQRS and OWA, we have performed an
experiment on the Spambase dataset[5]. This dataset has 4601 objects, 57 conditional attributes and two decision classes.

Specifically, we apply the hillclimbing procedure from Section 4.1 using γ,
γ^{VQRS} and γ^{OWA}; we use $\mathcal{I}(x,y) = \min(1, 1 - x + y)$ in (5), $Q_l = Q_{(0.2,1)}$ in
(8), $W_l = W_{Q_{(0.2,1)}}$ in (13) and $\mathcal{T}(x,y) = \max(0, x + y - 1)$ in (20). At each
step in the hillclimbing algorithm, the subsets obtained so far are evaluated by
means of their classification accuracy using the Nearest Neighbour classifier IBk
in the Weka toolkit [8]. As a baseline, we also consider the approach in which
attributes are added in random order.

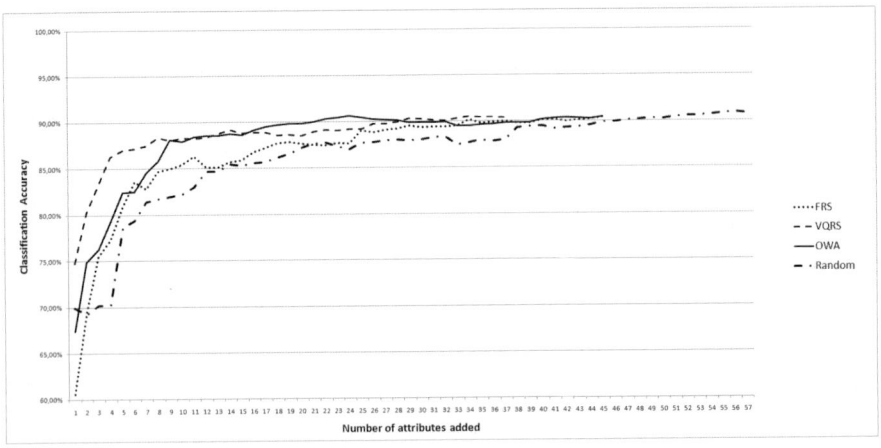

Fig. 1. Evaluation of feature selection approaches on Spambase data

The results can be found in Figure 1. The X axis contains the number of
attributes added so far, while the classification accuracy obtained with IBk for
the corresponding attribute subsets can be read on the Y axis. From this figure,
we can see that all fuzzy-rough approaches yield better subsets than those randomly generated, and that VQRS and OWA both outperform the fuzzy-rough
approach along most of the range. Initially, the VQRS approach selects higher-
quality attributes than those obtained with OWA, but roles switch after the
sixteenth attribute. It is also interesting to note that OWA obtains the highest
overall accuracy, viz. 90.59% using 24 attributes.

5   Conclusion and Future Work

In this paper, we have presented a new fuzzy-rough hybridization which relaxes
the traditional property that lower and upper approximations are determined

---

[5] Available from http://archive.ics.uci.edu/ml/datasets/Spambase

by means of a single object. As such, it provides an alternative to the previously introduced VQRS model; from a theoretical point of view, the new model is better-founded because it respects monotonicity w.r.t. the used fuzzy indiscernibility relation. As another advantage, it adheres more closely to the traditional fuzzy-rough approach, to which it converges as the andness (resp., orness) of the OWA-based lower (resp., upper) approximation nears 1. Initial experimentation has pointed out that the OWA-based model can outperform the traditional fuzzy-rough approach, and that it can compete with the VQRS approach.

More extensive experiments are needed, however, to fully investigate its performance; in particular, in order to study its noise-handling potential, we will evaluate the quality of subsets obtained on data that is deliberately contaminated with noise. Furthermore, we also plan to investigate various construction methods for the OWA weight vectors, and their optimization in function of the particular dataset used. Finally, apart from feature selection, the OWA-based model may also be applied to other data analysis tasks in which fuzzy-rough methods have been used, such as rule induction and instance selection.

References

1. Cornelis, C., De Cock, M., Radzikowska, A.M.: Vaguely quantified rough sets. In: An, A., Stefanowski, J., Ramanna, S., Butz, C.J., Pedrycz, W., Wang, G. (eds.) RSFDGrC 2007. LNCS (LNAI), vol. 4482, pp. 87–94. Springer, Heidelberg (2007)
2. Cornelis, C., De Cock, M., Radzikowska, A.M.: Fuzzy rough sets: from theory into practice. In: Pedrycz, W., Skowron, A., Kreinovich, V. (eds.) Handbook of Granular Computing, pp. 533–552. John Wiley and Sons, Chichester (2008)
3. Cornelis, C., Jensen, R.: A noise tolerant approach to fuzzy-rough feature selection. In: Proceedings of the 2008 IEEE International Conference on Fuzzy Systems (FUZZ-IEEE 2008), pp. 1598–1605 (2008)
4. Cornelis, C., Jensen, R., Hurtado Martín, G., Slezak, D.: Attribute selection with fuzzy decision reducts. Information Sciences 180(2), 209–224 (2010)
5. Dubois, D., Prade, H.: Rough fuzzy sets and fuzzy rough sets. International Journal of General Systems 17, 91–209 (1990)
6. Pawlak, Z.: Rough sets. International Journal of Computer and Information Sciences 11(5), 341–356 (1982)
7. Radzikowska, A.M., Kerre, E.E.: A comparative study of fuzzy rough sets. Fuzzy Sets and Systems 126, 137–156 (2002)
8. Witten, I.H., Frank, E.: Data Mining: Practical machine learning tools and techniques, 2nd edn. Morgan Kaufmann, San Francisco (2005)
9. Yager, R.R.: On ordered weighted averaging aggregation operators in multicriteria decision making. IEEE Transactions on Systems, Man, and Cybernetics 18, 183–190 (1988)
10. Yager, R.R.: Families of OWA operators. Fuzzy Sets and Systems 59(2), 125–148 (1993)

On Attribute Reduction of Rough Set Based on Pruning Rules

Hongyuan Shen[1,2], Shuren Yang[1,2], and Jianxun Liu[1]

[1] Key laboratory of knowledge processing and networked manufacturing,
college of hunan province, Xiangtan, 411201, China
[2] Institute of Information and Electrical Engineering,
Hunan University of Science and technology, Xiangtan, 411201, China
sryang86@gmail.com

Abstract. Combining the concept of attribute dependence and attribute similarity in rough sets, the pruning ideas in the attribute reduction was proposed, the estimate method and fitness function in the processing of reduction was designed, and a new reduction algorithm based on pruning rules was developed, the complexity was analyzed, furthermore, many examples was given. The experimental results demonstrate that the developed algorithm can got the simplest reduction.

Keywords: Rough Sets Theory, Reduction of Attribute, Pruning Rules, Information System, Completeness.

1 Introduction

The Rough Set theory proposed by Z.Pawlak in 1982 is a new and powerful mathematical approach to deal with imprecision,uncertainty and vagueness knowledge [1, 2]. A very rapid development and application of rough set theory have been witnessed in recent 30 years, the theory of rough set has been successfully applied to pattern recognition, knowledge discovery, machine learning, decision analysis, etc [3, 4, 5, 6]. Attribute reduction, also called feature subset selection, is the core issue of rough set theory. The main concept of attribute reduction can be described as follows: For a given information system, an attribute reduction is a minimal subset of attributes that supplies the equivalent discernibility power as all of the attributes [7], that is to say, anyone attribute in a reduction is jointly sufficient and individually necessary, and will be not redundant with each other. The same impact and conclusion by selecting a smaller and simpler set of attributes could be obtianed, as a result, the more ordinary and easier rules can be got naturally [8].

Different algorithms methods and strategies have been extensively studied and proposed for finding the attribute reduction [3, 7, 8, 9, 10]. In the general case, for a given information system, there is more than one attribute reduction which could be found. Unfortunately, the process of finding reduction has been proven to be an NP-Hard problem [11]. In consequence, many heuristic methods for exploring the optimal or sub-optimal reduction have also been investigated and

J. Yu et al. (Eds.): RSKT 2010, LNAI 6401, pp. 86–93, 2010.

developed [12, 4, 14, 15], at the same time, various software have been developed and applied with attribute reduction. Such as ROSE and RSES [16, 17].

2   Basic Knowledge

2.1   Rough Set

In the general case, for Rough Set Attributes Reduction(RSAS), the core issue is the concepts of indiscernibility. Let $S = (U, A, V, f)$ will be an information system, it is also a two dimension decision table actually, which is consist of discrete-valued datasets, the definitions of these notions are as follows:where U is a finite set of objects,Q is a finite set of the total attributes,$V = \bigcup_{q \in Q} V_q$ and V_q is a domain of the attribute q, and $f : U \times A \rightarrow V$ is an information function,if $\forall a \in A, x \in U$, then $f(x, a) \in V$, Sometimes, information system S can be written as $S = (U, A)$ briefly [1].

Definition 1. Let $S = (U, A, V, f)$ be an information system, for an attribute set $P \subseteq A$ there is associated an equivalence relation $IND_A(P)$:

$$IND_A(P) = \{(x, y) \in U^2 | \forall a \in P, x \neq y, f(a, x) = f(a, y)\} \tag{1}$$

$IND_A(P)$,or denoted $IND(P)$ too, is called the P-indiscernibility relation, its classes are denoted by $[x]_P$. By $X|P$ giving the partition of U defined by the indiscernibility relation $IND(P)$.

Definition 2. Let $P \subset A$ and let $Q \subset A$ be the equivalence relation over U, the positive region of P on Q, defined as $POS_P(Q)$, the form is

$$POS_P(Q) = \cup P_-(X), X \in U|Q \tag{2}$$

Where $P_-(X)$ is the P-lower approximation of X set. In terms of the indiscernibility relation $U|P$, the positive region contains all objects of U that can be classified to blocks of the partition $U|Q$ employing the knowledge in attributes set P.

Definition 3. Let $S = (U, A, V, f)$ be an information system, with $P, Q \subseteq A$, the dependency between the subset P and Q can be written as

$$k = \Upsilon(P, Q) = |POS_P(Q)|/|U| \tag{3}$$

Where we define that parameter $k(0 \leq k \leq 1)$ is dependency of P on Q, denoted $P \Rightarrow_k Q$ or $\Upsilon(P, Q)$, if $k = 1$ Q depends entirely on P, if $0 < k < 1$ Q depends partially on P , generally, called degree k, and if $k = 0$ Q does not depend on P completely.

Definition 4. Let $S = (U, A, V, f)$ be an information system, with $P, Q \subseteq A$, if $\forall r \in C$, the similarity between odd condition attribute r and decision attribute D can be defined as:

$$R(D, r) = \frac{GD(D \cup \{r\})}{\sqrt{GD(\{r\})} \times \sqrt{GD(D)}} \tag{4}$$

Where, $GD(r)$ express the granularity of r, can be written the following formulation:

$$GD(r) = \sum_{i=1}^{n} |r_i|^2 \tag{5}$$

Where, r_i is the different values of the different attributes, can be gotten by the partition commonly. The more values of the Eq.(5), the more similarity between the odd attribute r and decision D, if the value of the $R(D, r)$ is 1, then r and D is similar completely, conversely, if it is 0, then r and D is far from similar [9].

2.2   Pruning Thought

The proposed pruning thought is based on the fact that the pruning rules will be given in the next section 2.3. For a given information system, in order to find the minimal or simplest attribute reduction, pruning thought works as follows. It starts the search from the Level-Top set of attributes and works its way to smaller attributes sets through the defined pruning rules. In each process of pruning, the algorithm will prune an irrelevant attribute from the candidate set.

Let $S = (U, A, V, f)$ be an information system, suppose object attributes set A be composed of the set $\{a, b, c, d, e, f\}$, where the condition attributes set be $\{a, b, c, d, e\}$, denoted by C, the decision attributes set be, presented by D is $\{f\}$. In general, the all potential attribute reduction of an information system are contained by the all subsets of condition attribute C, as such, the simplest or minimum attribute reduction also are included in the all subset of condition attribute C. Thereby, a Level-Figure will be defined, which is shown in Fig. 1, shows the search space for attribute reduction, and the Level-Figure represents all possible nonempty combinations of the five condition attribute [21].

The pruning though in the process of finding the attribute reduction for Rough Set is as follows: If condition attributes subset $\{c, d\}$ is a attribute reduction, is located at Level-2, first, at Level-Top, with Top is 5, the condition attributes set $\{a, b, c, d, e\}$ is a attribute reduction too, that is to say, If given an information system, being the attribute reduction, the condition attribute C must be an potential attribute reduction. Then, at the next Level-4, the subset of the above Level-5 which contains the set $\{c, d\}$ is also a attribute reduction, for example, the subset $\{a, b, c, d\}$. In the same way, at the next Level-3, the set $\{a, c, d\}$ contains the set $\{c, d\}$, which is also a attribute reduction. At the following Level-2, among the subset of the set $\{a, c, d\}$, can obtain $\{c, d\}$ is a attribute reduction, At last, at the Level-Bottom, since the subset of the above level $\{c, b\}$ is not the attribute reduction, it can infer that the set $\{c, d\}$ is the final attribute reduction for the information system. The pruning and searching process is showed by the broken line in Fig.1.

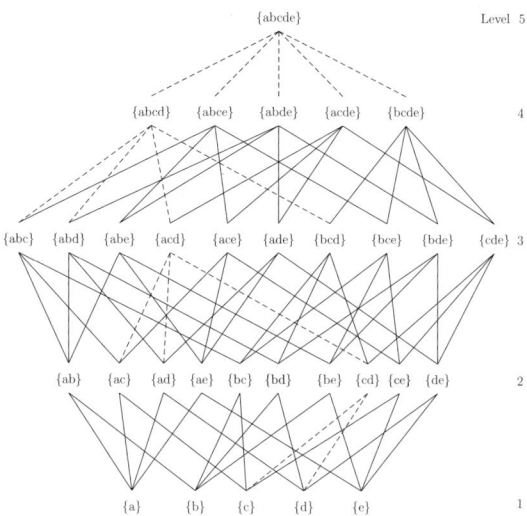

Fig. 1. Level-Figure of attribute reduction for searching space

2.3   Pruning Rules

In the pruning processing, the irrelevant condition attribute will be pruned, and need evaluate the superiority of the different subset of the candidate. For that reason, a fitness function which used to evaluate the superiority will be required, where, a new fitness function is defined by combining the formulation (4) and (7), which is devoted to measure the superiority of the candidate, the formulation (4) express the dependency of condition attribute over decision attribute, which used to check whether or not the candidate is potential reduction, the candidate is potential reduction if and only if the value of the formulation (4) is equal to 1.But, the Eq.(4) don't ensure the obtained reduction which one is more superior than the other. Thus, we introduce the formulation (7) to resolve the above problem, which is the similarity of the decision attribute D toward to the odd attribute, it can be used to measure which one is more superior than the other to decision attribute D. For example, in the Fig.1, for given an information system, with five condition attributes, $C = \{a, b, c, d, e\}$, and only one decision attribute $D = \{f\}$. Suppose that $\{c, b\}$ was the attribute reduction for the information system. At first, let the candidate be $\{a, b, c, d, e\}$, there exist three subset of the candidate make the value of the Eq.(4) is 1, that is to say, these subsets are the attribute reduction for the information system. However, it is impossible to confirm which one is more than the other, there, the formulation (7) can be used to check. Since there is only one different attribute element among them, by introducing the notion of similarity into the fitness function in rough set model, the final optimal reduction from many potential reductions can be selected by the above strategy.

3   Attribute Reduction of Rough Set Based on Pruning Rules(ARRSBPR)

3.1   Algorithm Description

In general, the ARRSBPR algorithm is a searching method based on level strategy. Figure .1 shows a Level-Figure in terms of the number of each subset includes condition attribute, there is only one different element among them in the same Level. The present searching scale of ARRSBPR is the length of the arbitrary set in the above Level. The idea of searching rules is to select an optimal potential reduction from the above level as the candidate of the below level, which called superset. The core issue of ARRSBPR is how to select an optimal relative reduction from the all reduction of same Level by the formulation (4) and (7).

ARRSBPR Algorithm:

Inputs: A two-dimensional relation table $R(S)$ over information system $S = (U, A, V, f)$, the row be used to express the objects U ,and the column stands for object attribute A.

Ouputs: A reduce over information system $S = (U, A, V, f)$.

Prune_Reduce(R(S))

```
1. Candidate=C; Dec=D; R=Length(C); K=1; Reduce=;
2. Simplest_Reduce =Calculate_POS(Candidate,Dec,K); //Eq.(4)
3. If(Simplest_Reduce=){
4. Reduce=Calculate_POS(Candidate,Dec,R);
5. If(Reduce){
6. While(R>1)
7. {R=R-1;
8. Level_R_All_Reduce=Calculate_POS(Candidate,Dec,R);
9. If(Level_R_All_Reduce){
10.Level_R_Reduce=Max(Calculate_R(Level_R_All_Reduce,Dec));//Eq.(7)
11.R_Candidate =Prune(Candidate,Level_R_Reduce);
      Candidate=R_Candidate;}
12.else{Reduce= Candidate, Output(Reduce)}}}
13.else{Output(Reduce){The R(S) exists not the Reduce}}
14.else{Reduce=Max(Caculate_R(Simplest_Reduce,Dec));
        Output(Reduce);}}
```

Line 1 initializes some necessary parameter. The function $Calculate\_POS()$, which is shown in Line 2 ,Line 4 and Line 8, calculates the attribute dependency degree that represents to what degree Dec depends on the subset of Candidate. The paremeter R is the number of attributes of the above level superset, and note that the function $Calculate\_POS()$ only return the subset, which the dependency degre is 1 for the Eq.(4). The function $Calculate\_R()$, which is shown in Line 10 and Line 14, uses the formulation (7) to calculate the similarity of different condition attribute among all possible reduction in the Level-R and decision attribute respectively, then, the function $Max()$ selects which one reduction from these possible reductions is better than the other, the first parameter is the all possible reduction in Level-R, the second is decision attribute. The function *Prune*(), which is shown in Line 11, prunes the redundant odd condition attribute from superset candidate step by step, and obtain the smaller superset candidate, the second parameter is the optimal reduction in Level-R so far. The function *Output*(), which is shown in Line 12, Line 13 and Line 14,outputs the final reduction for an given information system. In addition, in order to improve the efficiency and practicality of the algorithm, we can judge whether or not the final reduction only includes one condition attribute firstly, described in Line 2, Line 3 and Line14. Line 4 estimates whether there has a reduction or not in an given information system. Line 8 generates in all subset of the candidate all possible reduction for from Level R to Level R-1. An optimal reduction in Line 10 from these potential generated reduction can be selected in Line 8. Followed by, a new Candidate will be produced in Line 11 by the pruning rules.

3.2   Algorithm Analysis

ARRSBPR, the proposed algorithm for attribute reduction follows the basic concepts of rough set, its complexity is mainly lie on the calculation of the above function $Calculate\_POS()$ and $Calculate\_R()$. These are some same properties between them as follows:1) Doing partition the information system with the different attributes; 2) Calculate the dependency and similarity between the conditional attribute and decision attribute. We use the sorting method based on cardinal developed in the paper [22] to obtain the partition, and the complexity is $O(|T||U|)$. where T denotes the partition attribute set, and U is the domain. When the cardinal sorting used to calculate the dependency showed in formulation, the worst complexity is $O(|C| \cdot |U|, O(|C|^2 \cdot |U/C|))$. Suppose given an information system including g conditional attributes, and the number of the attribute reduction is e ultimately. Before not pruned, the number of all possible nonempty combinations of attributes is $2^g - 1$, and also is the searching space, by contrast, the searching scale will be $\sum_e^g i$ after pruned, the performance is visible. Since the fitness function in the prune processing of ARRSBPR is designed by the notion of attribute dependency degree and similarity, the developed ARRSBPR is complete.

4   Examples and Its Analysis

the proposed approach will be applied to several data sets from the UCI repository of machine learning databases and related actual data in the table of the paper [18],with different numbers of attributes and objects, and there is only one decision attribute included these datasets. So as to evaluate and verify the proposed approach. The ARRSBPR is implemented in Matlab environment and run it on Intel 2.8GHz CPU and 1GB of RAM PC.

Notes: O.N-the number of objects, C.N-the number of conditional attributes, E.R-whether the reduction exits or not, R.N-the odd attribute number in reduction, N.P.S-the original searching space, P.S-the pruned searching space.

Table 1. Summary of the experiment data sets

Data set	O.N	C.N	E.R	R.N	N.P.S	P.S
The table 1[22]	36	5	Y	2	31	14
Zoo	101	16	Y	5	65535	126
Ionosphere	351	34	Y	3	$2^{34} - 1$	592
Connectionist Bench	208	60	Y	1	$2^{60} - 1$	60
Cancer-Wisconsin	699	9	Y	4	511	39
Abalone	4177	7	Y	3	255	25
Waveform	5000	21	Y	3	2097151	228
Optical Recognition	3823	64	Y	7	$2^{64} - 1$	2059
Libras Movement	360	90	Y	2	$2^{90} - 1$	4094
Statlog(Shuttle)	58000	9	Y	4	511	39

5   Discussion and Future Work

A novel Pruning Rules-based algorithm, called ARRSBPR, is proposed firstly, to deal with attribute reduction in rough set theory. The main work is as follows. (a) Some important properties and conception of rough set are derived and analyzed, laying the theoretical foundation for our algorithms. (b) The pruning thought and pruning rules are developed in terms of rough set for the first time, and defines a novel fitness function which to measure the superiority of the candidate.(c) Using Level-Searching and Level-Pruning techniques, an algorithm for attribute reduction in rough set is presented based on the pruning rules. The program code is listed in detail, and the rigorous and detailed proofs of the algorithm's validity, complexity and completeness is provided. (d) Numerical experiments indicate that the algorithm is efficient and complete. The experiments results are consistent with the theoretical analysis. In sum, the algorithm is an effective means of attribute reduction for rough set, especially large ones in object or attribute. Future research will concentrate on developing the theoretical foundation necessary for analysis of efficient and completeness of algorithm in decision systems. More generally, as for large datasets ,our research strives to enrich and improve the proposed algorithm in this paper and present others effective and efficient approaches to attribute reduction in rough set.

Acknowledgements

This work was supported by the grant 90818004(the Natural Science Foundation of China), grant 09k085(the Education Department Foundation of Hunan Provincial.

References

1. Pawlak, Z.: Rough sets. International Journal of Parallel Programming 11(5), 341–356 (1982)
2. Pawlak, Z.: Rough sets: Theoretical aspects of reasoning about data. Springer, Heidelberg (1991)

3. Hoa, S.N., Son, H.N.: Some efficient algorithms for rough set methods. In: The sixth international conference, Information Procesing and Management of Uncertainty in Knowledge-Based Systems (IPMU), Granada, Spain, pp. 1451–1456 (1996)
4. Pawlak, Z.: Rough set theory and its applications. Cybernetics and Systems 29, 661–688 (2002)
5. Huang, J., Liu, C., Ou, C., et al.: Attribute reduction of rough sets in mining market value functions. In: Proceedings of the 2003 IEEE/WIC International Conference on Web Intelligence, pp. 470–473. IEEE Computer Society, Washington (2003)
6. Bhatt, R., Gopal, M.: On fuzzy-rough sets approach to feature selection. Pattern Recognition Letters 26(7), 965–975 (2005)
7. Meng, Z., Shi, Z.: A fast approach to attribute reduction in incomplete decision systems with tolerance relation-based rough sets. Information Sciences 179(16), 2774–2793 (2009)
8. Yamaguchi, D.: Attribute dependency functions considering data efficiency. International Journal of Approximate Reasoning 51, 89–98 (2009)
9. Xia, K.W., Liu, M.X., Zhang, Z.W.: An Approach to Attribute Reduction Based on Attribute Similarity. Journal of Hebei Unviersity of Technology 34(4), 20–23 (2005)
10. Hu, F., Wang, G.: Quick reduction algorithm based on attribute order. Chinese Journal of Computers 30(8), 1429–1435 (2007)
11. Wong, S., Ziarko, W.: On optimal decision rules in decision tables. Bulletin of Polish Academy of Sciences 33(11-12), 693–696 (1985)
12. Yu, H., Wang, G., Lan, F.: Solving the Attribute Reduction Problem with Ant Colony Optimization, pp. 242–251. Springer, Heidelberg (2008)
13. Hu, Q., Xie, Z., Yu, D.: Hybrid attribute reduction based on a novel fuzzy-rough model and information granulation. Pattern Recognition 40(12), 3509–3521 (2007)
14. xun, Z., Gu, D., Yang, B.: Attribute Reduction Algorithm Based on Genetic Algorithm. In: Intelligent Computation Technology and Automation, ICICTA (2009)
15. Ye, D., Chen, Z., Liao, J.: A new algorithm for minimum attribute reduction based on binary particle swarm optimization with vaccination. In: Advances in Knowledge Discovery and Data Mining, pp. 1029–1036 (2007)
16. ROSE, http://www.fizyka.umk.pl/~duch/software.html
17. RSES, http://logic.mimuw.edu.pl/~rses/
18. Liu, S.H., Sheng, Q.G., Wu, B.: Researh of Efficient Algorithms for Rough Set Methods. Chinese Journal of Computers 26(5), 524–529 (2003)
19. Zeng, H.L.: Rough set theory and its application(Revision). Chongqing University Press (1998)
20. Pawlak, Z.: Rough set. Communication of the ACM 11(38), 89–95 (1995)
21. Yao, H., Hamilton, H.: Mining functional dependencies from data. Data Mining and Knowledge Discovery 16(2), 197–219 (2008)
22. Xu, Z.y., Liu, Z.p., et al.: A Quick Attribute Reduction Algorithm with Complexity of $\max(O(|C||U|), O(|C|^2|U/C|))$. Chinese Journal of Computers 29, 17–23 (2006)

Set-Theoretic Models of Granular Structures

Yiyu Yao[1], Duoqian Miao[2], Nan Zhang[1,2], and Feifei Xu[2]

[1] Department of Computer Science, University of Regina,
Regina, Saskatchewan, Canada S4S 0A2
`yyao@cs.uregina.ca`
[2] Department of Computer Science and Technology, Tongji University,
Shanghai, China 201804
`{miaoduoqian,zhangnan0851}@163.com, xufeifei1983@hotmail.com`

Abstract. Granular structure is one of the fundamental concepts in granular computing. Different granular structures reflect multiple aspects of knowledge and information, and depict the different characteristics of data. This paper investigates a family of set-theoretic models of different granular structures. The proposed models are particularly useful for concept formulation and learning. Some of them can be used in formal concept analysis, rough set analysis and knowledge spaces. This unified study of granular structures provides a common framework integrating these theories of granular computing.

1 Introduction

Granular computing provides a general methodology for problem solving and information processing [1,4,6,7,8,11]. It simplifies complex real world problems by considering different levels of granularity. The triarchic theory of granular computing focuses on an understanding in three perspectives, granular computing as structured thinking, as structured problem solving, and as structured information processing [16,17,18,19]. One of its central notions is hierarchical multilevel granular structures defined by granules and levels. Each granule represents a focal point, or a unit of discussion on a particular level; each level is populated with granules of similar grain-sizes (i.e., granularity) or similar features; all levels are (partially) ordered according to their granularity. Problem solving can be approached as top-down, bottom-up or middle-out [12] processes based a granular structure.

The formulation and interpretation of granular structures are much application dependent. Different vocabulary, terminology or language may be used for description at different levels. Nevertheless, one may study some mathematical models independent of specific applications. The main objective of this paper is to investigate a family of set-theoretic models of granular structures. These models are particularly useful for concept formulation and learning, as the intension and extension of a concept are commonly represented as a pair of a set of properties and a set of instances [20].

A basic model of a granular structure is given by a poset $(G, \subseteq)$, where G is a family of subsets of a universal set U and $\subseteq$ is the set-theoretic inclusion relation.

J. Yu et al. (Eds.): RSKT 2010, LNAI 6401, pp. 94–101, 2010.

By imposing different sets of conditions on G, we derive seven sub-models of granular structures. Some of them are used in formal concept analysis [13], rough set analysis [9,10,14], and knowledge spaces [3,21]. This unified model of granular structures is an important step for integrating the three theories in a common framework with a common language. The discussion and classification of models of granular structures is restricted to a set-theoretic setting in this paper. A more detailed and comprehensive study of the taxonomy of granular computing models can be found in Kent's Ph.D. dissertation [5].

2 Overview of the Unified Model

Two key notions of granular computing are granules and a hierarchical granular structure formed by a family of granules. In constructing a unified set-theoretic model, we assume that a granule is represented by a subset of a universal set and granular structure is constructed based on the standard set-inclusion relation on a family of subsets of the universal set. Depending on the properties of the family of granules, specific models of granular computing can be obtained.

Definition 1. *Let U denote a finite nonempty universal set. A subset $g \in 2^U$ is called a granule, where 2^U is the power set of U.*

The power set 2^U consists of all possible granules formed from a universal set U. The standard set-inclusion relation $\subseteq$ defines a partial order on 2^U, which leads to sub-super relationship between granules.

Definition 2. *For $g, g' \in 2^U$, if $g \subseteq g'$, we call g a sub-granule of g' and g' a super-granule of g.*

Under the partial order $\subseteq$, the empty set $\emptyset$ is the smallest granule and the universe U is the largest granule. When constructing a granular structure, we may consider a family G of subsets of U and an order relation on G.

Definition 3. *Suppose $G \subseteq 2^U$ is a nonempty family of subsets of U. The poset $(G, \subseteq)$ is called a granular structure, where $\subseteq$ is the set-inclusion relation.*

By the relation $\subseteq$, we can arrange granules in G into a hierarchical multilevel granular structure. The relation $\subseteq$ is an example of partial orders. In general, one may consider any partial order on G and the corresponding poset $(G, \preceq)$. For simplicity, we consider only the poset $(G, \subseteq)$, but the argument can be easily applied to any poset.

The structure $(G, \subseteq)$ gives rise to the weakest set-theoretic model in which a granule is a subset of a universe, and a granular structure is a family of subsets of the universe. We denote this basic model by the pair $\mathcal{M}_0 = (U, G)$. In constructing the basic model, we only assume that $G \neq \emptyset$ and there are no other constraints. The family G does not have to be closed with respect to any set-theoretic operations. The structure of G is only a partial order defined by $\subseteq$.

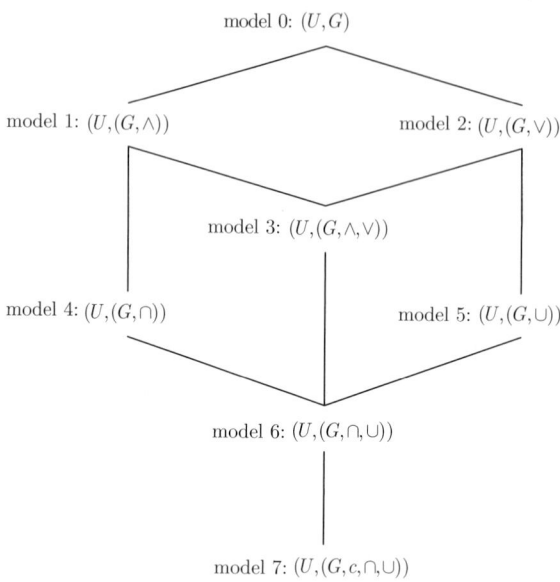

Fig. 1. Models of granular structures

The granular structure $(G, \subseteq)$ of model $\mathcal{M}_0$ is a substructure of $(2^U, \subseteq)$. Each granule in G represents a focal point of our discussion. The family G represents all focal points of our discussion. The construction and operation of granules depend on particular applications. By imposing extra conditions on G, we can derive more specific models from the basic model. Fig.1 summarizes eight models of granular structures.

A line connecting two models in Fig.1 indicates the sub-model relationship and relations that can be obtained from the transitivity are not explicitly drawn. For example, $\mathcal{M}_1$ is the sub-model of $\mathcal{M}_0$ and $\mathcal{M}_4$ is a sub-model of $\mathcal{M}_1$. It follows that $\mathcal{M}_4$ is the sub-model of $\mathcal{M}_0$.

The three models $\mathcal{M}_1$, $\mathcal{M}_2$ and $\mathcal{M}_3$ may be viewed as lattice-based models. They correspond to meet-semilattice, join-semilattice and lattice, respectively, where the symbols $\wedge$ and $\vee$ are lattice meet and joint operations. The meet $\wedge$ and join $\vee$ may not necessarily coincide with the set-theoretic operations $\cap$ and $\cup$. When they are in fact set intersection ($\cap$) and union ($\cup$), we have the three set-theoretic models $\mathcal{M}_4$, $\mathcal{M}_5$ and $\mathcal{M}_6$. The most specific model is the one in which G is closed under all three set-theoretic operations, where c denotes set complement.

The models in Fig.1 represent hierarchical granular structures that are commonly used in many studies. A mixture of models $\mathcal{M}_2$ and $\mathcal{M}_4$ are used in formal concept analysis [13], where a granular structure $(U, (G, \cap, \vee))$ is used; while the meet is given by the set intersection, the join is defined differently. Model $\mathcal{M}_5$ is used in the study of knowledge spaces [3]. Model $\mathcal{M}_7$ is used in

Pawlak rough set analysis [9]. All these models are considered in the generalized rough set models [15].

A detailed examination of the family of set-theoretic models of granular computing structure will be presented in the next section.

3   Models of Granular Structures

For the convenience of discussion, we divide models in Fig.1 into two groups. One group is lattice-based models of granular structures, the other group is set-based models of granular structures.

3.1   Lattice-Based Models of Granular Structures

In a granular structure $(G, \subseteq)$, where $G \subseteq 2^U$ and $\subseteq$ is the set-inclusion relation, the relation $\subseteq$ is a partial order (i.e., $\subseteq$ is reflexive, antisymmetric and transitive). One can derive three lattice-based models with respect to the partial order.

Definition 4. *Suppose* $(G, \subseteq)$ *is a granule spaces. For a pair of granules* $a, b \in G$, *a granule* $l \in G$ *is called a lower bound of* a *and* b *if* $l \subseteq a$ *and* $l \subseteq b$; *a granule* $u \in G$ *is called an upper bound of* a *and* b *if* $a \subseteq u$ *and* $b \subseteq u$. *In addition,* l *is called the greatest lower bound (glb) of* a *and* b, *if* $k \subseteq l$ *for any lower bound of* k *of* a *and* b; u *is called least upper bound (lub) of* a *and* b *if* $u \subseteq k$ *for any upper bound* k *of* a *and* b.

For an arbitrary pair of granules in G, their lower bounds, the greatest lower bound, upper bounds or the least upper bound may not exist in G. If the greatest lower bound exists, it is unique and is denoted by $a \wedge b$; if the least upper bound exists, it is unique and is denoted by $a \vee b$. Based on these notions, we immediately obtain two models of granular structures.

Definition 5. *A granular structure* $(G, \subseteq)$ *is a meet-semilattice, denoted by* $(G, \wedge)$, *if the greatest lower bound always exists in* G *for any pair of granules in* G; *a granular structure* $(G, \subseteq)$ *is a join-semilattice, denoted by* $(G, \vee)$, *if the least upper bound always exists in* G *for any pair of granules in* G.

In model $\mathcal{M}_1$, a granular structure $(G, \subseteq)$ is a meet-semilattice $(G, \wedge)$, $a \wedge b$ is the largest granule contained by both a and b. Since G is not necessarily closed under set intersection, $\wedge$ is not necessarily the same as $\cap$. Similarly, in model $\mathcal{M}_2$ a granular structure $(G, \subseteq)$ is a join-semilattice $(G, \vee)$, $a \vee b$ is the smallest granule in G that contains both a and b. Again, $\vee$ is not necessarily the same as $\cup$.

In a meet-semilattice granular structure $(G, \wedge)$, for a pair of granules p and g, if $p \cap g \in G$, then $p \wedge g = p \cap g$. In a join-semilattice granular structure $(G, \vee)$, for a pair of granules p and g, if $p \cup g \in G$, then $p \vee g = p \cup g$.

Example 1. Suppose $\mathcal{M}_1 = (U, (G_1, \wedge))$, where $U = \{a, b, c, d, e\}$, $G_1 = \{\{c\}, \{a, b, c\}, \{b, c, d\}\}$. It can be easily verified that $(G_1, \wedge)$ is a meet-semilattice. Consider two granules, $p = \{a, b, c\}$ and $q = \{b, c, d\} \in G_1$, The glb of p and g is $p \wedge q = \{c\}$. On the other hand, the intersection of p and g is $p \cap q = \{b, c\}$, which is not in G_1.

Example 2. Suppose $\mathcal{M}_2 = (U, (G_2, \vee))$, where $U = \{a, b, c, d, e\}$, $G_2 = \{\{a, b\},$ $\{c, d\}, \{a, b, c, d, e\}\}$. The granular structure $(G_2, \vee)$ is a join-semilattice. Consider two granules, $p = \{a, b\}$ and $q = \{c, d\} \in G_2$, The lub of p and g is $p \vee q = \{a, b, c, d, e\}$. On the other hand, the union of p and g is $p \cup q = \{a, b, c, d\}$, which is not in G_2.

The operators $\wedge$ and $\vee$ defined based on the partial order $\subseteq$ are referred to as the meet and join operators of semilattices. If both $\wedge$ and $\vee$ are defined for a granular structure, one can derive a lattice.

Definition 6. *A granular structure $(G, \subseteq)$ is a lattice, denoted by $(G, \wedge, \vee)$, if both the greatest lower bound and the least upper bound always exist in G for any pair of granules in G.*

In $\mathcal{M}_3$, a granular structure is a lattice, in which $\wedge$ and $\vee$ is not necessarily the same as set intersection $\cap$ and $\cup$, respectively.

Example 3. Suppose $\mathcal{M}_3 = (U, (G_3, \wedge, \vee))$, where $U = \{a, b, c, d, e\}$, $G_3 = \{\{c\},$ $\{a, b, c\}, \{b, c, d\}, \{a, b, c, d, e\}\}$. The granular structure $(G_3, \wedge, \vee)$ is a lattice. Consider two granules, $p = \{a, b, c\}$ and $q = \{b, c, d\} \in G_3$, the glb of p and g is $p \wedge q = \{c\}$. The intersection of p and g is $p \cap q = \{b, c\}$, which is not in G_3. The lub of p and g is $p \vee q = \{a, b, c, d, e\}$. The union of p and g is $p \cup q = \{a, b, c, d\}$, which is not in G_3.

3.2   Set-Based Models of Granular Structures

Set-based models of granular structures are special cases of lattice-based models, where the lattice meet $\wedge$ coincides with set intersection $\cap$ and lattice join $\vee$ coincides with set union $\cup$. In other words, G is closed under set intersection and union, respectively. We immediately obtain the set-based model $\mathcal{M}_4$, $\mathcal{M}_5$ and $\mathcal{M}_6$.

Definition 7. *A granular structure $(G, \subseteq)$ is a $\cap$-closed granular structure, denoted by $(G, \cap)$, if the intersection of any pair of granules of G is in G; a granular structure $(G, \subseteq)$ is a $\cup$-closed granular structure, denoted by $(G, \cup)$, if the union of any pair of granules of G is in G; a granular structure $(G, \subseteq)$ is a $(\cap, \cup)$-closed granular structure, denoted by $(G, \cap, \cup)$, if both the intersection and union of any pair of granules of G are in G.*

Example 4. Suppose $\mathcal{M}_4 = (U, (G_4, \cap))$, where $U = \{a, b, c, d, e\}$, $G_4 = \{\{b, c\},$ $\{a, b, c\}, \{b, c, d\}\}$, $\mathcal{M}_4$ is a model of $\cap$-closed granular structure. Suppose $\mathcal{M}_5 = (U, (G_5, \cup))$, where $U = \{a, b, c, d, e\}$, $G_5 = \{\{a, b, c\}, \{b, c, d\}, \{a, b, c, d\}\}$, $\mathcal{M}_5$ is a model of $\cup$-closed granular structure. Suppose $\mathcal{M}_6 = (U, (G_6, \cap, \cup))$, where $U = \{a, b, c, d, e\}$, $G_6 = \{\{b, c\}, \{a, b, c\}, \{b, c, d\}, \{a, b, c, d\}\}$, $\mathcal{M}_6$ is a model of $(\cap, \cup)$-closed granular structure.

In the granular structure $(G, \cap)$, the largest granule U may not be in G. If $U \in G$, the granular structure $(G, \cap)$ is a closure system.

Definition 8. *A $\cap$-closed granular structure $(G, \cap)$ is a closure system, if the $U \in G$.*

A closure system that is closed under set union is referred to as a $\cup$-closure system [2]. Given a granular structure $(G, \cap)$ that is a closure system, its dual system $(G^c, \cup) = \{g^c \mid g \in G\}$ contains $\emptyset$ and is closed under set union.

Definition 9. *A granular structure $(G, \subseteq)$ is a Boolean algebra, denoted by $(G, c, \cap, \cup)$, if G is closed under set complement, intersection and union, respectively.*

The Pawlak rough set model uses a Boolean algebra $(G, c, \cap, \cup)$ whose elements are called definable sets.

3.3   Characterization of Models of Granular Structures

Different models of granular structures have different properties. To characterize and classify these models, we introduce the following list of axioms:

$$S_0 : \ G \neq \emptyset,$$
$$S_1 : \ (a \in G, b \in G) \Longrightarrow \mathrm{glb}(a, b) = a \wedge b \text{ exists in } G,$$
$$S_2 : \ (a \in G, b \in G) \Longrightarrow \mathrm{lub}(a, b) = a \vee b \text{ exists in } G,$$
$$S_3 : \ (a \in G, b \in G) \Longrightarrow a \cap b \in G,$$
$$S_4 : \ (a \in G, b \in G) \Longrightarrow a \cup b \in G,$$
$$S_5 : \ a \in G \Longrightarrow a^c \in G,$$
$$S_6 : \ \emptyset \in G$$
$$S_7 : \ U \in G.$$

Axioms S_1 and S_2 define lattice-based structures; axioms S_3 and S_4 define set-based structures; axioms S_3 and S_4 are the special cases of S_1 and S_2, respectively. These axioms are not independent. For example, $(S_5, S_6) \Longrightarrow S_7$, $(S_5, S_7) \Longrightarrow S_6$, $(S_4, S_5) \Longrightarrow S_3$, and $(S_3, S_5) \Longrightarrow S_4$.

The family of models in Fig.1 is characterized by these axioms as follows:

$$\begin{array}{ll} \mathcal{M}_0 : S_0; & \mathcal{M}_1 : S_1; \\ \mathcal{M}_2 : S_2; & \mathcal{M}_3 : S_1, S_2; \\ \mathcal{M}_4 : S_3; & \mathcal{M}_5 : S_4; \\ \mathcal{M}_6 : S_3, S_4; & \mathcal{M}_7 : S_3, S_4, S_5. \end{array}$$

Additional models can also be obtained. For example, a closure system is defined by S_3 and S_7. A $\cup$-closure system is defined by S_3, S_4 and S_7, and the formal concept lattice is defined by S_2 and S_3. Some of the models discussed in this section have been studied in the context of rough set theory as generalized rough set models [15].

Axioms (S_1) and (S_2) may be considered as dual axioms in the sense that if a granular structure $(G, \subseteq)$ satisfies axiom (S_1), its dual structure $(G^c, \subseteq)$,

$G^c = \{g^c \mid g \in G\}$, satisfies axiom (S_2), and vice versa. Similarly, axioms (S_3) and (S_4) are dual axioms, and so are axioms (S_6) and (S_7). By the duality of axioms, we can discuss the duality of methods. In Fig. 1, we have two pairs of dual models, namely, models $\mathcal{M}_1$ and $\mathcal{M}_2$, and models $\mathcal{M}_3$ and $\mathcal{M}_4$.

4   Conclusion

Granular structures play a fundamental role in granular computing. A special type of granular structures is introduced based on a set-theoretic interpretation of granules. A granule is a subset of a universal set and a granular structure is a family of subsets, namely, granules, of the universal set equipped with the standard set-inclusion relation. Seven specific models are derived from this basic model by imposing extra conditions. These derived models can be formally defined and classified by a set of axioms.

Some of the models examined in this paper have been used in many studies, such as formal concept analysis, rough set theory and knowledge spaces. As future research, we will investigate concrete examples of these models and their real world applications.

Acknowledgements

This work is partially supported by a Discovery Grant from NSERC Canada and Grant No. 60970061 from the National Natural Science Foundation of China.

References

1. Bargiela, A., Pedrycz, W.: Granular Computing: An Introduction. Kluwer Academic Publishers, Boston (2002)
2. Caspard, N., Monjardet, B.: Some lattices of closure systems on a finite set. Discrete Mathematics and Theoretical Computer Science 6, 163–190 (2004)
3. Doignon, J.P., Falmagne, J.C.: Knowledge Spaces. Springer, Berlin (1999)
4. Hobbs, J.R.: Granularity. In: Joshi, A. (ed.) Proceedings of the 9th International Joint Conference on Artificial Intelligence, pp. 432–453. IEEE Computer Society Press, Los Angeles (1985)
5. Keet, C.M.: A Formal Theory of Granularity, PhD Thesis, KRDB Research Centre, Faculty of Computer Science, Free University of Bozen-Bolzano, Italy (2008), http://www.meteck.org/files/AFormalTheoryOfGranularity_CMK08.pdf (accessed June 8, 2008)
6. Lin, T.Y., Yao, Y.Y., Zedah, L.A. (eds.): Data Mining, Rough Set and Granular Computing. Physica-Verlag, Heidelberg (2002)
7. Miao, D.Q., Fan, S.D.: The calculation of knowledge granulation and its application. System Engeering-Theory and Practice 1, 48–56 (2002)
8. Miao, D.Q., Wang, G.Y., Liu, Q., Lin, T.Y., Yao, Y.Y. (eds.): Granular Computing: Past, Present and Prospect. Tsinghua University Press, Beijing (2007)
9. Pawlak, Z.: Rough sets. International Journal of Computer and Information Sciences 11, 341–356 (1982)

10. Pawlak, Z.: Rough Sets-Theoretical Aspects of Reasoning About Data. Kluwer Academic Publishers, Boston (1991)
11. Pedrycz, W., Skowron, A., Kreinovich, V. (eds.): Handbook of Granular Computing. Wiley Interscience, New York (2008)
12. Shiu, L.P., Sin, C.Y.: Top-down, middle-out, and bottom-up processes: a cognitive perspective of teaching and learning economics. International Review of Economics Education 5, 60–72 (2006)
13. Wille, R.: Concept lattices and conceptual knowledge systems. Computers Mathematics with Applications 23, 493–515 (1992)
14. Xu, F.F., Yao, Y.Y., Miao, D.Q.: Rough set approximations in formal concept analysis and knowledge spaces. In: An, A., Matwin, S., Raś, Z.W., Ślęzak, D. (eds.) ISMIS 2008. LNCS (LNAI), vol. 4994, pp. 319–328. Springer, Heidelberg (2008)
15. Yao, Y.Y.: On generalizing Pawlak approximation operators. In: Polkowski, L., Skowron, A. (eds.) RSCTC 1998. LNCS (LNAI), vol. 1424, pp. 298–307. Springer, Heidelberg (1998)
16. Yao, Y.Y.: A comparative study of formal concept analysis and rough set theory in data analysis. In: Tsumoto, S., Słowiński, R., Komorowski, J., Grzymała-Busse, J.W. (eds.) RSCTC 2004. LNCS (LNAI), vol. 3066, pp. 59–68. Springer, Heidelberg (2004)
17. Yao, Y.Y.: Perspectives of granular computing. In: Hu, X.H., Liu, Q., Skowron, A., Lin, T.Y., Yager, R.R., Zhang, B. (eds.) Proceedings of 2005 IEEE International Conference on Granular Computing (GrC 2005), pp. 85–90. IEEE Computer Society Press, Los Angles (2005)
18. Yao, Y.Y.: Three perspectives of granular computing. Journal of Nanchang Institute of Technology 25, 16–21 (2006)
19. Yao, Y.Y.: Granular computing: past, present and future. In: Hu, X.H., Hata, Y., Slowinski, R., Liu, Q. (eds.) Proceedings of 2008 IEEE International Conference on Granular Computing (GrC 2008), pp. 80–85. IEEE Computer Society Press, Los Angles (2008)
20. Yao, Y.Y.: Interpreting concept learning in cognitive informatics and granular computing. IEEE Transactions on Systems, Man, and Cybernetics (Part B) 4, 855–866 (2009)
21. Yao, Y.Y., Miao, D.Q., Xu, F.F.: Granular Structures and Approximations in Rough Sets and Knowledge Spaces. In: Ajith, A., Rafael, F., Rafael, B. (eds.) Rough Set Theory: A True Landmark in Data Analysis. Springer, Berlin (2009)

A Robust Fuzzy Rough Set Model Based on Minimum Enclosing Ball

Shuang An, Qinghua Hu, and Daren Yu

Harbin Institute of Technology, Harbin 150001, P.R. China
anshuang_001@163.com,
{huqinghua,yudaren}@hit.edu.cn

Abstract. Fuzzy rough set theory was introduced as a useful mathematical tool to handle real-valued data. Unluckily, its sensitivity to noise has a great impact on the application in real world. So it is necessary to enhance the robustness of fuzzy rough sets. In this work, based on the minimum enclosing ball problem we introduce a robust model of fuzzy rough sets. In addition, we define a robust fuzzy dependency function and apply it to evaluate features corrupted by noise. Experimental results show that the new model is robust to noise.

1 Introduction

Rough sets was originally introduced by Pawlak [9] as a mathematical approach to handle imprecision, vagueness and uncertainty in data analysis. It has been successfully applied in attribute reduction [11]. Unfortunately, the classic rough set model could not work efficiently on the real-valued data sets. Many extensions of rough sets, such as neighborhood rough sets [6], fuzzy rough sets [3] and covering rough sets [21], have been introduced to address this problem. Especially, fuzzy rough sets can effectively deal with both fuzziness and vagueness of the data sets with continuous features.

It was shown that fuzzy rough sets are sensitive to noise [10,20]. Actually, data are usually corrupted by noise (outliers or mislabeled samples) in real-world. So fuzzy rough sets can not accurately tackle the data sets containing noise, which has a great influence on the application and popularization of fuzzy rough sets in practice.

Consequently, some noise-tolerance models of rough sets were developed, such as decision-theoretic rough set model [19] and information-theoretic rough set model [18]. These models can get rid of the influence of some noise in handling the data. Besides, in [10] and [20], the authors showed the models of variable precision fuzzy rough sets and fuzzy variable precision rough sets to enhance the robustness of fuzzy rough sets, respectively. Unfortunately, we find that the model in [20] is still sensitive to mislabeled samples. Although there are some other robust models of fuzzy rough sets, it seems that handling noise is still an open problem in the rough set theory.

Minimum enclosing ball (MEB) problem, a classic problem in computational geometry [4], plays an important role in novelty detection, classification etc.

J. Yu et al. (Eds.): RSKT 2010, LNAI 6401, pp. 102–109, 2010.

in machine learning and pattern recognition in recent years [13,14]. Given a data set embedded into a feature space with mapping function $\phi(\cdot)$, the MEB problem is to find the smallest hypersphere to contain all the objects in the feature space. Considering that outliers have a great impact on the solution of the problem, which may result in a not robust pattern analysis system, the soft minimum enclosing ball is used to address this problem [12,15]. It makes most of the objects contained in the MEB and balances the loss incurred by missing a small number of objects with the reduction in radius, which can potentially give rise to robust pattern analysis systems.

Our contributions are presented as follows. Firstly, we introduce the robust minimum enclosing ball (RMEB) with the spatial median of a data as the center. Secondly, based on the RMEB we introduce a robust model of fuzzy rough sets. The new model changes the way of computing the fuzzy memberships of a sample to the lower and upper approximations of a set. As to classification, the fuzzy memberships are determined by the nearest sample in the RMEB of the set consisted of the samples from other classes (or the same class), but not by all the samples from other classes (or the same class). Besides, we define a new fuzzy dependency and apply it to evaluate features. Experimental results show that the new dependency is robust to noise.

The rest of this paper is organized as follows. Section 2 introduces the related work of fuzzy rough set theory and the MEB problem. In Section 3 we introduce a RMEB. Based on the RMEB problem we present a robust fuzzy rough set model and define a robust fuzzy dependency function. Next, the robustness of the new dependency function in measuring single feature is tested in Section 4 and some conclusions are provided in Section 5.

2   Related Work

2.1   Fuzzy Rough Sets and Robustness Analysis

Given a nonempty universe U, R is a fuzzy binary relation on U. If R satisfies reflexivity, symmetry and sup-min transitivity, we say R is a fuzzy equivalence relation that is used to measure the similarity between objects characterized with continuous features. The fuzzy equivalence class $[x]_R$ associated with x and R is a fuzzy set, where $[x]_R(y) = R(x, y)$ for all $y \in U$. Based on fuzzy equivalence relations fuzzy rough sets were defined as follows.

Definition 1. *Let U be a nonempty universe, R be a fuzzy similarity relation on U and $F(U)$ be the fuzzy power set of U. Given a fuzzy set $A \in F(U)$, the lower and upper approximations are defined as*

$$\begin{cases} \underline{R}A(x) = \inf_{y \in U} \max\{1 - R(x, y), A(y)\}, \\ \overline{R}A(x) = \sup_{y \in U} \min\{R(x, y), A(y)\}. \end{cases} \tag{1}$$

In [16] and [17], the authors studied the lower and upper approximations in (1) in detail from the constructive and axiomatic approaches. Recently, many researches were focusing on approximation operators. Morsi and Yakout replaced fuzzy equivalence relation with a T-equivalence relation and built an axiom system of the model [8]. Besides, based on the negator operator δ and implicator operator θ, Radzikowska and Kerre defined the lower and upper approximations of a fuzzy set $F \in F(U)$ [7].

If A is a crisp set, the lower and upper approximations in (1) degenerate into the following formulae

$$\begin{cases} \underline{R}A(x) = \inf_{A(y)=0} \{1 - R(x,y)\} \\ \overline{R}A(x) = \sup_{A(y)=1} \{R(x,y)\} \end{cases} . \tag{2}$$

If we take

$$R(x,y) = \exp(\frac{- \parallel x - y \parallel^2}{\delta}) \tag{3}$$

as a similarity measure, then $1 - R(x,y)$ can be considered as a distance measure. Then $\underline{R}A(x)$ is the minimal distance between x and the sample $y \in U - A$ and $\overline{R}A(x)$ is the minimal distance between x and the sample $y \in A$. As we known, the statistics of minimum is not robust. One noisy sample completely alters the lower approximation of a class. Therefore, fuzzy rough sets are not robust to noise.

2.2   Minimum Enclosing Ball

Given a sample set $X = \{\mathbf{x}_1, \mathbf{x}_2, ..., \mathbf{x}_N\}$ with an associated embedding $\phi(\cdot)$ in an Euclidean feature space F with an associated kernel k satisfying

$$k(\mathbf{x}, \mathbf{z}) = < \phi(\mathbf{x}), \phi(\mathbf{z}) >, \tag{4}$$

the MEB of X, denoted by $\text{MEB}(X)$, is the smallest hypersphere

$$B_X(\mathbf{c}, r) = \{\mathbf{x} \in X | \parallel \phi(\mathbf{x}) - c \parallel \le r\}, \tag{5}$$

where the center $\mathbf{c}^*$ of $\text{MEB}(X)$ is the point resolved with

$$\mathbf{c}^* = \arg\min_{\mathbf{c}} \max_{1 \le i \le N} \parallel \phi(\mathbf{x}_i) - \mathbf{c} \parallel \tag{6}$$

and the radius r is the value of the expression at the optimum. The MEB problem can be described as the following optimization problem:

$$\begin{aligned} &\min_{\mathbf{c}, r} \quad r^2 \\ &s.t. \quad \parallel \phi(\mathbf{x}_i) - \mathbf{c} \parallel^2 = (\phi(\mathbf{x}_i) - \mathbf{c})^T(\phi(\mathbf{x}_i) - \mathbf{c}) \le r^2, \quad \forall \mathbf{x}_i \in X. \end{aligned} \tag{7}$$

As the radius r is determined by the distance from $\mathbf{c}$ to the furthest point, the solution of formula (7) is sensitive to outliers. As we know, data usually contain noise (outliers or mislabeled samples) in practice. Then the MEB may result in a non-robust novelty detection or classification system in real-world.

Soft minimum enclosing ball (SMEB) is used to tackle the above problem. It can be formulated as the following constrained optimization problem:

$$\begin{aligned}
\min_{\mathbf{c}, r, \xi} \quad & r^2 + C \parallel \xi \parallel_1 \\
s.t. \quad & \parallel \phi(x_i) - \mathbf{c} \parallel^2 = (\phi(x_i) - \mathbf{c})^T (\phi(x_i) - \mathbf{c}) \le r^2 + \xi_i, \\
& \xi_i \ge 0, \quad i = 1, ..., N,
\end{aligned} \tag{8}$$

where $\xi_i = (\parallel c - \phi(x_i) \parallel^2 - r^2)_+$ is a slack variable and the parameter C is a penalty factor controlling the trade-off between the radius and ξ.

We can see that SMEB does not require all the points contained in the ball and can balances the loss incurred by missing a small number of points with the reduction in radius, which can potentially give rise to robust novelty detectors and classifiers.

3   MEB-Based Robust Fuzzy Rough Set Model

3.1   Robust Minimum Enclosing Ball

Given a data set X embedded in a feature space, resolving the problem of SMEB(X) is NP-hard [12]. Hence, we can only get the approximation of the optimal solution. In this section, we present a method for finding the suboptimal solution of the SMEB problem.

As median is a robust estimation of the center of a data set, we use the spatial median of a multidimensional data set as the center of the SMEB, where spatial median is described as follows.

Let $S: \mathbb{R}^n \longrightarrow \mathbb{R}^n$ be the spatial sign function i.e.

$$S(\mathbf{x}) = \begin{cases} \frac{\mathbf{x}}{\parallel \mathbf{x} \parallel}, \mathbf{x} \ne \mathbf{0}, \\ \mathbf{0}, \mathbf{x} = \mathbf{0}, \end{cases} \tag{9}$$

where $\mathbf{x} \in \mathbb{R}^n$, $\parallel \mathbf{x} \parallel = \sqrt{\mathbf{x}^T \mathbf{x}}$ and $\mathbf{0}$ is zero vector in $\mathbb{R}^n$. Then the spatial median $\mathbf{c}$ of $X = \{\mathbf{x}_1, \mathbf{x}_2, ..., \mathbf{x}_N\} \subset \mathbb{R}^n$ satisfies

$$\left\| \sum_{\mathbf{x}_i \in X} S(\mathbf{x}_i - \mathbf{c}) \right\| = 0. \tag{10}$$

By this way, the radius of the SMEB is determined by

$$r = \arg\min_r r^2 + C \parallel \xi \parallel, \tag{11}$$

where $\xi_i = (\parallel \phi(\mathbf{x}_i) - \mathbf{c} \parallel^2 - r^2)_+$. The hypersphere, with the spatial median as the center and r as the radius, is named robust minimum enclosing ball of X (RMEB(X)) denoted by $RB_X(\mathbf{c}, r)$.

3.2   A Robust Rough Set Model

Definition 2. *Let U be a nonempty universe, R be a fuzzy equivalence relation on U and $F(U)$ be the fuzzy power set of U. Given a fuzzy set $A \in F(U)$, the robust lower and upper approximations are defined as*

$$\begin{cases} \underline{R^R}A(\mathbf{x}) = \inf\limits_{\mathbf{y} \in RB_{Y_L}(\mathbf{c},r)} \{(1 - R(\mathbf{x}, \mathbf{y})\}, \\ \overline{R^R}A(\mathbf{x}) = \sup\limits_{\mathbf{y} \in RB_{Y_U}(\mathbf{c}',r')} \{R(\mathbf{x}, \mathbf{y})\}. \end{cases} \tag{12}$$

where

$$\begin{cases} Y_L = \{\mathbf{y} \in U | A(\mathbf{y}) \leq A(\mathbf{y}_L)\}, \mathbf{y}_L = \arg \inf\limits_{\mathbf{y} \in U} \max\{1 - R(\mathbf{x}, \mathbf{y}), A(\mathbf{y})\}, \\ Y_U = \{\mathbf{y} \in U | A(\mathbf{y}) \geq A(\mathbf{y}_U)\}, \mathbf{y}_U = \arg \sup\limits_{\mathbf{y} \in U} \min\{R(\mathbf{x}, \mathbf{y})\}. \end{cases} \tag{13}$$

If A is crisp set, (13) degenerates as

$$\begin{cases} Y_L = \{\mathbf{y} \in U | A(\mathbf{y}) = 0\}, \\ Y_U = \{\mathbf{y} \in U | A(\mathbf{y}) = 1\}. \end{cases} \tag{14}$$

In brief, $\underline{R^R}A(x)$ is computed with the samples in the RMEB composed of samples having different class labels, and $\overline{R^R}A(x)$ is computed with the samples in the RMEB composed of samples coming from the same class.

Consequently, the membership of an object $\mathbf{x} \in U$ belonging to the robust positive region of the decision D on feature subset B is defined as

$$POS_{R_B}^R(D)(\mathbf{x}) = \sup\limits_{X \in U/D} \underline{R_B^R}(X)(\mathbf{x}). \tag{15}$$

Then the robust fuzzy dependency (RFD) of D on B is defined as

$$\gamma_{R_B}^R(D) = \frac{\sum_{\mathbf{x} \in U} POS_{R_B}^R(D)(\mathbf{x})}{|U|}. \tag{16}$$

As the new model of fuzzy rough sets is more robust than fuzzy rough sets, the new dependency is more robust than fuzzy dependency in evaluating features, which would be validated in the next section.

4   Experiments

The experiments are conducted on the noisy data created from nine real-world data sets coming from UCI [1] to verify the finding of this study. We concentrate on the robustness of robust fuzzy dependency.

In this work, the noisy data are created as follows. We take the nine real-world data sets as the raw data, respectively. Firstly, we randomly draw $i\%$ ($i = 2, 4, 6, 8, 10$) samples, and then randomly give them labels that are distinct from

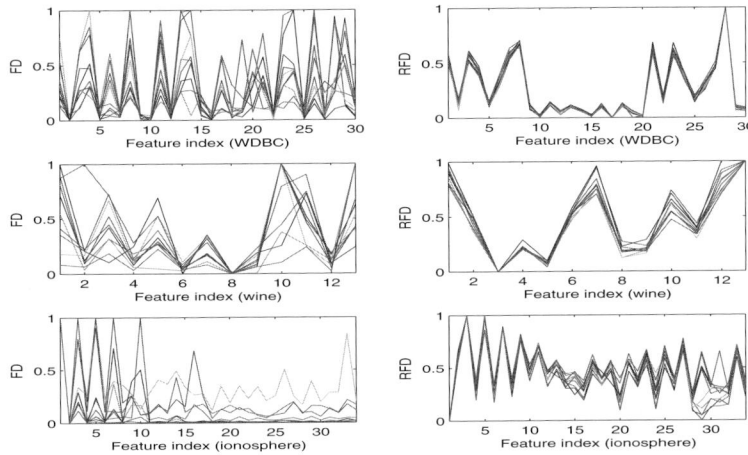

Fig. 1. Robustness comparison between FD and RFD

Table 1. Similarity between raw dependency and noisy dependency

Data	FD					RFD				
	2%	4%	6%	8%	10%	2%	4%	6%	8%	10%
glass	0.79	0.99	0.16	0.65	0.74	1.00	1.00	1.00	0.99	1.00
musk	0.92	0.59	0.46	0.34	0.51	0.99	0.96	0.99	0.99	0.98
diabetes	1.00	0.68	0.69	0.68	0.75	0.99	1.00	0.99	1.00	0.98
yeast	1.00	1.00	0.99	0.98	1.00	1.00	1.00	1.00	1.00	1.00
WDBC	0.73	0.19	0.58	0.53	0.28	1.00	1.00	1.00	0.99	0.99
WPBC	0.58	0.21	0.90	0.35	0.45	0.99	0.99	0.98	0.99	0.99
ionosphere	0.96	0.24	0.12	0.21	0.17	1.00	0.99	0.97	0.93	0.91
sonar	0.99	0.87	0.93	0.92	0.85	0.99	0.93	0.93	0.96	0.96
wine	0.98	0.72	0.76	0.49	0.41	1.00	0.99	0.99	0.99	0.98
average	0.88	0.61	0.62	0.57	0.57	1.00	0.98	0.98	0.98	0.97

their original labels. By this way, we get noisy data sets with $i\%$ ($i = 2, 4, 6, 8, 10$) noise levels.

First, we compute the RFD of single feature with a raw data and 10 noisy data sets. And we get 11 RFD values for each feature. If the 11 values are same or very similar, we consider RFD function is robust. Otherwise, RFD is sensitive to noise.

Fig.1 shows the evaluation results of features with FD and RFD. For each subgraph, x axis is the feature index, y is the dependency values and the 11 curves represent 11 evaluation results of features. It shows that the 11 values calculated with RFD for each feature are more similar than that with FD, which shows RFD is more robust than FD.

Now we use Pearson's correlation coefficient to compute the similarity between the raw dependency vector (computed with the raw data) and 10 noisy dependency vectors (computed with the 10 noisy data), respectively. The larger the similarity is, the more robust the measure is. Table 1 shows the results. It shows the similarity between raw RFD and noisy RFD is larger and more stable along with the increasing of noise levels than that between raw FD and noisy FD. So we conclude that RFD is more robust than FD in evaluating single feature.

In order to evaluate the robustness of all the evaluation results, we take the method in [5] to compare the robustness of similarity matrices computed with FD and RFD. The smaller the evaluation value is, the more robust the measure is. Table 2 shows the evaluation results. We can see that the values for RFD is much smaller than that for FD. This also shows RFD is more robust than FD.

Table 2. Robustness on similarity matrix

Data	glass	musk	diabetes	yeast	WDBC	WPBC	ionosphere	sonar	wine
FD	0.44	0.61	0.15	0.24	0.63	0.37	1.00	0.18	0.50
RFD	0.00	0.02	0.01	0.00	0.00	0.01	0.05	0.05	0.01

5   Conclusions

Fuzzy rough set model is the generalization of the classic rough set model to deal with both fuzziness and vagueness in data analysis. Due to its sensitivity to noise, the application and popularization of fuzzy rough sets are restricted in practice. In this work, based on the minimum enclosing ball, we present a robust fuzzy rough set model and define a robust fuzzy dependency function. And then we test the robustness of the robust fuzzy dependency in measuring features with experiments. The results show that the new dependency is more robust to noise than fuzzy dependency.

Acknowledgments

This work is partially supported by National Natural Science Foundation of China under Grants 60703013, 10978011 and The Hong Kong Polytechnic University (G-YX3B). and Prof. Yu is supported by National Science Fund for Distinguished Young Scholars under Grant 50925625.

References

1. Blake, C.L., Merz, C.J.: UCI Repository of Machine Learning Databases (1998), http://www.ics.uci.edu/mlearn/MLRepository.html
2. Chen, Y.X., Dang, X., Peng, H.X., Bart Jr., H.L.: Outlier detection with the kernelized spatial depth function. IEEE Transactions on Pattern Analysis and Machine Intelligence 31, 288–305 (2009)
3. Dubois, D., Prade, H.: Rough fuzzy sets and fuzzy rough sets. General systems 17, 191–209 (1990)

4. Fischer, K.: The smallest enclosing ball of balls: combinatorial structure and algorithms. In: Proceedings of the Nineteenth Annual Symposium on Computational Geometry, USA, pp. 292–301 (2003)
5. Hu, Q.H., Liu, J.F., Yu, D.R.: Stability analysis on rough set based feature evaluation. In: Wang, G., Li, T., Grzymala-Busse, J.W., Miao, D., Skowron, A., Yao, Y. (eds.) RSKT 2008. LNCS (LNAI), vol. 5009, pp. 88–96. Springer, Heidelberg (2008)
6. Hu, Q.H., Yu, D.R., Liu, J.F., Wu, C.X.: Neighborhood rough set based heterogeneous feature subset selection. Information Sciences 178, 3577–3594 (2008)
7. Radzikowska, A.M., Kerre, E.E.: A comparative study of fuzzy rough sets. Fuzzy Sets and Systems 126, 137–155 (2002)
8. Morsi, N.N., Yakout, M.M.: Axiomatics for fuzzy rough sets. Fuzzy Sets and Systems 100, 327–342 (1998)
9. Pawlak, Z.: Rough sets. International Journal of Computer and Information Sciences 11, 341–356 (1982)
10. Rolka, A.M., Rolka, L.: Variable precision fuzzy rough sets. In: Peters, J.F., Skowron, A., Grzymała-Busse, J.W., Kostek, B.z., Świniarski, R.W., Szczuka, M.S. (eds.) Transactions on Rough Sets I. LNCS, vol. 3100, pp. 144–160. Springer, Heidelberg (2004)
11. Skowron, A., Polkowski, L.: Rough sets in knowledge discovery. Springer, Berlin (1998)
12. Taylor, J.S., Cristianini, N.: Kernel methods for pattern analysis. Cambridge University Press, Cambridge (2004)
13. Tsang, I.W., Kwok, J.T., Cheung, P.M.: Core vector machines: Fast SVMtraining on large data sets. Journal of Machine Learning Research 6, 363–392 (2005)
14. Tsang, I.W., Kwok, J.T., Zurada, J.M.: Generalized core vector machines. IEEE Transactions on Neural Networks 17, 1126–1140 (2006)
15. Wu, M.R., Ye, J.P.: A Small Sphere and Large Margin Approach for Novelty Detection Using Training Data with Outliers. IEEE Transactions on Pattern Analysis and Machine Intelligence 11, 2088–2092 (2009)
16. Wu, W.-Z., Zhang, W.-X.: Constructive and axiomatic approaches of fuzzy approximation operators. Information Sciences 159, 233–254 (2004)
17. Wu, W.-Z., Mi, J.-S., Zhang, W.-X.: Generalized fuzzy rough sets. Information Sciences 151, 263–282 (2003)
18. Xu, F.F., Miao, D.Q., Wei, L.: An approach for fuzzy-rough sets attribute reduction via mutual information. In: Proceedings of the Fourth International Conference on Fuzzy Systems and Knowledge Discovery,USA, pp. 107–112 (2007)
19. Yao, Y.Y., Wong, S.K.M., Lingras, P.: A decision-theoretic rough set model. Methodologies for Intelligent Systems 5, 17–24 (1990)
20. Zhao, S.Y., Tsang, E.C.C., Chen, D.G.: The model of fuzzy variable precision rough sets. IEEE Transactions on Fuzzy Systems 17, 451–467 (2009)
21. Zhu, W., Wang, F.-Y.: Reduction and axiomization of covering generalized rough sets. Information Sciences 152, 217–230 (2003)

Indiscernibility and Similarity in an Incomplete Information Table

Renpu Li[1,2] and Yiyu Yao[2]

[1] School of Information Science and Technology,
Ludong University, Yantai, China, 264025
[2] Department of Computer Science, University of Regina, Regina,
Saskatchewan, Canada S4S 0A2

Abstract. A new method is proposed for interpreting and constructing relationships between objects in an incomplete information table. An incomplete information table is expressed as a family of complete tables. One can define an indiscernibility relation for each complete table and then get a family of indiscernibility relations of all complete tables. A pair of an indiscernibility and a similarity relation is constructed by the intersection and union of this family of indiscernibility relation. It provides a clear semantic interpretation for relationship between objects of an incomplete information table. In fact, it is a pair of bounds of the actual indiscernibility relation if all values in the incomplete table were known.

Keywords: Incomplete information table, completion, indiscernibility relation, similarity relation.

1   Introduction

For incomplete information tables, how to interpret the semantic of unknown values is a primary problem. In this paper, a basic assumption is that any object on any attribute possesses only one value, and if it is unknown its actual value is one known value of the value domain. Under this assumption, an incomplete information table is expressed as a family of compete tables, one of which, called actual complete table, contains the actual values of incomplete information table.

For each complete table, one can define an indiscernibility relation on all objects. All indiscernibility relations on the family of complete tables can be used to described the relation between objects in incomplete information tables. For example, through the union and intersection operations on the family of indiscernibility relations, a pair of an indiscernibility and a similarity relations can be constructed for an incomplete information table. They include the pairs of objects which certainly or possibly belong to the indiscernibility relation of the actual complete table.

This method of constructing relations in an incomplete information table provides a clear semantic interpretation for the relationship between objects. For the actual indiscernibility relation of an incomplete information table, traditional

J. Yu et al. (Eds.): RSKT 2010, LNAI 6401, pp. 110–117, 2010.

similarity relations only provide its upper bound while the pair of indiscernibility and similarity relations is a pair of bound of it.

2 Incomplete Information Table and Its Interpretations

2.1 Information Tables

Definition 1. *An information table T is expressed as the tuple $T = (U, At, V, f)$ where U is a finite nonempty set of objects, At is a finite nonempty set of attributes, $V = \{V_a | a \in At\}$ and V_a is a nonempty set of values for an attribute $a \in At$, $f = \{f_a | a \in At\}$ and $f_a : U \rightarrow V_a \cup P(V_a)$ is an information function.*

A single-valued information table is characterized by an information function $f_a : U \rightarrow V_a$ and a set-valued table by $f_a : U \rightarrow P(V_a)$. Our definition of an information table combines the standard definitions of single-valued and set-valued information tables [10][11].

   If $f_a(x) = \emptyset$, we denote that the attribute a is not applicable to x. For simplicity, we do not consider this case by assuming that information function is of the form $f_a : U \rightarrow V_a \cup (P(V_a) - \{\emptyset\})$.

   In fact, some authors use an information function of the form $I_a : U \longrightarrow P(V_a)$ to represent both types of tables [1][10][13]. In this study, we make a distinction by representing complete information with values from V_a and incomplete information with values from $P(V_a) - \{\emptyset\}$. The separation explicitly shows that two semantically different categories of values are used in a table representing either complete or incomplete information.

   The set-valued approach to incomplete information is more flexible and general than the well-known null-valued approach. The use of a null value restricts the discussion of incomplete information into the two extreme cases [8], that is, known (i.e., represented by a value from V_a) or unknown (i.e., represented by the null value). In set-valued approaches, a null value may be interpreted as our total ignorance on the value of x and is expressed by the entire set V_a while a subset $\emptyset \neq f_a(x) \neq V_a$ may be used to represent the fact that the value of x is partially known.

Definition 2. *Let $T = (U, At, V, f)$ be an information table. If for all $a \in At$ and $x \in U$, $f_a(x) \in V_a$, it is called a complete table, otherwise it is an incomplete table.*

Example 1. Table 1 is an example of an incomplete information table $IT = (U, At, V, f)$, where $U = \{o_1, o_2, o_3, o_4\}$, $At = \{a, b\}$, $V_a = \{0, 1, 2\}$, $V_b = \{0, 1\}$.

2.2 Interpretation of an Incomplete Information Table

In an incomplete information table, a set of values is used to indicate all possibilities of the actual value of an object. Many studies focus on selecting one value from the set of values to represent the actual value [2]. A main disadvantage

Table 1. An incomplete information table $T1$

U	a	b
o_1	0	0
o_2	0	1
o_3	{0,1}	0
o_4	{1,2}	0

is that each selection method is based on certain assumption and has its own bias. The resulting complete table may not necessarily be meaningful. To avoid this problem, some authors focus on the family of all complete tables that are consistent with an incomplete table [5][8][9].

Definition 3. *Let $T = (U, At, V, f)$ be an incomplete information table. A complete table $T' = (U, At, V, f')$ is called a completion of T if and only if for all $a \in At$ and $x \in U$, $f_a(x) \in V_a$ implies $f'_a(x) = f_a(x)$ and $f_a(x) \in P(V_a)$ implies $f'_a(x) \in f_a(x)$. The set of all completions of T is denoted as $\mathrm{COMP}(T)$.*

Example 2. Completions of Table 1 are given in Fig. 1.

Fig. 1. Completions of incomplete Table $T1$

T_1			T_2			T_3			T_4		
U	a	b	U	a	b	U	a	b	U	a	b
o_1	0	0	o_1	0	0	o_1	0	0	o_1	0	0
o_2	0	1	o_2	0	1	o_2	0	1	o_2	0	1
o_3	0	0	o_3	0	0	o_3	1	0	o_3	1	0
o_4	1	0	o_4	2	0	o_4	1	0	o_4	2	0

3   A Pair of Relations in an Incomplete Information Table

The interpretation of an incomplete table as a family of complete tables provides a new way to define and interpret relationships between objects in an incomplete information table. Based on the earlier related studies by Zhao [6], Lipski [8] and Nakamura [9], in this section we introduce a general method for constructing a pair of indiscernibility and similarity relations for an incomplete information table.

A central notion of Pawlak rough sets is an indiscernibility relation defined by a set of attributes in a complete table [11].

Definition 4. *Suppose $T = (U, At, V, f)$ is a complete information table. For a subset of attributes $A \subseteq At$, a binary indiscernibility relation $IND(T, A)$ on U is defined as follows:*

$$\mathrm{IND}(T, A) = \{(x, y) \mid (x, y) \in U \times U, \forall a \in A[f_a(x) = f_a(y)]\}$$

$$= \bigcap_{a \in A} \mathrm{IND}(T, \{a\}). \tag{1}$$

The relation $\text{IND}(T, A)$ is an equivalence relation. The equivalence class containing x is given by $[x]_{\text{IND}(T,A)} = \{y \mid y \in U, x\text{IND}(T, A)y\}$.

For an incomplete table, it is impossible to define an equivalence relation by using Definition 4 directly. Many studies generalize Definition 4 by modifying the condition $f_a(x) = f_a(y)$. We suggest an alternative method by using Definition 4 indirectly. Since each table T' in $\text{COMP}(T)$ is a complete table, one can directly use Definition 4 to define an equivalence relation in T', then we obtain a family of equivalence relations for an incomplete table. Based on this family, new relations on U can be constructed.

Definition 5. *Suppose $T = (U, At, V, f)$ is an incomplete information table, $A \subseteq At$ and $T' \in \text{COMP}(T)$. A pair of indiscernibility and similarity relations on U can be defined respectively as follows:*

$$\underline{S}(T, A) = \bigcap_{T' \in \text{COMP}(T)} \text{IND}(T', A),$$

$$\overline{S}(T, A) = \bigcup_{T' \in \text{COMP}(T)} \text{IND}(T', A). \tag{2}$$

The pair of relations can be equivalently defined by:

$$x\underline{S}(T, A)y \iff \forall T' \in \text{COMP}(T)[x\text{IND}(T', A)y],$$
$$x\overline{S}(T, A)y \iff \exists T' \in \text{COMP}(T)[x\text{IND}(T', A)y]. \tag{3}$$

By definition, for $\forall T' \in \text{COMP}(T)$ we have

$$\underline{S}(T, A) \subseteq \text{IND}(T', A) \subseteq \overline{S}(T, A). \tag{4}$$

The pair of relations $(\underline{S}(T, A), \overline{S}(T, A))$ represents the two extremes in defining relationship between objects in an incomplete table T. Since the actual table is one of the tables in $\text{COMP}(T)$, the formulation provides a pair of lower and upper bounds within which the actual indiscernibility relation lies. In other words, we can conclude with certainty that a pair of objects in $\underline{S}(T, A)$ must be indiscernible in the actual table, while it is only possible that a pair of objects in $\overline{S}(T, A)$ is indiscernible in the actual table.

It can be verified that $\underline{S}(T, A)$ is an equivalence relation and $\overline{S}(T, A)$ is a similarity relation. Let $[x]_{\underline{S}(T,A)}$ and $(x)_{\overline{S}(T,A)}$ denote the equivalence class and similarity class of x on $\underline{S}(T, A)$ and $\overline{S}(T, A)$ respectively, which are given by:

$$[x]_{\underline{S}(T,A)} = \{y \mid y \in U, x\underline{S}(T, A)y\},$$
$$(x)_{\overline{S}(T,A)} = \{y \mid y \in U, x\overline{S}(T, A)y\}. \tag{5}$$

The pair of relations can be equivalently defined by:

$$[x]_{\underline{S}(T,A)} = \bigcap_{T' \in \text{COMP}(T)} [x]_{\text{IND}(T',A)},$$

$$(x)_{\overline{S}(T,A)} = \bigcup_{T' \in \text{COMP}(T)} [x]_{\text{IND}(T',A)}, \tag{6}$$

where $[x]_{\text{IND}(T',A)}$ is the equivalence class of x in a complete table T'.

For $\forall T' \in \text{COMP}(T)$ we have

$$[x]_{\underline{S}(T,A)} \subseteq [x]_{\text{IND}(T',A)} \subseteq (x)_{\overline{S}(T,A)}. \tag{7}$$

Example 3. (Continued from Example 2). We have $\text{COMP}(T) = \{T_1, T_2, T_3, T_4\}$, the classes containing o_3 can be computed respectively by:

$$[o_3]_{\underline{S}(T,A)} = \{o_1, o_3\} \cap \{o_1, o_3\} \cap \{o_3, o_4\} \cap \{o_3\} = \{o_3\};$$
$$(o_3)_{\overline{S}(T,A)} = \{o_1, o_3\} \cup \{o_1, o_3\} \cup \{o_3, o_4\} \cup \{o_3\} = \{o_1, o_3, o_4\}.$$

The classes of all objects can be obtained as follows:

$$[o_1]_{\underline{S}(T,A)} = \{o_1\}, (o_1)_{\overline{S}(T,A)} = \{o_1, o_3\}$$
$$[o_2]_{\underline{S}(T,A)} = \{o_2\}, (o_2)_{\overline{S}(T,A)} = \{o_2\}$$
$$[o_3]_{\underline{S}(T,A)} = \{o_3\}, (o_3)_{\overline{S}(T,A)} = \{o_1, o_3, o_4\}$$
$$[o_4]_{\underline{S}(T,A)} = \{o_4\}, (o_4)_{\overline{S}(T,A)} = \{o_3, o_4\}.$$

4   Similarity Relations and the Pair of Relations

Similarity relations for incomplete information tables have been studied by many authors [4][10]. We examine two such examples, one is based on set-valued information tables [1][10][13], and the other is based on incomplete tables with missing values, denoted by special symbols such as "*", "?", "-" and "ϵ" [3][4][5][7][9][12].

Based on the interpretation of an incomplete information table as a family of complete tables and the corresponding construction method for defining the pair of relations, two existing similarity relations are proved to be equivalent to the similarity relation of the pair of relations, and a new indiscernibility relation is built on the objects of an incomplete information table directly.

4.1   Similarity Relation on Set-Valued Information Tables

Definition 6. *[10] Suppose $T = (U, At, V, f)$ is a set-valued information table, where $f_a : U \longrightarrow P(V_a) - \{\emptyset\}$ for $a \in At$. For $A \subseteq At$, a similarity relation on U is defined as:*

$$\text{TOR}(T, A) = \{(x, y) \mid (x, y) \in U \times U, \forall a \in A[f_a(x) \cap f_a(y) \neq \emptyset]\}. \tag{8}$$

By treating a singleton subset of attribute values as the value itself, one can easily transform between information tables characterized by $f_a : U \longrightarrow P(V_a) - \{\emptyset\}$ and $f_a : U \longrightarrow V_a \cup (P(V_a) - \{\emptyset\})$, respectively. It follows that the completion based interpretation can be immediately applied to a set-valued table with $f_a : U \longrightarrow P(V_a) - \{\emptyset\}$. The similarity relation $TOR(T, A)$ can be obtained under such an interpretation.

Theorem 1. $\text{TOR}(T, A) = \overline{S}(T, A).$

Proof. Suppose $(x, y) \in \text{TOR}(T, A)$. According to Definition 6, we have $f_a(x) \cap f_a(y) \neq \emptyset$ for all $a \in A$. Thus, we can select a value $v_a \in f_a(x) \cap f_a(y)$ for all $a \in A$. By setting $f'_a(x) = f'_a(y) = v_a$ for all $a \in A$, we can construct a completion $T' \in \text{COMP}(T)$. It follows that $(x, y) \in \text{IND}(T', A)$. By Definition 5, we have $\text{IND}(T', A) \subseteq \overline{S}(T, A)$, and hence $(x, y) \in \overline{S}(T, A)$ holds.

Now suppose $(x, y) \in \overline{S}(T, A)$. According to Definition 5, there exists a $T' \in \text{COMP}(T)$ such that $(x, y) \in \text{IND}(T', A)$. Thus, for all $a \in A$, we have $f'_a(x) = f'_a(y)$. Since T' is a completion of T, for all $a \in A$, we have $f'_a(x) \in f_a(x)$ and $f'_a(y) \in f_a(y)$. It means that $f'_a(x) = f'_a(y) \in (f_a(x) \cap f_a(y)) \neq \emptyset$ for all $a \in A$. According to Definition 6, $(x, y) \in \text{TOR}(T, A)$ holds.

$\text{TOR}(T, A) = \overline{S}(T, A)$ is proved.

4.2 Similarity Relation on Information Tables with Missing Values

Definition 7. *[4] Suppose $T = (U, At, V, f)$ is an incomplete information table, where $f_a : U \longrightarrow V_a \cup \{*\}$ for $a \in At$. For $A \subseteq At$, a similarity relation on U can be defined as:*

$$\text{TOR2}(T, A) = \{(x, y) \mid (x, y) \in U \times U,$$
$$\forall a \in A[f_a(x) = f_a(y) \vee f_a(x) = * \vee f_a(y) = *]\}.$$

In fact, the information tables with $f_a : U \longrightarrow V_a \cup \{*\}$ can be thought as a special case of the ones with $f_a : U \longrightarrow (V_a \cup (P(V_a) - \{\emptyset\}))$ because the symbol "*" on attribute a is equivalent to V_a, that means, in the table with $f_a : U \longrightarrow (V_a \cup \{*\})$, $f_a(x)$ is either a single value or V_a. Hence, the following corollary can be obtained based on Theorem 1.

Corollary 1. $\text{TOR2}(T, A) = \overline{S}(T, A)$.

4.3 An Indiscernibility Relation for Incomplete Information Tables

For an incomplete information table, above two similarity relations both provide the upper bound for the equivalence of the actual completion and equal to the similarity relation derived in Definition 5.

Naturally, from the lower bound, we can point out those object pairs which surely belong to the equivalence relation of the actual table by constructing a binary relation as follows:

Definition 8. *Suppose $T = (U, At, V, f)$ and $A \subseteq At$. A binary relation called incomplete indiscernibility relation on T can be defined as follows:* $\text{IIND}(T, A) = \{(x, y) \mid (x, y) \in U \times U, (x = y) \vee (\forall a \in A[f_a(x) = f_a(y) \in V_a])\}$.

Obviously incomplete indiscernibility relation is also an equivalence relation. If a pair of objects $(x, y) \in \text{IIND}(T, A)$ or $x\text{IIND}(T, A)y$, there are two cases for the relationship between x and y. The first case, if $\exists a \in A$ such that $f_a(x)$ is set-valued, then $y = x$, that is, x has the incomplete indiscernibility relation only with itself. The second case, if values of x on any $a \in A$ are all single-valued, then values of y on any $a \in A$ are also single-valued and equal to the ones of x on corresponding attributes of A.

Theorem 2. $\text{IIND}(T, A) = \underline{S}(T, A)$.

Proof. Suppose $(x, y) \in \text{IIND}(T, A)$. If $\exists a \in A$ such that $f_a(x)$ is set-valued, then $y = x$, according to the reflexivity $(x, x) \in \underline{S}(T, A)$. If all values of x on A are all single-valued, by Definition 8 we have $f_a(x) = f_a(y) \in V_a$ for all $a \in A$. So according to the definition of completion for any $T' \in COMP(T) = COMP(T)$, we have $f'_a(x) = f_a(x) = f_a(y) = f'_a(y)$ for all $a \in A$, then $(x, y) \in \text{IND}(T', A)$. According to Definition 5, $(x, y) \in \underline{S}(T, A)$ holds.

Now suppose $(x, y) \in \underline{S}(T, A)$. If x and y are the same object, according to the reflexivity $(x, y) \in \text{IIND}(T, A)$. If x and y are two different object, according to Definition 5, we can assure that $(x, y) \in \text{IND}(T', A)$ for all $T' \in COMP(T)$. That means for any $T' \in COMP(T)$, we have $f'_a(x) = f'_a(y)$ for all $a \in A$. According to the Definition 3, there is only two cases for the relationship between $f'_a(x)$ and $f_a(x)$: $f'_a(x) = f_a(x)$ or $f'_a(x) \in f_a(x)$.

Assume $f'_a(x) \in f_a(x)$. We can find another value t such that $t \in f_a(x) \wedge t \neq f'_a(x)$, and then by replacing $f'_a(x)$ with t we can construct another completion T^* such that $T^* \in COMP(T)$, and hence we have $(x, y)\overline{\in}\text{IND}(T^*, A)$ for $f^*_a(x) = t \neq f'_a(x) = f'_a(y)$. It conflicts with the prior conclusion that $(x, y) \in \text{IND}(T', A)$ for all $T' \in COMP(T)$. So we have $f'_a(x) = f_a(x)$ and $f_a(x)$ is a single value.

Similarly, it can be obtained that $f'_a(y) = f_a(y)$ and $f_a(y)$ is a single value, and hence $f_a(x) = f_a(y) \in V_a$ for all $a \in A$. Therefore by Definition 8 $(x, y) \in \text{IIND}(T, A)$ holds.

So $\text{IIND}(T, A) = \underline{S}(T, A)$ is proved.

5   Conclusion

An incomplete information table can be interpreted as a set of completions. This interpretation provides a clear semantic for incomplete data and is more suitable to modeling incomplete information. By applying the union and intersection operations on a family of equivalence relations of the set of completions, a pair of indiscernibility and similarity relations can be naturally obtained to approximate the discernibility relation of the actual completion from both upper and lower directions.

Comparatively, traditional similarity relations by extending equivalence relation directly for incomplete information tables are more abstract on its semantical interpretation and only provide upper bound for the discernibility of the actual completion.

This paper provide a new view for reasoning in incomplete information table. It has been proved that existing similarity relations can be included in the pair of relations and new results are easy to obtained from the new view.

Acknowledgements

This work is supported by the Shandong Provincial Scientific Research Foundation for Excellent Young Scientists No. 2008BS01014 and the Innovative Team Construction of Ludong University under No.08-CXB006.

References

1. Guan, Y.: Set-valued information systems. Information Sciences 176, 2507–2525 (2006)
2. Grzymala-Busse, J.W., Hu, M.: A comparison of several approaches to missing attribute values in data mining. In: Ziarko, W., Yao, Y. (eds.) Proceedings of the Second International Conference on Rough Sets and Current Trends in Computing, pp. 378–385. Physica-Verlag, Heidelberg (2001)
3. Jaworski, W.: Generalized indiscernibility relations: applications for missing values and analysis of structural objects. In: Peters, J.F., Skowron, A. (eds.) Transactions on Rough Sets VIII. LNCS, vol. 5084, pp. 116–145. Springer, Heidelberg (2008)
4. Kryszkiewicz, M.: Rough set approach to incomplete information systems. Information Sciences 112, 39–49 (1998)
5. Kryszkiewicz, M.: Rules in incomplete information systems. Information Sciences 113, 271–292 (1999)
6. Zhao, Y., Yao, Y.Y., Luo, F.: Data analysis based on discernibility and indiscernibility. Information Sciences 177, 4959–4976 (2007)
7. Leung, Y., Li, D.: Maximal consistent block technique for rule acquisition in incomplete information systems. Information Sciences 153, 85–106 (2003)
8. Witold Lipski, J.R.: On semantic issues connected with incomplete information databases. ACM Transactions on Database Systems 4, 262–296 (1979)
9. Nakamura, A.: A rough logic based on incomplete information and its applicaton. International Journal of Approximate Reasoning 15, 367–378 (1996)
10. Orlowska, E.: Introduction: what you always wanted to know about rough sets. In: Orlowska, E. (ed.) Incomplete Information: Rough Set Analysis, pp. 1–20. Physica-Verlag, Heidelberg (1998)
11. Pawlak, Z.: Information systems theoretical foundations. Information Systems 6, 205–218 (1981)
12. Wang, G.: Extension of rough set under incomplete information systems. Journal of Computer Research and Development 39, 1238–1243 (2002)
13. Zhang, W.: Incomplete information system and its optimal selections. Computers and Mathematics with Applications 48, 691–698 (2004)

A New Fitness Function for Solving Minimum Attribute Reduction Problem

Dongyi Ye, Zhaojiong Chen, and Shenglan Ma

College of Mathematics and Computer, Fuzhou University, Fuzhou 350108, China
yiedy@fzu.edu.cn

Abstract. The problem of minimum attribute reduction is formally a nonlinearly constrained combinatorial optimization problem and has been proved to be NP-hard. A most commonly used approach for dealing with the problem is to first transform it into a fitness maximization problem over a Boolean space and then to solve the latter via stochastic optimization methods. However, existing fitness functions either fail to assure in theory the optimality equivalence between the original problem and the fitness maximization problem or are not quite adequate in terms of fitness evaluation. In this paper, a new fitness function that overcomes the drawbacks of the existing fitness functions is given. The optimality equivalence using the proposed fitness function is theoretically proved. Comparisons are made experimentally between the proposed fitness function and two other typical fitness functions by using two recent attribute reduction algorithms. Experimental results show that for each algorithm in test, the proposed fitness function outperforms the other two in terms of both the probability of finding a minimum reduct and the average length of the output reducts.

1 Introduction

It is well known that the problem of finding a minimum attribute reduction in the context of rough set theory is NP-hard [1]. Formally, the minimum attribute reduction(or MAR for short) problem, is a nonlinearly constrained combinatorial optimization problem. So far, the most commonly used approach for dealing with this problem in the literature is to use randomized search methods. Actually, there have been such attribute reduction algorithms(see [2]-[10]). In these algorithms, the problem is first transformed into a problem of maximizing some fitness function over a multi-dimensional Boolean space, and then some randomized optimization method is applied to solve the fitness maximization problem. To get good performance, fitness functions should meet the requirements that the fitness evaluation of a candidate solution is appropriate and the optimality equivalence is guaranteed. Here, the optimality equivalence means that the optimal solution of the fitness maximization problem corresponds to a minimum attribute reduction. Unfortunately, the existing fitness functions do not well meet the above mentioned requirements and consequently affect the performance of

J. Yu et al. (Eds.): RSKT 2010, LNAI 6401, pp. 118–125, 2010.

the related algorithms. Hence, it is of significance to get a more adequate fitness function for the solution of the MAR problem. This motivated our study reported in this paper.

The main contribution of this paper is the design of a new fitness function that proves to meet the requirements better than existing fitness functions. Compared with other fitness functions, the proposed one takes into account more factors, including the size of the underlying decision table as well as some estimated lower bound on approximation quality differences. The equivalence of optimality is theoretically justified for the new fitness function. Experimentally, two existing search algorithms were implemented, each using three different fitness functions including the proposed one. Both algorithms were tested on a number of data sets. The experimental results show that for each of the two tested algorithms, the use of the proposed fitness function has a higher probability of finding a minimum reduction than the use of the other two functions.

The rest of the paper is organized as follows. Section 2 presents some basic concepts in rough sets and discusses related works on the fitness design for the MAR problem. In section 3, a new fitness function is given and some of its properties are presented. In section 4, experiments and comparison analysis are made to show the betterment of the proposed fitness function and finally, in section 5, we conclude.

2 Some Basic Concepts and Related Works

A decision table can be represented as a quadruple $L = \{U, A, V, f\}$ [11], where $U = \{x_1, \cdots, x_n\}$ is a non-empty finite set of objects called universe of discourse, A is a union of condition attributes set C and decision attributes set D, V is the domains of attributes belonging to A, and $f : U \times A \longmapsto V$ is a function assigning attribute values to objects belonging to U. Assume that C contains m condition attributes $a_1, \cdots, a_m$ and without loss of generality that D contains only one decision attribute which takes $k(> 1)$ distinct values. For a subset $P \subseteq A$, $IND(P)$ represents the indiscernibility relation induced by the attributes belonging to P. The partition induced by $IND(P)$ is called a knowledge base and denoted by $U/IND(P)$. In particular, $U/IND(D) = \{Y_1, \cdots, Y_k\}$ is the knowledge base of decision classes. Let

$$U/IND(C) = \{X_1, X_2, \cdots, X_N\}, N \leq n \qquad (1)$$

where the equivalence classes X_i are numbered such that $|X_1| \leq |X_2| \leq \cdots |X_N|$.

Let $X \subseteq U$ and $R \subseteq C$. The R–lower approximation of X is defined as $\underline{R}X = \{x \in U : [x]_R \subseteq X\}$, where $[x]_R$ denotes an equivalence class of $IND(R)$ determined by object x. The notation $POS_R(U, D)$ refers to the R-positive region which is given by $POS_R(U, D) = \bigcup_{i=1}^{k} \underline{R}Y_i$ or equivalently by $POS_R(U, D) = \bigcup\{X \in U/IND(R) : \exists i, X \subseteq Y_i, Y_i \in U/IND(D)\}$. The R–approximation quality with respect to decisions is defined as $\gamma_R = \sum_{i=1}^{k} \frac{|\underline{R}Y_i|}{|U|}$, where $|\cdot|$ denotes the cardinality of a set.

Definition 1. *Let $R \subseteq C$. If $\gamma_R = \gamma_C$ holds, then R is said to be a consistent set of C; if R is a minimal consistent set of C, then R is said to be a relative reduct of C or simply a reduct.*

By definition, a consistent set of attributes either is a reduct itself or contains a reduct. In general, there can be more than one reduct. A minimum reduct is a reduct that contains the least number of attributes. Now let $\{0,1\}^m$ be an m-dimensional Boolean space and ξ be a mapping from $\{0,1\}^m$ to 2^C, the power set of C, such that:

$$x_i = 1 \Leftrightarrow a_i \in \xi(x), i = 1, \cdots, m, a_i \in C.$$

Then, the MAR problem can be formulated as the following nonlinearly constrained binary optimization problem:

$$\min \ S(x)$$

$$s.t. \begin{cases} x \in \{0,1\}^m \\ \gamma_{\xi(x)} = \gamma_C \\ \forall q \in \xi(x), \gamma_{\xi(x)\backslash\{q\}} < \gamma_{\xi(x)} \end{cases} \tag{2}$$

where $0 \leq S(x) = \sum_{i=1}^{m} x_i \leq m$.

Given a vector $x \in \{0,1\}^m$, if it is a feasible solution to Problem (2), then its corresponding subset of attributes $\xi(x)$ is a reduct. Furthermore, if it is an optimal solution to Problem (2), then $\xi(x)$ is a minimum reduct. Here we do not consider the trivial case when $\gamma_C = 0$. In the trivial case, any single attribute is a reduct and also a minimum reduct. For this reason, we assume that the decision table under consideration is not trivial.

So far, a most commonly used approach for solving Problem (2) is to transform it into the following unconstrained maximization problem:

$$\max_{x \in \{0,1\}^m} \ F(x) \tag{3}$$

where the objective function F, usually referred to as fitness function, is defined by enforcing the constraints into the objective function of the original constrained problem.

A number of algorithms using this approach with different definitions of fitness evaluation have been proposed [2]-[10]. For instance, Jensen and Shen [4] used the following multiplicative fitness function:

$$F1(x) = \gamma_{\xi(x)} * \frac{m - S(x)}{m}$$

In some recent works [6][7], the following additive weighted fitness function was employed:

$$F2(x) = \alpha \frac{m - S(x)}{m} + \beta\gamma_{\xi(x)} \tag{4}$$

where $\alpha > 0, \beta = 1 - \alpha > 0$ are parameters to balance the importance of the two terms involved. In the literature, the two parameters are often set to be $\alpha = 0.1$

and $\beta = 0.9$ respectively to emphasize the importance of the approximation quality which is related to the constraints of the original problem.

The problem with the above two fitness functions is that a boolean vector solving the fitness maximization problem may not correspond to a minimum reduct. Actually, given a decision table, if the two terms in $F2$ are not suitably balanced, a solution corresponding to a small-sized non-consistent set may maximize $F2$. That is to say, the optimal solution to the fitness maximization problem may not correspond to a reduct, not to mention a minimum reduct. The same observation holds true for $F1$. In other words, for these two types of fitness evaluation, the equivalence of optimality between the MAR problem and the fitness maximization problem can not be guaranteed in theory.

Another way of defining the fitness is to use a piecewise function by giving some bonus to consistent sets of attributes. For instance, [10] proposed the use of the function:

$$F3(x) = \begin{cases} \frac{m-S(x)}{m} + \gamma_{\xi(x)}, & \gamma_{\xi(x)} < \gamma_C; \\ \frac{m-S(x)}{m} + \gamma_C + 1, & \gamma_{\xi(x)} = \gamma_C. \end{cases} \tag{5}$$

This amounts to giving 1.0 as the bonus to any consistent set of attributes. The equivalence of optimality is proved in [10]. However, the way of giving a bonus to consistent sets of attributes has turned out to have an obvious drawback. With the bonus, one consistent set of attributes, whatever its size, prevails over any non-consistent set of attributes in terms of fitness values, even though the non-consistent set, short in size, is a subset of some minimum reduct and can be a good candidate leading to the minimum reduct in the search space. This is obviously not appropriate as far as the fitness evaluation is concerned.

3   A New Fitness Function and Its Properties

In this section, we present a new fitness function that not only guarantees the equivalence of optimality, but also overcomes the drawback of using bonuses in the fitness evaluation as mentioned earlier. The new fitness function is defined as

$$F(x) = \frac{m - S(x)}{m} + \frac{n}{\Delta}\gamma_{\xi(x)}$$

where n is the size of the underlying decision table, $\Delta = |X_1|$ with X_1 being defined in (1).

Lemma 1. *For any non-empty subset $R \subseteq C$, if $\gamma_R < \gamma_C$, then $\gamma_R \leq \gamma_C - \frac{\Delta}{n}$.*

Proof. By definition of positive regions, if $\gamma_R < \gamma_C$, then at least one equivalence class of $IND(C)$, say X_i, which is inside $POS_C(U, D)$, is outside $POS_R(U, D)$. By (1), $|POS_R(U, D)| \leq |POS_C(U, D)| - |X_i| \leq |POS_C(U, D)| - |X_1|$. Hence, we have $\gamma_R \leq \gamma_C - \frac{\Delta}{n}$.

Theorem 1. *If x^* is an optimal solution to the MAR problem(2), then x^* is also an optimal solution to the fitness maximization problem(3).*

Proof. Let $P = \xi(x^*)$. By hypothesis, we have $\gamma_P = \gamma_C$ and hence $F(x^*) = \frac{m-S(x^*)}{m} + \frac{n}{\Delta}\gamma_C$. For any non-zero vector $x \in \{0, 1\}^m$, let $R = \xi(x)$. If $\gamma_R = \gamma_C$, then R either contains or is itself a reduct. By the optimality of x^*, $S(x^*) \le S(x)$ and thus $F(x) \le F(x^*)$. Consider the case $\gamma_R < \gamma_C$. By Lemma 1, $\gamma_R \le \gamma_C - \frac{\Delta}{n}$. Keeping in mind that $|S(x*) - S(x)| < m$, we have

$$F(x) \le \frac{m - S(x)}{m} + \frac{n}{\Delta}(\gamma_C - \frac{\Delta}{n})$$

$$\le \frac{m - S(x^*)}{m} + \frac{n}{\Delta}\gamma_C + \frac{S(x^*) - S(x)}{m} - 1$$

$$< \frac{m - S(x^*)}{m} + \frac{n}{\Delta}\gamma_C = F(x^*)$$

The proof is thus complete.

Theorem 2. *Suppose that x^* is an optimal solution to the maximization problem (3). Let $P = \xi(x^*)$. Then, x^* is also an optimal solution to the MAR problem (2), or P is a minimum reduct.*

Proof. First, we use proof by contradiction to show P is a consistent set of attributes, i.e., $\gamma_P = \gamma_C$. Assume that $\gamma_P < \gamma_C$. Let $\hat{x} = (1, 1, \cdots, 1)^T \in \{0, 1\}^m$. Then, $C = \xi(\hat{x})$ and $S(\hat{x}) = m$. By Lemma 1, $\gamma_P \le \gamma_C - \frac{\Delta}{n}$, then

$$F(x^*) \le \frac{m - S(x^*)}{m} + \frac{n}{\Delta}(\gamma_C - \frac{\Delta}{n})$$

$$= \frac{m - S(x^*)}{m} + \frac{n}{\Delta}\gamma_C - 1$$

$$= \frac{m - S(x^*)}{m} + \frac{n}{\Delta}\gamma_{\xi(\hat{x})} - \frac{S(\hat{x})}{m}$$

$$< \frac{m - S(\hat{x})}{m} + \frac{n}{\Delta}\gamma_{\xi(\hat{x})} = F(\hat{x})$$

contradicting the optimality of x^*. Hence, $\gamma_P = \gamma_C$. Next, assume that there exists P', a subset of P, such that $\gamma_{P'} = \gamma_C$. Let $P' = \xi(x')$. Then, $S(x') < S(x^*)$ and by definition

$$F(x') = \frac{m - S(x')}{m} + \frac{n}{\Delta}\gamma_C$$

$$> \frac{m - S(x^*)}{m} + \frac{n}{\Delta}\gamma_C = F(x^*)$$

again contradicting the optimality of x^*. We have thus shown that P is a reduct and x^* is a feasible solution of Problem (2).

Now, for any feasible solution x of Problem (2), we have $\gamma_R = \gamma_C$ with $R = \xi(x)$. By the optimality of x^*, we have

$$F(x) = \frac{m - S(x)}{m} + \frac{n}{\Delta}\gamma_C$$

$$\leq F(x^*) = \frac{m - S(x^*)}{m} + \frac{n}{\Delta}\gamma_C,$$

leading to $S(x^*) \leq S(x)$. This implies that x^* is also an optimal solution to Problem (2). The proof is thus done.

4   Performance Comparison between Different Fitness Functions

To evaluate the effectiveness of the proposed fitness function, we carried out experiments in the following way. Two different types of stochastic search based attribute reduction algorithms were used. One is the particle swarm optimization(PSO in short) based algorithm proposed in [7], denoted here by PSOAR, and the other is a tabu search based algorithm given in [9], denoted by TSAR. Both algorithms were then tested on a number of data sets using three different fitness functions, namely, the weighted fitness function as shown in (4), the bonus giving based fitness function as shown in (5), and the fitness proposed in this paper. For simplicity, these three fitness functions are respectively denoted as W-fitness, B-fitness and N-fitness. For W-fitness, we set the parameters $\alpha = 0.1$ and $\beta = 0.9$ which are commonly used in the literature. A PC running Windows XP with 2.8GHz CPU and 1GB of main memory was employed to run both algorithms. The data sets for test were obtained from the UCI machine learning repository and 12 data sets were chosen.

For each of the three fitness functions, both algorithms were independently run 20 times on each of the data sets in the experiment with the same setting of the parameters involved. For each run, if the output solution is a minimum reduct, then the run is said to be successful. Three values were reported, including the shortest length and the average length of the output reducts during the 20 runs, denoted respectively by STL and AVL, and the ratio of successful runs, denoted by RSR. The parameter settings for both algorithms are given in Table 1. And the computational results of PSOAR and TSAR are reported in Table 2 and Table 3 respectively.

In Table 1, population refers to the size of particle population in PSOAR, Imax is the maximum allowed number of iterations, Vmax is the upper bound on velocity needed in PSOAR algorithm, while TL is a tabu list whose length is set to 7. The list is used in TSAR to record the 5 most recently visited locations in the search space together with the zero vector and the vector with its components all equal to 1.

From Table 2, we see that for the algorithm PSOAR, the use of the proposed fitness function has the same performance as the use of the other two functions

Table 1. Parameter Settings for Both Algorithms

algorithm	population	Vmax	Imax	$|TL|$
PSOAR	20	7	100	–
TSAR	1	–	100	7

Table 2. Performance of PSOAR using different fitness functions

dataset	W- fitness			B- fitness			N- fitness		
	STL	AVL	RSR	STL	AVL	RSR	STL	AVL	RSR
Bupa	3	3	20/20	3	3	20/20	3	3	20/20
Breast-cancer	8	8	20/20	8	8	20/20	8	8	20/20
Corral	4	4	20/20	4	4	20/20	4	4	20/20
Soybean-small	2	2	20/20	2	2	20/20	2	2	20/20
lymphography	5	5.5	9/20	5	5.6	9/20	5	5.5	9/20
Soybean-large	9	9.6	10/20	9	9.2	18/20	9	9.1	18/20
splice	9	9.5	11/20	9	9.6	10/20	9	9.45	12/20
mushroom	4	4.1	18/20	4	4.05	19/20	4	4.05	19/20
Led24	11	11.85	3/20	11	12.09	2/20	11	11.5	10/20
Sponge	8	8.4	13/20	8	8.3	14/20	8	8.3	14/20
DNA-nominal	9	9.6	9/20	9	9.4	12/20	9	9.3	14/20
Vote	8	9.1	7/20	8	8.8	16/20	8	8.1	19/20

Table 3. Performance of TSAR using different fitness functions

dataset	W- fitness			B- fitness			N- fitness		
	STL	AVL	RSR	STL	AVL	RSR	STL	AVL	RSR
Bupa	3	3.05	19/20	3	3.1	18/20	3	3.05	19/20
Breast-cancer	8	8	20/20	8	8	20/20	8	8	20/20
Corral	4	4.2	16/20	4	4.3	14/20	4	4.1	18/20
Soybean-small	2	2.8	12/20	2	2.85	11/20	2	2.2	16/20
lymphography	5	5.6	7/20	5	5.6	7/20	5	5.55	8/20
Soybean-large	9	9.9	8/20	9	9.6	14/20	9	9.3	16/20
splice	9	9.6	10/20	9	9.8	8/20	9	9.5	11/20
mushroom	4	4.6	12/20	4	4.6	12/20	4	4.4	14/20
Led24	11	13.5	1/20	11	13.2	2/20	11	12.35	8/20
Sponge	8	8.75	7/20	8	8.9	6/20	8	8.65	8/20
DNA-nominal	9	9.6	9/20	9	9.4	12/20	9	9.3	14/20
Vote	8	8.2	16/20	8	8.3	14/20	8	8.2	18/20

on the first four data sets in the table, but outperforms the use of the other two fitness functions on the other 8 data sets in terms of both the average length of output solutions and the ratio of successful runs. This means that the use of the proposed fitness function has a higher probability of finding a minimum reduct and yields on average better solution sets than the use of two others.

Table 3 gives the computational results by using the algorithm TSAR. Almost the same observation as in the case of PSOAR can be obtained for the proposed fitness function.

Summing up the experiment results, we see that the proposed fitness function is more adequate than the other two typical fitness functions on the data sets for test.

5   Conclusions

We discussed in this paper how to design an adequate fitness function for the minimum attribute reduction problem. The drawbacks of existing fitness functions were pointed out and a new fitness function that takes into account the influence of difference between approximation qualities was presented. Both theoretical analysis and experimental results show that the proposed fitness function, assuring the optimality equivalence, is more adequate than the existing ones for solving the minimum attribute reduction problem.

Acknowledgement. This work was partly funded by National Science Foundation of China(No.60805042) and by Provincial Science Foundation of Fujian of China(No.2010J01329).

References

1. Wong, S.K.M., Ziarko, W.: On optimal decision rules in decision tables. Bulletin of Polish Academy of Science 33, 693–696 (1985)
2. Bello, R., Gomez, Y., Garcia, M.M., et al.: Two-Step Particle Swarm Optimization to Solve the Feature Selection Problem. In: Proceedings of Seventh International Conference on Intelligent Systems Design and Applications, pp. 691–696 (2007)
3. Dai, J.H.: Particle swarm algorithm for minimal attribute reduction of decision tables. In: Proceedings of IMSCCS 2006, vol. 2, pp. 572–575. IEEE Computer Society Press, Los Alamitos (2006)
4. Jensen, R., Shen, Q.: Finding Rough Set Reducts with Ant Colony Optimization. In: Proceedings of the 2003 UK workshop on computational intelligence, pp. 15–22 (2003)
5. Ke, L.J., Feng, Z.R., Ren, Z.G.: An efficient ant colony optimization approach to attribute reduction in rough set theory. Pattern Recognition Letters 29, 1351–1357 (2008)
6. Wang, X.Y., Yang, J., Peng, N.S., et al.: Finding Minimal Rough Set Reducts with Particle Swarm Optimization. In: Ślęzak, D., Wang, G., Szczuka, M.S., Düntsch, I., Yao, Y. (eds.) RSFDGrC 2005. LNCS (LNAI), vol. 3641, pp. 451–460. Springer, Heidelberg (2005)
7. Wang, X.Y., Yang, J., Teng, X., Xia, W., Jensen, R.: Feature selection based on rough sets and particle swarm optimization. Pattern Recognition Letters 28, 459–471 (2007)
8. Wroblewski, J.: Finding mininmal reducts using genetic algorithm. In: Proceedings of the Second Annual Joint Conference on Information Sciences, Wrightsville Beach, NC, pp. 186–189 (1995)
9. Hedar, A.R., Wang, J., Fukushima, M.: Tabu search for attribute reduction in rough set theory. Soft Computing 12(9), 909–918 (2008)
10. Ye, D.Y., Chen, Z.J., Liao, J.K.: A New Algorithm for Minimum Attribute Reduction Based on Binary Particle Swarm Optimization with Vaccination. In: Zhou, Z.-H., Li, H., Yang, Q. (eds.) PAKDD 2007. LNCS (LNAI), vol. 4426, pp. 1029–1036. Springer, Heidelberg (2007)
11. Pawlak, Z., Slowinski, R.: Rough set approach to multi-attribute decision analysis. European J. of Operational Research 72, 443–459 (1994)

Temporal Dynamics in Rough Sets Based on Coverings

Davide Ciucci

Dipartimento Di Informatica, Sistemistica e Comunicazione
Università di Milano – Bicocca, Viale Sarca 336 – U14, I–20126 Milano, Italia
ciucci@disco.unimib.it

Abstract. Given a covering of a universe, we study how time evolution of the covering influences rough approximations and covering reduction. The definition of time evolution considers two cases: new objects are added to the system or the granules are changed.

1 Introduction

In [2] we studied different kinds of dynamics in rough sets and divided them in two main classes: knowledge evolves in time (diachronic dynamics) and knowledge changes from a point of view to another (synchronic dynamics) [3]. Each of these two broad situations is then sub-classified according to three mainstreams: information tables, approximation spaces and coverings. The subsequent analysis of the existing works on dynamics pointed out that there are no works which deal with dynamics in coverings, an important topic in rough sets since many generalization directly start from a covering or lead to a covering instead of a partition. We intend here to fill this gap. Thus, at first, we will recall the basic definitions of dynamics in coverings given in [2] and also introduce new ones. Then, we will study how the temporal (synchronic) dynamics influences the approximations (lower and upper) and the notion of covering reduct [7].

2 The Classification of Covering Dynamics

Let us first recall the definition of covering on a given universe X.

Definition 2.1. *Let X be a non empty set, a covering $\mathbb{C}(X)$ of X is a collection of sets $C_i \subseteq \mathcal{P}(X)$ such that $\bigcup C_i = X$.*

As already anticipated in the introduction, when dealing with dynamic classification, the main distinction concerns the simultaneity or not of the events under investigation. These two branches are called by Pagliani [3] synchronic and diachronic dynamics. Here we will treat only the second case, i.e., the time evolution. However, as it will be clear later, we need to introduce also the notion of *compatibility* among covering, useful in synchronic dynamics.

J. Yu et al. (Eds.): RSKT 2010, LNAI 6401, pp. 126–133, 2010.

Definition 2.2. *A collection of coverings* $\mathbb{C}_i(X_i)$ *is said to be:*

- object compatible *if the objects are the same for all* $\mathbb{C}_i$: $\mathbb{C}_i(X)$;
- granule compatible *if for all common objects* $Y = \bigcap X_i$, *it happens that for all* $x, y \in Y$, *if* x, y *belong to the same set* $\mathbb{C}_j$ *for one covering* $\mathbb{C}_i(X_i)$, *then* x, y *belong to the same set in all coverings. More formally:* $\forall x, y \in Y$ *if* $\exists \gamma_i \in \mathbb{C}_i$ *such that* $x, y \in \gamma_i$ *then* $\forall \mathbb{C}_j, \exists \gamma_j : x, y \in \gamma_j$.

Let us note that the notion of compatibility is intended to point out those multi-source or multi-agent situations with a common background. In one case, the objects are the same. In the other (granule compatibility) the classification of two common objects is similar. We note that this form of granule compatibility is very weak since it deals only with pairs of objects: two coverings can be granule compatible even if they share only few characteristics.

Example 2.1. Let $X_1 = \{1, 2, 3, 4\}$ and $X_2 = \{1, 2, 3, 4, 5\}$ be the supports of the two coverings $\mathbb{C}_1(X_1) = \{\{1, 2, 4\}, \{1, 3\}, \{2, 3, 4\}\}$ and $\mathbb{C}_2(X_2) = \{\{1, 2\}, \{2, 4, 5\}, \{1, 3, 4\}, \{2, 3\}\}$. According to the above definition, the two coverings are granule compatible.

Now, let us consider the changes that occur in time. Since we are dealing with coverings, these changes can be of two types: new objects enter into the system under investigation or the granulation of the universe changes.

Of course, these changes can appear simultaneously. However, we will consider only situations where at each time step only one event can happen. We can reduce to this situation by splitting a complex event into simpler ones.

We try now to recognize the cases where there is a monotone increase of knowledge with respect to the two dynamics above outlined: new objects are added, granules are changed. To this aim, we need to introduce an order relation among coverings. However, to define an order relation on coverings is not so trivial, since definitions of orderings which are equivalent on partitions differ when generalized to coverings (see [1] for an overview). So, in the general case, we assume to have a notion of ordering or quasi-ordering available, let call it $\leq$, and we define monotonicity with respect to this ordering as follows.

Definition 2.3. *Given two coverings* $\mathbb{C}^{(t_1)}(X_1), \mathbb{C}^{(t_2)}(X_2)$ *with* $t_1, t_2 \in \mathbb{R}$, $t_1 \leq t_2$, *we have an* increase of knowledge in coverings

- wrt objects *of type 1 if* $X_1 \subseteq X_2$, $\mathbb{C}^{(t_1)}(X_1) \subseteq \mathbb{C}^{(t_2)}(X_2)$ *and all* $\gamma \in \mathbb{C}^{(t_2)}(X_2) \backslash \mathbb{C}^{(t_1)}(X_1)$ *contains at least one element* $x \in X_2 \backslash X_1$;
- wrt objects *of type 2 if* $X_1 \subseteq X_2$ *and the new objects are added to an existing class or form a new class. That is, for all the granules* $\gamma_2 \in \mathbb{C}^{(t_2)}(X_2)$ *either* $\gamma_2 = \gamma_1$ *with* $\gamma_1 \in \mathbb{C}^{(t_1)}(X_1)$ *or* $\gamma_2 = \gamma_1 \cup \{x_1, \ldots, x_n | x_i \in X_2 \backslash X_1\}$ *(and in this case* $\gamma_1 \notin \mathbb{C}^{(t_2)}(X_2)$*) or* $\gamma_2 = \{x_1, \ldots, x_m | x_i \in X_2 \backslash X_1\}$;
- wrt granules *if* $\mathbb{C}^{(t_1)}$ *and* $\mathbb{C}^{(t_2)}$ *are object compatible and* $\mathbb{C}^{(t_1)}(X) \leq \mathbb{C}^{(t_2)}(X)$.

Note that this definition is slightly different from the original one given in [2] where the increase of knowledge wrt objects just required granule compatibility. Here, we used two stronger notions which imply granule compatibility (the proof is straightforward). In the case of monotonicity of type 1 all the granules at time t_1 remains at time t_2 and new ones are added which contains at least a new element. This evolution is more "faithful" to approximation spaces dynamics in case the coverings are obtained from an approximation space [6,2].

In the monotonicity of type 2, the new elements either constitute a new class or are added to an existing one. As we will see this evolution gives a sufficient condition to obtain the conservation of covering reduction.

Example 2.2. Consider the universe $X = \{a, b, c, d\}$ and the covering $\{\{a, d\}, \{b, c\}\}$. At a following time a new object e can enter the system. Then, we have an increase of knowledge in coverings wrt objects of type 1 if the covering becomes, for instance, $\{\{a, d\}, \{b, c\}, \{d, e\}, \{e\}\}$. On the other hand, we have an increase of knowledge in coverings wrt objects of type 2 if the new covering is $\{\{a, d\}, \{b, c, e\}, \{e\}\}$. Then, if this last system is updated such that the new covering is $\{\{a, d\}, \{b, c, e\}, \{c, e\}, \{e\}\}$ we have an increase of knowledge with respect to granules if the quasi ordering of equation 4 is considered.

3   The Partition Case

Let us first consider the simpler case in which the covering is indeed a partition, i.e., for any two $\gamma_1, \gamma_2 \in \mathbb{C}(X)$, $\gamma_1 \cap \gamma_2 = \emptyset$. In this case we will use the symbol π to denote the partition instead of $\mathbb{C}$. Now, given a partition $\pi(X)$ and a set $A \subseteq X$, the lower and upper approximations of A are respectively defined as:

$$L(A) = \{\gamma \in \pi(X) | \gamma \subseteq A\} \tag{1}$$
$$U(A) = \{\gamma \in \pi(X) | \gamma \cap A \neq \emptyset\} \tag{2}$$

We now analyze how the approximations evolve in case of time evolution of the underlying partition. First, let us consider the monotone increase of knowledge wrt granules. In the case of partitions, the usual ordering (which can be expressed in several different ways) is:

$$\pi_1(X) \leq \pi_2(X) \quad \text{iff} \quad \forall C \in \pi_1(X), \exists D \in \pi_2(X) \quad \text{such that} \quad C \subseteq D \tag{3}$$

In this case it is trivial to prove that the approximations at time t_2 are better than the ones at time t_1, in the sense of the following proposition.

Proposition 3.1. *Let $\pi^{(t_1)}(X)$ and $\pi^{(t_2)}(X)$ two partitions of X such that there is a monotone increase of knowledge wrt granules going from $\pi^{(t_1)}$ to $\pi^{(t_2)}$. Then,*

$$\forall A \subseteq X \quad L^{t_2}(A) \subseteq L^{t_1}(A) \subseteq A \subseteq U^{t_1}(A) \subseteq U^{t_2}(A)$$

In case we add new objects to the system, we have two different behaviours according to type 1 or 2 increase of knowledge. When considering a monotone increase of knowledge wrt objects of type 1, we have that the approximations do not change.

Proposition 3.2. *Let $\pi_1(X_1)$ and $\pi_2(X_2)$ be two partitions such that there is a monotone increase of knowledge wrt objects of type 1 going from π_1 to π_2. Then,*

$$\forall A \subseteq X_1 \quad L^{t_2}(A) = L^{t_1}(A) \subseteq A \subseteq U^{t_1}(A) = U^{t_2}(A)$$

Proof. In the partition case, the definition of monotone increase of type 1 reduces to the fact that the new objects can only generate new classes with no older objects in them. Thus, they do not influence the approximations of $A \subseteq X_1$.

Finally, if we have a monotone increase of knowledge with respect to objects of type 2 the following holds.

Proposition 3.3. *Let $\pi^{(t_1)}(X_1)$ and $\pi^{(t_2)}(X_2)$ be two partitions such that there is a monotone increase of knowledge wrt objects of type 2 going from $\pi^{(t_1)}$ to $\pi^{(t_2)}$. Then,*

$$\forall A \subseteq X_1 \quad L^{t_2}(A) \subseteq L^{t_1}(A) \subseteq A \subseteq U^{t_1}(A) \subseteq U^{t_2}(A)$$

Proof. We can have several cases:

1. if the new objects generate new classes then the approximations do not change: $L^{t_2}(A) = L^{t_1}(A)$ and $U^{t_1}(A) = U^{t_2}(A)$;
2. if there is a new object x such that $x \in [y]_2$ with $y \in L^{t_1}(A)$ then it follows that $x \notin A$ and consequently $[y]_{t_2} \not\subseteq A$, i.e., $L^{t_2}(A) \subset L^{t_1}(A)$. In case of the upper approximation, we have that $x \notin U^{t_1}(A)$ but $x \in U^{t_2}(A)$;
3. finally, there is a new object x such that $x \in [y]_{t_2}$ with $y \notin L^{t_1}(A)$ but $[y]_{t_2} \cap A \neq \emptyset$ then it follows that $L^{t_1}(A) = L^{t_2}(A)$ but $U^{t_1}(A) \subseteq U^{t_2}(A)$.

Thus, we can see that when we add more objects, we have less precision in the approximations.

Example 3.1. Let $X_1 = \{a, b, c, d\}$ and $\pi^{(t_1)}(X_1) = \{\{a, b\}, \{c, d\}\}$. The approximations of the set $A = \{a, b, c\}$ are $L^{(t_1)}(A) = \{a, b\}$ and $U^{(t_1)}(A) = \{a, b, c, d\}$. Now, let us suppose that the object e enters the system. The only way to have a monotone increase of knowledge wrt objects of type 1 is to add the new class $\{e\}$, and the approximations of A do not change. We have an increase of knowledge wrt objects of type 2 if, for instance, the new partition is $\pi^{(t_2)}(X_2) = \{\{a, b, e\}, \{c, d\}\}$. In this case, $L^{t_2}(A) = \emptyset \subset L^{t_1}(A)$ and $U^{t_1}(A) \subset U^{t_2}(A) = \{a, b, c, d, e\}$.

4   The Covering Case: Approximations

We are now going to consider the general covering case. Our analysis will be based on the following two quasi-orderings (reflexive, transitive relations). Let $\mathbb{C}_1$ and $\mathbb{C}_2$ be two coverings of the same universe X, then $\mathbb{C}_1(X) \preceq \mathbb{C}_2(X)$ iff

$$\forall C_i \in \mathbb{C}_1(X) \; \exists D_j \in \mathbb{C}_2(X) \quad \text{such that} \quad C_i \subseteq D_j \tag{4}$$

and $\mathbb{C}_1(X) << \mathbb{C}_2(X)$ iff

$$\forall D \in \mathbb{C}_2(X) \, \exists \{C_1, C_2, \ldots, C_n\} \subseteq \mathbb{C}_1(X) \quad \text{such that} \quad D = \bigcup_{i=1}^{n} C_i \quad (5)$$

Both the two orderings generalize equation (3) and they are independent. That is, it may happen that $\mathbb{C}_1 \preceq \mathbb{C}_2$ and not $\mathbb{C}_1 << \mathbb{C}_2$ and vice versa.

We have now to define a lower and an upper approximation on coverings. There are more than twenty ways to do this (see for instance [4]). Here we will consider the most immediate (and widely used) ones which can be obtained by a generalization of equations 1:

$$L_1(A) = \{\gamma \in \mathbb{C}(X) | \gamma \subseteq A\} \qquad U_1(A) = L_1^c(A^c) \qquad (6)$$
$$L_2(A) = U_2^c(A^c) \qquad\qquad U_2(A) = \{\gamma \in \mathbb{C}(X) | \gamma \cap A \neq \emptyset\} \quad (7)$$

4.1   Increase wrt Granules

Let us first consider the approximations L_1 and U_1 of equations (6).
In case of order (4) there is no increase nor decrease of the quality of approximations, as can be seen in the following example.

Example 4.1. Let $\mathbb{C}^{t_1}(X) = \{\{a\}, \{b\}, \{c, d, e\}\}$ and $\mathbb{C}^{t_2}(X) = \{\{a, b, c\}, \{c, d, e\}\}$. Then we have that $\mathbb{C}^{t_1}(X) \preceq \mathbb{C}^{t_2}(X)$. However, if we consider the set $A = \{a, b, c\}$ then $L_1^{t_1}(A) = \{a, b\}$ and $L_1^{t_2}(A) = \{a, b, c\}$ with $L_1^{t_1}(A) \subset L_1^{t_2}(A)$. Vice versa, for the set $B = \{a, b\}$ it holds $\emptyset = L_1^{t_2}(A) \subset L_1^{t_1}(A) = \{a, b\}$.

On the contrary, in case of order (5) it holds the following monotonic behaviour.

Proposition 4.1. *Let $\mathbb{C}^{t_1}(X) << \mathbb{C}^{t_2}(X)$ then for all $A \subseteq X$, it holds*

$$L_1^{t_2}(A) \subseteq L_1^{t_1}(A) \subseteq A \subseteq U_1^{t_1}(A) \subseteq U_1^{t_2}(A) \, .$$

Proof. Trivial for L_1, by duality for U_1.

In case of the approximations (7) we have the opposite situation: order (4) has a monotonic behaviour, whereas order (5) is impredictable.

Proposition 4.2. *Let $\mathbb{C}^{t_1}(X) \preceq \mathbb{C}^{t_2}(X)$ then for all $A \subseteq X$, it holds*

$$L_2^{t_2}(A) \subseteq L_2^{t_1}(A) \subseteq A \subseteq U_2^{t_1}(A) \subseteq U_2^{t_2}(A) \, .$$

Proof. Trivial for U_2, by duality for L_2.

Example 4.2. Let $\mathbb{C}^{t_1}(X) = \{\{a, b\}, \{c\}, \{c, d\}, \{a, d\}\}$ and $\mathbb{C}^{t_2}(X) = \{\{a, b, c\}, \{c, d\}\}$. Then we have that $\mathbb{C}^{t_1}(X) << \mathbb{C}^{t_2}(X)$. However, if we consider the set $A = \{a, b\}$ then $U_2^{t_1}(A) = \{a, b, d\}$ and $U_2^{t_2}(A) = \{a, b, c\}$ with the two approximations which are not comparable.

4.2   Increase wrt Objects

Let us now consider the cases where new objects enter into the system. In case of type 1 increase of knowledge and contrary to the partition case, only the approximations of equation (6) do not change.

Proposition 4.3. *Let* $\mathbb{C}_1(X_1)$ *and* $\mathbb{C}_2(X_2)$ *be two coverings of* X *such that there is a monotone increase of knowledge wrt objects of type 1 going from* $\mathbb{C}_1$ *to* $\mathbb{C}_2$. *Then,*

$$\forall A \subseteq X \quad L_1^{t_2}(A) = L_1^{t_1}(A) \subseteq A \subseteq U_1^{t_1}(A) = U_1^{t_2}(A)$$
$$\forall A \subseteq X \quad L_2^{t_2}(A) \subseteq L_2^{t_1}(A) \subseteq A \subseteq U_2^{t_1}(A) \subseteq U_2^{t_2}(A)$$

Proof. In case of approximations L_2, U_2 (7), it can happen that there exists a new granule with a non-empty intersection with A, thus $U_2^{t_1}(A) \subseteq U_2^{t_2}(A)$. By duality, we obtain the symmetric relation with respect to the lower approximation L_2.

Proposition 4.4. *Let* $\mathbb{C}_1(X_1)$ *and* $\mathbb{C}_2(X_2)$ *be two partitions of* X *such that there is a monotone increase of knowledge wrt objects of type 2 going from* $\mathbb{C}_1$ *to* $\mathbb{C}_2$. *Then,*

$$\forall A \subseteq X \quad L_1^{t_2}(A) \subseteq L_1^{t_1}(A) \subseteq A \subseteq U_1^{t_1}(A) \subseteq U_1^{t_2}(A)$$
$$\forall A \subseteq X \quad L_2^{t_2}(A) \subseteq L_2^{t_1}(A) \subseteq A \subseteq U_2^{t_1}(A) \subseteq U_2^{t_2}(A)$$

Proof. Similar to the partition case of Proposition 3.3.

5   The Covering Case: Reduction

A question which we can pose when dealing with a covering is if it can be somehow simplified in order to not change the approximations. This leads to the definition of covering reduction as first investigated in [7].

Definition 5.1. *Let* $\mathbb{C}(X)$ *be a covering of* X. *A granule* $\gamma \in \mathbb{C}(X)$ *is reducible if it is the union of others granules, otherwise it is irreducible. A covering* $\mathbb{C}(X)$ *is irreducible if all* $\gamma \in \mathbb{C}(X)$ *are irreducible.*

Remark 5.1. We consider a covering $\mathbb{C}(X)$ as a set and not as a multi-set. So, it is not possible to have inside the same covering two granules which are equal.

Proposition 5.1. *[7] Let* $\gamma \in \mathbb{C}(X)$ *be a reducible granule. Then the lower approximation* L_1 *does not change if computed with respect to* $\mathbb{C}(X)$ *and* $\mathbb{C}(X) \backslash \gamma$.

By duality of U_1 with respect to L_1 we can prove the same result for the upper approximation.

Corollary 5.1. *Let* $\gamma \in \mathbb{C}(X)$ *be a reducible granule. Then the upper approximation* U_1 *does not change if computed with respect to* $\mathbb{C}(X)$ *and* $\mathbb{C}(X) \backslash \gamma$.

Thus, we can simplify a covering by deleting the reducible elements, without modifying the approximations L_1, U_1. However, using definition 5.1 we cannot obtain a result similar to proposition 5.1 for the approximations L_2, U_2 . This drawback has been pointed out also in [5] (see example 4.5) where a more general solution to the reduction problem has been proposed. According to their approach, approximation U_2 corresponds to their "second" approximation. In this case, the only reduction which maintains the approximation is deleting equal classes, every other change will modify also the approximations. Considering remark 5.1, we cannot have two equal granules in the same covering, so in this case every covering is irreducible with respect to this weaker definition. Thus, all the following discussion will be based on the reduction of definition 5.1 and it makes sense only with approximations L_1, U_1.

Now, let us suppose that $C^{(t_1)}(X_1)$ is not reducible (according to definition 5.1). We wonder what happens at time t_2 when there is a monotone increase of knowledge.

In case of a new object and a type 1 increase of knowledge, the new covering is not guaranteed to be irreducible.

Example 5.1. Let $X_1 = \{1, 2, 3, 4\}$ and $C^{(t_1)}(X_1) = \{\{1, 2\}, \{2, 3\}, \{3, 4\}\}$. As can be seen this covering is not reducible. If at time t_2 we add the object 5 and the new covering is $C^{(t_2)} = C^{(t_1)} \cup \{\{1, 2, 5\}, \{5\}\}$ then the new covering $C^{(t_2)}$ is reducible.

On the other hand, if we have a monotone increase of objects of type 2, then also the new covering is reducible.

Proposition 5.2. *Let $C^{(t_1)}(X_1)$ be a reducible covering. If we add a new object x and obtain a covering $C^{(t_2)}(X_2)$ in order to have a monotone increase of knowledge wrt objects of type 2, then, also $C^{(t_2)}(X_2)$ is irreducible.*

Proof. Let us suppose that $C^{(t_2)}(X_2)$ is reducible, that is there exists a $\gamma_2 \in C^{(t_2)}(X_2)$ which is reducible. Then we get a contradiction. Indeed, according to the definition of increase of knowledge wrt objects of type 2 we can have three cases:

1. $\gamma_2 = \gamma_1$ for a $\gamma_1 \in \mathbb{C}^{(t_1)}(X_1)$. In this case also γ_1 is reducible in $\mathbb{C}^{(t_1)}$, which is in contradiction with the hypothesis.
2. $\gamma_2 = \gamma_1 \cup \{x\}$. Also in this case we obtain that γ_1 is reducible in $\mathbb{C}^{(t_1)}$, a contradiction.
3. $\gamma_2 = \{x\}$, this case is obvious.

Of course if we add more than one object at a time, case 3 of the proof doe not hold. As an example, let us consider to add two objects x, y. If we add the three granules $\{x\}, \{y\}, \{x, y\}$, then, of course, this new covering is reducible. Further, we have that the condition of increase of knowledge wrt objects of type 2 is sufficient to have that the new covering is not reducible, but not necessary. That is, we can have a covering at time t_2 with more objects than the corresponding one at time t_1 such that both $C^{(t_1)}$ and $C^{(t_2)}$ are reducible but they are not linked by a monotone increase of knowledge.

Example 5.2. Let $\mathbb{C}^{(t_1)}(X_1) = \{\{1, 2, 3\}, \{2, 3\}, \{3, 4\}\}$ and $\mathbb{C}^{(t_2)}(X_2) = \mathbb{C}^{(t_1)} \cup \{2, 3, 5\}$. Then X_2 has one more element than X_1, both coverings are not reducible, but they are not linked by an increase of knowledge of type 2.

We leave as an open problem to find a sufficient and necessary condition.

Finally, if we consider an increase of knowledge with respect to granules, then also in this case the new covering can become reducible with respect to both considered orders.

Example 5.3. Let us consider the following two coverings of the set $X = \{1, 2, 3, 4\}$: $C^{(t_1)}(X) = \{\{1, 2, 3\}, \{2, 3\}, \{3, 4\}\}$ and $C^{(t_2)}(X) = \{\{1, 2\}, \{1, 2, 3\}, \{2, 3\}, \{3, 4\}\}$. Then, it holds that $C^{(t_1)}(X) \preceq C^{(t_2)}(X)$ and $C^{(t_1)}(X) << C^{(t_2)}(X)$. That is, there is a monotone increase of knowledge with respect to granules and orders $\preceq$ and $<<$. However, $C^{(t_1)}(X)$ is not reducible whereas $C^{(t_2)}(X)$ is.

Also in this case we have the open problem if it does exist an order which preserves irreducibility.

6 Conclusions

Given a covering, we have studied how approximations and reducibility change during the time evolution of the covering. In particular, we considered two specific definitions of lower and upper approximations, and two order relations on coverings. Of course the results depends on which approximation and which kind of evolution we consider. In some cases, we can establish a general result on the behaviour of approximations and on reduction, in some others this is not possible. An immediate generalization of this study is to consider other definitions of approximations or other order relations. Further, we have seen that in the time evolution approximations can become better or worst, we can think to evaluate the amount of the changes. This leads to evaluate indexes on the quality of rough sets (another dimension of investigation outlined in [2]). Finally, the analysis can be extended to synchronic dynamics, i.e., to the multi-source or multi-agent situation.

References

1. Bianucci, D., Cattaneo, G.: Information entropy and granulation co-entropy of partitions and coverings: A summary. Transactions on Rough Sets 10, 15–66 (2009)
2. Ciucci, D.: Classification of rough sets dynamics. In: Szczuka, M. (ed.) RSCTC 2010. LNCS (LNAI), vol. 6086, pp. 257–266. Springer, Heidelberg (2010)
3. Pagliani, P.: Pretopologies and dynamic spaces. Fundamenta Informaticae 59(2-3), 221–239 (2004)
4. Samanta, P., Chakraborty, M.K.: Covering based approaches to rough sets and implication lattices. In: Sakai, H., Chakraborty, M.K., Hassanien, A.E., Ślęzak, D., Zhu, W. (eds.) RSFDGrC 2009. LNCS, vol. 5908, pp. 127–134. Springer, Heidelberg (2009)
5. Yang, T., Li, Q.: Reduction about approximation spaces of covering generalized rough sets. International journal of approximate reasoning 51, 335–345 (2010)
6. Yao, Y.: Relational interpretations of neighborhood operators and rough set approximation operators. Information Sciences 111, 239–259 (1998)
7. Zhu, W., Wang, F.Y.: Reduction and axiomatization of covering generalized rough sets. Information Sciences 152, 217–230 (2003)

Data Classification Using Rough Sets and Naïve Bayes

Khadija Al-Aidaroos, Azuraliza Abu Bakar, and Zalinda Othman

Center for Artificial Intelligence Technology (CAIT)
Faculty of Information and Science Technology
Universiti Kebangsaan Malaysia, 43600 Bangi, Selangor, Malaysia
`bintalameen@gmail.com`, `{aab,zalinda}@ftsm.ukm.my`

Abstract. Naïve Bayesian classifier is one of the most effective and efficient classification algorithms. The elegant simplicity and apparent accuracy of naive Bayes (NB) even when the independence assumption is violated, fosters the on-going interest in the model. Rough Sets Theory has been used for different tasks in knowledge discovery and successfully applied in many real-life problems. In this study we make use of rough sets ability, in discovering attributes dependencies, to overcome the NB un-practical assumption. We propose a new algorithm called Rough-Naïve Bayes (RNB) that is expected to outperform other current NB variants. RNB is based on adjusting attributes' weights based on their dependencies and contribution to the final decision. Experimental results show that RNB can achieve better performance than NB classifier.

Keywords: Classification, Naïve Bayes (NB) , NB variants, Rough Sets (RS), attribute dependency , weighted NB.

1 Introduction

Naïve Bayes classifier is a straightforward, frequently used method for supervised learning. Data classification with naïve Bayes is the task of predicting the class of an instance from a set of attributes describing that instance and assume that all the attributes are conditionally independent given the class. It has proven its effective application, in text classification, medical diagnosis and systems performance management [1, 2]. A variety of adaptations to NB in the literature have been studied in order to improve upon its good performance while maintaining its efficiency and simplicity, for a review on different variants of NB classifier please refer to [3].

In recent years, Rough Set Theory (RST) witnessed a rapid growth of interest by many researchers worldwide and it has been applied successfully almost in all the areas [4, 5]. It has proven to be a good technique for data analysis and can combine with other complementary techniques, such as soft computing, statistics, intelligent systems and pattern recognition [4]. Among the different functions of RST in KDD is its use for discovering attribute relationships and dependencies.

NB have been widely criticized due to its unrealistic assumption of independence and many of the research aiming to improve NB is based on the idea of removing or relaxing the independence assumption. In this paper we investigate the use of rough sets in naïve Bayes to propose a new classifier, called Rough-Naïve Bayes (RNB).

J. Yu et al. (Eds.): RSKT 2010, LNAI 6401, pp. 134–142, 2010.

This paper is organized as follows. The background of rough sets is presented in Section 2. Naïve Bayes and hybrid models including the variants of NB are reviewed in Section 3. In Section 4 is the proposed RNB algorithm and Section 5 presents the experimental analysis. Finally is the conclusion and future work which is given in Section 6.

2 Rough Set

Rough Set Theory (RST) introduced by Pawlak in 1982, is a mathematical tool to deal with imprecision, vagueness and uncertainty of information [4, 6-8]. It is a formal approximation of a crisp set in terms of a pair of sets which give the lower and upper approximation of the original set. RST has been used for the discovery of data dependencies, evaluating the importance of features, reducing redundant objects and features, discovering the patterns of data, and extracting rules which functions as a classifier for unseen data sets [4, 8]. However, in this study we are more interested in the use of RST for detecting attributes' dependency and significance, and how both measures can be used in the proposed RNB algorithm. In the following we present some basic concepts of rough set theory along with the key definitions used in this paper.

2.1 Basic Concepts of Rough Set

An *information system* (IS) is a pair of the form IS = (U, A), where U = { $x_1, x_2, ..., x_m$} is a non-empty finite set of objects called the universe; A = {$a_1, a_2, ..., a_n$} is a non-empty finite set of attributes. In a special case of information systems, called *Decision System*, we have $A = C \cup D$ and $C \cap D = \emptyset$ where C is a non-empty set of condition attributes and D is a non-empty set of decision attributes.

For every set of attributes $B \subseteq A$, an equivalence relation, denoted by Ind_B and called *B-indiscernibility* relation, is defined by:

$$Ind_B = \{(x, y) \in U \times U : \forall_{a \in B} \; a(x) = a(y)\} \qquad (1)$$

For every $x \in U$, there is an equivalence class in the partition of U defined by Ind_B, where two objects (x, y) are members of the same equivalence class if and only if they cannot be discerned from each other on the basis of the set of attributes B. The equivalence classes of the *B-indiscernibility* relation are denoted by $[x]_B$.

The indiscernibility relation is used to define basic concepts of rough set theory. In the following we present the concepts used in this paper with their formal definitions.

Rough Set Approximations and Positive Region. A rough set approximates traditional set using a pair of sets named the lower and upper approximation of the set. Given a set $B \subseteq A$, the lower and upper approximations of a set $X \subseteq U$, denoted $\underline{B}X$ and $\bar{B}X$ respectively, are defined by:

$$\underline{B}X = \{x | [x]_B \subseteq X\} \qquad (2)$$

$$\bar{B}X = \{x | [x]_B \cap X \neq \emptyset\} \qquad (3)$$

Where the lower approximation consists of all objects which surely belong to the concept X and the upper approximation consists of all objects which possibly belong to the concept X.

The positive region of the partition U/Ind_D with respect to B, is defined by:

$$POS_B(D) = \bigcup_{X \in U/Ind_D} \underline{B}X \tag{4}$$

Where $POS_B(D)$ is the set of all elements of U that can be uniquely classified to blocks of the partition U/Ind_D by B.

Dependency of Attributes. Formally, dependency can be defined in the following way. Let D and C be subsets of A, then we say that D depends on C in a degree k ($0 \le k \le 1$), such that:

$$k = \gamma(C, D) = \frac{|POS_C(D)|}{|U|} \tag{5}$$

The coefficient k is called the *degree of dependency*; if $k = 1$ we say that D depends *totally* on C, and if $k < 1$ we say that D depends *partially* (in a degree k) on C.

Significance of Attributes. Significance of attributes enables us to evaluate features by assigning a real number from the closed interval $[0,1]$, expressing how important is a feature in an information table. Given a decision table $DT = (U, C \cup D)$, the significance of an attribute $a \in C$ can be evaluated by measuring the effect of removing attribute a from the attribute set C on the positive region defined by the table DT. Significance of the attribute a can be defined as:

$$\sigma_{(C,D)}(a) = \frac{\left(\gamma(C,D) - \gamma(C - \{a\}, D)\right)}{\gamma(C,D)} = 1 - \frac{\gamma(C - \{a\}, D)}{\gamma(C,D)} \tag{6}$$

3 Naïve Bayes

Naïve Bayesian classifier, or simply naive Bayes (NB), is a simple probabilistic classifier based on applying Bayes' theorem with strong (naive) independence assumptions. Given a set of training instances with class labels and a test case E represented by n attribute values ($a_1, a_2... a_n$), naive Bayesian classifier uses the following equation to classify E:

$$c_{NB}(E) = \arg_{c \in C} \max\ P(c) \prod_{i=1}^{n} P(a_i|c) \tag{7}$$

where $c_{NB}(E)$ denotes the classification given by NB on test case E.

Although independence is generally a poor assumption, in practice NB often competes well with much more sophisticated techniques. Naïve Bayes has a lot of features which attracted researchers to improve upon its performance, such as, simplicity, efficiency, transparency and understandability. Moreover, NB is naturally robust to missing values and noise data [3].

3.1 Hybrid Models and Variants of NB

The need of a hybrid approach is widely recognized by the data mining community, and much current work in data mining tends to hybridize diverse methods [10, 11]. Both naïve Bayes and rough set approaches have been hybridized with other approaches to produce more effective models [3, 12]. However, since our proposed method is considered a variant of NB, we list in the following some of NB's variants and highlight the most related approaches.

The basic independent NB model has been modified in various ways and new algorithms have been proposed in attempts to improve its performance. For example, NBTree [13], TAN (Tree Augmented Naïve Bayes) [1], LBR(Lazy Bayesian Rule) [14], SBC (Selective Bayesian Classifier) [15], LWNB (Locally Weighted NB) [16], AODE (Averaged One-Dependence Estimators) [17], RoughTree [18] and AWNB (Attribute Weighted NB) [19]. The different studies have shown that it is possible to improve upon the general performance of NB classifier which is extended using different approaches and combining different techniques.

Among the different mentioned variants, RoughTree and AWNB are two recent studies which are most related to our proposed method. RoughTree [18] is a hybrid classifier in decision tree representation and it uses the attribute dependence detecting measure in rough set. It splits the dataset into subspaces according to the selected attributes which hold the maximum values by the attribute dependence measure. On the other hand, AWNB (Attribute Weighted Naïve Bayes) [19] is based on setting attribute weights, and therefore it is considered a WNB. AWNB used decision tree to estimate the weights of attributes based on the minimum depth at which each attribute is tested in the tree.

For improving NB, based on the reviewed variants in [3], many researchers have paid considerable attention to investigate the use of decision tree and k-NN. Other techniques, such as rough set theory, still need more investigation. Based on the literature, at least up to our knowledge, RoughTree is the only Bayesian classifier which makes use of rough sets to improve NB and this encourages for more research in this area. Thus investigating the use of rough set to improve NB's performance is the main goal of our proposed RNB method.

To summarize the above discussions, we emphasize on the following points. Rough sets can combine with other complementary techniques and it has been used for discovering attributes' significance and dependencies. On the other hand, NB have been widely criticized due to its unrealistic assumption of independence and many of the research aiming to improve NB is based on the idea of removing or relaxing the independence assumption. Therefore, there is a real need to integrate the advantages of both approaches to overcome the NB un-practical assumption.

4   The Proposed RNB Algorithm

The proposed algorithm aims to enhance NB by integrating rough sets with a Weighted Naïve Bayes (WNB), which is a NB model with attributes weighted differently [20]. In the following is the formal definition of Weighted Naïve Bayes (WNB):

$$c_{WNB}(E) = arg_{c \in C} \max P(c) \prod_{i=1}^{n} P(a_i|c)^{w_i} \qquad (8)$$

Where $c_{WNB}(E)$ denotes the classification given by the WNB for the test instance E, and w_i is the weight of attribute A_i.

The proposed method enhances NB by weighting the predictive attributes according to the degree of their dependency on the values of other attributes. To alleviate the effect of independence assumption of NB, we will assign lower weights to those attributes that have many dependencies.

In order to estimate the degree to which an attribute depends on others, we will use the *degree of dependency* (γ) measure based on rough set theory (discussed in section 2.1). Since the weights assigned to each attribute should be inversely related to the degree of dependency and because $0 \leq \gamma \leq 1$, we set the weight for each attribute as, $w(a_i) = 1 - k(a_i)$, where $k(a_i) = \gamma(C_i, D_i)$, $C_i = \{a_i, d\}$ and $D_i = \{a_j: a_j \in Att, \forall j \in \{1,2,\ldots,n\} \& j \neq i\}$.

The main difference between AWNB [19] and the proposed method is the methodology used to learn the weight vector. AWNB used decision trees to estimate the weights of attributes while our method used rough sets for weights estimation.

RoughTree [18] on the other hand, uses the attribute dependence detecting measure in rough set. However, unlike our proposed method which is a NB variant, RoughTree is a classifier in decision tree representation that uses the attribute dependence measure to split the dataset into subspaces according to the selected attributes while our approach uses this measure to assign weights to the set of conditional attributes.

4.1   RNB Algorithm

In the following (Fig. 1) are the main steps of the proposed RNB algorithm. The algorithm starts with NB setting, that is, all weights are initially assigned to the value one. The calculation of $\sigma(a_i)$ for all attributes (step 2) is necessary to make the decision which attribute should be deleted in step 4.3. We adjust the weights for each attribute a_i, based on its dependency value (steps 3-6). Only highly correlated attributes (step 4.3) and totally dependent (if any based on step 6) are removed (step 8). The remaining attributes will get their weights adjusted to reflect their contribution toward the classification task. Finally, based on the calculated weights we produce the final model (step 9).

We hypothesize that if a feature is relevant and can contribute to the final decision taken by the classifier, it is better to keep it and assign it a small weight. By this method we avoid loss of information caused by attribute deletion and at the same time we reduce its effect, so only attributes with high dependent values are removed.

Input *Att*: set of attributes, *n*: number of attributes and θ: Threshold

Output RNB classifier

BEGIN

1. Set weights of all attributes to value 1. ($w(a_i) = 1 \ \forall \ a_i$ in *Att*).

2. Calculate significance $\sigma(a_i)$ for all attributes using eq. (in sec 2.1.3)

3. For each a_i in *Att* (where $w(a_i) \neq 0$) DO

4.   For each $j = i+1$ To n DO

        4.1   Calculate dependency measure, $k(a_i, a_j)$

        4.2   If both attributes are independent check next a_j, i.e. go to step 4.

        4.3   If both attributes are highly correlated, i.e. ($\theta \leq k \leq 1$), remove the less significant

           attribute.

5.   NEXT j

6.   Set weight for a_i, $w(a_i) = 1 - \gamma(C_i, D_i)$ where $C_i = \{a_i, d\}$ and $D_i = Att - C_i$

7. NEXT i (step 3)

8. Remove all attributes with zero weights.

9. Learn a weighted NB using the final attribute weights.

END

Fig. 1. Rough Naive Bayes (RNB) algorithm

5   Experimental Analysis

The proposed RNB algorithm is implemented to evaluate its performance and compare it with naïve Bayes (NB), rough sets (RS) and NBTree (which is one of the popular NB hybrids). In this paper, for NB and NBTree implementation, we used WEKA version 3.7.0 [21], and both algorithms were employed based on their default parameters of the WEKA application. For RS results we used the ROSETTA toolkit [22] and for the proposed RNB we implemented it based on the steps listed previously with threshold value θ set to 0.6 (based on several runs of RNB with different threshold values). In our experiment, 10 real-world problems from the UCI machine learning repository [23] were selected for evaluating the performance of the four methods. Table 1 gives a summary of the characteristics of these data sets. However, since our algorithm deals with discrete data, we selected data sets with only nominal attributes to avoid the effect of different discretization methods on the final result produced by each algorithm.

5.1   Experiment Setup

We used 10-fold cross-validation to minimize the bias associated with the random sampling of the training and holdout data samples [24]. Each data set is divided into 10 mutually exclusive subsets (or folds) and each fold is used once to test the performance of the classifier that is generated from the combined data of the remaining nine folds, leading to 10 independent performance estimates for each algorithm.

Table 1. Data sets for the experiment

Data set	Instances	Attr	Classes	Data set	Instances	Attr	Classes
Breast cancer	286	9	2	Postoperative patient	90	8	3
Contact lenses	24	4	3	Primary tumor	339	17	21
House votes	435	16	2	Soybean	683	35	19
Lung cancer	32	56	3	Spect_all	267	22	2
Lymph	148	18	4	Zoo	101	17	7

For evaluating the performance of the proposed RNB against the other three methods, we used the prediction accuracy as the main criteria for their performance. All
accuracy estimates were obtained by averaging the results from 10 runs of 10-fold
cross-validation.

5.2   Results and Discussions

We applied NB, RS, NBTree and RNB to the ten selected data sets, and based on the
conducted experiments the predictive accuracy results for each of them is presented in
Table2. The entries which are marked as bold indicate the highest value achieved for
each data set.

Table 2. Classification Accuracy (%)

No.	Data set	NB	RS	NBTree	RNB
1	Breast cancer	72.70	**73.18**	70.99	72.73
2	Contact lenses	76.17	60.00	76.17	**77.67**
3	House votes	90.02	92.18	**94.87**	90.02
4	Lung cancer	53.25	46.00	53.25	**56.83**
5	Lymph	84.97	81.69	82.20	**85.64**
6	Postoperative patient	**68.11**	67.77	65.56	**68.11**
7	Primary tumor	49.71	39.84	47.50	**49.77**
8	Soybean	**92.94**	85.97	92.87	91.33
9	Spect_all	78.68	**82.73**	80.01	78.68
10	Zoo	93.69	88.09	94.55	**95.25**
	Wins	2/10	2/10	1/10	6/10

The wins given at the bottom of Table 2., shows that RNB got best accuracy results, outperforming the other algorithms in 6 out of 10 data sets. RS and NB are the
second best algorithms with 2 wins followed by NBTree which outperformed the
others in only one data set.

However, by considering the performance of NB against RNB, the achieved results show that RNB is comparable to NB in four data sets, performed better than NB in five data sets and being outperformed by NB in only one data set. Compared to RS and NBTree, the proposed method is better in seven data sets and worse in only three data sets. This indicates that our proposed method did successfully improve upon the results obtained by NB and performed better than both RS and NBTree.

6   Conclusions and Future Work

Many classification methods have been proposed by researchers aiming to improve the performance of NB classifier and resolve its main weakness of independence assumption. In this paper we proposed a new algorithm which combines rough sets with NB approach resulting in a new variant of NB called RNB. The algorithm we propose is motivated by rough set's ability in discovering attributes' dependency and therefore utilized rough set theory to minimize the effect of dependent attributes through weighting approach.

Based on the conducted experiments we emphasize that the use of rough set theory provides us with better results for classifying new objects as compared to NB, RS and NBTree. However, the proposed algorithm still needs further improvements to compete well with other hybrid methods; and this is our main direction for future work. Possible improvements involve refining the calculated dependency value and modifying the positive region by allowing for approximate satisfaction as proposed, e.g., in the variable precision rough set (VPRS) model [25]. Our future work will also involve comparing the performance of RNB with existing hybrid techniques and investigating the use of different discretization methods with RNB algorithm to be able to recommend the best method for preprocessing of data.

References

1. Friedman, N., Geiger, D., Goldszmidt, M.: Bayesian Network Classifiers. Machine Learning 29, 131–163 (1997)
2. Rish, I.: An Empirical Study of the Naïve Bayes Classifier. In: Proceedings of the Int. Joint Conf. on Artificial Intelligence, Workshop on Empirical Methods in AI (2001)
3. Al-Aidaroos, K.M., Bakar, A.A., Othman, Z.: Naïve Bayes Variants in Classification Learning. In: Int. conf. on Information Retrieval and Knowledge Management, pp. 276–281 (2010)
4. Pawlak, Z.: Rough Sets and Intelligent Data Analysis. Information Sciences 147, 1–12 (2002)
5. Thangavel, K., Pethalakshmi, A.: Dimensionality Reduction based on Rough Set Theory: A Review. Applied Soft Computing 9, 1–12 (2009)
6. Pawlak, Z.: Rough Set Approach to Knowledge-based Decision Support. European Journal of Operational Research 99, 48–57 (1997)
7. Pawlak, Z., Skowron, A.: Rudiment of Rough Sets. Information Sciences 177, 3–27 (2007)
8. Skowron, A.: Rough Sets in KDD (2000)
9. Domingos, P., Pazzani, M.: On the Optimality of the Simple Bayesian Classifier under Zero-One Loss. Machine Learning 29, 103–130 (1997)

10. Hassan, S.Z., Verma, B.: A Hybrid Data Mining Approach for Knowledge Extraction and Classification in Medical Databases. In: IEEE Seventh International Conference on Intelligent Systems Design and Applications, pp. 503–508 (2007)

11. Pattaraintakorn, P., Cercone, N.: Integrating Rough Set Theory and Medical Applications. Applied Mathematics Letters 21, 400–403 (2008)

12. Li, R., Zhao, Y., Zhang, F., Song, L.: Rough Sets in Hybrid Soft Computing Systems. In: Alhajj, R., Gao, H., Li, X., Li, J., Zaïane, O.R. (eds.) ADMA 2007. LNCS (LNAI), vol. 4632, pp. 35–44. Springer, Heidelberg (2007)

13. Kohavi, R.: Scaling Up the Accuracy of Naive-Bayes Classifiers: A Decision-Tree Hybrid. In: Proceedings of the Second International Conference on Knowledge Discovery and Data Mining (KDD 1996), pp. 202–207 (1996)

14. Zheng, Z., Webb, G.I.: Lazy Learning of Bayesian Rules. Machine Learning 41, 53–84 (2000)

15. Ratanamahatana, C.A., Gunopulos, D.: Scaling up the Naive Bayesian Classifier: Using Decision Trees for Feature Selection. In: proceedings of Workshop on Data Cleaning and Preprocessing (DCAP 2002), at IEEE International Conference on Data Mining, ICDM 2002 (2002)

16. Frank, E., Hall, M., Pfahringer, B.: Locally Weighted Naive Bayes. In: Proceedings of the Conference on Uncertainty in Artificial Intelligence, pp. 249–256. Morgan Kaufmann, Seattle (2003)

17. Webb, G.I., Boughton, J., Wang, Z.: Not so Naive Bayes: Aggregating One-Dependence Estimators. Machine Learning 58, 5–24 (2005)

18. Ji, Y., Shang, L.: RoughTree: A Classifier with Naïve-Bayes and Rough Sets Hybrid in Decision Tree Representation. In: 2007 IEEE International Conference on Granular Computing, pp. 221–226 (2007)

19. Hall, M.: A Decision Tree-Based Attribute Weighting Filtering for Naïve Bayes. Knowledge-Based Systems 20, 120–126 (2007)

20. Zhang, H., Sheng, S.: Learning Weighted Naïve Bayes with Accurate Ranking. In: Proceedings of the Fourth IEEE International Conference on Data Mining, ICDM 2004 (2004)

21. Hall, M., Frank, E., Holmes, G., Pfahringer, B., Reutemann, P., Witten, I.H.: The WEKA Data Mining Software: An Update. SIGKDD Explorations 11(1) (2009)

22. The ROSETTA rough set toolkit, http://www.lcb.uu.se/tools/rosetta/

23. Asuncion, A., Newman, D.J.: UCI Machine Learning Repository. University of California, Department of Information and Computer Science, Irvine (2007), http://www.ics.uci.edu/~mlearn/MLRepository.html

24. Kohavi, R.: A Study of Cross-validation and Bootstrap for Accuracy Estimation and Model Selection. In: International joint Conference on artificial intelligence, pp. 1137–1145 (1995)

25. Ziarko, W.: Variable Precision Rough Set Model. Journal of Computer and System Sciences 46, 39–59 (1993)

A Heuristic Reduction Algorithm in IIS Based on Binary Matrix

Huaxiong Li[1,2], Xianzhong Zhou[1,2], and Meimei Zhu[1]

[1] School of Management and Engineering, Nanjing University,
Nanjing, Jiangsu, 210093, P.R. China
[2] State Key Laboratory for Novel Software Technology, Nanjing University,
Nanjing, Jiangsu, 210093, P.R. China
huaxiongli@gmail.com, zhouxz@nju.edu.cn, zmeimei84@gmail.com

Abstract. A binary discernibility matrix is presented in this paper, upon which a binary matrix-based heuristic reduction algorithm in incomplete information system(IIS) is proposed. In the proposed algorithm, the problem of finding an attribute reduction is converted to the problem of searching a set of binary matrices that can cover the objective binary matrix. The heuristic function in the proposed heuristic reduction algorithm is defined by a discernibility matrix associated with each condition attribute, which denotes the classification significance of the condition attribute. In the proposed heuristic reduction algorithm, attribute reduct is constructed by adding attributes in the sequence of attribute significance. An example of incomplete information system is presented to illustrate the algorithm and its validity.The algorithm is proved to be effective based on an illustrative example.

Keywords: rough set, incomplete information system, reduction, binary matrix, heuristic.

1 Introduction

Rough set theory, introduced by Pawlak Z. in 1982, provides a mathematical tool to deal with vague or imprecise information [6,7,8]. In recent decades, rough set theory has aroused wild attention of researchers on data mining and knowledge discovery, and witnessed great success in practical applications on intelligence information process, knowledge reduction and decision making support system.

However, classical rough set theory mainly concerns complete system, in which all the attribute values of each object are known, while in practice, many information systems are incomplete information systems(IIS). One of the methods to deal with this problem is to extend the equivalence relation to looser binary relations such as tolerance relation, similarity relation, etc. and then the incomplete information system can be analyzed [2,3,4,10,12]. This article introduces how a rough set model can be used in incomplete information system.

A fundamental notion in rough set theory is reduct, which has been extensively studied in many literatures. A reduct is a subset of attributes that is jointly

J. Yu et al. (Eds.): RSKT 2010, LNAI 6401, pp. 143–150, 2010.
© Springer-Verlag Berlin Heidelberg 2010

sufficient and individually necessary for preserving the same information under consideration as provided by the entire set of attributes. Discernibility matrix is a basic tool to search all reduct based on the construction of disjunctive form of attributes [9], which is widely used in the analysis on reduction [1,14,15,16]. However, the problem of transforming the conjunctive form of boolean function to the disjunctive form is NP-hard, which is not applicable in practice, thus a heuristic strategy is necessary, and many heuristic algorithm for reduct construction have been developed [5,11,13]. In this paper, we present a new form of discernibility matrix, which can be used to present a heuristic algorithm to construct reduct in incomplete information system.

For this purpose, a binary discernibility matrix with respect to each attribute is defined, in which the entry of the matrix is either 0 or 1, depends on whether this single attribute can discern the corresponding objects. By using the binary matrix, the problem of finding an attribute reduction can be converted to the problem of searching a set of binary matrices, in which the union of the binary matrices can cover the objective binary matrix of the decision attribute. In the proposed algorithm, a heuristic function is defined according to the intersection of the attribute binary matrix and the objective matrix, which indicates the distance between the attribute set and reduct. Based on the proposed heuristic function, a new heuristic algorithm is presented to construct the attribute reduct in an incomplete information system.

2   Basic Notions

Information system(IS) is a basic concept in rough set theory [6]. Let us briefly review the basic notions on information system and incomplete information system(IIS).

Denote an information system as $S =< U, A = C \cup D, V, f >$, where U and A are respectively called universe and the set of attributes. A is a finite set of attributes which can be divided into two sets: the condition attributes C and the decision attributes D. For each attribute $a \in A$, there is a mapping $f : U \times A \to V$ connecting elements of U and elements of V_a, and the value of an object $x \in U$ on an attribute $a \in A$ is denoted by $f(x, a)$. An information system is called an decision table if condition attributes and decision attributes are distinguished. Commonly in complete decision table, where the attribute values of each object are known, it is assumed that the mapping $f(x, a)$ is single-valued. While in incomplete information system, there exist objects whose attribute values are unknown or missing. In this case, the value of an object $x \in U$ on an attribute $a \in At$ i.e. $f(x, a)$ can be expressed by $*$, namely $f(x, a) = *$.

An equivalence relation defined on a complete information system induces an indiscernibility relation, which can be described as [6]:

$$IND(P) = \{(x, y) \in U \times U | \forall a \in P, f(x, a) = f(y, a)\},$$

where $P \subseteq C$ is a condition attribute set. While in an incomplete information system, the equivalence relation can be extended to looser binary relation such as tolerance relation so that it is available in semantics. A typical tolerance relation induced from an attribute set P is defined as [2]:

$$SIM(P) = \{(x,y) \in U \times U | \forall a \in P, f(x,a) = f(y,a) \vee f(x,a) = * \vee f(y,a) = *\},$$

Attribute reduction is an important problem of rough set theory. The notion of a reduct plays an essential role in analyzing an information table. A reduct is a minimum subset of attributes that provides the same descriptive or classification ability as the entire set of attributes. In other words, attributes in a reduct are jointly sufficient and individually necessary. Many methods have been proposed and examined for finding the set of reduct. An attribute a is dispensable in $P \subseteq C$ if $IND(P) = INP(P - \{a\})$. Otherwise, a is indispensable. Clearly, a dispensable attribute does not improve the classification ability thus it can be deleted. The intersection of all reduct set of P is called as a core of P, denoted as $CORE(P)$.

Discernibility matrix is a basic method to search all the reduct for an information system. The classical discernibility matrix is described as a $U \times U$ matrix $M = M(x,y)$, in which the element $M(x,y)$ for an object pair (x,y) is defined by $M(x,y) = \{a \in A | f(x,a) \neq f(y,a)\}$, i.e., it comprise all the attributes that can discern (x,y). For an incomplete decision table, discernibility matrix is described as a $U \times U$ matrix $M = M(x,y)$, in which the element $M(x,y)$ for an object pair (x,y) is defined by $M(x,y) = \{a \in A | f(x,a) \neq f(y,a) \wedge f(x,a) \neq * \wedge f(y,a) \neq *\}$. By using discernibility matrix, all reduct can be induced based on transforming the conjunctive form of boolean function $\bigwedge \bigvee (M(x,y))$ to a disjunctive form.

3   Binary Discernibility Matrix

In this section, we present a new form of discernibility matrix for incomplete information system based on binary matrix, which can be used to propose a new heuristic algorithm for attribute reduction.

Definition 1. *Let $S = (U, At = C \cup D, \{V_a \mid a \in At\}, \{I_a \mid a \in At\})$ be an incomplete information system, for $\forall a \in A$, the binary discernibility matrix of attribute a is a $n \times n$ binary matrix, denoted by M_a. Each entry $M_a(x_i, x_j)$ in the matrix M_a equals to either 0 or 1, which is determined by:*

$$M_a(x_i, x_j) = \begin{cases} 0, & I_a(x_i) = I_a(x_j) \vee I_a(x_i) = * \vee I_a(x_j) = *, \\ 1, & I_a(x_i) \neq I_a(x_j). \end{cases}$$

According to Definition 1, the binary discernibility matrix of an attribute a describes the discernible ability of the attribute on any two objects in universe. $M_a(x_i, x_j)$ equals to 1 when x_i and x_j are discernible on attribute a, i.e., $I_a(x_i) \neq I_a(x_j)$, and $M_a(x_i, x_j)$ equals to 0 when x_i and x_j are indiscernible on attribute a, i.e., $I_a(x_i) = I_a(x_j)$ or $I_a(x_i) = *$ or $I_a(x_j) = *$. It should be noted that x_i and x_j are indiscernible if there exist missing values on any of the two objects since the values of these two objects are not exactly unequal. Similarly, the binary discernibility matrix of an attribute set can be presented follows.

Definition 2. *Let* $S = (U, At = C \cup D, \{V_a \mid a \in At\}, \{I_a \mid a \in At\})$ *be an incomplete information system, for* $B \subset C$, *the binary discernibility matrix of attribute set* B *is a* $n \times n$ *binary matrix, denoted by* M_B. *Each entry* $M_B(x_i, x_j)$ *in the matrix* M_B *equals to either 0 or 1, which is determined by:*

$$M_B(x_i, x_j) = \begin{cases} 0, & \forall a \in B, I_a(x_i) = I_a(x_j) \vee I_a(x_i) = * \vee I_a(x_j) = *, \\ 1, & \exists b \in B, I_b(x_i) \neq I_b(x_j). \end{cases}$$

According to Definition 1 and Definition 2, the relationship between attribute discernibility matrix and attribute set discernibility matrix can be presented, which may be described as following proposition.

Proposition 1. *For an incomplete information system* $S = (U, At = C \cup D, \{V_a \mid a \in At\}, \{I_a \mid a \in At\})$, *suppose* B *is a subset of* C, *where* $B = \{b_1, b_2, \cdots, b_m\}$, *then it holds that:*

$$M_B = \bigcup_{b \in B} M_b = M_{b_1} \cup M_{b_2} \cup \cdots \cup M_{b_m},$$

where $\cup$ *denotes the union of binary matrices. The union of two* $n \times n$ *binary matrices* M_a *and* M_b *is a* $n \times n$ *binary matrix* T, *in which the ith row jth column entry* T_{ij} *is defined by:*

$$T_{ij} = \begin{cases} 0, & M_a(x_i, x_j) = 0 \wedge M_b(x_i, x_j) = 0, \\ 1, & M_a(x_i, x_j) = 1 \vee M_b(x_i, x_j) = 1. \end{cases}$$

Proposition 1 indicates that the discernable ability of attribute set B can be presented by the discernable ability union of all single attributes in B. In general, more attributes have stronger discernable ability on any two objects in universe, whereas less attributes have weaker discernable ability. The discernable ability relationship between two attribute sets B_1 and B_2 can be described by the respectively corresponding binary matrix, which is presented as follows.

Definition 3. *Suppose* M_{B_1} *and* M_{B_2} *are respectively the binary discernibility matrix of* B_1 *and* B_2. *The discernable ability of* B_1 *is called weaker than that of* B_2 *if for* $\forall i, j (i, j = 1, 2, \cdots, n)$ *it holds that:* $M_{B_1}(x_i, x_j) = 0$ *or* $M_{B_1}(x_i, x_j) = 1 \wedge M_{B_2}(x_i, x_j) = 1$, *and we denote it by* $M_{B_1} \subseteq M_{B_2}$, *which is also called that* M_{B_2} *covers* M_{B_1}. *Further, the discernable ability of* B_1 *is called strictly weaker than that of* B_2 *if* $M_{B_1} \subseteq M_{B_2}$ *and* $\exists i, j$ *satisfying* $M_{B_1}(x_i, x_j) = 0$ *and* $M_{B_2}(x_i, x_j) = 1$, *and we denote it by* $M_{B_1} \subset M_{B_2}$, *which is also called that* M_{B_2} *strictly covers* M_{B_1}.

Proposition 2. *Suppose* S *is an incomplete information system,* $S = (U, At = C \cup D, \{V_a \mid a \in At\}, \{I_a \mid a \in At\})$, B *is an absolute reduct of attribute set* C *if following two statements hold:(i)* $M_C = M_B$, *(ii)* $\forall a \in B$, M_C *strictly covers* $M_{B-\{a\}}$, *i.e.,* $M_{B-\{a\}} \subset M_C$.

Proposition 3. *Suppose S is an incomplete information system, $S = (U, At = C \cup D, \{V_a \mid a \in At\}, \{I_a \mid a \in At\})$, $B \subseteq C$. Let M^* be a $n \times n$ binary matrix, in which the ith row jth column entry $M^*(x_i, x_j)$ is defined by:*

$$M^*(x_i, x_j) = \begin{cases} 1, & x_i \in POS_C(D) \wedge x_j \notin POS_C(D) \\ & or \; x_i \notin POS_C(D) \wedge x_j \in POS_C(D) \\ & or \; x_i, x_j \in POS_C(D) \wedge I_D(x_i) \neq I_D(x_j), \\ \\ 0, & otherwise. \end{cases}$$

According to literature [9], for a positive region reduct, the objective binary matrix M^* can be calculated based on Proposition 3, and for a absolute reduct, the objective binary matrix M^* is equal to M_C [9]. Based on Proposition 2 and Proposition 3, we may construct a reduct by searching a set of binary matrix M_{a_i} that can cover the objective binary matrix M^*. Therefore, the problem of searching an attribute reduction is converted to the problem of searching a covering of matrix.

4 A Heuristic Reduction Algorithm Based on Binary Matrix

In this section, we present a heuristic algorithm for the construction of absolute reduct. The main idea is to convert the problem of searching an attribute reduct to the problem of searching a set of binary matrices, in which the union of all binary matrices may cover the objective binary matrix. However, it is a NP-hard problem to search a minimum cover for a matrix. Therefore, in order to find out the set of matrices that may cover the objective matrix, a heuristic function is presented to determine the sequence of attributes in the process of constructing a reduct. The heuristic function is defined according to the intersection of the attribute binary matrix and the objective matrix, which indicates the distance between the attribute set and reduct. For absolute reduct, the objective matrix is , and for relative reduct, the objective matrix is M_C, and for relative reduct, the objective matrix is M^*, which is presented in section 3. For simpleness, we denote the objective matrix uniformly as M.

Definition 4. *Suppose S is an incomplete information system, $S = (U, At = C \cup D, \{V_a \mid a \in At\}, \{I_a \mid a \in At\})$, M is the objective binary matrix. $M_a^* = M_a \cap M$, where the ith row jth column entry of $M_a \cap M$ is defined by:*

$$(M_a^*)_{ij} = (M_a \cap M)_{ij} = \begin{cases} 1, & M_a(x_i, x_j) = M(x_i, x_j) = 1, \\ 0, & otherwise. \end{cases}$$

The covering significance of attribute $a \in C$ under objective binary matrix M is defined by $\gamma(a) = \sum_{i,j} (M_a^*)_{ij}$.

In definition 4, the covering significance is the total number of the common entries equal to 1, which evaluates the importance of an attribute with regard to the objective binary matrix. In the proposed heuristic algorithm, attribute with highest covering significance is selected firstly to add in the attribute set, then the attribute with the second highest covering significance is taken into account. The process will continue until the union of binary discernibility matrix of attributes set cover the objective binary matrix. The algorithm is described as follows:

Input: an incomplete decision table S;
Output: an attribute reduct B.
Step 1: For each $a \in C$, calculate binary discernibility matrix M_a;
Step 2: Calculate the objective binary matrix M based on:
(1)$M = M_C$ if B is an absolute reduct,
      where M_C is calculated based on Definition 2;
(2)$M = M^*$ if B is a relative reduct,
      where M^* is calculated based on Proposition 3;
Step 3: Let $B = \emptyset$, $E = C$, repeat following loop:
(I)For each $a \in E$, calculate $M_a^* = M_a \cap M$, $\gamma(a) = \sum_{i,j}(M_a^*)_{ij}$;
(II)Select the attribute a with the highest covering significance $\gamma(a)$,
      and add a to attribute set B: $B = B \cup \{a\}$;
(III) if $M_B = M_C$, then go to **Step 4**;
(IV) Let $E = E - \{a\}$, set all entries which are covered by
      M_a to 0, update M, go to (I);
Step 4: Delete redundant attributes in B.
Step 5: Output a reduct B.

5   An Illustrative Example

In this section, we present an example of an incomplete information system to illustrate the algorithm and its validity.

Example. Table 1 is an incomplete information system, where $U = \{x_1, x_2, x_3, x_4\}$, $C = \{a_1, a_2, a_3, a_4, a_5\}$.
    Suppose we construct an absolute reduct of attribute set C. Based on the proposed algorithm, the whole process is described as follows:

Step 1: Calculate all binary discernibility matrices M_a, and we get three matrices:

Table 1. An Incomplete Information Table

U	a_1	a_2	a_3	a_4	a_5
x_1	1	2	1	*	1
x_2	1	*	2	1	2
x_3	1	1	3	*	2
x_4	2	*	1	3	1

$$M_{a_1} = \begin{pmatrix} 0\,0\,0\,1 \\ 0\,0\,0\,1 \\ 0\,0\,0\,1 \\ 1\,1\,1\,0 \end{pmatrix}, M_{a_2} = \begin{pmatrix} 0\,0\,1\,0 \\ 0\,0\,0\,0 \\ 1\,0\,0\,0 \\ 0\,0\,0\,0 \end{pmatrix}, M_{a_3} = \begin{pmatrix} 0\,1\,1\,0 \\ 1\,0\,1\,1 \\ 1\,1\,0\,1 \\ 0\,1\,1\,0 \end{pmatrix},$$

$$M_{a_4} = \begin{pmatrix} 0\,0\,0\,0 \\ 0\,0\,0\,1 \\ 0\,0\,0\,0 \\ 0\,1\,0\,0 \end{pmatrix}, M_{a_5} = \begin{pmatrix} 0\,1\,1\,0 \\ 1\,0\,0\,1 \\ 1\,0\,0\,1 \\ 0\,1\,1\,0 \end{pmatrix};$$

Step 2: Calculate M_C, set objective matrix $M = M_C$, then we have:

$$M = M_C = \begin{pmatrix} 0\,1\,1\,1 \\ 1\,0\,1\,1 \\ 1\,1\,0\,1 \\ 1\,1\,1\,0 \end{pmatrix};$$

Step 3: Set $B = \emptyset$, $C = \{a_1, a_2, a_3, a_4, a_5\}$, calculate all $\gamma(a)$:$\gamma(a_1) = 6$,$\gamma(a_2) = 2$,$\gamma(a_3) = 10$,$\gamma(a_4) = 2$,$\gamma(a_5) = 8$, and the attribute with the maximum $\gamma(a)$ is attribute a_3, then we add a_3 to B, thus $B = \{a_3\}$. Clearly, $M_B \neq M$, it is necessary to add more attributes to B. Let $E = C - \{a_3\} = \{a_1, a_2, a_4, a_5\}$, and update the objective matrix M:

$$M = M_C = \begin{pmatrix} 0\,0\,0\,1 \\ 0\,0\,0\,0 \\ 0\,0\,0\,0 \\ 1\,0\,0\,0 \end{pmatrix};$$

and calculate the covering significance of each attribute in $E = \{a_1, a_3\}$ with regard to the updated M: $\gamma(a_1) = 2$,$\gamma(a_2) = 0$,$\gamma(a_4) = 0$,$\gamma(a_5) = 0$. The attribute with the maximum $\gamma(a)$ is attribute a_1, then we add a_1 to B, and we have $B = \{a_1, a_3\}$. For the updated attribute set B, it holds that $M_B = M$.

Step 4: it is necessary to check the redundancy of $B = \{a_1, a_3\}$. It can be testified that B is independent because $M_{B-\{a_1\}} \subset M_C$ and $M_{B-\{a_3\}} \subset M_C$. Therefore, $B = \{a_1, a_3\}$ is an absolute reduct of attribute set C. We may testify that $B = \{a_1, a_3\}$ is an absolute reduct of C according to the definition of absolute reduct in literature [9].

6   Conclusion

Attribute reduction in incomplete information system is an important topics in rough set theory. It has been proven that the construction of minimal reduct is a NP-hard problem. In this paper, we present a new heuristic function based on binary discernibility matrix. The heuristic function is defined by the covering significance, which is the total number of the common entries equal to 1, which presents an evaluation for the importance of an attribute with regard to the objective binary matrix and the distance between the attribute set and reduct. Based on the proposed heuristic function, a new heuristic algorithm is presented.

Acknowledgments

This research is partially supported by the National Natural Science Foundation of China under grant No.70971062,70571032, and the open foundation of State Key Laboratory for Novel Software Technology.

References

1. Hu, X.H., Cercone, N.: Learning in relational databases: a rough set approach. Computational Intelligence 11, 323–338 (1995)
2. Kryszkiewicz, M.: Rough Set Approach to Incomplete Information Systems. Information Sciences 112, 39–49 (1998)
3. Kryszkiewicz, M.: Rules in Incomplete Information Systems. Information Sciences 113, 271–292 (1999)
4. Li, H.X., Zhou, X.Z., Huang, B.: Method to determine α in rough set model based on connection degree. Journal of Systems Engineering and Electronics 20(1), 98–105 (2009)
5. Miao, D.Q., Hu, G.R.: A heuristic algorithm for reduction of knowledge. Journal of Computer Research and Development 36(6), 681–684 (1999) (in Chinese)
6. Pawlak, Z.: Rough sets. International Journal of Computer and Information Science 11, 341–356 (1982)
7. Pawlak, Z.: Rough classification. International Journal of Man-Machine Studies 20, 469–483 (1984)
8. Pawlak, Z., Skowron, A.: Rudiments of rough sets. Information Sciences 177(1), 3–27 (2007)
9. Skowron, A., Rauszer, C.: The discernibility matrices andfunctions in information systems. In: Slowiski, R. (ed.) Intelligent Decision Support, Handbook of Applications and Advances of the Rough Sets Theory, pp. 331–362. Kluwer, Dordrecht (1992)
10. Stefamowski, J., Tsoukeas, A.: On the Extension of Rough Sets under Incomplete Information. International Journal of Intelligent System 16, 29–38 (1999)
11. Wang, G.Y.: Calculation methods for core attributes of decision table. Chinese Journal of Computers 26(5), 611–615 (2003)
12. Wang, G.Y.: Extension of rough set under incomplete information systems. Journal of computer research and development 39(10), 1238–1243 (2002) (in Chinese)
13. Xu, Z.Y., Liu, Z.P., Yang, B.R., et al.: A quick attribute reduction algorithm with complexity of $\max(O(|C||U|), O(|C|^2|U/C|))$. Chinese Journal of Computers 29(3), 391–399 (2006)
14. Yang, M.: An incremental updating algorithm for attribute reduction based on improved discernibility matrix. Chinese Journal of Computers 29(3), 407–413 (2006)
15. Yao, Y.Y., Zhao, Y.: Discernibility matrix simplification for constructing attribute reducts. Information Sciences 179(5), 867–882 (2009)
16. Ye, D.Y., Chen, Z.J.: A new discernibility matrix and the computation of a core. Acta Electronica Sinica 30(7), 1086–1088 (2002) (in Chinese)

Generalized Distribution Reduction in Inconsistent Decision Systems Based on Dominance Relations

Yan Li[1], Jin Zhao[1], Na-Xin Sun[1], and Sankar Kumar Pal[2]

[1] Faculty of Mathematics and Computer Science, Hebei University, Baoding 071002,
Hebei Province, China
[2] Machine Intelligence Unit, Indian Statistical Institute, Kolkata, 700 035, India
ly@hbu.cn, {zhaojinjin111,sunnaxin}@126.com, sankar@isical.ac.in

Abstract. By incorporating dominance principle in inconsistent decision systems based on dominance relations, two new types of distribution reductions are proposed, i.e., generalized distribution reduction and generalized maximum distribution reduction, and their properties and relationship are also discussed. The corresponding generalized distribution discernibility matrix is then defined to provide a convenient computation method to obtain the generalized distribution reductions. The validation of this method is showed by both theoretical proofs and illustrative examples.

Keywords: Rough set, Decision table, Dominance Relation, Distribution function, Generalized distribution reduction.

1   Introduction

Rough set theory [1] proposed by Pawlak is used to handle uncertainty and ambiguity in data. Knowledge reduction is one of the most important problems in data mining, and is also an essential task of rough set theory. Based on equivalence relation, the concept of reduct is introduced in rough set theory to remove the indispensable attributes without losing discernibility ability of information systems. Up to date, researchers have done extensive studies in this aspect, and many useful results have been reported in [2-7].

In some practical problems, particularly in multi-criteria decision analysis, there are some attributes with ordered values, such as attribute "score" taking values of high, medium and low. This type of attributes is often used to give an evaluation of each object in the universe, e.g., judging a student according to his scores of several subjects. In this case, the preference-orders of attribute values should not be ignored. The concept of equivalence relation cannot reflect this kind of order, and therefore the generated rules (knowledge) cannot convey this kind of information. To address this problem, Greco [8] firstly developed the framework of dominance relation-based rough set theory, in which the original equivalence relation is substituted by dominance relation. Different

J. Yu et al. (Eds.): RSKT 2010, LNAI 6401, pp. 151–158, 2010.

types of reductions are then introduced based on dominance relations[8-13]. In [10], by defining the distribution function for each object in universe, distribution reduction and maximum distribution reduction are proposed and their judgment theorem is given. [11] has pointed out that the judgment theorem in [10] is wrong, and defined a new maximum distribution reduction. Another two types of reduction,positive region reduction and low approximation reduction are given in [12, 13]. The distribution function in [10, 11] is defined by computing the overlaps of each object's condition dominance class and all the decision dominance classes of the information system. It is difficult to have an intuitive understanding of this definition. It seems that, the distribution function of an object describes the distribution of condition dominance class of this object in every decision dominance class, which cannot reflect the idea of Greco's dominance principle [8]. Here, the dominance principle means if condition attributes of object x are not worse than those of object y, then the decision attributes of object x should not be worse than those of y. In present paper, incorporating dominance principle, we introduce a new type of distribution function, namely, the generalized distribution function and correspondingly define the generalized distribution reduction and generalized maximum distribution reduction.

Some basic concepts are given the Section 2. Section 3 defines the concepts of generalized distribution reduction and generalized maximum distribution reduction in inconsistent systems based on dominance relations, and some propositions about their relationships are also given. In Section 4, the generalized distribution discernibility matrix is obtained based on generalized distribution function, which provides an approach to knowledge reduction in inconsistent systems based on dominance relations. Finally, some conclusions are given in Section 5.

2   Basic Concepts

In this section, some basic concepts are introduced for information systems based on dominance relations.

Definition 1. A 5-tuple $DS = (U, A, F, D, G)$ is referred to as a target information system [10], where

$U = (x_1, x_2, \cdots, x_n)$ is a non-empty finite set of objects;

$A = (a_1, a_2, \cdots, a_n)$ is a finite set of condition attributes;

$D = (d_1, d_2, \cdots, d_n)$ is a finite set of decision attributes;

$F = \{f_k(x): U \rightarrow V_k, k \leq p\}$, $f_k(x)$ is the value of a_k, $x \in U$, V_k is the finite domain of a_k, a_k;

$G = \{g_k(x): U \rightarrow V'_k, k \leq q\}$, $g_k(x)$ is the value of d_k, $x \in U$, V'_k is the finite domain of d_k, d_k.

Actually, a target information system is a decision system (table). In the following, without any further statement, we mean a target information system by decision system.

Definition 2. Let $DS = (U, A, F, D, G)$ be a target information system, $B \subseteq A$, the dominance relations induce by B and D of DS are respectively denoted as

$R_B^{\leq}=\{(x_i, x_j) \in U \times U : f_l(x_i) \leq f_l(x_j), \forall a_l \in B\}$

$R_D^{\leq}=\{(x_i, x_j) \in U \times U: g_m(x_i) \leq g_m(x_j), \forall d_m \in B\}$

Based on $R_B^{\leq}$ and $R_D^{\leq}$, B-dominance relation class and D-dominance class of an object x can be defined as follows:

$[x_i]_{\overline{B}}^{\leq}=\{x_j \in U: (x_i, x_j) \in R_B^{\leq}\}=\{x_j \in U: f_l(x_i) \leq f_l(x_j), \forall a_l \in B\}$

$[x_i]_{\overline{D}}^{\leq}=\{x_j \in U: (x_i, x_j) \in R_D^{\leq}\}=\{x_j \in U: g_m(x_i) \leq g_m(x_j), \forall d_m \in D\}$

Followed these definitions, the properties of dominance relation can be directly obtained.

Proposition 1. Let $R_B^{\leq}$ be a dominance relation.

(1) $R_B^{\leq}$ is reflexive and transitive, but not necessarily symmetric.

(2) If $P \subseteq Q \subseteq A$, then $R_A^{\leq} \subseteq R_Q^{\leq} \subseteq R_P^{\leq}$.

(3) If $P \subseteq Q \subseteq A$, then $\forall x \in U$, $[x]_A^{\leq} \subseteq [x]_Q^{\leq} \subseteq [x]_P^{\leq}$

(4) If $y \in [x]_B^{\leq}$, then $[y]_B^{\leq} \subseteq [x]_B^{\leq}$

From (1), a dominance relation is not necessarily an equivalence relation.

Definition 3. Let $DS = (U, A, F, D, G)$ be a target information system based on dominance relations. DS is called as a consistent target information system, if $R_A^{\leq} \subseteq R_D^{\leq}$; Otherwise, if $R_A^{\leq} \not\subseteq R_D^{\leq}$, it is inconsistent.

Since we focus on reductions in inconsistent decision systems, an example is given for further understanding of this type of systems based on dominance relations.

Example 1. Consider the target information system I shown in Table 1.

Table 1. Generalized distribution discernibility matrix

U	a_1	a_2	a_3	d
x_1	1	2	1	3
x_2	3	2	2	2
x_3	1	1	2	1
x_4	2	1	3	2
x_5	3	3	2	3
x_6	3	2	3	1

Here, $U=\{x_1, x_2, x_3, x_4, x_5, x_6\}$, $A=\{a_1, a_2, a_3\}$, $D=\{d\}$. According to definition 2, we have

$[x_1]_A^{\leq}=\{x_1, x_2, x_5, x_6\}$, $[x_2]_A^{\leq}=\{x_2, x_5, x_6\}$

$[x_3]_A^{\leq}=\{x_2, x_3, x_4, x_5, x_6\}$, $[x_4]_A^{\leq}=\{x_4, x_6\}$

$[x_5]_A^{\leq}=\{x_5\}$, $[x_6]_A^{\leq}=\{x_6\}$

$D_1=[x_1]_{\overline{D}}^{\leq}=\{x_1, x_5\}$, $D_2=[x_2]_{\overline{D}}^{\leq}=\{x_2, x_5, x_6\}$

$D_3=[x_3]_{\overline{D}}^{\leq}=\{x_1, x_2, x_3, x_4, x_5, x_6, \}$, $D_4=[x_4]_{\overline{D}}^{\leq}=\{x_1, x_2, x_4, x_5\}$

$D_5=[x_5]_{\overline{D}}^{\leq}=\{x_1, x_5\}$, $D_6=[x_6]_{\overline{D}}^{\leq}=\{x_1, x_2, x_3, x_4, x_5, x_6\}$

Obviously, $R_A^{\leq} \not\subseteq R_D^{\leq}$. For example, $[x_1]_A^{\leq} \not\subseteq [x_1]_D^{\leq}$, which means there are objects whose condition attributes are not worse than those of x_1, but their decision attribute is worse than that of x_1. Therefore, the information system in Table 1 is inconsistent in the sense of Greco's dominance principle.

3   Generalized Distribution Reductions

In order to incorporate Greco's dominance principle, we generalize the concepts of distribution function and maximum distribution function defined in [10, 11]. In this section, without further statement, DS denotes an inconsistent decision system based on dominance relations.

Definition 4. (Generalized distribution function)

Let $DS = (U, A, F, D, G)$, $B \subseteq A$, $x \in U$. Denote $U/R_B^{\leq} = \{[x_i]_B^{\leq} | x \in U\}$, $D_i = [x_i]_D^{\leq}$.

Let $\mu_B^{\leq}(x_i) = \frac{|D_i \cap [x_i]_B^{\leq}|}{|U|} = \frac{|[x_i]_D^{\leq} \cap [x_i]_B^{\leq}|}{|U|}$

The generalized distribution function of U is defined as:

$\rho_B(U) = \{ \frac{|D_1 \cap [x_1]_B^{\leq}|}{|U|}, \frac{|D_2 \cap [x_2]_B^{\leq}|}{|U|}, \cdots, \frac{|D_n \cap [x_n]_B^{\leq}|}{|U|} \}$

$= \{ \mu_B^{\leq}(x_1), \mu_B^{\leq}(x_2), \cdots, \mu_B^{\leq}(x_n) \}$

The maximum distribution function of U is:

$\lambda_B(U) = max\{ \frac{|D_1 \cap [x_1]_B^{\leq}|}{|U|}, \frac{|D_2 \cap [x_2]_B^{\leq}|}{|U|}, \cdots, \frac{|D_n \cap [x_n]_B^{\leq}|}{|U|} \}$

$= max(\mu_B(x_1), \mu_B(x_2), \cdots, \mu_B(x_n))$

It should be noted that, the generalized distribution function is defined on U, $\mu_B(x_i)$ represents the proportion of such objects in U that both of their condition attributes and decision attributes are not worse than x_i. While in [1, 5], a distribution function is defined for each $x \in U$ as follows:

$\mu_B(x) = \{ \frac{|D_1' \cap [x_1]_B^{\leq}|}{|U|}, \frac{|D_2' \cap [x_2]_B^{\leq}|}{|U|}, \cdots, \frac{|D_r' \cap [x_r]_B^{\leq}|}{|U|} \}$,

where $D_1', D_2', \cdots, D_r'$ are different decision dominance classes of DS. Obviously, definition 4 is much simpler in form. Moreover, the generalized definition can be used to reflect the consistent degree of the whole decision system in the sense of dominance principle. While the distribution function in [10, 11] describes the overlaps of B-condition dominance class of x_i and every different decision dominance class of DS, which cannot describe the consistency of DS.

Definition 5. (Generalized distribution consistent set, generalized distribution reduct)

Given $DS = (U, A, F, D, G)$. For $B \subseteq A$, B is said to be a generalized distribution consistent set if $\rho_B(U) = \rho_A(U)$. If no proper subset of B is generalized distribution consistent set, then is called a generalized distribution reduction of DS. B is said to be a generalized maximum distribution consistent set if $\lambda_B(U) = \lambda_A(U)$; If no proper subset of B is generalized distribution consistent set, then B is called a generalized maximum distribution reduction of DS.

From Definition 5, we can directly obtain Proposition 2-3.

Proposition 2. Let $DS = (U, A, F, D, G)$, $B \subseteq A$. If B is a generalized distribution consistent set, B must be a generalized maximum distribution consistent set.

Proposition 3. Let $DS = (U, A, F, D, G)$, $B \subseteq A$. B is a generalized distribution consistent set if only and only if for any x, $\mu_B^{\leq}(x_i) = \mu_B^{\leq}(x_j)$.

Example 2. Consider the inconsistent decision system in example 1.

According to definition 4-5, we can compute

$\rho_A(U) = \{ \frac{1}{3}, \frac{1}{3}, \frac{5}{6}, \frac{1}{6}, \frac{1}{6}, \frac{1}{6} \}$

Let $B_1=\{a_1, a_2\}$, $B_2=\{a_1, a_3\}$, $B_3=\{a_2, a_3\}$, their generalized distribution functions are:

$\rho_{B_1}(U)=\{\frac{1}{3}, \frac{1}{3}, 1, \frac{1}{2}, \frac{1}{6}, \frac{1}{3}\}$, $\rho_{B_2}(U)=\{\frac{1}{3}, \frac{1}{3}, \frac{5}{6}, \frac{1}{6}, \frac{1}{6}, \frac{1}{6}\}$, $\rho_{B_3}(U)=\{\frac{1}{3}, \frac{1}{3}, 1, \frac{5}{6}, \frac{1}{6}, \frac{1}{6}\}$

Obviously,

$\rho_A(U)\neq\rho_{B_1}(U)$, $\lambda_A(U)\neq\lambda_{B_1}(U)$, $\rho_{B_2}(U)=\rho_{B_3}(U)=\rho_A(U)$,
$\lambda_{B_2}(U)=\lambda_{B_3}(U)=\lambda_A(U)$.

Thus, B_2 and B_3 is two generalized distribution consistent sets and generalized maximum distribution consistent sets. B_1 is not a generalized distribution consistent set and generalized maximum distribution consistent set.

Furthermore, it is easy to verify that $\{a_1, a_3\}$, $\{a_1\}$, $\{a_2\}$ is not generalized distribution consistent sets and generalized maximum distribution consistent sets. $\{a_3\}$ is not a generalized distribution consistent set, but it is a generalized maximum distribution reduction. So $B_2=\{a_1, a_3\}$ and $B_3=\{a_2, a_3\}$ is generalized distribution reductions. $\{a_3\}$ is the generalized maximum distribution reduction.

Next, before we give the judgment theorem for generalized distribution consistent set, the definition of vector comparisons is introduced, which will be used in the proof of the theorem.

Definition 6. Let $\alpha = \{\alpha_1, \alpha_2, \cdots, \alpha_n\}^T$, $\beta = \{\beta_1, \beta_2, \cdots, \beta_n\}^T$ are two n-dimensional vectors. α is equal to vector $\beta(\alpha=\beta)$, if $\alpha_i=\beta_i(i=1, 2, \cdots, n)$; α is greater than vector β, $(\alpha\geq\beta)$, if $\alpha_i\geq\beta_i(i=1, 2, \cdots, n)$.

Theorem 1. $DS = (U, A, F, D, G)$, $B\subseteq A$, x_i, $x_j\in U$. When $\mu_{\overline{A}}^{\leq}(x_i) < \mu_{\overline{A}}^{\leq}(x_j)$, $D_i \supseteq D_j$, B is a generalized distribution consistent set if only and only if $\exists b\in B$ such that $f_b(x_i)>f_b(x_j)$.

Proof : "$\Rightarrow$"The proof is by contradiction.

Assume when $\mu_{\overline{A}}^{\leq}(x_i) < \mu_{\overline{A}}^{\leq}(x_j)$, $D_i\supseteq D_j$, for $\forall b\in B$, $f_b(x_i)>f_b(x_j)$ doesn't hold, i.e., $f_b(x_i)\leq f_b(x_j)$. So, $\exists x_j\in[x_i]_{\overline{B}}^{\leq}$, and from proposition 1(4), $[x_i]_{\overline{B}}^{\leq} \supseteq [x_j]_{\overline{B}}^{\leq}$. It follows that, when $D_i\supseteq D_j$, $\mu_{\overline{B}}^{\leq}(x_i) \geq \mu_{\overline{B}}^{\leq}(x_j)$ holds.

On the other hand, since B is a generalized distribution consistent set, we have $\forall x_i\in U$, $\mu_{\overline{B}}^{\leq}(x_i) = \mu_{\overline{A}}^{\leq}(x_i)$. By proposition 3 and according to proposition1(3), when $B\subseteq A$, $\mu_{\overline{A}}^{\leq}(x_j) \leq \mu_{\overline{A}}^{\leq}(x_j)$.

Therefore, when $\mu_{\overline{A}}^{\leq}(x_i) < \mu_{\overline{A}}^{\leq}(x_j)$, we have $\mu_{\overline{B}}^{\leq}(x_i) = \mu_{\overline{A}}^{\leq}(x_i) < \mu_{\overline{A}}^{\leq}(x_j) \leq \mu_{\overline{B}}^{\leq}(x_j)$ which is in contradiction to $\mu_{\overline{B}}^{\leq}(x_i) \geq \mu_{\overline{B}}^{\leq}(x_j)$.

"$\Leftarrow$" The proof is also by contradiction.

Assume when $\mu_{\overline{A}}^{\leq}(x_i) < \mu_{\overline{A}}^{\leq}(x_j)$, $D_i\supseteq D_j$, $\exists b\in B$, such that $f_b(x_i)>f_b(x_j)$, but B is not a generalized distribution consistent set.

By proposition 3, there exists x_i, such that $\mu_{\overline{B}}^{\leq}(x_i)\neq\mu_{\overline{A}}^{\leq}(x_i)$. Since $[x_i]_{\overline{A}}^{\leq} \subseteq [x_i]_{\overline{B}}^{\leq}$, i.e., $\mu_{\overline{A}}^{\leq}(x_i)=\frac{|D_i\cap[x_i]_{\overline{A}}^{\leq}|}{|U|}\leq\frac{|D_i\cap[x_i]_{\overline{B}}^{\leq}|}{|U|}=\mu_{\overline{B}}^{\leq}(x_i)$. Then $\mu_{\overline{B}}^{\leq}(x_i)\geq \mu_{\overline{A}}^{\leq}(x_i)$. It follows that, $| D_i \cap [x_i]_{\overline{B}}^{\leq} |>| D_i \cap [x_i]_{\overline{A}}^{\leq} |$, which implies $D_i\cap[x_i]_{\overline{B}}^{\leq} \supset D_i\cap[x_i]_{\overline{A}}^{\leq}$, so there exist $x_j\in D_i\cap[x_i]_{\overline{B}}^{\leq}$ and $x_j\notin D_i\cap[x_i]_{\overline{A}}^{\leq}$. That is, $x_j\in D_i$, $x_j\in[x_i]_{\overline{B}}^{\leq}$, and $x_j\notin[x_i]_{\overline{A}}^{\leq}$.

Therefore, $\forall b\in B$, $f_b(x_i)\leq f_b(x_j)$. However, according to the conditions of this theorem, when $\mu_{\overline{A}}^{\leq}(x_i) < \mu_{\overline{A}}^{\leq}(x_j)$, $D_i \supseteq D_j$, there exists $b\in B$ such that $f_b(x_i) > f_b(x_j)$, which is in contradiction with $x_j\in[x_i]_{\overline{B}}^{\leq}$.

Thus, B is a generalized distribution consistent set.

4 Discernibility Matrix-Based Approach of Generalized Distribution Reductions

This section provides a convenient computation approach of generalized distribution reduction for inconsistent decision systems based on dominance relations.

Definition 7. Consider $DS = (U, A, F, D, G)$, denote

$$D^*{}_\mu = \{(x_i, x_j) \mid \mu_{\overline{A}}^{\leq}(x_i) < \mu_{\overline{A}}^{\leq}(x_j),\ D_i \supseteq D_j\}$$

$$D_\mu(x_i, x_j) = \{a_k \in A, f_{a_k}(x_i) > f_{b_k}(x_j), (x_i, x_j) \in D^*{}_\mu; \emptyset, (x_i, x_j) \notin D^*{}_\mu.$$

where $D_\mu(x_i, x_j)$ is called the generalized distribution discernibility attribute set of x_i and x_j.

And $M_\mu = (D_\mu(x_i, x_j), x_i, x_j \in U)$ is referred as the generalized distribution discernibility matrix of DS.

Theorem 2. Let $DS = (U, A, F, D, G)$, $B \subseteq A$. B is a generalized distribution consistent set if only and only if for $\forall (x_i, x_j) \in D^*{}_\mu$, $B \cap D_\mu(x_i, x_j) \neq \emptyset$.

Proof : "$\Rightarrow$" From Definition 7, $\forall (x_i, x_j) \in D^*{}_\mu$, we have $\mu_{\overline{A}}^{\leq}(x_i) < \mu_{\overline{A}}^{\leq}(x_j)$, $D_i \supseteq D_j$. Since B is a generalized distribution consistent set, by Theorem 1, there exists $b \in B$ such that $f_b(x_i) > f_b(x_j)$. Therefore, $b \in D^*{}_\mu$. This means that, if B is generalized distribution consistent set, for $\forall (x_i, x_j) \in D^*{}_\mu$, we have $B \cap D_\mu(x_i, x_j) \neq \emptyset$.

"$\Leftarrow$" For $\forall (x_i, x_j) \in D^*{}_\mu$, $B \cap D_\mu(x_i, x_j) \neq \emptyset$. So $\exists a_k \in B$, $a_k \in D_\mu(x_i, x_j)$, we have $f_{a_k}(x_i) > f_{a_k}(x_j)$, and $\mu_{\overline{A}}^{\leq}(x_i) < \mu_{\overline{A}}^{\leq}(x_j)$, $D_i \supseteq D_j$. By Theorem 1, B is generalized distribution consistent set.

Definition 8. Let $DS = (U, A, F, D, G)$ and M_μ be distribution discernibility matrix of DS. F_μ is called generalized distribution discernibility function,

$$F_\mu = \wedge \{\vee \{a_k:\ a_k \in D_\mu(x_i, x_j)\},\ x_i, x_i \in U\} = \wedge \{\vee \{a_k:\ a_k \in D_\mu(x_i,\ x_j)\},\ x_i, x_i \in D^*{}_\mu\}.$$

Theorem 3. Denote the minimal disjunctive normal form of generalized distribution discernibility matrix by

$$F_\mu = \bigwedge_{k=1}^{p} \left(\bigvee_{s=1}^{q_k} a_k \right), a_k \in D_\mu(x_i, x_j), x_i, x_i \in U$$

and $B^k{}_\mu = \{a_s, s=1, 2, \cdots\}$. $B^k{}_\mu$ is then a generalized distribution reduction.

Proof : It can be directly obtained from Theorem 2 and the definition of minimal disjunctive normal of the generalized discernibility function.

Theorem 2 provides a convenient way to find generalized distribution reduction of inconsistent decision systems based on dominance relations. An example is given to illustrate the computation process of finding a reduction based on the concept of generalized distribution function.

Example 3. Consider the DS showed in Table 1. According to Definition 8, the generalized distribution discernibility matrix is given in Table 2.

Consequently, we have

$$F_\mu = a_3 \wedge (a_1 \vee a_2) = (a_1 \wedge a_3) \vee (a_2 \wedge a_3)$$

Therefore, by Theorem 3, we obtain that both $\{a_1, a_3\}$ and $\{a_2, a_3\}$ are generalized distribution reducts of DS in Table 1, which is consistent with the result of Example 2.

Table 2. Generalized distribution discernibility matrix

U	x_1	x_2	x_3	x_4	x_5	x_6
x_1	$\emptyset$	$\emptyset$	$\emptyset$	$\emptyset$	$\emptyset$	$\emptyset$
x_2	$\emptyset$	$\emptyset$	$\emptyset$	$\{a_1,a_2\}$	$\emptyset$	$\emptyset$
x_3	$\{a_3\}$	$\emptyset$	$\emptyset$	$\emptyset$	$\emptyset$	$\emptyset$
x_4	$\emptyset$	$\emptyset$	$\emptyset$	$\emptyset$	$\emptyset$	$\emptyset$
x_5	$\emptyset$	$\emptyset$	$\emptyset$	$\emptyset$	$\emptyset$	$\emptyset$
x_6	$\emptyset$	$\emptyset$	$\emptyset$	$\emptyset$	$\emptyset$	$\emptyset$

It should be noted that, by theorem 1 and definition 7, when computing the generalized distribution reductions, there are conditions to be satisfied, i.e., $\mu_{\bar{A}}^{\leq}(x_i) < \mu_{\bar{A}}^{\leq}(x_j)$, $D_i \supseteq D_j$. Therefore, Theorem 3 can not guarantee that all the reductions can be found.

5   Conclusion

In multi-criteria problems, there are often inconsistent decision systems with ordered attribute values. Based on dominance relations and incorporating Greco's dominance principle, we introduce a generalized distribution function which is simple in form and directly reflect the consistent degree of the given decision systems. Then two new types of distribution reductions are correspondingly defined, and their relationships are also discussed. Finally, the computation method based on generalized discernibility matrix is given for finding generalized distribution reductions. This provides an approach to knowledge reduction in inconsistent systems based on dominance relations.

Acknowledgments. This work is supported by NSFC (No. 60903088), Natural Science Foundation of Hebei Province (No. F2009000227), 100-Talent Programme of Hebei Province, and Natural Science Foundation of Education Department of Hebei Province (No. 2007105).

References

1. Pawlak, Z.: Theoretical aspects of reasoning about data. Kluwer Academic Publishers, Boston (1991)
2. Liu, Q., Liu, S.H., Zheng, F.: Rough Logic and its Applicationsin Data Reduction. Journal of Software 12, 415–419 (2001) (in Chinese)
3. Zhang, W.X., Mi, J.S., Wu, W.Z.: Approaches to Knowledge Reductions in Inconsistent Systems. Chinese Journal of Computers 26, 12–18 (2003) (in Chinese)
4. Slowinski, R. (ed.): Intelligent Decision Support: Handbook of Applications and Advances of the Rough sets Theory. Kluwer academic Publishers, Boston (1992)
5. Kryszkiewicz, M.: Rough Set Approach to Incomplete Information System. Information Sciences 112, 39–49 (1998)
6. Slezak, D.: Approximate Reducts in Decision Tables.Granada. In: Procedings of IMPU 1996, vol. 3, pp. 1159–1164 (1996)

7. Kryszkiewicz, M.: Comparative studies of alternative of knowledge reduction in inconsistent systems. Intelligent Systems 1, 105–120 (2001)
8. Greco, S., Matarazzo, B., Slowingski, R.: Rough approximation of a preference relation by dominance relation. European Journal of Operation Research 117, 63–83 (1999)
9. Shao, M.W., Zhang, W.X.: Dominance relation and rules in an incomplete ordered information system. International Journal of Intelligent Systems 20, 13–27 (2005) (in Chinese)
10. Xu, W.H., Zhang, W.X.: Distribution reduction in Inconsisten information systems based on dominance relations. Fuzzy Systems and Mathematics 21(4), 122–131 (2007) (in Chinese)
11. Gui, X.C., Peng, H.: Research on distribution reduction and maximum distribution reduction by dominance relations. Computer Engineering and Applications 45(2), 150–153 (2009) (in Chinese)
12. Chen, J., Wang, G.Y., Hu, J.: Positive Domain Reduction Based on Dominance Relation in Inconsistent System. Computer Science 35(3), 216–218, 217 (2008) (in Chinese)
13. Xu, W.H., Zhang, X.Y., Zhang, W.X.: Lower approximation reduction in inconsistent information systems based on dominance relations. Computer Engineering and Applications 45(16), 66–68 (2009) (in Chinese)

Towards Multi-adjoint Property-Oriented Concept Lattices

Jesús Medina*

Department of Mathematics, University of Cádiz
jesus.medina@uca.es

Abstract. In this paper we present some properties related to adjoint triples when we consider dual supports. These results are used in order to generalize the classical property oriented concept lattices, which itself embeds rough set theory. Specifically, we define a fuzzy environment based on the philosophy of the multi-adjoint paradigm, which is related to the multi-adjoint concept lattice. As a consequence, we can move the properties from one to another.

1 Introduction

Rough set theory was originally proposed by Pawlak [22] as a formal tool for modelling and processing incomplete information in information systems. This theory was extended by Düntsch and Gediga in [10, 11] in order to consider two different sets, the set of objects and the set of attributes. This extension is called *property-oriented concept lattice* [6].

The relation between fuzzy sets and rough sets has been studied and some fuzzy extensions of rough set theory and property-oriented concept lattice have been presented in order to represent and analyze both incomplete information and imprecise information. In [24] the authors introduce the $(\mathcal{I}, \mathcal{T})$-fuzzy rough sets, where $\mathcal{I}$ is a S-, R- or QL-implication satisfying some properties and $\mathcal{T}$ is a t-norm, which extend the fuzzy rough set framework given in [9]. This was embedded, in the residuated case, by the fuzzy framework of the property-oriented concept lattice presented in [12, 15].

On the other hand, formal concept analysis, introduced by Wille in [26], arise as other useful tool for qualitative data analysis, which has become an important and appealing research topic both from a theoretical perspective and from an applicative one.

Both formal concept analysis and rough set theory have been related [16, 18, 28, 29]. As a consequence, we can apply the results presented in a formal concept analysis framework to a rough set theory [6, 25].

A number of different fuzzy extensions of formal concept analysis have been presented. To the best of our knowledge, the first one was given in [4], although they did not advance much beyond the basic definitions, probably due to the fact that they did not use residuated implications. Later, in [2, 23] the authors independently used complete residuated lattices as structures for the truth degrees; for this approach, a representation theorem was proved directly in a fuzzy framework, setting the basis of most of the subsequent direct proofs.

* Partially supported by the Spanish Science Ministry under grant TIN2009-14562-C05-03 and by Junta de Andalucía under grant P09-FQM-5233.

J. Yu et al. (Eds.): RSKT 2010, LNAI 6401, pp. 159–166, 2010.

Multi-adjoint concept lattices were introduced [19] as a new general approach to formal concept analysis, in which the philosophy of the multi-adjoint paradigm to formal concept analysis was applied (see [13, 21] for more information), with the idea of providing a general framework in which the different approaches stated above could be conveniently accommodated. The authors worked in a general non-commutative environment; and this naturally led to the consideration of several adjoint triples, also called implication triples [1] as the main building blocks of a multi-adjoint concept lattice. As a consequence, different degrees of preference can be easily established on the set of objects or attributes.

In this paper, we present some properties about adjoint triples obtained when we consider dual lattices. These results are used in order to generalize the classical property-oriented concept lattices [6] to a fuzzy environment based on the philosophy of the multi-ajoint paradigm [20].

Moreover, these definitions generalize the fuzzy definitions given in [12, 15, 24], if we consider the classical equality; and we can apply the theory developed from one kind of concept lattice to another one. For instance, we could change the Fundamental Theorem of the multi-adjoint concept lattice [19], in order to obtain a Fundamental Theorem, to multi-adjoint property-oriented concept lattice.

2   Formal Concept Analysis: Derivation Operators

In classical formal concept analysis, we consider a set of **attributes** A, a set of **objects** B and a crisp relation between them $R \colon A \times B \to \{0, 1\}$, where, for each $a \in A$ and $b \in B$, we have that $R(a, b) = 1$, if a and b are related, or $R(a, b) = 0$, otherwise. We can write aRb when $R(a, b) = 1$. The triple (A, B, R) is called a *context* and the mappings $^{\triangle} \colon 2^B \to 2^A$, $^{\triangle} \colon 2^A \to 2^B$, are defined, for each $X \subseteq B$ and $Y \subseteq A$, as:

$$X^{\triangle} = \{a \in A \mid \text{for all } b \in X, aRb\} = \{a \in A \mid \text{if } x \in X, \text{ then } aRb\} \qquad (1)$$

$$Y^{\triangle} = \{b \in B \mid \text{for all } a \in Y, aRb\} = \{b \in B \mid \text{if } a \in Y, \text{ then } aRb\} \qquad (2)$$

A *concept* in the context (A, B, R) is defined as a pair (X, Y), where $X \subseteq B, Y \subseteq A$, and satisfies $X^{\triangle} = Y$ and $Y^{\triangle} = X$. The element X of the concept (X, Y) is the *extent* and Y the *intent*.

The set of concepts in a context (A, B, R) is denoted as $\mathcal{B}(A, B, R)$ and it is a complete lattice [7, 26], with the order: $(X_1, Y_1) \leq (X_2, Y_2)$ if $X_1 \subseteq X_2$ (or, equivalently, $Y_2 \subseteq Y_1$), for all $(X_1, Y_1), (X_2, Y_2) \in \mathcal{B}(A, B, R)$.

An important fact is that the extent and intent mappings form a Galois connection [7]. There are two dual versions of this definition. The one we adopt here is the most famous Galois connection of all, that which was discovered by Galois, where the maps are order-reversing, which will be properly called *Galois connection*, and the other in which the maps are order-preserving, will be called *isotone Galois connection*. There are arguments for both versions, although, at a theoretical level, the difference is not significant since we can pass from one to another substituting a lattice by its dual, for example, 2^B by its dual[1] $(2^B)^{\partial}$.

---

[1] The definition of dual set will be remembered later.

In order to make this contribution self-contained, we recall its formal definitions:

Let $(P_1, \leq_1)$ and $(P_2, \leq_2)$ be posets, and $\downarrow\colon P_1 \to P_2$, $\uparrow\colon P_2 \to P_1$ mappings, the pair $(\uparrow, \downarrow)$ forms a *Galois connection* between P_1 and P_2 if and only if: $\uparrow$ and $\downarrow$ are order-reversing; $x \leq_1 x^{\downarrow\uparrow}$, for all $x \in P_1$, and $y \leq_2 y^{\uparrow\downarrow}$, for all $y \in P_2$.

The dual definition is given as: Let $(P_1, \leq_1)$ and $(P_2, \leq_2)$ be posets, and $\downarrow\colon P_1 \to P_2$, $\uparrow\colon P_2 \to P_1$ mappings, the pair $(\uparrow, \downarrow)$ forms an *isotone Galois connection* between P_1 and P_2 if and only if: $\uparrow$ and $\downarrow$ are order-preserving; $x \leq_1 x^{\downarrow\uparrow}$, for all $x \in P_1$, and $y^{\uparrow\downarrow} \leq_2 y$, for all $y \in P_2$.

This last definition arises from the notion of residuated mappings; that is, the mappings of an isotone Galois connection are residuated mappings of each other.

Before continuing with the comments, we need to recall the definition of an opposite order. Given a set P and an order relation, $\leq$, on P, the *opposite* order (also *dual, inverse*, or *converse*, etc.) of $\leq$ is the relation $\leq^{op}$, defined as $x_1 \leq^{op} x_2$ if and only if $x_2 \leq x_1$, for all $x_1, x_2 \in P$. Usually, we will write P instead of the partially ordered set $(P, \leq)$, P^∂ instead of $(P, \leq^{op})$, and we will say that P^∂ is the *dual* of P.

Now, as we observed above, the definition of isotone Galois connection follows from the original one by considering P_2^∂ instead of P_2. Hence, an isotone Galois connection $(\uparrow, \downarrow)$ on P_1 and P_2 is a Galois connection on P_1 and P_2^∂, and we can transform the properties of Galois connections to isotone Galois connections.

A direct consequence of Galois connection definition is that $x^\downarrow = x^{\downarrow\uparrow\downarrow}$, $y^\uparrow = y^{\uparrow\downarrow\uparrow}$, for all $x \in P_1$ and $y \in P_2$. Hence, all concepts in $\mathcal{B}(A, B, R)$ are pairs $(X^{\triangle\triangle}, X^\triangle)$, where $X \subseteq B$, or $(Y^\triangle, Y^{\triangle\triangle})$, where $Y \subseteq A$, that is:

$$\mathcal{B}(A, B, R) = \{(X^{\triangle\triangle}, X^\triangle) \mid X \subseteq B\} = \{(Y^\triangle, Y^{\triangle\triangle}) \mid Y \subseteq A\}$$

The above definition of extent and intent operators is one of the modal-style operators [10,11]. There are three more definitions considered in several frameworks: qualitative data analysis [10,11], crisp rough set theory [29] and fuzzy rough set theory [5,17]. Some extra motivations about these operators are also introduced in [8,12,15,24,27].

Given the sets A, B, and a crisp relation $R\colon A \times B \to \{0, 1\}$, we have the mappings $^\pi\colon 2^B \to 2^A$, $^N\colon 2^B \to 2^A$, $^\nabla\colon 2^B \to 2^A$ defined, for each $X \subseteq B$, as:

$$X^\pi = \{a \in A \mid \text{ there is } b \in X, \text{ such that } aRb\} = \{a \in A \mid aR \cap X \neq \varnothing\} \quad (3)$$
$$X^N = \{a \in A \mid \text{ for all } b \in B, \text{ if } aRb, \text{ then } b \in X\} = \{a \in A \mid aR \subseteq X\} \quad (4)$$
$$X^\nabla = \{a \in A \mid \text{ there exists } b \in X^c, \text{ such that } aR^c b\} \quad (5)$$

where $aR = \{b \in B \mid aRb\}$, and X^c, R^c are the complement of X and the complement relation of R, respectively.

Analogously, we can define the mappings: $^\pi\colon 2^A \to 2^B$, $^N\colon 2^A \to 2^B$ and $^\nabla\colon 2^A \to 2^B$.

These operators are called *possibility, necessity* and *dual sufficiency operators*, respectively; the classical one is called *sufficient operator*. They are composed to form Galois connections and, hence, new concept lattices: *classical formal concept lattice, dual formal concept lattice, object-oriented concept lattice* and *property-oriented concept lattice* [6]; this last framework is a generalization of rough set theory [11].

Clearly, the dual sufficiency operator satisfies that: $X^\nabla = ((X^c)^\triangle)^c$, for each $X \subseteq B$, therefore these operators are not independent and the concept lattice given from them are related. Specifically, we can obtain one from the other. Moreover, the necessity and possibility operators are related to the sufficient operators. For details, see [15].

3 Adjoint Triples and Multi-adjoint Concept Lattices

Assuming non-commutativity on the conjunctor, directly provides two different ways of generalising the well-known adjoint property between a t-norm and its residuated implication, depending on which argument is fixed.

Definition 1. *Let* $(P_1, \leq_1)$, $(P_2, \leq_2)$, $(P_3, \leq_3)$ *be posets and* $\&\colon P_1 \times P_2 \to P_3$, $\swarrow\colon P_3 \times P_2 \to P_1$, $\nwarrow\colon P_3 \times P_1 \to P_2$ *be mappings, then* $(\&, \swarrow, \nwarrow)$ *is an* adjoint triple *with respect to* P_1, P_2, P_3 *if:*

1. $\&$ *is order-preserving in both arguments, and* $\swarrow$, $\nwarrow$ *are order-preserving on the first argument and order-reversing on the second argument.*
2. $x \leq_1 z \swarrow y$ *iff* $x \& y \leq_3 z$ *iff* $y \leq_2 z \nwarrow x$, *where* $x \in P_1$, $y \in P_2$ *and* $z \in P_3$.

Now, we introduce an interesting result about adjoint triples that we will use later. This result studies what happens if we change a lattice by its dual.

Lemma 1. *Given the posets* $(P_1, \leq_1)$, $(P_2, \leq_2)$, $(P_3, \leq_3)$ *and an adjoint triple with respect them* $(\&, \swarrow, \nwarrow)$, *we obtain that* $(\swarrow, \&, \nwarrow_{op})$ *is an adjoint triple with respect to* P_3^∂, P_2, P_1^∂.

Proof. (2). Since $(\&, \swarrow, \nwarrow)$ is an adjoint triple with respect to P_1, P_2 and P_3, we have that: $\&\colon P_1 \times P_2 \to P_3$, is increasing on both arguments, and $\swarrow\colon P_3 \times P_2 \to P_1$, $\nwarrow\colon P_3 \times P_1 \to P_2$, are increasing on the first argument and decreasing on the second. Hence, we conclude that $\swarrow\colon P_3^\partial \times P_2 \to P_1^\partial$, is increasing on both arguments, and $\&\colon P_1^\partial \times P_2 \to P_3^\partial$, $\nwarrow_{op}\colon P_1^\partial \times P_3^\partial \to L_2$, are increasing on the first argument and decreasing on the second. Furthermore, for all $x \in P_1$, $y \in P_2$ and $z \in P_3$, we have:

$$x \leq_1 z \swarrow y \quad \text{iff} \quad x \& y \leq_3 z \quad \text{iff} \quad y \leq_2 z \nwarrow x$$

which is equivalent to: $z \swarrow y \leq_1^{op} x$ iff $z \leq_3^{op} x \& y$ iff $y \leq_2 x \nwarrow_{op} z$ that is:

$$z \leq_3^{op} x \& y \quad \text{iff} \quad z \swarrow y \leq_1^{op} x \quad \text{iff} \quad y \leq_2 x \nwarrow_{op} z$$

which allows us to prove that $(\swarrow, \&, \nwarrow_{op})$ is an adjoint triple w.r.t. P_3^∂, P_2, P_1^∂.

There are many other possibilities, following a similar idea, in order to define new adjoint triples from another one, but we will only use one in this paper.

In the following definition we will present the basic structure which allows the existence of several adjoint triples for a given triplet of lattices.

A *multi-adjoint frame* $\mathcal{L}$ is a tuple

$$(L_1, L_2, P, \preceq_1, \preceq_2, \leq, \&_1, \swarrow^1, \nwarrow_1, \ldots, \&_n, \swarrow^n, \nwarrow_n)$$

where $(L_1, \preceq_1)$ and $(L_2, \preceq_2)$ are complete lattices, $(P, \leq)$ is a poset and, for all $i = 1, \ldots, n$, $(\&_i, \swarrow^i, \nwarrow_i)$ is an adjoint triple with respect to L_1, L_2, P.

Multi-adjoint frames are denoted as $(L_1, L_2, L, \&_1, \dots, \&_n)$.

Let $(L_1, L_2, P, \&_1, \dots, \&_n)$ be a multi-adjoint frame, a *context* is a tuple (A, B, R, σ) such that A and B are non-empty sets (usually interpreted as attributes and objects, respectively), R is a P-fuzzy relation $R \colon A \times B \to P$ and $\sigma \colon B \to \{1, \dots, n\}$ is a map which associates any element in B with some particular adjoint triple in the frame.[2]

Considering a multi-adjoint frame and a context for that frame, we can define the following mappings $\uparrow^\sigma \colon L_2^B \longrightarrow L_1^A$ and $\downarrow^\sigma \colon L_1^A \longrightarrow L_2^B$ which generalize the classical definitions (Eq. (1), (2)), and the given ones in [3, 14]:

$$g^{\uparrow_\sigma}(a) = \inf\{R(a, b) \swarrow^{\sigma(b)} g(b) \mid b \in B\} \tag{6}$$

$$f^{\downarrow^\sigma}(b) = \inf\{R(a, b) \nwarrow_{\sigma(b)} f(a) \mid a \in A\} \tag{7}$$

It is not difficult to show that these two arrows generate a Galois connection [19].

The notion of concept is defined as usual: a *multi-adjoint concept* is a pair $\langle g, f \rangle$ satisfying that $g \in L_2^B$, $f \in L_1^A$ and that $g^{\uparrow_\sigma} = f$ and $f^{\downarrow^\sigma} = g$; with $(\uparrow_\sigma, \downarrow^\sigma)$ being the Galois connection defined above.

Finally, the definition of concept lattice in this framework is defined [19].

Definition 2. *Let* $(L_1, L_2, P, \&_1, \dots, \&_n)$ *be a multi-adjoint frame and* (A, B, R, σ) *a context. So, a multi-adjoint concept lattice is the set*

$$\mathcal{M} = \{\langle g, f \rangle \mid g \in L_2^B, f \in L_1^A \text{ and } g^{\uparrow_\sigma} = f, f^{\downarrow^\sigma} = g\}$$

in which the ordering is defined by $\langle g_1, f_1 \rangle \preceq \langle g_2, f_2 \rangle$ *iff* $g_1 \preceq_2 g_2$ *(or* $f_2 \preceq_1 f_1$*).*

In [19], the authors proved that the ordering defined above provides $\mathcal{M}$ with the structure of a complete lattice. Furthermore, a representation theorem to multi-adjoint concept lattices is proved, which generalizes the classical one and some other fuzzy generalizations. The details can be seen in [19].

4   Multi-adjoint Property-Oriented Concept Lattices

In this section we will generalize the definitions of the necessity and possibility operators to a fuzzy environment, in a similar way that the fuzzy definitions of the sufficient operators were given.

From now on, a multi-adjoint frame $(L_1, L_2, P, \&_1, \dots, \&_n)$ and context (A, B, R, σ) will be fixed. Moreover, to improve readability, we will write $(\uparrow, \downarrow)$ instead of $(\uparrow_\sigma, \downarrow^\sigma)$ and $\swarrow^b$, $\nwarrow_b$ instead of $\swarrow^{\sigma(b)}$, $\nwarrow_{\sigma(b)}$. Now, we define: $\uparrow^N \colon L_2^B \to L_1^A$, $\uparrow^\pi \colon L_2^B \to L_1^A$:

$$g^{\uparrow^\pi}(a) = \sup\{R(a, b) \&_b g(b) \mid b \in B\} \tag{8}$$

$$g^{\uparrow^N}(a) = \inf\{g(b) \nwarrow_b R(a, b) \mid b \in B\} \tag{9}$$

Analogously, we can define the mappings: $\downarrow^N \colon L_2^A \to L_1^B$, $\downarrow^\pi \colon L_2^A \to L_1^B$:

$$f^{\downarrow^\pi}(b) = \sup\{R(a, b) \&_b f(a) \mid a \in A\} \tag{10}$$

$$f^{\downarrow^N}(b) = \inf\{f(a) \nwarrow_b R(a, b) \mid a \in A\} \tag{11}$$

---

[2] A similar theory could be developed by considering a mapping $\tau \colon A \to \{1, \dots, n\}$ which associates any element in A with some particular adjoint triple in the frame.

Clearly, these definitions are generalizations of the classical and fuzzy possibility and necessity operators [11, 12] and [15] with the usual equality. Additionally, the operators (8) and (9) can be combined with (10) and (11) in order to consider new concept lattices which generalize the given one in [6].

Now, we will prove that the mappings $^{\uparrow_\pi}: L_2^B \rightarrow L_1^A$, $^{\downarrow^N}: L_2^A \rightarrow L_1^B$ will allow us to obtain a fuzzy generalization of the concept lattice introduced in [6]. Moreover, the properties given in several paper, as in [6, 11], can be presented in this general framework.

First, as usual, we introduce the notion of concept, in this environment, as a pair of mappings $\langle g, f \rangle$, with $g \in L^B$, $f \in L^A$, such that $g^{\uparrow_\pi} = f$ and $f^{\downarrow^N} = g$, which will be called *multi-adjoint property-oriented concept*.

Definition 3. *A multi-adjoint property-oriented concept lattice is the set*

$$\mathcal{M}_{\pi N} = \{\langle g, f \rangle \mid g \in L_2^B, f \in L_1^A \text{ and } g^{\uparrow_\pi} = f, f^{\downarrow^N} = g\}$$

in which the ordering is defined by $\langle g_1, f_1 \rangle \preceq \langle g_2, f_2 \rangle$ *iff* $g_1 \preceq_2 g_2$ *(or* $f_2 \preceq_1 f_1$*).*

As a consequence of the following result, $(M_{\pi N}, \preceq)$ is a complete lattice.

Lemma 2. *If the posets* L_1^∂, L_2 *and* P^∂ *are considered instead of* L_1, L_2 *and* P, *then we conclude that the pair* $(^{\uparrow_\pi}, {}^{\downarrow^N})$ *is equal to the pair* $(^\uparrow, {}^\downarrow)$ *given by Eqs. (6) and (7).*

Proof. From Lemma 1 we obtain, for each adjoint triple $(\&_i, \nearrow^i, \nwarrow_i)$ with respect to L_1, L_2 and P, the triple $(\nearrow^i, \&_i, \nwarrow_{i,op})$ is an adjoint triple with respect to L_1^∂, L_2 and P^∂, where $\nwarrow_{i,op}$ is the opposite operator of $\nwarrow_i$. Hence, the mappings $(^{\uparrow_\pi}, {}^{\downarrow^N})$ are defined, for all $a \in A$ and $b \in B$, as:

$$g^{\uparrow_\pi}(a) = \sup\{R(a, b) \,\&_b\, g(b) \mid b \in B\} = g^\uparrow(a)$$
$$f^{\downarrow^N}(b) = \inf\{f(a) \nwarrow_b R(a, b) \mid a \in A\} = \inf\{R(a, b) \nwarrow_{b,op} f(a) \mid a \in A\} = f^\downarrow(b)$$

the equalities are given since $\&$ and $\nwarrow_{b,op}$ are the second and third operators of the adjoint triple, respectively.

Theorem 1. *The multi-adjoint property-oriented concept lattice* $(M_{\pi N}, \leq)$ *is, indeed, a complete lattice where*

$$\inf\{\langle g_i, f_i \rangle \mid i \in I\} = \langle \inf_2\{g_i \mid i \in I\}, (\inf_1\{f_i \mid i \in I\})^{\downarrow^N \uparrow_\pi}\rangle$$
$$\sup\{\langle g_i, f_i \rangle \mid i \in I\} = \langle (\sup_2\{g_i \mid i \in I\})^{\uparrow_\pi \downarrow^N}, \sup_1\{f_i \mid i \in I\}\rangle$$

Proof. From Lemma 2 we understand that $(^{\uparrow_\pi}, {}^{\downarrow^N})$ is a Galois connection on L_1^∂ and L_2, hence, applying the result given in [19], we have the set $M_{\pi N}$ form a complete lattice with the order $\leq^{op}$ defined as: $\langle g_1, f_1 \rangle \leq^{op} \langle g_2, f_2 \rangle$ if and only if $g_1 \preceq_2 g_2$, or equivalently, if and only if $f_2 \preceq_1^{op} f_1$, where:

$$\inf\{\langle g_i, f_i \rangle \mid i \in I\} = \langle \inf_2\{g_i \mid i \in I\}, (\sup_{1,op}\{f_i \mid i \in I\})^{\downarrow^N \uparrow_\pi}\rangle$$
$$\sup\{\langle g_i, f_i \rangle \mid i \in I\} = \langle (\sup_2\{g_i \mid i \in I\})^{\uparrow_\pi \downarrow^N}, \inf_{1,op}\{f_i \mid i \in I\}\rangle$$

such that $\sup_{1,op}$ and $\inf_{1,op}$ are the supremum and infimum on L_1^∂, respectively. Thus, by changing L_1 by L_1^∂ we obtain the result.

As a consequence of Lemma 2, returning to the original lattices, we also conclude that the composition mapping $\uparrow_\pi \downarrow^N : L_2 \to L_2$ is a closure operator and $\downarrow^N \uparrow_\pi : L_1 \to L_1$ is an interior operator, which is very important to obtain the elements of $(M_{\pi N}, \leq)$.

Similarly, we may introduce the multi-adjoint object-oriented concept lattice, which generalizes the classical one. Furthermore, some generalization of the dual sufficient operator could define a new concept lattice. These new frameworks will be studied as future work.

5   Conclusions and Future Work

We have generalized the classical property oriented concept lattices [6], which itself embeds rough set theory [10]. In order to have done that, we have used the philosophy of the multi-adjoint paradigm [13, 21] to formal concept analysis. With the idea of providing a general framework in which a general non-commutative environment may be considered; and this naturally leads to the consideration of several adjoint triples defined on non-linear sets, which could allow us to easily establish different degrees of preference on the set of objects or attributes.

To check that we have first proved some properties about adjoint triples, and, as a consequence, we relate the new lattices with the multi-adjoint concept lattice [19], which allow us to translate the properties from one to the other.

As future work, we will present useful examples where we can check the potential of multi-adjoint property-oriented concept lattices. Note that, in the multi-adjoint concept lattices, considering several adjoint triples allows us to obtain preference among objects or attributes. This fact is very important but could not be included by the lack of space. Additionally, we will obtain some important properties to the new framework from the already given one.

References

1. Abdel-Hamid, A., Morsi, N.: Associatively tied implicacions. Fuzzy Sets and Systems 136(3), 291–311 (2003)
2. Bělohlávek, R.: Fuzzy concepts and conceptual structures: induced similarities. In: Joint Conference on Information Sciences, pp. 179–182 (1998)
3. Bělohlávek, R.: Concept lattices and order in fuzzy logic. Annals of Pure and Applied Logic 128, 277–298 (2004)
4. Burusco, A., Fuentes-González, R.: The study of L-fuzzy concept lattice. Mathware & Soft Computing 3, 209–218 (1994)
5. Chen, X., Li, Q.: Construction of rough approximations in fuzzy setting. Fuzzy Sets and Systems 158(23), 2641–2653 (2007)
6. Chen, Y., Yao, Y.: A multiview approach for intelligent data analysis based on data operators. Information Sciences 178(1), 1–20 (2008)
7. Davey, B., Priestley, H.: Introduction to Lattices and Order, 2nd edn. Cambridge University Press, Cambridge (2002)
8. Dubois, D., de Saint-Cyr, F.D., Prade, H.: A possibility-theoretic view of formal concept analysis. Fundamenta Informaticae 75(1-4), 195–213 (2007)
9. Dubois, D., Prade, H.: Putting fuzzy sets and rough sets together. In: Slowiński, R. (ed.) Intelligent Decision Support, pp. 203–232. Kluwer Academic, Dordrecht (2004)

10. Düntsch, I., Gediga, G.: Approximation operators in qualitative data analysis. In: de Swart, H., Orłowska, E., Schmidt, G., Roubens, M. (eds.) Theory and Applications of Relational Structures as Knowledge Instruments. LNCS, vol. 2929, pp. 214–230. Springer, Heidelberg (2003)

11. Gediga, G., Düntsch, I.: Modal-style operators in qualitative data analysis. In: Proc. IEEE Int. Conf. on Data Mining, pp. 155–162 (2002)

12. Georgescu, G., Popescu, A.: Non-dual fuzzy connections. Arch. Math. Log. 43(8), 1009–1039 (2004)

13. Julian, P., Moreno, G., Penabad, J.: On fuzzy unfolding: A multi-adjoint approach. Fuzzy Sets and Systems 154(1), 16–33 (2005)

14. Krajci, S.: A generalized concept lattice. Logic Journal of IGPL 13(5), 543–550 (2005)

15. Lai, H., Zhang, D.: Concept lattices of fuzzy contexts: Formal concept analysis vs. rough set theory. International Journal of Approximate Reasoning 50(5), 695–707 (2009)

16. Lei, Y., Luo, M.: Rough concept lattices and domains. Annals of Pure and Applied Logic 159(3), 333–340 (2009)

17. Liu, G.L.: Construction of rough approximations in fuzzy setting. Information Sciences 178(6), 1651–1662 (2008)

18. Liu, M., Shao, M., Zhang, W., Wu, C.: Reduction method for concept lattices based on rough set theory and its application. Computers & Mathematics with Applications 53(9), 1390–1410 (2007)

19. Medina, J., Ojeda-Aciego, M., Ruiz-Calviño, J.: Formal concept analysis via multi-adjoint concept lattices. Fuzzy Sets and Systems 160(2), 130–144 (2009)

20. Medina, J., Ojeda-Aciego, M., Vojtáš, P.: Multi-adjoint logic programming with continuous semantics. In: Eiter, T., Faber, W., Truszczyński, M. (eds.) LPNMR 2001. LNCS (LNAI), vol. 2173, pp. 351–364. Springer, Heidelberg (2001)

21. Medina, J., Ojeda-Aciego, M., Vojtáš., P.: Similarity-based unification: a multi-adjoint approach. Fuzzy Sets and Systems 146, 43–62 (2004)

22. Pawlak, Z.: Rough sets. International Journal of Computer and Information Science 11, 341–356 (1982)

23. Pollandt, S.: Fuzzy Begriffe. Springer, Berlin (1997)

24. Radzikowska, A.M., Kerre, E.E.: A comparative study of fuzzy rough sets. Fuzzy Sets and Systems 126(2), 137–155 (2002)

25. Wang, L., Liu, X.: Concept analysis via rough set and afs algebra. Information Sciences 178(21), 4125–4137 (2008)

26. Wille, R.: Restructuring lattice theory: an approach based on hierarchies of concepts. In: Rival, I. (ed.) Ordered Sets, pp. 445–470. Reidel (1982)

27. Yao, Y.: A comparative study of formal concept analysis and rough set theory in data analysis. In: Tsumoto, S., Słowiński, R., Komorowski, J., Grzymała-Busse, J.W. (eds.) RSCTC 2004. LNCS (LNAI), vol. 3066, pp. 59–68. Springer, Heidelberg (2004)

28. Yao, Y.Y.: Concept lattices in rough set theory. In: Proceedings of Annual Meeting of the North American Fuzzy Information Processing Society (NAFIPS 2004), pp. 796–801 (2004)

29. Yao, Y.Y., Chen, Y.: Rough set approximations in formal concept analysis. In: Peters, J.F., Skowron, A. (eds.) Transactions on Rough Sets V. LNCS, vol. 4100, pp. 285–305. Springer, Heidelberg (2006)

Extension of Covering Approximation Space and Its Application in Attribute Reduction*

Guoyin Wang[1] and Jun Hu[1,2]

[1] Institute of Computer Science and Technology,
Chongqing University of Posts and Telecommunications,
Chongqing, 400065, P.R. China
[2] School of Electronic Engineering, XiDian University,
Xi'an, Shaanxi, 710071, P.R. China
{hujun,wanggy}@cqupt.edu.cn

Abstract. The concept of the complement of a covering is introduced firstly, and then the complement space and extended space of a covering approximation space is defined based on it. It is proved that a covering approximation space will generate the same covering lower and upper approximations as its complement space and extended space if the covering is degenerated to a partition. Moreover, the extended space of a covering approximation space often generate a bigger covering lower approximation or smaller covering upper approximation than itself. Through extending each covering in a covering decision system, the classification ability of each covering is improved. Thus, a heuristic reduction algorithm is developed to eliminate some coverings in a covering decision system without decreasing the classification ability of the system for decision. Theoretic analysis and example illustration indicate that this algorithm can get shorter reduction than other algorithms.

Keywords: covering, rough set, covering decision system, attribute reduction.

1 Introduction

Rough set theory was proposed by Pawlak in 1982 to deal with uncertainty, incompleteness, and vagueness[1]. Because of its advantage of not depending on prior knowledge, it has achieved great success in many fields in the past years, e.g., data mining, machine leaning and pattern recognition, etc. The classical rough set theory is based on an indiscernibility relation or a partition on a universe, which is too restricted to apply it in many real problems. To address this issue, several interesting and meaningful extensions of rough set model have

---

* This paper is supported by the National Natural Science Foundation of P. R. China (No.60773113), the Science & Technology Research Program of Chongqing Education Committee of P. R. China (No.KJ090512), the Natural Science Foundation of Chongqing of P. R. China (No.2008BA2017), the Science Fund for Distinguished Young Scholars of Chongqing of P. R. China (No.2008BA2041).

J. Yu et al. (Eds.): RSKT 2010, LNAI 6401, pp. 167–174, 2010.

been proposed, such as tolerance relation based rough set model[2], similarity relation based rough set model[3], limited tolerance relation based rough set model[4], and others[5].

Particularly, Zakowski used covering of a universe to establish an extension called covering generalized rough set model[6]. Since then, covering generalized rough set theory has wan the wide attention. In view of the definition of the lower and upper approximation of a set in a covering approximation space, several different approximation operators have been developed[7][8][9][10], and the relationship among them has been studied[11]. Zhu studied the problem in which case two different coverings generate the same covering lower and upper approximations, and proposed the concept of reductions of coverings, which provides a method to get rid of redundancy in a covering approximation space[12]. For a given decision, Hu found that a covering can be reduced on the condition of not decreasing the classification ability of it[13]. Rough entropy and fuzziness were proposed by Huang and Xu respectively for the uncertainty measure of rough set induced by a covering[14][15].

Attribute reduction is an important application of rough set theory. In the past, researchers have developed many attribute reduction algorithms from the perspective of algebra and information[16], but all of them can not be used to deal with covering decision systems. In order to solve this problem, Chen applied covering generalized rough set theory to it, and developed a reduction method based on discernibility matrix for consistent and inconsistent covering decision systems respectively[17]. Based on information theory, another approach for reduction of covering decision system was studied by Li[18]. The experiment results show that covering generalized rough set theory has advantage in processing covering decision systems.

The key idea of reduction of covering decision system is to eliminate some of coverings from a covering decision system but not reduce the classification ability of it. Generally, in order to simplify the system or reduce the cost of computation, we wish the reduction to be as short as possible. Can the reductions gotten by the existing methods be further reduced? How to get shorter reduction than those gotten by the existing methods? We will study these problems in this paper. The remaining parts of this paper are arranged as follows. In section 2, some basic concepts about covering generalized rough sets are introduced. Section 3 proposes the concepts of complement space and extended space of a covering approximation space, and their relationship on approximation ability is analyzed. Section 4 develops a heuristic reduction algorithm of a covering decision system. Section 5 uses an example to illustrate the idea of the algorithm developed in section 4. This paper concludes in section 6.

2   Preliminaries

For the convenience of discussion, some relevant concepts of covering generalized rough sets will be introduced in this section[19].

Definition 1. *Let U be a universe of discourse, C be a covering of U. We call the ordered pair (U, C) a covering approximation space, or covering space for short.*

Definition 2. *Let (U, C) be a covering approximation space, x be an object of U, then $C_x = \bigcap\{K \in C | x \in K\}$ is called the neighbor of x in C.*

Definition 3. *Let (U, C) be a covering approximation space, for a subset X of U, the covering generalized rough set $C(X) = (\underline{C}(X), \overline{C}(X))$ is defined as follows:*

$$\underline{C}(X) = \{x \in U | C_x \subseteq X\} \tag{1}$$

$$\overline{C}(X) = \{x \in U | C_x \cap X \neq \emptyset\} \tag{2}$$

where $\underline{C}(X)$ is the covering lower approximation, and $\overline{C}(X)$ is the covering upper approximation. If $\underline{C}(X) = \overline{C}(X)$, then X is called exact in C. Otherwise, X is called inexact in C.

It can be proofed that the covering lower and upper approximations will degenerate to Pawlak's lower and upper approximations respectively if the covering is degenerated to a partition. That is, Pawlak's rough set is a special case of covering generalized rough set.

3   Extension of Covering Approximation Space

At first, we introduce the concept of the complement of a covering. And then, the complement space and extended space of a covering approximation space are defined based on it.

Definition 4. *Let U be a universe of discourse, C be a covering of U, then $C^\sim = \{\sim K | K \in C\}$ is called the complement of C.*

In real problems, if an element of a covering includes all the objects in a universe, it is clear that this kind of element has no use for problem solving. Therefore, we will not discuss coverings with this kind of element in this paper.

Proposition 1. *Let U be a universe of discourse, C be a covering of U, then $C^\sim$ is a covering of U if and only if $\cap\{K | K \in C\} = \emptyset$.*

Definition 5. *Let (U, C) be a covering approximation space. If $\cap\{K | K \in C\} = \emptyset$, then the ordered pair $(U, C^\sim)$ is called the complement space of (U, C).*

Proposition 2. *Let (U, C) be a covering approximation space, X be a subset of U. If C is degenerated to a partition, then $\underline{C}(X) = \underline{C^\sim}(X)$ and $\overline{C}(X) = \overline{C^\sim}(X)$.*

Proof. Let $C = \{K_1, K_2, ..., K_n\}$, then $C^\sim = \{\sim K_1, \sim K_2, ..., \sim K_n\}$. Suppose $x \in K_i$, then $x \in \sim K_j, j \neq i$. Hence,

$$C_x = K_i, \ (C^\sim)_x = \cap_{1 \leq j \leq i \lor i \leq j \leq n} \sim K_j.$$

Since $K_i = \sim (\cup_{1 \leq j \leq i \lor i \leq j \leq n} K_j) = \cap_{1 \leq j \leq i \lor i \leq j \leq n} \sim K_j$, we have $C_x = (C^\sim)_x$. Thus, by definition 3, we have $\underline{C}(X) = \underline{C^\sim}(X)$ and $\overline{C}(X) = \overline{C^\sim}(X)$ for $X \subseteq U$.   □

Proposition 2 indicates that a covering approximation space and its complement space generate the same covering lower and upper approximations. That is, they are equal in approximation ability.

Definition 6. *Let (U, C) be a covering approximation space. The ordered pair $(U, C^\sharp)$ is called the extended space of (U, C), where $C^\sharp = C \cup C^\sim$ is the extension of C.*

Proposition 3. *Let (U, C) be a covering approximation space, X be a subset of U. If C is degenerated to a partition, then $\underline{C}(X) = \underline{C^\sharp}(X)$ and $\overline{C}(X) = \overline{C^\sharp}(X)$.*

Proposition 3 tells us that a covering approximation space and its extended space generate the same covering lower and upper approximations. That is, the extended space of a covering approximation space have the same approximation ability as itself.

Proposition 4. *Let (U, C) be a covering approximation space, X be a subset of U. If C is degenerated to a partition, then $\underline{C^\sim}(X) = \underline{C^\sharp}(X)$ and $\overline{C^\sim}(X) = \overline{C^\sharp}(X)$.*

We can conclude that a covering approximation space, its complement space and its extended space are equal in approximation ability if the covering is degenerated to a partition.

Proposition 5. *Let (U, C) be a covering approximation space, X be a subset of U, then $\underline{C}(X) \subseteq \underline{C^\sharp}(X)$ and $\overline{C}(X) \supseteq \overline{C^\sharp}(X)$.*

Proof. Since $C^\sharp = C \cup C^\sim$, then $C_x \supseteq (C^\sharp)_x$. For $X \subseteq U$,

If $x \in \underline{C}(X)$, then $C_x \subseteq X$. By $C_x \supseteq (C^\sharp)_x$, we have $(C^\sharp)_x \subseteq X$, and then $x \in \underline{C^\sharp}(X)$. Therefore, $\underline{C}(X) \subseteq \underline{C^\sharp}(X)$.

If $x \in \overline{C^\sharp}(X)$, then $(C^\sharp)_x \cap X \neq \emptyset$. By $C_x \supseteq (C^\sharp)_x$, we have $C_x \cap X \neq \emptyset$, and then $x \in \overline{C}(X)$. Therefore, $\overline{C}(X) \supseteq \overline{C^\sharp}(X)$. $\square$

By Proposition 5, we know that the extended space of a covering approximation space generate more exact approximations than itself generally.

4   A Heuristic Reduction Algorithm of Covering Decision System

A covering decision system can be denoted as an ordered triple (U, Δ, D), where U is a universe of discourse, Δ is a family of coverings of U, D is a decision. For $x \in U$, the neighbor of x in Δ is defined as follows.

$$\Delta_x = \cap\{(C_i)_x | C_i \in \Delta\}. \tag{3}$$

Thus, the covering lower and upper approximations of $d_i \in D$ is defined as follows[17].

$$\underline{\Delta}(d_i) = \{x | \Delta_x \subseteq d_i\} \tag{4}$$

$$\overline{\Delta}(d_i) = \{x | \Delta_x \cap d_i \neq \emptyset\} \tag{5}$$

And then, the positive region of D with respect to Δ is defined as follows[17].

$$POS_\Delta(D) = \bigcup_{d_i \in D} \underline{\Delta}(d_i) \tag{6}$$

The positive region of D with respect to Δ is made up of the objects which can be classified into D definitely by Δ. If $POS_\Delta(D) = U$, we say that D is consistent in Δ. Otherwise, we say that D is inconsistent in Δ.

Proposition 6. *Let (U, Δ, D) be a covering decision system, C_i be a covering of Δ, then $POS_\Delta(D) \supseteq POS_{\Delta - \{C_i\}}(D)$.*

Proposition 6 indicates that the more covering a covering decision system has, the stronger approximation ability it has, whereas the much it will cost in computation. In order to reduce the complexity of computation, we wish to eliminate some coverings from a covering decision system but not reduce its classification ability. That is, the reduction of covering decision system is to get the same classification ability as all coverings with part of them.

For this purpose, Chen and Li, based on discernibility matrix and information theory respectively, proposed two reduction methods[17][18]. Generally, we wish the reduction to be as short as possible. How to get shorter reduction than those gotten by Chen's and Li's methods? We will try to address this problem in this section.

Definition 7. *Let $\Delta = \{C_1, C_2, ..., C_m\}$ be a family of coverings of a universe, then $\Delta_E = \{(C_1)^\sharp, (C_2)^\sharp, ..., (C_m)^\sharp\}$ is called the extension of Δ.*

Proposition 7. *Let (U, Δ, D) be a covering decision system, then $POS_\Delta(D) \subseteq POS_{\Delta_E}(D)$.*

Proposition 7 tells us that the classification ability of a covering decision system maybe improve by extending each covering in it.

Definition 8. *Let (U, Δ, D) be a covering decision system. For a covering C_i in Δ, if $POS_{\Delta_E}(D) = POS_{(\Delta - \{C_i\})_E}(D)$, then C_i is reducible in Δ with respect to D. Otherwise, C_i is irreducible in Δ with respect to D. Moreover, all coverings, which are irreducible in Δ with respect to D, make up of the core of Δ with respect to D, denoted as $core_D(\Delta)$.*

Definition 9. *Let (U, Δ, D) be a covering decision system, Θ be a subset of Δ. If both of the following conditions are satisfied, then Θ is a reduction of Δ with respect to D.*

(1) $\forall_{C_i \in \Theta}(POS_{\Theta_E}(D) \neq POS_{(\Theta - \{C_i\})_E}(D))$,
(2) $POS_{\Theta_E}(D) = POS_{\Delta_E}(D)$.

Definition 10. *Let (U, Δ, D) be a covering decision system. For a covering C_i in Δ, its significance with respect to Δ is defined as follows.*

$$Sig_\Delta(C_i) = \gamma_\Delta(D) - \gamma_{\Delta - \{C_i\}}(D) \tag{7}$$

where $\gamma_\Delta(D) = |POS_{\Delta_E}(D)|/|U|$ is the approximation classification quality of Δ with respect to D.

Proposition 8. *Let (U, Δ, D) be a covering decision system. For a covering C_i in Δ, C_i is irreducible in Δ with respect to D if and only if $Sig_\Delta(C_i) > 0$.*

Proposition 9. *Let (U, Δ, D) be a covering decision system, then $core_D(\Delta) = \{C_i \in \Delta | Sig_\Delta(C_i) > 0\}$.*

Algorithm 1. A heuristic reduction of covering decision system

(1) let $core_D(\Delta) = \emptyset$
(2) for each $C_i \in \Delta$
(3)   calculate $Sig_\Delta(C_i)$
(4)   if $Sig_\Delta(C_i) > 0$ then
(5) $core_D(\Delta) = core_D(\Delta) \cup \{C_i\}$
(6)   end if
(7) end for
(8) if $POS_{core_D(\Delta)}(D) = POS_\Delta(D)$ then
(9)   return $core_D(\Delta)$
(10) else
(11)   let $red = core_D(\Delta)$
(12)   for each $C_j \in \{\Delta - red\}$
(13)   calculate $Sig_{red \cup \{C_j\}}(C_j)$
(14)   end for
(15) $Sig_{red \cup \{C_k\}}(C_k) = max_{C_j \in \{\Delta - red\}} Sig_{red \cup \{C_j\}}(C_j)$
(16) $red = red \cup C_k$
(17)   if $POS_{red}(D) = POS_\Delta(D)$ then
(18)   return red
(19)   else
(20)   goto (12)
(21)   end if
(22) end if

Let $|\Delta| = m$ and $|U| = n$, then the time complexity of Algorithm 1 is $O(m^3 n^2)$.

5   Example Illustration

In this section, we will use the example, which has been used in [17], to illustrate our idea presented in the previous sections.

Suppose $U = \{x_1, x_2, ..., x_{10}\}$ to be a set of ten houses, and $E = \{price, structure, color, surrounding\}$ to be a set of attributes. According to the evaluation results on E given by four specialists, we can get a family of four coverings $\Delta = \{C_1, C_2, C_3, C_4\}$ of U. They are as follows:

$C_1 = \{\{x_1, x_2, x_3, x_4, x_6, x_7, x_8, x_9, x_{10}\}, \{x_3, x_4, x_6, x_7\}, \{x_3, x_4, x_5, x_6, x_7\}\}$
$C_2 = \{\{x_1, x_2, x_3, x_4, x_5, x_6, x_7\}, \{x_6, x_7, x_8, x_9\}, \{x_{10}\}\}$
$C_3 = \{\{x_1, x_2, x_3, x_6, x_8, x_9, x_{10}\}, \{x_2, x_3, x_4, x_5, x_6, x_7, x_9\}\}$
$C_4 = \{\{x_1, x_2, x_3, x_6\}, \{x_2, x_3, x_4, x_5, x_6, x_7\}, \{x_6, x_8, x_9, x_{10}\}, \{x_6, x_7, x_9\}\}$

where C_1, C_2, C_3, and C_4 are induced by *price, structure, color, and surrounding* respectively. Suppose the final decision D includes *sale, further evaluation* and *reject*, which partition the universe into three classes:

$D = \{\{x_1, x_2, x_3, x_6\}, \{x_4, x_5, x_7\}, \{x_8, x_9, x_{10}\}\}$

Thus, the positive region of D with respect to Δ_E is $POS_{\Delta_E}(D) = U$. That is, the decision D is consistent in Δ_E.

The significance of each covering in Δ with respect to D is:

$Sig_\Delta(C_1) = Sig_\Delta(C_2) = Sig_\Delta(C_3) = Sig_\Delta(C_4) = 0.$

Therefore, $core_D(\Delta) = \emptyset$. Let $red = core_D(\Delta) = \emptyset$, then

$Sig_{red\cup\{C_1\}}(C_1) = 1/10,\ Sig_{red\cup\{C_2\}}(C_2) = 3/10,$
$Sig_{red\cup\{C_3\}}(C_3) = 3/10,\ Sig_{red\cup\{C_4\}}(C_4) = 1.$

Select C_4 with the largest significance, and let

$red = red \cup \{C_4\} = \{C_4\}.$

Because $POS_{red_E}(D) = POS_{\Delta_E} = U$, we get a reduction $\{C_4\}$. Comparing with the reduction $\{\{C_2, C_4\}, \{C_2, C_3\}\}$ gotten by the existing methods, the reduction gotten by our method needs less attributes. Moreover, our method can get a consistent decision, while the existing methods can not.

6 Conclusions

By defining the complement of a covering, we defined the concepts of complement space and extended space. It is found that the complement space and extended space of a covering approximation space will generate the same covering lower and upper approximation as itself if the covering is degenerated to a partition. Moreover, the extended space of a covering approximation space generate bigger covering lower approximation or smaller covering upper approximation than it in general. That is, the approximation ability of a covering approximation space maybe improve by extending. Through extending each covering in covering decision systems, the classification ability of each covering maybe improve. Based on the extended covering decision system, a heuristic reduction algorithm of covering decision system is developed. The example illustration shows that this algorithm can get shorter reduction than existing methods. It will be our future work to apply our method in more real problems.

References

1. Pawlak, Z.: Rough set. International Journal of Computer and Information Sciences 11, 341–356 (1982)
2. Kryszkiewicz, M.: Rough set approach to incomplete information systems. Information Science 112, 39–49 (1998)
3. Slowinski, R., Vsnderpooten, D.: A generalized definition of rough approximations based on similarity. IEEE Tansactions on Knowledge and Data Engineering 12, 331–326 (2000)
4. Wang, G.Y.: Extension of rough set under incomplete information systems. Journal of Computer Research and Development 39, 1238–1243 (2002)
5. Zhu, W.: Generalized rough sets based on relations. Information Sciences 177, 4997–5011 (2007)
6. Zakowski, W.: Approximation in the space (U, Π). Demonstratio Mathematica 16, 761–769 (1983)
7. Bonikowski, Z., Bryniarski, E., Wybraniec, U.: Extensions and intentions in the rough set theory. Information Sciences 107, 149–167 (1998)
8. Tsang, E.C.C., Chen, D.G., Lee, J.W.T., Yeung, D.S.: On the upper approximations of covering generalized rough sets. In: 3rd International Conference Machine Learning and Cybermetics, Shanghai, China, pp. 4200–4203 (2004)
9. Zhu, W., Wang, F.Y.: A new type of covering rough set. In: 3rd International IEEE Conference Intelligent Systems, London, pp. 444–449 (2006)
10. Zhu, W.: Topological approaches to covering rough sets. Information Sciences 177, 1499–1508 (2007)
11. Zhu, W., Wang, F.Y.: On Three Types of Covering-Based Rough Sets. IEEE Transactions on Knowledge and Data Engineering 19, 1131–1144 (2007)
12. Zhu, W., Wang, F.Y.: Reduction and axiomization of covering generalized rough sets. Information Sciences 152, 217–230 (2003)
13. Hu, J., Wang, G.Y.: knowledge reduction of covering approximation space. Transction on Computer Science 5540, 69–80 (2009)
14. Huang, B., He, X., Zhou, X.Z.: Rough entropy based on generalized rough sets covering reduction. Journal of Software 15, 215–220 (2004)
15. Xu, W.H., Zhang, W.X.: Measuring roughness of generalized rough sets induced by a covering. Fuzzy sets and systems 158, 2443–2455 (2007)
16. Wang, G.Y.: Rough set theory and knowledge acquisition. Xi'an Jiaotong University Press, Xi'an (2001)
17. Chen, D.G., Wang, C.Z., Hu, Q.H.: A new approach to attribute reduction of consistent and inconsistent covering decision systems with covering rough sets. Information Sciences 177, 3500–3518 (2007)
18. Li, F., Yin, Y.Q.: Approaches to knowledge reduction of covering decision systems based on information theory. Information Sciences 179, 1694–1704 (2009)
19. Zhu, W.: Relationship between generalized rough sets based on binary relation and covering. Information Sciences 179, 210–225 (2009)

A New Extended Dominance Relation Approach Based on Probabilistic Rough Set Theory

Decui Liang[1], Simon X. Yang[2], Chaozhe Jiang[3], Xiangui Zheng[4], and Dun Liu[1]

[1] School of Economics and Management, Southwest Jiaotong University,
Cheng du 610031, China
decuiliang@126.com, newton83@163.com
[2] School of Engineering, University of Guelph, Ontario N1G 2W1, Canada
syang@uoguelph.ca
[3] College of Traffic and Transportaton, Southwest Jiaotong University,
Cheng du 610031, China
jiangchaozhe@163.com
[4] Sichuan Higher Institute of Cuisine, Cheng du 610031, China
zhengxiangui@yahoo.cn

Abstract. A new extended dominance relation is proposed for multi-attribute decision making problems with preference and incomplete information. First, the concept of new extended dominance relation based on the possibility of dominance relation between two objects is proposed. It substitutes the possibility of dominance relation for the conditions of limited extended dominance relation. In addition, it restricts that attribute values of two objects are missing in the same attribute. Through using probabilistic rough set theory, the threshold for the new extended dominance relation is estimated. It is decided by the loss regarding the risk or cost of those classification actions with respect to different states. Finally, the effectiveness of the new extended dominance relation is validated by experimental studies.

Keywords: Extended dominance relation, Bayesian decision theory, incomplete information system, rough sets.

1 Introduction

In the economic management's activities, some multi-attribute decision making problems (MADM)are often involved, such as the evaluation of suppliers, investment analysis, and project evaluation. The rough set theory, first proposed by Pawlak in 1982 [9], provides a new idea to MADM. Its main advantage is that it does not require any prior knowledge besides the data itself [10]. Due to the environmental complexity and uncertainty, and the decision maker's preference, incomplete information and the preference are prominent features of actual decision system. Therefore, it is necessary to extend classical rough set model to better suit practical applications.

Greco et al [1] proposed an extension of rough sets theory which is based on substitution of the indiscernibility relation by a dominance relation.

J. Yu et al. (Eds.): RSKT 2010, LNAI 6401, pp. 175–180, 2010.
© Springer-Verlag Berlin Heidelberg 2010

The dominance relation gives a new way to solve MADM with preference. In the aspect of incomplete preference decision system, He and Hu [3] proposed an extended dominance relation, based on the study of the paper [1]. But this model has a hypothesis in which missing attribute value can be regarded as any value: it can dominate any attribute value, and also be dominated by any attribute value. It is easy to misjudge different classes of objects into the same class. Then limited extended dominance relation [4] and generalized extended relation [5] are proposed. Although the two kinds of extended dominance relation limit the number of missing attribute, they do not solve this defect existed in [3]. In order to solve this problem, a new limited extended dominance relation is proposed in [7]. Meanwhile, a limited similarity dominance relation is defined in [8]. In [7] and [8], they take into account attribute values have preference-ordered, but just discuss the most preference value and the worst value, ignore the possibility of dominance relation for other attribute values. Besides this, attribute values of two objects may appear missing in the same attribute is not restricted. The purpose of this paper is to solve these problems. The concept of a new extended dominance relation based on the possibility of dominance relation between two objects is presented, while using probabilistic rough set theory to require the threshold of the new dominance relation.

The remaining of the paper is organized as follows. Some basic concepts are reviewed in Section 2. A new extended dominance relation based on probabilistic rough set theory is proposed in Section 3. In Section 4, the effectiveness of the new extended dominance relation are validated by experimental studies. Section 5 concludes the research work of this paper.

2   Preliminaries

The basic concepts, correlative notations of incomplete preference decision system and several extended dominance relations are reviewed [1,3,7].

Supposed a decision system $S = (U, A, V, f)$, where U is a non-empty finite set of objects, $A = C \bigcup D$ is a non-empty finite set of attributes, C denotes the set of condition attributes and D denotes the set of decision attributes, $C \bigcap D = \phi$. $V = \bigcup_{\alpha \in A} V_\alpha$ and V_α is a domain of the attribute α, and V_C is the domain of the condition attributes, V_D is the domain of the decision attributes and $* \notin V_D$. $f : U \times A \to V$ is an information function such that $f(x, \alpha) \in V_\alpha$ for every $x \in U$, $\alpha \in A$. All attributes are preference-ordered in this system. If there exists $x \in U$ and $q \in C$ where the value of $f(x, q)$ is missing, denoted by $''*''$, the decision system is an incomplete preference decision system. $\forall x \in U$, there at least exists $a \in C$ such that $f(x, \alpha) \neq *$.

To solve MADM with preference and incomplete information, let us review extended dominance relation [3] and limited extended dominance relation [7].

Definition 1. *Supposed a decision system $S = (U, A, V, f)$,$P \subseteq A$, $x, y \in U$, extended dominance relation can be defined as:*

$$ED(P) = \{(x, y) \in U \times U | \forall q \in P, (f(y, q) \geq f(x, q)) \vee (f(x, q) = *)$$
$$\vee (f(y, q) = *)\} \qquad (1)$$

If y dominates x under this relation, denoted by $yD_p^E x$. This model has a hypothesis that missing attribute value can be regarded as any value. It can dominate any attribute value, and be dominated by any attribute value. It's easy to misjudge objects of different classes into the same class. In order to solve this problem, the paper [7] proposed limited extended dominance relation, using attribute values' preference-ordered.

Definition 2. *Supposed a decision system $S = (U, A, V, f)$, $P \subseteq A$, $x, y \in U$, limited extended dominance relation can be defined as:*

$$LED(P) = \{(x, y) \in U \times U | \forall q \in P, (f(y, q) \geq f(x, q)) \vee (f(x, q) = *$$
$$\wedge f(y, q) = maxV_q) \vee (f(y, q) = * \wedge f(x, q) = minV_q)$$
$$\vee (f(x, q) = f(y, q) = *)\} \qquad (2)$$

where $maxV_q$ denotes the most preference value and $minV_q$ denotes the worst value. If y dominates x under this relation, denoted by yD_P^{LE}.

3   A New Extended Dominance Relation Based on Probabilistic Rough Set Theory

For the limited extended dominance relation, it takes into account attribute values have preference-ordered, but just discusses the most preference value and the worst value and ignores the possibility of dominance relation for other attribute values. In addition, attribute values of two objects may appear missing in the same attribute, the aforementioned models do not restrict this case. Thus we propose a new extended dominance relation to solve these problems. With reference to valued tolerance relation [2,11], the possibility of dominance relation in pairs of objects are defined as follow.

Definition 3. *Supposed a decision system $S = (U, A, V, f)$, $P \subseteq A$, $x, y \in U$, $c \in P$, $P_c(y, x)$ is the possibility of object y dominates object x in the attribute c, while $P_P(y, x)$ denotes the possibility of object y dominates object x in the set of attributes P. $P_c(y, x)$ and $P_P(y, x)$ can be defined as follow, respectively:*

$$P_c(y, x) = \begin{cases} 0, if (f(x, c) \neq *) \wedge (f(y, c) \neq *) \wedge (f(x, c) > f(y, c)) \\ |\{V | V \in V_c, V_c \geq f(x, c)\}| / |V_c|, if (f(x, c) \neq *) \wedge (f(y, c) = *) \\ |\{V | V \in V_c, f(y, c) \geq V_c\}| / |V_c|, if (f(x, c) = *) \wedge (f(y, c) \neq *) \\ 1/2 + 1/(2|V_c|), if (f(x, c) = *) \wedge (f(y, c) = *) \\ 1, if (f(x, c) \neq *) \wedge (f(y, c) \neq *) \wedge (f(y, c) \geq f(x, c)) \end{cases} \quad (3)$$

$$P_P(y, x) = \prod_{c \in P} P_c(y, x) \qquad (4)$$

where $|\cdot|$ stands for the number of elements in sets.

The possibility of dominance relation includes all conditions of the extended dominance relation and the limited extended dominance relation. The new extended dominance relation is defined as follow.

Definition 4. *Supposed a decision system* $S = (U, A, V, f)$, $P \subseteq A$, $x, y \in U$, *the new extended dominance relation can be defined as:*

$$VED(P) = \{(x, y) \in U \times U | P_P(y, x) \geq \alpha\} \bigcup I_U \tag{5}$$

Where α *denotes the threshold,* $I_U = \{(x, x) : x \in U\}$. *If* y *dominates* x *under this relation, denoted by* $y D_P^{VE} x$.

Obviously, the new extended dominance relation is reflexive and transferable, but not symmetric. It takes into account all attribute values have preference-ordered, and limits the case that attribute values of two objects may appear missing in the same attribute.

Definition 5. *Given* $P \subseteq A$ *and* $x \in U$, *let* $D_P^{+VE}(x) = \{y \in U : y D_P^{VE} x\}$, *so-called,* P−*new extended dominate set with respect to* x.

The decision-theoretic rough set model(DTRS) [12,13], introduces a way to estimate the threshold parameter in probabilistic rough set models, e.g. variable precision rough set model, on the basis of the loss regarding the risk or cost of those classification actions with respect to different states . According to the idea of DTRS, the threshold of the new extended dominance relation can be estimated by Bayesian decision theory. The set of states is given by $St = \{V, V^C\}$ indicating that an element is in $D_P^{+VE}(x)$ and not in $D_P^{+VE}(x)$. The set of actions is $A = \{a_1, a_2\}$, where a_1, a_2 represent the two actions in classifying an object, deciding in $D_P^{+VE}(x)$ and not in $D_P^{+VE}(x)$. Supposed the loss of the same action to different objects is the same, denoted as $\lambda_{i1} = \lambda(a_i | V)$, $\lambda_{i2} = \lambda(a_i | V^C)$. Let $P_P(y, x)$ be the possibility of object y dominates object x with respect to set of criteria $P \subseteq A$. Given an object $y \in U$, the expected losses $R(y \in D_P^{+VE}(x) | y)$ and $R(y \notin D_P^{+VE}(x) | y)$ can be expressed as:

$$R(y \in D_P^{+VE}(x) | y) = \lambda_{11} P_P(y, x) + \lambda_{12}(1 - P_P(y, x))$$

$$R(y \notin D_P^{+VE}(x) | y) = \lambda_{21} P_P(y, x) + \lambda_{22}(1 - P_P(y, x))$$

By applying Bayesian decision theory, we obtain the following minimum risk decision rules:

If $R(y \in D_P^{+VE}(x) | y) \leq R(y \notin D_P^{+VE} | y)$, $y \in D_P^{+VE}(x)$.
If $R(y \in D_P^{+VE}(x) | y) > R(y \notin D_P^{+VE} | y)$, $y \notin D_P^{+VE}(x)$.

It is quite natural to assume that $\lambda_{11} < \lambda_{21}$, $\lambda_{22} < \lambda_{12}$, so the above decision rules can be expressed as:

If $P_P(y, x) \geq \alpha$, decide $y \in D_P^{+VE}(x)$.
If $P_P(y, x) < \alpha$, decide $y \notin D_P^{+VE}(x)$.

Where $\alpha = \dfrac{\lambda_{12} - \lambda_{22}}{(\lambda_{21} - \lambda_{11}) + (\lambda_{12} - \lambda_{22})} = (\dfrac{\lambda_{21} - \lambda_{11}}{\lambda_{12} - \lambda_{22}} + 1)^{-1}$. Thus this step gives us the formula to estimate the threshold α of Definition 4.

4   Experimental Evaluation

In this section, some experiments are used to validate the effectiveness of the new extended dominance relation. The Car Evaluation database in UCI machine learning repository (http://archive.ics.uci.edu/ml/) is chosen. First, 300 objects are sampled from this database by random function. Missing values in sample are also created by random function according to the proportion of the amount of sample: 10%, 20%, 30%, 40%, 50%, 60% and 70%, then seven incomplete test databases are obtained. Second, all attributes with decreasing preference are converted into increasing. There, the comparability of test databases under some different extended dominance relations are defined with reference to the comparability in [6]. Finally, we test the five incomplete databases with different extended dominance relations.

Definition 6. *Let U be test database, $P \subseteq A$, $D_P^+(x)$ be the dominate set of $x \in U$ under dominance relation in complete information systems, while $D_P^{+E}(x)$ be the dominate set of x under different extended dominance relations in incomplete information systems. The comparability of test databases that extended dominance relation compare with dominance relation can be defined as:*

$$R^+ = \frac{\sum_{x \in U} \frac{|D_P^+(x) \bigcap D_P^{+E}(x)|}{|D_P^+(x)| + |D_P^{+E}(x)| - |D_P^+(x) \bigcap D_P^{+E}(x)|}}{|U|} \quad (6)$$

Obviously, the bigger R^+, the better the effectiveness of classification. Let $\frac{\lambda_{21} - \lambda_{11}}{\lambda_{12} - \lambda_{22}} = 7/3$, the threshold α is 0.3. The comparability of test databases in different extended dominance relations(i.e. extended dominance relation(EDR), limited extended dominance relation(LEDR), new extended dominance relation(NEDR)) are shown in Table 1.

In Tabel 1, it is obviously that the new extended dominance relation has better effectiveness than the other extended dominance relation in the same test database when α is 0.3.

Table 1. The comparability of test databases in different extended dominance relations

	EDR	LEDR	NEDR
10%	0.4721	0.5531	0.6570
20%	0.3007	0.3610	0.4646
30%	0.2473	0.3009	0.3903
40%	0.1670	0.2209	0.3012
50%	0.1425	0.1776	0.2484
60%	0.5340	0.6202	0.6683
70%	0.4794	0.5582	0.6608

5   Conclusions

In this paper, after analyzing several present models based on extended dominance relation in incomplete information systems, a new extended dominance relation is proposed. It takes into account all attribute values have preference-ordered, and limits the case that attribute values for two objects may appear missing in the same attribute. On the other hand, the threshold is estimated by the loss regarding the risk or cost of those classification actions with respect to different states. The effectiveness of the new extended dominance relation is validated by experimental studies.

References

1. Greco, S., Matarazzo, B., Slowinski, R.: Rough sets theory for multicriteria decision analysis. European Journal of Operational Research 129, 1–47 (2001)
2. Gao, Y., Zheng, G.L.: Extended rough sets based on variable-precision tolerance relation. Systems Engineering and Electronics 30, 477–480 (2008) (in Chinese)
3. He, Y.Q., Hu, S.S.: Rough analysis method of multi-attribute decision making with incomplete information. Journal of Systems Engineering 19, 117–120 (2004) (in Chinese)
4. Hu, M.L., Liu, S.F.: An analysis method of rough multi-attribute decision making based on limited extended dominance relation. Systems engineering 24, 106–110 (2006) (in Chinese)
5. Hu, M.L., Liu, S.F.: Rough analysis method of multi-attribute decision making based on generalized extended dominance relation. Control and Decision 22, 1347–1351 (2007) (in Chinese)
6. Ji, X., Li, L.S., Chen, S.B.: A new extended rough set model based on the dynamic tolerance relation and improved tolerance degree calculating method. In: Information Processing, APCIP 2009, pp. 94–96. IEEE Press, Los Alamitos (2009)
7. Luo, G.Z., Yang, X.J., Zhou, D.Q.: Rough analysis model of multi-attribute decision making based on limited extended dominance relation. Journal of Systems & Management 18, 391–396 (2009) (in Chinese)
8. Luo, G.Z., Yang, X.J.: Rough analysis model of multi-attribute decision making based on limited similarity dominance relation. Systems Engineering-Theory & Practice 29, 134–140 (2009) (in Chinese)
9. Pawlak, Z.: Rough Sets. International Journal of Computer and Information Sciences 11, 341–356 (1982)
10. Pawlak, Z.: Rudiments of rough sets. Information Sciences 177, 3–27 (2007)
11. Stefanowski, J., Tsoukias, A.: On the Extension of Rough Sets under Incomplete Information. In: Zhong, N., Skowron, A., Ohsuga, S. (eds.) RSFDGrC 1999. LNCS (LNAI), vol. 1711, pp. 73–82. Springer, Heidelberg (1999)
12. Yao, Y.Y., Wong, S.K.M.: A decision theoretic framework for approximating concepts. International Journal of Man-machine Studies 37(6), 793–809 (1992)
13. Yao, Y.Y.: Probabilistic rough set approximations. International Journal of Approximate Reasoning 49(2), 255–271 (2008)

An Equivalent Form of Rough Logic System RSL

Yingchao Shao[1], Zhongmin Xie[2], and Keyun Qin[1]

[1] School of Mathematics Southwest jiaotong university Chengdu,
Sichuan, 610031, China
[2] School of Communication and Information Engineer,
University of Electronic Science
and Technology of China Chengdu, Sichuan, 611731, China

Abstract. Firstly, based on regular double Stone algebra, a new kind of rough logic system is established, and obtain some properties in this paper.Secondly,we prove that it is one equivalent form of rough logic system RSL.

Keywords: fuzzy logic systems, regular double Stone algebras, rough logic systems RSL.

1 Introduction

An argument was risen on the application of the fuzzy logic after Elken made a report titled as the paradoxical success of fuzzy logic in the 11th American Congress and School on Artificial Intelligence. By this argument, a lot of scholars had been absorbed to take their attention on the fuzzy logic, which had made it rapidly progress, a lot of logical theory and logical methods had been risen.

In order to seek a reliable logical theory foundation of fuzzy theory, professor Wang proposed a new formal deductive system L^* [9]. Daowu Pei in[11]proposed a simplified axioms for both systems IMTL and NM which only contain six and seven axioms respectively and proved the independence of the simplified systems from the systems IMTL.

With the development of the research, more and more interest is turned to the connection of fuzzy logic and rough logic. Zhang Xiaohong proposed the rough logic system RSL in[12]. It is a rough logic system based on Stone algebra.In this paper, we will give a new rough fuzzy logic system RSL^*, which describes the rough fuzzy logic system from the other view. At last,we prove that it is equivalent to the system RSL.

2 Preliminaries

Definition 1. *[12] A double Stone algebra is an algebraic structure$(L, \vee, \wedge, ^*, ^+, 0, 1)$ of type(2,2,1,1,0,0) such that:*

1. *$(L, \vee, \wedge, 0, 1)$is a bounded distributive lattice.*
2. *For all $a \in L, a^*$ is the pseudo-complement of a,i.e., $a \wedge x = 0 \Longleftrightarrow x \leq a^*$.*

J. Yu et al. (Eds.): RSKT 2010, LNAI 6401, pp. 181–186, 2010.

3. *For all $a \in L, a^+$ is the dual pseudo-complement of a, i.e., $a \vee x = 1 \Longleftrightarrow$*
   $a^+ \leq x$.
4. *For all $a \in L, a^* \vee a^{**} = 1, a^+ \wedge a^{++} = 0$.*

Example 1.
(1)Let $(L; \vee, \wedge, ', 0, 1)$be a Boole-algebra ,then $(L; \vee, \wedge, ^*, ^+, 0, 1)$ is a double Stone algebra,where,for all $a \in L, a^* = a^+ = a'$,
(2) Let $(L; \vee, \wedge, 0, 1)$ be a finite distributive lattice,then$(L, \vee, \wedge, ^*, ^+, 0, 1)$is a double Stone algebra.

Definition 2. *[2,12] A double Stone algebra$(L, \vee, \wedge, ^*, ^+, 0, 1)$ is called regular if one of the following terms,which are equivalent to each other,is satisfying:*

1. *(RDSA) $\forall a, b \in L, (a^* = b^*, a^+ = b^+) \Rightarrow a = b$;*
2. *(RDSA1) $\forall a, b \in L, a \leq b \Leftrightarrow (a^{**} \leq b^{**}, a^{++} \leq b^{++})$;*
3. *(RDSA2) $\forall a, b \in L, a \wedge a^+ \leq b \vee b^*$.*

Definition 3. *[12,14] Let $S=\{P_1, P_2, \ldots\}$ be a set of propositional variables .A free algebra $F(S)$generated by S is a structure $(^*, ^+, \wedge, \rightarrow), \perp$is a propositional constant. The axioms in the formed system RSL are consisted of the following formulas ,for all $\alpha, \beta \in F(S)$:*

(AX1) $(\varphi \rightarrow \psi) \rightarrow ((\psi \rightarrow \chi) \rightarrow (\varphi \rightarrow \chi))$;
(AX2) $(\varphi \wedge \psi) \rightarrow \varphi$;
(AX3)$(\varphi \wedge \psi) \rightarrow (\psi \wedge \varphi)$;
(AX4)$\varphi \rightarrow ((\varphi \rightarrow \psi) \rightarrow (\varphi \wedge \psi))$;
(AX5)$((\varphi \rightarrow \psi) \rightarrow \chi) \rightarrow (((\psi \rightarrow \varphi) \rightarrow \chi) \rightarrow \chi)$;
(AX6)$(\neg\varphi \rightarrow \neg\varphi) \rightarrow (\psi \rightarrow \varphi)$;
(AX7)$\varphi \wedge (\varphi \wedge \psi)^* \rightarrow \varphi \wedge \psi^*, \varphi \wedge \psi^* \rightarrow \varphi \wedge (\varphi \wedge \psi)^*$;
(AX8)$\varphi \vee (\varphi \vee \psi)^+ \rightarrow \varphi \vee \psi^+, \varphi \vee \psi^+ \rightarrow \varphi \vee (\varphi \vee \psi)^+$;
(AX9)$\varphi \rightarrow \perp^*, \top^+ \rightarrow \varphi$;
(AX10)$\perp^{**} \rightarrow \perp, \top \rightarrow \top^{++}$;
(AX11)$\top \rightarrow \varphi^* \vee \varphi^{**}, \varphi^+ \wedge \varphi^{++} \rightarrow \perp$;
(AX12)$\varphi \wedge \varphi^+ \rightarrow \varphi \vee \varphi^*$;
(AX13)$(\varphi \rightarrow \psi) \rightarrow (\varphi^+ \vee \psi) \wedge (\varphi^+ \vee \psi^{*+}) \wedge (\varphi \vee \varphi^* \vee \psi^{*+})$;
$(\varphi^+ \vee \psi) \wedge (\varphi^+ \vee \psi^{*+}) \wedge (\varphi \vee \varphi^* \vee \psi^{*+}) \rightarrow (\varphi \rightarrow \psi)$.
The deduction rule in RSL is MP.Where, $\varphi \vee \psi = \neg(\neg\varphi \wedge \neg\psi); \neg\varphi = \varphi \rightarrow \perp; \top = \perp \rightarrow \perp$.Moreover,$\varphi \& \psi$is used to denote $\neg(\varphi \rightarrow \neg\psi)$.

Theorem 1. *[12] The following formulas are theorems in RSL:*

1. *$(\varphi \wedge \psi) \wedge \chi \rightarrow \varphi \wedge (\psi \wedge \chi)$;*
2. *$\varphi \rightarrow \varphi \wedge (\varphi \vee \psi), \varphi \wedge (\varphi \vee \psi) \rightarrow \varphi$;*
3. *$\varphi \rightarrow \top$;*
4. *$\varphi \wedge (\psi \vee \chi) \rightarrow (\varphi \wedge \psi) \vee (\varphi \wedge \chi), (\varphi \wedge \psi) \vee (\varphi \wedge \chi) \rightarrow \varphi \wedge (\psi \vee \chi)$;*
5. *$(\varphi \wedge \psi \rightarrow \perp) \rightarrow ((\psi \rightarrow \varphi^*) \rightarrow \perp), (\top \rightarrow \varphi \vee \psi) \rightarrow (\psi^+ \rightarrow \varphi)$;*
   $((\psi \rightarrow \varphi^*) \rightarrow \perp) \rightarrow (\varphi \wedge \psi), (\psi^+ \rightarrow \varphi) \rightarrow (\top \rightarrow \varphi \vee \psi)$;
6. *$\varphi \wedge (\varphi \rightarrow \psi) \rightarrow \varphi \wedge \psi$.*

Theorem 2. *[12] The following deduction rules hold in RSL:*

1. $\{\varphi \to \psi, \psi \to \chi\} \vdash \varphi \to \chi$;
2. $\{\varphi, \psi\} \vdash \varphi \wedge \psi$;
3. $\{\varphi \to (\psi \to \chi), \chi \to \xi\} \vdash \varphi \to (\psi \to \xi)$.

3　The Rough Fuzzy Logic System RSL^*

Definition 4. *Let $S=\{P_1, P_2, \ldots\}$ be a propositional variables set ,a free algebra $F(s)$generated by S is a structure such that$(*, +, \wedge, \to), \perp$ is a propositional
constant.The logic system consisted of the following formulas and the deduction
rule MP is called RSL^*:*

1. $\varphi \wedge \psi \to \psi \wedge \varphi, \varphi \vee \psi \to \psi \vee \varphi$;
2. $\varphi \wedge (\psi \wedge \chi) \to (\varphi \wedge \psi) \wedge \chi, (\varphi \vee \psi) \vee \chi \to \varphi \vee (\psi \vee \chi)$;
3. $\varphi \wedge (\varphi \vee \psi) \equiv \varphi, \varphi \vee (\varphi \wedge \psi) \equiv \varphi$;
4. $\varphi \to \psi \equiv \top \Leftrightarrow \varphi \wedge \psi \equiv \varphi \Leftrightarrow \varphi \vee \psi \equiv \psi$;
5. $\varphi \to ((\varphi \to \psi) \to \varphi \wedge \psi)$;
6. $\varphi \to \top, \perp \to \varphi$;
7. $\varphi \wedge (\psi \vee \chi) \to (\varphi \wedge \psi) \vee (\varphi \wedge \chi)$;
8. $(\varphi \wedge \psi \to \perp) \to (\top \to (\psi \to \varphi^*)), (\top \to \varphi \vee \psi) \to (\psi^+ \to \varphi)$,
   $((\psi \to \varphi^*) \to \perp) \to (\top \to \varphi \wedge \psi), (\psi^+ \to \varphi) \to (\top \to \varphi \vee \psi)$;
9. $\varphi \wedge \varphi^+ \to \psi \vee \psi^*$;
10. $(\varphi \to \psi) \to (\varphi^+ \vee \psi) \wedge (\varphi \vee \varphi^* \vee \psi^{*+}), (\varphi^+ \vee \psi) \wedge (\varphi \vee \varphi^* \vee \psi^{*+}) \to (\varphi \to \psi)$.

*where,$\varphi \vee \psi = \neg(\neg\varphi \wedge \neg\psi); \neg\varphi = \varphi \to \perp; \top = \perp \to \perp; \varphi \equiv \psi = (\varphi \to \psi) \wedge (\psi \to
\varphi)$.Moreover,$\varphi \& \psi$is used to denote $\neg(\varphi \to \neg\psi)$.*

Proposition 1. *The following formulas hold in RSL^* :*

1. $\varphi \wedge \perp \equiv \perp, \varphi \vee \perp \equiv \varphi, \varphi \wedge \top \equiv \varphi, \varphi \vee \top \equiv \top$;
2. $\varphi \wedge \varphi \equiv \varphi; \varphi \vee \varphi \equiv \varphi$;
3. If $\varphi \to \psi, \xi \to \chi$,then $\varphi \wedge \xi \to \psi \wedge \chi, \varphi \vee \xi \to \psi \vee \chi$;
4. $\varphi \wedge \varphi^* \equiv \perp, \varphi \vee \varphi^+ \equiv \top, \varphi^* \vee \varphi^{**} \equiv \top$,
   $\varphi^+ \wedge \varphi^{++} \equiv \perp, \perp^* \equiv \perp^+ \equiv \top, \top^* \equiv \top^+ \equiv \perp$;
5. If $\varphi \to \psi$,then $\psi^* \to \varphi^*, \psi^+ \to \varphi^+$;
6. $\varphi \to \varphi^{**}, \varphi^* \equiv \varphi^{***}; \varphi^{++} \to \varphi, \varphi^+ \equiv \varphi^{+++}$;
7. $\varphi \wedge (\varphi \wedge \psi)^* \equiv \varphi \wedge \psi^*, \varphi \vee (\varphi \vee \psi)^+ \equiv \varphi \vee \psi^+$;
8. $(\varphi \vee \psi)^* \equiv \varphi^* \wedge \psi^*, (\varphi \wedge \psi)^+ \equiv \varphi^+ \vee \psi^+$;
9. $(\varphi \wedge \psi)^{**} \equiv \varphi^{**} \wedge \psi^{**}, (\varphi \vee \psi)^{++} \equiv \varphi^{++} \vee \psi^{++}$;
10. $(\neg\varphi)^* \equiv \varphi^{+*}, \neg(\varphi^+) \equiv \varphi^{++}, \neg(\varphi^*) \equiv \varphi^{*+}, (\neg\varphi)^+ \equiv \varphi^{**}$;
11. $(\varphi \to \psi) \wedge (\varphi \to \chi) \equiv (\varphi \to \psi \wedge \chi); (\varphi \to \psi) \vee (\varphi \to \chi) \equiv (\varphi \to \psi \vee \chi)$;
    $(\varphi \to \chi) \wedge (\psi \to \chi) \equiv (\varphi \vee \psi \to \chi); (\varphi \to \chi) \vee (\psi \to \chi) \equiv (\varphi \wedge \psi \to \chi)$;
12. $\varphi \to (\psi \to \chi) \equiv \varphi \& \psi \to \chi$.

Proof

1. By Definition4(4)and(6), we can obtain them immediately.
2. This is immediate from Definition4(3) by $\psi \equiv \perp$.

3. By Definition4(4),we have $\varphi \to \psi \Leftrightarrow \varphi \wedge \psi \equiv \varphi, \xi \to \chi \Leftrightarrow \xi \wedge \chi \equiv \xi$,so,$(\varphi \wedge \xi) \wedge (\psi \wedge \chi) \equiv (\varphi \wedge \psi) \wedge (\xi \wedge \chi) \equiv \varphi \wedge \xi$.Thus, $\varphi \wedge \xi \to \psi \wedge \chi$.
   Similarly,we can prove the second formula.

4. This is immediate from Definition4(8).

5. It follows by Definition4(8)and (10).

6. Since $\varphi \wedge \varphi^* \equiv \bot$ we have $\varphi \to \varphi^{**}$; by(4),we get $\varphi^{***} \to \varphi^*$.Likewise,since $\varphi^* \wedge \varphi^{**} \equiv \bot$,we have $\varphi^* \to \varphi^{***}$.Thus, $\varphi^* \equiv \varphi^{***}$.
   Similarly,we can prove the other two formulas.

7. It is clear that $\varphi \wedge \psi^* \to \varphi \wedge (\varphi \wedge \psi)^*$.Likewise,since $(\varphi \wedge \psi) \wedge (\varphi \wedge \psi)^* \equiv \bot$we can obtain $\varphi \wedge (\varphi \wedge \psi)^* \to \psi^*$,so $\varphi \wedge (\varphi \wedge \psi)^* \to \varphi \wedge \psi^*$,Thus, $\varphi \wedge (\varphi \wedge \psi)^* \equiv \varphi \wedge \psi^*$.
   Similarly,we can get the other formula.

8. By Definition4(9),we have $(\varphi^* \wedge \psi^*) \wedge (\varphi \vee \psi) \equiv \bot \vee \bot \equiv \bot$,so, $(\varphi^* \wedge \psi^*) \equiv (\varphi \vee \psi)^*$.
   Similarly,we can get the other formula.

9. It is clear that $(\varphi \wedge \psi)^{**} \to (\varphi^{**} \wedge \psi^{**})$.
   Since $\varphi \wedge \varphi^* \equiv \bot$,we have $(\varphi \wedge \psi) \wedge (\varphi \wedge \psi)^* \equiv \bot$.Likewise,by $\psi^* \equiv \psi^{***}$,we have $\varphi \wedge (\varphi \wedge \psi)^* \to \psi^{***}$,so,$(\varphi \wedge (\varphi \wedge \psi)^*) \wedge \psi^{**} \to \psi^{***} \wedge \psi^{**}$, i.e., $(\varphi \wedge (\varphi \wedge \psi)^*) \wedge \psi^{**} \to \bot$,and then,$(\varphi \wedge (\varphi \wedge \psi)^*) \wedge \psi^{**} \equiv \bot$,thus,we have $(\varphi \wedge \psi)^* \wedge \psi^{**} \to \varphi^*$,so, $(\varphi \wedge \psi)^* \wedge \psi^{**} \to \varphi^{***}$and then $((\varphi \wedge \psi)^* \wedge \psi^{**}) \wedge \varphi^{**} \to \varphi^{***} \wedge \varphi^{**}$, i.e., $((\varphi \wedge \psi)^* \wedge \psi^{**}) \wedge \varphi^{**} \equiv \bot$.i.e., $(\varphi \wedge \psi)^* \wedge (\psi^{**} \wedge \varphi^{**}) \equiv \bot$.Thus we get $(\varphi^{**} \wedge \psi^{**}) \to (\varphi \wedge \psi)^{**}$.
   Similarly,we can get the other formula.

10. $(\neg\varphi)^* \equiv (\varphi^+ \wedge (\varphi \vee \varphi^*))^* \equiv \varphi^{+*} \vee (\varphi^* \wedge \varphi^{**}) \equiv \varphi^{+*} \vee \bot \equiv \varphi^{+*}$.
    $\neg(\varphi^+) \equiv \varphi^{++} \wedge (\varphi^+ \vee \varphi^{+*}) \equiv \varphi^{++} \wedge (\varphi^+ \vee \varphi^{++}) \equiv \varphi^{++} \wedge \top \equiv \varphi^{++}$.
    $\neg(\varphi^*) \equiv \varphi^{*+} \wedge (\varphi^* \vee \varphi^{**}) \equiv \varphi^{*+} \wedge \top \equiv \varphi^{*+}$.
    $(\neg\varphi)^+ \equiv (\varphi^+ \wedge (\varphi \vee \varphi^*))^+ \equiv \varphi^{++} \vee (\varphi^+ \wedge \varphi^{*+}) \equiv (\varphi^{++} \vee \varphi^+) \wedge (\varphi^{++} \vee \varphi^{*+}) \equiv \varphi^{**}$.

11. It follows immediately from Definition4(10).

12. $\varphi \to (\psi \to \chi) \equiv \varphi \to (\psi^+ \vee \chi) \wedge (\neg\psi \vee \chi^{*+})$
    $\equiv \{\varphi^+ \vee ((\psi^+ \vee \chi) \wedge (\neg\psi \vee \chi^{*+}))\} \wedge \{\neg\varphi \vee ((\psi^+ \vee \chi) \wedge (\neg\psi \vee \chi^{*+}))^{*+}\}$
    $\equiv \{(\varphi^+ \vee \psi^+ \vee \chi) \wedge (\varphi^+ \vee \neg\psi \vee \chi^{*+})\} \wedge \{\neg\varphi \vee ((\psi^{+*+} \vee \chi^{*+}) \wedge ((\neg\psi)^{*+} \vee \chi^{*+*+}))\}$
    $\equiv \{(\varphi^+ \vee \psi^+ \vee \chi) \wedge (\varphi^+ \vee \neg\psi \vee \chi^{*+})\} \wedge \{\neg\varphi \vee ((\psi^{+*+} \vee \chi^{*+}) \wedge (\psi^{+*+} \vee \chi^{*+}))\}$
    $\equiv \{(\varphi^+ \vee \psi^+ \vee \chi) \wedge (\varphi^+ \vee \neg\psi \vee \chi^{*+})\} \wedge \{\neg\varphi \vee ((\psi^+ \vee \chi^{*+}) \wedge (\psi^+ \vee \chi^{*+}))\}$
    $\equiv (\varphi^+ \vee \psi^+ \vee \chi) \wedge (\varphi^+ \vee \neg\psi \vee \chi^{*+}) \wedge (\neg\varphi \vee \psi^+ \vee \chi^{*+})$.
    moreover,
    $\varphi\&\psi \to \chi \equiv \neg(\varphi \to \neg\psi) \to \chi$
    $\equiv \neg\{(\varphi^+ \vee \neg\psi) \wedge (\neg\varphi \vee (\neg\psi)^{*+})\} \to \chi$
    $\equiv \{(\neg(\varphi)^+ \wedge \neg\neg\psi) \vee (\neg\neg\varphi \wedge \neg(\neg\psi)^{*+})\} \to \chi$
    $\equiv (\phi^{++} \wedge \psi) \vee (\varphi \wedge \psi^{++}) \to \chi$
    $\equiv \{((\varphi^{++} \wedge \psi) \vee (\varphi \wedge \psi^{++}))^+ \vee \chi\} \wedge \{\neg((\varphi^{++} \wedge \psi) \vee (\varphi \wedge \psi^{++})) \vee \chi^{*+}\}$
    $\equiv \{((\varphi^{+++} \vee \psi^+) \wedge (\varphi^+ \vee \psi^{+++})) \vee \chi\} \wedge \{(\neg\varphi^{++} \vee \neg\psi) \wedge (\neg\varphi \vee \neg\psi^{++})) \vee \chi^{*+}\}$
    $\equiv (\varphi^+ \vee \psi^+ \vee \chi) \wedge \{((\varphi^+ \vee \neg\psi) \wedge (\neg\varphi \vee \psi^+)) \vee \chi^{*+}\}$
    $\equiv (\varphi^+ \vee \psi^+ \vee \chi) \wedge (\varphi^+ \vee \neg\psi \vee \chi^{*+}) \wedge (\neg\varphi \vee \psi^+ \vee \chi^{*+})$.

Thus the formula holds.

Corollary 1

1. If $\varphi \to \chi$, then $\varphi \wedge \chi \to \psi \wedge \chi, \varphi \vee \chi \to \psi \vee \chi$.
2. $\varphi \wedge \psi \to \varphi, \varphi \to \varphi \vee \psi$.
3. $\neg\neg\varphi \equiv \varphi, \neg(\varphi \vee \psi) \equiv \neg\varphi \wedge \neg\psi, \neg(\varphi \wedge \psi) \equiv \neg\varphi \vee \neg\psi$.
4. $\varphi \& (\varphi \to \psi) \to \psi$.

Theorem 3. *The logic system RSL^* is equivalent to RSL.*

Proof. *At first we prove that $RSL \Rightarrow RSL^*$:*

By observing it is easy to get that in the system $RSL^*,(1),(2),(5),(6),(11),(12),$ (13)are just like $(AX2),(AX3),(AX4),(AX2),(AX11),(AX12),(AX13)$in RSL^*, respectively;and $(3),(4),(7),(8),(9),(10)$are exactly $(1),(2),(4),(3),(4),(5)$in Theorem1,respectively.
We prove that $RSL^* \Rightarrow RSL$in the following:

As is shown in the above that in the $RSL^*,(AX3),(AX4),(AX9),(AX10),$ $(AX12),(AX13)$are just like $(1),(5),(6),(9),(10),(11)$in the system RSL and $(AX7),(AX8)$ are exactly (6)in the proposition1.So it is only essential to prove $(AX1),(AX5)$ and $(AX6)$.

$(AX1)\ (\varphi \to \psi) \to [(\psi \to \chi) \to (\varphi \to \chi)]$.

Since $\varphi \& (\varphi \to \psi) \to \psi$ we have $\varphi \& (\varphi \to \psi) \& (\psi \to \chi) \to \psi \& (\psi \to \chi)$, that is,$\varphi \& (\varphi \to \psi) \& (\psi \to \chi) \to \chi$, so we get $(\varphi \to \psi) \& (\psi \to \chi) \to (\varphi \to \chi)$,thus,$(\varphi \to \psi) \to ((\varphi \to \psi) \to (\varphi \to \chi))$. $(AX5)\ ((\varphi \to \psi) \to \chi) \to (((\psi \to \varphi) \to \chi) \to \chi)$.

By Definition4(4) we have $\top \equiv \varphi \wedge \psi \to \varphi \vee \psi \equiv (\varphi \wedge \psi \to \varphi) \vee (\varphi \wedge \psi \to \psi) \equiv (\varphi \to \varphi) \vee (\psi \to \varphi) \vee (\varphi \to \psi) \vee (\psi \to \psi) \equiv (\psi \to \varphi) \vee (\varphi \to \psi)$.
moreover,
$(\top \to \chi) \to \chi \equiv ((\psi \to \varphi) \vee (\varphi \to \psi) \to \chi) \to \chi$
$\equiv ((\psi \to \varphi) \to \chi) \wedge (((\varphi \to \psi) \to \chi) \to \chi)$
$\equiv ((\psi \to \varphi) \to \chi) \& (((\varphi \to \psi) \to \chi) \to ((\psi \to \varphi) \to \chi) \wedge ((\varphi \to \psi) \to \chi))$
$\equiv ((\psi \to \varphi) \to \chi) \& ((\varphi \to \psi) \to \chi) \to \chi$
$\equiv ((\psi \to \varphi) \to \chi) \to \{((\varphi \to \psi) \to \chi) \to \chi\}$.
$(AX6)(\neg\varphi \to \neg\varphi) \to (\psi \to \varphi)$.
$\neg\varphi \to \neg\psi \equiv ((\neg\varphi)^+ \vee \neg\psi) \wedge (\varphi \vee (\neg\psi)^{*+})$
$\equiv (\varphi^{*+} \vee \neg\psi) \wedge (\varphi \vee \psi^{+*+})$
$\equiv (\varphi \vee \psi^+) \wedge (\varphi^{**} \vee (\psi^+ \wedge (\psi \vee \psi^*)))$
$\equiv \psi \to \varphi$.
Therefore,we get the equivalence of the two logic systems.

Corollary 2. $\{\varphi \to \psi, \psi \to \chi\} \vdash \varphi \to \chi$.

Acknowledgments. The authors acknowledge gratefully for the support from the National Natural Science Foundation of China (No. 60875034).

References

1. Duntsch, I.: Rough relation algebras. Fundamenta Informaticae 21, 321–331 (1994)
2. Düntsch, I., Winter, M.: Rough Relation Algebras Revisited. Fundamenta Informaticae 74, 283–300 (2006)
3. Xu, Y., Ruan, D., Qin, K.Y., et al.: Lattice-Valued logic. Springer, Heidelberg (2003)
4. Blyth, T.S.: Lattices and Ordered Algebraic structures. Springer, New York (2005)
5. Pei, D.: The Natural Deductive System of Fuzzy Logic. Joural of Engineering Mathematics 19(3), 95–100 (2002) (in Chinese)
6. Esteva, F., Godo, L.: Monoidal t-norm based logic: towards a logic for left-continuous t-norms. Fuzzy Sets and Systems 124, 271–288 (2001)
7. Radzikowska, A.M., Kerre, E.E.: A comparative study off fuzzy rough sets. Fuzzy Sets and Systems 126, 137–155 (2002)
8. Pei, D.: On equivalent forms of fuzzy logic systems NM and IMTL. Fuzzy Sets and Systems 138, 187–195 (2003)
9. Wang, G.J.: On the logic foundation of fuzzy reasoning. Inform Sci. 117(1), 47–88 (1999)
10. Qin, K.Y., Pei, Z.: On the topological properties of fuzzy rough sets. Fuzzy sets and systems 114, 601–613 (2005)
11. Pei, D.: Simplification and independence of axioms of fuzzy logic systems IMTL and NM. Fuzzy Sets and Systems 152, 303–320 (2005)
12. Zhang, X.H.: Fuzzy logic and its algebraic analysis. Science Press (2008) (in Chinese)
13. Beazer, R.: A note on injective double Stone algebras. Algebra Univ. 5, 239–241 (1975)
14. Changliu, H., Zhenming, S.: The foundation of Lattice Theory. Henan University Press (1991)

Conceptual Reduction of Fuzzy Dual Concept Lattices

Xiao-Xue Song[1], Wen-Xiu Zhang[2], and Qiang Zhao[1]

[1] Department of Computer, Xianyang Normal College, Xianyang,
Shaan'xi, 712000, P.R. China
[2] Institute for Information and System Sciences, Faculty of Science
Xi'an Jiaotong University, Xi'an, Shaan'xi, 710049, P.R. China
`sxx1669@163.com`, `wxzhang@mail.xjtu.edu.cn`

Abstract. In this paper we discuss the conceptual reduction of fuzzy dual concept lattices. Three pairs of operators in a fuzzy formal context are introduced. Based on the proposed operators, we present three types of variable threshold dual concept lattices. The properties and the relations of them are discussed. The result shows that the number of concepts in variable threshold dual concept lattices is less than that in fuzzy dual concept lattices, and the important concepts are preserved.

Keywords: Galois connection, fuzzy formal context, fuzzy dual concept lattice, conceptual reduction.

1 Introduction

The theory of concept lattices, called also the formal concept analysis (FCA), was introduced by R.Wille in 1982 [7]. The basis of FCA is formal concepts and concept lattices. As an effective tool for data analysis and knowledge discovery, FCA has been applied to various fields such as data mining, decision making, information retrieval, etc.

In two-valued formal concept analysis, attributes are assumed to be binary, i.e. table entries are 1 or 0 according to whether an attribute are applies to an object or not. But in many practical applications, the relation between objects and attributes is a fuzzy set, that is, each table entry contains a truth degree to which an attribute applies to an object. For this fuzzy binary relation, several generalizations to formal concept can be found in the existent literatures [1,3,4,8,9]. The main problem of this generalization is that the number of fuzzy formal concepts extracted from a fuzzy formal context will be very large. So the reduction for fuzzy concepts has become a hotspot in fuzzy concept lattices. Zhang et al [10] introduced the notions of variable threshold concept lattices for reducing the numbers of fuzzy formal concepts. Belohlavek et al proposed several approaches which can decrease the size of fuzzy concepts [1,2]. The notion of the dual concept is proposed by [7]. In this paper, we present reduction approaches for fuzzy dual concept lattices. Three kinds of variable threshold dual concept lattices and their properties and relations are discussed. The number of concepts

J. Yu et al. (Eds.): RSKT 2010, LNAI 6401, pp. 187–194, 2010.

in variable threshold dual concept lattices is less than that in fuzzy dual concept lattices.

The paper is organized as follows. In Section 2 we recall basic properties of complete residuated lattices and Galois connection. Approaches of conceptual reduction for fuzzy dual concept lattices, i.e., three types of variable threshold dual concept lattices are defined in Sections 3. The properties and relationships of different types of variable threshold dual concept lattices are given in Section 4. A final section contains conclusions.

2   Preliminaries

In this section, we recall some notions of fuzzy logic which will be used in this paper. We first review the notion of a residuated lattice which provides a very general truth structure for fuzzy logic and fuzzy set theory.

2.1   Residuated Lattices and Fuzzy Formal Contexts

Definition 1. [1] *A residuated lattice is a structure* $\mathbf{L} = (L, \vee, \wedge, \otimes, \rightarrow, 0, 1)$, *where*

(1) $(L, \vee, \wedge, 0, 1)$ is a lattice with the least element 0 and the greatest element 1;

(2) $(L, \otimes, 1)$ is a commutative monoid;

(3) $(\otimes, \rightarrow)$ is a residuated pair in L, i.e. $a \otimes b \leq c \Leftrightarrow a \leq b \rightarrow c, a, b, c \in L$.

A residuated lattice is called complete if $(L, \vee, \wedge, 0, 1)$ is a complete lattice. A complete residuated lattice $(L, \vee, \wedge, \otimes, \rightarrow, 0, 1)$ is involutive if $a = a^{cc}$ for all $a \in L$ (where $a^c = a \rightarrow 0$).

Each residuated lattice satisfies the following properties: $\forall x, y, z \in L, x_i, y_i \in L, i \in I$, where I is an index set,

$$(1)\ x \rightarrow x = 1, x \rightarrow 1 = 1, 0 \rightarrow x = 1, 1 \rightarrow x = x, x \otimes 0 = 0,$$
$$(2)\ x \leq y \text{ iff } x \rightarrow y = 1,$$
$$(3)\ x \otimes (x \rightarrow y) \leq y, y \leq x \rightarrow (x \otimes y), x \leq (x \rightarrow y) \rightarrow y,$$
$$(4)\ \bigvee_{i \in I} x_i \rightarrow y = \bigwedge_{i \in I} (x_i \rightarrow y), x \rightarrow \bigwedge_{i \in I} y_i = \bigwedge_{i \in I} (x \rightarrow y_i),$$
$$(5)\ x \otimes \bigvee_{i \in I} y_i = \bigvee_{i \in I} (x \otimes y_i), x \otimes \bigwedge_{i \in I} y_i \leq \bigwedge_{i \in I} (x \otimes y_i).$$

Moreover, if a residuated lattice is involutive, then it also satisfies the following properties:

$$(1)\ x \rightarrow y = (x \otimes y^c)^c,$$
$$(2)\ (\bigwedge_{i \in I} x_i)^c = \bigvee_{i \in I} x_i^c,$$
$$(3)\ x \rightarrow y = y^c \rightarrow x^c, x^c \rightarrow y = y^c \rightarrow x.$$

Let $\mathbf{L} = (L, \vee, \wedge, \otimes, \rightarrow, 0, 1)$ be a complete residuated lattice. A *L*-fuzzy set X in a universe U is a map $X : U \rightarrow L, X(x)$ is interpreted as the truth degree of the fact "x belong to X". The set of all L-sets in U is denoted as L^U. For all $X_1, X_2 \in L^U$, $X_1 \subseteq X_2$ if $X_1 \leq X_2$ for all $x \in U$. Operations $\vee$ and $\wedge$ on L^U are defined as follows: $(X_1 \vee X_2)(x) = X_1(x) \vee X_2(x)$, $(X_1 \wedge X_2)(x) = X_1(x) \wedge X_2(x)$. Let $\mathbf{L} = (L, \vee, \wedge, \otimes, \to, 0, 1)$ be a complete residuated lattice. A L-fuzzy formal context (see [1,10]) is a triple $(U, A, \tilde{I})$, where U is a set of objects, and A is a set of attributes, $\tilde{I}$ is a L-fuzzy relation between U and A, i.e. $\tilde{I} : U \times A \to L$. A L-fuzzy formal context is called *involutive* if $\mathbf{L}$ is an involutive residuated lattice.

In what follows, we suppose $\mathbf{L} = (L, \vee, \wedge, \otimes, \to, 0, 1)$ be a complete involutive residuated lattice.

2.2   Galois Connection

Let U be an universe of discourse. $(L^U, \subseteq)$ is a poset. The dual of $(L^U, \subseteq)$ is denoted by $(L^U, \subseteq)^\partial$(Also written $(L^U)^\partial$ for short).

Definition 2. [5] Let $(L_1^U, \subseteq)$ and $(L_2^A, \subseteq)$ be two posets. The pair (f, g) of mapping $f : L_1^U \to L_2^A$ and $g : L_2^A \to L_1^U$ is called *a Galois Connection* between L_1^U and L_2^A if and only if $X \subseteq g(B) \Leftrightarrow f(X) \subseteq B$, where $X \in L_1^U, B \in L_2^A$(both X and B can be either a crisp or a L-fuzzy set).

Definition 3. [6] Let $(L_1^U, \subseteq)$ and $(L_2^A, \subseteq)$ be two posets, and $f : L_1^U \to L_2^A$, $g : L_2^A \to L_1^U$ be two mappings. A pair $(X, B), X \in L_1^U, B \in L_2^A$, is called *a formal concept* if $f(X) = B, g(B) = X$. X and B are respectively referred to as the *extension* and the *intention* of (X, B).

Theorem 1. [6] Let $(L_1^U, \subseteq)$ and $(L_2^A, \subseteq)$ be two posets, and (f, g) be a Galois connection between L_1^U and L_2^A. Then the set of all formal concepts forms a complete lattice in which the meet and join are defined as follows:

$$\wedge_{i \in I}(X_i, B_i) = (\cap_{i \in I} X_i, f \circ g(\cap_{i \in I} B_i)),$$
$$\vee_{i \in I}(X_i, B_i) = (g \circ f(\cup_{i \in I} X_i), \cup_{i \in I} B_i).$$

To converse the order in the Galois connection, we can obtain other forms of Galois connection. In the followings we give the notion of a complementary formal concept and its properties.

Theorem 2. Let $(L_1^U, \subseteq)$ and $(L_2^A, \subseteq)$ be two posets, and $f : L_1^U \to L_2^A$, $g : L_2^A \to L_1^U$ be two mappings. Then for any $X, X_1, X_2 \in L_1^U, B, B_1, B_2 \in L_2^A$, the following statements are equivalent: where X^c denotes complement of the set X, and $(f(X))^c$ is simply denoted by $f^c(X)$,

(1) $X^c \subseteq g(B) \Leftrightarrow B^c \subseteq f(X)$;
(2) $X_1^c \subseteq X_2 \Rightarrow f(X_2^c) \subseteq f(X_1)$, $B_1^c \subseteq B_2 \Rightarrow g(B_2^c) \subseteq g(B_1)$,
      $X^c \subseteq g \circ f^c(X)$, $B^c \subseteq f \circ g^c(B)$;
(3) $f(X_1 \cap X_2) = f(X_1) \cap f(X_2)$, $g(B_1 \cap B_2) = g(B_1) \cap g(B_2)$,
      $X^c \subseteq g \circ f^c(X)$, $B^c \subseteq f \circ g^c(B)$.

Definition 4. Let $(L_1^U, \subseteq)$ and $(L_2^A, \subseteq)$ be two posets, $f : L_1^U \rightarrow L_2^A$, $g : L_2^A \rightarrow L_1^U$ be two mappings. A pair (X, B), $X \in L_1^U$, $B \in L_2^A$, is called *a complementary formal concept* if $f(X^c) = B, g(B^c) = X$. X and B are respectively referred to as the extension and the intention of the complementary formal concept (X, B).

Theorem 3. Let $(L_1^U, \subseteq)$ and $(L_2^A, \subseteq)$ be two posets, $f : L_1^U \rightarrow L_2^A$, $g : L_2^U \rightarrow L_1^A$ satisfy $X^c \subseteq g(B) \Leftrightarrow B^c \subseteq f(X)$. Then the set of all complementary formal concepts forms a complete lattice, in which the meet and join are given by:

$$\wedge_{i \in I}(X_i, B_i) = (\cap_{i \in I} X_i, f \circ g^c(\cap_{i \in I} B_i^c)),$$
$$\vee_{i \in I}(X_i, B_i) = (g \circ f^c(\cap_{i \in I} X_i^c), \cap_{i \in I} B_i).$$

For reasons of clarity, in what follows, the power set of U is denoted by $\mathcal{P}(U)$ and the element of $\mathcal{P}(U)$ is denoted by X, i.e., $X \in \mathcal{P}(U)$. A fuzzy set in U is denoted by $\tilde{X} : U \rightarrow L$.

3   Approaches of Conceptual Reduction for Fuzzy Dual Concept Lattices

Definition 5. [6] Let $(U, A, \tilde{I})$ be a fuzzy formal context. For $\tilde{X} \in L^U, \tilde{B} \in L^A$, two operators $(\sharp, \sharp)$ are defined as follows:

$$\tilde{X}^\sharp(a) = \bigvee_{x \in U} (\tilde{I}^c(x, a) \otimes \tilde{X}^c(x)),$$
$$\tilde{B}^\sharp(x) = \bigvee_{a \in A} (\tilde{I}^c(x, a) \otimes \tilde{B}^c(a)). \tag{1}$$

The operators $(\sharp, \sharp)$ in Definition 5. is a Galois connection between L^U and L^A, i.e., $\tilde{X}^\sharp \subseteq \tilde{B} \Leftrightarrow \tilde{B}^\sharp \subseteq \tilde{X}$.

The set of all formal concepts generated by the operators $(\sharp, \sharp)$ is a complete lattice called *a fuzzy dual concept lattice*, denoted by $L_d(\tilde{U}, \tilde{A}, \tilde{I})$. The sets of all extensions and intentions are respectively denoted by $Ext_d(\tilde{U}, \tilde{A}, \tilde{I})$ and $Int_d(\tilde{U}, \tilde{A}, \tilde{I})$.

Example 1. A fuzzy formal context $(U, A, \tilde{I})$ is presented in Table 1, which $U = \{1, 2, 3\}, A = \{a, b, c, d\}$.

We choose the Lukasiewicz operators, i.e.

$$a \rightarrow_L b = \begin{cases} 1, & a \leq b, \\ 1 - a + b, & a > b. \end{cases} \qquad a \otimes_L b = (a + b - 1) \vee 0.$$

Let $L = [0, 1]$. Fig.1. describes all fuzzy dual concepts induced from Table 1.

Table 1. A fuzzy formal context

U	a	b	c	d
1	1.0	0.3	0.7	0.1
2	0.5	0.0	0.4	0.2
3	0.7	0.1	0.2	0.2

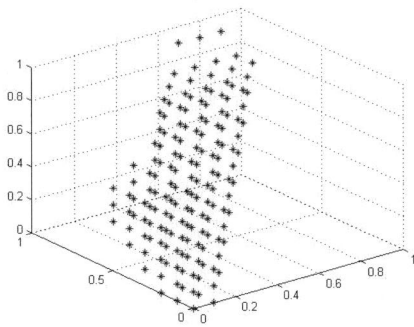

Fig. 1. Fuzzy dual concepts induced from Table 1

In the followings we give three type approaches of conceptual reduction for fuzzy dual concept lattices $L_d(\tilde{U}, \tilde{A}, \tilde{I})$.

Definition 6. Let $(U, A, \tilde{I})$ be a fuzzy formal context. For $X \in \mathcal{P}(U)$, $B \in \mathcal{P}(A)$, and $0 < \delta \le 1$, two operators $\sharp_1, \sharp_1$ are defined as follows:

$$
\begin{aligned}
X^{\sharp_1} &= \{a \in A | \bigvee_{x \in X^c} (\tilde{I}^c(x, a) \otimes X^c(x)) \le \delta^c\}, \\
B^{\sharp_1} &= \{x \in U | \bigvee_{a \in B^c} (\tilde{I}^c(x, a) \otimes B^c(a)) \le \delta^c\}.
\end{aligned}
\tag{2}
$$

Theorem 4. The pair $(\sharp_1, \sharp_1)$ in Definition 6. satisfies the following property:

$$X^c \subseteq B^{\sharp_1} \Leftrightarrow B^c \subseteq X^{\sharp_1}.$$

Proof: Since $X \in \mathcal{P}(U)$, $B \in \mathcal{P}(A)$, we have $\forall x \in X^c, X^c(x) = 1$ and $\forall a \in B^c, B^c(x) = 1$. Then

$$
\begin{aligned}
X^c \subseteq B^{\sharp_1} &\Leftrightarrow \forall x \in X^c, \forall a \in B^c, \tilde{I}^c(x, a) \otimes B^c(a) \le \delta^c \\
&\Leftrightarrow \forall a \in B^c, \forall x \in X^c, \tilde{I}^c(x, a) \otimes X^c(x) \le \delta^c \\
&\Leftrightarrow \forall a \in B^c, \bigvee_{x \in X^c} (\tilde{I}^c(x, a) \otimes X^c(x)) \le \delta^c \\
&\Leftrightarrow \forall a \in B^c, a \in X^{\sharp_1} \Leftrightarrow B^c \subseteq X^{\sharp_1}. \qquad \square
\end{aligned}
$$

From Theorem 3., the set of all formal concepts generated by the operators $(\sharp_1, \sharp_1)$ is a complete lattice called *a crisp-crisp variable threshold dual concept lattice*, denoted by $L_d^\delta(U, A, \tilde{I})$. The sets of all extensions and intentions are respectively denoted by $Ext_d^\delta(U, A, \tilde{I})$ and $Int_d^\delta(U, A, \tilde{I})$.

Definition 7. Let $(U, A, \tilde{I})$ be a fuzzy formal context. $\forall X \in \mathcal{P}(U)$, $\tilde{B} \in L^A$, $0 < \delta \le 1$, two operators $\sharp_2, \sharp_2$ are defined as follows:

$$
\begin{aligned}
X^{\sharp_2}(a) &= \delta \otimes \bigvee_{x \in X} \tilde{I}^c(x, a), \\
\tilde{B}^{\sharp_2} &= \{x \in U | \bigvee_{a \in A} (\tilde{I}^c(x, a) \otimes \tilde{B}^c(a)) \le \delta^c\}.
\end{aligned}
\tag{3}
$$

Theorem 5. The pair $(\sharp_2, \natural_2)$ in Definition 7. is a Galois connection between $\mathcal{P}(U)$ and L^A, i.e., $X \subseteq \tilde{B}^{\natural_2} \Leftrightarrow X^{\sharp_2} \subseteq \tilde{B}$.

From Theorem 1., the set of all formal concepts generated by the operators $(\sharp_2, \natural_2)$ is a complete lattice called *a crisp-fuzzy variable threshold dual concept lattice*, denoted by $L_d^\delta(U, \tilde{A}, \tilde{I})$. The sets of all extensions and intentions are respectively denoted by $Ext_d^\delta(U, \tilde{A}, \tilde{I})$ and $Int_d^\delta(U, \tilde{A}, \tilde{I})$.

Definition 8. Let $(U, A, \tilde{I})$ be a fuzzy formal context. $\forall \tilde{X} \in L^U$, $B \in \mathcal{P}(A)$, $0 < \delta \leq 1$, two operators $\sharp_3, \natural_3$ are defined as follows:

$$\tilde{X}^{\sharp_3} = \{a \in A | \bigvee_{x \in U} (\tilde{I}^c(x, a) \otimes \tilde{X}^c(x)) \leq \delta^c\},$$

$$B^{\natural_3}(x) = \bigvee_{a \in B} \tilde{I}^c(x, a) \otimes \delta. \tag{4}$$

Theorem 6. The pair $(\sharp_3, \natural_3)$ in Definition 8. is a Galois connection between L^U and $\mathcal{P}(A)$, i.e., $B \subseteq \tilde{X}^{\sharp_3} \Leftrightarrow B^{\natural_3} \subseteq \tilde{X}$.

From Theorem 1. and properties of Galois connection, the set of all formal concepts generated by the operators $(\sharp_3, \natural_3)$ is a complete lattice called *a fuzzy-crisp variable threshold dual concept lattice*, denoted by $L_d^\delta(\tilde{U}, A, \tilde{I})$. The sets of all extensions and intentions are respectively denoted by $Ext_d^\delta(\tilde{U}, A, \tilde{I})$ and $Int_d^\delta(\tilde{U}, A, \tilde{I})$.

4   Properties and Relationships of Variable Threshold Dual Concept Lattices

In this section, we write "$(X^{\sharp_1})_{\delta_1}$" instead of "$X^{\sharp_1}$" for $\delta = \delta_1$. Other operators are similar.

Theorem 7. Let $L_d^\delta(U, A, \tilde{I}), L_d^\delta(U, \tilde{A}, \tilde{I}), L_d^\delta(\tilde{U}, A, \tilde{I})$ be a crisp-crisp, a crisp-fuzzy, a fuzzy-crisp variable threshold dual concept lattice, respectively. Then for $1 \geq \delta_1 \geq \delta_2 > 0$, the following properties are concluded:

(1) $\forall X \in \mathcal{P}(U)$, $B \in \mathcal{P}(A)$, we have $(X^{\sharp_1})_{\delta_1} \subseteq (X^{\sharp_1})_{\delta_2}$, $(B^{\natural_1})_{\delta_1} \subseteq (B^{\natural_1})_{\delta_2}$.
(2) $\forall X \in \mathcal{P}(U)$, $\tilde{B} \in L^A$, we have $(X^{\sharp_2})_{\delta_1} \supseteq (X^{\sharp_2})_{\delta_2}$, $(\tilde{B}^{\natural_2})_{\delta_1} \subseteq (\tilde{B}^{\natural_2})_{\delta_2}$, and $Ext_d^{\delta_2}(U, \tilde{A}, \tilde{I}) \subseteq Ext_d^{\delta_1}(U, \tilde{A}, \tilde{I})$.
(3) $\forall \tilde{X} \in L^U$, $\forall B \in \mathcal{P}(A)$, we have $(\tilde{X}^{\sharp_3})_{\delta_1} \subseteq (\tilde{X}^{\sharp_3})_{\delta_2}$, $(B^{\natural_3})_{\delta_1} \supseteq (B^{\natural_3})_{\delta_2}$, and $Int_d^{\delta_2}(\tilde{U}, A, \tilde{I}) \subseteq Int_d^{\delta_1}(\tilde{U}, A, \tilde{I})$.

Theorem 8. Let $L_d^\delta(U, A, \tilde{I}), L_d^\delta(U, \tilde{A}, \tilde{I}), L_d^\delta(\tilde{U}, A, \tilde{I})$ be a crisp-crisp, a crisp-fuzzy, a fuzzy-crisp variable threshold dual concept lattice, respectively. $L_d(\tilde{U}, \tilde{A}, \tilde{I})$ be a fuzzy dual concept lattice. Then for $1 \geq \delta > 0$,

$$Ext_d^\delta(U, A, \tilde{I}) \subseteq Ext_d^\delta(U, \tilde{A}, \tilde{I}), \ Int_d^\delta(U, \tilde{A}, \tilde{I}) \subseteq Int_d(\tilde{U}, \tilde{A}, \tilde{I}), \tag{5}$$
$$Int_d^\delta(U, A, \tilde{I}) \subseteq Int_d^\delta(\tilde{U}, A, \tilde{I}), \ Ext_d^\delta(\tilde{U}, A, \tilde{I}) \subseteq Ext_d(\tilde{U}, \tilde{A}, \tilde{I}). \tag{6}$$

Theorem 9. Let $L_d^\delta(U, A, \tilde{I}), L_d^\delta(U, \tilde{A}, \tilde{I}), L_d^\delta(\tilde{U}, A, \tilde{I})$ be a crisp-crisp, a crisp-fuzzy, a fuzzy-crisp variable threshold dual concept lattice, respectively. $L_d(\tilde{U}, \tilde{A}, \tilde{I})$ be a fuzzy dual concept lattice. For $1 \geq \delta > 0$, denote

$$[\tilde{X}_i] = \{x \in U | \tilde{X}_i(x) \leq 0\}, \quad [\tilde{B}_i] = \{a \in A | \tilde{B}_i(a) \leq 0\},$$
$$[\tilde{X}_i]_\delta = \{x \in U | \tilde{X}_i(x) \leq \delta^c\}, \quad [\tilde{B}_i]_\delta = \{a \in A | \tilde{B}_i(a) \leq \delta^c\}.$$

Then

$$L_d^\delta(U, A, \tilde{I}) = \{(X, B) | X = \cup \{X_i | [\tilde{B}_i] = B\}, (X_i, \tilde{B}_i) \in L_d^\delta(U, \tilde{A}, \tilde{I})\}, \tag{7}$$
$$L_d^\delta(U, A, \tilde{I}) = \{(X, B) | B = \cup \{B_i | [\tilde{X}_i] = X\}, (\tilde{X}_i, B_i) \in L_d^\delta(\tilde{U}, A, \tilde{I})\}, \tag{8}$$
$$L_d^\delta(\tilde{U}, A, \tilde{I}) = \{(\tilde{X}, B) | \tilde{X} = \cap \{\tilde{X}_i | [\tilde{B}_i]_\delta = B\}, (\tilde{X}_i, \tilde{B}_i) \in L_d(\tilde{U}, \tilde{A}, \tilde{I})\}, \tag{9}$$
$$L_d^\delta(U, \tilde{A}, \tilde{I}) = \{(X, \tilde{B}) | \tilde{B} = \cap \{\tilde{B}_i | [\tilde{X}_i]_\delta = X\}, (\tilde{X}_i, \tilde{B}_i) \in L_d(\tilde{U}, \tilde{A}, \tilde{I})\}. \tag{10}$$

Example 2. There are two concepts in crisp-crisp variable threshold dual concept lattice for $\delta = 1$ induced from Table 1: $(\emptyset, \emptyset)$ and (U, A). The concepts of the other two variable threshold dual concept lattices induced from Table 1 are listed in Table 2-3.

Table 2. The concepts of crisp-fuzzy variable threshold dual concept lattice for $\delta = 1$

X	a	b	c	d
$\emptyset$	0.0	0.0	0.0	0.0
$\{1\}$	0.0	0.7	0.3	0.9
$\{2\}$	0.5	1.0	0.6	0.8
$\{3\}$	0.3	0.9	0.8	0.8
$\{1,2\}$	0.5	1.0	0.6	0.9
$\{1,3\}$	0.3	0.9	0.8	0.9
$\{2,3\}$	0.5	1.0	0.8	0.8
$\{1,2,3\}$	0.5	1.0	0.8	0.9

Table 3. The concepts of fuzzy-crisp variable threshold dual concept lattice for $\delta = 1$

B	1	2	3
$\emptyset$	0.0	0.0	0.0
$\{a\}$	0.0	0.5	0.3
$\{a, c\}$	0.3	0.6	0.8
$\{a, b, c\}$	0.7	1.0	0.9
$\{a, c, d\}$	0.9	0.8	0.8
$\{a, b, c, d\}$	0.9	1.0	0.9

5   Conclusion

Formal concept analysis (FCA), as a useful tool to deal with data, has recently received wide attention on the research areas. Because of the limits of two-value formal contexts, fuzzy set has been introduced into FCA. However, the number of formal concepts extracted from fuzzy formal contexts is too large. In this paper we discuss the conceptual reduction of fuzzy dual concept lattices. Three kinds of variable threshold concept lattices, i.e., crisp-crisp, crisp-fuzzy, fuzzy-crisp variable threshold dual concept lattices, and the properties and the relations of them are discussed. The results shows that the number of concepts in variable threshold dual concept lattices is far less than that in fuzzy dual concept lattices, and the important concepts are preserved.

Acknowledgment. The paper was supported by the Natural Science Foundation of the Science and technology Department of Shaanxi Province in China (No.2009JM8021), and the Scientific Research Project of the Education Department of Shaanxi Province in China (No.07JK422, No.09JK811).

References

1. Belohlavek, R., Vychodil, V.: Reducing the size of fuzzy concept lattices by hedges. In: The IEEE International Conference on Fuzzy Systems, Nevada, USA, pp. 663–668 (2005)
2. Belohlavek, R., Dvorák, J., Outrata, J.: Fast factorization by similarity in formal concept analysis of data with fuzzy attributes. Journal of Computer and System Sciences 73, 1012–1022 (2007)
3. Burusco, A., Gonzalez, R.F.: The study of the L-fuzzy concept lattice. Mathware and Soft Computing 1(3), 209–218 (1994)
4. Elloumi, S., Jaam, J., Hasnah, A., Jaoua, A., Nafkha, I.: A multi-level conceptual data reduction approach based on the Lukasiewicz implication. Information Sciences 163, 253–262 (2004)
5. Erné, M., Koslowski, J., Melton, A., Strecker, G.E.: A primer on Galois connections. In: Papers on General Topology and Applications Seventh Summer Conference Annals of the New York Academy of Sciences, vol. 704, pp. 103–125 (1993)
6. Fan, S.Q., Zhang, W.X., Xu, W.: Fuzzy Inference based on fuzzy concept lattice. Fuzzy sets and systems 157, 3177–3187 (2006)
7. Ganter, B., Wille, R.: Formal Concept Analysis, Mathematical Foundations. Springer, Berlin (1999)
8. Georgescu, G., Popescu, A.: Non-dual fuzzy connections. Archive Math Logic 43(8), 1009–1039 (2004)
9. Popescu, A.: A general approach to fuzzy concept. Math. Logic Quaterly 50(3), 1–17 (2001)
10. Zhang, W.X., Ma, J.M., Fan, S.Q.: Variable threshold concept lattice. Information Sciences 177(22), 4883–4892 (2007)

Qualitative Approximations of Fuzzy Sets and Non-classical Three-Valued Logics (I)

Xiaohong Zhang[1,*], Yiyu Yao[2], and Yan Zhao[2]

[1] Department of Mathematics, Ningbo University
Ningbo 315211, Zhejiang Province, P.R. China
[2] Department of Computer Science, University of Regina
Regina, Saskatchewan, S4S 0A2, Canada

Abstract. $(0,1)$-Qualitative approximations of fuzzy sets are studied by using the core and support of a fuzzy set. This setting naturally leads to three disjoint regions and an analysis based on a three-valued logic. This study combines both an algebra view and a logic view. From the algebra view, the mathematical definition of a $(0,1)$-approximation of fuzzy sets are given, and algebraic operations based on various t-norms and fuzzy implications are established. From the logic view, a non-classical three-valued logic is introduced. Corresponding to this new non-classical three-valued logic, the related origins of t-norms and fuzzy implications are examined.

Keywords: Fuzzy set, $(0,1)$-approximation, non-truth functional logic, three-valued logic.

1 Introduction

Crisp set approximations of a fuzzy set have been studied extensively in fuzzy set theory [1,2,3,9,12,15,17,18,21]. From the existing studies, we can identify two types of approximations: one uses a single crisp set and the other uses a pair of crisp sets. All crisp set approximations are defined using either an α-cut or a strong α-cut.

Yao [27] considered a pair of crisp set approximations, called the (α, β)-approximations, of a fuzzy set. Similar to other crisp set approximations of a fuzzy set, such as interval sets [21], (α, β)-rough sets [1], and shadowed sets [17,18], (α, β)-approximations are defined by an α-cut and a strong β-cut. By the (α, β)-approximations, a set is divided into three disjoint regions. Based on the three regions, the common qualitative features of fuzzy sets and rough sets are established.

Based on [27], the main objective of this paper is to study approximations of a fuzzy set in detail from both an algebra view and a logic view. From the

* This work is supported by National Natural Science Foundation of China (Grant No. 60775038) and K.C.Wong Magna Fund at Ningbo University. The author acknowledges the support of the Department of Computer Science, University of Regina, Canada, during a visit in which the paper was finalized.

J. Yu et al. (Eds.): RSKT 2010, LNAI 6401, pp. 195–203, 2010.

algebra view, we give the mathematical definitions of the $(0,1)$-approximation, and discuss its algebraic properties from various t-norms and residuated implications. From the logic view, we introduce a non-classical three-valued logics, and examine the related origins of t-norms and fuzzy implications corresponding to this new non-classical three-valued logic.

2   Fuzzy Set, t-Norm and Fuzzy Implication

In this section, we review some key notions of fuzzy sets, fuzzy set-theoretic operations and fuzzy implications.

2.1   Fuzzy Sets

Let U be a non-empty set called universe. A fuzzy set $\mathcal{X} \subseteq U$ is defined by a membership function:

$$\mu_{\mathcal{X}} : U \longrightarrow [0,1].$$

We can denote $\mu_{\emptyset}$ and μ_U as the membership functions of $\emptyset$ and U, respectively, namely, $\mu_{\emptyset}(a) = 0$ for all $a \in U$ and $\mu_U(a) = 1$ for all $a \in U$.

A fuzzy set $\mu_{\mathcal{X}}$ is a subset of another fuzzy set $\mu_{\mathcal{Y}}$, written $\mu_{\mathcal{X}} \subseteq \mu_{\mathcal{Y}}$, if and only if $\mu_{\mathcal{X}}(a) \leq \mu_{\mathcal{Y}}(a)$ for all $a \in U$.

The core of a fuzzy set $\mu_{\mathcal{X}}$ is a crisp subset of U, consisting of those elements with full membership:

$$\mathrm{core}(\mu_{\mathcal{X}}) = \{a \in U \mid \mu_{\mathcal{X}}(a) = 1\}. \tag{1}$$

The support of a fuzzy set $\mu_{\mathcal{X}}$ is a crisp subset of U consists of those elements with non-zero membership:

$$\mathrm{supp}(\mu_{\mathcal{X}}) = \{a \in U \mid \mu_{\mathcal{X}}(a) > 0\}. \tag{2}$$

2.2   Fuzzy Set-Theoretic Operations

Let $\mathcal{F}(U)$ denote the set of all fuzzy sets on U, we can define fuzzy set-theoretic operations on the set $\mathcal{F}(U)$. The fuzzy set-theoretic operations can be defined through functions on the unit interval [4,12]. Let $^c, \sqcap, \sqcup : \mathcal{F}(U) \longrightarrow \mathcal{F}(U)$ denote the fuzzy set complement, intersection and union, respectively. They can be defined by negation functions, t-norms and t-conorms, respectively.

Definition 1. *A negation function, $N : [0,1] \longrightarrow [0,1]$, is defined by the following properties: for all $a, b \in [0,1]$,*

(N1) Boundary conditions: $N(0) = 1, N(1) = 0$;
(N2) Monotonicity: $a \leq b \Longrightarrow N(b) \leq N(a)$.

The standard negation function is defined as follows. For any $a \in [0, 1]$,

$$N_S(a) = 1 - a.$$

The H-negation function N_H is defined as follows. For any $a \in [0, 1]$,

$$N_H(a) = \begin{cases} 1, & a = 0; \\ 0, & otherwise. \end{cases}$$

Definition 2. *A t-norm is a function* $T : [0, 1] \times [0, 1] \to [0, 1]$ *satisfying the following conditions. For any* $a, b, c \in [0, 1]$,

    (T1) Boundary conditions: $T(0, 0) = 0, T(1, a) = T(a, 1) = a$;
    (T2) Monotonicity: $(a \leq c, b \leq d) \Longrightarrow T(a, b) \leq T(c, d)$;
    (T3) Symmetry: $T(a, b) = T(b, a)$;
    (T4) Associativity: $T(a, T(b, c)) = T(T(a, b), c))$.

Definition 3. *A T-conorm is a function* $S : [0, 1] \times [0, 1] \to [0, 1]$ *satisfying the following conditions. For any* $a, b, c \in [0, 1]$,

    (S1) Boundary conditions: $S(1, 1) = 1, S(a, 0) = S(0, a) = a$;
    (S2) Monotonicity: $(a \leq c, b \leq d) \Longrightarrow S(a, b) \leq S(c, d)$;
    (S3) Symmetry: $S(a, b) = S(b, a)$;
    (S4) Associativity: $S(a, S(b, c)) = S(S(a, b), c))$.

Suppose $N : [0, 1] \longrightarrow [0, 1]$ is a negation function. If N satisfies properties (N1), (N2), and the property of involution:

$$(N3) \quad Involution : \quad N(N(a)) = a,$$

then N is called an involutive negation function. With respect to an involutive negation function, we have a pair of dual T-norm and T-conorm stated as:

$$S(a, b) = N(T(N(a), N(b))), \text{ and}$$
$$T(a, b) = N(S(N(a), N(b))).$$

Example 1. The following are the four basic t-norms, namely, the minimum t-norm $t_{\mathbf{M}}$, the product t-norm $T_{\mathbf{P}}$, the Lukasiewicz t-norm $T_{\mathbf{L}}$, and the drastic product t-norm $T_{\mathbf{D}}$. For all $a, b \in [0, 1]$,

$$T_{\mathbf{M}}(a, b) = \min(a, b);$$
$$T_{\mathbf{P}}(a, b) = a \cdot b;$$
$$T_{\mathbf{L}}(a, b) = \max(0, a + b - 1);$$
$$T_{\mathbf{D}}(a, b) = \begin{cases} 0, & if \ (a, b) \in [0, 1)^2, \\ \min(a, b), & otherwise. \end{cases}$$

According to the standard negation function $N_S(a) = 1 - a$, the dual t-conorms of these t-norms are defined as follows. For all $a, b \in [0, 1]$,

$$S_{\mathbf{M}}(a, b) = \max(a, b);$$

$$S_{\mathbf{P}}(a, b) = a + b - a \cdot b;$$
$$S_{\mathbf{L}}(a, b) = \min(a + b, 1);$$
$$S_{\mathbf{D}}(a, b) = \begin{cases} 1, & if\ (a, b) \in (0, 1]^2, \\ \max(a, b), & otherwise. \end{cases}$$

Definition 4. *Let N be a negation function, and T and S be t-norm and t-conorm, respectively. We define fuzzy set complement, intersection and union as follows. For all $a \in [0, 1]$,*

$$(\mu_X)^c(a) = \mu_{X^c}(a) = N(\mu_X(a)),$$
$$(\mu_X \sqcap \mu_Y)(a) = \mu_{X \cap Y}(a) = T(\mu_X(a), \mu_Y(a)),$$
$$(\mu_X \sqcup \mu_Y)(a) = \mu_{X \cup Y}(a) = S(\mu_X(a), \mu_Y(a)). \qquad (3)$$

2.3   Fuzzy Implication

Definition 5. *[7] A binary function $I : [0, 1] \times [0, 1] \longrightarrow [0, 1]$ is a fuzzy implication if it is an extension of the Boolean implication, i.e., for any $a, b \in [0, 1]$, $I(a, b) \in [0, 1]$ and*

$$I(0, 0) = I(0, 1) = I(1, 1) = 1,\ I(1, 0) = 0.$$

For example, $I_{\mathbf{KD}}(a, b) = \max(1 - a, b)$ is a fuzzy implication, and is called Kleene-Dienes implication.

Example 2. The following binary functions are fuzzy implication. For any $a, b \in [0, 1]$,

$$I_1(a, b) = \begin{cases} b, & if\ a = 1, \\ a, & if\ a \in (0, 1)\ and\ b = 0, \\ 1, & otherwise. \end{cases}$$

$$I_2(a, b) = \begin{cases} b, & if\ a = 1, \\ 0, & if\ a \in (0, 1)\ and\ b = 0, \\ 1, & otherwise. \end{cases}$$

Left-continuity is the necessary and sufficient condition for a t-norm T and its residuum I, defined as follows[5,28], where the supremum is denoted by sup:

$$I(a, b) = \sup\{c \in [0, 1] \mid t(a, c) \le b\}. \qquad (4)$$

We can verify the residuation property:

$$T(a, b) \le c \Longleftrightarrow a \le I(b, c). \qquad (5)$$

In this case, (T, I) is called a residuated pair.

Example 3. The following are the residual implication of the four basic t-norms, denoted as $I_{\mathbf{M}}$, $I_{\mathbf{P}}$, $I_{\mathbf{L}}$, and $I_{\mathbf{D}}$, respectively. For any $a, b \in [0, 1]$,

$$I_{\mathbf{M}}(a, b) = \begin{cases} 1, & if\ a \leq b, \\ b, & otherwise; \end{cases}$$

$$I_{\mathbf{P}}(a, b) = \begin{cases} 1, & if\ a = 0, \\ \min(1, b/a), & otherwise; \end{cases}$$

$$I_{\mathbf{L}}(a, b) = \min(1, 1 - a + b);$$

$$I_{\mathbf{D}}(a, b) = \begin{cases} b, & if\ a = 1, \\ 1, & otherwise. \end{cases}$$

Example 4. Given an involutive negation N in $[0, 1]$, the corresponding weak nilpotent minimum t-norm $T_{\mathbf{NM}}$ is defined as [5,6] follows. For any $a, b \in [0, 1]$,

$$T_{\mathbf{NM}}(a, b) = \begin{cases} 0, & a \leq N(b), \\ \min(a, b), & otherwise. \end{cases}$$

The corresponding residual implication $I_{\mathbf{NM}}$ is given by: for any $a, b \in [0, 1]$,

$$I_{\mathbf{NM}}(a, b) = \begin{cases} 1, & a \leq b, \\ \max(N(a), b), & otherwise. \end{cases}$$

The symbols $T_{\mathbf{NM}}$ and $I_{\mathbf{NM}}$ are restricted in standard form, that is, N is defined as N_S.

Definition 6. *Let I be a fuzzy implication. We define fuzzy implication by:*

$$(\mu_{\mathcal{X}} \to \mu_{\mathcal{Y}})(a) = \mu_{\mathcal{X} \to \mathcal{Y}}(a) = I(\mu_{\mathcal{X}}(a), \mu_{\mathcal{Y}}(a)). \tag{6}$$

3   (0,1)-Approximations and TP$_1$ Logic

According to the definitions of core and support of a fuzzy set we can define the positive, boundary and negative regions of a fuzzy set as follows.

$$\begin{aligned}
\mathrm{POS}(\mu_{\mathcal{X}}) &= \mathrm{core}(\mu_{\mathcal{X}}) = \{a \in U \mid \mu_{\mathcal{X}}(a) = 1\}, \\
\mathrm{BND}(\mu_{\mathcal{X}}) &= \mathrm{supp}(\mu_{\mathcal{X}}) - \mathrm{core}(\mu_{\mathcal{X}}) = \{a \in U \mid 0 < \mu_{\mathcal{X}}(a) < 1\}, \\
\mathrm{NEG}(\mu_{\mathcal{X}}) &= (\mathrm{supp}(\mu_{\mathcal{X}}))^c = \{a \in U \mid \mu_{\mathcal{X}}(a) = 0\}.
\end{aligned} \tag{7}$$

The three regions are pairwise disjoint and form a partition

$$\{\mathrm{POS}(\mu_{\mathcal{X}}), \mathrm{BND}(\mu_{\mathcal{X}}), \mathrm{NEG}(\mu_{\mathcal{X}})\}$$

of the universe. To characterize the three regions, we can introduce three new membership values, namely, $\{0, M, 1\}$, where 0 indicates a non-membership (i.e., the negative region), M a partial membership (i.e., the boundary region or middle region), and 1 a full membership (i.e., the positive region). In this way, three regions can be used to define a three-valued set as an alternative representation of the core and support.

Definition 7. *Let $L3 = \{0, M, 1\}$ be a lattice with the relation $0 \leq M \leq 1$. A $(0, 1)$-approximation of a fuzzy set $\mu_{\mathcal{X}}$ is defined by a mapping $Q_{\mathcal{X}} : U \longrightarrow L3$ as follows. For any $a \in U$,*

$$Q_{\mathcal{X}}(a) = \begin{cases} 1, & a \in \text{POS}(\mu_{\mathcal{X}}), \\ M, & a \in \text{BND}(\mu_{\mathcal{X}}), \\ 0, & a \in \text{NEG}(\mu_{\mathcal{X}}). \end{cases} \tag{8}$$

Definition 8. *Let $Q(U)$ denote the set of all $(0, 1)$-approximations of fuzzy sets in $\mathcal{F}(U)$, ^{c}, $\sqcap$, $\sqcup$, and $\rightarrow$ be fuzzy set-theoretic operations on $\mathcal{F}(U)$. We define $(0, 1)$-approximation algebraic operations as follows (where $\tilde{\neg}$, $\tilde{\sqcap}$, $\tilde{\sqcup}$ and $\tilde{\rightarrow}$ denote the negation, join, meet and implication of a $(0, 1)$-approximation):*

$$\tilde{\neg}(Q_{\mathcal{X}}) = Q_{\mathcal{X}^c},$$
$$Q_{\mathcal{X}} \tilde{\sqcap} Q_{\mathcal{Y}} = Q_{\mathcal{X} \sqcap \mathcal{Y}},$$
$$Q_{\mathcal{X}} \tilde{\sqcup} Q_{\mathcal{Y}} = Q_{\mathcal{X} \sqcup \mathcal{Y}},$$
$$Q_{\mathcal{X}} \tilde{\rightarrow} Q_{\mathcal{Y}} = Q_{\mathcal{X} \rightarrow \mathcal{Y}}. \tag{9}$$

The $(0, 1)$-approximation algebraic operations are computed in two steps. For example, $Q_{\mathcal{X}} \tilde{\sqcap} Q_{\mathcal{Y}}$, is obtained by first computing intersection of two fuzzy sets $\mu_{\mathcal{X}}$ and $\mu_{\mathcal{Y}}$, namely, $\mu_{\mathcal{X}} \sqcap \mu_{\mathcal{Y}}$, and then computing the membership values using Equation (8). Although $\sqcap$ and $\sqcup$ are truth-functional, the new $(0, 1)$-approximation algebraic operations $\tilde{\sqcap}$ and $\tilde{\sqcup}$ are non-truth-functional.

Definition 9. *Let $Q(U)$ denote the set of all $(0, 1)$-approximations of fuzzy sets in $\mathcal{F}(U)$, $L3 = \{0, M, 1\}$, ^{c}, $\sqcap$, $\sqcup$ and $\rightarrow$ be fuzzy set-theoretic operations on $\mathcal{F}(U)$. We define negation, meet, join and implication operations on $L3$ as follows. For any $p, q \in L3$,*

(i) $\neg p = \{\tilde{\neg} Q_{\mathcal{X}}(a) \mid \exists \mathcal{X} \in \mathcal{F}(U) \text{ and } a \in U \text{ such that } p = Q_{\mathcal{X}}(a)\}$;

(ii) $p \sqcap q = \{(Q_{\mathcal{X}} \tilde{\sqcap} Q_{\mathcal{Y}})(a) \mid \exists \mathcal{X}, \mathcal{Y} \in \mathcal{F}(U) \text{ and } a \in U \text{ such that } p = Q_{\mathcal{X}}(a), q = Q_{\mathcal{Y}}(a)\}$;

(iii) $p \sqcup q = \{(Q_{\mathcal{X}} \tilde{\sqcup} Q_{\mathcal{Y}})(a) \mid \exists \mathcal{X}, \mathcal{Y} \in \mathcal{F}(U) \text{ and } a \in U \text{ such that } p = Q_{\mathcal{X}}(a), q = Q_{\mathcal{Y}}(a)\}$;

(iv) $p \rightarrow q = \{(Q_{\mathcal{X}} \tilde{\rightarrow} Q_{\mathcal{Y}})(a) \mid \exists \mathcal{X}, \mathcal{Y} \in \mathcal{F}(U) \text{ and } a \in U \text{ such that } p = Q_{\mathcal{X}}(a), q = Q_{\mathcal{Y}}(a)\}$.

We call $(L3, \neg, \sqcap, \sqcup, \rightarrow)$ a three-valued partial algebra (TP-algebra for short) induced by $(\mathcal{F}(U), ^c, \sqcap, \sqcup, \rightarrow)$.

Remark. (1) If $\neg p$ (or $p \sqcap q, p \sqcup q, p \rightarrow q$) is a singleton subset $\{a\}$, then we denote it by a. (2) The operator $\neg$ (or $\sqcap, \sqcup, \rightarrow$) on $L3$ may be not a classical algebraic operator (the uniqueness may not hold), since $\neg p$ (or $p \sqcap q, p \sqcup q, p \rightarrow q$) may be not a singleton. For this reason, we call it "partial algebra" (it is similar to quantum effect algebras [29]).

It is easy to prove the following theorems.

Theorem 1. *Let $^c, \sqcap, \sqcup$ and $\rightarrow$ be the standard negation N_S, the minimum t-norm $T_\mathbf{M}$, the minimum t-conorm $S_\mathbf{M}$ and the fuzzy implication I_1, respectively. Then the three-valued partial algebra $(L3, \neg, \cap, \cup, \rightarrow)$ induced by $(\mathcal{F}(U), ^c, \sqcap, \sqcup, \rightarrow)$ is isomorphic to Lukasiewicz three-valued algebra.*

Theorem 2. *Let $^c, \sqcap, \sqcup$ and $\rightarrow$ be the standard negation N_S, the minimum t-norm $T_\mathbf{M}$, the minimum t-conorm $S_\mathbf{M}$ and the Kleene-Dienes implication $I_{\mathbf{KD}}$, respectively. Then the three-valued partial algebra $(L3, \neg, \cap, \cup, \rightarrow)$ induced by $(\mathcal{F}(U), ^c, \sqcap, \sqcup, \rightarrow)$ is isomorphic to Kleene three-valued algebra.*

Theorem 3. *Let $^c, \sqcap, \sqcup$ and $\rightarrow$ be the H-negation N_H, the minimum t-norm $T_\mathbf{M}$, the minimum t-conorm $S_\mathbf{M}$ and the fuzzy implication I_2, respectively. Then the three-valued partial algebra $(L3, \neg, \cap, \cup, \rightarrow)$ induced by $(\mathcal{F}(U), ^c, \sqcap, \sqcup, \rightarrow)$ is isomorphic to Heyting three-valued algebra.*

Example 5. Let $^c, \sqcap, \sqcup$ and $\rightarrow$ be the standard negation N_S, the standard nilpotent minimum t-norm $T_\mathbf{NM}$, the standard nilpotent minimum t-conorm $S_\mathbf{NM}$ and the residual implication $I_\mathbf{NM}$, respectively. Denote $L3 = \{0, M, 1\}$. Putting $\mu_{\mathcal{X}_1}, \mu_{\mathcal{X}_2}, \mu_{\mathcal{Y}_1}, \mu_{\mathcal{Y}_2} \in \mathcal{F}(U)$ as follows. For all $a \in U$,

$$\mu_{\mathcal{X}_1}(a) = 0.2, \ \mu_{\mathcal{Y}_1}(a) = 0.1, \ \mu_{\mathcal{X}_2}(a) = 0.6, \ \mu_{\mathcal{Y}_2}(a) = 0.5.$$

By Definition 8, we have

$$(Q_{\mathcal{X}_1} \tilde{\cap} Q_{\mathcal{Y}_1})(a) = Q_{\mathcal{X}_1 \cap \mathcal{Y}_1}(a) = 0,$$
$$(Q_{\mathcal{X}_2} \tilde{\cap} Q_{\mathcal{Y}_2})(a) = Q_{\mathcal{X}_2 \cap \mathcal{Y}_2}(a) = M.$$

By Definition 9, we get $M \cap M = \{0, M\}$ in $L3$. Similarly, we have $M \cup M = \{M, 1\}, M \rightarrow M = \{M, 1\}$ in $L3$.

From the above example, we introduce a new notion of $\mathbf{TP}_1$ logic.

Definition 10. *The language of $\mathbf{TP}_1$ logic is $\mathcal{L} = \mathcal{L}_{\{\neg, \rightarrow, \cup, \cap\}}$. The logical connectives $\neg, \rightarrow, \cup, \cap$ of $\mathbf{TP}_1$ are defined as the following tables in the set $\{\bot, m, \top\}$, where $\bot < m < \top$, $\bot$ represents approximately false, $\top$ represents approximately true, and m represents medium or middle:*

$\neg$	$\bot$	m	$\top$
	$\top$	m	$\bot$

$\cup$	$\bot$	m	$\top$
$\bot$	$\bot$	m	$\top$
m	m	$\{m, \top\}$	$\top$
$\top$	$\top$	$\top$	$\top$

$\cap$	$\bot$	m	$\top$
$\bot$	$\bot$	$\bot$	$\bot$
m	$\bot$	$\{\bot, m\}$	m
$\top$	$\bot$	m	$\top$

$\rightarrow$	$\bot$	m	$\top$
$\bot$	$\top$	$\top$	$\top$
m	m	$\{m, \top\}$	$\top$
$\top$	$\bot$	m	$\top$

We call $(\{\bot, m, \top\}, \neg, \cap, \cup, \rightarrow)$ *the* TP_1 *three-valued partial algebra, where* $\neg, \cap, \cup, \rightarrow$ *are defined by above tables, respectively.*

Remark. For logic operators of $\mathbf{TP}_1$ logic, some operation result are undetermined, this means that the operation result of the same pair of logic constant may be selected in several deference value. That is, $\mathbf{TP}_1$ logic is non-truth functional logic, it is a kind of non-classical three-valued logic.

We can prove the following theorem.

Theorem 4. *Let* $^c, \sqcap, \sqcup$ *and* $\rightarrow$ *be an arbitrary involutive negation* N, *a* t-norm T, *its dual* t-conorm S *with respect to* N *and the residual implication* $\rightarrow$ *of* T, *respectively. Then the three-valued partial algebra* $(L3, \neg, \cap, \cup, \rightarrow)$ *induced by* $(\mathcal{F}(U),\ ^c, \sqcap, \sqcup, \rightarrow)$ *is isomorphic to the* TP_1 *three-valued partial algebra.*

4   Conclusion

In order to express clearly the approximations of fuzzy sets, we introduce the mathematical notion of a $(0, 1)$-approximation. By inducing approximation algebraic operators, we discuss a kind of non-classical three-valued partial algebras that some logical operations may not have unique results, nor are non-truth functional. From these results, we can see that approximations of fuzzy sets have some new properties that are differ from classical set theory and traditional fuzzy set theory.

For future study, the approximations of rough sets and related non-classical three-valued logics should be investigated. Moreover, the applications of these non-truth functional three-valued logics are emerged as an important research direction.

References

1. Chakrabarty, K., Biswas, R., Nanda, S.: Nearest ordinary set of a fuzzy set: a rough theoretic construction. Bulletin of the Polish Academy of Sciences, Technical Sciences 46, 105–114 (1998)
2. Chakrabarty, K., Biswas, R., Nanda, S.: Fuzziness in rough sets. Fuzzy Sets and Systems 110, 247–251 (2000)
3. Chanas, S.: On the interval approximation of a fuzzy number. Fuzzy Sets and Systems 122, 353–356 (2001)
4. Dubois, D., Prade, P.: A class of fuzzy measures based on triangular norms. International Journal of General Systems 8, 43–61 (1982)
5. Esteva, F., Godo, L.: Monoidal t-norm-based logic: towards a logic for left-continuous t-norms. Fuzzy Sets and Systems 124, 271–288 (2001)
6. Fodor, J.: Nilpotent minimum and related connectives for fuzzy logic. In: Proceedings of FUZZ-IEEE 1995, pp. 2077–2082 (1995)
7. Fodor, J., Keresztfalvi, T.: Nonstandard conjunctions and implications in fuzzy logic. International Journal of Approximate Reasoning 12, 69–84 (1995)
8. Giles, R.: The concept of grade of membership. Fuzzy Sets and Systems 25, 297–323 (1988)

9. Grzegorzewski, P.: Nearest interval approximation of a fuzzy number. Fuzzy Sets and Systems 130, 321–330 (2002)
10. Klement, E.P., Mesiar, R., Pap, E.: Triangular Norms. Kluwer Academic Publishers, Dordrecht (2000)
11. Klement, E.P., Mesiar, R., Pap, E.: Triangular norms. Position paper I: basic analytical and algebraic properties. Fuzzy Sets and Systems 143, 5–26 (2004)
12. Klir, G.J., Yuan, B.: Fuzzy Sets and Fuzzy Logic: Theory and Applications. Prentice Hall, New Jersey (1995)
13. Liu, G., Zhu, W.: The algebraic structures of generalized rough set theory. Information Sciences 178, 4105–4113 (2008)
14. Mi, J.S., Leung, Y., Zhao, H.Y., Feng, T.: Generalized fuzzy rough sets determined by a triangular norm. Information Sciences 178, 3203–3213 (2008)
15. Nguyen, H.T., Pedrycz, W., Kreinovich, V.: On approximation of fuzzy sets by crisp sets: from continuous control-oriented defuzzification to discrete decision making. In: Proceedings of International Conference on Intelligent Technologies, pp. 254–260 (2000)
16. Pawlak, Z., Skowron, A.: Rough membership functions, in: Yager, R.R., Fedrizzi, M. and Kacprzyk, J (Eds.). In: Advances in the Dempster-Shafer Theory of Evidence, pp. 251–271. John Wiley and Sons, New York (1994)
17. Pedrycz, W.: Shadowed sets: representing and processing fuzzy sets. IEEE Transactions on System, Man, and Cybernetics, Part B 28, 103–109 (1998)
18. Pedrycz, W.: Shadowed sets: bridging fuzzy and rough sets. In: Pal, S.K., Skowron, A. (eds.) Rough Fuzzy Hybridization a New Trend in Decision-making, pp. 179–199. Springer, Singapore (1999)
19. Pedrycz, W., Vukovich, G.: Investigating a relevance of fuzzy mappings. IEEE Transactions on System, Man, and Cybernetics, Part B 30, 249–262 (2000)
20. Wasilewska, A.: An Introduction to Classical and Non-Classical Logics. Suny, Stony Brook (2007)
21. Yao, Y.Y.: Interval-set algebra for qualitative knowledge representation. In: Proceedings of the Fifth International Conference on Computing and Information, pp. 370–374 (1993)
22. Yao, Y.Y.: Combination of rough and fuzzy sets based on α-level sets. In: Lin, T.Y., Cercone, N. (eds.) Rough Sets and Data Mining: Analysis for Imprecise Data, pp. 301–321. Kluwer Academic Publishers, Boston (1997)
23. Yao, Y.Y.: A comparative study of fuzzy sets and rough sets. Information Sciences 109, 227–242 (1998)
24. Yao, Y.Y.: Decision-theoretic rough set models. In: Yao, J., Lingras, P., Wu, W.-Z., Szczuka, M.S., Cercone, N.J., Ślęzak, D. (eds.) RSKT 2007. LNCS (LNAI), vol. 4481, pp. 1–12. Springer, Heidelberg (2007)
25. Yao, Y.Y.: Probabilistic rough set approximations. International Journal of Approximation Reasoning 49, 255–271 (2008)
26. Yao, Y.Y., Zhao, Y.: Attribute reduction in decision-theoretic rough set models. Information Sciences 178, 3356–3373 (2008)
27. Yao, Y.Y.: Crisp set approximations of fuzzy sets: a decision-theoretic rough sets perspective. Fuzzy Sets and Systems (2008)
28. Zhang, X.H.: Fuzzy Logics and Algebraic Analysis. Science Press, Beijing (2008)
29. Zhang, X.H., Fan, X.S.: Pseudo-BL algebras and pseudo-effect algebras. Fuzzy Sets and Systems 159(1), 95–106 (2008)
30. Zhu, W.: Relationship between generalized rough sets based on binary relation and covering. Information Sciences (2008)

Qualitative Approximations of Fuzzy Sets and Non-classical Three-Valued Logics (II)

Xiaohong Zhang[1,*], Yiyu Yao[2], and Yan Zhao[2]

[1] Department of Mathematics, Ningbo University
Ningbo 315211, Zhejiang Province, P.R. China
[2] Department of Computer Science, University of Regina
Regina, Saskatchewan, S4S 0A2, Canada

Abstract. (α, β)-qualitative approximations of fuzzy sets are studied by using a pair of subsets defined by an α-cut and a strong β-cut. This setting naturally leads to three disjoint regions and an analysis based on a three-valued logic. This study combines both an algebra view and a logic view. From the algebra view, the mathematical definition of an (α, β)-approximation of fuzzy sets is given, and algebraic operations based on various t-norms and fuzzy implications are established. From the logic view, two non-classical three-valued logics are introduced. Corresponding to these new non-classical three-valued logics, the related origins of t-norms and fuzzy implications are examined.

Keywords: Fuzzy set, (α, β)-approximation, non-truth functional logic, three-valued logic.

1 Introduction and Some Classical Three-Valued Logics

This paper is a continuation of our paper [4]. The motivation, basic concepts and some symbols, please see [4].

Now, we recall some classical three-valued logics, i.e, Łukasiewicz, Kleene and Heyting logics. These content are quoted from the textbook by Wasilewska [1].

1.1 Łukasiewicz Logic Ł

Łukasiewicz developed his semantics (called logic) to deal with future contingent statements. The language of Łukasiewicz logic, $\mathcal{L} = \mathcal{L}_{\{\neg, \rightarrow, \cup, \cap\}}$, is closed under $\neg$ negation, $\cup$ join, $\cap$ meet and $\rightarrow$ implication, where the logical connectives are defined as Table 1 on the set $\{F, \bot, T\}$. T and F are for true and false, respectively, and $\bot$ means that the statement cannot be assigned a truth value. It is not simply that we do not have sufficient information to decide the truth value but rather the statement does not have one. We further define that $F < \bot < T$.

* This work is supported by National Natural Science Foundation of China (Grant No. 60775038) and K.C.Wong Magna Fund at Ningbo University. The author acknowledges the support of the Department of Computer Science, University of Regina, Canada, during a visit in which the paper was finalized.

J. Yu et al. (Eds.): RSKT 2010, LNAI 6401, pp. 204–211, 2010.

Table 1. Truth tables of negation, meet, join and implication for Łukasiewicz Logic

$\neg_{\mathbf{L}}$	F	$\perp$	T
	T	$\perp$	F

$\cup_{\mathbf{L}}$	F	$\perp$	T
F	F	$\perp$	T
$\perp$	$\perp$	$\perp$	T
T	T	T	T

$\cap_{\mathbf{L}}$	F	$\perp$	T
F	F	F	F
$\perp$	F	$\perp$	$\perp$
T	F	$\perp$	T

$\rightarrow_{\mathbf{L}}$	F	$\perp$	T
F	T	T	T
$\perp$	$\perp$	T	T
T	F	$\perp$	T

$(\{F, \perp, T\}, \neg, \cap, \cup, \rightarrow)$ is called Łukasiewicz three-valued algebra if and only if $\neg, \cap, \cup, \rightarrow$ are defined by Łukasiewicz logic.

1.2   Kleene Logic K

Kleene logic was originally conceived to accommodate undecided mathematical statements. The language of Kleene logic also is $\mathcal{L} = \mathcal{L}_{\{\neg, \rightarrow, \cup, \cap\}}$. In Kleene's semantics, the logical value $\perp$ represents undecided. Its purpose is to signal a state of partial ignorance. A sentence is assigned a value $\perp$ just in case it is not known to be either true of false. The logical connectives of Kleene logic are defined on the set $\{F, \perp, T\}$. The definitions of the operators $\neg, \cup$ and $\cap$ of Kleene logic are the same as the ones of Łukasiewicz logic, while the definition of implication of Kleene logic is different, and is defined as Table 2.

Table 2. Truth table of implication for Kleene logic

$\rightarrow_{\mathbf{K}}$	F	$\perp$	T
F	T	T	T
$\perp$	$\perp$	$\perp$	T
T	F	$\perp$	T

In fact, the logical connectives $\cup, \cap$ of both Łukasiewicz logic and Kleene logic can be defined as follows. For any $a, b \in \{F, \perp, T\}$,

$$a \cup_{\mathbf{L}} b = a \cup_{\mathbf{K}} b = \max\{a, b\},$$
$$a \cap_{\mathbf{L}} b = a \cap_{\mathbf{K}} b = \min\{a, b\}.$$

The implication in Łukasiewicz and Kleene logic can be defined as follows. For any $a, b \in \{F, \perp, T\}$,

$$a \rightarrow_{\mathbf{L}} b = \begin{cases} \neg a \cup b, & \textit{if } a > b, \\ T, & \textit{otherwise.} \end{cases}$$
$$a \rightarrow_{\mathbf{K}} b = \neg a \cup b.$$

$(\{F, \perp, T\}, \neg, \cap, \cup, \rightarrow)$ is called Kleene three-valued algebra if and only if $\neg, \cap,$ $\cup, \rightarrow$ are defined by Kleene logic.

1.3   Heyting Logic H

The language of Heyting logic is the same as the one for previous cases, i.e., $\mathcal{L} = \mathcal{L}_{\{\neg, \rightarrow, \cup, \cap\}}$. The logical connectives of Heyting logic are defined on the set $\{F, \perp, T\}$. The definitions of the operations $\cup$ and $\cap$ of Heyting logic are the same as the ones of Łukasiewicz and Kleene logics, while the definitions of implication and negation of Heyting logic are different, and are defined as Table 3.

Table 3. Truth tables of negation and implication for Heyting logic

$\neg_{\mathbf{H}}$	F	$\perp$	T
	T	F	F

$\rightarrow_{\mathbf{H}}$	F	$\perp$	T
F	T	T	T
$\perp$	F	T	T
T	F	$\perp$	T

In fact, the logical connectives $\rightarrow_{\mathbf{H}}, \neg_{\mathbf{H}}$ of Heyting logic can be defined as follows. For any $a, b \in \{F, \perp, T\}$,

$$a \rightarrow_{\mathbf{H}} b = \begin{cases} T, & \text{if } a \leq b, \\ b, & \text{otherwise.} \end{cases}$$

$$\neg_{\mathbf{H}} a = a \rightarrow_{\mathbf{H}} F.$$

$(\{F, \perp, T\}, \neg, \cap, \cup, \rightarrow)$ is called Heyting three-valued algebra if and only if $\neg, \cap,$ $\cup, \rightarrow$ are defined by Heyting logic. Heyting algebra provides an algebraic model for the intuitionistic logic.

2 (α, β)-Approximations and TP$_1$ Logic

Given a number $\alpha \in [0, 1]$, an α-cut of a fuzzy set μ_χ is defined by:

$$\alpha(\mu_\chi) = \{a \in U \mid \mu_\chi(a) \geq \alpha\}, \qquad (1)$$

which is a subset of U. A strong α-cut is defined by:

$$\alpha^+(\mu_\chi) = \{a \in U \mid \mu_\chi(a) > \alpha\}. \qquad (2)$$

Through either an α-cut or a strong α-cut, a fuzzy set determines a family of nested subsets of U. By definition, we have:

$$\text{core}(\mu_\chi) = 1(\mu_\chi),$$
$$\text{supp}(\mu_\chi) = 0^+(\mu_\chi). \qquad (3)$$

By generalizing these relations, for a pair of parameters $0 \leq \beta < \alpha \leq 1$, Yao [2] introduce an (α, β)-core and an (α, β)-support as follows:

$$\text{core}_{(\alpha,\beta)}(\mu_{\mathcal{X}}) = \alpha(\mu_{\mathcal{X}}) = \{a \in U \mid \mu_{\mathcal{X}}(a) \geq \alpha\},$$
$$\text{supp}_{(\alpha,\beta)}(\mu_{\mathcal{X}}) = \beta^+(\mu_{\mathcal{X}}) = \{a \in U \mid \mu_{\mathcal{X}}(a) > \beta\}. \tag{4}$$

For a pair of numbers $0 \leq \beta < \alpha \leq 1$, we derive three regions based on the (α, β)-core and (α, β)-support:

$$\text{POS}_{(\alpha,\beta)}(\mu_{\mathcal{X}}) = \text{core}_{\alpha}(\mu_{\mathcal{X}}) = \{a \in U \mid \mu_{\mathcal{X}}(a) \geq \alpha\},$$
$$\text{BND}_{(\alpha,\beta)}(\mu_{\mathcal{X}}) = \text{supp}_{\beta}(\mu_{\mathcal{X}}) - \text{core}_{\alpha}(\mu_{\mathcal{X}}) = \{a \in U \mid \beta < \mu_{\mathcal{X}}(a) < \alpha\},$$
$$\text{NEG}_{(\alpha,\beta)}(\mu_{\mathcal{X}}) = (\text{supp}_{\beta}(\mu_{\mathcal{X}}))^c = \{a \in U \mid \mu_{\mathcal{X}}(a) \leq \beta\}. \tag{5}$$

They are pairwise disjoint and form a partition

$$\{\text{POS}_{(\alpha,\beta)}(\mu_{\mathcal{X}}), \text{BND}_{(\alpha,\beta)}(\mu_{\mathcal{X}}), \text{NEG}_{(\alpha,\beta)}(\mu_{\mathcal{X}})\}$$

of the universe.

Definition 1. *Let $L3 = \{0, M, 1\}$ be a lattice with the relation $0 \leq M \leq 1$. An (α, β)-approximation of a fuzzy set $\mu_{\mathcal{X}}$ is defined by a mapping $Q_{\mathcal{X}}^{(\alpha,\beta)} : U \longrightarrow L3$ as follows,*

$$Q_{\mathcal{X}}^{(\alpha,\beta)}(a) = \begin{cases} 1, & a \in \text{POS}_{(\alpha,\beta)}(\mu_{\mathcal{X}}), \\ M, & a \in \text{BND}_{(\alpha,\beta)}(\mu_{\mathcal{X}}), \\ 0, & a \in \text{NEG}_{(\alpha,\beta)}(\mu_{\mathcal{X}}). \end{cases} \tag{6}$$

When $\beta = 1 - \alpha$ and $\alpha > 0.5$, (α, β)-approximation is called an (α)-approximation, and we can simply denote $Q_{\mathcal{X}}^{(\alpha,\beta)}$ by $Q_{\mathcal{X}}^{(\alpha)}$.

The three values represent another kind of qualitative properties of elements of U with respect to their memberships of $\mu_{\mathcal{X}}$. That is, all elements with membership values that are greater than or equal to α are considered to be qualitatively equivalent. The same interpretation also applied to other two classes. Thus, we qualitatively characterize a fuzzy set by three crisp sets.

Definition 2. *Let $Q^{(\alpha,\beta)}(U)$ denote the set of all (α, β)-approximations of a fuzzy set in $\mathcal{F}(U)$, ^{c}, $\sqcap$, and $\sqcup$, and $\rightarrow$ be fuzzy set-theoretic operations on $\mathcal{F}(U)$. We define (α, β)-approximation algebraic operations as follows (where $\tilde{\neg}$, $\tilde{\sqcap}$, $\tilde{\sqcup}$ and $\tilde{\rightarrow}$ denote the negation, meet, join and implication of an (α, β)-approximation):*

$$\tilde{\neg}(Q_{\mathcal{X}}^{(\alpha,\beta)}) = Q_{\mathcal{X}^c}^{(\alpha,\beta)},$$
$$Q_{\mathcal{X}}^{(\alpha,\beta)} \tilde{\sqcap} Q_{\mathcal{Y}}^{(\alpha,\beta)} = Q_{\mathcal{X} \sqcap \mathcal{Y}}^{(\alpha,\beta)},$$
$$Q_{\mathcal{X}}^{(\alpha,\beta)} \tilde{\sqcup} Q_{\mathcal{Y}}^{(\alpha,\beta)} = Q_{\mathcal{X} \sqcup \mathcal{Y}}^{(\alpha,\beta)},$$
$$Q_{\mathcal{X}}^{(\alpha,\beta)} \tilde{\rightarrow} Q_{\mathcal{Y}}^{(\alpha,\beta)} = Q_{\mathcal{X} \rightarrow \mathcal{Y}}^{(\alpha,\beta)}. \tag{7}$$

They are obtained by first taking fuzzy set-theoretic operations and then computing from generalized fuzzy core and support by Equation (6).

Definition 3. *Let $Q^{(\alpha,\beta)}(U)$ denote the set of all (α,β)-approximations of fuzzy sets in $\mathcal{F}(U)$, $L3 = \{0, M, 1\}$, and ^{c}, $\sqcap$, $\sqcup$ and $\rightarrow$ be fuzzy set-theoretic operations on $\mathcal{F}(U)$. We define negation, meet, join and implication operations on $L3$ as follows. For any $p, q \in L3$,*

(i) $\neg p = \{\tilde{\neg}Q_{\mathcal{X}}^{(\alpha,\beta)}(a) \mid \exists \mathcal{X} \in \mathcal{F}(U) \text{ and } a \in U \text{ such that } p = Q_{\mathcal{X}}^{(\alpha,\beta)}(a)\}$;

(ii) $p \cap q = \{(Q_{\mathcal{X}}^{(\alpha,\beta)}\tilde{\sqcap}Q_{\mathcal{Y}}^{(\alpha,\beta)})(a) \mid \exists \mathcal{X}, \mathcal{Y} \in \mathcal{F}(U) \text{ and } a \in U \text{ such that}$
$\qquad p = Q_{\mathcal{X}}^{(\alpha,\beta)}(a), q = Q_{\mathcal{Y}}^{(\alpha,\beta)}(a)\}$;

(iii) $p \cup q = \{(Q_{\mathcal{X}}^{(\alpha,\beta)}\tilde{\sqcup}Q_{\mathcal{Y}}^{(\alpha,\beta)})(a) \mid \exists \mathcal{X}, \mathcal{Y} \in \mathcal{F}(U) \text{ and } a \in U \text{ such that}$
$\qquad p = Q_{\mathcal{X}}^{(\alpha,\beta)}(a), q = Q_{\mathcal{Y}}^{(\alpha,\beta)}(a)\}$;

(iv) $p \rightarrow q = \{(Q_{\mathcal{X}}^{(\alpha,\beta)}\tilde{\rightarrow}Q_{\mathcal{Y}}^{(\alpha,\beta)})(a) \mid \exists \mathcal{X}, \mathcal{Y} \in \mathcal{F}(U) \text{ and } a \in U \text{ such that}$
$\qquad p = Q_{\mathcal{X}}^{(\alpha,\beta)}(a), q = Q_{\mathcal{Y}}^{(\alpha,\beta)}(a)\}$.

We call $(L3, \neg, \cap, \cup, \rightarrow)$ a three-valued (α,β)-partial algebra (for short, $TP^{(\alpha,\beta)}$-algebra) induced by $(\mathcal{F}(U),^c, \sqcap, \sqcup, \rightarrow)$. When $\beta = 1 - \alpha$ and $\alpha > 0.5$, $TP^{(\alpha,\beta)}$-algebra is called $TP^{(\alpha)}$-algebra.

We can prove the following theorems.

Theorem 1. *Let $^c, \sqcap, \sqcup$ and $\rightarrow$ be the standard negation N_S, the minimum t-norm $T_\mathbf{M}$, the minimum t-conorm $S_\mathbf{M}$ and the fuzzy implication I_1^α, respectively. If $\alpha > 0.5$ and $\beta = 1-\alpha$, then the three-valued (α)-partial algebra $(L3, \neg, \cap, \cup, \rightarrow)$ induced by $(\mathcal{F}(U),^c, \sqcap, \sqcup, \rightarrow)$ is isomorphic to Łukasiewicz three-valued algebra, where the fuzzy implication I_1^α is defined as follows. For any $a, b \in [0, 1]$,*

$$I_1^\alpha(a, b) = \begin{cases} b, & if \ a \geq \alpha, \\ a, & if \ a \in (1 - \alpha, \alpha) b = 0, \\ 1, & otherwise. \end{cases}$$

Theorem 2. *Let $^c, \sqcap, \sqcup$ and $\rightarrow$ be the standard negation N_S, the minimum t-norm $T_\mathbf{M}$, the minimum t-conorm $S_\mathbf{M}$ and the Kleene-Dienes implication $I_{\mathbf{KD}}$, respectively. If $\alpha > 0.5$ and $\beta = 1 - \alpha$, then the three-valued (α)-partial algebra $(L3, \neg, \cap, \cup, \rightarrow)$ induced by $(\mathcal{F}(U),\ ^c, \sqcap, \sqcup, \rightarrow)$ is isomorphic to Kleene three-valued algebra.*

Theorem 3. *Let $^c, \sqcap, \sqcup$ and $\rightarrow$ be the H-negation N_H, the minimum t-norm $T_\mathbf{M}$, the minimum t-conorm $S_\mathbf{M}$ and the fuzzy implication I_2^α, respectively. If $\alpha > 0.5$ and $\beta = 1 - \alpha$, then the three-valued (α)-partial algebra $(L3, \neg, \cap, \cup, \rightarrow)$ induced by $(\mathcal{F}(U),\ ^c, \sqcap, \sqcup, \rightarrow)$ is isomorphic to Heyting three-valued algebra, where the fuzzy implication I_2^α is defined as follows. For any $a, b \in [0, 1]$,*

$$I_2^\alpha(a, b) = \begin{cases} b, & if \ a \geq \alpha, \\ 0, & if \ a \in (1 - \alpha, \alpha) b = 0, \\ 1, & otherwise. \end{cases}$$

Theorem 4. *Let $^c, \sqcap, \sqcup$ and $\rightarrow$ be the standard negation N_S, the standard nilpotent minimum t-norm $T_{\mathbf{NM}}$, its dual t-conorm $S_{\mathbf{NM}}$ and the residual implication $I_{\mathbf{NM}}$, respectively. If $\alpha > 0.5$ and $\beta = 1 - \alpha$, then the three-valued (α)-partial algebra $(L3, \neg, \cap, \cup, \rightarrow)$ induced by $(\mathcal{F}(U),\ ^c, \sqcap, \sqcup, \rightarrow)$ is isomorphic to the TP_1 three-valued partial algebra.*

3 (α, β)-Approximations and $\mathbf{TP_2}/\mathbf{TP_3}$ Logic

Example 1. Let $^c, \sqcap, \sqcup$ and $\rightarrow$ be the standard negation N_S, the product t-norm $T_{\mathbf{P}}$, its dual t-conorm $S_{\mathbf{P}}$ and the residual implication $I_{\mathbf{P}}$, respectively. Denote $L3 = \{0, M, 1\}$. Putting $\mu_{\mathcal{X}_1}, \mu_{\mathcal{X}_2}, \mu_{\mathcal{Y}_1}, \mu_{\mathcal{Y}_2} \in \mathcal{F}(U)$ as follows. For all $a \in U$,

$$\mu_{\mathcal{X}_1}(a) = 0.35,\ \mu_{\mathcal{Y}_1}(a) = 0.8,\ \mu_{\mathcal{Z}_1}(a) = 0.95,$$
$$\mu_{\mathcal{X}_2}(a) = 0.85,\ \mu_{\mathcal{Y}_2}(a) = 0.9,\ \mu_{\mathcal{Z}_2}(a) = 0.98.$$

Let $\alpha = 0.85, \beta = 0.3 \neq 1 - \alpha$, by Definition 12 we have

$$(Q_{\mathcal{X}_1}^{(\alpha,\beta)} \tilde{\sqcap} Q_{\mathcal{Y}_1}^{(\alpha,\beta)})(a) = Q_{\mathcal{X}_1 \sqcap \mathcal{Y}_1}^{(\alpha,\beta)}(a) = 0,$$
$$(Q_{\mathcal{Y}_1}^{(\alpha,\beta)} \tilde{\sqcap} Q_{\mathcal{Z}_1}^{(\alpha,\beta)})(a) = Q_{\mathcal{Y}_1 \sqcap \mathcal{Z}_1}^{(\alpha,\beta)}(a) = M;$$
$$(Q_{\mathcal{X}_2}^{(\alpha,\beta)} \tilde{\sqcap} Q_{\mathcal{Y}_2}^{(\alpha,\beta)})(a) = Q_{\mathcal{X}_2 \sqcap \mathcal{Y}_2}^{(\alpha,\beta)}(a) = M,$$
$$(Q_{\mathcal{Y}_2}^{(\alpha,\beta)} \tilde{\sqcap} Q_{\mathcal{Z}_2}^{(\alpha,\beta)})(a) = Q_{\mathcal{Y}_2 \sqcap \mathcal{Z}_2}^{(\alpha,\beta)}(a) = 1;$$
$$(Q_{\mathcal{X}_1}^{(\alpha,\beta)} \tilde{\rightarrow} Q_{\mathcal{Y}_1}^{(\alpha,\beta)})(a) = Q_{\mathcal{X}_1 \rightarrow \mathcal{Y}_1}^{(\alpha,\beta)}(a) = 1,$$
$$(Q_{\mathcal{Y}_1}^{(\alpha,\beta)} \tilde{\rightarrow} Q_{\mathcal{X}_1}^{(\alpha,\beta)})(a) = Q_{\mathcal{Y}_1 \rightarrow \mathcal{X}_1}^{(\alpha,\beta)}(a) = M.$$

By Definition 13, we get $M \cap M = \{0, M\}, 1 \cap 1 = \{M, 1\}$ and $M \rightarrow M = \{M, 1\}$ in $L3$. Similarly, we have

$\neg 0 = \{M, 1\},\ \neg M = \{0, M\},\ \neg 1 = 0;$
$0 \cup 0 = \{0, M\},\ 0 \cup M = \{M, 1\},\ M \cup M = \{M, 1\},\ 1 \cup M = 1,\ 1 \cup 1 = 1;$
$0 \cap 0 = 0,\ 0 \cap M = 0,\ M \cap M = \{0, M\},\ 1 \cap M = \{0, M\},\ 1 \cap 1 = \{M, 1\};$
$0 \rightarrow 0 = \{0, M, 1\},\ 0 \rightarrow M = 1,\ M \rightarrow 0 = \{0, M, 1\};$
$M \rightarrow M = \{M, 1\},\ 1 \rightarrow 0 = \{0, M\},\ 1 \rightarrow M = \{M, 1\}.$

From the above example, we introduce a new notion of $\mathbf{TP_2}$ logic.

Definition 4. *The language of $\mathbf{TP_2}$ logic is $\mathcal{L} = \mathcal{L}_{\{\neg, \rightarrow, \cup, \cap\}}$. The logical connectives $\neg, \rightarrow, \cup, \cap$ of $\mathbf{TP_2}$ are defined as the following tables in the set $\{\bot, m, \top\}$, where $\bot < m < \top$, $\bot$ represents approximately false, $\top$ represents approximately true, and m represents medium or middle:*

$\neg$	$\bot$	m	$\top$
	$\{m, \top\}$	$\{\bot, m\}$	$\bot$

$\cup$	$\bot$	m	$\top$
$\bot$	$\{\bot, m, \top\}$	$\{m, \top\}$	$\top$
m	$\{m, \top\}$	$\{m, \top\}$	$\top$
$\top$	$\top$	$\top$	$\top$

$\cap$	$\bot$	m	$\top$
$\bot$	$\bot$	$\bot$	$\bot$
m	$\bot$	$\{\bot,m\}$	$\{m,\top\}$
$\top$	$\bot$	$\{m,\top\}$	$\{\bot,m,\top\}$

$\to$	$\bot$	m	$\top$
$\bot$	$\{\bot,m,\top\}$	$\top$	$\top$
m	$\{\bot,m,\top\}$	$\{m,\top\}$	$\top$
$\top$	$\{\bot,m,\top\}$	$\{m,\top\}$	$\top$

We call $(\{\bot,m,\top\},\neg,\cap,\cup,\to)$ the TP_2 three-valued partial algebra, where $\neg,\cap,\cup,\to$ are defined by above tables, respectively.

Remark. Though we have $1 \to 0 = \{0,M\}$ in Example 6, but we define $\top \to \bot = \{\bot,m,\top\}$ in Definition 14. Because when we select other α and β, $1 \to 0$ maybe include 1. For example, putting $\alpha = 0.46$ and $\beta = 0.56$, then $0 \le \beta < 0.5 < \alpha \le 1$, $\alpha + \beta > 1$, and

$$(Q_{\mathcal{X}}^{(\alpha,\beta)} \tilde{\to} Q_{\mathcal{Y}}^{(\alpha,\beta)})(a) = Q_{\mathcal{X} \to \mathcal{Y}}^{(\alpha,\beta)}(a) = 1,$$

where $\mu_{\mathcal{X}}(a) = 0.6$, $\mu_{\mathcal{Y}}(a) = 0.45, \forall a \in [0,1]$, and $\to$ is a residual implication I_P. Similarly, we define $\bot \cup \bot = \{\bot,m,\top\}$, $\top \cap \top = \{\bot,m,\top\}$ and so on.

We can prove the following theorem.

Theorem 5. *Let $^c,\cap,\sqcup$ and $\to$ be the standard negation N_S, the standard t-norm T, its dual t-conorm S with respect to N and the residual implication $\to$ of T, respectively. If $0 \le \beta < 0.5 < \alpha \le 1$ and $\alpha + \beta > 1$, then the three-valued (α,β)-partial algebra $(L3,\neg,\cap,\cup,\to)$ induced by $(\mathcal{F}(U),\ ^c,\cap,\sqcup,\to)$ is partial isomorphism to the TP_2 three-valued partial algebra.*

Remark. Here, "partial isomorphism" means that there exists bijection f such that

$$f(\neg p) \subseteq \neg(f(p));$$
$$f(p \cap q) \subseteq f(p) \cap f(q);$$
$$f(p \cup q) \subseteq f(p) \cup f(q);$$
$$f(p \to q) \subseteq f(p) \to f(q).$$

Similarly, we can introduce the notion of $\mathbf{TP_3}$ logic.

Definition 5. *The language of $\mathbf{TP_3}$ logic is $\mathcal{L} = \mathcal{L}_{\{\neg,\to,\cup,\cap\}}$. The logical connectives $\neg, \to, \cup, \cap$ of $\mathbf{TP_3}$ are defined as the following tables in the set $\{\bot,m,\top\}$, where $\bot < m < \top$, $\bot$ represents approximately false, $\top$ represents approximately true, and m represents medium or middle:*

$\neg$	$\bot$	m	$\top$
	$\{m,\top\}$	$\{m,\top\}$	$\{\bot,m\}$

$\cup$	$\bot$	m	$\top$
$\bot$	$\{\bot,m,\top\}$	$\{m,\top\}$	$\top$
m	$\{m,\top\}$	$\{m,\top\}$	$\top$
$\top$	$\top$	$\top$	$\top$

$\cap$	$\bot$	m	$\top$
$\bot$	$\bot$	$\bot$	$\bot$
m	$\bot$	$\{\bot, m\}$	$\{m, \top\}$
$\top$	$\bot$	$\{m, \top\}$	$\{\bot, m, \top\}$

$\rightarrow$	$\bot$	m	$\top$
$\bot$	$\{\bot, m, \top\}$	$\top$	$\top$
m	$\{\bot, m, \top\}$	$\{m, \top\}$	$\top$
$\top$	$\{\bot, m, \top\}$	$\{\bot, m\}$	$\top$

We call $(\{\bot, m, \top\}, \neg, \cap, \cup, \rightarrow)$ *the* TP_3 *three-valued partial algebra, where* $\neg, \cap, \cup, \rightarrow$ *are defined by above tables, respectively.*

We can prove the following theorem.

Theorem 6. *Let* $^c, \sqcap, \sqcup$ *and* $\rightarrow$ *be the standard negation* N_S, *the standard t-norm* T, *its dual t-conorm* S *with respect to* n *and the residual implication* $\rightarrow$ *of* T, *respectively. If* $0 \leq \beta < 0.5 < \alpha \leq 1$ *and* $\alpha + \beta < 1$, *then the three-valued* (α, β)*-partial algebra* $(L3, \neg, \cap, \cup, \rightarrow)$ *induced by* $(\mathcal{F}(U), {}^c, \sqcap, \sqcup, \rightarrow)$ *is partial isomorphic to the* TP_3 *three-valued partial algebra.*

4   Conclusion

In order to express clearly the approximations of fuzzy sets, we introduce the mathematical notion of an (α, β)-approximation. By inducing approximation algebraic operators, we discuss two kinds of non-classical three-valued partial algebras that some logical operations may not have unique results, nor are non-truth functional. From these results, we can see that approximations of fuzzy sets have some new properties that are differ from classical set theory and traditional fuzzy set theory. For future study, the approximations of rough sets and related non-classical three-valued logics should be investigated. Moreover, the applications of these non-truth functional three-valued logics are emerged as an important research direction.

References

1. Wasilewska, A.: An Introduction to Classical and Non-Classical Logics. Suny, Stony Brook (2007)
2. Yao, Y.Y.: Crisp set approximations of fuzzy sets: a decision-theoretic rough sets perspective. Fuzzy Sets and systems (2008)
3. Zhang, X.H.: Fuzzy Logics and Algebraic Analysis. Science Press, Beijing (2008)
4. Zhang, X., Yao, Y., Zhao, Y.: Qualitative Approximations of Fuzzy Sets and Non-classical Three-Valued Logics (I). In: Yu, J., Greco, S., Lingras, P., Wang, G., Skowron, A. (eds.) RSKT 2010. LNCS(LNAI), vol. 6401, pp. 195–203. Springer, Heidelberg (2010)

Implication Operator of Linguistic Truth-Valued Intuitionistic Fuzzy Lattice

Chunying Guo[1], Fengmei Zhang[2], Li Zou[2], and Kaiqi Zou[3]

[1] Vocational College, Liaoning Technical University, Fu Xin 123000, P.R. China
[2] School of Computer and Information Technology,Liaoning Normal University,
Dalian 116029, P.R. China
[3] College of Information Engineering, University Key Lab of Information Science and
Engineering, Dalian University, Dalian 116622, China
zoulicn@163.com

Abstract. We construct linguistic truth-valued intuitionistic fuzzy lattice b based on linguistic truth-valued lattice implication algebra to deal with linguistic truth-valued. The proposed system can better express both comparable and incomparable information. Also it can deal with both positive and negative evidences which are represented by linguistic truth values at the same time during information processing system. This paper discusses the intuitionistic fuzzy implication operator on the linguistic truth-valued intuitionistic fuzzy lattice and its properties.

Keywords: Linguistic truth-valued lattice implication algebra, Linguistic truth-valued intuitionistic fuzzy lattice, Intuitionistic fuzzy implication.

1 Introduction

Natural languages can be formalized by Zadeh's linguistic variables [7], in which, a linguistic value is consist of an atomic linguistic value and a linguistic hedge, *e.g.*, very true (true is the atomic linguistic value and very is the linguistic hedge). In computing with words (CWW), very true is expressed by a fuzzy set on $[0, 1]$, *i.e.*, semantics of very true. Information processing corresponding to linguistic values is translated to their semantics, and fuzzy set theory becomes a main tool for CWW. Nowadays, there exist many alternative methods to deal with linguistic valued of intelligent information processing[2], *e.g.*, Huynh proposed a new model for parametric representation of linguistic truth-values[3]. Ho discussed the ordering structure of linguistic hedges, and proposed hedge algebra for CWW[4]. Linguistic truth-valued lattice implication algebra proposed by Xu, et al. to deal with linguistic truth inference [5]. A framework of linguistic truth-valued propositional logic was developed by Zou [8,9,10] proposed the reasoning method of linguistic truth-valued logic system.

Intuitionistic fuzzy sets (A-IFSs) introduced by Atanassov is a powerful tool to deal with uncertainty [6]. A-IFSs concentrate on expressing advantages and disadvantages, pros and cons. Inspired by linguistic truth-valued lattice implication algebra and A-IFSs, we will put the linguistic truth-values into intuitionistic

J. Yu et al. (Eds.): RSKT 2010, LNAI 6401, pp. 212–219, 2010.

fuzzy logic. The truth values of the intuitionistic fuzzy logic are linguistic truth-valued intuitionistic fuzzy sets insteading of number.

The organization of this paper is as follows: In Section 2, we discuss a framework of the linguistic truth-valued lattice implication algebra. We analysis implication operator on linguistic truth-valued intuitionistic fuzzy lattice in Section 3. We conclude the paper in Section 4.

2   Linguistic Truth-Valued Lattice Implication Algebra

Definition 1. *Let* $L_n = \{d_1, d_2, \cdots, d_n\}$, $d_1 < d_2 < \cdots < d_n$, $L_2 = \{b_1, b_2\}$, $b_1 < b_2$, $(L_n, \vee_{(L_n)}, \wedge_{(L_n)},'^{(L_n)}, \rightarrow_{(L_n)}, d_1, d_n)$ *and* $(L_2, \vee_{(L_2)}, \wedge_{(L_2)},'^{(L_2)}, \rightarrow_{(L_2)}, b_1, b_2)$ *be two Lukasiewicz implication algebra. For any* $(d_i, b_j), (d_k, b_m) \in L_n \times L_2$, *if*

$$(d_i, b_j) \vee (d_k, b_m) = (d_i \vee_{(L_n)} d_k, b_j \vee_{(L_2)} b_m), \tag{1}$$

$$(d_i, b_j) \wedge (d_k, b_m) = (d_i \wedge_{(L_n)} d_k, b_j \wedge_{(L_2)} b_m), \tag{2}$$

$$(d_i, b_j)' = (d_i'^{(L_n)}, b_j'^{(L_2)}), \tag{3}$$

$$(d_i, b_j) \rightarrow (d_k, b_m) = (d_i \rightarrow_{(L_n)} d_k, b_j \rightarrow_{(L_2)} b_m). \tag{4}$$

Then $(L_n \times L_2, \vee, \wedge,', \rightarrow, (d_1, b_1), (d_n, b_2))$ *is a lattice implication algebra, denote as* $\mathcal{L}_n \times \mathcal{L}_2$ *(Figure.1).*

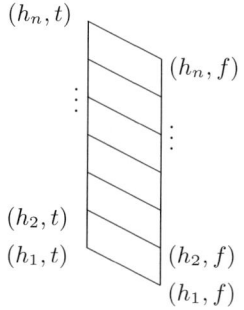

(h_n, t)

(h_n, f)

(h_2, t)
(h_1, t)

(h_2, f)
(h_1, f)

Fig. 1. Hasse Diagrams of $\mathcal{L}_n \times \mathcal{L}_2$

Let $AD_n = \{a_1, a_2, \cdots, a_n\}$ be a set of n hedge operators and $a_1 < a_2 < \cdots < a_n$, $MT = \{f, t\}$ be meta truth values and $f < t$, denote it by $L_{V(n \times 2)} = AD_n \times MT$.

Define the mapping g: $L_{V(n \times 2)} \longrightarrow \mathcal{L}_n \times \mathcal{L}_2$,

$$g((a_i, mt)) = \begin{cases} (d_i', b_1), & mt = f, \\ (d_i, b_2), & mt = t. \end{cases} \tag{5}$$

Then g is bejuction, its inverse mapping is g^{-1}. For any $x, y \in L_{V(n \times 2)}$, define

$$x \vee y = g^{-1}(g(x) \vee g(y)), \tag{6}$$
$$x \wedge y = g^{-1}(g(x) \wedge g(y)), \tag{7}$$
$$x' = g^{-1}((g(x))'), \tag{8}$$
$$x \to y = g^{-1}(g(x) \to g(y)). \tag{9}$$

Then $\mathcal{L}_{V(n \times 2)} = (L_{V(n \times 2)}, \vee, \wedge, ', \to, (a_n, f), (a_n, t))$ is called linguistic truth-valued lattice implication algebra from AD_n and MT (figure.2). g is an isomorphic mapping from $(L_{V(n \times 2)}, \vee, \wedge, ', \to, (a_n, f), (a_n, t))$ to $\mathcal{L}_n \times \mathcal{L}_2$.

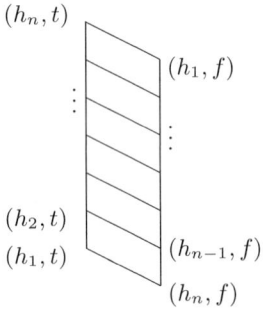

Fig. 2. Hasse Diagrams of $\mathcal{L}_{V(n \times 2)}$

Based on $2n$ linguistic truth-valued lattice implication algebra $\mathcal{L}_{V(n \times 2)}$, we can construct linguistic truth-valued intuitionistic fuzzy lattice. Formally, denote $\mathcal{L I}_{2n} = (LI_{2n}, \cup, \cap)$ as linguistic truth-valued intuitionistic fuzzy lattice, where $((h_n, t), (h_n, f))$ and $((h_1, t), (h_1, f))$ are the greatest element and the least element of $\mathcal{L I}_{2n}$ respectively.

3   Linguistic Truth-Valued Intuitionistic Fuzzy Lattice $\mathcal{L I}_{2n}$

Definition 2. *In the linguistic truth-valued intuitionistic fuzzy lattice* $\mathcal{L I}_{2n} = (LI_{2n}, \cup, \cap)$ *(Figure.3), for any* $((h_i, t), (h_j, f)), ((h_k, t), (h_l, f)) \in LI_{2n}$, $((h_i, t), (h_j, f)) \leq ((h_k, t), (h_l, f))$ *if and only if* $i \leq k$ *and* $j \leq l$, *also*

$$((h_i, t), (h_j, f)) \cup ((h_k, t), (h_l, f)) = ((h_{max(i,k)}, t), (h_{max(j,l)}, f)), \tag{10}$$
$$((h_i, t), (h_j, f)) \cap ((h_k, t), (h_l, f)) = ((h_{min(i,k)}, t), (h_{min(j,l)}, f)). \tag{11}$$

We can prove the following propeties in $\mathcal{L I}_{2n}$.

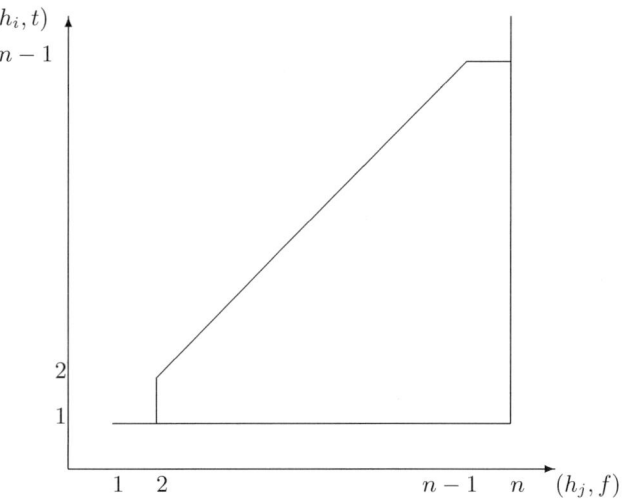

Fig. 3. Structure Diagrams of $\mathcal{LI}_{2n}$

Corollary 1. *For any* $((h_i,t),(h_j,f)),((h_k,t),(h_l,f)),((h_m,t),(h_s,f)) \in LI_{2n}$,

1. *Idempotent:* $((h_i,t),(h_j,f)) \cup ((h_i,t),(h_j,f)) = ((h_i,t),(h_j,f))$, $((h_i,t),(h_j,f)) \cap ((h_i,t),(h_j,f)) = ((h_i,t),(h_j,f))$;
2. *Commutative:* $((h_i,t),(h_j,f)) \cup ((h_k,t),(h_l,f)) = ((h_k,t),(h_l,f)) \cup ((h_i,t),(h_j,f))$, $((h_i,t),(h_j,f)) \cap ((h_k,t),(h_l,f)) = ((h_k,t),(h_l,f)) \cap ((h_i,t),(h_j,f))$;
3. *Associative:* $((h_i,t),(h_j,f)) \cup (((h_k,t),(h_l,f)) \cup ((h_m,t),(h_s,f))) = (((h_i,t),(h_j,f)) \cup ((h_k,t),(h_l,f))) \cup ((h_m,t),(h_s,f))$, $((h_i,t),(h_j,f)) \cap (((h_k,t),(h_l,f)) \cap ((h_m,t),(h_s,f))) = (((h_i,t),(h_j,f)) \cap ((h_k,t),(h_l,f))) \cap ((h_m,t),(h_s,f))$;
4. *Absorption:* $((h_i,t),(h_j,f)) \cup (((h_i,t),(h_j,f)) \cap ((h_k,t),(h_l,f))) = ((h_i,t),(h_j,f))$, $((h_i,t),(h_j,f)) \cap (((h_i,t),(h_j,f)) \cup ((h_k,t),(h_l,f))) = ((h_i,t),(h_j,f))$
5. *Distributive:* $((h_m,t),(h_s,f)) \cap (((h_i,t),(h_j,f)) \cup ((h_k,t),(h_l,f))) = (((h_m,t),(h_s,f)) \cap ((h_i,t),(h_j,f))) \cup (((h_m,t),(h_s,f)) \cap ((h_k,t),(h_l,f)))$, $((h_m,t),(h_s,f)) \cup (((h_i,t),(h_j,f)) \cap ((h_k,t),(h_l,f))) = (((h_m,t),(h_s,f)) \cup ((h_i,t),(h_j,f))) \cap (((h_m,t),(h_s,f)) \cup ((h_k,t),(h_l,f)))$.

Corollary 2. *For any* $((h_i,t),(h_j,f)),((h_k,t),(h_l,f)) \in LI_{2n}$, $((h_i,t),(h_j,f)) \leq ((h_k,t),(h_l,f))$ *if and only if* $(h_i,t) \leq (h_k,t)$ *and* $(h_j,f) \geq (h_l,f)$ *if and only if* $i \leq k$ *and* $j \leq l$.

Theorem 1. *In the linguistic truth-valued intuitionistic fuzzy lattice* $\mathcal{LI}_{2n}$,

1. $((h_i,t),(h_i,f))(i \in \{2,3,\cdots,n\})$ *are* $\vee-$*irreducible elements of* $\mathcal{LI}_{2n}$, *denote as* $J_1 = \{((h_i,t),(h_i,f))|i = 2,3,\cdots,n\}$;

2. $((h_1, t), (h_i, f))(i \in \{2, 3, \cdots, n\})$ are $\vee-irreducible$ elements of $\mathcal{LI}_{2n}$, denote as $J_2 = \{((h_1, t), (h_i, f)) | i = 2, 3, \cdots, n\}$.

Corollary 3. For any $((h_k, t), (h_l, f)) \in LI_{2n} - \{((h_1, t), (h_1, f))\}$, $((h_k, t),$ $(h_l, f))$ can be represented as the union of the two elements in $J_1 \cup J_2$, i.e,

$$((h_k, t), (h_l, f)) = ((h_k, t), (h_k, f)) \cup ((h_1, t), (h_l, f)).$$

Corollary 4. $(LI_{2n} - \{((h_1, t), (h_1, f))\}, \cup, \cap)$ is isomorphic with $(2^{J_1 \cup J_2}, \vee, \wedge)$.

From the above corollary, we consider the following implication operator on $J_1 \cup J_2 \cup \{((h_1, t), (h_1, f))\}$, where denote it as $\mathcal{J}_1 = J_1 \cup \{((h_1, t), (h_1, f))\}$, $\mathcal{J}_2 = J_2 \cup \{((h_1, t), (h_1, f))\}$.

Definition 3. For any $((h_i, t), (h_j, f))$, $((h_k, t), (h_l, f)) \in J_1 \cup J_2 \cup \{((h_1, t),$ $(h_1, f))\}$,

1. If $((h_i, t), (h_i, f)), ((h_k, t), (h_k, f)) \in \mathcal{J}_1$,

$$((h_i, t), (h_i, f)) \rightarrow ((h_k, t), (h_k, f)) = ((h_{min(n, n-i+k)}, t), (h_{min(n, n-i+k)}, f)).$$

2. If $((h_1, t), (h_j, f)), ((h_1, t), (h_l, f)) \in \mathcal{J}_2$,

$$((h_1, t), (h_j, f)) \rightarrow ((h_1, t), (h_l, f)) = ((h_{min(n, n-j+l)}, t), (h_n, f)).$$

3. If $((h_i, t), (h_i, f)) \in \mathcal{J}_1$ and $((h_1, t), (h_l, f)) \in \mathcal{J}_2$,

$$((h_i, t), (h_i, f)) \rightarrow ((h_1, t), (h_l, f)) = ((h_{min(n, n-i+1)}, t), (h_{min(n, n-i+l)}, f)),$$
$$((h_1, t), (h_l, f)) \rightarrow ((h_i, t), (h_i, f)) = ((h_{min(n, n-l+i)}, t), (h_n, f)).$$

Now we extend the implication operator from $J_1 \cup J_2 \cup \{((h_1, t), (h_1, f))\}$ to LI_{2n}.

Definition 4. For any $((h_i, t), (h_j, f)), ((h_k, t), (h_l, f)) \in LI_{2n}$, define

$$
\begin{aligned}
&((h_i, t), (h_j, f)) \rightarrow ((h_k, t), (h_l, f)) \\
&= (((h_i, t), (h_i, f)) \cup ((h_1, t), (h_j, f))) \rightarrow (((h_k, t), (h_k, f)) \cup ((h_1, t), (h_l, f))) \\
&= ((((h_i, t), (h_i, f)) \rightarrow (((h_k, t), (h_k, f)) \cup ((h_1, t), (h_l, f)))) \cap ((((h_1, t), (h_j, f)) \rightarrow (((h_k, t), \\
&\quad (h_k, f)) \cup ((h_1, t), (h_l, f)))) \\
&= ((((h_i, t), (h_i, f)) \rightarrow ((h_k, t), (h_k, f))) \cup (((h_i, t), (h_i, f)) \rightarrow ((h_1, t), (h_l, f)))) \cap \\
&\quad ((((h_1, t), (h_j, f)) \rightarrow ((h_k, t), (h_k, f))) \cup (((h_1, t), (h_j, f)) \rightarrow ((h_1, t), (h_l, f)))) \\
&= (((h_{min(n, n-i+k)}, t), (h_{min(n, n-i+k)}, f)) \cup ((h_{min(n, n-i+1)}, t), (h_{min(n, n-i+l)}, f)))) \cap \\
&\quad (((h_{min(n, n-j+k)}, t), (h_n, f)) \cup ((h_{min(n, n-j+l)}, t), (h_n, f))) \\
&= ((h_{max(min(n, n-i+k), min(n, n-i+1))}, t), (h_{max(min(n, n-i+k), min(n, n-i+l))}, f)) \cap \\
&\quad ((h_{max(min(n, n-j+k), min(n, n-j+l))}, t), (h_n, f)) \\
&= ((h_{min(n, n-i+k)}, t), (h_{min(n, n-i+l)}, f)) \cap ((h_{min(n, n-j+l)}, t), (h_n, f)) \\
&= ((h_{min(n, n-i+k, n-j+l)}, t), (h_{min(n, n-i+l)}, f)). \qquad (12)
\end{aligned}
$$

From the above we obtain the following properties.

Corollary 5. *For any* $((h_i, t), (h_j, f)), ((h_k, t), (h_l, f)), ((h_m, t), (h_s, f)) \in LI_{2n}$,

1. $(((h_i, t), (h_j, f)) \cup ((h_k, t), (h_l, f))) \rightarrow ((h_m, t), (h_s, f)) = (((h_i, t), (h_j, f)) \rightarrow ((h_m, t), (h_s, f))) \cap (((h_k, t), (h_l, f)) \rightarrow ((h_m, t), (h_s, f)));$
2. $((h_m, t), (h_s, f)) \rightarrow (((h_i, t), (h_j, f)) \cup ((h_k, t), (h_l, f))) = (((h_m, t), (h_s, f)) \rightarrow ((h_i, t), (h_j, f))) \cup (((h_m, t), (h_s, f)) \rightarrow ((h_k, t), (h_l, f)));$
3. $(((h_i, t), (h_j, f)) \cap ((h_k, t), (h_l, f))) \rightarrow ((h_m, t), (h_s, f))) = (((h_i, t), (h_j, f)) \rightarrow ((h_m, t), (h_s, f))) \cup (((h_k, t), (h_l, f)) \rightarrow ((h_m, t), (h_s, f)));$
4. $((h_m, t), (h_s, f)) \rightarrow (((h_i, t), (h_j, f)) \cap ((h_k, t), (h_l, f))) = (((h_m, t), (h_s, f)) \rightarrow ((h_i, t), (h_j, f))) \cap (((h_m, t), (h_s, f)) \rightarrow ((h_k, t), (h_l, f))).$

In fact, for any $((h_i, t), (h_j, f)), ((h_k, t), (h_l, f)), ((h_m, t), (h_s, f)) \in LI_{2n}$,

$$(((h_i, t), (h_j, f)) \cup ((h_k, t), (h_l, f))) \rightarrow ((h_m, t), (h_s, f))$$
$$= ((h_{max(i,k)}, t), (h_{max(j,l)}, f)) \rightarrow ((h_m, t), (h_s, f))$$
$$= ((h_{min(n, n-max(i,k)+m, n-max(j,l)+s)}, t), (h_{min(n, n-max(i,k)+s)}, f))$$
$$= ((h_{min(n, min(n-i+m, n-k+m), min(n-j+s, n-l+s))}, t), (h_{min(n, min(n-i+s, n-k+s))}, f))$$
$$= ((h_{min(n, n-i+m, n-j+s)}, t), (h_{min(n, n-i+s)}, f)) \cap ((h_{min(n, n-k+m, n-l+s)}, t),$$
$$(h_{min(n, n-k+s)}, f))$$
$$= (((h_i, t), (h_j, f)) \rightarrow ((h_m, t), (h_s, f))) \cap (((h_k, t), (h_l, f)) \rightarrow ((h_m, t), (h_s, f))).$$

The others equations can be proved similar.

Theorem 2. *For any* $((h_i, t), (h_j, f)), ((h_k, t), (h_l, f)), ((h_m, t), (h_s, f)) \in LI_{2n}$,

1. $((h_1, t), (h_1, f)) \rightarrow ((h_k, t), (h_l, f)) = ((h_n, t), (h_n, f));$
2. $((h_n, t), (h_n, f)) \rightarrow ((h_k, t), (h_l, f)) = ((h_k, t), (h_l, f));$
3. $((h_i, t), (h_j, f)) \rightarrow ((h_k, t), (h_l, f))) = ((h_n, t), (h_n, f))$ *if and only if* $((h_i, t), (h_j, f)) \leq ((h_k, t), (h_l, f));$
4. $((h_k, t), (h_l, f)) \leq ((h_i, t), (h_j, f)) \rightarrow ((h_k, t), (h_l, f));$
5. *If* $((h_i, t), (h_j, f)) \leq ((h_k, t), (h_l, f))$, *then*

$$((h_m, t), (h_s, f)) \rightarrow ((h_i, t), (h_j, f)) \leq ((h_m, t), (h_s, f)) \rightarrow ((h_k, t), (h_l, f)),$$

$$((h_i, t), (h_j, f)) \rightarrow ((h_m, t), (h_s, f)) \geq ((h_k, t), (h_l, f)) \rightarrow ((h_m, t), (h_s, f)).$$

Prove.   According to above results,
  (1)

$$((h_1, t), (h_1, f)) \rightarrow ((h_k, t), (h_l, f))$$
$$= ((h_{min(n, n-1+k, n-1+l)}, t), (h_{min(n, n-1+l)}, f))$$
$$= ((h_n, t), (h_n, f)).$$
$$((h_n, t), (h_n, f)) \rightarrow ((h_k, t), (h_l, f))$$
$$= ((h_{min(n, n-n+k, n-n+l)}, t), (h_{min(n, n-n+l)}, f))$$
$$= ((h_k, t), (h_l, f)).$$

(2) If $((h_i, t), (h_j, f)) \geq ((h_k, t), (h_l, f))$, then $i \leq k \leq l$ and $j \leq l$,

$$((h_i, t), (h_j, f)) \rightarrow ((h_k, t), (h_l, f)))$$
$$= ((h_{min(n, n-i+k, n-j+l)}, t), (h_{min(n, n-i+l)}, f))$$
$$= ((h_n, t), (h_n, f)).$$

On the contrary, If $((h_i, t), (h_j, f)) \rightarrow ((h_k, t), (h_l, f))) = ((h_n, t), (h_n, f))$, then $min(n, n-i+k, n-j+l) = n$, i.e., $n-i+k \geq n$ and $n-j+l \geq n$, Hence, $i \leq k$ and $j \leq l$, $((h_i, t), (h_j, f)) \leq ((h_k, t), (h_l, f))$.

3) Since $k \leq n-i+k$ and $k \leq l \leq n-j+l$, hence $k \leq min(n, n-i+k, n-j+l)$. As the same case, we get $l \leq n-i+l$, $l \leq min(n, n-i+l)$, i.e.,

$$((h_k, t), (h_l, f)) \leq ((h_i, t), (h_j, f)) \rightarrow ((h_k, t), (h_l, f)).$$

4) Since $((h_i, t), (h_j, f)) \leq ((h_k, t), (h_l, f))$, then$i \leq k$ and $j \leq l$. And

$$((h_m, t), (h_s, f)) \rightarrow ((h_i, t), (h_j, f)) = ((h_{min(n, n-m+i, n-s+j)}, t), (h_{min(n, n-m+j)}, f)),$$
$$((h_m, t), (h_s, f)) \rightarrow ((h_k, t), (h_l, f)) = ((h_{min(n, n-m+k, n-s+l)}, t), (h_{min(n, n-m+l)}, f)),$$

Hence $min(n, n-m+i, n-s+j) \leq min(n, n-m+k, n-s+l)$ and $min(n, n-m+j) \leq min(n, n-m+l)$, i.e.,

$$((h_m, t), (h_s, f)) \rightarrow ((h_i, t), (h_j, f)) \leq ((h_m, t), (h_s, f)) \rightarrow ((h_k, t), (h_l, f)).$$

As the same case, since

$$((h_i, t), (h_j, f)) \rightarrow ((h_m, t), (h_s, f)) = ((h_{min(n, n-i+m, n-j+s)}, t), (h_{min(n, n-i+s)}, f)),$$
$$((h_k, t), (h_l, f)) \rightarrow ((h_m, t), (h_s, f)) = ((h_{min(n, n-k+m, n-l+s)}, t), (h_{min(n, n-k+s)}, f)),$$

$min(n, n-i+m, n-j+s) \geq min(n, n-k+m, n-l+s)min(n, n-i+s) \geq min(n, n-k+s)$. Hence,

$$((h_i, t), (h_j, f)) \rightarrow ((h_m, t), (h_s, f)) \geq ((h_k, t), (h_l, f)) \rightarrow ((h_m, t), (h_s, f)).$$

Theorem 3. *For any* $((h_i, t), (h_j, f)), ((h_k, t), (h_l, f)), ((h_m, t), (h_s, f)) \in LI_{2n}$,

1. $((h_i, t), (h_j, f)) \rightarrow (((h_k, t), (h_l, f)) \rightarrow ((h_m, t), (h_s, f))) = ((h_k, t), (h_l, f)) \rightarrow (((h_i, t), (h_j, f)) \rightarrow ((h_m, t), (h_s, f)));$
2. $((h_i, t), (h_j, f)) \leq (((h_i, t), (h_j, f)) \rightarrow ((h_k, t), (h_l, f))) \rightarrow ((h_k, t), (h_l, f)).$

4   Conclusions

According to the linguistic truth-valued lattice implication algebra, we constructed a linguistic truth-valued intuitionistic fuzzy lattice. From the framework of the linguistic truth-valued intuitionistic fuzzy algebra, we consider the implication operator on the linguistic truth-valued intuitionistic fuzzy lattice. The linguistic truth-valued intuitionistic fuzzy lattice can better express both the comparable information and incomparable information. Also we can use this approach to deal with both positive evidence and negative evidence which are represented by linguistic truth values at the same time during information processing.

The further work is to find some operation properties for linguistic truth-valued intuitionistic fuzzy logic reasoning. The new method is expected to use into the field of decision making, risk analysis and so on.

Acknowlegments

This work is partly supported by National Nature Science Foundation of China (Grant no.60875034,60873042), the Specialized Research Fund for the Doctoral Program of Higher Education of China (Grant no. 20060613007).

References

1. Herrera, F., Martnez, L.: The 2-Tuple linguistic computational model. Advantages of its linguistic description, accuracy and consistency. International Journal of Uncertainty, Fuzziness and Knowledge-Based Systems 9, 33048 (2001)
2. Xu, Z.S.: Intuitionistic fuzzy aggregation operators. IEEE Transactions on Fuzzy Systems 15(11), 1179–1187 (2007)
3. Huynh, V.N., Ho, T.B., Nakamori, Y.: A Parametric Representation of Linguistic Hedges in Zadeh's Fuzzy Logic. International Journal of Approximate Reasoning 30, 203–223 (2002)
4. Ho, N.C.: A topological completion of refined hedge algebras and a model of fuzziness of linguistic terms and hedges. Fuzzy Sets and Systems 158, 436–451 (2007)
5. Xu, Y., Liu, J., Ruan, D., Lee, T.T.: On the consistency of rule bases based on lattice-valued first-order logic $LF(X)$. International Journal of Intelligent Systems 21, 399–424 (2006)
6. Atanassov, K.T.: Answer to D. Dubois, S. Gottwald, P. Hajek, J. Kacprzyk and H. Prade's paper Terminological difficulties in fuzzy set theory the case of 'Intuitionistic Fuzzy Sets'. Fuzzy Sets and Systems 156, 496–499 (2005)
7. Zadeh, L.A.: The comcept of linguistic variable and its application to approximate reasoning (I)(II). Information Sciences 8, 199–249, 310–357 (1975)
8. Zou, L., Ruan, D., Pei, Z., Xu, Y.: A linguistic truth-valued reasoning approach in decision making with incomparable information. Journal of Intelligent and Fuzzy Systems 19(4-5), 335–343 (2008)
9. Zou, L., Pei, Z., Liu, X., Xu, Y.: Semantic of Linguistic Truth-Valued Intuitionistic Fuzzy Proposition Calculus. International Journal of Innovative Computing, Information and Control 5(12), 4745–4752 (2009)
10. Zou, L., Liu, X., Wu, Z., Xu, Y.: A Uniform Approach of Linguistic Truth Values in Sensor Evaluation. International Journal of Fuzzy Optimization and Decision Making 7(4), 387–397 (2008)
11. Zou, L., Ma, J., Xu, Y.: A Framework of Linguistic Truth-valued Propositional Logic Based on Lattice Implication Algebra. In: Proceedings of 2006 IEEE International Conference on Granular Computing, pp. 574–577 (2006)
12. Wang, D.G., Song, W.Y., Li, H.X.: Unified forms of Fuzzy Similarity Inference Method for Fuzzy Reasoning and Fuzzy Systems. International Journal of Innovative Computing Information and Control 4(10), 2285–2294 (2008)

Robust Granular Neural Networks, Fuzzy Granules and Classification

G. Avatharam and Sankar K. Pal

Center for Soft Computing Research,
Indian Statistical Institute,
Kolkata - 700108, India
avatharg@yahoo.co.in, sankarpal@yahoo.com

Abstract. We introduce a robust granular neural network (RGNN) model based on the multilayer perceptron using back-propagation algorithm for fuzzy classification of patterns. We provide a development strategy of the network mainly based upon the input vector, linguistic connection weights and target vector. While the input vector is described in terms of fuzzy granules, the target vector is defined in terms of class membership values and zeros. The connection weights among nodes of RGNN are in terms of linguistic variables, whose values are updated by adding two linguistic hedges. The updated linguistic variables are called generalized linguistic variables. The node functions of RGNN are defined in terms of linguistic arithmetic operations. We present the experimental results on several real life data sets. Our results show that the classification performance of RGNN is superior to other similar type of networks.

Keywords: granular computing, linguistic variable, linguistic arithmetic, granular neural networks, fuzzy classification.

1   Introduction

Granular Computing (GC) is a new information-processing paradigm developing in the past few years, which recognizes that precision, is sometimes an expensive goal in modeling and controlling complex systems. In particular, GC recognizes that human thought does not appear to operate at a numeric level of precision, but at a more abstract level of detail. GC is a framework for expressing that abstraction within computational processes, which has recently drawn extensive attention. While seminal research in GC has occurred in fuzzy systems community, GC includes Pawlak's rough sets, including set theory (power algebra, interval number algebra, interval set algebra) [1].

In granular computing, the atomic units of computation are groups of objects. Objects from some collection, such as real numbers, are grouped together based on their similarity, equality, equivalence or functionality, and then all computations are performed on these groups rather than individual objects. These groups are the information granules manipulated in a granular world. The formal definition of granule is a clump of objects defined by a generalized constraint

J. Yu et al. (Eds.): RSKT 2010, LNAI 6401, pp. 220–227, 2010.

[2]. Random set constraint is one of the several classes of constraints that can be defined by granules. It is the combination of probabilistic and possibilistic constraints.

It may be mentioned that human reasoning somewhat fuzzy in nature. The utility of granules (fuzzy sets) lies in their ability to model uncertain data that is so often encountered in real life. Hence, to enable a system to tackle real life situations in a manner more akin to humans, one may incorporate the concept of granules into the neural networks. A number of authors have discussed a variety of granular neural network architectures in the past few years. Zhang, Fraser, Gagliano, and Kandal [12] described granular neural networks for passing granulated input data to ensemble of numeric neural networks. Vasilakos and Stathakis described granular neural networks for land use classification for remote sensing images [11]. Zhang, Jin and Tang built a granular neural network using fuzzy sets as their formalism and an evolutionary training algorithm [13].

We describe here a new application of GC, namely, the development of a neural network architecture and inductive learning rule that uses linguistic connection weights and neurons with logistic activation functions. The proposed networks, called robust granular neural networks (RGNN), store their knowledge as a pattern of connection weights between simple processing units and learn new knowledge in a granular form. While the input features are presented in the form fuzzy granules with π membership function, the desired vector is given class membership values and zeros. This provides a scope for incorporating linguistic information in both the training and testing phases of RGNN and increases its robustness in tackling imprecise or uncertain input specifications. If the granular computing premise is correct, the RGNN will be able to learn input-output relationships just like a standard neural network.

2 Representations of Weights in Linguistic Form

The connection weights of RGNN are in the form generalized linguistic variables based on linguistic arithmetic developed in [3]. For the reader's convenience, we describe the main concepts behind the linguistic arithmetic and provide the definitions of the operators in linguistic arithmetic.

2.1 Generalized Linguistic Variables

A linguistic variable is a 5-tuple (X, U, T, S, M) where X is the name of the variable, U is the universe of discourse for X, T is the term set for X, S is a syntactic rule for generating elements of T, and M is a semantic rule. In its most general form, S is a context free grammar that contains two types of terminals which are the atomic terms and linguistic hedges. Elements of the term set T are composed of single atomic term, and one or more hedges as specified by the context free grammar S. The semantic rule M associates each atomic term with a single fuzzy subset F of U, and each hedge with a function h : [0,1] → [0,1] that modifies the membership function of F. The hedge function 'h' will be applied to F in the order indicated by the parse tree for linguistic term, derived from S. The fundamental principal for updating connection weights is to add hedges to atomic terms in a structured manner. The change of connection weights based on the less than (LT) and greater than (GT) type hedges has been explained in [3].

A **generalized linguistic variable (GLV)** is a 6-tuple (X, U, τ, S, M, χ) where X is the name of the variable, U is the universe of discourse for X, τ is the term set, S a syntactic rule for generating new atomic terms (a phrase structure grammar), M is a semantic rule that associates with each atomic term generated by S, a fuzzy subset of U, and χ is the crossover limit. The crossover limit χ represents the maximum number of greater than or less than type hedges that can be added to an atomic term before crossover operation occurs. For example, the syntactic rule S for a GLV with atomic terms N, Z, P and $\chi = \mathbf{2}$ is given in [4]. The total ordering of linguistic terms in GLV, based on an ordering of the fuzzy sets associated with them by the semantic rule M, is given in [5].

2.2   Linguistic Arithmetic

The linguistic arithmetic operations are operated on GLV based on ordering of atomic terms in a standard sequence σ, an analog of standard vector in [6]. The standard sequence is a list of atomic terms of a linguistic variable, where the atomic terms are arranged in some reasonable order according to some rules followed in [3]. It is proved in [6] that the standard sequence σ, represents a countable infinite number of atomic terms, due to the existence of a bijection function Γ between τ and the set of integers. Given two terms $t1, t2 \in \tau$, the difference between these two terms is defined as the number of intervening terms between $t1$ and $t2$ plus 1. The value is positive if $t1 < t2$ and negative if $t1 > t2$ and the term t1 should be a zero atomic term.

Definition 2.2.1. The operations of linguistic arithmetic: (i) linguistic sum, (ii) linguistic difference, (iii) scalar product, and (iv) linguistic product.

1. The linguistic sum is a mapping
   $\oplus : \tau \times \tau \to \tau$
   $t1 \oplus t2 = \Gamma^{-1}(\Gamma(t1) + \Gamma(t2))$
2. The linguistic difference is a mapping
   $\ominus : \tau \times \tau \to \tau$
   $t1 \ominus t2 = \Gamma^{-1}(\Gamma(t1) - \Gamma(t2))$
3. The linguistic scalar product is a mapping
   $\otimes : \mathbb{R} \times \tau \to \tau$
   $s \otimes t1 = \Gamma^{-1}(round(s * \Gamma(t1)))$
4. The linguistic product is a mapping
   $\bullet : \tau \times \tau \to \tau$
   $t1 \bullet t2 = \Gamma^{-1}(round((\Gamma(t1) * \Gamma(t2))/\chi))$

From the above equations, τ is the set of composite terms for a GLV, $t1, t2 \in \tau$, $s \in \mathbb{R}$ is a real value, '+' is an integer addition, '-' is an integer subtraction,

'*' is an integer multiplication, '/' is real division, and round(x) is the nearest integer to the real value x. $1/\chi$ is the scaling factor in the linguistic product.

Example 1. Consider three atomic terms=$\{NS, AZ, PS\}$, and crossover χ is 2 for a GLV. The crossover rule executed only between adjacent atomic terms. The two hedges less than (LT) and greater than (GT) type are used to generate a new linguistic connection weight by syntactic rule [4].

The linguistic product LT LT AZ $\bullet$ GT AZ is computed for the GLV as follows.

$$=\Gamma^{-1}(\text{round}((\Gamma(\text{LT LT AZ })*\Gamma(\text{GT AZ}))/2))$$
$$=\Gamma^{-1}(\text{round}((\text{-2 * 1})/2))$$
$$=\Gamma^{-1}(\text{round(-1)})$$
$$=\Gamma^{-1}(\text{-1})$$
$$=\text{LT AZ}$$

2.3   RGNN Back Propagation Algorithm

Input

D, a data set consisting of the training tuples in the form of random set granules and their associated target vector in terms of membership value and zeros.

η, the learning rate

network, a granular feed-forward network.

Output

A trained neural network

Method:

1. Initialize all linguistic weights and biases in network;
2. While terminating condition is not satisfied{
3. for each training tuple
   Propagate the inputs forward:
4. for each input layer unit j{
5. $O_j = I_j$; here, the output of an input unit is its actual input value.
6. for each hidden or output layer unit j, compute the net input of unit j with respect to the previous layer, i{
7. $I_j = \sum_i w_{ij} \bullet O_i \oplus b_j$; }
8. Apply logistic activation function to compute the output of each unit j.
9. $\Phi(O_j) = \frac{1}{1+e^{-\alpha \Xi(I_j)}}$

   where $\Xi(I_j) = \frac{\Gamma(I_j)}{(2.\chi+1).((n-1)/2)}$; } in which χ is the crossover rule and n is the number of atomic terms.
   Linguistic back propagation:
10. for each unit j in the output layer, compute the error {
11. $Error_j = ((\phi(O_j).(1 - \phi(O_j))).(T_j - \phi(O_j))) \bullet sup(w)$;
    w is least upper bound of the standard vector of atomic terms [4].
12. for each unit j in the hidden layers, from the last to the first hidden layer, compute the error with respect to the next higher layer.

13. $Error_j = (\phi(O_j).(1 - \phi(O_j))) \bullet (\sum_k Error_k \otimes w_{jk})$; }
14. for each weight w_{ij} in network {
15. $\Delta w_{ij} = (\eta.x_i) \bullet \gamma_j$;
    where η is network wise learning rate parameter.
16. $\Delta w_{ij} = \lfloor (\eta.x_i(k)) \bullet \gamma_j(k) \rfloor \oplus \lfloor \alpha \bullet \Delta w_{ij}(k-1) \rfloor$; }
    }

where α is a momentum parameter used to escape local minima in weight space, and k denotes an epoch (i.e. k-1 denotes the previous epoch), which is clearly, the number of times training data will be presented to the network for updating the weight matrix w_{ij}. The trained network is used to classifying the test patterns.

3   Input Pattern Representation in Granular Form

In general, human minds cannot perform a wide variety of physical and mental tasks without any measurement or computation. Familiar examples of such tasks are parking a car, driving in heavy traffic, understanding speech. Based on such tasks perceptions of size, distance, weight, speed, time, direction, smell, color, shape force etc occur. But a fundamental difference between such measurements on one hand and perception on the other, is that, the measurements are crisp numbers whereas perceptions are fuzzy numbers or more generally, fuzzy granules [2].

A formal definition of fuzzy granule is a group of objects defined by the generalized constraint form "X   isr   R" where 'R' is constrained relation, 'r' is a random set constraint, which is a combination of probabilistic and posibilistic constraints, 'X' is a fuzzy set valued random variable which takes the values low, medium, high. Using fuzzy-set theoretic techniques [7], [8] a pattern point x, belonging to the universe U, may be assigned a grade of membership with the membership function $\mu_A(x)$ to a fuzzy set A. This is defined as

$$A = \{(\mu_A(x), x)\}, \quad x \in U, \quad \mu_A(x) \in [0,1]. \tag{1}$$

The π membership function, with range [0,1] and $x \in \mathbb{R}^n$, is defined as

$$\pi(x, c, \lambda) = \begin{cases} 2(1 - \frac{\|x-c\|}{\lambda})^2 & for \quad \frac{\lambda}{2} \leq \| x - c \| \leq \lambda \\ 1 - 2(\frac{\|x-c\|}{\lambda})^2 & for \quad 0 \leq \| x - c \| \leq \frac{\lambda}{2} \\ 0 & otherwise \end{cases} \tag{2}$$

where $\lambda > 0$ is the radius of the π function with c as the central point, and $\| \ \|$ denotes Euclidian norm. Each input feature F_j can be expressed in terms of membership values to each of the three linguistic properties low, medium, high as granules. Therefore, an n-dimensional pattern can be represented as a 3n-dimensional vector

$$\overrightarrow{F}_i = [\mu_{low(F_{i1})}(\overrightarrow{F}_i), \mu_{medium(F_{i1})}(\overrightarrow{F}_i), \mu_{high(F_{i1})}(\overrightarrow{F}_i)....\mu_{high(F_{in})}(\overrightarrow{F}_i)] \tag{3}$$

3.1   Choice of Parameters of π Functions for Numerical Features

When the input feature is numerical, we use π fuzzy set of (2) with appropriate parameters, center c and λ, chosen as explained in the article of Pal and Mitra [9]. These are used to express features of each input pattern in terms of membership values to each of three fuzzy granules low, medium, and high.

3.2   Class Memberships as Output Vectors

The membership of the ith pattern to class c_k is defined as,

$$\mu_k(\overrightarrow{F}_i) = \frac{1}{1 + (\frac{Z_{ik}}{f_d})^{f_e}} \qquad (4)$$

where Z_{ik} is the weighted distance (eqn. 5), and f_d, f_e are the denominational and exponential fuzzy generators controlling the amount of fuzziness in the class membership set. Obviously class membership lies in $[0, 1]$. We use

$$Z_{ik} = \sqrt{\sum_{j=1}^{n}\left[\sum_{p=1}^{3}\frac{1}{3}(\mu_p(F_{ij}) - \mu_p(o_{kj})^2)\right]} \qquad for \quad k = 1, 2, ..., l. \qquad (5)$$

where o_{kj} is the center of jth feature vector which belongs to k^{th} class. In the fuzziest case, we may use fuzzy modifier contrast internification (INT) [9] to enhance contrast with in the class membership to decrease the ambiguity in taking a decision.

3.3   Applying the Membership Concept to Target Vector

The target vector at the output layer is defined by membership values and zeros (eqn.6). For the patterns belonging to a particular class, the desired vectors of those patterns at the corresponding class node were assigned membership values, and the rest of the class nodes were assigned zeros. For i^{th} input pattern we define the desired output of the jth output node as

$$d_j = \begin{cases} \mu_{INT(j)}(\overrightarrow{F_i}) & for \quad i^{th} pattern \quad in \quad the \quad fuzziest \quad case \\ 0 & otherwise \end{cases} \qquad (6)$$

4   Experimental Results

We have determined the classification accuracy, incorporating fuzzy concepts at various levels, of the best RGNN classifier on several real life data sets [10]. The proposed model has been tested for different numbers of hidden layers, and with different numbers of neurons in each such layer. During learning, we have used a 10-fold cross validation design with stratified sampling. In all our experiments, for each dataset, the parameters f_d and f_e in (6) are chosen as 5 and 1 respectively, and 'fdenorm' as 0.8. The parameter α is put equal to 1.0,

Table 1. Experimental Results

Dataset Name	Features	Classes	Algorithm	Min Accuracy	Max Accuracy	Average Accuracy
Telugu vowel	3	6	RGNN	79.31	94.25	86.55
			GNN	78.16	85.06	81.38
Balance Scale	4	3	RGNN	96.78	100.00	98.07
			GNN	85.48	96.78	93.71
Thyroid	5	3	RGNN	90.91	100.00	96.82
			GNN	85.71	100.00	94.29
Zoo	16	7	RGNN	90.00	100.00	98.00
			GNN	80.00	100.00	91.00
Square2	2	4	RGNN	100.00	100.00	100.00
			GNN	82.50	100.00	91.63
Lympography	18	4	RGNN	83.33	88.89	86.11
			GNN	77.77	88.88	81.25
Wine	13	3	RGNN	90.00	100.00	97.16
			GNN	82.25	86.36	85.56
Breast cancer	9	2	RGNN	92.64	99.58	95.21
			GNN	90.58	98.32	94.84

each connection weight in a GLV with 7 atomic terms, and crossover limit of 7. In our all experiments, The parameter η traverses a range of values from **0** to **1** and finally we choose $0.25 < \eta < 0.39$ for good results. In the case of other similar network, the datasets are normalized to all the variables between 0 and 1. We have reported minimum, maximum, average accuracies across all 10folds, for both the RGNN (our proposed model) and other similar network on the real life data sets. The results are given in Table 1. Examining the above experimental results, in the 10-fold cross validation, for Telugu vowel dataset, GNN accuracy is seen to vary between 78.16% and 85.06% with an average of 81.38%. In contrast, the RGNN varied between 79.31% and 94.25% with average 86.55% accuracy. For Balance Scale dataset, GNN accuracy varied between 85.48% and 96.78% with an average of 93.71%. In contrast, the RGNN varied between 96.78% and 100% with average 98.07% accuracy. Similar observations can also be made from the remaining part of the experimental results for other data sets in Table 1. This indicates that the difference between RGNN and the other similar network, as stated above, is likely to be statistical significant and RGNN is superior to GNN. The experimental results shows that the time-consuming of datasets in Table 1 in RGNN is almost similar to GNN. However, the performance of RGNN is always better than GNN.

5   Conclusion

In this paper, we presented a new granular neural network model based on the multilayer perceptron using back propagation algorithm. The linguistic connection weights, treating atomic terms as granules, were used in the form of GLV throughout the network. The weights were mapped to set of integers, which describe the location of that weight, between a particular interval of integers, defined by atomic terms, to perform general arithmetic operations during the training and testing phases. The model converts numerical inputs in terms of fuzzy granules low, medium, high and provides output decisions in terms of class membership values and zeros.

The experimental results show that the RGNN architecture is a useful application of granular computing to real world classification problems. Future investigation includes study of more efficient hybrid granular computing methods and learning algorithms for complex applications such as bioinformatics, security and Web intelligence.

Acknowledgments. The work was done while Prof. S. K. Pal was a J.C. Bose Fellow of the Government of India.

References

1. Yao, Y.Y.: Granular Computing: basic issues and possible solutions. In: 5th Joint Conference on Information Sciences, pp. 186–189 (2001)
2. Zadeh, L.A.: From computing with numbers to computing with words - from manipulation of measurements to manipulation of perceptions. IEEE Trans. Circuits Syst. 1-45, 105–119 (1999)
3. Dick, S., Schenkar, A., Pedrycz, W., Kandal, A.: A Granular Algorithm for Re-Granulating a Fuzzy Rule base. Information Sciences 177, 408–435 (2007)
4. Dick, S., Tappenden, A., Badke, C., Olarewaju, O.: A Novel Granular Neural Network Architecture. In: Annual Meeting of the North American Fuzzy Information Processing Society - NAFIS, North America, pp. 42–47 (2007)
5. Kocyz, L.T., Hirota, K.: Left Ordering, distance and closeness of fuzzy sets. Fuzzy Sets and Systems 59, 281–293 (1993)
6. Liu, C., Shindhelm, A., Li, D., Jin, K.: A numerical approach to linguistic variables and linguistic space. In: IEEE Int. C. Fuzzy Systems, pp. 954–959 (1996)
7. Klir, G.J., Folger, T.: Fuzzy sets, Uncertainty and Information. Addison Wesley, Reading (1989)
8. Pal, S.K., Dutta Majumder, D.: Fuzzy Mathematical Approach to Pattern Recognition. Wiley Halsted Press, New York (1986)
9. Pal, S.K., Mitra, S.: Multilayer Perceptron, Fuzzy Sets, and Classification. IEEE Transactions on Neural Networks 3(5) (1992)
10. Newman, D.J., Hettich, S., Blake, C.L., Merz, C.J.: UCI repository of machine learning databases. University of California, Department of Information and Computer Science, Irvine, CA (1998), http://archive.ics.uci.edu/ml/
11. Vasilakos, A., Stathakis, D.: Granular neural networks for land use classification. Soft Computing 9, 332–340 (2005)
12. Zhang, Y.-Q., Fraser, M.D., Gagliano, R.A., Kandel, A.: Granular neural networks for numerical-linguistic data fusion and knowledge discovery. IEEE Transactions on Neural Networks 11, 658–667 (2000)
13. Zhang, Y.-Q., Jin, B., Tang, Y.: Granular neural networks with evolutionary interval learning. IEEE Transactions on Fuzzy Systems 16-2, 309–319 (2008)

Perturbed Iterative Approximation of Common Fixed Points on Nonlinear Fuzzy and Crisp Mixed Family Operator Equation Couples in Menger PN-Spaces

Heng-you Lan[1], Tian-xiu Lu[1], Huang-lin Zeng[2], and Xiao-hong Ren[2]

[1] Department of Mathematics, Sichuan University of Science & Engineering
Zigong, Sichuan 643000, People's Republic of China
[2] School of Automation and Electronic Information
Sichuan University of Science & Engineering
Zigong, Sichuan 643000, P.R. China

Abstract. In this paper, based on the concept of probabilistic (φ, ψ)-contractor couple introduced by Mihet, a new class of nonlinear operator equation couples with a mixed family of fuzzy and crisp operator equations in Menger probabilistic normed spaces (briefly, Menger PN-spaces) is introduced and studied. Further, some new iterative algorithms are constructed, and the existence of solutions for the nonlinear operator equation couples and the convergence of iterative sequences generated by the algorithms under a larger class of t-norms and joint orbitally complete conditions are discussed.

Keywords: Common fixed point, probabilistic (φ, ψ)-contractor, nonlinear fuzzy and crisp mixed family operator equation couple, joint orbitally complete condition, iterative algorithm and convergence.

1 Introduction

In this paper, we shall consider the following nonlinear operator equation couple:

$$\widetilde{S}_{ix}(u) \geq d(x), \quad \widetilde{S}_{jx}(u) \geq e(x), \tag{1}$$

where $\widetilde{S}_i, \widetilde{S}_j : \mathbb{X} \to W(\mathbb{Y})$ are two fuzzy operators for some $i, j \in \mathbb{N}$, $\mathbb{X}$ and $\mathbb{Y}$ are two separable real vector spaces, $W(\mathbb{Y})$ denotes the collection of all fuzzy sets in $\mathbb{Y}$, $d, e : \mathbb{X} \to (0, 1]$ are two operators. Obviously, if for some $i, j \in \mathbb{N}$, S_i and S_j are the set-valued operators induced by the fuzzy operators $\widetilde{S}_i$ and $\widetilde{S}_j$, respectively, then (1) is equivalent to the following nonlinear equation couple for set-valued operators:

$$u \in S_i(x), \qquad u \in S_j(x). \tag{2}$$

Moreover, (2) becomes to the following nonlinear operator equation couple:

$$u = f(x), \qquad u = g(x), \tag{3}$$

J. Yu et al. (Eds.): RSKT 2010, LNAI 6401, pp. 228–233, 2010.

when $f, g : \mathbb{X} \to \mathbb{X}$ are two nonlinear operators satisfying $S_i(x) \subset f(\mathbb{X})$ and $S_j(x) \subset g(\mathbb{X})$ for any $x \in \mathbb{X}$ and some $i, j \in \mathbb{N}$.

Without loss of generality, we can suppose that $u = \theta$. In fact, if $u \neq \theta$, and $Q_i, Q_j : \mathbb{X} \to \Omega_\mathbb{Y}$ are two set-valued operators with $Q_i(x) = S_i(x) - u$ and $Q_j(x) = S_j(x) - u$ for all $x \in \mathbb{X}$ and some $i, j \in \mathbb{N}$, where $\Omega_\mathbb{Y}$ denotes the collection of all nonempty closed probabilistically bounded subsets of $\mathbb{Y}$, then (2) is equivalent to $\theta \in Q_i(x)$ and $\theta \in Q_j(x)$. Thus, in order to discuss equation couples (1) or (2) or (3), we can turn to discuss the following equation couples:

$$\begin{cases} \widetilde{S}_{ix}(\theta) \geq d(x), \\ \widetilde{S}_{jx}(\theta) \geq e(x), \end{cases} \text{or,} \quad \begin{cases} \theta \in S_i(x), \\ \theta \in S_j(x), \end{cases} \text{or,} \quad \begin{cases} \theta = f(x), \\ \theta = g(x). \end{cases} \quad (4)$$

The concept of probabilistic contractors in probabilistic normed spaces has been introduced by Chang [1] based on Altman's theory of contractors on normed spaces ([2]). Recently, by using the theory of countable extension of t-norms ([3]) and the concept of probabilistic Ψ-contractor, many results for the more general classes of t-norms and for nonlinear operator equations with set-valued operator or a mixed family of fuzzy and crisp operators in probabilistic normed spaces have been discussed. See, for example, [4]-[7] and the references therein.

Motivated and inspired by the above works, the purpose of this paper is to introduce and study a new class of nonlinear operator equation couples with a mixed family of fuzzy and crisp operators in Menger PN-spaces by using the concept of probabilistic (φ, ψ)-contractor couple introduced by Mihet [7]. Moreover, we shall construct some new perturbed iterative algorithms and prove the existence of random solutions for the nonlinear operator equation couples and the convergence of random iterative sequences generated by the algorithms under a larger class of t-norms and joint orbitally complete conditions.

2   Preliminaries

Let Δ^+ denote the set of all distribution functions F with $F(0) = 0$ (F is a nondecreasing, left continuous operator from $\mathbb{R}$ into $[0, 1]$ with $\sup_{x \in \mathbb{R}} F(x) = 1$). The special distribution function H is defined by $H(t) = \begin{cases} 1, t > 0, \\ 0, t \leq 0. \end{cases}$

A fuzzy set A in $\mathbb{X}$ is a function from $\mathbb{X}$ into $[0, 1]$. If $x \in \mathbb{X}$, then the function value $A(x)$ is called the grade of membership of x in A. The α-level set of A, denoted by A_α, is defined by $A_q = \{x : A(x) \geq q\}$ for all $q \in (0, 1]$. Let $W(\mathbb{X})$ denote the collection of all fuzzy sets A in $\mathbb{X}$ such that A_q is compact and convex for all $q \in (0, 1]$ and $\sup_{x \in \mathbb{X}} A(x) = 1$. For any $A, B \in W(\mathbb{X})$, $A \subset B$ means $A(x) \leq B(x)$ for all $x \in \mathbb{X}$. Let M be an arbitrary set and $\mathbb{X}$ be any linear metric space. A function $\widetilde{S} : M \to W(\mathbb{X})$ is called fuzzy operator.

Definition 1. Let f, g be two operators from $\mathbb{X}$ into itself and $\{\widetilde{S}_n\}_{n=1}^\infty$ a sequence of fuzzy operators from $\mathbb{X}$ into $W(\mathbb{X})$. If, for some $x_0 \in \mathbb{X}$, there exist sequences $\{x_n\}$ and $\{y_n\}$ in $\mathbb{X}$ such that for all $n \geq 0$,

$$\{y_{2n+1}\} = \{f(x_{2n+1})\} \subset \widetilde{S}_{2n+1}(x_{2n}), \quad \{y_{2n+2}\} = \{g(x_{2n+2})\} \subset \widetilde{S}_{2n+2}(x_{2n+1}). (5)$$

then $\vartheta(\widetilde{S}_n, f, g, x_0) = \{y_n : n \in \mathbb{N}\}$ is called an orbit for the mixed operators $(\widetilde{S}_n, f, g)$.

Definition 2. $\mathbb{X}$ is called x_0-joint orbitally complete if every Cauchy sequence of each orbit at x_0 is convergent in $\mathbb{X}$.

Remark 1. ([8]) Clearly, if $\mathbb{X}$ is a any complete space and $x_0 \in \mathbb{X}$, then $\mathbb{X}$ is x_0-joint orbitally complete, while the converse is not necessarily true.

3   Iterative Algorithm and Convergence

Let $(\mathbb{X}, \widehat{\mathcal{F}}, T)$ and $(\mathbb{Y}, \mathcal{F}, T_\mathbb{Y})$ be two Menger PN-space, and $\widetilde{S}_i, \widetilde{S}_j : \mathbb{X} \to W(\mathbb{Y})$ be two random fuzzy operators satisfying the following condition (I):

(I) There exist two operators $a, b : \mathbb{X} \to (0, 1]$ such that, for all $x \in \mathbb{X}$ and some $i, j \in \mathbb{N}$, the set $(\widetilde{S}_{i\,x})_{a(x)} \in \Omega_\mathbb{Y}$ and $(\widetilde{S}_{j\,x})_{b(x)} \in \Omega_\mathbb{Y}$.

We note that $(\widetilde{S}_{ix})_{a(x)} = \{y| \widetilde{S}_{ix}(y) \ge a(x)\} \in \Omega_\mathbb{Y}$, where $a(x) \in (0, 1]$ is a real number and $\widetilde{S}_{ix} \in W(\mathbb{Y})$ is a fuzzy set in $\mathbb{Y}$ decided by the fuzzy operator $\widetilde{S}_i$ at $x \in \mathbb{X}$. By using each pair of fuzzy operators $\widetilde{S}_i$ and $\widetilde{S}_j$, we can define two set-valued operators S_i and S_j as follows:

$$S_i : \mathbb{X} \to \Omega_\mathbb{Y}, \ x \mapsto (\widetilde{S}_{ix})_{a(x)}, \ \forall\, x \in \mathbb{X}; \ S_j : \mathbb{X} \to \Omega_\mathbb{Y}, \ x \mapsto (\widetilde{S}_{jx})_{b(x)}, \ \forall\, x \in \mathbb{X}.$$

In the sequel, for some $i, j \in \mathbb{N}$, S_i and S_j are called the set-valued operators induced by the fuzzy operators $\widetilde{S}_i$ and $\widetilde{S}_j$, respectively. We need the following definitions:

Definition 3. Let $(\mathbb{X}, \widehat{\mathcal{F}}, T)$ and $(\mathbb{Y}, \mathcal{F}, T)$ be two Menger PN-spaces. A set-valued operator $P : D(P) \subset \mathbb{X} \to \Omega_\mathbb{Y}$ is said to be τ-closed if, for any $x_n \in D(P)$ and $y_n \in P(x_n)$, whenever $x_n \xrightarrow{\tau_\mathbb{X}} x$ and $y_n \xrightarrow{\tau_\mathbb{Y}} y$, we have $x \in D(P)$ and $y \in P(x)$.

Let Φ denote the class of all functions $\varphi : [0, \infty) \to [0, \infty)$ and Ψ be the class of all functions $\psi : [0, 1] \to [0, 1]$. Further, let $\overline{\Phi}$ be the class of all functions φ in Φ which are bijective, nondecreasing and such that $\sum_{n=1}^{\infty} \varphi^n(t) < \infty$, for all $t > 0$, where φ^n denotes the n-th iteration of φ, and let $\overline{\Psi}$ be the collection of all nondecreasing functions $\psi \in \Psi$ such that $\lim_{n \to \infty} \Psi^n(t) = 1$, for all $t \in [0, 1]$. It is easy to see that if $\varphi \in \overline{\Phi}$ then $\varphi(t) < t$ and $\lim_{n \to \infty} (\varphi^{-1})^n(t) = \infty$ for all $t > 0$ and if $\psi \in \overline{\Psi}$, then $\psi(t) > t$ for $t \in [0, 1)$.

Definition 4. Let $(\mathbb{X}, \widehat{\mathcal{F}}, T)$, $(\mathbb{Y}, \mathcal{F}, T)$ be two Menger PN-spaces and $P : D(P) \subset \mathbb{X} \to 2^\mathbb{Y}$, $Q : D(Q) \subset \mathbb{X} \to 2^\mathbb{Y}$ be two set-valued operators, $\Gamma_1, \Gamma_2 : \mathbb{X} \to L(\mathbb{Y}, \mathbb{X})$ be two operators, where $L(\mathbb{Y}, \mathbb{X})$ denotes the space of all linear operators from $\mathbb{Y}$ to $\mathbb{X}$, and $\varphi \in \Phi, \psi \in \Psi$. The pair (Γ_1, Γ_2) is called a probabilistic (φ, ψ)-contractor couple of P and Q if $F_{P(x+\Gamma_1(x)y), Q(x)+y}(\varphi(t)) \ge \psi(\min\{F_y(t), F_{Q(x)}(t), F_{P(x+\Gamma_1(x)y)}(t)\})$ for all $x \in D(Q)$, $y \in \{y \in \mathbb{Y} : x + \Gamma_1(x)y \in D(P)\}$, $t \ge 0$, and $F_{Q(x+\Gamma_2(x)y), P(x)+y}(\varphi(t)) \ge \psi(\min\{F_y(t), F_{P(x)}(t), F_{Q(x+\Gamma_2(x)y)}(t)\})$ for all $x \in D(P)$, $y \in \{y \in \mathbb{Y} : x + \Gamma_2(x)y \in D(Q)\}$, $t \ge 0$.

Remark 2. If $\varphi(t) = t$ for $t \in [0, \infty)$, (Γ_1, Γ_2) is simply called a probabilistic ψ-contractor couple of P and Q (see [6]).

Now, by the results in [4], we give the following algorithms for our main results:

Algorithm 1. Let $(\mathbb{X}, \widehat{\mathcal{F}}, T)$ be a complete Menger PN-space with a t-norm T and $(\mathbb{Y}, \mathcal{F}, T_{\mathbb{Y}})$ be a Menger PN-space with a t-norm $T_{\mathbb{Y}}$. Let f, g be two operators from $\mathbb{X}$ into itself, $\{\widetilde{S}_n\}_{n=1}^{\infty}$ be a sequence of fuzzy operators from $\mathbb{X}$ into $W(\mathbb{X})$ satisfying the condition (I) and S_n be the τ-closed set-valued operators induced by the fuzzy operators $\widetilde{S}_n$ for all $n \in \mathbb{N}$. Let Γ_1, $\Gamma_2 : \mathbb{X} \to L(\mathbb{Y}, \mathbb{X})$ and $\varphi \in \overline{\Phi}, \psi \in \overline{\Psi}$. Suppose that the following conditions (i)-(iv) are satisfied: (i) $S_i(x) \subset f(\mathbb{X})$ and $S_j(x) \subset g(\mathbb{X})$ for all $x \in \mathbb{X}$; (ii) $x + \Gamma_1(x)y \in D(S_i)$ for all $x \in D(S_j)$ and $y \in \mathbb{Y}$, $x + \Gamma_2(x)y \in D(S_j)$ for all $x \in D(S_i)$ and $y \in \mathbb{Y}$; (iii) (Γ_1, Γ_2) is a probabilistic (φ, ψ)-contractor couple of S_i and S_j; (iv) for all $x \in D(S_j)$ and $y \in S_j(x)$, there exists $v \in S_i(x + \Gamma_1(x)y)$ such that $F_v(t) \geq F_{S_i(x+\Gamma_1(x)y), S_j(x)-y}(t)$ for $t \geq 0$, and for every $x \in D(S_i)$ and $y \in S_i(x)$, there exists $w \in S_j(x + \Gamma_2(x)y)$ such that $F_w(t) \geq F_{S_j(x+\Gamma_2(x)y), S_i(x)-y}(t)$ for all $t \geq 0$. For any $x_0 \in D(S_j)$ and $y_0 \in S_j(x_0)$, we define one sequence $\{x_n\}$ in $\mathbb{X}$ by:

$$x_{2n+1} = (1 - \alpha_{2n})x_{2n} + \alpha_{2n}(x_{2n} - \Gamma_1(x_{2n})y_{2n}),$$
$$x_{2n+2} = (1 - \alpha_{2n+1})x_{2n+1} + \alpha_{2n+1}(x_{2n+1} - \Gamma_2(x_{2n+1})y_{2n+1}), \tag{6}$$

where $\{\alpha_n\}$ is a real monotone decreasing sequence in $(0, 1]$ and $\alpha_n \to \alpha \in (0, 1]$ as $n \to \infty$, the sequence $\{y_n\}$ in $\mathbb{Y}$ is defined by (5) and satisfies the following:

$$F_{y_n}(\varphi^n(t)) \geq \psi^n(F_{y_0}(t)), \quad \forall t > 0. \tag{7}$$

Algorithm 2. Let $(\mathbb{X}, \widehat{\mathcal{F}}, T)$ be a complete Menger PN-space with a t-norm T, $(\mathbb{Y}, \mathcal{F}, T_{\mathbb{Y}})$ be a Menger PN-space with a t-norm $T_{\mathbb{Y}}$ and $A, V : \mathbb{X} \to \mathbb{X}$ be two operators. Let $\{\widetilde{S}_n\}_{n=1}^{\infty}$, S_n, Γ_1, Γ_2, φ and ψ be the same as in Algorithm 1. Suppose that the conditions (ii)-(iv) in Algorithm 1 are satisfied. If

(II) $S_i(x) \subset \mathbb{X} - A(\mathbb{X})$ and $S_j(x) \subset \mathbb{X} - V(\mathbb{X})$ for all $x \in \mathbb{X}$ and some $i, j \in \mathbb{N}$, then, for any $x_0 \in D(S_j)$ and $y_0 \in S_j(x_0)$, we have the sequence $\{x_n\}$ in $\mathbb{X}$ defined by

$$x_{2n+1} = x_{2n} - \Gamma_1(x_{2n})y_{2n}, \; x_{2n+2} = x_{2n+1} - \Gamma_2(x_{2n+1})y_{2n+1},$$

where the sequence $\{y_n\}$ in $\mathbb{Y}$ is defined by (5).

Theorem 1. Let $(\mathbb{X}, \widehat{\mathcal{F}}, T)$, $(\mathbb{Y}, \mathcal{F}, T_{\mathbb{Y}})$, f, g, $\{\widetilde{S}_n\}_{n=1}^{\infty}$, S_n, Γ_1, Γ_2, φ and ψ be the same as in Algorithm 1. Suppose that the conditions (i)-(iv) in Algorithm 1 hold and the following conditions (v)-(vii) are satisfied: (v) $g(\mathbb{X})$ is x_0-joint orbitally complete for some $x_0 \in \mathbb{X}$; (vi) there exists a constant $M > 0$ such that, for any constant $\lambda_1 > \lambda_2 > 0$, $\widehat{F}_{\lambda_1 \Gamma_1(x)y}(\frac{t}{\lambda_2 M}) \geq F_y(t)$ for all $x \in D(S_j), y \in \mathbb{Y}, t \geq 0$ and $\widehat{F}_{\lambda_1 \Gamma_2(x)y}(\frac{t}{\lambda_2 M}) \geq F_y(t)$ for all $x \in D(S_i), y \in \mathbb{Y}, t \geq 0$; (vii) $\lim_{n \to \infty} T_{i=n}^{\infty} \psi^i(s) = 1$ for all $s \in [0, 1]$. Then the nonlinear operator equation couples (4) has a solution z such that $\{f(z)\} = \{g(z)\} \subset \bigcap_{i=1}^{\infty} S_i(z)$. Further, $\{x_n\}$ τ-converges to a solution of (4) and $\{y_n\}$ τ-converges to θ, where $\{x_n\}$ in $\mathbb{X}$ and $\{y_n\}$ in $\mathbb{Y}$ are two sequences generated by Algorithm 1.

Proof. By (6), (7) and the assumption (vi), since $\{\alpha_n\} \subset (0,1]$ is a monotone decreasing sequence with $\alpha_n \to \alpha \in (0,1]$, we have

$$\widehat{F}_{x_{2n+1}-x_{2n}}(\alpha M \varphi^{2n}(t)) \geq F_{y_{2n}}(\varphi^{2n}(t)) \geq \psi^{2n}(F_{y_0}(t)), \forall n \in \mathbb{N}, \, t \geq 0,$$

$$\widehat{F}_{x_{2n+2}-x_{2n+1}}(\alpha M \varphi^{2n+1}(t)) \geq F_{y_{2n+1}}(\varphi^{2n+1}(t)) \geq \psi^{2n+1}(F_{y_0}(t)), \forall n \in \mathbb{N}, \, t \geq 0,$$

which imply that $\widehat{F}_{x_{n+1}-x_n}(\alpha M \varphi^n(t)) \geq \psi^n(F_{y_0}(t))$ for all $n \in \mathbb{N}$ and $t \geq 0$. Hence, we know that $\{x_n\}$ is a τ-Cauchy sequence in $\mathbb{X}$. Since $g(\mathbb{X})$ is x_0-joint orbitally complete, we can assume that $x_n \xrightarrow{\tau_{\mathbb{X}}} z \in \mathbb{X}$. Moreover, by (7), it is easy to see that $\lim_{n \to \infty} F_{y_n}(t) = 1$ for all $t > 0$ and so $y_n \xrightarrow{\tau_{\mathbb{Y}}} \theta$. Since S_i and S_j are τ-closed, it follows from (6) and the assumption (i) that $\theta \in S_i(z)$, $\theta \in S_j(z)$ for all $i, j \in \mathbb{N}$, i.e., z is a solution of (4) and $\{f(z)\} = \{g(z)\} \subset \bigcap_{i=1}^{\infty} S_i(z)$. This completes the proof.   □

From Theorem 1 and the results in [3], we have the following results:

Corollary 1. Let $(\mathbb{X}, \widehat{\mathcal{F}}, T)$, $(\mathbb{Y}, \mathcal{F}, T_{\mathbb{Y}})$, f, g, $\{\widetilde{S}_n\}_{n=1}^{\infty}$, S_n, Γ_1, Γ_2, φ and ψ be the same as in Algorithm 1. Suppose that the conditions (i)-(iv) in Algorithm 1 and (v)-(vi) in Theorem 1 are satisfied. Then the conclusions of Theorem 1 still hold if one of the following conditions is satisfied:

(i) t-norm T is of H-type.

(ii) There exist $x_0 \in D(S_j)$ and $y_0 \in S_j(x_0)$ for some $j \in \mathbb{N}$ such that $\sum_{n=1}^{\infty}(1 - \psi^n(F_{y_0}(\frac{t}{\mu M})))^{\lambda} < \infty$ for all $t \geq 0$, where $\mu > 0$ is a constant.

(iii) There exist $x_0 \in D(S_j)$ and $y_0 \in S_j(x_0)$ for some $j \in \mathbb{N}$ such that $\sum_{n=1}^{\infty}(1 - \psi^n(F_{y_0}(\frac{t}{\mu M}))) < \infty$ for all $t \geq 0$, where $\mu > 0$ is a constant.

(iv) There exist $x_0 \in D(S_j)$ and $y_0 \in S_j(x_0)$ for some $j \geq 1$ such that $\sum_{n=1}^{\infty}(1 - \psi^n(F_{y_0}(\frac{t}{\mu M})))^{\lambda} < \infty$ for all $t \geq 0$, where $\mu > 0$ is a constant.

Remark 3. Since $T_L \in \bigcup_{\lambda \in (-1, \infty)} T_{\lambda}^{SW}$, it is easy to see that Corollary 1 is a generalization of the corresponding result in Fang [6].

Corollary 2. Let $(\mathbb{X}, \widehat{\mathcal{F}}, T)$ be a complete Menger PN-space and $L : \mathbb{X} \to \Omega_{\mathbb{X}}$ satisfy $F_{Lx, Ly}(t) \geq \psi(\min\{F_{x-y}(t), F_{x-L(x)}(t), F_{y-L(y)}(t)\})$ for all $t \geq 0$ and $x, y \in \mathbb{X}$, where operator $\psi \in \overline{\Phi}$. Suppose that the conditions (i) in Algorithm 1 and (v) in Theorem 1 are satisfied and there exists $x_0 \in \mathbb{X}$ and $y_0 \in x_0 - L(x_0)$ such that t-norm T satisfies $\lim_{n \to \infty} T_{i=n}^{\infty} \psi^i(F_{y_0}(t)) = 1$ for all $t \geq 0$, and for all $x \in \mathbb{X}$ and $y \in x - L(x)$, there exists $v \in x + y - L(x + y)$ such that $F_v(t) \geq F_{x+y-L(x+y), x-L(x)-y}(t)$ for all $t \geq 0$. Then there exists $x^* \in \mathbb{X}$ such that $x^* \in Lx^*$, i.e., x^* is a fixed point of L.

Theorem 2. Let $(\mathbb{X}, \widehat{\mathcal{F}}, T)$, $(\mathbb{Y}, \mathcal{F}, T_{\mathbb{Y}})$, A, V, $\{\widetilde{S}_n\}_{n=1}^{\infty}$, S_n, Γ_1, Γ_2, φ and ψ be the same as in Algorithm 2. Suppose that the conditions (ii)-(iv) in Algorithm 1 and (II) in Algorithm 2 are satisfied. If (i) $\mathbb{X} - V(\mathbb{X})$ is x_0-joint orbitally complete for some $x_0 \in \mathbb{X}$, (ii) there exists a constant $M > 0$ such that $\widehat{F}_{\Gamma_1(x)y}(\frac{t}{M}) \geq F_y(t)$ for all $x \in D(S_j)$, $y \in \mathbb{Y}$, $t \geq 0$ and $\widehat{F}_{\Gamma_2(x)y}(\frac{t}{M}) \geq F_y(t)$ for all $x \in D(S_i)$, $y \in \mathbb{Y}$, $t \geq 0$, (iii) $\lim_{n \to \infty} T_{i=n}^{\infty} \psi^i(s) = 1$ for all $s \in [0, 1]$, then the following system of nonlinear operator equations: $x = A(x)$, $x = V(x)$ has a solution z such that $\{z - A(z)\} = \{z - V(z)\} \subset \bigcap_{i=1}^{\infty} S_i(z)$. Further, $\{x_n\}$ τ-converges to a solution of z and $\{y_n\}$ τ-converges to θ, where the sequences $\{x_n\}$ in $\mathbb{X}$ and $\{y_n\}$ in $\mathbb{Y}$ are defined by Algorithm 2.

4 Conclusions

In this paper, based on the concept of probabilistic (φ, ψ)-contractor couple introduced by Mihet [7], we introduce and study a new class of nonlinear fuzzy and crisp mixed family operator equation couples in Menger PN-spaces. Further, by using the results in [4], we construct some new perturbed iterative algorithms and prove the existence of solutions for this kind of nonlinear operator equation couples and the convergence of iterative sequences generated by the algorithms under a larger class of t-norms and joint orbitally complete conditions. Our results improve and generalize corresponding results in the literature.

Acknowledgments

We would like to express our thanks to the referees for their valuable suggestions. This work was supported by the Sichuan Youth Science and Technology Foundation (08ZQ026-008), the Open Foundation of Artificial Intelligence of Key Laboratory of Sichuan Province (2009RZ001) and the Scientific Research Fund of Sichuan University of Science & Engineering (2009XJKYL007).

References

1. Chang, S.S.: Probabilistic contractor and the solutions for nonlinear equations in PN-spaces. Chinese Sci. Bull. 35, 1451–1454 (1990) (in Chinese)
2. Altman, M.: Contractors and Contractor Directions Theory and Applications. Marcer Dekker, New York (1977)
3. Hadžić, O., Pap, E.: Fixed Point Theory in Probabilistic Metric Spaces: Theory in Probabilistic Metric Spaces. Kluwer Academic Publishers, Dordrecht (2001)
4. Chang, S.S., Cho, Y.J., Kang, S.M.: Nonlinear Operator Theory in Probabilistic Metric Spaces. Nova Science Publ. Inc., New York (2001)
5. Cho, Y.J., Lan, H.Y., Huang, N.J.: A system of nonlinear operator equations for a mixed family of fuzzy and crisp operators in probabilistic normed spaces. J. Inequal. Appl., Art. ID 152978, 16p. (2010)
6. Fang, J.X.: On nonlinear equations for fuzzy mappings in probabilistic normed spaces. Fuzzy Sets and Systems 131, 357–364 (2002)
7. Mihet, D.: On set-valued nonlinear equations in Menger probabilistic normed spaces. Fuzzy Sets and Systems 158(16), 1823–1831 (2007)
8. Sharma, B.K., Sahu, D.R., Bounias, M.: Common fixed point theorems for a mixed family of fuzzy and crisp mappings. Fuzzy Sets and Systems 125, 261–268 (2002)

Improving the Learning of Recurring Concepts through High-Level Fuzzy Contexts

João Bártolo Gomes[1],[*], Ernestina Menasalvas[1],[**], and Pedro A.C. Sousa[2]

[1] Facultad de Informática - Universidad Politécnica Madrid, Spain
joao.bartolo.gomes@alumnos.upm.es,
emenasalvas@fi.upm.es
[2] Faculdade de Ciências e Tecnologia, Universidade Nova de Lisboa, Portugal
pas@fct.unl.pt

Abstract. In data stream classification the problem of recurring concepts is a special case of concept drift where the underlying concepts may reappear. Several methods have been proposed to learn in the presence of concept drift, but few consider recurring concepts and context integration. To address these issues, we presented a method that stores previously learned models along with context information of that learning period. When concepts recur, the appropriate model is reused, avoiding relearning a previously seen concept. In this work, in order to model the vagueness and uncertainty associated with context, we propose the inference of high-level fuzzy contexts from fuzzy logic rules, where the conditions result from fuzzified context inputs. We also present the changes required for our method to deal with this new representation, extending the approach to handle uncertain contexts.

Keywords: Data Stream Mining, Concept Drift, Recurring Concepts, Context-awareness, Fuzzy Logic, Ubiquitous Knowledge Discovery.

1 Introduction and Related Work

Learning from data streams in ubiquitous devices where the data distributions and target concepts may change over time is a challenging problem, known as concept drift[7]. In real world classification problems it is common for previously seen concepts to reappear[8]. This represents a particular type of concept drift[7], known as concept recurrence[8,2,9,1].

Prediction models usually change over time, as in product recommendations where customer interests change due to fashion, economy or other *hidden context*[8,4]. Several methods have been proposed to detect and adapt to concept drift[7,3]. The usual approach is to use a forgetting mechanism and learn a new decision model when drift is detected[3]. A possible solution to exploit recurrence

---

[*] The work of J.P. Bártolo Gomes is supported by a Phd Grant of the Portuguese Foundation for Science and Technology (FCT).
[**] The research is partially financed by project TIN2008-05924 of Spanish Ministry of Science and Innovation.

J. Yu et al. (Eds.): RSKT 2010, LNAI 6401, pp. 234–239, 2010.

is to store previously learned models that represent observed concepts and thus avoid relearning a previously learned one when it reappears[2,9,1].

The works presented in[9,2] are the most similar to our approach[1] for this problem, as both use drift detection and store past models in order to adapt to concept drift and recurrence. The main difference lies in the mechanism used to store and retrieve past models. The context-aware approach proposed in[1] resembles and shares the motivation with the one presented in[4], where the method infers the periods where *hidden context* is stable from available context features, which are described as contextual clusters. We explore the relation between concepts and context to improve the adaptation to drift in the presence of recurring concepts. The main difference in relation to [4] is that we use an online approach, which does not require the partition of the dataset into batches, since the concepts can have arbitrary size as determined by the drift detection method[3].

When using context in the proposed adaptation mechanism[1], it is important to represent the uncertainty and vagueness observed in physical world situations from context input attributes and in the high level context inferred from those attributes. In this work, we propose the extension of the mechanism presented in[1] by integrating fuzzy set theory in context modeling, high level context inference and reasoning. The usage of fuzzy logic in context-aware systems is not new and has been proposed in[6] as a mechanism to handle uncertainty regarding context reasoning.

This paper is organized as follows. Section 2 describes the concept recurrence problem and summarizes fundamentals presented in[1]. The proposed extension to deal with uncertainty of context is found in section 3 where we describe fuzzy context and model storage/retrieval extensions according to it. Finally we provide some concluding remarks and outline future research work in section 4.

2   Learning Process Fundamentals

Let us assume a system learning from a stream of records where the target concept can change over time, based on an incremental version of the Naive Bayes algorithm. The performance of the learning algorithm is monitored using a drift detection mechanism[3] that triggers an event when drift is detected. In previous work[1] we presented a mechanism, assuming the same learning system, that keeps learned decision models along with context information, improving adaptation to drift in scenarios where concepts associated with context reappear.

If the records distribution is stationary the classifier error-rate decreases as the number of training records grows. This assumption is shared by most approaches dealing with drift[7], as it is a natural property in real world problems where periods of stable concepts are observed followed by change to a new period of stability (with a different target concept). Proposed in[3] a method for drift detection uses this assumption to find drift events. Two levels are defined, a warning level and a drift level, based on the error-rate of the learning algorithm. Note that an increase in the error-rate reaching the warning level but followed immediately by a decrease is considered a false alarm.

The continuous learning process presented in[1], consists of the following steps:

1. process the incoming data stream records using an incremental learning algorithm (base learner) to obtain a decision model that represents the underlying concept.
2. the drift detection method[3] monitors the error-rate of the learning algorithm.
3. when the error-rate goes up the drift detection method indicates,
   - **warning level:** store the incoming records into a warning window and prepare a new learner that processes incoming records while the warning level is signaled.
   - **drift level:** store the current model and its associated context into a model repository; use the model repository to find a past model with a context similar to the occurring context that performs well with the new data records (i.e. represents the current concept, as presented by the authors in[1]). Reuse the model from the repository as base learner to continue the learning process as in point 1. If no model is found use the learner that was initiated during warning level period as base learner.
   - **false alarm (normal level after warning):** the warning window is cleared and the learner used during the warning period is discarded. The learning process continues as in point 1.

In this work we propose a modification of our method to deal with the problem of ambiguity and uncertainty in the inference and usage of context. The requirements for such system are: i) adapt the classification model to concept drift ; ii) recognize and use past models from previously seen concepts when these reappear ; iii) handle vagueness and uncertainty in context information ; iv) use contextual information associated with stored models to improve adaptation to drift.

3   Fuzzy Based Approach to Deal with Context Integration and Recurring Concepts

The main challenges of the proposed extension are related with: 1) context inference and usage 2) model storage and retrieval. The models for reuse are stored in a model repository with the associated context as described in subsection 3.2. This context is the one observed during the corresponding learning period of the stored model. In this work, we integrate the use of fuzzy context as described in section 3.1. In the case of model retrieval, the repository is searched for a model that represents the current data and with context similar to the currently occurring one. The metrics used for this task are described in subsection 3.3.

3.1   Fuzzy Context

Context attributes depend on the problem domain. In this work, in order to deal with vagueness and uncertainty in context attributes and high level context inference (i.e. context inferred from context attributes), the previous context representation is extended using fuzzy set theory[5]. The previous context representation defined each context state C_i as a tuple of N attribute-values $C_i = (a_1^i, ..., a_n^i)$, where a_n^i represents the value of attribute a_n for context state C_i. A context space defines the regions of acceptable values for these attributes.

In fuzzy context representation the crisp context attributes are fuzzified into degrees of membership for linguistic terms of fuzzy sets, according to the problem domain semantics and defined by experts, such as 'high', 'good', or 'cold'. High level contexts are defined using fuzzy rules[5]. Each high level context fuzzy rule R_i consists of a conjunction of conditions and a condition can itself be expressed by a disjunction of conditions,

$$R_i : \textbf{If } a_1 \textit{ is } A_{i1} \text{ ... } \textbf{and } a_n \textit{ is } A_{in} \textbf{ then } c \textit{ is } hlc_i$$

In R_i conditions represent the membership degree of a context attribute a_n to a particular fuzzy set A_{in}. The output variable c represents the high level context and hlc_i is one output fuzzy set, such as 'summer' or 'havingHeartAttack'. For each rule a weight is assigned to each condition with a value ranging between 0 and 1, where the sum of the weights is 1 per rule. The weights w_j represent the importance of each condition j in the inference of the high level context. Note that when evaluating rules where disjunctions appear, the fuzzy set A_{in} with maximum degree is used. The membership degree of high level context hlc_i inferred from rule R_i is defined as,

$$hlc_i = \sum_{j=1}^{n} w_j * A_{in}(a_j)$$

where n is the number of conditions in R_i.

3.2   Model Storage

In the Naive Bayes learning algorithm, let P_C be the table storing estimation of $P(C)$ as the observed frequency for each class C, and P_A be the vector that stores estimation of $P(A_n|C)$ as the observed frequency of each feature A_n given class C. In the assumed learning system, storing a decision model M also requires storing:

- The context frequency C_f representing the frequency of each high level context during the learning period of M. This implies storing a vector with the average membership value $\bar{hlc_i}$ for each high level context hlc_i inferred from the context input C_{input} observed during M learning period. We calculate the average degree $\bar{hlc_i}$ using n as the number of context records seen in the learning period using the formulation, $\bar{hlc_i} = \frac{1}{n}\sum_{j=1}^{n} hlc_i(C_{input_j})$. The context frequency C_f is expressed as $C_f = (\bar{hlc_1}, ..., \bar{hlc_n})$
- The accuracy Acc_k of M, representing the accuracy value obtained during the learning period k, with $numCRecords_k$ being the number of correctly classified records by M and $numRecords_k$ being the total number of records processed during k. The accuracy value is updated if M is reused. This is formulated as $Acc_k = \frac{numCRecords_k}{numRecords_k}$
- T is the timestamp that records the time when model M was stored.

Consequently each decision model M stored in the model repository is defined as the tuple:

$$M = \{P_C, P_A, C_f, Acc_k, T\}$$

3.3   Model Retrieval

The main goal of the model retrieval procedure is to find the model that best represents the underlying concept and with context similar to the occurring one. This will be a high accurate model for the current underlying concept. Our approach searches the model repository and calculates the error of the stored models on the new data, using the records available in the warning window. If the number of records is lower than a specified threshold no model is returned, as we are unable to estimate the performance of past models. The same happens when the repository is empty or the performance of the stored models is too low.

The error prediction of the past models is calculated by the mean square error MSE_i of model M_i, using the warning window W_n of n records.

The other important aspect of model retrieval is to minimize the similarity between the occurring context and the context stored in previously learned models, as this indicates that we are observing a similar context. This is achieved by means of a distance function $Dis(Context_i, Context_o)$, where $Context_i$ is the context vector with the high level context membership values (as described in section 3.2) associated with model M_i and $Context_o$ is the current context vector. The distance used is the Euclidean distance expressed as:

$$Dis(Context_i, Context_o) = \sqrt{\sum_{K=1}^{N} (hlc_k^i - hlc_k^o)^2}$$

Thus the selection criterion resulting from the two metrics is the utility function defined as:

$$u(MSE_i, Dis(Context_i, Context_o)) = \frac{w_m}{1 + MSE_i} + \frac{w_c}{1 + Dis(Context_i, Context_o)}$$

To maximize the utility value, both metrics should be minimal. This means that the selected model will be the one with the lowest error and with a context similar to the occurring one. Weights w_c and w_m are assigned to the context and learner error components of the utility function. The model with the highest utility value is selected. Note that if ties are found between stored models, the accuracy Acc_k and the timestamp T are compared, giving preference to the model with the highest past accuracy and timestamp value. If no model in the repository reaches the utility threshold value, no model is returned and the new model that was processing records during the warning period is used instead of a stored one.

4 Conclusions and Future Work

In this work, we have presented an extension of an existing method for the problem of data stream classification with concept recurrence. This method associates context information with learned models, improving adaptation to drift. The main contribution of the work presented in this paper consists in the use of fuzzy contexts to represent uncertainty, mapping crisp context attributes into fuzzy sets and the inference of high level context from fuzzy rules using the fuzzified context inputs. This context is stored along with previously learned models to track recurrence when drift occurs. A description of the solution, with the modifications in model storage and retrieval was presented.

As future work we would like to integrate our mechanism in the drift detection method as it does not integrate context and only considers the learning algorithm performance as an indicator of drift. It would also be interesting to study the accuracy gains of such extensions on recurring concept problems.

References

1. Bartolo Gomes, J.P., Menasalvas, E., Sousa, P.: Tracking recurrent concepts using context. In: Rough Sets and Current Trends in Computing, Proceedings of the Seventh International Conference RSCTC 2010 (2010) (to appear)
2. Gama, J., Kosina, P.: Tracking Recurring Concepts with Meta-learners. In: Lopes, L.S., Lau, N., Mariano, P., Rocha, L.M. (eds.) EPIA 2009. LNCS, vol. 5816, p. 423. Springer, Heidelberg (2009)
3. Gama, J., Medas, P., Castillo, G., Rodrigues, P.: Learning with drift detection. LNCS, pp. 286–295. Springer, Heidelberg (2004)
4. Harries, M.B., Sammut, C., Horn, K.: Extracting hidden context. Machine Learning 32(2), 101–126 (1998)
5. Konar, A. (ed.): Computational Intelligence: Principles, Techniques and Applications. Springer, New York (2005)
6. Ranganathan, A., Al-Muhtadi, J., Campbell, R.H.: Reasoning about uncertain contexts in pervasive computing environments. IEEE Pervasive Computing 3(2), 62–70 (2004)
7. Tsymbal, A.: The problem of concept drift: definitions and related work. Computer Science Department, Trinity College Dublin (2004)
8. Widmer, G., Kubat, M.: Learning in the presence of concept drift and hidden contexts. Machine learning 23(1), 69–101 (1996)
9. Yang, Y., Wu, X., Zhu, X.: Combining proactive and reactive predictions for data streams. In: Proceedings of the eleventh ACM SIGKDD international conference on Knowledge discovery in data mining, p. 715. ACM, New York (2005)

A Frequent Pattern Mining Method for Finding Planted (l, d)-motifs of Unknown Length

Caiyan Jia[1], Ruqian Lu[2,3], and Lusheng Chen[2]

[1] Department of Computer Science, Beijing Jiaotong University,
Beijing 100044, China
cyjia@bjtu.edu.cn

[2] Shanghai Key Lab of Intelligent Information Processing & Department of
Computer Science and Engineering, Fudan University, Shanghai 200433, China
lschen@fudan.edu.cn

[3] Institute of Mathematics, Chinese Academy of Sciences, Beijing 100080, China
rqlu@math.ac.cn

Abstract. Identification and characterization of gene regulatory binding motifs is one of the fundamental tasks toward systematically understanding the molecular mechanisms of transcriptional regulation. Recently, the problem has been abstracted as the challenge planted (l, d)-motif problem. Previous studies have developed numerous methods to solve the problem. But most of methods need to specify the length l of a motif in advance. In this study, we present an exact and efficient algorithm, called Apriori-Motif, without given l. The algorithm uses breadth first search and prunes the search space quickly by the downward closure property used in Apriori, a classical algorithm of frequent pattern mining. Empirical study shows that Apriori-Motif is better than some existing methods.

1   Introduction

Finding regulatory binding motifs in DNA sequences is an important step to decipher expression regulation of thousands of annotated genes in genomes. In the statistical sense, a motif in DNA sequences is not exactly identical but presents mutations. It is often a short conserved subsequence in the midst of a great amount of statistical noise, and hard to be discriminated from random signals [1,2,3]. In the literature, Prevzner & Sze formalize the problem as the planted (l, d)-motif problem (PMP for short) [4].

Given a set of strings $S=\{s_1, s_2, \ldots, s_N\}$ over a symbol set $\Sigma=\{A,C,G,T\}$ such that $|s_i| \leq L$, $1 \leq i \leq N$ and positive integers l and d such that $1 \leq l \leq L$ and $0 \leq d < l$, the planted (l, d)-motif problem is to find a string $t \in \Sigma^l$ such that for every string s_i in S, there exists a substring (consecutive symbol string in a sequence) t_i in s_i such that $d(t, t_i) = d$, where $d(t, t_i)$ means the Hamming distance between t and t_i. In other words, a conserved motif at length l with just d mutations is planted into each background sequence with length L. We need to find the motif from a set of N sequences just given l and d.

J. Yu et al. (Eds.): RSKT 2010, LNAI 6401, pp. 240–248, 2010.

Previous studies have developed numerous methods in order to find planted motifs, including SPELLER [5], WINNOWER [4], SP-STAR [4], MITRA [6], PROJECTION [7], MULTI-PROFILE [8], PatternBranching [9], CENSUS [10], WEEDER [11], PMS [12], Voting [13,14], RISOTTO [15], etc.

By large, tow kinds of strategies are used. One is local optimal strategy, e.g. Gibbs Sampling method [7,16]. The kind of algorithms are based on PWM (Position-Specific Weighted Matrix, also called profile) model, can find motifs of any specified length very efficiently. But they are inevitably to slump into a local optimal. The other is heuristic enumeration strategy based on consensus model (more accurately, mismatch model) [5,10,11]. The kind of algorithms can find all exact solutions by searching the whole pattern space, but they are limited by the time or the space complexity of the algorithms [17].

In general, the heuristic enumeration algorithms have the following properties. 1. The depth first search is used to enumerate all potentials. 2. Most of algorithms need to know the length l of a motif in advance. 3. Most of algorithms suppose that there exists an occurrence of a motif in each sequence. 4. Half of methods only deal with the case that the allowed mismatches are just d.

In fact, a researcher may not know the size l of a planted motif a priori. And he is more likely to be interested in motifs that are d or fewer mutations rather than exact d. Moreover, as experimental data are commonly rife with noise, it is likely that some sequences may contain no motifs at all [18,19]. Thus, in this study, we intend to solve PMP under the conditions that l is unknown in advance, there are at least q $(q \leq N)$ sequences containing a planted signal, and the allowed mismatches are at most d.

Based on the previous study [17], at the inspiration of frequent pattern mining techniques, we give a new, efficient algorithm Apriori-Motif. It makes a trade-off between the time and the space complexities to identify planted subtle motifs exactly and efficiently in terms of the down closure property used in Apriori [20]. Similar with Apriori, Apriori-Motif uses the breadth first search to scan the whole patten space indexed by the consensus tree [17]. The empirical studies have shown that Apriori-Motif can solve PMP under the conditions specified above at reasonable time and low main memory.

The rest of paper is organized as follows. Section 2 introduces some definitions used in the paper. Section 3 presents the algorithm Aprior-Motif. Section 4 gives some experimental results. Section 5 compares Apriori-Motif with some related methods. Section 6 concludes the paper.

2 Preliminaries

Definition 1. Given $d \geq 0$ as the number of maximally allowed mutations (or mismatches), any string b with $d(a, b) = x \leq d$ is called an x-mutated copy (or simply mutated copy) of string a and vice versa, where $d(a, b)$ means the Hamming distance between a and b. A zero mutated copy is also called an exact copy. All x-mutated copies of a, where $x \leq d$, form the d-neighborhood of a. a is called the center of this neighborhood.

Given a center a, the size of d-neighborhood of a is bounded by $v(d, l)$, where

$$v(d, l) = \sum_{i=0}^{d} \binom{l}{i} (|\Sigma| - 1)^i.$$

And the number of all d-mutated copies of a is at most

$$v'(d, l) = \binom{l}{d} (|\Sigma| - 1)^d.$$

Thus, PMP with at most d mutations is harder than that with exact d mutations since the former has larger search space than the latter.

Definition 2. A set with property P is said to satisfy downward closure property if all nonempty subsets of the set also have the property P.

It's easy to know that motifs satisfy the downward closure property. For an example, if TCTGAC satisfies the constraint that their x-mutated copies $(x \le d)$ appears in at least q $(q \le N)$ sequences of a sample, then its any substring, e.g. TCTGA and CTGAC, satisfies this constraint. Thus, for finding a planted motif, we can search the pattern space from short strings to longer strings since if one of TCTGA and CTGAC does not satisfy the constraint, TCTGAC does not satisfy the constraint definitely, can be pruned from the search space. The idea is just the essence of Apriori [20], will be used in Apriori-Motif.

Definition 3. A tree-like structure is called consensus tree if it is used to storage all possible motifs in the searching process (see [17]). And a consensus tree has the following properties.

1. There is one and only one path corresponding to a potential motif. The length of the path is just that of the motif.

2. A full consensus tree is a complete $|\Sigma|$ branches tree. But in real applications, only nodes (corresponding motifs) with x-mutated copies in at least q inputting sequences are allowed to grow in the tree, where $x \le d$ and $q \le N$.

3. A candidate signal spelled by the path ending at the node might not occur in sequences at all although its x-mutated copies $(x \le d)$ occur more than q times in at least q difference sequences.

4. All nodes in a tree can be divided into two classes. One is real nodes representing the real signals contained in sequences. They are allowed to grow in the next level of the tree. The other is virtual nodes representing the candidate but false signals. They are prohibited to grow in the tree.

5. For a non-root node, there is a link (pointer) between the son of the node and the brother of the node, where the son and the brother have the same node content, a symbol in Σ. Thus, the branches of a node are as the same as the active branches (their nodes stand for real signals) of its uncle. We call it heredity property of a consensus tree. The property just comes from the down closure property of a motif. Thus, only real nodes need the link structure.

A typical example of a consensus tree is shown in Fig. 1. It is a four-branch tree. And the content of a node is one of the four nucleotides A, C, G, T. The

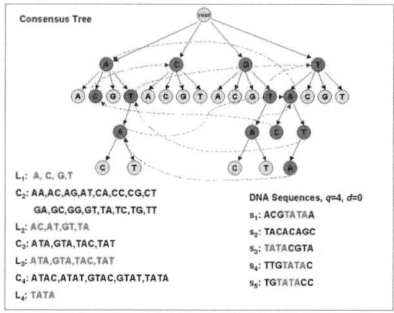

Fig. 1. Consensus tree of S=s1, s2, $\cdots$, s5 when $q = 4$, $d = 0$

path from the root to a node represents a candidate motif. All candidate motifs make up the entire search space. And each candidate motif has one and only one path in the tree. This is a compressed compact structure for representing the search space. In general, the longest signal is just the planted motif.

In Fig. 1, the red nodes represent all real nodes. The green nodes stand for all virtual nodes. L_i denotes all real signals and C_i denotes all candidate signals in the i-th level where $L_i \subseteq C_i$. The longest signal contained in the sequences is TATA. The branch TAT can grow up one branch TATA since it uncle AT has a red branch ATA. It's just an example of heredity property of the consensus tree.

3   Apriori-Motif

For the problem of motif finding, we generally need to repeatedly search DNA substrings in a set of sequences multiple times. A brute force string search is going to be terrible and inefficient. Thus, similar with SPELLER [5] and WEEDER [11], an index structure, suffix tree, is used in our algorithm.

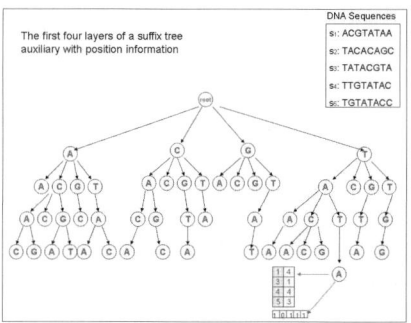

Fig. 2. The first four levels of the suffix tree

In Apriori-Motif, we only build the first $l + 1$ levels of the suffix tree for a sample of sequences since only the first $l + 1$ levels are useful for detecting whether a candidate signal is real or not, where l is the maximal length of the real signals. And we attach (j, k) tuples of all exact copies of a candidate signal and an N-bit string to the corresponding node of the tree, where (j, k) stands for a substring starting at the k-th position of the j-th sequence and the i-th bit of an N-bit string is set to 1 if the candidate motif spelled by the path ending at the node occurs in the i-th sequence, otherwise it is 0. (j, k) tuples show positions of a signal appeared in a sample and the N-bit string can be used to count the number of occurrences of the signal directly. The first four levels of the suffix tree are shown in Fig. 2 for the example in Fig. 1. Due to space limitations, we only draw the (j, k) tuples and the 5-bit string of 'TATA'.

Since the first d levels of a consensus tree is a full $|\Sigma|$ branches tree, we build a tree from the $(d + 1)$-th level. The process of motif finding is just that of constructing the consensus tree for a sample shown in the **Algorithm Apriori-Motif**.

Algorithm Apriori-Motif

---

1. Let $i = d + 1$.
2. Build the i-th level of the suffix tree for a DNA sample.
3. Generate a node of the consensus tree in terms of its uncle with breadth first manner in the i-th level.
4. Count the number (denoted by t) of occurrences of x-mutated copies of the node from the constructed suffix tree, where $x \leq d$.
   4.1 If $t \geq q$, the node is permitted to grow in the next level.
   4.2 if $t < q$, the node is forbidden to grow in the next level.
5. $i = i + 1$, go to the step 2 until there is no real or virtual node in the current level of the consensus tree.

According to the heredity property of a consensus tree, a node will grow out a new branch in the next level only if it is a real node and its link node (uncle of the node) has the active branch in the current level. With the growth of the suffix tree, the consensus tree is built up layer by layer, synchronously. What we should do just count the number q of occurrences of a candidate.

The space complexity of Apriori-Motif is bound by

$$O_{s-Apriori-Motif} = O(N \times L \times v(d, l)).$$

And the time complexity of Apriori-Motif is

$$O_{t-Apriori-Motif} = O(N^2 \times L^2 \times l \times v(d, l)).$$

But the time and space complexities at worst case analysis are heavily overestimated in real world since the number of nodes in a consensus tree is far less than the theoretical bound $N \times L \times v(d, l)$ in the level i ($i \leq l$). For an example, the time used by Apriori-Motif is almost linear to the maximal length L of background sequences in all the real experiments tested in the next section while it is quadratic time with L in theory.

4   Experimental Results

We test the performance of Apriori-Motif on some benchmark synthetic samples for PMP under the conditions concerned in the study. And we compare the algorithm with some other algorithms including brute force algorithm, classical PWM based algorithm PROJECTION [7], Gemoda algorithm [18,19] which aims to identify a planted motif also without given the length l (but the algorithm needs to specify the window size l' of allowed mutations, $l' \leq l$). All experiments are performed on an Intel computer with 2 GHz processor and 2GB main memory. The operating system is Windows XP.

Following previous work, we generate the benchmark testing (l, d) samples by the method in [7,17]. In all of our experiments, if not specified, $N = 20$ and $L = 600$ and the real mumble of mismatches is at most d.

The results of Apriori-Motif, brute force, PROJECTION and Gemoda are shown in the table 1, where s means second, min means minute, h denotes hour, and mon denotes month. And the accuracy of the algorithms (i.e. acc in table 1) is measure by the motif level accuracy in [1]. Among the four algorithms, the length l of the planted motif for a sample is given to brute force algorithm and PROJECTION a priori. And Gemoda needs to known the mutated window size l' ($l' = l$, $q = N$ for all cases in the table 1). But we only know the number of maximally allowed mismatches d and $q = N$ for Apriori-Motif.

Table 1. The Performance Comparison on a Range of (l, d)[1]

(l, d)	Brute force		PROJECTION		Gemoda		Apriori-Motif	
	acc	time	acc	time	acc	time	acc	time
$(10, 2)$	1.00	$72min$	0.82	$161.1s$	1.00	$8min$	1.00	$60.109s$
$(11, 2)$	-	-	0.95	$12.5s$	1.00	$<1min$	1.00	$60.235s$
$(12, 3)$	-	-	0.71	$8.7min$	1.00	$10.5h$	1.00	$15.9min$
$(13, 3)$	-	-	0.94	$46.0s$	1.00	$10min$	1.00	$15.6min$
$(14, 4)$	-	-	0.65	$15.4min$	1.00	$>3mon$	1.00	$3.368h$
$(15, 4)$	-	-	0.90	$129.0s$	1.00	$6h$	1.00	$3.134\ h$

The table 1 shows the speed of PROJECTION is very fast, but the algorithm can only find the approximate answers. It is just the pros and the cons of a local optimal method. And it is will known that PMP is NP-hard. The brute force algorithm, which enumerates all potential patterns without any pruning strategy, will work only when l and d are both small. Compared with Gemoda which is designed for solving PMP with a specified mutation window of a planted motif, Apriori-Motif is much more efficient while it is designed for solving PMP without given any information both on the length of a motif and the window size of mutations.

Moreover, in all of the experiments, the longest predicted motifs are just the planted ones except for the (14, 4) sample. For the (14, 4) sample, Apriori-Motif

---

[1] The data of Brute force, PROJECTION and Gemoda comes from [14] and [18].

reports two predicted motifs with length 14. One is false positive, the other is just the planted one. But the false positive and the real signal have long-range overlaps. What's more, the larger the parameter d, the harder the problem.

And we test the influence of the length L of background sequences on Apriori-Motif. The time complexity is almost linear with L. Moreover, the main memory used by all of the experiments is no more than 150MB. Thus, the algorithm is suitable for running on the current personal computers. Also, the parameter q has strong influence on the hardness of the problem. We do not show it since the result is similar with that in [17]. So does the parameter N.

5   Related Researches and Comparison

Firstly, Apriori-Motif is a breadth first search method. It finds planted motifs exactly without given the length l while almost all of the existing enumerative algorithms use the depth first search, e.g. SPELLER [5] and WEEDER [11], and need to specify the length l a priori. Although the depth first search can save the main memory, the breadth first search can speed up the algorithm if it is used well.

Secondly, WEEDER introduces ϵ to narrow down the pattern space (see [11]). Similar with WEEDER, we introduce a parameter e to Apriori-Motif for speeding up the algorithm, where e is the number of maximally allowed consecutive mismatches between a planted occurrence and the planted motif. The testing results can be found in the table 2, where e varies from 1 to d. The table 2 shows the smaller e, the faster but the less accuracy the algorithm.

Table 2. The Influence of e on Apriori-Motif

(l, d)	$e=1$		$e=2$		$e=3$		$e=4$	
	acc	time	acc	time	acc	time	acc	time
$(10, 2)$	0.7	47.140s	1.0	60.109s	-	-	-	-
$(11, 2)$	1.0	50.157s	1.0	60.235s	-	-	-	-
$(12, 3)$	0.66667	5.169min	0.58333	12.765min	1.00	15.908min	-	-
$(13, 3)$	0.76923	4.954min	1.0	14.320min	1.00	15.612min	-	-
$(14, 4)$	0.64285	12.948min	0.92857	1.820h	1.00	2.840	1.00	3.368h
$(15, 4)$	0.60	10.643min	0.80	1.945h	1.00	2.914h	1.00	3.134 h

Thirdly, Apriori-Motif is different from Bpriori series algorithms in [17]. In fact, it is an improved version of Bprori algorithms.

6   Conclusion

In the paper, we present an algorithm Apriori-Motif for solving PMP under the conditions that l is not given in advance, $q \leq N$ and mismatches is at most d. It uses the breadth first search to prune the pattern space. Both theoretical analysis and experimental tests show the good performance of the algorithm.

Acknowledgements. This work was supported in part by NSFC (Grant No. 90820013, 60875031 and 60905029), 973 Project (Grant No. 2007CB311002).

References

1. Tompa, M.: Assessing computational tools for the discovery of transcription factor binding sites. Nature Biotechnology 23, 137–144 (2005)
2. Hu, J., Li, B., Kihara, D.: Limitations and potentials of current motif discovery algorithms. Nucleic Acids Research 33(15), 4899–4913 (2005)
3. Das, M.K., Dai, H.K.: A survey of DNA motif finding algorithms. BMC Bioinformatics 8(suppl. 7) (2007)
4. Pevzner, P., Sze, S.: Combinatorial approaches to finding subtle signals in DNA sequences. In: Proceedings of the Eighth International Conference on Intelligent Systems for Molecular Biology, California, USA, pp. 269–278 (2000)
5. Sagot, M.F.: Spelling approximate repeated or common motifs using a suffix tree. In: Lucchesi, C.L., Moura, A.V. (eds.) LATIN 1998. LNCS, vol. 1380, pp. 111–127. Springer, Heidelberg (1998)
6. Eskin, E., Pevzner, P.: Finding composite regulatory patterns in DNA sequences. Bioinformatics, 354–363 (2002)
7. Buhler, J., Tompa, M.: Finding motifs using random projections. In: Proceeding of the Fifth Annual Internal Conference Computational Molecular Biology, Canada. ACM Press, New York (2001)
8. Keich, U., Pevzner, P.A.: Subtle motif: defining the limits of finding algorithms. Bioinformatics 18(10), 1382–1390 (2002)
9. Price, A., Ramabhadran, S., Pevzner, P.A.: Finding subtle motifs by branching from sample string. Bioinformatics 2, 1–7 (2003)
10. Evans, P.A., Smith, A.D.: Toward optimal motif enumeration. In: Proceedings of Algorithms and Data Structures, 8th International Workshop, pp. 47–58 (2003)
11. Pavesi, G., Mauri, G., Pesole, G.: An algorithm for finding signals of unknown length in DNA sequences. Bioinformatics 17, 207–214 (2001)
12. Davila, J., Balla, S., Rajasekaran, S.: Fast and practical algorithms for planted (l, d) motif search. IEEE/ACM Trans. on Computational Biology and Bioinformatics 4, 544–552 (2007)
13. Chin, Y.L., Leung, C.M.: Voting algorithms for discovering long motifs. In: Proceedings of the Third Asia-Pacific Bioinformatics Conference, Singapore, pp. 261–271 (2005)
14. Leung, C.M., Chin, Y.L.: An efficient algorithm for the extended (l, d)-motif problem with unknown number of binding sites. In: Proceedings of the Fifth IEEE Symposium on Bioinformatics and Bioengineering, pp. 11–18 (2005)
15. Pisanti, N., Carvalho, A.M., et al.: RISOTTO: fast extraction of motifs with mismatches. In: Proceeding of the Seventh Latin Am. Theoretical Informatics Symp., pp. 757–768 (2006)
16. Lawrence, C.E., Altschul, S.F., et al.: Detecting subtle sequence signals: A Gibbs sampling strategy for multiple alignment. Science 262, 208–214 (1993)
17. Lu, R.Q., Jia, C.Y., et al.: An exact data mining method for finding center strings and all their instances. IEEE Trans. on Knowledge and Data Engineering 19(4), 509–522 (2007)

18. Styczynski, M.P., Jensen, K.L.: An extension and novel solution to the (l, d)-motif challenge problem. Genome Informatics 15, 63–71 (2004)
19. Jensen, K.L., Styczynski, M.P.: et al: A generic motif discovery algorithm for sequential data. Bioinformatics 22, 21–28 (2006)
20. Agrawal, R., Srikant, R.: Fast algorithms for mining association rules. In: Proceedings of the 20th International Conference on Very Large Data Bases, Santiago de Chile, Chile, pp. 487–499 (1994)

A Quick Incremental Updating Algorithm for Computing Core Attributes

Hao Ge[1], Chuanjian Yang[2], and Wanlian Yuan[3],*

[1] Department of Electronic and Information Engineering, Chuzhou University
[2] Department of Computer Science, Chuzhou University
[3] Department of Mathematics, Chuzhou University,
Chuzhou 239012, P.R. China
{togehao,cjy474,yuanwanlian}@126.com

Abstract. Computing core attributes is one of key problems of rough set theory. Many researchers proposed lots of algorithms for computing core. Unfortunately, most of them are designed for static databases. However, many real datasets are dynamic. In this paper, a quick incremental updating core algorithm is proposed, which only allies on the updating parts of discernibility matrix and does not need to store, re-compute and re-analyze discernibility matrix, when new objects are added to decision table. Both of theoretical analysis and experimental results show that the algorithm is effective and efficient.

Keywords: rough sets, discernibility matrix, incremental updating, core attributes.

1 Introduction

Rough set theory[1], introduced by Z.Pawlak in 1982, has been used in a lot of domains, such as machine learning, pattern recognition, decision support system and expert systems and so on. Computing core is one of the most important issues of Rough Set theory.

HU[2] proposed the algorithm for computing core based on Skowron's discernibility matrix[3]. Its time and space complexity are $O(|C||U|^2)$ and $O(|C||U|^2)$ respectively. Ye[4] pointed out core attributes calculated by Hu's method is different to that found by algorithm based on positive region and present a new algorithm based on new discernibility matrix,which time and space complexity are also $O(|C||U|^2)$ and $O(|C||U|^2)$ respectively. Unfortunately, Ye did not found the inconsistency of the decision table led to false results[5][6]. In the paper[7]-[9], some improved approaches based on discernibility matrix are presented, which need to construct discernibility matrix firstly, and then compute core attributes. However,The time and space complexity of these algorithms are all not ideal.

Above the algorithm for computing core attributes are almost based on static decision table. However, real databases are always dynamic. Incremental arithmetic for computing core has been developed, but there are few incremental algorithms about incremental updating core. Yang[10] has proposed an incremental

---

* Corresponding author.

J. Yu et al. (Eds.): RSKT 2010, LNAI 6401, pp. 249–256, 2010.

updating algorithm of the computation of core based on improved discernibility matrix, which avoid repeating discernibility matrix. When a new object is dynamically added, the algorithm needs to update discernibility matrix, and then compute the core. Unfortunately, the time and space complexity of the algorithm are $O(|C|(5|U_1| + 3|U_2'|))$ and $O(|C||U_1|(|U_1| + |U_2'|))$ respectively (Yang don't consider the cardinality value of condition attributes C). The space cost of Yang's algorithm is too high, so it don't adapt to the large database.

Therefore, in this paper, we put forward the notion of simplified discerniblity matrix and the method for computing core attributes based on the simplified discerniblity matrix, which can well deal with inconsistent decision table. On the basis of that, we present a quick algorithm for incremental updating core attributes, which only need to analyze the updating parts of discernibility matrix and doesn't need to store and re-calculate discerniblily matrix, when a new object is added. Its time and space complexity are $O(2|C||U'|)$ and $O(|C|)$ respectively.

2   Core Equivalence and Algorithm for Computing Core

A decision table is defined as $S = (U, A, V, f)$, where U is a non-empty finite set of objects, called universe; A is a non-empty finite set of attributes, $A = C \cup D$ and $C \cap D = \phi$; C is the set of condition attributes and D is the set of decision attributes; $V = \bigcup_{a \in A} V_a$, and V_a is the domain of the attribute a; $f : U \times R \to V$ is an information function such that $f(x_i, a) \in V_a$ for every $a \in A, x_i \in U$. A decision table $S = (U, C \cup D, V, f)$ is inconsistent, if there have $f(x_i, C) = f(x_j, C)$ and $f(x_i, D) \neq f(x_j, D)$, where $x_i, x_j \in U, i \neq j$. And x_i and x_j are called inconsistent objects (Namely x_i conflicts with x_j). Otherwise, S is consistent.

Property 1. Given a decision table $S = (U, C \cup D, V, f), \forall P \subseteq C$, we have $POS_P(D) = \{[x_i]_P | x_j \in [x_i]_P (i \neq j),$ have $f(x_i, P) = f(x_j, P) \land f(x_i, D) = f(x_j, D)\}$.

Theorem 1. Given a decision table $S = (U, C \cup D, V, f), a \in C, a \in Core(C)$ if and only if $POS_{C-a}(D) \neq POS_C(D)$ [6].

2.1   Algorithm for Computing Core Based on the Simplified Decision Table

Definition 1. Given a decision table $S = (U, C \cup D, V, f)$, the compatible decision table is defined as $S_1' = (U, C \cup D, V', f')$ with:

$$f'(x_i, C) = f(x_i, C)$$

$$f'(x_i, D) = \begin{cases} V_\xi & \exists x_j, f(x_i, C) = f(x_j, C) \land f(x_i, D) \neq f(x_j, D) \\ f(x_i, D) & otherwise \end{cases}$$

Where, V_ξ is a decision value which is different to the other object's decision values. Even if new objects are added, there is not exist with the same as V_ξ.

Definition 2. For decision table $S_1' = (U, C \cup D, V', f')$, denote $U/C = \{[x_1]_C, [x_2]_C, \ldots, [x_l]_C, [x_{l+1}]_C, \cdots, [x_t]_C\}.x_i, x_j \in [x_r]_C (i \neq j, 1 \leq r \leq l)$, there has $f'(x_i, D) = f'(x_j, D)$. $x_i, x_j \in [x_s]_C (i \neq j, l < s \leq t)$, there is $f'(x_i, D) \neq f'(x_j, D)$. where $U_1' = \{x_1, x_2, \ldots, x_l,\}, U_2' = \{x_{l+1}, \ldots, x_t\}, U' = U_1' + U_2' = \{x_1, x_2, \ldots, x_t\}$, then $S_1' = (U', C \cup D, V', f')$ is called simplified decision table.

Definition 3. For decision table $S_1' = (U', C \cup D, V', f')$, then we define simplified discernibility matrix as $SM = \{m_{ij}\}$, and m_{ij} satisfies:

$$m_{ij} = \begin{cases} a & a \in C, f'(x_i, a) = f'(x_j, a) \wedge f'(x_i, D) \neq f'(x_j, D), j > i \\ \phi & otherwise \end{cases}$$

Definition 4. For decision table $S_1' = (U', C \cup D, V', f')$ and simplified discernibility matrix SM, $MCore(C)$ is called the core of S_1', where $MCore(C) = \{m_{ij} || m_{ij}| = 1, m_{ij} \in SM\}$.

Theorem 2. For decision table S, we have $MCore(C) = Core(C)$[11].

2.2 Algorithm for Computing Core Based on the Simplified Discernibility Matrix

The algorithm for computing core of literature[10] needs to create the discernibility matrix, which limits its suitability with large datasets. Thus we propose a computation approach based on the simplified decision table S_1' to find the core attribute in the section. The method does not need to store the discernibility matrix, and only carry on calculating the element of the discernibility matrix. For every core attribute, we add a counter count to record the number of the core attributes in accordance with the literature [10] method. Thus the core attribute set can be denoted $Core(C) = \{(a, count)\}$.

Algorithm 1. Computing Core Based on the Simplified Decision Table

Input: The decision table $S_1' = (U', C \cup D, V', f')$;
Ouput: $Core(C)$.
Step1:$Core(C) = \phi$;
Step2:for $(i = 1; i <= |U_1'|; i++)$
        for $(j = i + 1; j <= |U'|; j++)$
    Step2.1: if $(f'(x_i, D) \neq f'(x_j, D))$
        Step2.1.1: $m = \phi, len = 0$;
        Step2.1.2: For $a_k \in C$, if $(f'(x_i, a_k) \neq f'(x_j, a_k))$ $\{m = m \cup \{a_k\}, len++;\}$
    Step2.2: if $(len == 1)Core(C) = Core(C) + (m, 1)$;
Step3:Output $Core(C)$.

In the Algorithm 1, the time complexity of the algorithm is $O(|C||U'|^2)$. Because it do not store discernibility matrix, its space complexity is only $O(|C|)$.

3   Incremental Updating Algorithm of the Computation Core

3.1   Method of Incremental Updating Core

In the process of incremental computing core, analyzing the variation regularities of discernibility matrix is the key, when a new object is added. For the convenience of description, We can divide the decision table S_1' into $|U_1'/D| + 1$ parts according to decision attribute values, that is $\{Y_1, Y_2, \cdots, Y_{|U_1'/D|}, Y_\xi\}$. There exists four cases, when a new object x is added to the S_1'. $case(1)$: x is the same as one of the objects of U_1', or x conflict with one of the objects of U_2'; $case(2)$: x does not conflict with any objects, and x belongs to the class existed, let's $Y_t (1 \le t \le |U_1'/D|)$; $case(3)$: x does not conflict with any objects, and x is a new class; $case(4)$: x conflicts with the object y which belongs to the class $Y_t (1 \le t \le |U_1'/D|)$.

Definition 4. For decision table S_1', the new decision table is $S_2' = (U' \cup x, C \cup D, V', f')$, when a new object x is added to the S_1'. The new discernibility matrix of S_2' is M_{new}, which satisfies $M_{new} = SM + M_{add} - M_{del}$. The Properties of M_{add} and M_{del} will be described as follows, which together lead to Theorem 3.

Property 2. For decision table S_1', the discernibility matrix SM of S_1' and a new object x, if x belongs to $case(1)$, we have $M_{new} = SM$.

Proof. According to definition 1 and 2, it is obvious to $M_{new} = SM$, if x is the same as one of the objects of the S_1'. If x conflicts with the object y of U_2', x will merged with y. And only one of them holds in the U_2'. So SM holds the line, that is $M_{new} = SM$.

Property 3. For decision table S_1' , the discernibility matrix SM of S_1' and a new object x, if x belongs to $case(2)$, we have $M_{new} = SM + M_{add}$, where $M_{add} = m$ and $m = \{a \in C, f'(x_i, a) \ne f'(x, a) \wedge x_i \in (U' - Y_t)\}$.

Proof. Since x doesn't collide with anyone of U', and x belongs to Y_t. According to the symmetry of the discernibility matrix and definition 3, when x is added into the decision table, there will add new part which is $M_{add} = \{m\}$, where $m = \{a \in C, f'(x_i, a) \ne f'(x, a) \wedge x_i \in (U' - Y_t)\}$. Thus $M_{new} = SM + M_{add}$.

Property 4. For decision table S_1', the discernibility matrix SM of S_1' and a new object x, if x belongs to $case(3)$, we have $M_{new} = SM + M_{add}$, where $M_{add} = \{m\}$ and $m = \{a \in C, f'(x_i, a) \ne f'(x, a) \wedge x_i \in U'\}$.

Proof. Since x doesn't collide with anyone of U', and x is a new decision class. According to the symmetry of the discernibility matrix and definition 3, when x is added into the decision table, there will add new part which is $M_{add} = \{m\}$, where $m = \{a \in C, f'(x_i, a) \ne f'(x, a) \wedge x_i \in U'\}$. Thus $M_{new} = SM + M_{add}$.

Property 5. For decision table S'_1 , the discernibility matrix SM of S'_1 and a new object x, if x belongs to $case(4)$, there have $M_{new} = SM + M_{add} - M_{del}$, where $M_{add} = \{m\}$, $m = \{a \in C, f'(x_i, a) \neq f'(x, a) \wedge x_i \in Y_t\}$; $M_{del} = \{m'\}$, $m' = \{a \in C, f'(x_i, a) \neq f'(x, a) \wedge x_i \in V_\xi\}$.

Proof. According to definition 1 and 2, $y \in U'_1$, y conflicts with x, so y will be transferred to Y_ξ from Y_t. Therefore, when y is transferred form Y_t, there will add new part which is $M_{add} = \{m\}$,where $m = \{a \in C, f(x_i, a) \neq f(x, a) \wedge x_i \in Y_t\}$. In addition to, when y is moved into Y_ξ, there will make deleted part, which is $M_{del} = \{m'\}$, where $m' = \{a \in C, f'(x_i, a) \neq f'(x, a) \wedge x_i \in Y_\xi\}$.Thus the new discernibility matrix is $M_{new} = SM + M_{add} - M_{del}$.

Theorem 3. Given core attributes $Core(C)$ of S'_1 and a new object x suppose $Core'(C)$ is the core attributes of $S'_2 = (U' \cup \{x\}, C \cup D, V', f')$, then there have $Core'(C) = Core(C) + Core(C)_{add} - Core(C)_{del}$, where $Core(C)_{add} = \{(m, count) | |m| = 1, m \in M_{add}\}$, $Core(C)_{del} = \{(m, count) | |m| = 1, m \in M_{del}\}$.

By Theorem 3, on the basis of the $Core(C)$, the incremental core computation only needs to analyze the updating parts of discernibility matrix.

3.2   Incremental Core Computing Algorithm

Based above, the incremental algorithm for computing core can be summarized as follows:

Algorithm 2. Incremental Updating Algorithm of the Computation Core.

Input: $S1' = (U', C \cup D, V', f')$, $Core(C)$ of S'_1 and new object x.
Ouput: New $Core(C)$.
Step1: Travel the S'_1 and analyze x;
Step2: if $(x \in case(1))$ goto Step6;
Step3: if $(x \in case(2))$ // Suppose x belong to Y_t;
    Step3.1: for $i = 1$ to $|U' - Y_t|$ Do
        Step3.1.1: $m = \phi, len = 0$;
        Step3.1.2: For $a_k \in C$, if $(f'(x_i, a_k) \neq f'(x, a_k))$ $\{m = m \cup \{a_k\}, len++;\}$
        Step3.1.3: if $(len == 1) Core(C) = Core(C) + (\{m\}, 1)$;
    Step3.2: $Y_t = Y_t + x$;
Step4: if $(x \in case(3))$
    Step4.1: for $i = 1$ to $|U'|$ Do
        Step4.1.1: $m = \phi, len = 0$;
        Step4.1.2: For $a_k \in C$, if $(f'(x_i, a_k) \neq f'(x, a_k))$ $\{m = m \cup \{a_k\}, len++;\}$
        Step4.1.3: if $(len == 1) Core(C) = Core(C) + (\{m\}, 1)$;
    Step4.2: $U'_1 = U'_1 + x$;
Step5: if $(x \in case(4))$ // $y \in U'_1$, y is inconsistent to x, which belongs to Y_t;
    Step5.1: for $i = 1$ to $|Y_t|$ Do
        Step5.1.1: $m = \phi, len = 0$;
        Step5.1.2: For $a_k \in C$, if $(f'(x_i, a_k) \neq f'(x, a_k))$ $\{m = m \cup \{a_k\}, len++;\}$
        Step5.1.3: if $(len == 1) Core(C) = Core(C) + (\{m\}, 1)$;

Step5.2: for $i = 1$ to $|Y_\xi|$ Do
  Step5.2.1: $m = \phi, len = 0$;
  Step5.2.2: For $a_k \in C$, if $(f'(x_i, a_k) \neq f'(x, a_k))$ $\{m = m \cup \{a_k\}, len + +;\}$
  Step5.2.3: if $(len == 1)Core(C) = Core(C) - (\{m\}, 1)$;
Step5.3: $Y_t = Y_t - y; U_2' = U_2' + y$;
Step6: Output $Core(C)$.

3.3   Algorithm Complexity Analysis

1) Time complexity analysis

The time complexity of Step 1 is $O(|C||U'|)$.

For $case(1)$, the time complexity of Step2 is $O(1)$,so the time complexity of this case is $O(|C||U'|) + O(1) = O(|C||U'|)$.

For $case(2)$, the time complexity of Step3 is $O(|C||U' - Y_t|)$,so the time complexity of this case is $O(|C||U'|) + O(|C||U' - Y_t|) < 2O(|C||U'|)$.

For $case(3)$, the time complexity of Step4 is $O(|C||U'|)$,so the time complexity of this case is $O(|C||U'|) + O(|C||U'|) = 2O(|C||U'|)$

For $case(4)$, the time complexity of Step5 is $O(|C||Y_t|) + O(|C||Y_\xi|)$so the time complexity of this case is $O(|C||U'|) + O(|C||Y_t|) + O(|C||Y_\xi|) < 2O(|C||U'|)$.

Thus, the worst time complexity of the algorithm 2 is $2O(|C||U'|)$, which is less than literature [10] $O(|C|(5|U_1| + 3|U_2'|))$.

2) Space complexity analysis

The algorithm doesn't store the discernibility matrix, and the space consumption is m and $Core(C)$. Therefore, the space complexity is $O(|C|)$, which is far less than literature [10] $O(|C||U_1|(|U_1| + |U_2'|))$.

Clearly, both of the time complexity and space complexity of algorithm are superior to the algorithm 2 of literature [10].

4   Example and Experiment

4.1   Example Verification

Given a decision tables $S, \{a, b, c, d\}$ are condition attributes C, D is decision attribute and $\{x_1, x_2, \cdots, x_7\}$ are the objects of the decision table.

It is obvious that S is an inconsistent decision table. By definition 1, we can get decision table S_1',which has $Y_1 = \{x_2, x_4\}, Y_2 = \{x_3, x_5\}, Y_\xi = \{x_2\}, U_1' = \{x_1, x_4, x_3, x_5\}, U_2' = \{x_2\}$.

According to algorithm 1, we can attain $Core(C) = (b, 1)$.

Now we illustrate $case(2), case(3)$ and $case(4)$ of Algorithm 2 by adding different objects to S_1':

1) The new object x is $(1,1,1,1,2)$.

Analysis shows that x belongs to case (2), then $M_{del} = \phi$ and $M_{add} \neq \phi$. We have $M_{add} = \begin{matrix} x_1 & x_4 & x_2 \\ x \begin{bmatrix} d & c & bc \end{bmatrix} \end{matrix}$, then $Core(C)_{add} = \{(c, 1), (d, 1)\}$. Thus, $Core'(C) = Core(C) + Core(C)_{add} = \{(b, 1), (c, 1), (d, 1)\}$.

Table 1. Decision table S

U	a	b	c	d	D
x_1	1	1	1	0	1
x_2	1	0	1	0	1
x_3	0	1	0	0	2
x_4	1	1	0	1	1
x_5	0	1	1	1	2
x_6	1	0	1	0	2
x_7	0	1	0	0	2

Table 2. Decision table S_1'

U_i'	Y_i	U	a	b	c	d	D
	Y_1	x_1	1	1	1	0	1
U_1'		x_4	1	1	0	1	1
	Y_2	x_3	0	1	0	0	2
		x_5	0	1	1	1	2
U_2'	Y_ξ	x_2	1	0	1	0	V_ξ

2) The new object y is (1,1,1,1,4).

Analysis shows that y belongs to $case(3)$, then $M_{del} = \phi$ and $M_{add} \neq \phi$. We have

$$M_{add} = \frac{x_1\ x_4\ x_3\ x_5\ x_2}{y\ [\ d\quad c\quad acd\quad a\quad bd\ ]}, \text{ then } Core(C)_{add} = \{(a,1),(c,1),(d,1)\}. \text{Thus,}$$

$Core'(C) = Core(C) + Core(C)_{add} = \{(a,1),(b,1),(c,1),(d,1)\}$.

3) The new object z is (1,1,1,0,2).

Analysis shows that z belongs to $case(4)$, and z conflicts with the object x_1.

Then $M_{del} \neq \phi$ and $M_{add} \neq \phi$. We can respectively get: $M_{add} = \dfrac{x_3\ x_5}{z\ [\ ac\ ad\ ]}$

and $M_{del} = \dfrac{x_4}{z\ [\ b\ ]}$, then $Core(C)_{add} = \phi$ and $Core(C)_{del} = \{(b,1)\}$. Thus,

$Core'(C) = Core(C) + Core(C)_{add} - Core(C)_{del} = \phi$.

4.2 Experimental Results

In order to test the validity of Algorithm 2, we select five datasets of UCI database. For every example, there choose 90% objects as the basic dataset, and the rest of objects use as incremental objects. The experiments make on Pentium IV 2.8GHZ, 512MB and Windows XP. The results of experiments are shown in Table 3. Alg_A is algorithm 2 of documentation [10], Alg_B represents the algorithm 1 of the paper and Alg_C represents the algorithm 2 of the paper.

From Table 3, we can find that the running time of our incremental algorithm is much less than non-incremental Alg_B and is also less than Alg_A.

Table 3. The comparison of three algorithms for computing core(Time:ms)

Datasets	$|U|$	$|C|$	$|U'|$	Alg_A	Alg_B	Alg_C
Monks data	432	6	432	1.54	3.03	0.52
Monkey data	556	17	432	7.95	4.57	3.87
Car data	1728	6	972	4.98	10.25	4.31
Mushroom data	8124	23	8124	512.89	1263.61	391.68
Nursery data	12960	8	12960	101.07	438.22	40.18

5   Conclusion

In the paper, firstly, a new algorithm computing core is present based on simplified discernibility matrix, which don't store the discernibility matrix. The core based on simplified discernibility matrix is equivalent to the core based on positive region. And then, we propose a quick incremental core computing algorithm, which only needs to analyze the updating parts of the discernibility matrix, and don't store and ally on the original discernibility matrix, when a new object is added. Its time and space complexity are $O(2|C||U'|)$ and $O(|C|)$ respectively. Finally, both of theoretical analysis and experimental results show that the algorithm is effiective and efficient.

Acknowledgments. This research work is supported by the Research Foundation for Outstanding Younger Talents in Higher Education Institutions of Anhui Province (No.2010SQRL138),the Natural Science Foundation of Education of Anhui Province of China (No.KJ2010B137, KJ2009B237Z) and the Natural Science Foundation of Chuzhou University (No.2008KJ013B).

References

1. Pawlak, Z.: Rough set. International Journal of Information and Computer Science 11(5), 341–356 (1982)
2. Hu, X.H., Cercone, N.: Learning in Relational Databases: A Rough Set Approach. International Journal of Computational Intelligence 11(2), 323–338 (1995)
3. Skowron, A., Rauszer, C.: The Discernibility Matrices and Functions in Information Systems. In: Slowinski (ed.) Intelligent Decision Support-Handbook of Applications and Sdvances of the Rough Sets Theory, pp. 331–362. Kluwer Academic Publisher, Dordrecht (1991)
4. Ye, D.Y., Chen, Z.J.: A New Discernibility Matrix and the Computation of a Core. Acta Electronica Sinica 30(7), 1086–1088 (2002)
5. Wang, G.Y.: Rough Reduction in Algebra View and Information View. International Journal of Intelligent System 18(6), 679–688 (2003)
6. Wang, G.Y., Zhao, J., An, J.J., Wu, Y.: A comparative study of algebra viewpoint and information viewpoint in attribute reduction. Fundamenta Informaticae 68(6), 289–301 (2005)
7. Zhang, W.X., Wu, W.Z., Liang, J.Y.: Theory and Method of Rough Set, p. 7. Science Press, Beijing (2001)
8. Liu, W.J., Gu, Y.D., Feng, Y.B.: An Improved Attribute Reduction Algorithm of Decision Table. PR&AI 17(1), 119–123 (2004)
9. Yang, M., Sun, Z.H.: Improvement of Discernibility Matrix and the Computation of a Core. Journal of Fudan University(Natural Science) 43(5), 865–868 (2004)
10. Yang, M.: An incremental updating Algorithm of the computation of a core based on the improved discernibility matrix. Chinese Journal of Computers 29(3), 407–413 (2006)
11. Yang, C.J., Ge, H., Yao, G.S., Ma, L.S.: Core and Attribute Reduction Algorithms Based on Compatible Discernibility Matrix. In: The International Conference on Computational Intelligence and Natural Computing, CINC 2009, Wuhan, China, vol. 2, pp. 103–106 (June 2009)

Using Lexical Ontology for Semi-automatic Logical Data Warehouse Design

Mior Nasir Mior Nazri[1], Shahrul Azman Noah[2], and Zarinah Hamid[1]

[1] International Islamic University Malaysia, Kuala Lumpur, Malaysia
[2] Universiti Kebangsaan Malaysia, Bangi, Malaysia
miornasir@iiu.edu.my, samn@ftsm.ukm.my, inahumkc@iiu.edu.my

Abstract. Data Warehouse has become a very important tool for data analysis but the conversion from Operational Database to Data Warehouse is very complex. This paper presents a new method to design a data warehouse multidimensional model based on supply driven framework. We propose a semi-automatic tool using lexical ontology as the knowledge domain. The tool is able to identify fact and dimensional tables using the knowledge domain. Once fact table is identified, the following step is to generate Data Warehouse multidimensional model with minimal user interaction. The feasibility of the proposed method is demonstrated by a prototype using WordNet as knowledge domain.

Keywords: Data Warehouse, Multidimensional Model, Ontology, WordNet.

1 Introduction

Many different methods and approaches have been discussed to design DW in the literatures [1,2]. However, these approaches are being carried out manually by designers with the involvement of respective users. Several attempts have been made by researchers to automate the DW design process [3,4]. Most of existing DW design tools employ direct transformation of input into corresponding designs and rely on the users to identify suitable entities to be modeled as fact tables. Moody and Kortink [5] suggest classification of entities in order to select appropriate fact tables. Currently existing tools are unable to detect such entities and leave them to users for decision. However, such classifications can be derived and manipulated from existing semantic knowledge sources or ontology such as the WordNet [6] and ConceptNet [7].

This paper presents a methodology to develop a logical dimensional DW model by using lexical ontology as the knowledge domain. WordNet database contains more than 100,000 words representing nouns, verbs, adjectives and adverbs [6]. Given entity name as input, WordNet is able to produce semantic definition, syntactic category and logical groupings of the input word. With the semantic knowledge stored in WordNet, the system is able to classify entity into different categories for the process of selecting facts table, dimensions and hierarchies. A

J. Yu et al. (Eds.): RSKT 2010, LNAI 6401, pp. 257–264, 2010.

prototype is developed and tested in order to demonstrate the feasibility of the project.

Section 2 discusses the related work from wide varieties of areas starting with development approaches, DW design methods and knowledge domain. Section 3 describes the foundation for fact table selection. Section 4 presents and demonstrates the automated approach to develop dimensional model for DW. Section 5 concludes and discusses about future research.

2   Related Work

According to List et al. [8] DW development methods can be divided into three different approaches such as supply-driven, goal-driven and demand-driven. Supply-driven approaches begin with the study of the existing operational database which in turn reengineered to produce a multidimensional model. Several researchers manage to semi-automate the DW design process within the supply driven framework [3,9]. On the other hand, goal-driven approaches integrate corporate strategy and business objective as the requirement for DW design. Giorgini et al. [10] adopt organizational and decisional modelling while Guo et al. [11] use Key Performance Indicators as the guideline toward the DW design. Lastly, demand-driven approaches compel users to play very critical role in shaping and contributing to DW design [12]. Our research deploys designing multidimensional model using an enterprise logical model as input.

Many literatures have indicated that identifying fact is the most important step in DW design process, and most of the time pointing fact is usually done manually [4,5]. Moody and Kortink [5] claim that the most suitable entity for fact table is the transaction type entity. Few researchers use numeric attributes as the criteria to determine fact table [3,9]. Phipps and Davis [3] evaluate candidates schema utilizing demand-driven process through user queries. Romero and Abelló [4] recommend that entity with many-to-one relationship is most likely to be a suitable fact table hence this entity is the middle of the operation related to other supporting entity. We incorporate those suggestions for selecting fact table with details written in section 3.

Noah and Williams [13] indicate that design automation is more intelligent with the ability to tap business knowledge about semantics of the application domain. Romero and Abelló [4] propose a semi-automatic method to generate DW multidimensional concept using domain ontology. The ontology is used to discover business multidimensional concept buried in the business domain. Sugumaran and Storey [14] use domain ontology to represent the domain knowledge in order to assist database designer.

It is obvious that ontologies play an important role in automating the process of data related design such as data-base and data warehouse as exhibited from the aforementioned literatures. Lassila and McGuinness[15] provides a broad categorization ontology, of which can be categorized as lightweight and heavyweight ontology. Lexical ontology such as the WordNet belongs to the category of lightweight ontology. While many research have been focusing on exploiting domain specific ontologies such as business and finance ontology, none has make an effort to tap the potential of lexical ontologies. This paper therefore, aims at exploring the potential of lexical ontologies (i.e. WordNet) for semi-automated data warehouse design. Subsequently, we use WordNet to extract semantic knowledge from entity name and attribute name in order to identify the fact table.

3   Method Foundation

The purpose of this research is to semi-automate the design of multidimensional model for DW by exploiting lexical ontologies. The basic elements for multidimensional model are fact, dimension and measure[5]. Fact represents the vital data for the organization business transaction that can be used for analysis such as *Sales*. Measure is the numeric attribute for fact which can be aggregated. Dimension provides business perspective on how to analyze fact such as *Sales* based on *Geography*[2]. Dimension can be broken down into hierarchies such as *Country, State* and *City*. We propose to automate the process of selecting fact tables based on the following conditions:

[C1] **Entity type:** Entities can be divided into three different categories such as transaction, component and classification [5]. Transaction entity captured the event happened in business such as *orders* and *sales*. This entity is the most suitable candidate for fact table because it relates to activity or business process. On the other hand, component entity describes the situation related to the transaction such as person involve, product being exchange, time and location. Component entity is best candidate for dimension table.

[C2] **Numeric attribute:** Entity with numeric attributes can represent measures related to business event; hence entity with highest count of numeric attributes is most likely to be a potential fact table [3]. Based on our observation, some numeric attributes are not suitable for measure because they act as descriptor for example *telephone number*. To solve this problem, we utilize knowledge stored in WordNet to identify numeric attributes type; hence only transaction attributes are counted.

[C3] **Relationship:** Fact table usually has many-to-one relationship with the surrounding dimensional tables [4]. High cardinality of a relationships corresponding to other entities shown the importance of the entity in the schema [3]. With this premise, we set up a merit algorithm to calculate entity with many-to-one relationship as potentially good candidate for fact table.

In our method, a merit algorithm calculates score for entity evaluated by each of the criteria. Only entity that meets at least two criteria will be selected as potential fact table. The controlling mechanism is to limit the maximum score for each criterion not to exceed the minimum score required for the entity to be declared as potential fact table.

4   Proposed Method

This section presents the details of our proposed method. Our method uses the supply-driven approach discussed earlier. As depicted in Fig. 1, there are three major stages to transform logical model into a multidimensional model.

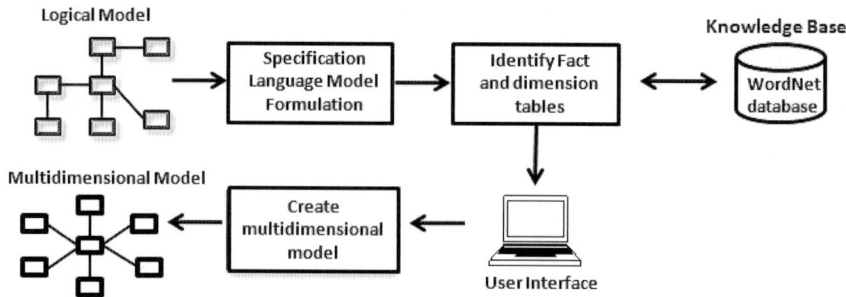

Fig. 1. Method Overview

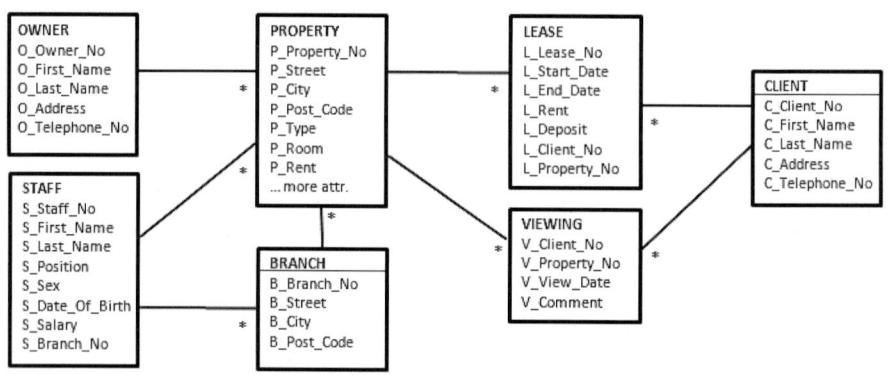

Fig. 2. Dream House Logical Model

The process starts with the conversion of an enterprise logical model into a specification language model as input. The next stage involves identifying candidates for fact and dimensional tables with the help from the WordNet. The final stage is to generate the logical multidimensional model. To demonstrate the feasibility of the proposed method, a prototype has been developed. We discuss our proposed method in the following section using the *dream house* example provided in [16] (see Fig. 2).

4.1   Specification Language Formulation

In stage 1, the input model is translated manually from UML notation into a specification language model where the core information such as entity, attribute, attribute type, identifier, and relationship are captured as text input file. The specification language is very similar to [9].

The prototype reads the input file and processes each record. At this stage diagnostic process is performed to capture mistakes such as duplicate entity. Captured errors are reported to the user for correction. Once the input file has been validated, it is stored in the systems memory in the form of entities and attributes.

4.2   Identify Fact and Dimension tables

In stage 2, the proposed method selects suitable entities as candidates for fact and dimension table based on the condition of [C1], [C2] and [C3] as previously described. This process is divided into four steps.

Step 1: Classify entity type according to [C1] - Database analyst uses natural language to define entity name which is usually derived from English word noun or verb. Our task is to identify the semantic of the entity name by extracting the word definition from knowledge stored in the WordNet such as synonyms and lexicographer file numbers (Lex No). For example entity *LEASE* has several meaning represented by seven sense index files (see Fig. 3). The Lex numbers are matched against transaction code table and component code table (see Fig. 3). The point for each types of description has been generated by analyzing the probability of occurrence of Lex. No in WordNet. Due to limited space, it is beyond the scope of this paper to discuss it.

The score for measuring transaction entity is using the following equation where x is the total number of matched Lex no.

$$Score(E_{C1}) = \frac{1}{x} \sum_{i=1}^{n} Point(L_i); L_i \text{ is the point for all Lex. No of entity } E$$

As a result, the entity *LEASE* is classified as transaction type with total score of 2.7. Since *LEASE* is already defined as transaction type, the score for component code is ignored. If an entity was defined as component type, the next two steps will be skipped.

Step 2: Process numeric attributes according to [C2] - Similarly with entity name, attribute name usually defined from noun or verb. In this step, the method extracts the word definition from WordNet for numeric attribute names only. The score is calculated using the following equation, where α some non-negative value, n_1 total entities with numeric attributes, n_2 is total number of numeric attributes and n_3 is total transaction type numeric attributes of entity E.

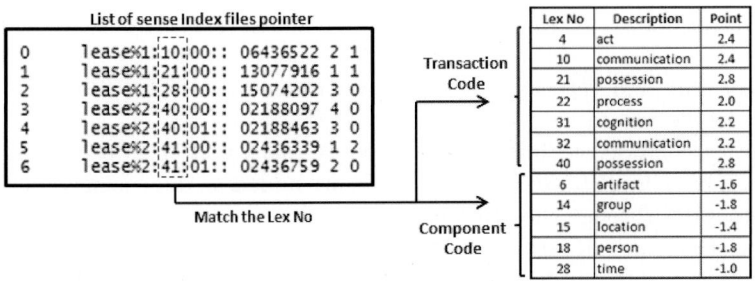

Fig. 3. Classification of entity *LEASE*

$$Score(E_{C2}) = \alpha + \frac{n_1}{n_2}n_3; \; Score(E_{C2}) = 0, \; \text{if } n_2 = 0 \text{ or } n_3 = 0;$$

Step 3: Process many-to-one relationship [C3] - In this step, we counts the number of many-to-one relationship that the entity participates. Score is given based on the following equation, where β is some non-negative value, r_1 is number of entities with M:1 relations, r_2 is number of M:1 relations and r_3 is total M:1 relations of entity E.

$$Score(E_{C3}) = \beta + \frac{r_1}{r_2}r_3; \; Score(E_{C3}) = 0, \; \text{if } r_2 = 0 \text{ or } r_3 = 0;$$

Step 4: Identify potential fact and dimension table - The final step is to sum total score from the previous three processes. Entity that has total score which exceeds the minimum requirement is defined as transaction type. Entity with negative score is defined as component type. Other entity that does not fall under the two categories is defined as unknown type. The score is calculated using the following equation.

$$Score(E) = \begin{cases} Score(E_{C1}) + Score(E_{C2}) + Score(E_{C3}), & \text{if } E = \text{transaction} \\ Score(E_{C1}), & \text{if } E = \text{component} \end{cases}$$

For example the $Score(Lease) = 6.9$: $Score(Lease_{C1}) = 2.7$; $Score(Lease_{C2}) = 2.2$ and $Score(Lease_{C3}) = 2.0$. Using the aformentioned approach to the remaining entities i.e. *STAFF, BRANCH, PROPERTY, OWNER, CLIENT* and *VIEWING* give the scores of -1.8, -1.8, 6.9, -1.8, -1.7 and 4.4 respectively. Therefore the entities *PROPERTY, LEASE* and *VIEWING* are selected as the potential fact table.

4.3   Generate Multidimensional Schema

Stage 3 is the creation of multidimensional model similar to [3]. Fact table is created from entity confirmed by the user as fact table in stage 2. Our method first creates fact table name *F_PROPERTY*. The numeric attributes *room* and *rent* are added as measures. Since *PROPERTY* does not have any temporal attributes, system generated temporal attributes are added as fact attributes. Next the primary key *property_no* and foreign keys *owner_no, staff_no* and *branch_no* are added to the fact table. Other non-numeric attributes *street_no, city, post_code* and *type* are stored in dimension table call *PROPERTY* link directly to the fact table.

All other component dimension with one-to-many relationship with fact table such as *OWNER, BRANCH* and *STAFF* are added as dimension tables. Next step is to walks through the component entity for other entities that have one-to-many relationship and add those entities as hierarchies dimension until there is no more entity with one-to-many relationship. Next process is to add entity with many-to-one relationship with the fact table. In this case we include *VIEWING* and *LEASE* to the fact table. The final result will be the creation of star or snow

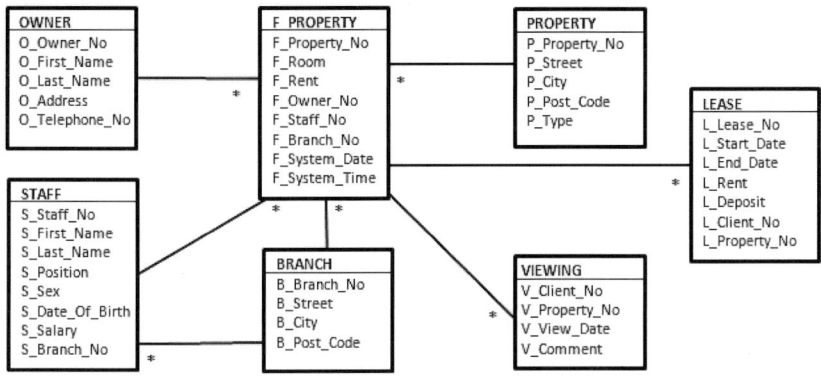

Fig. 4. Property Fact Table

flake dimensional schema. The same process will be repeated for other fact table candidates. The final multidimensional model for *F_PROPERTY* fact table is displayed in Fig. 4.

5   Discussion and Conclusion

We have described a semi-automated process to generate a DW multidimensional model. In our approach, we use lexical ontology represented by WordNet database to assist us in identifying potential fact and dimensional tables with minimum user interaction. Once the fact tables are identified, the prototype is able to generate a star like multidimensional model from the selected potential fact table. The paper has demonstrated the viability of exploiting the simplest form of ontology in assisting the process of automated data warehouse design by making intelligent decision for selecting potential facts and dimensional tables. Ontology is not claimed to be a panacea, however exploiting lightweight ontologies such as WordNet is seen able to suggest the correct entities for potential fact tables which will remain unidentified for novice human designer. The work and method presented in this paper, however, require further investigation. Additional testing and evaluation are required involving multiples case studies in order to ascertain the real findings presented in this paper.

References

1. Husemann, B., Lechtenbörger, J., Vossen, G.: Conceptual Data Warehouses Design. In: DMDW 2000: Proceedings of the third International Workshop on Design and Management of Data Warehouses (2000)
2. Ponniah, P.: Data Warehousing Fundamentals. John Wiley & Sons, Inc., New York (2001)
3. Phipps, C., Davis, K.C.: Automating Data Warehouse Conceptual Schema Design and Evaluation. In: DMDW 2002: Proceedings of the International Workshop on Design and Management of Data Warehouses, Canada (2002)

4. Romero, O., Abelló, A.: Automating Multidimensional Design from Ontologies. In: DOLAP 2007: Proceedings of the ACM tenth International Workshop on Data Warehousing and OLAP, pp. 1–8. ACM Press, New York (2007)
5. Moody, D.L., Kortink, M.A.R.: From Enterprise Models to Dim. Models: A Methodology for DW and Data Mart Design. In: DMDW 2000: Proceedings of the International Workshop on Design and Management of Data Warehouses, p. 5.1–5.12 (2000)
6. Miller, G.A., Beckwith, R., Fellbaum, C., Gross, D., Miller, K.: WordNet: An online lexical database. International Journal of Lexicography 3 (1990)
7. Singh, P., Liu, H.: ConceptNet - a practical commonsense reasoning tool-kit. BT Technology Journal 22, 211–226 (2004)
8. List, B., Bruckner, R.M., Machaczek, K., Schiefer, J.: A Comparison of Data Warehouse Development Methodologies Case Study of the Process Warehouse. In: Hameurlain, A., Cicchetti, R., Traunmüller, R. (eds.) DEXA 2002. LNCS, vol. 2453, p. 203. Springer, Heidelberg (2002)
9. Sitompul, O.S., Noah, S.A.: A Transformation-oriented Methodology to Knowledge-based Conceptual Data Warehouse Design. Journal of Computer Science 2, 460–465 (2006)
10. Giorgini, P., Rizzi, S., Garzetti, M.: Goal-Oriented Requirement Analysis for Data Warehouse Design. In: DOLAP 2005: Proceedings of the 8th ACM International Workshop on Data Warehousing and OLAP. ACM Press, Bremen (2005)
11. Guo, Y., Tang, S., Tong, Y., Yang, D.: Triple-driven Data Modeling Methodology in Data Warehousing: a case study. In: DOLAP 2006: Proceedings of the 9th ACM International Workshop on Data Warehousing and OLAP, pp. 59–66. ACM Press, Arlington (2006)
12. Winter, R., Strauch, B.: A Method for Demand-driven Information Requirements Analysis in Data Warehousing Projects. In: HICSS 2003: Proceedings of the 36th Hawaii International Conference on System Sciences (2003)
13. Noah, S.A., Williams, M.D.: Intelligent Database Design Diagnosis: Performance Assessment with the Provision of Domain Knowledge. Artificial Intelligence Review 21, 57–84 (2004)
14. Sugumaran, V., Storey, V.C.: The Role of Domain Ontologies in DB Design: An Ontology Mgmt & Conceptual Modeling Environment. ACM Transactions on Database Systems 31, 1064–1094 (2006)
15. Lassila, O., McGuinness, D.: The Role of Frame-Based Representation on the Semantic Web. Technical Report KSL-01-02. Knowledge System Laboratory. Stanford University, Stanford, California (2001)
16. Connolly, T., Begg, C.: Database System - A Practical Approach to Design, Implementation and Management. Addison Wesley Longman Publishing Co., Inc., Redwood City (2005)

Likelihood-Based Sampling from Databases for Rule Induction Methods

Shusaku Tsumoto, Shoji Hirano, and Hidenao Abe

Department of Medical Informatics, Faculty of Medicine,
Shimane University
89-1 Enya-cho Izumo 693-8501 Japan
{tsumoto,hirano,abe}@med.shimane-u.ac.jp

Abstract. This paper introduces the idea of log-likelihood ratio to measure the similarity between generated training samples and original tracing samples. The ratio is used as a test statistic to determine whether the statistical information of generated training samples(S_k) is almost equivalent to that of original training samples(S_0), denoted by $S_0 \simeq S_k$. If the test statistic obtained rejects the hypothesis $S_0 \simeq S_k$, then these samples are abandoned. Otherwise, the generated samples are accepted and rule induction methods or statistical methods are applied. This method was evaluated to three medical domains. The results show that the proposed method selects training samples which reflect the statistical characteristics of the original training samples although the performance with small samples is not so good.

1 Introduction

One of the most important problems in data mining is how to manage a large amount of data and to extract efficient knowledge from large databases. Although many machine learning methods and statistical methods have been proposed to solve this problem, they are not powerful when we have more than 1000 samples, since the computational complexity of these algorithms is larger than or approximately equal to n^2.

Thus, it is very important to introduce sampling technique in order to solve the computational complexity problems. However, for this purpose, we have to check whether sampled examples are not deviated from the original training samples: if we only apply random sampling to generate training samples, then it is probable that the statistics of the generated samples are completely different from those of the original training samples. For example, let original training samples be equal to {1,2,3,4,5,6}, whose information is given in Table 1. Then, let us consider a case when we generate training samples whose cardinality is equal to the original training samples. When generated samples are equal to {1,1,2,2,6,6}, we cannot induce rules from this generated dataset. Or, when generated samples are {1,1,1,3,3,3}, the induced rules are different from those rules obtained by the original training samples.

J. Yu et al. (Eds.): RSKT 2010, LNAI 6401, pp. 265–272, 2010.

Table 1. An Small Database on Headache

No.	age	location	nature	severity	history	jolt	prod	nau	M1	M2	class
1	50-59	occular	persistent	weak	persistent	yes	yes	yes	no	no	m.c.h.
2	40-49	whole	persistent	strong	persistent	yes	yes	yes	no	no	m.c.h.
3	40-49	occular	throbbing	strong	persistent	no	no	no	yes	yes	migraine
4	40-49	whole	throbbing	weak	paroxysmal	no	no	no	yes	yes	migraine
5	40-49	whole	persistent	strong	paroxysmal	yes	yes	yes	no	no	m.c.h.
6	50-59	whole	persistent	weak	persistent	yes	yes	no	no	yes	m.c.h.

NOTATIONS. jolt: jolt headache, prod: prodrome, nau:nausea,
M1/M2: tenderness of M1/M2, m.c.h.: muscle contraction headache

Thus, we have to check the following two problems when we want to apply a sampling technique to save the computational complexity: whether generated training samples include all the class to be classified, and whether generated training samples reflects the statistical tendency of the original training samples. In this paper, we introduce the idea of log-likelihood ratio to measure the similarity between generated training samples and original tracing samples. Then, we use this ratio as a test statistic to determine whether the statistical information of generated training samples(S_k) is almost equivalent to that of original training samples(S_0), denoted by $S_0 \simeq S_k$. If the test statistic obtained rejects the hypothesis $S_0 \simeq S_k$, then these samples are abandoned. Otherwise, the generated samples are accepted and rule induction methods[2,5] or statistical methods[1] are applied. This method, called LISA(likelihood based sampling), was evaluated on three medical domains. The results show that the proposed method selects training samples which reflect the statistical characteristics of the original training samples although the performance with small samples is not so good.

The paper is organized as follows: Section 2 makes a brief description about log-likelihood ratio. Section 3 shows an algorithm for likelihood-based sampling(LISA) method. Section 4 gives experimental results. Section 5 discuss the problems of our work.

2   Likelihood Ratio

Likelihood is defined as a direct product of probabilities, which can be viewed as a statistic of the probabilistic situation[3].

Let $p_{a_i}(v_j)$ denote the probability or frequency when an attribute a_i take a value v_j. Then, this probability is defined as:

$$p_{a_i}(v_j) = \frac{|[x]_{[a_i=v_j]}|}{|[x]_{a_i}|} = \frac{|[x]_{[a_i=v_j]}|}{|\cup_{i,j}[x]_{[a_i=v_j]}|},$$

where $[x]_{[a_i=v_j]}$ denotes a set of samples which satisfy $[a_i = v_j]$, and where $|D|$ denotes the cardinality of a set $|D|$. For example, $p([age = 50 - 59])$ is equal to

$1/3$, since $|[x]_{[age=50-59]}| = 2$ and $|\cup [x]_{age}| = 6$. By the use of this definition, likelihood is defined as:

$$L(S_k) = \prod_{i=1}^{n} \prod_{j} p_{a_i}(v_j).$$

As discussed above, likelihood can be viewed as a statistic of a probabilistic situation, denoted by S_k in the above formula, which also means that likelihood is equivalent to a point in the real axis. Thus, we can compare the similarity between two situations S_k and S_l using $L(S_k)$ and $L(S_l)$. Although it is very clear, there are two kinds of comparison between two likelihood values: difference or ratio. In the statistical context, it has been stressed that a ratio $L(S_k)/L(S_l)$ is better than a difference $L(S_k) - L(S_l)$, since the approximate distribution of the log-likelihood ratio, $-2\log(L(S_k)/L(S_l))$ is equal to the chi-square distribution with n freedom, $\chi^2(n)$, where n is equal to the total number of attributes, if $L(S_k) < L(S_l)$[3]. That is,

$$-2\log \frac{L(S_k)}{L(S_l)} \sim \chi^2(n).$$

It is notable that the left side is equal to 0 when $L(S_k) = L(S_l)$, which is the minimum value. From the degree of freedom and the value of χ^2 statistic, the p-value can be calculated, which means the probability to accept the null hypothesis.

For example, let $L(S_0)$ denote the likelihood of the original training samples. Then,

$$
\begin{aligned}
L(s_0) = &\, p_{age}(40-59)p_{age}(50-59) \\
&\times p_{location}(occular)p_{location}(whole) \\
&\times p_{nature}(persistent)p_{nature}(throbbing) \\
&\times p_{severity}(weak)p_{severity}(strong) \\
&\times p_{history}(persistent)p_{history}(paroxysmal) \\
&\times p_{jolt}(yes)p_{jolt}(no) \\
&\times p_{prod}(yes)p_{prod}(no) \\
&\times p_{nau}(yes)p_{nau}(no) \\
&\times p_{M1}(yes)p_{M1}(no) \\
&\times p_{M2}(yes)p_{M2}(no) \\
&\times p_{class}(m.c.h)p_{class}(migraine) \\
= &\, 7.342 \times 10^{-8}.
\end{aligned}
$$

When selected samples S_1 consists of $\{1,1,1,1,1,1\}$, $L(S_1)$ is equal to 0, since the probability $p_{class}(migraine) = 0$. Thus, since the nominator of the likelihood ratio is equal to 0,

$$-2\log(L(S_0)/L(S_1)) = -2\log 0 \simeq \infty,$$

whose p-value is approximately equal to 0. Thus, the probability that the new training samples is equivalent to the original samples is approximately equal to 0. Thus, this hypothesis can be rejected. On the other hand, when S_2 is equal to $\{1,1,3,3,4,4\}$, $L(S_2)$ is obtained as 6.526×10^{-8}. Thus, the log-likelihood ratio is

$$-2\log(L(S_2)/L(S_1)) = -2\log(6.526/7.342) = 0.235634,$$

whose p-value is much larger than 0.999. Thus, this sampling can be accepted.

3 Sampling Design

Using the idea of log-likelihood ratio, we introduce the following sampling design, $LISA$ (Likelihood-based Sampling)to get training samples which reflect the statistical characteristics of the original training samples.

In this sampling design, we first set two parameters: a sampling size m, and a precision α for rejection of the null hypothesis. Then, $LISA$ first computes the likelihood of original training samples(S_0). Next, $LISA$ performs random sampling, and generate new training samples(S_k) whose cardinality is equal to m. After sampling, $LISA$ calculates the likelihood of S_k. Then, this algorithm computes the log-likelihood ratio $|2\log(S_k)/(S_0)|$, and checks its p-value. If the p-value obtained is larger than α, then $LISA$ accepts this sampling. Otherwise, $LISA$ rejects the generated samples. After accepting new training samples S_k, $LISA$ evokes a learning method or statistical method. These processes are summarized into Fig. 3. For the above example, let m and α be equal to 6 and 0.01. Then, S_0 is equal to $\{1,2,3,4,5,6\}$. When S_1 is equal to $\{1,1,1,1,1,1\}$, the

```
procedure Likelihood based Sampling (LISA)
    var
        m : integer; /* A cardinality of a sampling set */
        n : integer; /* A total number of attribute-value pairs */
        α : real; /* A precision for acceptance */
    begin
        Calculate the likelihood L(S₀);
        Generate training samples Sₖ randomly
            such that |Sₖ| = m.
        Calculate the likelihood L(Sₖ);
        Calculate the likelihood ratio R = | − 2L(Sₖ)/L(S₀)|;
        Calculate the threshold θ with the degree of
            freedom n and the precision α;
        if (R ≥ θ) then do
            Accept this sampling set S₁;
        else do
            Reject this sampling set S₁;
    end { Likelihood based Sampling }
```

Fig. 1. An Algorithm for Sampling

log-likelihood is ∞, much larger than 45.64, which is the value of χ^2-statistic with the degree of freedom 26 and the precision 0.01. Thus, $LISA$ rejects this sampling.

On the other hand, when S_2 is equal to $\{1,1,3,3,4,4\}$, $L(S_2)$ is obtained as 6.526×10^{-8}. Then, the log-likelihood ratio is $-2\log(6.526/7.342) = 0.235634$, whose p-value is much larger than 0.999. Thus, $LISA$ accepts this sampling.

4   Experimental Results

We apply $LISA$ to the following three medical domains: headache(RHINOS domain), whose training samples consist of 1477 samples, 10 classes, and 20 attributes, cerebulovasular diseases(CVD), whose training samples consist of 620 samples, 15 classes, and 25 attributes, and meningitis, whose training samples consists of 213 samples, 3 classes, and 27 attributes.

The experiments were performed by the following four procedures, which combines cross-validation, $LISA$ method and random sampling for efficient computation. First, original samples(S_0) were randomly split into new training samples(S_k) and new test samples(T_k) with a given ratio $r = m/N$, where m and N denotes $|S_k|$ and $|S_0|$, respectively. Second, learning methods were applied to S_k (random sampling), the predictive accuracy is estimated by T_k and is stored in a list L_r. Third, $LISA$ checking algorithm was applied. If the condition was satisfied, S_k would be accepted and the above test results were also stored in a list L_l. Otherwise, the test results would be abandoned. These procedures were repeated for 1000 times and all the estimators were averaged over 1000 trials, and acceptance ratio for $LISA$ is calculated. Finally, fourth, learning methods were applied to the original training samples, predictive accuracy was estimated using ten-fold cross-validation[1].

A ratio r is set to be 7/8, 3/4, 1/2, 1/4, and 1/8, and apply the above procedures, and the precision α was set to 0.01. For learning methods, C4.5[5] were used. Then, we compare the results of random sampling and $LISA$ sampling with t-test. The experimental results are shown in Table 2 to 5.

Table 2 shows an acceptance rate for each domain and for each setting of a ratio. The second to the sixth columns denotes the value of the acceptance ratio. Each row denotes the applied domain. As shown in this table, Each acceptance rate decreases rapidly when a sampling size is around 100, which degrades the

Table 2. Acceptance Rate

Domain	Ratio				
	7/8	3/4	1/2	1/4	1/8
Headache(1477)	97.6 %	96.5 %	84.2%	79.2%	75.2%
CVD(620)	94.4 %	89.2 %	75.7%	73.2%	65.1%
Meningitis(213)	98.2 %	81.6 %	69.9%	54.2%	43.9%

quality of this estimation method. Table 3 to 5 show comparison between random sampling and $LISA$. The second column gives the estimators obtained by ten-fold cross validation, and the third to the sixth column presents accuracy estimated by original training samples. The first row and the second row denotes the two sampling design: random sampling, LISA. The third column shows the results of statistical test between the first row and the second row.

When a size of training samples is large, the statistical difference between two methods is not significant. However, when the size decreases, $LISA$ outperforms random sampling.

Thus, when we have to take small samples because of some reasons, such as the cost for examinations, $LISA$ is much better tin induce clarification rules.

However, even when $LISA$ is applied, small samples cause the decrease of accuracy. For each domain, the statistical difference between estimators obtained by $LISA$ and estimators obtained by original training samples S_0 is significant, which suggests that even $LISA$ does not do well with small training samples, whose size is around 50, although the performance of this sampling is much better than that of random sampling.

Table 3. Comparison of Two Sampling Designs (C4.5): Headache

		Ratio				
Sampling Design	(9/10)	7/8	3/4	1/2	1/4	1/8
Random Sampling	85.4%	85.2 %	85.0 %	84.1%	69.3%	65.8%
LISA	–	84.9 %	84.5 %	84.2%	82.6%	81.5%
p-value		NS	NS	NS	< 0.01	< 0.01

DEFINITION: NS: not significant (9/10): ten-fold cross validation

Table 4. Comparison of Two Sampling Designs (C4.5): CVD

		Ratio				
Sampling Design	(9/10)	7/8	3/4	1/2	1/4	1/8
Random Sampling	73.2%	72.2 %	68.9 %	67.8%	59.3%	55.4%
LISA	–	71.9 %	69.7 %	68.4%	69.6%	69.5%
p-value		NS	NS	NS	< 0.01	< 0.01

DEFINITION: NS: not significant (9/10): ten-fold cross validation

Table 5. Comparison of Two Sampling Designs (C4.5): Meningitis

		Ratio				
Sampling Design	(9/10)	7/8	3/4	1/2	1/4	1/8
Random Sampling	84.3%	82.2 %	80.9 %	73.1%	69.2%	65.4%
LISA	–	82.9 %	80.7 %	80.4%	79.6%	67.5%
p-value		NS	NS	< 0.01	< 0.01	< 0.05

DEFINITION: NS: not significant (9/10): ten-fold cross validation

5   Discussion

Using $LISA$, we can obtain the results induced by generated samples whose performance is almost equal to the original training samples. Furthermore, this performance is conserved by around 100 samples.

One of the most important features in $LISA$ method is that the precision α determines the characteristics of sampling. If α is set to be very small, an acceptance test tends to be conservative. For example, when α is set to 0.0001, most of the generated training samples will be accepted. On the other hand, if α is set to be very large, most of the generated training samples will be rejected. Thus, it is very important to tune this parameter. Unfortunately, since α cannot be determined by original sampling method, we have to set α to a fixed value.

Another important feature is that likelihood imposes a constraint on the sampling algorithm: if at least one probability is equal to 0, then the sampling set will be rejected since its likelihood is equal to 0. Therefore, an attribute-value pair should have one entry in generated training samples. This knowledge is very reasonable, because if one attribute-value pair do not appear in new training samples, no rule which includes the pair will be induced: this may cause the overfitting. Thus, this likelihood test plays an important role in suppressing overfitting rules. However, this constraint may be sometimes too strict to select training samples from original training samples.

In order to loose this constraint, we can select a Laplace estimator $l_{a_i}(v_j)$ rather than a frequency defined in Section 2. For example, the simplest Laplace formula is:

$$l_{a_i}(v_j) = \frac{|[x]_{[a_i=v_j]}| + 1}{|\cup_j [x]_{[a_i=v_j]}| + 2}.$$

Then, even in the case of some probability, zero can be removed from the definition of the likelihood. Even if $[x]_{[a_i=v_j]}$ is equal to 0,

$$l_{a_i}(v_j) = \frac{1}{2 + |\cup_j [x]_{[a_i=v_j]}|} = \frac{1}{2 + N},$$

where N denotes the cardinality of original training samples. For example, let us consider a case when $S_1 = \{1,1,1,1,1,1\}$. the term whose value is equal to 0 will become $1/8$. Thus, $L(S_1)$ is equal to $(1/8)^1 1 = 2^{-33}$.

Then, if the precision α is set so that the log-likelihood ratio $|-2\log(S_1)/(S_0)|$ is smaller than the threshold, even the above sampling set S_1 will be accepted. However, too small precision causes the whole algorithm to be nearer to random sampling. Thus, this technique should be used when we have a preference about sampling. For example, in the case when "m.c.h." should be put an emphasis on inducing rules, the above scenario can be applied.

6   Conclusion

In this paper, we introduce the idea of log-likelihood ratio to measure the similarity between generated training samples and original tracing samples. Then, we use this ratio as a test statistic to determine whether the statistical information of generated training samples(S_1) is almost equivalent to that of original training samples(S_0), denoted by $S_0 \simeq S_1$. If the test statistic obtained rejects the hypothesis $S_0 \simeq S_1$, then these samples are abandoned. Otherwise, the generated samples are accepted and rule induction methods or statistical methods are applied. This system, called LISA, was applied to three medical domains. The results show that the proposed method selects training samples which reflect the statistical characteristics of the original training samples although the performance with small samples is not so good.

References

1. Breiman, L., Freidman, J., Olshen, R., Stone, C.: Classification And Regression Trees. Wadsworth International Group, Belmont (1984)
2. Clark, P., Niblett, T.: The CN2 Induction Algorithm. Machine Learning 3, 261–283 (1989)
3. Edwards, A.W.F.: Likelihood, expanded edition. Johns Hopkins University Press, Baltimore (1992)
4. Efron, B.: The Jackknife, the Bootstrap and Other Resampling Plans. Society for Industrial and Applied Mathematics, Philadelphia (1982)
5. Quinlan, J.R.: C4.5 - Programs for Machine Learning. Morgan Kaufmann, CA (1993)
6. Walker, M.G., Olshen, R.A.: Probability Estimation for Biomedical Classification Problems. In: Proceedings of the 16th SCAMC. McGrawHill, New York (1992)

Residual Analysis of Statistical Dependence in Multiway Contingency Tables

Shusaku Tsumoto and Shoji Hirano

Department of Medical Informatics, Faculty of Medicine,
Shimane University
89-1 Enya-cho Izumo 693-8501 Japan
{tsumoto,hirano}@med.shimane-u.ac.jp

Abstract. A Pearson residual is defined as the residual between actual values and expected ones of each cell in a contingency table. This paper shows that this residual is represented as linear sum of determinants of 2×2, which suggests that the geometrical nature of the residuals can be viewed from grasmmanian algebra.

1 Introduction

Statistical independence between two attributes is a very important concept in data mining and statistics. The definition $P(A, B) = P(A)P(B)$ show that the joint probability of A and B is the product of both probabilities. This gives several useful formula, such as $P(A|B) = P(A)$, $P(B|A) = P(B)$. In a data mining context, these formulae show that these two attributes may not be correlated with each other. Thus, when A or B is a classification target, the other attribute may not play an important role in its classification.

Although independence is a very important concept, it has not been fully and formally investigated as a relation between two attributes.

In this paper, a statistical independence in a contingency table is focused on from the viewpoint of granular computing and linear algebra. which is continuation of studies on contingency matrix theory in [2,4,5,6]. Tsumoto[2,6] discusses that a contingency table compares two attributes with respect to information granularity and shows that statistifcal independence in a contingency table is a special form of linear depedence of two attributes. Especially, when the table is viewed as a matrix, the above discussion shows that the rank of the matrix is equal to 1.0. Tsumoto[4] matrix algebra is a key point of analysis of this table. A contingency table can be viewed as a matrix and several operations and ideas of matrix theory are introduced into the analysis of the contingency table. On the other hand, in [5], Tsumoto shows that the Pearson residual of a contingency matrix is represented as linear sum of 2×2 submatrices, whose number is equal to the degree of freedom.

In this paper, we extend residual analysis to multidimensional cases. The results show that multidimensional residuals are also represented as linear sum

J. Yu et al. (Eds.): RSKT 2010, LNAI 6401, pp. 273–280, 2010.
© Springer-Verlag Berlin Heidelberg 2010

of determinants of 2×2 submatrices, whose number is equal to the degree of freedom in a given contingency table. Furthermore, linear sum includes several kinds of units for statistical independence/dependence, such as total independence and partial independence. This also suggests that 2×2 submatrices in a multidimensional data cube can be viewed as information granules for measuring statistical dependence.

The paper is organized as follows: Section 2 discusses the characteristics of contingency tables. Section 3 gives the results of residual analysis of multiway contingency tables. Section 4 shows that the number of the determinants of 2×2 submatrices is exactly the same as the degree of freedom in a three-way contingency table. Finally, Section 5 concludes this paper.

2   Multiway Contingency Table

Definition 1. *Let $R_1, R_2, \cdots, R_n$ denote $n(\in N)$ multinominal attributes in an attribute space A which have $m_1, m_2, \cdots, m_n$ values Let $|R_j = A_{j_i}|$ denote the set of data whose jth-attribute is equal to A_{j_i} (ith-partition of j). A contingency table $T(R_1, R_2, \cdots, R_n)$ is a table, each of whose cells can be defined as:*

$$x_{i_1 i_2 \cdots i_n} = \#\{x \in |R_1 = A_{i_1}| \wedge |R_2 = A_{i_2}| \cdots \wedge |R_n = A_{i_n}|\},$$

with these marginal sums. For example, in the two dimensional case, this table is arranged into the form shown in Table 1, where: $|[R_1 = A_j]_A| = \sum_{i=1}^{m} x_{1i} = x_{\cdot j}$, $|[R_2 = B_i]_A| = \sum_{j=1}^{n} x_{ji} = x_{i\cdot}$, $|[R_1 = A_j \wedge R_2 = B_i]_A| = x_{ij}$, $|U| = N = x_{\cdot\cdot}$ ($i = 1, 2, 3, \cdots, n$ and $j = 1, 2, 3, \cdots, m$).

Table 1. Contingency Table $(m \times n)$

	A_1	A_2	$\cdots$	A_n	Sum
B_1	x_{11}	x_{12}	$\cdots$	x_{1n}	$x_{1\cdot}$
B_2	x_{21}	x_{22}	$\cdots$	x_{2n}	$x_{2\cdot}$
$\cdots$	$\cdots$	$\cdots$	$\cdots$	$\cdots$	$\cdots$
B_m	x_{m1}	x_{m2}	$\cdots$	x_{mn}	$x_{m\cdot}$
Sum	$x_{\cdot 1}$	$x_{\cdot 2}$	$\cdots$	$x_{\cdot n}$	$x_{\cdot\cdot} = |U| = N$

Definition 2. *A multiway contigency matrix $M_{R_1, R_2, \cdots, R_n}(N)$ is defined as: which is composed of*

$$x_{i_1 i_2 \cdots i_n} = \#\{x \in |R_1 = A_{i_1}| \wedge |R_2 = A_{i_2}| \cdots \wedge |R_n = A_{i_n}|\},$$

where their marginal sums are not included as elements.   □

For simplicity, if we do not need to specify R_1 and R_2, we use $M(m, n, N)$ as a contingency matrix with m rows, n columns and N samples.

One of the important observations from granular computing is that a contingency table shows the relations between two attributes with respect to intersection of their supporting sets. When two attributes have different number of equivalence classes, the situation may be a little complicated. But, in this case, due to knowledge about linear algebra, we only have to consider the attribute which has a smaller number of equivalence classes. and the surplus number of equivalence classes of the attributes with larger number of equivalnce classes can be projected into other partitions. In other words, a $m \times n$ matrix or contingency table includes a projection from one attributes to the other one.

Residual of Multi-way Table

Tsumoto and Hirano[5] gives a formula for conditional probabilities of a multi-way contingency table.

Theorem 1. *If a m-way contingency table satisfy statistical independence, then the following equation should be satisfied for any k-th attribute i_k and j_k (k = $1, 2, \cdots, n$) where n is the number of attributes.*

$$\frac{x_{i_1 i_2 \cdots i_k \cdots i_n}}{x_{i_1 i_2 \cdots j_k \cdots i_n}} = \frac{x_{\bullet \bullet \cdots i_k \cdots \bullet}}{x_{\bullet \bullet \cdots j_k \cdots \bullet}}$$

Also, the following equation should be satisfied for any i_k:

$$x_{i_1 i_2 \cdots i_n} \times x_{\bullet \bullet \cdots \bullet}^{n-1} = x_{i_1 \bullet \cdots \bullet} x_{\bullet i_2 \cdots \bullet} \times \cdots \times x_{\bullet \bullet \cdots i_k \cdots \bullet} \times \cdots \times x_{\bullet \bullet \cdots \bullet i_n} \qquad \square$$

Thus, a Pearson residual, a difference between an observed value for each cell in a contingency table and an expected value, is defined as:

$$\sigma_{i_1 i_2 \cdots i_n} = x_{i_1 i_2 \cdots i_n} - \frac{x_{i_1 \bullet \cdots \bullet} x_{\bullet i_2 \cdots \bullet} \times \cdots \times x_{\bullet \bullet \cdots i_k \cdots \bullet} \times \cdots \times x_{\bullet \bullet \cdots \bullet i_n}}{x_{\bullet \bullet \cdots \bullet}^{n-1}} \qquad (1)$$

3   Information Granule for Contingency Matrices

3.1   Residual of Contingency Matrices

Tsumoto and Hirano [3,5] discusses the meaning of pearson residuals from the viewpoint of linear algebra.

From Equation (1), the residual is defined as:

$$\sigma_{ij} = x_{ij} - \frac{x_{i\bullet} \times x_{\bullet j}}{x_{\bullet\bullet}}.$$

And simple calculation leads to the following theorem.

Theorem 2. *The residual σ_{ij} of two-dimensional contigency matrix is obtained as:*

$$\sigma_{ij} = \frac{1}{x_{\bullet\bullet}} \{x_{ij}x_{\bullet\bullet} - x_{i\bullet} \times x_{\bullet j}\}$$

$$= \frac{1}{x_{\bullet\bullet}} \left\{ x_{ij} \sum_{k\neq i}\sum_{l\neq j} x_{kl} - \left(\sum_{l\neq j} x_{il}\right) \left(\sum_{k\neq i} x_{kj}\right) \right\}$$

$$= \frac{1}{x_{\bullet\bullet}} \sum_{\substack{k\neq i \\ l\neq j}} (x_{ij}x_{kl} - x_{kj}x_{il})$$

$$= \frac{1}{x_{\bullet\bullet}} \sum_{\substack{k\neq i \\ l\neq j}} \Delta_{k,l}^{i,j},$$

where $\Delta_{k,l}^{i,j}$ is the determinant of a 2×2 submatrix of the original contingency matrix with selection of i and k rows and j and l columns. Also, the sum takes over $k = m$ and $l = n$. Equivalently, the above formula can be represented as:

$$\sigma_{ij}x_{\bullet\bullet} = \sum_{\substack{k\neq i \\ l\neq j}} \Delta_{k,l}^{i,j},$$

where the sum takes over $k = m$ and $l = n$.   □

Especially, for $m = n = 2$, the residual is equal to the determinant of the orinal matrix.

$$\sigma_{ij}x_{\bullet\bullet} = \Delta_{k,l}^{i,j},$$

where $k \neq i$ and $l \neq j$. Thus, $\sigma_{11}x_{\bullet\bullet} = \Delta_{2,2}^{1,1}$,

In the case of 3×3 tables, the following formulas are obtained.

$$\sigma_{11} = \frac{1}{x_{..}} \left(\Delta_{2,2}^{1,1} + \Delta_{2,3}^{1,1} + \Delta_{3,2}^{1,1} + \Delta_{3,3}^{1,1} \right)$$

$$\sigma_{12} = \frac{1}{x_{..}} \left(\Delta_{2,1}^{1,2} + \Delta_{2,3}^{1,2} + \Delta_{3,1}^{1,2} + \Delta_{3,3}^{1,2} \right)$$

$$\sigma_{21} = \frac{1}{x_{..}} \left(\Delta_{1,2}^{2,1} + \Delta_{1,3}^{2,1} + \Delta_{3,2}^{2,1} + \Delta_{3,3}^{2,1} \right)$$

$$\sigma_{22} = \frac{1}{x_{..}} \left(\Delta_{1,1}^{2,2} + \Delta_{1,3}^{2,2} + \Delta_{3,1}^{2,2} + \Delta_{3,3}^{2,2} \right)$$

Thus, a 2×2 submatrix in a contingency table can be viewed as a information granule for statistical (in)dependence.

Can we generalize these results into statistical independence of multivariate cases ? This question is answered in this paper as follows.

3.2   Information Granule for Three-Way Contingency Tables

The residual for x_{ijk} is defined as:

$$\sigma_{ijk} = x_{ijk} - \frac{x_{i\bullet\bullet} \times x_{\bullet j\bullet} \times x_{\bullet\bullet k}}{x_{\bullet\bullet\bullet}^2}. \tag{2}$$

Here, we define "partial residuals" in which one of three attributes are summarized (marginalized) as follows:

$$\sigma_{\bullet jk} = x_{\bullet jk} - \frac{x_{\bullet j\bullet} \times x_{\bullet\bullet k}}{x_{\bullet\bullet\bullet}} \tag{3}$$

$$\sigma_{i\bullet k} = x_{i\bullet k} - \frac{x_{i\bullet\bullet} \times x_{\bullet\bullet k}}{x_{\bullet\bullet\bullet}} \tag{4}$$

$$\sigma_{ij\bullet} = x_{ij\bullet} - \frac{x_{i\bullet\bullet} \times x_{\bullet j\bullet}}{x_{\bullet\bullet\bullet}} \tag{5}$$

Then, by using (3), the residual (2) is reformualted as:

$$\sigma_{ijk} = x_{ijk} - \frac{x_{i\bullet\bullet} \times x_{\bullet j\bullet} \times x_{\bullet\bullet k}}{x_{\bullet\bullet\bullet}^2}$$

$$= x_{ijk} - \frac{x_{i\bullet\bullet}}{x_{\bullet\bullet\bullet}} \frac{x_{\bullet j\bullet} x_{\bullet\bullet k}}{x_{\bullet\bullet\bullet}}$$

$$= x_{ijk} - \frac{x_{i\bullet\bullet}}{x_{\bullet\bullet\bullet}} \left(x_{\bullet jk} - \sigma_{\bullet jk} \right)$$

$$= \frac{x_{i\bullet\bullet}}{x_{\bullet\bullet\bullet}} \sigma_{\bullet jk} + x_{ijk} - \frac{x_{i\bullet\bullet}}{x_{\bullet\bullet\bullet}} x_{\bullet jk}$$

$$= \frac{x_{i\bullet\bullet}}{x_{\bullet\bullet\bullet}} \sigma_{\bullet jk} + \frac{1}{x_{\bullet\bullet\bullet}} \left(x_{ijk} x_{\bullet\bullet\bullet} - x_{i\bullet\bullet} x_{\bullet jk} \right) \tag{6}$$

Since $\sigma_{\bullet jk}$ is equivalent to the residual of two-dimensional contigency matrix, it can be represented as linear sum of 2×2 submatrices. Thus, the latter part $x_{ijk} x_{\bullet\bullet\bullet} - x_{i\bullet\bullet} x_{\bullet jk}$ should be examined. Let us denote this part by σ_{jk}^i.

$$\sigma_{jk}^i = x_{ijk} x_{\bullet\bullet\bullet} - x_{i\bullet\bullet} x_{\bullet jk}$$

$$= x_{ijk} \left(x_{i\bullet\bullet} + \sum_{l \neq i} x_{l\bullet\bullet} \right) - x_{i\bullet\bullet} \left(x_{ijk} + \sum_{m \neq i} x_{mjk} \right)$$

$$= x_{ijk} \sum_{l \neq i} x_{l\bullet\bullet} - x_{i\bullet\bullet} \sum_{m \neq i} x_{mjk}$$

$$= \sum_{l \neq i} \left(x_{ijk} x_{l\bullet\bullet} - x_{i\bullet\bullet} x_{ljk} \right)$$

$$= \sum_{\substack{l \neq i \\ m \neq j \text{ or } n \neq k}} \left(x_{ijk} x_{lmn} - x_{imn} x_{ljk} \right)$$

$$= \sum_{\substack{l \neq i \\ n \neq k}} \Delta(j)_{ln}^{ik} + \sum_{\substack{l \neq i \\ m \neq j}} \Delta(k)_{lm}^{ij} + \sum_{\substack{l \neq i \\ m \neq j \\ n \neq k}} \left(x_{ijk} x_{lmn} - x_{imn} x_{ljk} \right), \tag{7}$$

where $\Delta(j)_{ln}^{ik}$ denotes the determinant of 2×2 matrix in which i and l rows and k and n columns are selected with $m = j$ fixed and $\Delta(k)_{lm}^{ij}$ denotes the determinant of 2×2 matrix in which i and l rows and j and m columns are selected with $n = k$ fixed. On the other hand, the last part is different from the former two sums. This sum is zero if the conditional independence are satisfied, that is, if $x_{ijk} = x_{i\bullet\bullet}x_{\bullet jk}$, whose meaning is i and j are statistical independence with k fixed. Thus, the following theorem is obtained.

Theorem 3. *The residual of three-dimensional contingency table is decomposed as:*

$$\sigma_{ijk} = \frac{x_{i\bullet\bullet}}{x_{\bullet\bullet\bullet}}\sigma_{\bullet jk} + \sigma_{jk}^i$$

$$= \frac{x_{i\bullet\bullet}}{x_{\bullet\bullet\bullet}}\sigma_{\bullet jk}$$

$$+ \frac{1}{x_{\bullet\bullet\bullet}}\left(\sum_{\substack{l\neq i \\ n\neq k}} \Delta(j)_{ln}^{ik} + \sum_{\substack{l\neq i \\ m\neq j}} \Delta(k)_{lm}^{ij} + \sum_{\substack{l\neq i \\ m\neq j \\ n\neq k}} (x_{ijk}x_{lmn} - x_{imn}x_{ljk}) \right). \quad (8)$$

□

3.3   Information Granule for Multi-Way Contingency Table

The structure of Equation (8) suggests that the residual can be defined in a recursive way.

Theorem 4. *Let n be a number of attributes in a contingency table (i.e. n-way contingency table. The residual of the cell$(i_1, i_2, \cdots, 1_n)$, $\sigma_{i_1 i_2 \cdots i_n}$ is decomposed as follows.*

$$\sigma_{i_1 i_2 \cdots i_n} = \frac{x_{i_1 \bullet \cdots \bullet}}{x_{\bullet\bullet\cdots\bullet}}\sigma_{\bullet i_2 \cdots i_n} + \frac{1}{x_{\bullet\bullet\cdots\bullet}}\sigma_{i_2 \cdots i_n}^{i_1}$$

$$= \frac{x_{i_1 \bullet \cdots \bullet}}{x_{\bullet\bullet\cdots\bullet}}\sigma_{\bullet i_2 \cdots i_n} + \frac{1}{x_{\bullet\bullet\cdots\bullet}}\left(x_{i_1 i_2 \cdots i_n}x_{\bullet\bullet\cdots\bullet} - x_{i_1 \bullet\cdots\bullet}x_{\bullet i_2 \cdots i_n} \right), \quad (9)$$

□

For example, in the case of a four-dimensional contingency table, a residual $\sigma_{i_1 i_2 i_3 i_4}$ is decomposed as:

$$\sigma_{i_1 i_2 i_3 i_4} = \frac{x_{i_1 \bullet\bullet\bullet}}{x_{\bullet\bullet\bullet\bullet}}\sigma_{\bullet i_2 i_3 i_4} + \frac{1}{x_{\bullet\bullet\bullet\bullet}}\sigma_{i_2 i_3 i_4}^{i_1}$$

$$= \frac{x_{i_1 \bullet\bullet\bullet}}{x_{\bullet\bullet\bullet\bullet}}\sigma_{\bullet i_2 i_3 i_4} + \frac{1}{x_{\bullet\bullet\bullet\bullet}}\left(x_{i_1 i_2 i_3 i_4}x_{\bullet\bullet\bullet\bullet} - x_{i_1 \bullet\bullet\bullet}x_{\bullet i_2 i_3 i_4} \right),$$

where $\sigma_{\bullet i_2 i_3 i_4}$ can be decomposed as shown in Equation (8):

$$\sigma_{\bullet i_2 i_3 i_4} = \frac{x_{\bullet i_2 \bullet \bullet}}{x_{\bullet \bullet \bullet \bullet}} \sigma_{\bullet \bullet i_3 i_4} + \frac{1}{x_{\bullet \bullet \bullet \bullet}} \sigma_{\bullet i_3 i_4}^{i_2}$$

$$= \frac{x_{\bullet i_2 \bullet \bullet}}{x_{\bullet \bullet \bullet \bullet}} \sigma_{\bullet \bullet i_3 i_4} + \frac{1}{x_{\bullet \bullet \bullet \bullet}} \left(x_{\bullet i_2 i_3 i_4} x_{\bullet \bullet \bullet \bullet} - x_{\bullet i_2 \bullet \bullet} x_{\bullet \bullet i_3 i_4} \right).$$

Then, $\sigma_{\bullet i_3 i_4}^{i_2}$ is reformulated as shown in Equation (7).

4   Discussion

In the above section, we mentioned that the number of derminants of 2×2 submatrices in a three-way contingency table is equal to the degree of freedom of χ^2 test statistic in contingency table analysis. This section shows that it is not accidental.

Everitt [1] shows that the degree of freedom of χ^2 test statistic is given as:

$$d.f. = rcl - (r-1) - (c-1) - (l-1) - 1 = rcl - r - c - l + 2 \qquad (10)$$

On the other hand, with the hypothesis of conditional independence on $(r-1)$ row probabilities and column $\times$ layer probabilities $(cl-1)$, the degree of freedom is given as:

$$d.f. = rcl - (cl-1) - (r-1) - 1 = rcl - cl - r + 1 \qquad (11)$$

Let us go back to Equation 9 in Theorem 3 and assume that indices i,j, and k correspond to row, column, and layer, respectively. The numbers of the derminants are summarized into:

$$
\begin{array}{cc}
\sigma_{\bullet jk} & (c-1)(l-1) \\
\Delta(j)_{ln}^{ik} & (r-1)(l-1) \\
\Delta(k)_{lm}^{ij} & (r-1)(c-1) \\
x_{ijk} x_{lmn} - x_{imn} x_{ljk} & (r-1)(c-1)(l-1)
\end{array}
$$

The total number of the determinants sum is:

$$
\begin{aligned}
sum &= (c-1)(l-1) + (r-1)(l-1) + (r-1)(c-1) + (r-1)(c-1)(l-1) \\
&= rcl - r - c - l + 2,
\end{aligned}
$$

which is equal to Equation 10. Thus, the following theorem is obtaied.

Theorem 5. *The total number of derminants of 2×2 submatrices in a three-way contingency table is equal to the degree of freedom of χ^2 test statistic in contingency table analysis.*

Furthermore, since the total number of the derminants of two-way contingency tables with c rows and l columns is $(c-1)(l-1)$, the difference between the number of the determinants of three-way tables and that of two tables, δ is equal to:

$$\delta = (c-1)(l-1) + (r-1)(l-1) + (r-1)(c-1) + (r-1)(c-1)(l-1)$$
$$-(c-1)(l-1)$$
$$= rcl - cl - r + 1,$$

which is equal to Equation 11. Thus,

Theorem 6. *The total number of derminants of 2×2 submatrices in a three-way contingency table can be decomposed into the sum of the total number of derminants in a two-way contingency table and that in a three-way contingency table with conditional independence.*

It will be our future work to generalize these results into multi-way contingency tables, whose dimension is larger than 4.

5   Conclusion

This paper focuses on statistical independence of three variables from the viewpoint of linear algebra and show that information granules of statistical independence of three or four variables are decomposed into linear sum of determinants of 2×2-submatrices. The analysis also shows that the residuals of a multiway contigency table can be defined in a recursive way. For example, the residuals of four-way tables are described as those of three-way and two-way tables.

Then, we focus on this recursive nature in the case of three-way contingency tables. The geometric characteristics of selected dertimants in a residual x_{ijk} show that the components of the sum give possible combinations of parallelograms which has one fixed edge with x_{ijk} as a vertex. Furthermore, The total number of derminants of 2×2 submatrices in a three-way contingency table is equal to the degree of freedom of χ^2 test statistic in contingency table analysis.

Thus, the derminants of 2×2 matrices are principal information granules to measure the degree of statistical dependence in a given contingency table.

References

1. Everitt, B.: The Analysis of Contingency Tables, 2nd edn. Chapman & Hall/CRC, Boca Raton (1992)
2. Tsumoto, S.: Contingency matrix theory: Statistical dependence in a contingency table. Inf. Sci. 179(11), 1615–1627 (2009)
3. Tsumoto, S., Hirano, S.: Meaning of pearson residuals - linear algebra view. In: Proceedings of IEEE GrC 2007. IEEE press, Los Alamitos (2007)
4. Tsumoto, S., Hirano, S.: Contingency matrix theory ii: Degree of dependence as granularity. Fundam. Inform. 90(4), 427–442 (2009)
5. Tsumoto, S., Hirano, S.: Dependency and granularity indata. In: Meyers, R.A. (ed.) Encyclopedia of Complexity and Systems Science, pp. 1864–1872. Springer, Heidelberg (2009)
6. Tsumoto, S., Hirano, S.: Statistical independence and determinants in a contingency table - interpretation of pearson residuals based on linear algebra. Fundam. Inform. 90(3), 251–267 (2009)

A Note on the Effect of Knowledge Refinement on Bag Structures

Kankana Chakrabarty

School of Science and Technology
University of New England
Armidale 2351, NSW
Australia

Abstract. In this paper, the author discusses the notion of bags and describes a conceptual model of an information system where the effect of knowledge refinement on bag structure can be clearly observed. In this context, the indiscernibility relation is defined over one or more attribute(s) of a collection of objects of the same type. Here the author further define attributive indiscernibility relations and their types.

1   Introduction

The notion of Bags [3], [7], [9], [2] are important in case of knowledge representation problems where definite collection types allow repeated occurrence of its members and leads to consequences where it is difficult to formulate the model. It is observed that in case of relational databases, bags play significant role since database languages and systems require data models with multiset semantics. Bags can be used as type constructors and can describe dependencies between attributes and can assist to evaluate significance of concerned attributes in the model. In case of any bag, the indiscernibility of objects plays a crucial role. For any given information system, some indiscernibility relation should be pre-defined based on the available knowledge regarding a set of objects depending on which any bag drawn from that set can be formed. In this paper, we are describing a conceptual model of an information system where the effect of knowledge refinement on bag structure can be clearly observed. We find that the indiscernibility relation can be defined over one or more attribute(s) of a collection of objects of the same type. Here we are further defining the attributive indiscernibility relations and their types.

2   Bags, An Overview

In this section, we briefly present the notion of bags [9],[3].

A bag (or a crisp bag) B drawn from a set X is represented by a function $Count_B$ or C_B defined as

$$C_B : X \longrightarrow N$$

where N represents the set of non-negative integers.

J. Yu et al. (Eds.): RSKT 2010, LNAI 6401, pp. 281–287, 2010.

Here $C_B(x)$ is the number of occurrences of the element x in the bag B. We represent the bag B drawn from the set $X = \{x_1, x_2, \ldots, x_m\}$ as

$$B = \{x_1/n_1, x_2/n_2, \ldots x_m/n_m\}$$

where n_i is the number of occurrence of the element x_i (i = 1,2,3,....m) in the bag B.

For any two bags B_1 and B_2 drawn from a set X, the following are defined:

- $B_1 = B_2$ if $C_{B_1}(x) = C_{B_2}(x) \; \forall \; x \in X$.
- $B_1 \sqsubseteq B_2$, i.e B_1 is a sub bag of B_2, if $C_{B_1}(x) \leq C_{B_2}(x) \; \forall \; x \in X$.
- $B = B_1 \oplus B_2$ if $C_B(x) = C_{B_1}(x) + C_{B_2}(x) \; \forall \; x \in X$.
- $B = B_1 \ominus B_2$ if $C_B(x) = max[C_{B_1}(x) - C_{B_2}(x), 0] \; \forall \; x \in X$,

where '$\oplus$' and '$\ominus$' represents the 'bag addition' and 'bag removal' operations respectively.

If B be a bag drawn from a set X, then the support set of B denoted as B^* is a subset of X with the characteristic function given by

$$\phi_{B^*}(x) = min[C_B(x), 1] \quad \forall \; x \in X.$$

A bag B is called an empty bag if for all $x \in X$, $C_B(x) = 0$. The support set of an empty bag is the null set.

The cardinality of a bag B drawn from a set X is denoted by $Card(B)$ and is defined as

$$Card(B) = \sum_{x \in X} C_B(x).$$

For a bag B drawn from a set X, $\max_{x \in X} C_B(x)$ is said to be the peak value of the bag. Any $x^* \in X$ satisfying

$$C_B(x^*) = \max_{x \in X} C_B(x)$$

is called to be a peak element of the bag B.

$$C_B(x^*) = \max_{x \in X} C_B(x)$$

is called to be a peak element of the bag B.

The union of two bags B_1 and B_2 drawn from a set X is a bag denoted by $B_1 \sqcup B_2$ such that $\forall x \in X$,

$$C_{B_1 \sqcup B_2}(x) = max[C_{B_1}(x), C_{B_2}(x)].$$

The intersection of B_1 and B_2 results in a bag denoted by $B_1 \sqcap B_2$ such that $\forall x \in X$,

$$C_{B_1 \sqcap B_2}(x) = min[C_{B_1}(x), C_{B_2}(x)].$$

For two bags A and B drawn from the sets X and Y respectively, their cartesian product denoted by $A \otimes B$ is a bag drawn from $X \times Y$ such that for all $(x, y) \in (X \times Y)$,

$$C_{A \otimes B}(x, y) = C_A(x).C_B(y).$$

3   Effect of Knowledge Refinement

In this section, the effect of knowledge refinement on bag structures is discussed.

A bag is perceived as a collection of objects where the indiscernibility of objects play a crucial role in the sense that we allow the occurrence of indiscernible objects in a bag.

For any given information system, an indiscernibility relation should be pre-defined based on the available knowledge regarding a set of objects depending on which any bag drawn from that set can be formed. Different indiscernibility relations can be defined on a given class of objects and these relations with their multitude of properties may give rise to different bag structures.

Considering the fact that each object O_k in the class $\{O_i\}_{i=1}^p$ posesses a set of properties, for any set of indiscernibility relations $\{I_{R_i}\}_{i=1}^n$, the meaning of each I_{R_i} is defined based on a set of criteria $\{C_j^i\}_{j=1}^m$.

We consider the conceptual model of an information system where

$$B_0 = \{\theta_1/4, \theta_2/2, \theta_3/1, \theta_4/2, \theta_5/2, \theta_6/3, \theta_7/1, \theta_8/1\}$$

is a bag drawn from X where $X = \{\theta_i\}_{i=1}^8$ represents a collection of objects such that X is a unary projection $R\{A_1\}$ of a relvar R with the set of attributes $\{A_i\}_{i=1}^6$.

In this case, B_0 is formed on the basis of the knowledge that for any $x_1, x_2 \in X$, $x_1 I_{R_0} x_2$ holds if and only if x_1 and x_2 have the same value for A_1, I_{R_0} being an indiscernibility relation (refer to Figure 1).

For this particular information system, the said indiscernibility relation can be defined by using one or more attribute(s) of a collection of objects of the

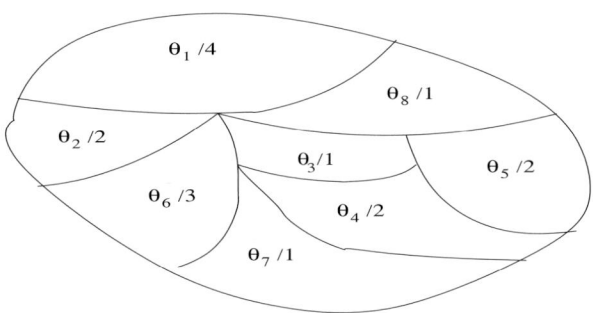

Fig. 1. Classes due to I_{R_0} relates to B_0

same type. Also further, this collection could be the instance of a relvar. Also any bag(s) drawn from this collection could also be regarded as relvar(s).

For a fixed information system, if any indiscernibility relation defined over a collection of objects involve only one attribute of that collection, then we call it unary attributive indiscernibility relation. Similarly if it involves two attributes of that collection, then we call it binary attributive indiscernibility relation and so on. In this information system, I_{R_0} represents a unary-attributive indiscernibility relation. This indicates that it is a bag representing the projection of a single attribute. By considering the tuples of a relvar of degree n, if there exists an n-ary attributive indiscernibility relation, then the relvar represents a bag of tuples. In this model we are considering the collections of objects as whole as well as the collections of certain attribute values of instantiation of objects. Any indiscernibility relation decomposes the object collection into disjoint classes. We call them indiscernibility classes. For the bag B_0, the indiscernibility classes consists of 4,2,1,2,2,3,1, and 1 elements respectively. This refers to an attribute level classification. It is important to note that if a unary projection of a relvar represents a bag, then it is not necessarily true that any instantiation of the relvar as whole will also represent a bag. We call the indiscernibility relation of type I_{R_0} the level-0 indiscernibility relation. After further refinement of knowledge, we can define a refined binary attributive indiscernibility relation called level-1 indiscernibility relation.

As an example, let I_{R_1} be a level-1 indiscernibility relation such that for any $x_1, x_2 \in X$, $x_1 I_{R_1} x_2$ holds if and only if x_1 and x_2 have the same value for A_1 and A_4. On the basis of this level of knowledge, a bag B_1 can be formed (refer to Figure 2).

$$B_1 = \{(\theta_1, 3)/1, (\theta_1, 4)/3, (\theta_2, 3)/2, (\theta_3, 3)/1, (\theta_4, 3)/1, (\theta_4, 4)/1,$$
$$(\theta_5, 5)/2, (\theta_6, 4)/2, (\theta_6, 5)/1, (\theta_7, 3)/1, (\theta_8, 4)/1\}.$$

Continuing the same way, we can further define a level-2 indiscernibility relation I_{R_2} such that for any $x_1, x_2 \in X$, $x_1 I_{R_2} x_2$ iff x_1 and x_2 have the same value for A_1, A_4, and A_5. On the basis of this level of knowledge, a bag B_1 can be formed.

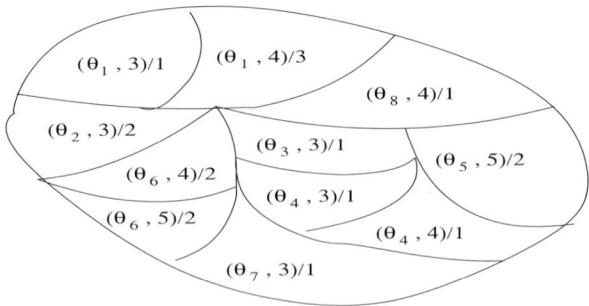

Fig. 2. Classes due to I_{R_1} relates to B_1

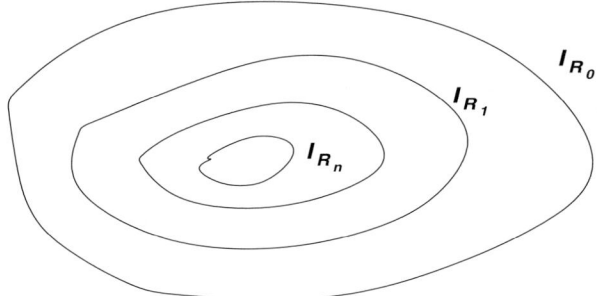

Fig. 3. Nested Indiscernibility Relations

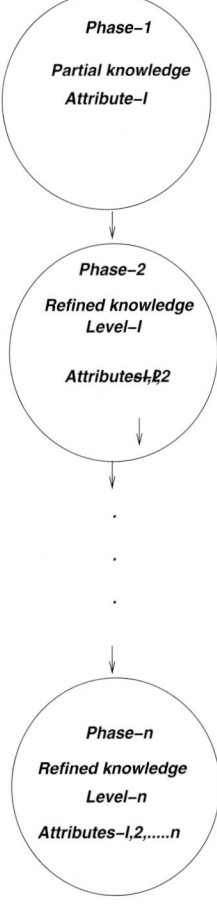

Fig. 4. Phases of Knowledge Refinement

It is to be noted that θ_3, θ_7, and θ_8 will not fall into a further refined class under any attribute level classification.

The number of attributes used for defining a level-2 indiscernibility relation is greater than the number used for defining a level-1 indiscernibility relation and similarly, the number of attributes used for defining a level-1 indiscernibility relation is greater than the number used for defining a level-0 indiscernibility relation.

Thus, $\{I_{R_0}, I_{R_1}, I_{R_2}, \ldots\ldots, I_{R_n}\}$ represents a set of nested indiscernibility relation (refer to Figure 3). These kinds of neted indiscernibility relations are very significant and well justified for knowledge representation under the frames of partial judgment.

It is clear that any indiscernibility relation above level-0 represents a conjunctive normal form expression. Hence, in case any conjunct is false, then the expression returns a false value. Since these are nested, hence, we have $\forall x, y \in X$,

$$x I_{R_n} y \Longrightarrow x I_{R_{n-1}} y \Longrightarrow x I_{R_{n-2}} y \Longrightarrow \ldots\ldots \Longrightarrow x I_{R_1} y \Longrightarrow x I_{R_0} y.$$

For any information system, incomplete or partial knowledge can give rise to a lower level indiscernibility relation. Partial knowledge tends to transform to

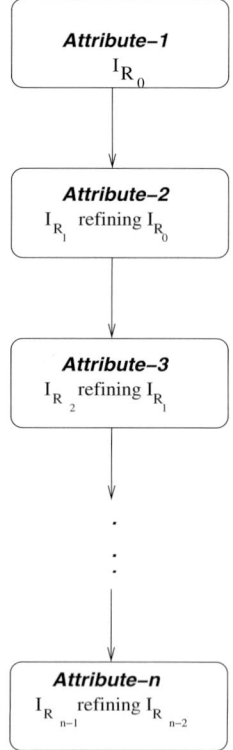

Fig. 5. Ordered Hierarchy of Attributes

complete knowledge after going through the process of gradual refinement by going through a series of phases (refer to Figure 4).

This can follow a hierarchical order and consequently, some subprocesses may be combined if enough knowledge can be obtained at a stage for refinement and hence reducing the number of phases. Order is of importance in case of refinement phases and hence it forms a list and not a set (refer to Figure 5). Thus, a bag can only be drawn from a set if and only if there exists atleast one basic indiscernibility relation defined on it. If the basic indiscernibility relation is refined further based on knowledge refinement, then with each refinement, a bag can be formed and the refinement can go through a hierarchical process forming disjoint object classes in each phase. In case of rough sets, we consider a set with an indiscernibility relation which plays the fundamental role in defining the lower and upper approximations. Hence our notions can be embedded into the system while incorporating the concepts of rough sets for decision modeling.

4   Conclusion

In this paper, the author briefly discussed the concept of bag structure, and observed the effect of knowledge refinement on bag structures. The conceptual model of an information system has been considered for studying this effect. Consequently, different levels of indiscernibility relations are defined and their significance are discussed. This paper is a short note on the preliminary study done by the author and more work on this area is in progress.

Acknowledgment. The author is thankful to the referees for their very valuable comments which helped in the modification of this paper.

References

1. Blizard, W.D.: Multiset Theory. Notre Dame Journal of Formal Logic 30, 36–66 (1989)
2. Chakrabarty, K., Despi, I.: n^k-bags. International Journal of Intelligent Systems 22(2), 223–236 (2007)
3. Chakrabarty, K., Biswas, R., Nanda, S.: On Yager's theory of Bags and Fuzzy bags. Computers and Artificial Intelligence 18(1), 1–17 (1999)
4. Chakrabarty, K.: Notion of Fuzzy IC-Bags. International Journal of Uncertainty, Fuzziness, and Knowledge-Based Systems 12(3), 327–345 (2004)
5. Chakrabarty, K.: Decision Analysis using IC-Bags. International Journal of Information Technology and Decision Making 3(1), 101–108 (2004)
6. Chakrabarty, K.: Bags with Interval Counts. Foundations of Computing and Decision Sciences 25(1), 23–36 (2000)
7. Chakrabarty, K.: On Bags and Fuzzy Bags. In: John, R., Birkenhead, R. (eds.) Advances in Soft Computing, Soft Computing Techniques and Applications, pp. 201–212. Physica-Verlag, Heidelberg (2000)
8. Date, C.J.: An Introduction to Database Systems. Addison-Wesley, San Francisco (2004)
9. Yager, R.R.: On the Theory of Bags. International Journal of General Systems 13, 23–37 (1986)

A Belief Structure for Reasoning about Knowledge

S.K.M. Wong[1] and Nasser Noroozi[2]

[1] Computer Science Department
University of Regina, Regina, SK, Canada
skmwong@rogers.com
[2] Computer Science Department
Lakehead University, ON, Canada
nnoroozi@lakeheadu.ca

Abstract. A logic-based belief structure is proposed for knowledge representation and reasoning. This structure is semantically different from the standard Kripke structure. It is demonstrated that such a representation of knowledge is particularly useful in a multi-agent environment.

The proposed model is also suitable for dealing with inconsistent and incomplete information and provides a natural measure of uncertainty for the knowledge modal operators.

1 Introduction

The Kripke structure in modal logic has been used in many reasoning systems [1,2]. The relationships between rough-sets [3] and modal logic were explored by many researchers [4,5]. In these studies, knowledge is represented by a relation on a set of possible worlds. Each possible world is described by a set of primitive propositions (attributes). The knowledge relation can be inferred from the properties of the primitive propositions (attribute values). In this case, the resulting relation would inevitably be an equivalence relation. We believe that such an approach may be too restrictive in some applications. Instead we suggest a *belief structure* in which the knowledge of an agent is represented by a family of subsets of possible worlds, not necessarily a partition. Moreover, in our model the semantics of the knowledge modal operators are different from those of the standard Kripke structure. In many reasoning systems, the emphasis is on a single "expert". Here we consider the situations involving the knowledge of a group of agents rather than that of a single agent. In a multi-agent environment, one needs to introduce additional modal operators representing the common and distributed knowledge. Common knowledge is a set of statements that every agent knows, and every agent knows that every agent knows, ... and so on. It is also useful to consider the knowledge distributed among the members of a group. That is, by pooling the knowledge of a group of agents, the group then knows a statement, although no member of the group individually knows it.

J. Yu et al. (Eds.): RSKT 2010, LNAI 6401, pp. 288–297, 2010.

In the real world, knowledge is often inconsistent and incomplete. Hence, it may be necessary to deal with uncertainty inherent in our knowledge. We show in our model that belief functions [6] provide an appropriate measure of uncertainty for the modal operators. Our results therefore support the claim that belief functions are indeed useful in their own right.

In Section 2, we introduce the key concepts of the proposed belief structure. We describe the event-based approach in Section 3, which is in fact equivalent to the logic-based approach. In Section 4, belief functions are introduced as a natural measure of probabilistic uncertainty of knowledge. The conclusion is presented in Section 5.

2   A Belief Structure

We first describe a language for the *belief structure*. We have a set Φ of *primitive propositions*, typically labeled $p, p', q, q', \dots$. These primitive propositions stand for basic facts about the individual *possible worlds* (states) in a universe S. Suppose there are n agents, named $1, 2, \dots, n$. We augment the language by the *knowledge (modal) operators* $K_1, K_2, \dots, K_n$. A statement like $K_i\varphi$ is then read "agent i knows φ". In addition, we introduce modal operators to describe *common knowledge* and *distributed knowledge* for a group $G \subseteq \{1, 2, \dots, n\}$ of agents. A statement like $E_G\varphi$ is read "every agent in group G knows φ". $C_G\varphi$ represents "φ is common knowledge among the agents in G" and $D_G\varphi$ denotes "φ is distributed knowledge among the agents in G".

Technically, a language is just a set of formulas. We start with the primitive propositions in Φ and form more complicated formulas by closing off under negation ($\neg$), conjunction ($\wedge$), and the modal operators $K_1, K_2, \dots, K_n, E_G^k, C_G$, and D_G. The operator $E_G^k\varphi$ is defined inductively based on the operator E_G as $E_G^1\varphi = E_G\varphi$ and $E_G^k\varphi = E_G(E_G^{k-1}\varphi)$ for $k > 1$. Thus, if φ and ψ are formulas, then so are $\neg\varphi, (\varphi \wedge \psi), K_i\varphi, E_G^k\varphi, C_G\varphi$, and $D_G\varphi$. We may omit the parentheses in formulas such as $(\varphi \wedge \psi)$, whenever it does not lead to confusion. We also use the standard abbreviations from propositional logic, such as $\varphi \vee \psi$ for $\neg(\neg\varphi \wedge \neg\psi), \varphi \rightarrow \psi$ for $\neg\varphi \vee \psi$, and $\varphi \leftrightarrow \psi$ for $(\varphi \rightarrow \psi) \wedge (\psi \rightarrow \varphi)$. We take "*true*" to be an abbreviation for some propositional tautology such as $p \vee \neg p$, and "*false*" to be an abbreviation for $\neg true$.

We have just described the syntax of our language (consisting the set of all well formed formulas). Next we need to define the semantics which will enable us to determine if a formula is *true* or *false*.

Traditionally, the knowledge of an agent is represented by a relation on a universe S of possible worlds. The Kripke [7] and Aumann [8,1] structures are typical examples of this approach. In particular, if the knowledge relation is *inferred* directly form the properties (attributes) of the individual possible worlds (e.g., in the theory of rough-sets [3]), then the resulting relation inevitably becomes an equivalence relation.

Here in our approach we assume that the *knowledge* of agent $i \in \{1, 2, \ldots, n\}$ is represented by a family F^i of subsets of S, namely,

$$F^i = \{f^i\} = \{f^i_1, f^i_2, \ldots, f^i_l\}, \text{ where } f^i \subseteq S.$$

Note that in general the f^i's are not necessarily disjoint. We call F^i the *focal set* of agent i and the f^i's are referred to as the *focal elements* of F^i. For simplicity, we may assume:

$$\bigcup_{1 \leq i \leq n} F^i = S.$$

The *belief structure* $\mathcal{B}$ for n agents over the set of primitive propositions $\Phi = \{p, p', q, q', \ldots\}$ is a tuple:

$$\mathcal{B} = (S, \pi, F^1, F^2, \ldots, F^n),$$

where S is a set of possible worlds, $\pi(s)$ is a truth assignment to the primitive propositions for each $s \in S$, that is,

$$\pi(s) : \Phi \to \{true, false\},$$

and F^i is the focal set representing the knowledge of agent i.

The truth assignment $\pi(s)(p)$ tells us whether the primitive proposition p is true or false in state s. We now define what it means for a formula φ to be true at a given world in belief structure $\mathcal{B}$. It is possible that a formula is true in one world and false in another. To capture this, we introduce the notion,

$$(\mathcal{B}, s) \models \varphi,$$

which can be read as "φ is true at $(\mathcal{B}, s)$" or "φ holds at $(\mathcal{B}, s)$"or "$(\mathcal{B}, s)$ satisfies φ". The relation $\models$ is defined by induction. The π component of the structure $\mathcal{B}$ gives us the information we need to deal with the base case, that is, when $p \in \Phi$ is a primitive proposition we have:

$$(\mathcal{B}, s) \models p \text{ iff } \pi(s)(p) = true.$$

We then work our way up to a more complicated formula φ, assuming $\models$ has been defined for all the subformulas of φ. For the conjunction and negation operators, we follow the standard treatment from propositional logic. A conjunction $\varphi \wedge \psi$ is true exactly if both conjuncts φ and ψ are true, while a negated formula $\neg\varphi$ is true exactly if φ is not true. That is,

$$(\mathcal{B}, s) \models \varphi \wedge \psi \text{ iff } (\mathcal{B}, s) \models \varphi \text{ and } (\mathcal{B}, s) \models \psi, \tag{1}$$

$$(\mathcal{B}, s) \models \neg\varphi \text{ iff } (\mathcal{B}, s) \not\models \varphi. \tag{2}$$

Next we define the meaning of the formula $K_i\varphi$. Here we try to capture the intuition that "agent i knows φ" in world s exactly if φ is true at all the possible worlds of some focal element $f \in F^i$ containing s. Thus, the semantics for the knowledge operator in the belief structure is formally defined as:

$$(\mathcal{B}, s) \models K_i \varphi \text{ iff } \exists f \in F^i \text{ such that } s \in f \text{ and } f \subseteq \{t \mid (\mathcal{B}, t) \models \varphi\}. \quad (3)$$

The formula $E_G \varphi$ is true at world $s \in S$ if everyone in the group G of agents knows φ at s, namely,

$$(\mathcal{B}, s) \models E_G \varphi \text{ iff } (\mathcal{B}, s) \models K_i \varphi \text{ for all } i \in G. \quad (4)$$

The formula $C_G \varphi$ is true at world s if everyone in G knows φ at s, everyone in G knows that everyone in G knows φ at s, and etc. Let $E_G^{k+1} \varphi$ be an abbreviation for $E_G(E_G^k \varphi)$. In particular, $E_G^1 \varphi$ denotes $E_G \varphi$. Then, the semantics for $C_G \varphi$ is defined as:

$$(\mathcal{B}, s) \models C_G \varphi \text{ iff } (\mathcal{B}, s) \models E_G^k \text{ for } k = 1, 2, 3, \dots .$$

This definition of common knowledge has a useful graph-theoretical interpretation, which will facilitate the construction of the focal set for C_G as being described in the following discussion.

We construct a *graph* as follows. The possible worlds are the nodes of the graph. Consider a group G of agents. There is an undirected edge (s, s') between two nodes s_1 and s_2, if both of those nodes belong to some focal element $f \in F^i$ and $i \in G$. A node t is said to be G-reachable from node s in k steps ($k \geq 1$), if there exists a sequence of edges $(s_0, s_1), (s_1, s_2), \dots (s_{k-1}, s_k)$ such that $s_0 = s$ and $s_k = t$.

We say t is G-reachable from s if t is G-reachable from s in k steps for some $k \geq 1$. That is, t is G-reachable form s exactly if there is a sequence of edges from s to t. When G is the set of all agents, we say simply that t is reachable from s. In fact, t is reachable from s exactly if s and t are in the same *connected component* of the graph.

Lemma 1

(a) $(\mathcal{B}, s) \models E_G^k \varphi$ *iff* $(\mathcal{B}, t) \models \varphi$ *for all t's that are G-reachable from s in k steps.*

(b) $(\mathcal{B}, s) \models C_G \varphi$ *iff* $(\mathcal{B}, t) \models \varphi$ *for all t's that are G-reachable from s.*

Proof. We shall prove part (a) by induction, while part (b) is an immediate consequence of part (a). Assume that part (a) is true for $k - 1$, which means, $(\mathcal{B}, u) \models E_G^k$ iff $(\mathcal{B}, t) \models \varphi$ for all t's that are G-reachable from u in $k-1$ steps. From definition 4,

$$(\mathcal{B}, s) \models E_G^k \varphi = E_G(E_G^{k-1} \varphi) \text{ iff } (\mathcal{B}, s) \models K_i(E_G^k \varphi) \text{ for all } i \in G.$$

Also from definition 3, we have:

$$(\mathcal{B}, s) \models K_i(E_G^k \varphi) \text{ iff } \exists f \in F^i \text{ such that } s \in f \text{ and } f \subseteq \{u \mid (\mathcal{B}, u) \models E_G^{k-1} \varphi\}.$$

This means that $(\mathcal{B}, s) \models K_i(E_G^k \varphi)$ for all the u's that are G-reachable from s in 1 step and $(\mathcal{B}, u) \models E_G^{k-1}\varphi$. By the inductive assumption, it immediately follows:

$(\mathcal{B}, s) \models E_G^k \varphi$ iff $(\mathcal{B}, t) \models \varphi$ for all t's that are G-reachable from s in k steps.   □

Based on the above lemma, we now show how to construct the focal sets of the knowledge operators E_G^k and C_G. Consider the undirected graphs $H_G^k = (N, \mathcal{E}^k)_G$, $k = 1, 2, \cdots$. Each node in N represents a possible world in S. There is an undirected edge in H_G^k between two nodes s and s', if s is G-reachable from s' in k steps. $\mathcal{E}^k$ is the set containing all such edges. A *clique* of H_G^k is a subset of nodes of N such that every pair of nodes in the subset forms an edge. A *maximal clique* in H_G^k is a clique which is not properly contained by any other clique in that graph.

According to Lemma 1(a), $(\mathcal{B}, s) \models E_G^1 \varphi$ iff $(\mathcal{B}, t) \models \varphi$ for all the t's that are G-reachable from s in 1 step, it follows that each focal element in the individual focal set F^{K_i}, $i \in G$, is a clique in the graph $H_G^1 = (N, \mathcal{E}^1)_G$. Thus, the focal set $F^{E_G^1}$ of the operator E_G^1 is the set of all maximal cliques in the following set:

$$F_G^1 = \bigcup_{i \in G} F^{K_i}.$$

Again by Lemma 1(a), the set of cliques in the graph $H_G^2 = (N, \mathcal{E}^2)$ can be computed from the focal set $F^{E_G^1}$ of the knowledge operator E_G^1. Let

$$F_G^2 = \{f = f_\alpha^{E_G^1} \cup f_\beta^{E_G^1} \mid f_\alpha^{E_G^1} \cap f_\beta^{E_G^1} \neq \emptyset\},$$

where $1 \leq \alpha, \beta \leq m$, and m is the number of focal elements in $F^{E_G^1}$. The focal set $F^{E_G^2}$ is then obtained from F_G^2 by removing all those cliques properly contained by some other cliques in the graph H_G^2.

Likewise, we can construct the focal set $F^{E_G^k}$ from the focal sets $F^{E_G^1}$ and $F^{E_G^{k-1}}$. The maximal cliques in the following set:

$$F_G^k = \{f = f_\alpha^{E_G^1} \cup f_\gamma^{E_G^{k-1}} \mid f_\alpha^{E_G^1} \cap f_\gamma^{E_G^{k-1}} \neq \emptyset\}, \tag{5}$$

where $1 \leq \gamma \leq r$, and r is the number of focal elements in $F^{E_G^{k-1}}$, are the focal elements of the focal set $F^{E_G^k}$.

Lemma 2. *Each maximally connected component of the graph $H_G^1 = (N, \mathcal{E}^1)_G$ is a focal element belonging to the focal set of F^{C_G} of the common knowledge operator C_G.*

Proof. The proof of this lemma follows directly from the above discussion.   □

The semantics of the operator C_G is defined as in equation 3, that is,

$$(\mathcal{B}, s) \models \varphi \text{ iff } \exists f \in F^{C_G} \text{ such that } s \in f \text{ and } f \subseteq \{t \mid (\mathcal{B}, t) \models \varphi\}.$$

A group G of agents has distributed knowledge of φ, if the combined knowledge of the agents implies φ. The combined knowledge is represented by those possible worlds that are contained by at least one focal element in the focal set F^{K_i} of every agent $i \in G$. That is, the focal set F^{D_G} of the distributed knowledge operator D_G is defined by:

$$F^{D_G} = \{f \mid f = (F^{K_\sigma} \cap F^{K_\delta} \cap \ldots \cap F^{K_\lambda}) \neq \emptyset\}, \tag{6}$$

where $G = \{\sigma, \delta, \ldots, \lambda\}$, $1 \leq \sigma, \delta, \ldots, \lambda \leq n$.

Again similar to equation 3, we have:

$$(\mathcal{B}, s) \models \varphi \text{ iff } \exists f \in F^{D_G} \text{ such that } s \in f \text{ and } f \subseteq \{t \mid (\mathcal{B}, t) \models \varphi\}$$

which defines the semantics of D_G.

3 An Event-Based Approach

In the previous section, we have discussed a logic-based language for modeling knowledge and reasoning. This language is based on a set of primitive propositions and is closed under logical operators, whereas knowledge is expressed syntactically by modal operators. Semantically, knowledge is defined by the notion of *focal sets*.

Here, we describe an equivalent language (an *event-based* approach) for reasoning, which is typically used in rough sets, probabilities, belief functions, game theory, etc.

As in the logic-based approach, we start out with a universe S of possible worlds. An *event* e is a set of possible worlds , i.e., $e \subseteq S$. The semantics for the belief structure $\mathcal{B}$ are defined in terms of formulas (see equation 3), whereas we define the semantics of the event-based approach in terms of events.

With a primitive proposition p, we associate the event,

$$e_\mathcal{B}(p) = \{s \mid (\mathcal{B}, s) \models p\}. \tag{7}$$

In general, the *intension* of a formula φ is defined by the set of possible worlds at which φ holds, namely,

$$e_\mathcal{B}(\varphi) = \{s \mid (\mathcal{B}, s) \models \varphi\}. \tag{8}$$

Let $\mathcal{O}$ denote a knowledge operator in the event-based approach and let O be its corresponding modal operator. Recall that such a modal operator in the belief structure $\mathcal{B}$ is characterized by a focal set F^O,

$$F^O = \{f^O\}.$$

We define a knowledge operator $\mathcal{O}$ as a function on events as follows:

$$\mathcal{O}(e_\mathcal{B}(\varphi)) = \bigcup_{f^O \subseteq e_\mathcal{B}(\varphi)} f^O,$$

where f^O (a subset of possible worlds) is a focal element in the focal set F^O. According to definitions 3 and 8, we can immediately conclude that

$$\mathcal{O}(e_{\mathcal{B}}(\varphi)) \equiv \bigcup_{f^O \subseteq e_{\mathcal{B}}(\varphi)} f^O = \{s \mid (\mathcal{B}, s) \models O\varphi\} \equiv e_{\mathcal{B}}(O\varphi).$$

The mapping defined by equations 7 and 8, namely,

$$e_{\mathcal{B}} : \{\varphi\} \longrightarrow 2^S,$$

satisfies the following properties:

$$e_{\mathcal{B}}(\varphi_1 \wedge \varphi_2) = e_{\mathcal{B}}(\varphi_1) \cap e_{\mathcal{B}}(\varphi_2),$$
$$e_{\mathcal{B}}(\neg\varphi) = \overline{e_{\mathcal{B}}(\varphi)},$$
$$e_{\mathcal{B}}(O\varphi) = \mathcal{O}(e_{\mathcal{B}}(\varphi)).$$

This means that the mapping $e_{\mathcal{B}}$ is an *isomorphism* between the logic-based structure $(\Psi, \wedge, \neg, O_1, O_2, \ldots, O_q)$ and $(2^S, \cap, -, \mathcal{O}_1, \mathcal{O}_2, \ldots, \mathcal{O}_q)$, the event-based structure.

It should be noted that if the focal set F^{K_i} of the individual agent is a partition on S (i.e., the focal elements in each of the F^{K_i}'s are pairwise disjoint), then the belief structure $\mathcal{B} = (S, \pi, F^{K_1}, F^{K_2}, \ldots, F^{K_n})$ is equivalent to a standard Kripke structure [7]. In this special case, for an event $e_{\mathcal{B}}(\varphi)$, the following subset of possible worlds:

$$\mathcal{K}_i(e_{\mathcal{B}}(\varphi)) = \bigcup_{f^{K_i} \subseteq e_{\mathcal{B}}(\varphi)} f^{K_i}$$

is in fact the well known *lower bound* (approximation) of $e_{\mathcal{B}}(\varphi)$ in the rough-sets model [3]. This observation indicates that the proposed belief structure has a wider application, particularly in the area of reasoning about knowledge. More importantly, perhaps, both the common and the distributed knowledge operators play a crucial role for reasoning in a multi-agent environment.

4   A Measure for Uncertainty

Consider a modal operator O in the belief structure $\mathcal{B}$ and assume that a *probability function* μ on S is given. We can define a *numeric measure* of uncertainty as:

$$B^O(A) = \sum_{f \in F^O, \ f \subseteq A \subseteq S} m^O(f), \qquad (9)$$

where F^O is the focal set of O, and for $f \in F^O$,

$$m^O(f) = \frac{\mu(f)}{\displaystyle\sum_{f' \in F^O} \mu(f')}. \qquad (10)$$

Note that $m^O(f)$ is referred to as a *basic probability assignment* and $B^O(A)$ is called a *belief function* [6].

Thus, for every knowledge operator O, we can associate a belief function B^O. For an event $e_B(\varphi)$, $B^O(e_B(\varphi))$ can be interpreted as the *degree of belief* that the truth lies in $e_B(\varphi)$. For example:

(a) $B^{K_i}(e_B(\varphi))$ is the degree of belief that "agent i knows $e_B(\varphi)$".
(b) $B^{E_G^1}(e_B(\varphi))$ is the degree of belief that "every agent in the group G knows $e_B(\varphi)$".
(c) $B^{C_G}(e_B(\varphi))$ is the degree of belief that "$e_B(\varphi)$ is the common knowledge among the agents in the group G".
(d) $B^{D_G}(e_B(\varphi))$ is the degree of belief that "$e_B(\varphi)$ is the distributed knowledge among the agents in the group G".

It is interesting to note that the computation of the distributed belief function is closely related to the *rule of combination* of belief functions [9,6]. According to equation 6 , the focal set F^{D_G} of the distributed knowledge operator D_G is defined by:

$$F^{D_G} = \{f = f_\alpha^{K_\sigma} \cap f_\beta^{K_\delta} \cap \ldots \cap f_\gamma^{K_\lambda} \neq \emptyset\}, \tag{11}$$

where $1 \leq \alpha \leq k_\sigma$, $1 \leq \beta \leq k_\delta$, $\ldots$, $1 \leq \gamma \leq k_\lambda$, and $k_\sigma, k_\delta, \ldots, k_\lambda$ are the numbers of focal elements in the focal sets $F^{K_\sigma}, F^{K_\delta}, \ldots, F^{K_\lambda}$, respectively. From equations 9,10, and 11, we can compute the belief function of the distributed knowledge operator:

$$B^{D_G}(A) = \sum_{f \in F^D, \, f \subseteq A \subseteq S} m^{D_G}(f),$$

where for $f \in F^{D_G}$,

$$m^{D_G}(f) = \frac{\mu(f)}{\displaystyle\sum_{f' \in F^{D_G}} \mu(f')}. \tag{12}$$

However if the probability function μ is not known, we can compute the distributed belief function from the belief functions of the individual agents in the group G. Let us first construct the following basic probability assignment for $f = (f_\alpha^{K_\sigma} \cap f_\beta^{K_\delta} \cap \ldots \cap f_\gamma^{K_\lambda}) \in F^{D_G}$,

$$m(f) = \frac{m^{K_\sigma}(f_\alpha^{K_\sigma}).m^{K_\delta}(f_\beta^{K_\delta}). \; \ldots \; .m^{K_\lambda}(f_\gamma^{K_\lambda})}{\displaystyle\sum_{(f_{\alpha'}^{K_\sigma} \cap f_{\beta'}^{K_\delta} \cap \ldots \cap f_{\gamma'}^{K_\lambda}) \in F^{D_G}} m^{K_\sigma}(f_{\alpha'}^{K_\sigma}).m^{K_\delta}(f_{\beta'}^{K_\delta}). \; \ldots \; .m^{K_\lambda}(f_{\gamma'}^{K_\lambda})}. \tag{13}$$

We call the belief function $\tilde{B}^{D_G}$ defined by:

$$\tilde{B}^{D_G}(A) = \sum_{f \in F^{D_G}, f \subseteq A \subseteq S} m(f) \ , \tag{14}$$

the *orthogonal sum* of $B^{K_1}, B^{K_2}, \ldots, B^{K_l}$, namely,

$$\tilde{B}^{D_G} = B^{K_\sigma} \oplus B^{K_\delta} \oplus \ldots \oplus B^{K_\lambda}.$$

We refer to the above method for computing the belief function $\tilde{B}^{D_G}$ as the Dempster's rule of combination [9,6]. Clearly, B^{D_G} is not necessarily the same as $\tilde{B}^{D_G}$.

There have been a lot of controversies about belief functions since it was first introduced by Shafer [6]. In particular, many researchers often questioned the usefulness of such an uncertainty measure (versus the well accepted probability function). Our results indicate that belief functions arise in a natural way as a measure of uncertainty for the knowledge modal operators. The necessity for introducing the combination rule in equations 13 and 14 to compute the distributed knowledge is primarily due to the lack of sufficient information.

5   Conclusion

In this paper we argued that using only equivalence relations to represent knowledge is rather restrictive in some situations, although such assumptions were successfully used in a variety of applications [1,3].

We showed that the proposed belief structure provides a more general framework than the standard Kripke structure for reasoning about knowledge in a multi-agent environment.

In our knowledge model, the notion of belief functions arises in a natural manner. Our exposition has clearly demonstrated the nature and usefulness of such a measure of uncertainty and its difference from the conventional probability function. Moreover, the Dempster's rule of combination is closely connected with the computation of distributed knowledge among a group of agents.

References

1. Fagin, R., Halpern, J., Moses, Y., Vardi, M.: Reasoning About Knowledge. MIT Press, Cambridge (1996)
2. Halpern, J.Y.: An analysis of First-order logics of probability. Artificial Intelligence (46), 311–350 (1990)
3. Pawlak, Z.: Rough sets. International Journal of Computer and Information Sciences (1982)
4. Orlowska, E.: Logic for reasoning about knowledge. Mathematical Logic Quarterly (1989)
5. Yao, Y., Lin, T.: Generalization of rough sets using modal logic. Intelligent and Automation and Soft Computing, an International Journal 2(2), 103–120 (1996)

6. Shafer, G.: A mathematical Theory of Evidence. Princeton University Press, Princeton (1979)
7. Kripke, S.: A semantical analysis of modal logic I: Normal modal propositional calculi. Zeitschriftfür Mathematische Logik und Grundlagen der Mathematik (1963)
8. Aumann, R.: Agreeing to disagree. Annals of Statistics (1976)
9. Dempster, A.: Construction and local computational aspects of network belief functions. Technical Report S-125, Department of Statistics. Harvard University, Cambridge, MA (1988)

Research on Mapping Mechanism of Learning Expression

Lili Zhou* and Fanzhang Li

School of Computer Science and Technology
Soochow University, Suzhou 215006, P.R. China
lfzh@suda.edu.com

Abstract. In this paper, we propose a description of Machine Learning System (MLS) which is based on the category theory in order to study mapping mechanism of the learning expression. And a detailed instance is provided. This is that Decision Tree Learning (DTL) can be denoted based on category theory. In addition, we prove that the machine learning system is indeed a category, and propose the learning expression and the mapping mechanism of the learning expression based on category theory. Also we describe the proof and verify the mechanism.

Keywords: Category, Category of Machine Learning, Learning Expression, Mapping Mechanism.

1 Introduction

The machine learning system composes of four modules which are performance system, critic, generalizer and experiment generator. Each module which can be expressed and operated independently has its own input and output that link these independent modules together into a loop system [1].

Thus it can be seen that regardless of experience training examples or target function, the representation is always a vital concept in the process of building a machine learning system. So far, there are many methods about representing, such as logical representation [7], generative representation [3], state space representation [3], semantic network representation [8], algebraic representation [9] and so on. In this paper we study the presentation of machine learning system based on category theory. It has two advantages that we use category theory to describe the representation of machine learning. First, the presentation of machine learning can be converted into mathematical problems, which can be concluded as some special mathematical objects and the research on the mapping between them. This belongs to the category theory. Second, in the field of machine learning, people have been finding the algebraic and geometry module of machine learning for a long time, and the category is a mathematical theory which uses the algebra geometry tools.

---

* This work was supported by the National Science Foundation of China under Grant NO.60775045;Tel:13962116494.

J. Yu et al. (Eds.): RSKT 2010, LNAI 6401, pp. 298–303, 2010.

2 The Representation of Machine Learning Based on Category Theory

Definition 2.1. Let $\mathcal{C} = (\mathrm{ob}\mathcal{C}, \mathrm{Mor}\mathcal{C})$ be the machine learning system, which consists of two parts as follows: (1)An object $\mathrm{ob}\mathcal{C}$, which members are called $\mathcal{C}$−objects; (2)For each pair (A, B) of $\mathrm{ob}\mathcal{C}$, a set $\mathcal{C}(A, B) \in \mathrm{Mor}\mathcal{C}$, which members are called $\mathcal{C} -$ morphisms from A to B; when $A \neq A'$ or $B \neq B'$, $\mathcal{C}(A, B)$ and $\mathcal{C}(A', B')$ is disjoin.

Note: (1) These objects correspond with the input/output set of the learning problem; (2) All of these objects compose a family of objects; (3) Morphism is an abstract process which keeps the structure between two objects. So we call $\mathcal{C}$ is a category.

Example 2.1. Next we will use category theory to describe the learning problem mentioned above. We all know that the solution set, the training sample set, the objective function set and the new problem set are all included by $\mathrm{ob}\mathcal{C}$. The functions of these four modules correspond to the morphisms between each set. Assume the solution set as object A, the training sample set as B, the objective function set as C, the new problem set as D, performance system as morphism f, critic as morphism g, generalizer as morphism h and experiment generator as morphism k, and we could get a conclusion that $g \in \mathcal{C}(A, B)$, $h \in \mathcal{C}(B, C)$, $k \in \mathcal{C}(C, D)$ and $f \in \mathcal{C}(D, A)$.

Based on theorem 2.1, there are several computation rules of the machine learning category $\mathcal{C}$.

(a) Composition low; i.e., for $A, B, C \in \mathrm{ob}\mathcal{C}$, $f \in \mathcal{C}(A, B)$, $g \in \mathcal{C}(B, C)$, then there is only a $gf \in \mathcal{C}(A, C)$ called the composite of f and g.

(b) Associativity; i.e., for $A, B, C, D \in \mathrm{ob}\mathcal{C}$, $f \in \mathcal{C}(A, B)$, $g \in \mathcal{C}(B, C)$, $h \in \mathcal{C}(C, D)$, the equation $h(gf) = (hg)f$ holds.

(c) Identity morphism; i.e., for each $1_A \in \mathcal{C}(A, A)$ and $f \in \mathcal{C}(A, B)$, $f1_A = f$ and $1_A g = g$.

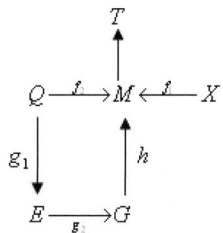

Fig. 1. The step of classifying data based on category theory. Let a morphism f_1 of $\mathcal{C} -$ morphism decide whether the leaf nodes are the same species.If the attribute of leaf nodes is null, we use a morphism f_2. we define a morphism g_1 to compute the entropy object E and g_2 to compute the information gain object G. Then choose the maximum of object G by defining a morphism h, finally update the decision tree T.

Example 2.2. Prove the feasibility of machine learning category by using the decision tree as an example.

ID3 learning algorithm is a learning category $\mathcal{C}$, and the morphism between data set object and sample set object is obvious. So the main algorithm will only be discussed. Let the object X be the whole sample set which consist of instances that is represented by attribute-value pair. The whole attribute set according to sample set is assumed as the object Q of learning category $\mathcal{C}$. Finally the decision tree is built as the object T of learning category $\mathcal{C}$.Consequently, the category representation of decision tree algorithm can be described by Fig. 1.

3   The Mapping between Learning Expressions

3.1   The Representation of Learning Expression Based on Category Theory

We could get the expression of learning problem from the standard description of learning problem gradually. A computer program is said to learn from experience E with respect to some class of tasks T and performance measure P , if its performance at tasks in T, as measured by P, improves with experience E[1]. The main goal of learning in this step is to discover an operation description of the perfect target function. Moreover, we could make it by using the learning expression.

Assume object A as a initial state set of training samples, object B as a intermediate and object as the final output state, so the learning process could be represented as follows:

$$f : A \mapsto B, g : B \mapsto C \tag{1}$$

f and g are $\mathcal{C}$ − morphism. Using composition law and associability of category theory, we could get this:

$$gf : A \mapsto C \tag{2}$$

In order to discuss the category representation of learning expression, let's take the learning progress of linear expression for example.

$$f : W \times X \mapsto B \tag{3}$$

$$g : B \mapsto M \tag{4}$$

$$h : B \times M \mapsto W \tag{5}$$

Note: we could get (4) and (5) through (1). Connect initial weight object W with initial state object X by using f , and we will get intermediate state object B ,from which the final output object M could be got by using the morphism g . In addition, through (4) and (5) we could get (6) which is to recalculate weight by using minimum error and update the weight object.

3.2   The Mapping between Expressions

From the previous section, we know that we can use learning expression category to describe learning process. Thus, the homomorphism relation between the different learning processes exists and forms a mapping mechanism between expressions. In other words, there is a mapping $\phi : X \rightarrow Y$ s.t.$\phi(u \cdot v) = \phi(u) \cdot \phi(v)$.

Definition 3.1. If $\mathcal{C}$ and $\mathcal{D}$ are learning expression category, then a learning expression function F from $\mathcal{C}$ to $\mathcal{D}$ is a function that assigns to each $\mathcal{C}$−object A a D - object $F(A)$,and to each $\mathcal{C}$ − morphism $f : A \rightarrow A'$ a $\mathcal{D}$ − morphism $F(f) : F(A) \rightarrow F(A')$, in such a way that
(1) F preserves composition; i.e., $F(f \circ g) = F(f) \circ F(g)$ whenever $f \circ g$ is defined;
(2) F preserves identity morphism; i.e., $F(1_A) = 1_{F(A)}$ for each $\mathcal{C}$ − object A.

Example 3.1. A mapping mechanism between learning expression categories.
Proof: Assume that $\mathcal{C}$ and $\mathcal{D}$ are learning expression category. $W, X, B, M \in$ ob$\mathcal{C}$,$W', X', B', M' \in$ ob$\mathcal{C}$,$f, g, h \in$ Mor$\mathcal{C}$,$f', g', h' \in$ Mor$\mathcal{C}$ (see details in 3.1).

The objects of learning expression category are dominated by the training sample set, so the following conclusion is obvious.

$$(W, X, B, M) \mapsto F(W', X', B', M')$$

If the operator between morphisms is , then we can define this:

$$F(f \bullet f') = F(f) \bullet F(f'), F : X \rightarrow Y$$

Therefore, the morphism of learning expression functor is equivalent to a homomorphism mapping.

Using the mapping mechanism of learning expression category, we could expand the classification condition about decision tree learning, allowing some attribute values of sample instances are absent. We will use priori probability and posterior probability of Bayesian theory and maximum hypothesis to sort the instance space, but training sample instances should have all data.

Firstly, we should calculate the information entropy and the information gain repeatedly to build a decision tree by using the classical decision-tree learning algorithm. Finally we can start classifying the data.

The classification algorithm includes two steps. First, build a decision tree. Second, classify data by using the decision tree. The details are as following: (1) Build a decision tree. You could find the specific building method in example 2.2. (2) Classify data by using the decision tree.

We should define the objects in the beginning. The object $Q = \theta \cup W$ is the attribute set of the testing samples, where θ is an unknown attribute set, and W is a known attribute set. The probability object P includes the priori probabilities and the posterior probabilities. Object A as the object of the attribute values has the maximum posterior probability. In addition, object R is the result object of classifying data.

We could obtain the posterior probability of the absent attributes by using object Q and object P, and we can build a morphism p where $p = P(\theta_i | w) = P(w|\theta_i) \cdot P(\theta_i) \Big/ \sum_j P(w|\theta_{ij}) \cdot P(\theta_{ij})$, and j is values of attributes.

Now we will use the PlayTennis data from the book of Machine Learning as the training data to build a decision tree.

Define the data set as object Q—the attribute set, and use a morphism g_1 that is to calculate entropy.

$$E \supset \begin{cases} E(S_{sunny}|out) = 0.971, E(S_{overcast}|out) = 0, E(S_{rain}|out) = 0.971, \\ E(out) = 0.693, E(S_{cool}|temp) = 0.811, E(S_{mild}|temp) = 0.918, \\ E(S_{hot}|temp) = 1.00, E(temp) = 0.911, E(S_{high}|humidity) = 0.985, \\ E(S_{normal}|humidity) = 0.591, E(humidity) = 0.789 \end{cases}$$

Use a morphism g_2 that is to calculate information gain, and we will obtain the elements of object $G \supset \begin{cases} G(S, out) = 0.246, G(S, temp) = 0.029, \\ G(S, hum) = 0.151, G(S, wind) = 0.048 \end{cases}$. Then we will obtain the root nodes $M \supset \{out\,(sunny, overcast, rain)\}$.

Now reselect three attributes of object Q. And there are three branches: outlook (sunny), outlook(overcast), outlook(rain). We could get other branch nodes and leaf nodes through repeating the process, and obtain the final decision tree in Fig. 2. Modify the training data at random in order to make some values of attributes absent. Take the new data as the testing set, and the eighth data for example. The value of Outlook is absent, and we will judge its classification.

Define an object $Q \supset \{\emptyset, Mild, High, Weak, ?\}$ as the attributes of the testing example data. We could obtain that:

$$P \supset \begin{cases} p(Sunny) = 1/3, p(Overcast) = 1/3, p(Rain) = 1/3, \\ p(N|Sunny) = 0.60, p(N|Overcast) = 0, p(N|Rain) = 0.40, \\ p(Y|Sunny) = 0.40, p(Y|Overcast) = 1.0, p(Y|Rain) = 0.60 \end{cases}$$

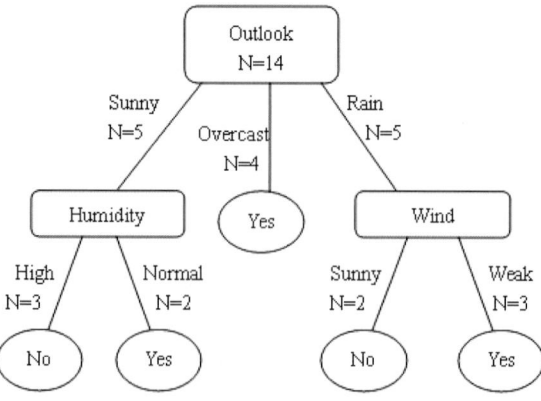

Fig. 2. The decision tree got by using data set. The leaf nods classify.

Calculate the posterior probabilities of the attribute values by using the morphism p, and update the probability object P,

$$P \supset \left\{ \begin{array}{l} p(Sunny\,|N\,) = 0.60, p(Sunny\,|Y\,) = 0.20, p(Overcast\,|N\,) = 0, \\ p(Overcast\,|Y\,) = 1.0, p(Rain\,|N\,) = 0.40, p(Rain\,|Y\,) = 0.30 \end{array} \right\}$$

And we will get the object $A \supset \{p(Sunny\,|N\,) = 0.60, outlook(Sunny)\}$ by using morphism f and object P. Morphism k updates the object Q, and the new data are as follows: $Q \supset \{Sunny, Mild, High, Weak, ?\}$. Finally, we can classify the data by using the decision-tree T.

The accuracy of this classification method is 85.714%. Although it is unsatisfactory, it could solve the classification problem that some values are absent.

4   Conclusion

In this paper, we find that we could use category theory to study the learning problem of the machine learning through studying the machine learning system and its modules and build a theory framework of the machine-learning category. In order to validate that the machine-learning category can represent the machine learning algorithms, we use one instance, decision tree learning. The aim of using category theory to describe the learning problem is to make the specific issues abstract and obtain a uniform form. We can analyze the specific problem by using the abstract form, which provides a theory for improving and optimizing algorithms in the future.

References

1. Mitchell, T.M.: Machine Learning. McGraw-Hill Science/Engineering/Math (1997)
2. Adamek, J., Herelich, H., Strecker, G.E.: Abstract and Concrete Categories. John Wiley and Sons, Inc., Chichester (1990)
3. Zixing, C., Guangyou, X.: Artificial Intelligence: Principles and Application. Tsinghua University press, Beijing (2004) (in chinese)
4. Fanzhang, L., Yu, K.: The study of machine learning theory frame based on Lie group. J. Journal of Yunnan Natioalities University 13, 251–255 (2004)
5. Zhigang, N., Shi, L., Lanfang, M., ShangPing, L.: Study and Application of Knowledge Expression. J. Application Research of Computers 24, 234–235 (2007)
6. Zhao, H.-W., Zhang, H.-l., Zang, X.-b.: Method of Semantic Nets Based on Mental Imagery. J. Jounrnal Of Beuing University Of Techology 35 (2009)
7. Xiaodong, L., Yuqing, G.: Algebra Operations and Optimization Techniques for XML Query. J. Computer Science 29, 57–63 (2002)
8. Yang, X.-d., He, N., Wu, L.-b.: Ontology Integration Description Based on Category Theory. J. Computer Engineering 35, 76–78 (2009)
9. Huan, X., Fanzhang, L.: Study on Lie Group Machine Learning. J. Journal of Computational Information System 1, 843–849 (2005)

Linking Open Spatiotemporal Data in the Data Clouds

He Hu[1,2] and Xiaoyong Du[1,2]

[1] School of Information, Renmin University of China, Beijing 100872, China
[2] Key Laboratory of Data Engineering and Knowledge Engineering, MOE, Beijing 100872, China
{hehu,duyong}@ruc.edu.cn

Abstract. The Linked Data Clouds are the best expression and realization of the Semantic Web vision to date. The Data Clouds can significantly benefit both the AI and Semantic Web communities by enabling new classes of tasks and enhancing reasoning, data mining and knowledge discovery applications. There are various forms of spatiotemporal data in the Linked Data Clouds. To efficiently utilize these spatiotemporal data, schema heterogeneity problem must be resolved. A hierarchical spatiotemporal model is proposed to give a conceptual description of different spatiotemporal dataset of Linked Data Clouds. The model can support schema level mappings and convey relationships between concepts of different datasets at the schema level. The hierarchical spatiotemporal model contains three layers: an meta level for abstract level spacetime knowledge; an schema level for well-known models in spatial and temporal reasoning - Allen's Interval Algebra in temporal reasoning and the RCC model in spatial reasoning; and an instantiations level to provide systematic mapping and formal descriptions of the various ground spatiotemporal statements.

Keywords: Spatiotemporal Data, Linked Data, SPARQL.

1 Introduction

The Linked Data Clouds [1] are one of the fastest growing data resources in the world with billions of RDF triples and millions of RDF links. Various kinds of data which are of interest and use to almost everyone are buried in this vast, quickly growing data collection; Linked Data Clouds can significantly benefit both the AI and Semantic Web communities by enabling new classes of tasks and enhancing reasoning, data mining and knowledge discovery applications. However, there is no formal description of diverse Data Clouds components and the links between them; this hampers seamlessly utilization of the vast number of facts present in the clouds.

There are various forms of spatiotemporal data in the Linked Data Clouds. Lots of spatial information and temporal information exists in DBpedia, GeoNames, and Freebase dataset. To efficiently utilize these spatiotemporal data,

J. Yu et al. (Eds.): RSKT 2010, LNAI 6401, pp. 304–309, 2010.

schema heterogeneity problem must be resolved. We propose a hierarchical spatiotemporal model to give a conceptual description of different spatiotemporal dataset of Linked Data Clouds. The model can support schema level mappings and convey relationships between concepts of different datasets at the schema level. The hierarchical spatiotemporal model contains three levels: an meta level for abstract level space-time knowledge; an schema level for well-known models in spatial and temporal reasoning - Allen's Interval Algebra in temporal reasoning and the RCC model in spatial reasoning; and an instantiations level to provide systematic mapping and formal descriptions of the various ground spatiotemporal statements.

1.1   The Linked Data Clouds

The Linked Data Clouds are the best expression and realization of the Semantic Web vision to date. It contains many large datasets contributed by people belonging to diverse communities such as geography, music, and life sciences.

The data formats used by different Data Clouds components are all RDF triples and these components may be overlapping by RDF links. But each Linked Data system is designed separately and there is no global knowledge on what information is available in diverse data sets; this may severely impact the usage and limit the applications that can be built using the Linked Data Clouds.

In Semantic Web communities, upper level ontologies have been used to integrate heterogeneous knowledge bases. Upper ontologies can been integrated with domain specific ontologies to provide advantages such as better data mining, knowledge discovery and reasoning performance.

1.2   Spatiotemporal Reasoning

Spatial and temporal information constitutes a most elementary part of our everyday life. The representation and reasoning of spatial and temporal knowledge remains an important field in artificial intelligence research, various models have been proposed in the last one or two decades, among them most famous are the two models: Allen's Interval Algebra [2] for temporal reasoning and Randell, Cui, and Cohn 's RCC model [3] for spatial reasoning. These models are fundamental for expressing spatiotemporal information in the Linked Data Clouds. However, there is schema heterogeneity exists in diverse spatiotemporal datasets. We need build an abstract layer to encapsulate various schemas.

Due to the diversity and the multitude of spatiotemporal data sources, the collaborative data usage becomes crucial. A spatiotemporal model for the Linked Data Clouds environment should abstract and support the the generally used spatiotemporal schemas in the Linked Data Clouds. The hierarchical spatiotemporal model should contains three levels: an meta level for abstract level space-time knowledge; an schema level for well-known models in spatial and temporal reasoning - Allen's Interval Algebra in temporal reasoning and the RCC model in spatial reasoning; and an instantiations level to provide systematic mapping and formal descriptions of the various ground spatiotemporal statements existing in the Linked Data Clouds.

1.3   Paper Organization

The paper is structured as follows: in section 2, we give a brief description of related research in spatiotemporal modeling and spatiotemporal ontologies; section 3 analyzes the Linked Data Clouds environment and gives the details of the proposed hierarchical spatiotemporal model; section 4 concludes the paper.

2   Related Work

Spatiotemporal information modeling and spatiotemporal ontologies have been well explored in GIS and Semantic Web studies. In LinkedGeoData project, Auer et. al. [4] studied the transformation and publication of the OpenStreetMap data according to the Linked Data principles to add a new dimension to the Data Web; spatial data can be retrieved and interlinked on an unprecedented level of granularity. This enhancement enables a variety of new Linked Data applications such as geo-data syndication or semantic-spatial searches. They also established mappings with DBpedia as the central interlinking hub in the Web of Data. DBpedia Ontology is a shallow, cross-domain ontology, which has been manually created based on the most commonly used infoboxes within Wikipedia [5]. The ontology currently covers over 205 classes which form a subsumption hierarchy and has 1,210 properties. With the DBpedia 3.2 release, DBpedia currently contains about 1,173,000 instances including 339,000 places, 119,000 organizations and 30,000 buildings. GeoNames is another geospatial data set in the Linked Data Clouds. It covers all countries and contains over eight million place names for more than 6.5 million unique features, describing about 2.2 million populated places and 1.8 million alternate names. Its features have a unique identifier, a name, alternative names (e.g. in different languages), part-of relations to administrative divisions and geo-spatial coordinates. All features are categorized into one out of 9 classes and further subcategorized into one of 645 codes.

In [6], Martins et. al. used standard information extraction techniques to find relevant geo-temporal references in the text. Following the identification of possible geo-temporal references, each of them is disambiguated into the corresponding gazetteer feature(s). The document is finally assigned to an encompassing geotemporal scope, determined with basis on the most general gazetteer features that combine the references in the text. In [7], Henson et al. proposed an ontological representation of time series observations (based on Observation and Measurement in OWL) to allow conversion between event-based and interval-based observations. Henson's work includes the ontological representation of observed properties, features of interests, sampling time, and observation location; however, a geo-process representation is still missing. Bittner et. al. studied spatiotemporal ontology for geographic information integration in [8]; they gave an axiomatic formalization of a theory of top-level relations between three categories of entities: individuals, universals, and collections. They dealt with a variety of relations between entities in these categories, including the sub-universal relation among universals and the parthood relation among individuals, as well as cross-categorical relations such as instantiation and membership.

3   A Hierarchical Spatiotemporal Model

In this section, we propose and discuss a hierarchical spatiotemporal model for the Linked Data Clouds environment. The hierarchical spatiotemporal model should contains three levels: an meta level for abstract level spacetime knowledge; an schema level for well-known models in spatial and temporal reasoning - Allen's Interval Algebra in temporal reasoning and the RCC model in spatial reasoning; and an instantiations level to provide systematic mapping and formal descriptions of the various ground spatiotemporal statements existing in the Linked Data Clouds.

3.1   The Meta Level Model

In [9], Grenon gave a formal definition for spatiotemporal ontology with first order predicate logic. The logic supports a bi-categorical reading and formalization which accounts for both endurants and continuants in time and a certain number of perdurants or occurrents in time. This is a very generous abstract view of spatiotemporal relationships. First order logic can not directly be applied in Semantic Web environment. It must be converted into description logic [10] and encoded in OWL ontology language; this will enable us to reuse known algorithms and results on the tractability of such logic systems. The closest Description Logic known in the literature is $\mathcal{SROIQ}(\mathcal{D})$.

The meta level model provides an abstract level of spatiotemporal knowledge, and plays a fundamental role for spatiotemporal information exchange with schema level and instantiations level entities.

3.2   The Schema Level Model

In temporal reasoning, of particular importance is the system of relations introduced by Allen [2] which describes the possible relationships between two convex intervals. Every interval can be represented using the two end points of the interval. When comparing the end points of two intervals according to the relations of the point algebra, two intervals can be related by the 13 different JEPD base relations: Before, Meets, Overlaps, Starts, During, Ends, converse relations: After, Met-by, Overlapped-by, Started-by, Includes, Ended-by, and the identity relation equal. The full algebra, known as the interval algebra, consists of the 2^{13} possible disjunctions of the base relations.

The Region Connection Calculus (RCC) is a topological approach to spatial representation and reasoning developed by Randell , Cui , and Cohn [3]. In this calculus, spatial regions are non-empty regular subsets of some topological space D. Spatial regions do no have to be internally connected, i.e., they might consist of multiple disconnected pieces. Since all spatial regions are regular subsets of the same topological space $\mathcal{D}$, all spatial regions have the same dimension, namely, the dimension of $\mathcal{D}$.

In this schema level of spatiotemporal model, we advocate two well-known models in spatial and temporal reasoning - Allen's Interval Algebra in temporal reasoning and the RCC model in spatial reasoning; these models can serve as underlying framework for various spatiotemporal data sets in the Linked Data Clouds. we experimented at schema level with linking spatiotemporal data from DBpedia to several other public Linked Data sources. Spatial and temporal data in DBpedia were linked both to their spatiotemporal counterparts and to their relative data in FOAF, OpenCyc, SKOS and Freebase. The spatiotemporal features in these datasets conform to the basic temporal and spatial models described above.

3.3   The Instantiation Level Model

The meta level spatiotemporal model captures various domains at the abstract level, while the schema level spatiotemporal model creates a bridge between the abstraction of the ontology and instantiations available in the Linked Data Clouds. The instantiation level spatiotemporal model gives a set of wrappers for various spatiotemporal datasets. For example, GeoNames, Freebase, DBpedia etc. This will help in providing systematic and formal descriptions and mapping of the various ground statements, the classes to which the instances belong, and for identifying schema level relationships.

3.4   The Overall Model and a Demo

The overall spatiotemporal model is illustrated in a demo system. For the sake of space, this demo system was not fully addressed in the paper. For more detailed information, or information on the latest version of the demonstration system, please visit $http://spatiotemporalweb.appspot.com$.

4   Conclusion

Research on spatiotemporal semantics plays an increasing role to support complex queries and reasoning tasks across heterogeneous data sources. We advocate a hierarchical spatiotemporal model for discovering and processing spatiotemporal information. The model contains a meta layer for abstract level spacetime knowledge, and schema level which supports Interval Algebra in temporal reasoning and the RCC model in spatial reasoning; and an instantiations level to provide systematic mapping and formal descriptions of the various ground spatiotemporal statements in the Linked Data Clouds.

Acknowledgments. The research is supported by NSFC with project number: 60773215.

References

1. Bizer, C., Heath, T., Ayers, D., Raimond, Y.: Interlinking Open Data on the Web. In Demonstrations Track. In: 4th European Semantic Web Conference, Innsbruck, Austria (2007)
2. Allen, J.F.: Maintaining Knowledge about Temporal Intervals. Communications of the ACM 26(11), 832–843 (1983)

3. Randell, D.A., Cui, Z., Cohn, A.G.: A Spatial Logic Based on Regions and Connection. In: Principles of Knowledge Representation and Reasoning: Proceedings of the 3rd International Conference, pp. 165–176 (1992)

4. Auer, S., Lehmann, J., Hellmann, S.: LinkedGeoData - adding a spatial dimension to the web of data. In: Bernstein, A., Karger, D.R., Heath, T., Feigenbaum, L., Maynard, D., Motta, E., Thirunarayan, K. (eds.) ISWC 2009. LNCS, vol. 5823, pp. 731–746. Springer, Heidelberg (2009)

5. Auer, S., Bizer, C., Kobilarov, G., Lehmann, J., Ives, Z.: Dbpedia: A nucleus for a web of open data. In: Aberer, K., Choi, K.-S., Noy, N., Allemang, D., Lee, K.-I., Nixon, L.J.B., Golbeck, J., Mika, P., Maynard, D., Mizoguchi, R., Schreiber, G., Cudré-Mauroux, P. (eds.) ASWC 2007 and ISWC 2007. LNCS, vol. 4825, pp. 11–15. Springer, Heidelberg (2007)

6. Martins, B., Manguinhas, H., Borbinha, J.: Extracting and exploring geo-temporal semantics of textual resources. In: Proceedings of the 2nd IEEE International Conference on Semantic Computing (2008)

7. Henson, C., Neuhaus, H., Sheth, A., Thirunarayan, K., Buyya, R.: An Ontological Representation of Time Series Observations on the Semantic Sensor Web. In: 1st International Workshop on the Semantic Sensor Web 2009, collocated with the 6th European Semantic Web Conference, ESWC (2009)

8. Bittner, T., Donnelly, M., Smith, B.: A Spatio-Temporal Ontology for Geographic Information Integration. International Journal of Geographical Information Science 23(6), 765–798 (2009)

9. Grenon, P.: The spatio-temporal ontology of reality and its formalization. In: AAAI Spring Symposium on Foundations and Applications of Spatio-Temporal Reasoning (FASTR) (2003)

10. Borgida, A.: On the relative expressiveness of description logics and predicate logics. Artificial Intelligence 82(1-2), 353–367 (1996)

11. Horrocks, I., Kutz, O., Sattler, U.: The even more irresistible SROIQ. In: KR 2006 (2006)

Review of Software Security Defects Taxonomy

Zhanwei Hui, Song Huang, Zhengping Ren, and Yi Yao

PLA Software Test and Evaluation Centre for Military Training
PLA University of Science and Technology
Nanjing, jiangsu province PRC
hzw_1983821@163.com,
hs0317@sohu.com,
rzp@hotmail.com,
yaoyi226@yahoo.com.cn

Abstract. an organized list of actual defects can be useful for software security test (SST). In order to target their technology on a rational basis, it would be useful for security testers to have available a taxonomy of software security defects organizing the problem space. Unfortunately, the only existing suitable taxonomies are mostly for tool-builders and software designers, or based on vulnerabilities and security errors, and do not adequately represent security defects that are found in modern software. In our work, we have reviewed the traditional software security errors or vulnerabilities taxonomies. Based on analyzing in its target, motivation and insufficiency, we have compared 9 kinds of taxonomies, which would be useful for defects based software security testing.

Keywords: software security defect, vulnerability, error, security test, taxonomy.

1 Introduction

The coming of the internet age should have led developers to collectively acknowledge that all applications need to be secure [1], but this has not been the case. Few software engineering would argue against the value of performing testing of security software, or testing security mechanisms in general software. However software security defects (SSD) have a nasty habit of cropping up in the most unlikely places. Although SSD may be less visible than those of the bridges he describes, they can be equally damaging. The history of software defects, apart from a few highly visible ones [3], [4], is relatively undocumented.

Traditional security error or vulnerability taxonomies [5, 6] have recently been developed for organizing data about software defects and vulnerabilities of all kinds. These taxonomies are primarily oriented toward collecting data during the software development process for the purpose of improving it.

1.1 Software Security Defects Definition

As different organizations have their own definitions to SSD and security errors. In fact, an inadvertently introduced security defects in a program is a bug. Generally, a security defect is a part of a program that can cause the system to violate its security

J. Yu et al. (Eds.): RSKT 2010, LNAI 6401, pp. 310–321, 2010.

requirements. Software security testing requirements vary according to the software under test and the application, so we cannot address them in detail here.

As normal, they concern identification and authentication of users, authorization of particular actions, and accountability for actions taken. So as a result, a SSD could result in software security mechanism incorrect.

The IEEE Standard Glossary of Software Engineering Terminology [7] includes definitions of

➢ error: human action that produces an incorrect result (such as software containing a fault),
➢ fault: an incorrect step, process, or data definition in a computer program, and
➢ failure: the inability of a system or component to perform its required functions within specified performance requirements.

We have tried to keep our use of the term "defect" intuitive without conflicting with standard terminology. A failure may be produced when a fault is encountered. This glossary lists bug as a synonym for both error and fault. We use defect as a synonym for bug, hence (in IEEE terms) as a synonym for fault, except that we include flaws that have been inserted into a system intentionally, as well as accidental ones.

Other researches are also pay attention to related definition of SSD. IFIP WG10.4, as an international researching organization, published a kind of taxonomy and definitions of terms [8] in this area. These define faults as the cause of errors that may lead to failures. A system fails when the delivered service no longer complies with the specification.

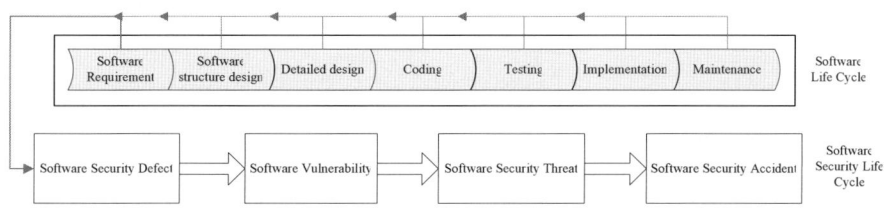

Fig. 1. The evolvement of SSD

This definition of "error" seems more consistent with its use in "error detection and correction" as applied to noisy communication channels or unreliable memory components than the IEEE one.

N. Davis et al. use a fairly broad definition of a SSD in the report [9]. Software security defects are any conditions or circumstances that can result in denial of service, unauthorized disclosure, unauthorized destruction of data, or unauthorized modification of data [11].

A software security defect is a security flaw in software which can result in a security policy violation. Software vulnerability is a kind of SSD representation as software function. The initial test bug reports are a kind of software vulnerabilities. As they exist in software system, then malicious users would make use of them to unpredictable operating. And software security accident will happen.

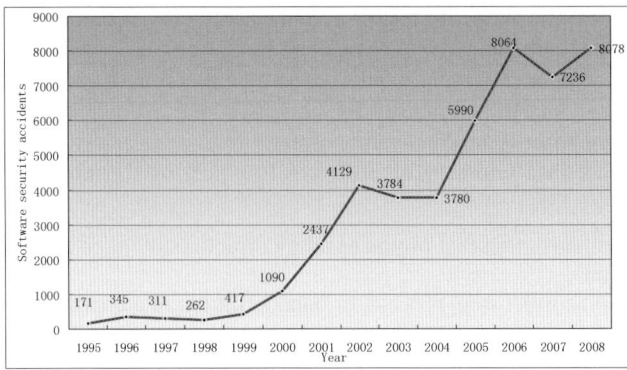

Fig. 2. CERT/CC security accident statistics

Tabel 1. 2006~2009 Top_10 List of Software Vulnerabilities

	2006	2007	2008	2009
A1	Cross Site Scripting (XSS)	Cross Site Scripting (XSS)	Cross Site Scripting (XSS)	Injection
A2	Injection Flaws	Injection Flaws	Injection Flaws	Cross Site Scripting (XSS)
A3	Insecure Remote File Include	Insecure Remote File Include	Insecure Remote File Include	Broken Authentication and Session Management
A4	Buffer Overflows	Insecure Direct Object Reference	Insecure Direct Object Reference	Insecure Direct Object References
A5	Insecure Direct Object Reference	Cross Site Request Forgery (CSRF)	Cross Site Request Forgery (CSRF)	Cross Site Request Forgery (CSRF)
A6	Information Leakage and Improper Error Handling	Information Leakage and Improper Error Handling	Information Leakage and Improper Error Handling	Security Misconfiguration No direct mappings;
A7	Unvalidated Input	Broken Authentication and Session Management	Broken Authentication and Session Management	Failure to Restrict URL Access
A8	Insecure Storage	Insecure Cryptographic Storage	Insecure Cryptographic Storage	Unvalidated Redirects and Forwards
A9	Broken authentication and session management	Insecure Communications	Insecure Communications	Insecure Cryptographic Storage
A10	Cross-site request forgery	Failure to Restrict URL Access	Failure to Restrict URL Access	Insufficient Transport Layer Protection

Our notion of SSD corresponds to that of a fault, with the possibility that the fault may be introduced either accidentally or maliciously. Software security defect will be different during software life cycle. Figure 1 gives out the evolvement and lifecycle of SSD.

1.2 Why Should We Catalog SSD?

As more and more researches reveal that, many security vulnerabilities result from defects that are unintentionally introduced into software during design and development. According to a preliminary analysis by the CERT/CC [10], most software vulnerabilities arise from common causes (same as SSD), and the top 10 cause account for about 75percent of all software vulnerabilities; more than 90 percent are caused by known SSD types. Figure 2 shows that the reported security accident has been increasing quickly, especially from 1999 to 2008. And table I give out the top 10 software security vulnerability list from 2006 to 2009[27].

Although most of the accidents happened at the customer of network, the underlying cause was the software security defect.

The rest of this paper is structured as follows: In Sec. 2 we analysis typical security vulnerability taxonomies in detail. In Sec. 3 we compare the current software security defects taxonomies in their target, motivation and their insufficiency. And then, in Sec. 4 we describe our planned further work.

2 Taxonomy Types and Previous Work

Reports of security violations due to software errors are becoming increasingly common. We refer to such errors as "software security errors" or "software security defects (SSD)". This has resulted in security related concerns among software developers and users. All stages of software development are motivated by the desire to make the product secure and invulnerable to malicious intentions of some users. Our work is concerned with the classifying of SSD with the goal of formal methods for SSD describing, and can help software testers detecting defects that might lead to security violations.

Most of the serious efforts to locate security flaws in computer programs through penetration exercises have used the Flaw Hypothesis Methodology developed in the early 1970s [12]. This method requires system developers first to become familiar with the details of the way the system works (its control structure), then to generate hypotheses as to where flaws might exist in a system; to use system documentation and tests to confirm the presence of a flaw; and finally to generalize the confirmed flaws and use this information to guide further efforts. Although Linde [12] provides lists of generic system functional flaws and generic operating system attacks, he does not provide a systematic organization for security flaws. Many published lists and taxonomies, such as [14~16], concern themselves with either vulnerabilities or attacks.

2.1 DI Based Vulnerabilities Taxonomy

DI based taxonomy is a typical method for software Vulnerabilities. Based on the phase of software defects induced during software life cycle (SLC), the way suggests that SUT threatened by two kinds of vulnerability: Design Vulnerabilities and Implementation Vulnerabilities (DI). [17]

The software design vulnerability is same as software defect, as well as software structure defect. It is induced at the design phase by design errors. And this kind of defects usually influences the security function of SUT. As programmers can not prevent mistakes, so the defect exists eventually.

And implementation vulnerabilities are induced by the programmer errors. This kind of vulnerabilities, just like the flaw exist in hardware when they were produced, no matter how perfect the design, will exist in software system. Furthermore, when the outside environment condition is satisfied, the slim flaw would result in the dangerous failure of the hardware.

In software systems, the defects are usually induced by slipshod coding behaviors most time. As incorrect definition the buffer size (buffer overflow defects), without checking up the format of backtrack code (format string defects), incorrect treating with non-expected input information (command injected defects), improper encoding or escaping of output (output domain defects), would result in software security failure.

2.2 DIO Based Vulnerabilities Taxonomy

On the base of DI taxonomy, list [17] is concerned that the incorrect operation and configuration would also induce security defects, beside design vulnerabilities and implementation vulnerabilities, which are operational vulnerabilities (DIO: Design vulnerability, Implementation vulnerability and Operation vulnerability). This kind of vulnerabilities is usually induced by incorrect development, improper configuration, and unsafe management. It will be not influenced by the code, but in the alternation of software with support environment.

2.3 RISOS Taxonomy and PA Taxonomy

The NIST (National Institute of Standards and Technology) and research organization made great effort to catalog all the software security defects. They collected information and accident reports from several organizations. (Include business institutions and government organizations) The Software Assurance Metrics and Tool Evaluation Project content the two flaw taxonomies, which were produced in the 1970's, were the result of the RISOS project [19] and the Protection Analysis project [20].

The former classified operating system flaws into seven categories:

- Incomplete parameter validation;
- Inconsistent parameter validation;
- Implicit sharing of privileged/confidential data;
- Asynchronous validation/inadequate serialization;
- Inadequate identification/authentication/authorization;
- Violable prohibition/limit;
- Exploitable logic error.

The goal of the Protection Analysis (PA) project was to collect error examples and abstract patterns from them that, it was hoped, would be useful in automating the search for flaws. According to the final report [20], more than 100 errors that could permit system penetrations were recorded from six different operating systems (GCOS, MULTICS, and Unix, in addition to those investigated under RISOS).

Unfortunately, this error database was never published and no longer exists [22]. However, the researchers published some examples, and they developed a classification scheme for software errors. Initially, they hypothesized 10 error categories; these were eventually reorganized into five "globa"l categories:

- Domain errors, including errors of exposed representation,
- Incomplete destruction of data within a deallocated object, or incomplete destruction of its context;
- Validation errors, including failure to validate operands or to handle boundary conditions properly in queue management;
- Naming errors, including aliasing and incomplete revocation of access to a deallocated object; and
- Serialization errors, including multiple reference errors and interrupted atomic operations.

Although the researchers felt that they had developed a very successful method for finding errors in operating systems, the technique resisted automation. Research attention shifted from finding flaws in systems to developing methods for building systems that would be free of such errors. Both pieces of work have received some criticism [23].

2.4 Landwehr's Three Dimension Taxonomy

Landwehr et al. [24] classified system security flaws in three dimensions:

- Flaw genesis (how it was introduced into the system);
- Time of flaws introduction (when in the development cycle it originated);
- Location of flaws.

The 'genesis' classification is the most pertinent to us. This classification starts by dividing flaws into those that are "intentional" or "inadvertent", where intentional flaws are further subdivided into those that are deliberately inserted into the system ("malicious"), versus being the unintended but direct result of design features ("non-malicious"). Further refinements yield thirteen ultimate categories.

As someone points out [16], Landwehr et al's inclusion of Trojan Horses and viruses into their taxonomy is incorrect: if there is a virus in a system, the flaw is not the virus itself, but rather whatever defect allowed the virus to be injected. From our point of view, the major problem with this taxonomy is that it is not fine enough and somewhat outdated: flaws that are now common are grouped into categories along with others that seem distinctly different or which lend themselves to different detection techniques.

2.5 Aslam's Security Flaw Taxonomy

Although, Landwehr's work is not perfect for software security defects taxonomy, it is well-thought-out enough to serve as a starting point. Aslam [18, 21] created a flaw taxonomy whose primary purpose was to serve as an organizational method for databases of flaws and to assist in static analysis efforts.

One of Aslam's criticisms of Landwehr's work, which also strongly influenced his taxonomy, was that Landwehr's classification of intentional flaws as either "malicious" or "nonmalicious" was inappropriate, because it requires a decision as to the motives of the programmer, which can be hard to determine. Although we agree that for Aslam's main purpose, database classification, decisions about people's motives are problematical, we strongly disagree with this with regards to static analysis efforts. For example, Ashcraft and Engler [25] describe how they've found many security flaws in code by means of a static analysis which verified whether user-supplied integers were range-checked against some values, not necessarily the correct ones. Clearly this tool can be trivially defeated by a programmer intent on inserting a Trojan Horse into the system, but has proven to be effective at finding accidental security flaws. This example demonstrates that the programmer's intent, even if impossible to determine ex post facto, is a strong determinator of what flaw detection techniques are applicable.

Aslam's taxonomy reflects this decision to omit intent from consideration, and suffers thereby. The main defect in this taxonomy, however, is that its divisions are not useful for static analysis. For example, "interaction errors between functionally correct modules", misunderstood exception handling and errors created by faulty compilers are all assigned to the same category, and the taxonomy admits of no coding faults that are neither synchronization errors nor condition validation errors. Aslam uses his taxonomy to state which flaws might be detected using static analysis. Unfortunately, this work has not aged well: both input validation errors and synchronization errors are now fruitful lines of static analysis research, even though Aslam stated that these are both "not viable."

2.6 Bishop and Bailey's Vulnerabilities Taxonomy

List [13] criticizes the Aslam's taxonomy for that each one should be unambiguously assignable into exactly one taxonomy category. They produced a kind of vulnerability taxonomy for Unix system and network related system in 1995. The method used 6 axes to catalog system vulnerabilities.

1) Character of vulnerabilities;
2) Time of vulnerabilities induced;
3) Utilized domain;
4) Affected domain;
5) Least components;
6) Source;

Lastly, we must take into account Bishop and Bailey's criticisms of taxonomies, especially because these criticisms could be applied to ours. Their major principle is that each item should be unambiguously assignable into exactly one taxonomy category. They point out two reasons why taxonomies fail to achieve this. First, a given flaw or vulnerability might have different causes depending upon whether one's viewpoint is that of an attacker, the flawed process, or the operating system. This criticism is not pertinent to our work: we always take the viewpoint of the flawed system, not the attacker, and we consider an operating system to be simply a low-level subcomponent of the system as a whole.

Bishop and Bailey's second general observation raises more serious issues. They observe that systems have various levels of abstraction, and that a single flaw might have different classifications, depending upon which abstraction level's viewpoint is taken. Although we agree with this observation, we disagree that this should be necessarily viewed as a problem with the taxonomy.

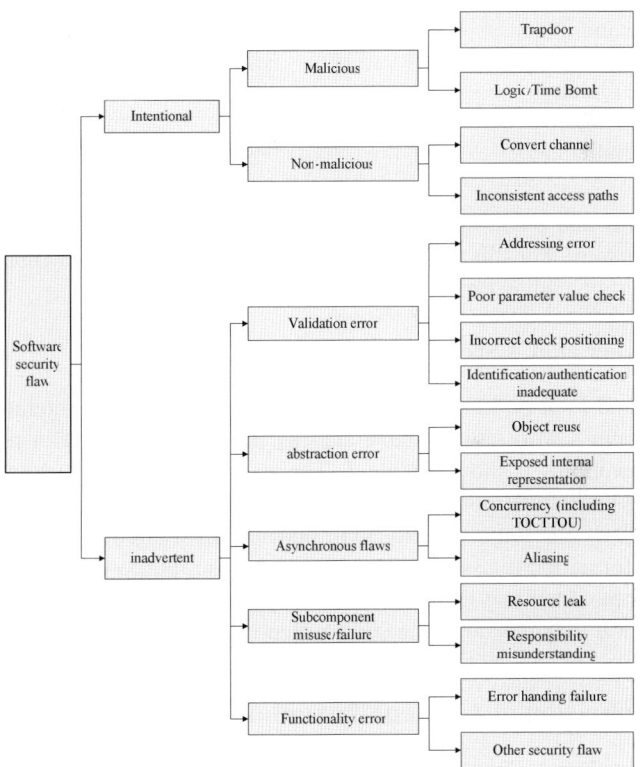

Fig. 3. S.Weber's Security Flaws Taxonomy

2.7   McGraw Coding Security Errors Taxonomy

The purpose of this taxonomy is to help software developers and security practitioners concerned about software understand common coding mistakes that impact security. This approach represents a striking alternative to taxonomies of attack patterns [2] or simple-minded collections of specific vulnerabilities [10]. The goal of this taxonomy is to educate and inform software developers so that they better understand the way their work affects the security of the systems they build. Developers who know this stuff (or at least use a tool that knows this stuff) will be better prepared to build security in than those who don't. The taxonomy is made up of two distinct kinds of sets which is called a phylum is a type or particular kind of coding error; for example, Illegal Pointer Value is a phylum. What is called a kingdom is a collection of phyla that share a common theme. That is, kingdoms are sets of phyla. The eight kingdoms include:

1) Input Validation and Representation
2) API Abuse
3) Security Features
4) Time and State
5) Error Handling
6) Code Quality
7) Encapsulation
8) Environment

Both kingdoms and phyla naturally emerge from a soup of coding rules relevant to enterprise software. For this reason, the taxonomy is likely to be incomplete and may be missing certain coding errors.

2.8 H. S.Weber et al.'s SecurityFlaws Taxonomy[26]

Sam Weber and Paul A. karger started by reviewing the security literature. They selected Landwehr's work as a suitable starting point, modifying it in response to its critics. The taxonomy is show in table I. The major design choice was that flaw categories should be defined positively, not negatively. This is most typical character of the method. It is for static security tool, and negatively-defined categories provide little to no information to assist tool designers. Another purpose of the taxonomy is to aid designers of code analysis tools, and clearly configuration errors are not visible at the code level. It only considers the coding security flaw.

3 A Comparision of Current Taxonomy

Despite the heroic contributions made by the authors of previous empirical studies, serious flaws remain and have detrimentally influenced our models for SSD Taxonomies.

Of course, such weaknesses exist in all scientific endeavors but if we are to improve scientific enquiry in software engineering we must first recognize past mistakes before suggesting ways forward. In Table II, we compare previous defects taxonomy from the target, the motivation and the insufficiency.

Table 2. Comparison of Researches on Defects Taxonomy

Taxonomy	Target	Motivation	Insufficiency
DI	Software vulnerabilities related to SLC	Security of SLC	Improper for defect
DIO	Software vulnerabilities	Security of SLC and support environment	Improper for defect
RISOS	Operating system flaws	Security management	Not integrated
PA	Operating system flaws	To produce flaw-detection tools	Not integrated
Landwehr's	Security flaws	Security development	Outdated for SST
Aslam's	Unix system security defects	An organizational method for databases of flaws and to assist in static analysis efforts.	Ambiguity
Bishopand Bailey's	Unix system and network related system vulnerabilities	To help the programmer improve software security	Professional
McGraw's	Coding security errors	To help software developers and security practitioners	Not distinguish code errors, attacks and security defects
S.Weber's	Coding security flaws	To aid designers of code analysis tools	Not include design security flaws

4 Conclusion and Future Work

In light of the current interest in software security test that detect software security problems, we've compared different taxonomies related to SSD, and showed how this taxonomy relates to available information on currently experienced attacks and vulnerabilities.

To be precise, there are a number of techniques which can be used to approach security problems, each with specific advantages and disadvantages. Instead of doing a cross-comparison of each technique against each taxonomy category, which is not feasible, we are trying to find useful factors among sets of analyses that affect difficulty, and relate these factors to the taxonomy.

We believe that SSD taxonomy would be essential to defects based software security test, and which will allow us to produce security test cases for each kind of defects, and illuminate which research directions are feasible.

Acknowledgment

This work was supported by National High Technology Research and Development Program of China (No: 2009AA01Z402). Resources of the PLA Software Test and Evaluation Centre for Military Training were used in this research.

References

[1] McGraw, G.: Software Security. IEEE Security & Privacy 2(2), 80–83 (2004)
[2] Potter, B., McGraw, G.: Software security testing. Security and Privacy Magazine IEEE 2(5), 81–85 (2004)
[3] Leveson, N., Turner, C.S.: An investigation of the Therac-25 accidents. UCI TR 92-108. Inf. and Comp. Sci. Dept., Univ. of Cal.-Irvine, Irvine, CA (1992)
[4] Spafford, E.H.: Crisis and aftermath. Comm. ACM 32, 678–687 (1989)
[5] Brehmer, C.L., Carl, J.R.: Incorporating IEEE Standard 1044 into your anomaly tracking process. CrossTalk. J. Defense Software Engineering 6, 9–16 (1993); Chillarege, R.
[6] Bhandari, I.S., Chaar, J.K., Halliday, M.J., Moebus, D.S., Ray, B.K., Wong, M.-Y.: Orthogonal defect classification—a concept for in-process measurements. IEEE Trans. on Software Engineering 18(11), 943–956 (1992)
[7] IEEE Computer Society 1990. Standard glossary of software engineering terminology. ANSI/IEEE Standard 610.12-1990. IEEE Press, New York (1990)
[8] Landwehr, C.E.: Dependability: Basic Concepts and Terminology. Dependable Computing and Fault-Tolerant Systems, vol. 6. Springer, New York (1992)
[9] Davis, N., Humphrey, W., Zibulski, G., Mcgraw, G.: Processes for producing secure software. IEEE Security& Privacy (2004)
[10] http://cwe.mitre.org/
[11] Landwehr, C.E.: Formal models for computer security. ACM Computing Surveys 13(3), 247–278 (1981)
[12] Linde, R.R.: Operating system penetration. In: AFIPS National Computer Conference, pp. 361–368 (1975)
[13] Bishop, M., Bailey, D.: A critical analysis of vulnerability taxonomies. Technical Report CSE-96-11, Department of Computer Science at the University of California at Davis (September 1996)
[14] Anderson, J.P.: Computer security technology planning study. Technical Report ESD–TR–73–51, Vols. I, II, James P. Anderson and Co., Fort Washington, PA, USA, HQ Electronic Systems Division, Hanscom AFB, MA, USA (October 1972)
[15] Lindquist, U., Jonsson, E.: How to systematically classify computer security intrusions. In: Proceedings of the IEEE Symposium on Security and Privacy, pp. 154–163 (1997)
[16] Howard, J.D.: An Analysis of Security Incidents on the Internet 1989 - 1995. PhD thesis, Carnegie Mellon University (April 1997)
[17] Wysopal, C., Nelson, L., Zovi, D.D., Dustin, E.: The Art of Software Security Testing. Symatec press
[18] Aslam, T., Krsul, I., Spafford, E.H.: Use of a taxonomy of security faults. In: Proc. 19th NIST-NCSC National Information Systems Security Conference, pp. 551–560 (1996)
[19] Abbott, R.P., Chin, J.S., Donnelley, J.E., Konigsford, W.L., Tukubo, S., Webb, D.A.: Security analysis and enhancements of computer operating systems. NBSIR 76-1041, The RISOS Project, Lawrence Livermore Laboratory, Livermore, CA, USA (April 1976)
[20] Bisbey, R., Hollingworth, D.: Protection analysis: Final report. Technical Report ISI/SR-78-13, Information Sciences Institute, University of Southern California, Marina del Rey, CA (May 1978)
[21] Aslam, T.: A taxonomy of security faults in the UNIX operating system. Master's thesis, Purdue University (August 1995)
[22] Bisbey, R.:Private communication (July 26, 1990)

[23] Weber, S., Karger, P.A., Paradkar, A.: A Software Flaw Taxonomy: Aiming Tools At Security. In: Software Engineering for Secure Systems – Building Trustworthy Applications (SESS 2005), St. Louis, Missouri, USA (2005)

[24] Landwehr, C.E., Bull, A.R., McDermott, J.P., Choi, W.S.: A taxonomy of computer program security flaws. ACM Computer Surveys 26(3), 211–254 (1994)

[25] Ashcraft, K., Engler, D.: Using programmer-written compiler extensions to catch security holes. In: IEEE Symposium on Security and Privacy, Oakland, California (May 2002)

[26] Weber, S., Karger, P.A., Paradkar, A.: A Software Flaw Taxonomy: Aiming Tools At Security. In: Software Engineering for Secure Systems – Building Trustworthy Applications (SESS 2005), St. Louis, Missouri, USA (2005)

[27] Hui, Z.: Research on the techniques of software security testing based on software security defects. Master thesis. PLA University of Science and Technology (2009)

A New Hybrid Method of Generation of Decision Rules Using the Constructive Induction Mechanism

Wiesław Paja, Krzysztof Pancerz, and Mariusz Wrzesień

Department of Artificial Intelligence and Expert Systems,
Institute of Biomedical Informatics,
University of Information Technology and Management in Rzeszów, Poland
{wpaja,kpancerz,mwrzesien}@wsiz.rzeszow.pl

Abstract. Our research is devoted to develop a new method of generation of a set of decision rules. This method is compiled using two different mechanisms. The first one is based on applying a new constructive induction algorithm to the investigated dataset. The belief networks are used in this algorithm. The aim is to find the most important descriptive attribute that is calculated on the basis of other attributes. The second part of the presented method constitutes the improvement algorithm that is used in an optimization process of a gathered rule set. The results of our research contain the comparison of classification efficiency using several datasets.

Keywords: decision rule, constructive induction, belief networks, hybrid method, classification process.

1 Introduction

The main task of data mining is to assist users in extracting useful information or knowledge, which is usually represented in the form of rules due to their easy understandability and interpretability, from the rapidly growing volumes of data [1]. Among various techniques in data mining, classification rule mining is one of the major and traditional techniques. An inductive learning algorithm, in general, performs well on data that have been pre-processed to reduce their complexity. However, they are not particularly effective in reducing data complexity while learning complicated cases [2]. The constructive induction is a useful methodology to reduce the complexity of an instance space for learning applications. The constructive induction usually uses operators to combine two or more features (descriptive attributes) to generate new features. For a classifier, besides the classification capability, its size is another vital aspect. In pursuit of high performance, many classifiers do not take into consideration their sizes and contain numerous both essential and significant rules. This, however, may bring an adverse situation for a classifier, because its efficiency will be put down greatly by redundant decision rules. So, it is necessary to eliminate those unwanted rules

J. Yu et al. (Eds.): RSKT 2010, LNAI 6401, pp. 322–327, 2010.

[3]. Inductive learning algorithms used commonly for the development of sets of decision rules can cause the appearance of some specific anomalies in learning models [4]. These anomalies can be devoted to *redundancy, consistency, reduction* and *completeness* of learning models in the form of a decision rule set [5], [6]. These anomalies are often investigated in post-processing operations [1], [7], and may be fixed (and sometimes removed) using some schemes generally known as verification and validation procedures [7]. The analysis of the literature [2] suggests expectation that application of the constructive induction mechanism over a source information database, prior to the development of rule sets, can lead to more effective learning models. There are numerous well-known CI algorithms and systems available in the literature cited in [2] like: BACON application, FRINGE system, decision-tree based CITRE algorithm, Explora and MIDOS systems, and also a group of AQ algorithms developed by Michalski [8]. Some new approaches to this issue like the Tertius algorithm and a genetic algorithm used by Smith and Bull should also be mentioned.

2   Examined Datasets

The six different datasets were used in our research. The first one concerns melanocytic skin lesion which is a very serious skin and lethal cancer. It is a disease of contemporary time, the number of melanoma cases is constantly increasing, due to, among other factors, sun exposure and a thinning layer of ozone around the Earth [9]. Statistical details on these data are given in [10]. Descriptive attributes were divided into four categories: *Asymmetry, Border, Color,* and *Diversity of structures*. This dataset consists of 548 diagnosed cases. All cases are assigned into four decision classes. The second dataset, with mental diseases cases, contains the description of patients that were examined using the Minnesota Multiphasic Personality Inventory (MMPI) from the psychical disturbances perspective [11]. The results of examination are presented in the form of profiles. Each profile is a data vector consisting of 13 attributes. This dataset consists of over 1000 cases classified by clinic psychologists. Each case is assigned to one of 20 classes. Each class corresponds to one of the nosological types. The next dataset is one of datasets available at the UCI Machine Learning Repository. It is the Heart Disease data set. This database contains 76 attributes, but in this research, like in most of published experiments, we refer to a subset of 14 of them. The decision field refers to the presence of heart disease in the patient. The next three datasets are smaller than those mentioned above. The Iris dataset is perhaps the best known database to be found in the pattern recognition literature. The data set contains 3 classes (three different varieties of Irises) of 50 instances each, where each class refers to a type of the iris plant. Next, the Lenses data set tries to predict whether people will need soft contact lenses, hard contact lenses or none contacts. It contains 24 cases described by 4 attributes and captured into 3 different classes. It is also a well-known data set in the literature. The last data set is devoted to recognition of types of fruits. It contains 23 learning cases, 5 descriptive attributes and 6 decision classes.

3   Hybrid Method of Rule Set Generation

A new hybrid method is proposed. It may be planned in four steps: (1) *Initial step*: a decision table is treated as a primary source of knowledge. The decision rule set is extracted and tested before an improvement and without application of a constructive induction process. (2) *Feature construction step*: the decision table is expanded by inclusion of a new, additional attribute obtained by a new constructive induction mechanism called CIBN (Constructive Induction-Based on Belief Networks) and described in the next section. (3) *Extraction step*: the decision rules are extracted from an extended decision table using the GTS (General-To-Specific) algorithm [12]. The decision rule set is extracted and tested after application of a constructive induction process but before improvement. (4) *Improvement step*: the set of generic operations is applied to the gathered rule set. The goal is to optimize and improve a form of decision rules. The decision rule set is tested after improvement of a learning model. All three learning models were evaluated via testing the classification accuracy of unseen cases. Our new constructive induction algorithm, called CIBN (Constructive Induction-Based on Belief Networks) creates a new descriptive attribute (a new column in the decision table). In this process, a new column-vector is created. Its all cells are filled with numerical values obtained in the following way: to create a new attribute, the information from the belief network [13] (generated for an investigated dataset) is used (see Figure 1). Descriptive attributes used in the original (source) decision table affect a course of the development cell-values of the new column-vector, according to the general formula: $Cell\_value_{n,constr.ind.} = factor_1 \cdot V_{A1,n} + factor_1 \cdot V_{A3,n} + factor_2 \cdot V_{A1,n} + factor_2 \cdot V_{A4,n} + factor_3 \cdot V_{A5,n}$, where: $factor_N$ is a coefficient related to the level of the structure of a belief network, $V_{AN,n}$ is a value of a cell in the n-th case for the N-th attribute. In Figure 2, an example of the CIBN algorithm application is shown (the Bayesian networks created for the data sets: Iris and Lenses). As it is stated in Figure 2a, two attributes on level L1: *petal_width* and *petal_length* have the greatest significance, direct influence on the decision attribute. Thus, these attributes have factors equal to 1 in the constructive induction process. Additionally, the *petal_length* attribute has indirect influence on a decision through

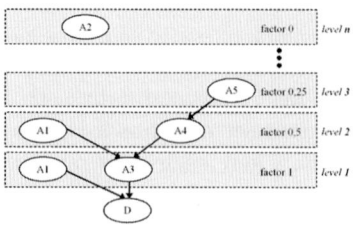

Fig. 1. The belief network scheme: A1, A2, ..., A5 - description attributes, D - decision attribute

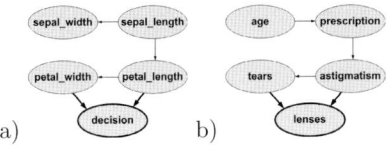

a)   b)

Fig. 2. Examples of the Bayesian networks for the data sets: (a) Iris, (b) Lenses

the *petal_width* attribute, so a factor increases to 1.5. Similarly, the next attribute *sepal_length*, on level L2, has indirect influence on a decision attribute. In that way, these attributes have factors equal to $0.5 + 0.25$, etc. The last attribute, *sepal_width* has no influence on a decision so a factor is equal to 0. Using this schema, a new attribute (called by us attribute CI) could be calculated: $CI = 0.75 \cdot sepal\_length + 0 \cdot sepal\_width + 1.5 \cdot petal\_length + 1 \cdot petal\_width$. Similarly, in case of two other (Fruits and Lenses) datasets, a new CI attribute could be calculated: $CI = 0 \cdot color + 0.5 \cdot size + 1.5 \cdot shape + 0 \cdot taste + 1 \cdot weight$, $CI = 0.375 \cdot age + 0.75 \cdot prescription + 1.5 \cdot astigmatism + 1 \cdot tears$. This algorithm was implemented in the PlaneSEEKER system [14], and next, analysis on selected information datasets was performed. After calculation of the attribute CI, the optimization procedures were applied. The main optimizing algorithm used in the research was implemented in the RuleSEEKER system [15], and was based on an exhaustive application of a collection of the following generic operations. *Finding and removing redundancy*: the data may be overdetermined, that is, some rules may explain the same cases. Here, redundant (excessive) rules were analyzed, and the redundant rule (or some of the redundant rules) was (were) removed. Providing this operation did not increase the error rate. *Finding and removing of incorporative rules*: another example when the data may be overdetermined. Here, some rule(s) being incorporated by another rule(s) was (were) analyzed, and the incorporative rule(s) was (were) removed. Providing this operation did not increase the error rate. *Merging rules*: in some circumstances, especially when continuous attributes were used for the description of objects being investigated, generated learning models contained rules that are more specific than they should be. In these cases, more general rules were applied, so that they cover the same investigated cases, without making any incorrect classifications. *Finding and removing of unnecessary rules*: sometimes rules developed by the systems were unnecessary, that is, there were no objects classified by these rules. Unnecessary rule(s) was (were) removed. Providing this operation did not increase the error rate. *Finding and removing of unnecessary conditions*: sometimes rules developed by the systems contained unnecessary conditions, that were removed. Providing this operation did not increase the error rate. *Creating of missing rules*: sometimes developed models did not classify all cases from a learning set. Missing rules were generated using a set of unclassified cases. *Discovering of hidden rules*: this operation generates a new rule by combination of similar rules, containing the same set of attributes and the same attribute values except one. *Rule specification*: some rules caused correct and incorrect classifications of selected cases. This operation divides a considered rule into a few rules by adding additional conditions. *Selecting of a final set of rules*: there were some rules that classify the same set of cases but have different composition, a simpler rule stayed in a set.

4   Results of Experiments

The results of improvement of learning models are gathered in Table 1. All experiments were performed using the well known n-fold cross validation method. Column 1 contains raw results for a learning model (in the form of a set of decision rules) developed for the source database. Column 2 presents results obtained after application of the constructive induction method, e.g., after adding the attribute CI. Column 3 contains results after adding CI and optimization of a learning model. All datasets were improved by decreasing of the number of rules and the error rate. The greatest influence of experiments is observed in the case of the Melanocytic skin lesions dataset. This model contained rather a large number of rules (153) and the error rate was on the level of 14%. Just after application of the constructive induction algorithm (column 2), the number of rules dropped by roughly 40%, and additionally, what is very interesting, also the error rate was distinctly decreased (from 14,29% to 7,69%). In the next step of the developed methodology (column 3), after using the optimization algorithm, the set of decision rules was smaller (92 rules vs. 48 rules). It means, that the decreasing number of rules was even larger, i.e., about 48%. In the case of the Mental disease and Iris datasets similar results are observed. A number of rules decreased by about 20%, and in the same time, the error rate decreased by nearly half. Similar results could be found in the case of other datasets.

Table 1. Results of experiments with learning models

Data set	Number of rules			Classification error [%]		
	1	2	3	1	2	3
Melanocytic skin lesions	153	92	48	14.29	7.69	7.14
Mental disease	508	497	385	39.24	27.33	22.85
Heart disease	125	131	88	44.44	32.94	30.74
Iris	32	32	17	22.00	20.00	14, 67
Fruits	13	13	11	21.74	13.04	13.04
Lenses	7	7	6	33.33	12.50	29.17

5   Conclusions

It should be stressed that the truncation of the learning model did not spoil its efficiency. The error rate is much smaller. Thus, it may be assumed that the combined algorithm of constructive induction, based on data taken from belief networks, and some improvement of decision rule sets was performed quite satisfactorily in classification of selected medical datasets. In the future, in the case of the melanocytic skin lesions dataset, the comparison of the calculated attribute CI with the well known in medicine parameter TDS is going to be done. Presumably, using this method, it would be possible to find some general parameter combined by means of other descriptive attributes for every investigated dataset.

References

1. Fayyad, U., Piatetsky-Shapiro, G., Smyth, P.: From data mining to knowledge discovery in databases. AI Magazine 17, 37–54 (1996)
2. Piramuthu, S., Sikora, R.: Iterative feature construction for improving inductive learning algorithms. Expert Systems with Application 36, 3401–3406 (2009)
3. Liu, H., Sun, J., Zhang, H.: Post-processing of associative classification rules using closed sets. Expert Systems with Application 36, 6659–6667 (2009)
4. Spreeuwenberg, S., Gerrits, R.: Requirements for successful verification in practice. In: Haller, S., Simmons, G. (eds.) Proc. of the 15th International Florida Artificial Intelligence Research Society Conference, Pensacola Beach, Florida, USA (2002)
5. Jensen, F.: Logical Foundations for Rule-Based Systems. Springer, Heidelberg (2006)
6. Lo, D., Khoo, S., Wong, L.: Non-redundant sequential rules – theory and algorithm. Information Systems 34(4-5), 438–453 (2009)
7. Gonzales, A., Barr, V.: Validation and verification of intelligent systems. Journal of Experimental & Theoretical Artificial Intelligence 12(2), 407–420 (2000)
8. Wnek, J., Michalski, R.: Hypothesis-driven constructive induction in AQ17-HCI: A method and experiments. Machine Learning 14(2), 139–168 (1994)
9. Friedman, R., Rigel, D., Kopf, A.: Early detection of malignant melanoma: the role of physician examination and self-examination of the skin. CA: A Cancer Journal for Clinicians 35, 130–151 (1985)
10. Hippe, Z., Bajcar, S., Blajdo, P., Grzymala-Busse, J., Grzymala-Busse, J., Knap, M., Paja, W., Wrzesien, M.: Diagnosing skin melanoma: Current versus future directions. TASK Quarterly 7(2), 289–293 (2003)
11. Duch, W., Kucharski, T., Gomuła, J., Adamczak, R.: Machine learning methods in analysis of psychometric data. Application to Multiphasic Personality Inventory MMPI-WISKAD, Toruń (1999) (in polish)
12. Hippe, Z.: Machine learning – a promising strategy for business information systems? In: Abramowicz, W. (ed.) Business Information Systems 1997, pp. 603–622. Academy of Economics, Poznan (1997)
13. Jensen, F.: Bayesian Networks and Decision Graphs. Springer, Heidelberg (2001)
14. Błajdo, P., Grzymała-Busse, J., Hippe, Z., Knap, M., Marek, T., Mroczek, T., Wrzesień, M.: A suite of machine learning programs for data mining: chemical applications. In: Debska, B., Fic, G. (eds.) Information Systems in Chemistry, vol. 2, pp. 7–14. University of Technology Editorial Office, Rzeszow (2004)
15. Paja, W.: RuleSEEKER – a new system to manage knowledge in form of decision rules. In: Tadeusiewicz, R., Ligeza, A., Szymkat, M. (eds.) Computer Methods and Systems. Ed. Office ONT, Cracow, pp. 367–370 (1997) (in polish)

An Effective Principal Curves Extraction Algorithm for Complex Distribution Dataset

Hongyun Zhang*, Duoqian Miao, Lijun Sun, and Ying Ye

The Key Laboratory of Embedded System and Service Computing,
Ministry of Education, China, Tongji University, Shanghai 201804, China
School of Electronic and Information Engineering, Tongji University, Shanghai
201804, China
Zhangongyun583@sina.com

Abstract. This paper proposes a new method for finding principal curves from complex distribution dataset. Motivated by solving the problem, which is that existing methods did not perform well on finding principal curve in complex distribution dataset with high curvature, high dispersion and self-intersecting, such as spiral-shaped curves, Firstly, rudimentary principal graph of data set is created based on the thinning algorithm, and then the contiguous vertices are merged. Finally the fitting-and-smoothing step introduced by Kégl is improved to optimize the principal graph, and Kégl's restructuring step is used to rectify imperfections of principal graph. Experimental results indicate the effectiveness of the proposed method on finding principal curves in complex distribution dataset.

Keywords: Principal curves, Complex distribution dataset, Thinning algorithm, Fitting-smoothing step, Image skeletonization.

1 Introduction

Hastie and Stuetzle introduced the notion of principal curves to solve the problems in traditional machine learning and multivariate data analysis in 1989 [1]. Principal curves are self-consistent smooth curves that pass through the middle of a data set. More specifically, we hope to find curves passing through the middle of the datasets, which can truly reflect the shape of the data. Principal curves are non-linear generalizations of principal components, the basic idea of which is to seek low-dimensional manifolds embedded in the high-dimensional space [2]. Due to all the satisfying properties and advantages, principal curves have gained its rapid development since 1990's with various definitions of principal curves having been proposed. Banfield and Raftery gave their principal curve definition called BR principal curve in 1992 [3]. Kégl proposed PL principal curve definition in 2000 [4]. Verbeek defined K-segment principal curve in 2002 [5], while Delicado introduced D principal curve in 2003 [8]. Currently, a

---

* Corresponding author, Mobile: 13917907676; Email: zhanghongyun583@sina.com

J. Yu et al. (Eds.): RSKT 2010, LNAI 6401, pp. 328–335, 2010.

great number of achievements have been reported concerning the applications of principal curves, such as shape detection [3], intelligent transportation analysis [2], speech recognition [7], image skeletonization, feature extraction [6] and Data Compression and Regression analysis [9].

Based on these definitions, ways of finding principal curves from data sets have been proposed one after the other. Hastie and Stuetzle offered an alternation between projecting data onto the curve and estimating conditional expectations on projectors by the scatter smoother or the spline smoother [1]. Banfield and Raftery modified the Hastie-Stuetzle method, using the projection residual of the data, instead of the data themselves, to estimate conditional expectations, for reducing both bias and variance [3]. Verbeek et al. proposed a k-segments algorithm which incrementally combines local line segments into the polygonal line to achieve an objective [5]. Kégl et al. presented the polygonal line algorithm which starts with an initial polygonal line, adds a new vertex to the polygonal line at each iteration, and updates the positions of all vertices so that the value of a penalized distance function is minimized [4]. Delicado found the principal oriented points one by one and orderly linked them to estimate principal curves [8].

However, for complex distribution dataset with high curvature, high dispersion and self-intersecting, such as spiral-shaped curves and spring-shaped curves, existing methods did not work well. Verbeek et al. attempted to solve this problem by combining line segments, which were optimized to minimize the total squared distance of all points to their closest segments into a polygonal line, but did not fit the curve well. All of these algorithms almost use the first principal component of all data as the initial estimation of the principal curve when lacking the prior knowledge. Unfortunately it is bad initialization for complex distribution dataset with high curvature, high dispersion and self-intersecting. So we need to consider the global structure feature of data set from the beginning. In this paper, an effective strategy is introduced to extract principal curves for complex distribution dataset. Instead of starting with a simple topology such as the first principal component and then increasing its complexity iteratively, we directly span the sufficient complex topology and then refine it iteratively. In our algorithm, instead of principal component analysis, thinning-based method is used to initialize data and create rudimentary principal graph of data set. Then, considering the large scale of dispersion and amount, we merging the contiguous vertices and improve the fitting-smoothing step of Kégl's principal curves algorithm [4]to optimize the principal graph. Finally, Kégl's restructuring step [4] is used to refine the graph.

2   Principal Curve

Loosely speaking, principal curves are smooth one-dimensional (1D) curves that pass through the "middle" of a set of p-dimensional data points[1]. The goal is to provide smooth and low dimensional summaries of the data. Here, a 1D curve in a p-dimensional space is a vector f of p functions indexed by one single variable s. The parameter s is the arc length along the curve. For any density h in R^p

with finite second moments, the curve f is a principal curve of h if the following self-consistent criterion is satisfied for almost every s :

$$f(s) = E[X|s_f(X) = s] \qquad (1)$$

And

$$s_f(X) = sup\{s : \|X - f(s)\| = inf\|Y - f(\tau)\|\} \qquad (2)$$

In the above X is a random vector from h, s_f is the projection index function which maps any value of $X = x$ to the value of s for which $f(x)$ is closest to x. The definition of a principal curve indicates that any point of a principal curve is the condition expectation of those points that project to this point, and a principal curve satisfies the property of self-consistent. Fig. 1 shows a first principal component line and a principal curve, Compared with corresponding first principal component, two obvious advantages of a principal curve can be observed: first, a principal curve can keep more information of data; and second, it can describe the outline of primitive information better.

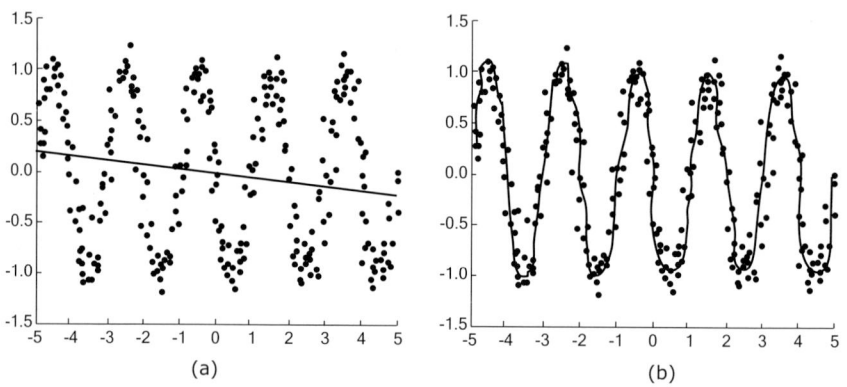

Fig. 1. Comparison between principal curve and first principal component

3   Principal Curves Extraction Algorithm

To overcome the ineffectiveness of the first principal component which is used to initialize data with high curvature, high dispersion and self-intersecting, such as spiral-shaped curves, thinning-based method is used to initialize data and create rudimentary principal graph of data set. Then, considering the large scale of dispersion and amount, we merging the contiguous vertices and improve the fitting-smoothing step of Kégl's principal curves algorithm [4] to optimize the principal graph. Finally, Kégl's restructuring step[4] is used to refine the graph. It contains the following steps:

STEP 1. Global Structure Extraction Step. Zhang-Suen thinning algorithm is adopted to generate sufficiently complex topology, that is, directly obtain the approximate initial skeleton of the original data template. However, it is not smooth and usually contains a number of spurious branches and inadequate structural elements. The skeleton is denoted by G_{vs}, which consists of V and S, where $V = \{v_1, \ldots, v_m\}$ is a set of *vertices*, and $S = \{(v_{i1}, v_{j1}), \ldots, (v_{ik}, v_{jk})\}, 1 \leq i1, j1, \ldots, ik, jk \leq m$ is a set of *edges*, while S_{ij} is a line segment that connects v_i and v_j.

STEP 2.Vertices-Merge Step. Considering two remarkable characteristics of complex distribution dataset, high level of dispersion and large quantity of data points, in this process, we merge the adjacent vertices in terms of distance and curvature. The distance criterion makes sure that at least a certain number of vertices are retained in a certain area coverage while the curvature criterion is set to reduce the number of vertices merged in areas with big curvature. Vertices-Merges step has two advantages: (a) complex distribution dataset includes thousands of vertices, and the Vertices-Merge step can effectively reduce the number of vertices which need to be adjusted. As a result, the efficiency of the algorithm is improved; (b) Vertices -Merge step increases the ratio of data points to skeleton vertices. In that case, more data points on average are involved in adjusting a single skeleton vertex so that the deviation of skeleton is controlled under a certain degree.

STEP 3. The Improved Fitting-Smoothing Step. Iteratively fit and smooth the skeleton by repeatedly projecting data point and optimizing vertex, until convergence is achieved while keeping the skeleton approximately equidistant from the contours of the dataset. The vertices smoothing and fitting process includes the following two steps.

STEP 3.1. The Improved Projection Step: The original Kégl algorithm points out: given a data set $X_n = \{x_1, \ldots, x_n\}$, scan the whole skeleton for every data point x_i , the data point x_i is partitioned into "the nearest neighbor regions" according to which segment or vertex it projects. This step is time-consuming, since thousands of scans are required. Considering the characteristic of complex distribution dataset, the Projection Step of original algorithm is improved. The step only scans certain areas of the vertices around the data point x_i instead of the whole skeleton. Through a large scale of samples-training, we set the width of 30 to 60 pixels which can lead to relatively good results both on effectiveness and time-costing.

STEP 3.2. The Improved Vertex Optimization Step: In original algorithm, every vertex v_i in the skeleton is optimized by using a gradient method to adjust the positions of vertices and segments for finding a local minimum of $E(G)$. The penalized distance function $E(G)$ is as follows:

$$E(G) = \frac{1}{n} \sum_{i=1}^{n} \Delta(x_i, G) + \lambda P(G) \tag{3}$$

$$P(G) = \frac{1}{m} \sum_{i=1}^{m} P_v(v_i) \tag{4}$$

$\frac{1}{n}\sum_{i=1}^{n}\Delta(x_i, G)$ is the average squared distance of all points from the G_{vs}. The smaller the value of $\frac{1}{n}\sum_{i=1}^{n}\Delta(x_i, G)$ is, the better G_{vs} fits the data. λ is a penalty coefficient that determines the trade-off between the accuracy of the approximation and smoothness of the curves. $P(G)$ is a penalty on the total curvature of the skeleton. The smaller the value of $P(G)$, the smoother G_{vs} is . n is the number of the data points. $\Delta(x_i, G)$ is the Euclidean squared distance between a point x_i and the nearest point of the skeleton to x_i. m is the number of vertices. $P_v(v_i)$ is the curvature penalty at vertex v_i.

According to the distribution of complex pattern data, since lots of triangle functions and branch structures are used in the original penalty function $P(G)$, the calculation process is really time-costing. We have figured out a way to solve this problem by replacing $P(G)$ with a new penalty function $D(G)$.

$$D(G) = \frac{1}{m}\sum_{i=1}^{m}\sum_{x \epsilon V_i \bigcup S_i}\Delta(x, v_i) \tag{5}$$

From the experimental results, we sum up that the function $D(G)$ has three advantages: (a) it helps the skeleton convergence much more; (b) it reduces the skeleton deviation; (c) since $D(G)$ only involves simple calculation of addition and average, it is more efficient than $P(G)$ which uses triangle functions.

Obviously, we redefine the penalized distance function $E(G)$:

$$E(G) = \Delta(G) + \lambda D(G) \tag{6}$$

STEP 3.3. The Judge Step: Judge if the adjusted skeleton meets the convergent condition, if true, goto STEP 4, else goto STEP 3.1.

STEP 4. The Restructuring Step: Rectify the structural imperfections of the skeleton graph by deleting short paths and small loops to get more accurate skeleton.

STEP 5. END.

4   Experimental Results and Analysis

We have carried on two typical experiments on complex distribution dataset. The first experiment tested the capability of the our algorithm for extracting principal curves from data distribution of simulated data sets, and the second one's aim was to test the effects of the proposed algorithm in the applications of image skeletonization with images of logo and fingerprint.

4.1   Simulated Data Sets

We generated data sets along some curves by the commonly used additive noise model.

$$X' = X + e \tag{7}$$

Where X is a data set uniformly distributed on a smooth curve which is called generating curve. $e = (c_1, c_2)$ is a bivariate additive noise which is independent of X, and X' is the generated data set.

We constructed various shaped curves, such as line, circle, half-circle, S-shaped, etc. Then we tested the proposed algorithm on these data sets. It turned out to have exciting results. In order to emphasize the algorithm's effectiveness for finding principal curves from complex distribution dataset with high curvature, high dispersion and self-intersecting, we select experimental result on the spiral-shaped curve and spring-shaped curve. Fig. 2 proves the effectiveness of our proposed algorithm.

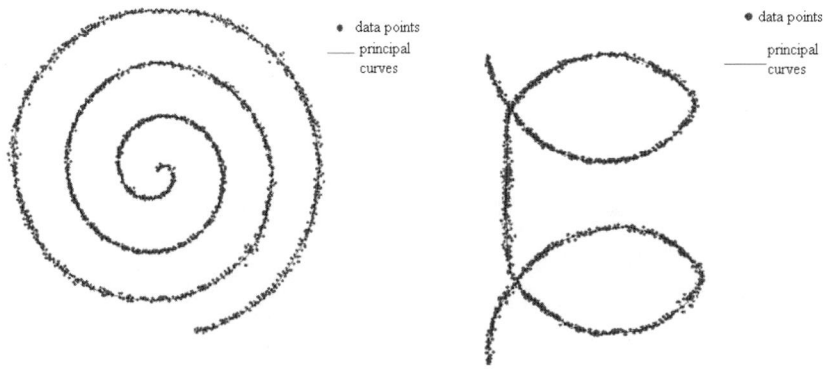

Fig. 2. Principal curves on simulated data set

Kégl et al. [4] pointed out the gradual degradation of their polygonal line algorithm on spiral-shaped generating curves of increasing length. Their algorithm performed well on spirals with one and a half, and two loops, but failed when the number of loops reached three. The Hastie-Stuetzle algorithm performed worse in the same experiments. Delicado and Hueta [8] also compared the performances of their algorithm with the PL algorithm and the Hastie-Stuetzle algorithm on the spiral with two loops and concluded that the Kégl's algorithm and theirs behave quite similarly which are both better than the Hastie-Stuetzle algorithm. Verbeek et al. [5] made further efforts, summed up the failures of different algorithms and successfully validated the effectiveness of their k-segment algorithm on the spiral with two and a half loops. The resultant polygonal line using twelve line segments by the k-segment algorithm recovered the basic shape of the spiral, but did not fit the curve well. In our experiment, the loops of spiral were more than that of all the above experiments. The number of loops reached four and the experimental result was almost indistinguishable from the generating spiral-shaped curve. In addition, our algorithm fitted the curve well.

4.2   Image Skeletonization

The self-consistency property of principal curves is quite similar to the equidistance property of medial axis of shapes. If foreground pixels of a shape are represented by a two dimensional data set, then the principal curves of this data set is the approximation to its skeleton of shape. Singh et al. and Kégl et al. have already applied the principal curves in image skeletonization. Now we also used the proposed algorithm to look for skeletons of images.

The performance of the algorithm was tested on three suites of images. The first one is the logo of Tongji university. The second one is the fingerprint pictures from FVC2002 fingerprint database [10]. Experimental results are listed as Fig. 3 Results confirmed the effectiveness of our algorithm on skeletonization of images whose dataset distribution meets complex distribution.

Fig. 3. Result of logo and fingerprint image skeletonization

5   Conclusions

Since the notion of principal curves was put forward, researchers have already introduced several methods of how to find principal curves from data sets. However, as for the complex distribution dataset such as high degree of dispersion, bending or self-intersecting, those existing methods are not ideal. To solve this problem, the paper proposes a new method. Firstly, it initializes a set of vertices to create the preliminary skeleton by using thinning algorithm after which adjacent vertices are merged. Then the fitting-smoothing step of Kgl's principal curves algorithm is improved to smooth vertices positions and construct the principal graph from these vertices through iteration. Finally, Kgl's restructuring step is used to refine the graph. Algorithm has been tested on simulated data sets and applied to image skeletonization. Experimental results confirm the effectiveness of the proposed method on finding principal curves from complex distribution dataset.

Acknowledgments. This work is supported by the National Natural Science Foundation of China (granted No. 60775036, 60970061), National 973 Program (granted No. 2003CB316902).

References

1. Hastie, T., Stuetzle, W.: Principal curves. Journal of the American Statistical Association 84(406), 502–516 (1989)
2. Zhang, J.P., Chen, D.W., Kruger, U.: Adaptive Constraint K-segment Principal Curves For Intelligent Transportation Systems. IEEE Transactions on Intelligent Transportation Systems 9(4), 666–677 (2008)
3. Banfield, J.D., Raftery, A.E.: Ice floe identification in satellite images using mathematical morphology and clustering about principal curves. Journal of the American Statistical Association 87(417), 7–16 (1992)
4. Kégl, B., Krzyzak, A., Linder, T.: Learning and Design of Principal Curves. IEEE Trans. on Distribution Analysis and Machine Intelligence 22(3), 281–297 (2000)
5. Verbeek, J.J., Vlassis, N., Krose, B.: A k-segments algorithm for finding principal curves. Distribution Recognition Letter 23(10), 1009–1017 (2002)
6. Zhang, H.Y., Miao, D.Q., Zhang, D.X.: Analysis and Extraction of Structural Features of Off -Line Handwritten Digits Based on Principal Curves. Journal of Computer Research and Development 42(8), 1344–1349 (2005)
7. Reinhard, K., Niranjan, M.: Parametric subspace modeling of speech transitions. Speech Communication 27(1), 19–42 (1999)
8. Delicado, P., Huetra, M.: Principal curves of oriented points.theoretical and computational improvements. Comput. Stat. 18(2), 293–315 (2003)
9. Jochen, E., Gerhard, T., Ludger, E.: Data Compression and Regression Based on Local Principal Curves. In: Advances in Data Analysis, Data Handling and Business Intelligence. Springer, Heidelberg (2009)
10. FVC2002 web site, http://bias.csr.unibo.it/fvc2000/download.asp

Parallel Reducts Based on Attribute Significance

Dayong Deng[1,2], Dianxun Yan[2], and Jiyi Wang[2]

[1] Xingzhi College, Zhejiang Normal University, Jinhua, China, 321004
dayongd@163.com
[2] College of Mathematics, Physics and Information Engineering,
Zhejiang Normal University, Jinhua, China, 321004

Abstract. In the paper, we focus on how to get parallel reducts. We present a new method based on matrix of attribute significance, by which we can get parallel reduct as well as dynamic reduct. We prove the validity of our method in theory. The time complex of our method is polynomial. Experiments show that our method has advantages of dynamic reducts.

Keywords: Rough Sets, Attribute Significance, Matrix, Parallel Reducts, Dynamic Reducts.

1 Introduction

Rough set theory is a valid mathematical tool, which deals with imprecise, vague and incomplete information [1,4]. There are many models of rough sets corresponding with their attribute reducts. Most of these models deal with data in a decision system, it is difficult for these models of rough sets to handle dynamic information or increase information. Some of researchers tried to solve this problem, in which a few of them extended rough set theory to a series of decision subsystems.

Deng and Huang[9,11] presented a method of discernibility matrix and function between two decision tables. The method can parallel compute, distributed compute and increasingly compute. Liu[6] introduced an algorithm of the smallest attribute reducts with increase data. Wang and Wang [7] proposed a distributed algorithm of attribute reduction based on discernibility matrix and function. In fact Liu's algorithm [6] and Wang's algorithm [7] are special cases of Deng's algorithm [9,11]. Zheng et al. [8] presented an incremental algorithm based on positive region. Kryszkiewicz and Rybinski[10] introduced an algorithm of attribute reducts in composed decision systems. Deng [13] presented a method of attribute reduction by voting in a series of decision subsystems.

Jan G. Bazan et al.[2,3] presented the concept of dynamic reducts to solve the problem of large amount of data or incremental data. They selected parts of data to process reduction, and then selected the intersection of all reducts as stable reducts. The method must count all of reducts in these decision subsystems. Its time complex is NP-Complete.

In[14,15,16,17] we extend rough set theory and introduce a new method to compute the stable reducts in a series of decision subsystems, which is called

J. Yu et al. (Eds.): RSKT 2010, LNAI 6401, pp. 336–343, 2010.

parallel reducts. The parallel reducts have all of advantages of dynamic reducts but avoid two their faults. The time complex of dynamic reducts is NP-complete, and the way to get dynamic reducts is not complete because the insection of all of Pawlak reducts in a series subsystems may be empty. The time complex of parallel reducts could be the same as that of the best algorithm of Pawlak reducts. We can always get parallel reducts in any cases. Just like dynamic reducts, parallel reducts could deal with increase data, dynamic data and tremenously large data, and we can also parallel compute to get parallel reducts.

In this paper we continue to investigate parallel reducts and its properties by attribute significance. We introduce a new method based on matrix of attribute significance to get parallel reduct. We have proved in theory that if the decision subsystems exist dynamic reduct, then the parallel reduct we got must be dynamic reduct. Experiments show that our method has advantages of the method of dynamic method.

2 Primary Knowledge

2.1 Rough Sets

We assume that readers are familiar with rough set theory. Therefore, we only introduce some primary knowledge of rough sets briefly.

In a decision system $DS = (U, A, d)$, where $\{d\} \cap A \neq \emptyset$, the decision attribute d divides the universe U into parts, denoted by $U/d = \{Y_1, Y_2, ..., Y_p\}$, where Y_i is an equivalence class. The positive region is defined as

$$POS_A(d) = \bigcup_{Y_i \in U/\{d\}} POS_A(Y_i)$$

Sometimes the positive region $POS_B(d)$ is also denoted by $POS_B(DS, d)$. In rough set theory there are various kinds of definitions of attribute reducts[1,4,12], the most popular definition is Pawlak reducts(reducts in short) in a decision system. It could be showed as follows:

Definition 1. *Given a decision system $DS = (U, A, d)$, $B \subseteq A$ is called the reduct of the decision system DS if B satisfies the following two conditions:*

(1) $POS_B(d) = POS_A(d)$,
(2) For any $S \subset B$, $POS_S(d) \neq POS_A(d)$.

All reducts of a decision system DS is denoted by $RED(DS)$.

Definition 2. *In a decision system $DS = (U, A, d)$ we will say d depends on A to a degree $h(0 \leq h \leq 1)$, if*

$$h = \gamma(A, d) = \frac{|POS_A(d)|}{|U|}$$

Where $|\cdot|$ denotes the cardinality of a set.

Definition 3. *The significance of an attribute a in a decision system DS =*
(U, A, d) is defined by

$$\sigma(a) = \frac{\gamma(A, d) - \gamma(A - \{d\}, d)}{\gamma(A, d)} = 1 - \frac{\gamma(A - \{d\}, d)}{\gamma(A, d)}.$$

2.2   Dynamic Reducts

The purpose of dynamic reducts is to get the stable reducts from decision subsystems. We will give their definitions in the following.

Definition 4. *If DS = (U, A, d) is a decision system, then any system DT =*
(U′, A, d) such that U′ ⊆ U is called a subsystem of DS. By P(DS) we denote
the set of all subsystems of DS. Let DS = (U, A, d) be a decision system and
F ⊆ P(DS). By DR(DS, F) we denote the set

$$RED(DT) \cap \bigcap_{DT \in F} RED(DT)$$

Any elements of DR(DS, F) is called an F-dynamic reduct of DS.

Definition 5. *Let DS = (U, A, d) be a decision system and F ⊆ P(DS). By*
GDR(DS, F) we denote the set

$$\bigcap_{DT \in F} RED(DT)$$

Any elements of GDR(DS, F) is called an F-generalized dynamic reduct of DS.

3   Parallel Reducts

In the sequel the formula $DS = (U, A, d)$ denotes a decision system. By $P(DS)$
we denote the set of all subsystems of DS. The symbol F denotes a nonempty
subset of $P(DS)$, which excludes the empty element ϕ, i.e. $\phi \notin F$.

Definition 6. *Let DS = (U, A, d) be a decision system, and P(DS) the set of*
all subsystems of DS, F ⊆ P(DS). B ⊆ A is called a parallel reduct of F (F-
parallel reduct in short) iff B satisfies the following two conditions:

(1) For any subsystem DT ∈ F it satisfies $POS_B(DT, d) = POS_A(DT, d)$.
(2) For any S ⊂ B, there exists at least a subsystem DT ∈ F such that
$POS_S(DT, d) \neq POS_A(DT, d)$.

Definition 7. *Let PRED be the set of parallel reducts of F, then the inter-*
section of elements of PRED is called the core of F-parallel reducts, denoted
by

$$PCORE = \bigcap PRED$$

We redefine the F-parallel reducts as follows by the dependence of attributes.

Definition 8. *Let $DS = (U, A, d)$ be a decision system, and $P(DS)$ the set of all subsystems of DS, $F \subseteq P(DS)$. $B \subseteq A$ is called a parallel reduct of F iff B satisfies the following two conditions:*

(1) For any subsystem $DT \in F$ it satisfies $\gamma(B, d) = \gamma(A, d)$.
(2) For any $S \subset B$, there exists at least a subsystem $DT \in F$ such that $\gamma(S, d) \neq \gamma(A, d)$.

From the primary knowledge of rough set theory, it is easy to know that the definition 8 is equivalence to the definition 6.

Definition 9. *The core of F-parallel reducts in the subsystems $F \subseteq P(DS)$ is the set of attributes which significance are bigger than zero for any $DT \in F$.*

4   Matrix of Attribute Significance

In this section we will define the matrix of attribute significance and investigate its properties.

Definition 10. *Let $DS = (U, A, d)$ be a decision system, and $P(DS)$ the set of all subsystems of DS, $F \subseteq P(DS)$, $B \subseteq A$, the matrix of attribute significance B relative to F is defined by*

$$M(B, F) = \begin{bmatrix} \sigma_{11} & \sigma_{12} & \cdots & \sigma_{1m} \\ \sigma_{21} & \sigma_{22} & \cdots & \sigma_{2m} \\ \cdots & \cdots & \cdots & \cdots \\ \sigma_{n1} & \sigma_{n2} & \cdots & \sigma_{nm} \end{bmatrix}.$$

Where $\sigma_{ij} = \gamma_i(B, d) - \gamma_i(B - \{a_j\}, d)$, $a_j \in B$, $(U_i, A, d) \in F$, $\gamma_i(B, d) = \frac{|POS_B(DT_i, d)|}{|U_i|}$, n denotes the number of decision tables in F, m denotes the number of condition attributes in B.

The matrix of attribute significance B relative to F is the attribute significance of any attribute in B relative to any decision table in F. Every elements in a row denote the attribute significance of various attributes in the same decision table. Every elements in a column denote the attribute significance of an condition attribute in various decision tables.

In the above formula the reason that we replace σ in the definition 3 with σ_{ij} is that $\gamma_i(B, d)$ may be equal to zero. This is to say, the positive region for the attribute set B may be empty. So we avoid using $\gamma_i(B, d)$ as a denominator. However, the attribute significance σ_{ij} could be the same effect as σ in the definition 3.

Definition 11. *Let $DS = (U, A, d)$ be a decision system, and $P(DS)$ the set of all subsystems of DS, $F \subseteq P(DS)$, $B \subseteq A$, the modified matrix of attribute significance B relative to F is defined by*

$$M'(B, F) = \begin{bmatrix} \sigma'_{11} & \sigma'_{12} & \cdots & \sigma'_{1m} \\ \sigma'_{21} & \sigma'_{22} & \cdots & \sigma'_{2m} \\ \cdots & \cdots & \cdots & \cdots \\ \sigma'_{n1} & \sigma'_{n2} & \cdots & \sigma'_{nm} \end{bmatrix}.$$

Where $\sigma'_{ij} = \gamma_i(B \cup \{a_j\}, d) - \gamma_i(B, d)$, $a_j \in A$, $(U_i, A, d) \in F$, $\gamma_i(B, d) = \frac{|POS_B(DT_i, d)|}{|U_i|}$, *n denotes the number of decision tables in F, m denotes the number of condition attributes in DS.*

It is easy to know that if $a_j \in B$, the element σ'_{ij} in the matrix $M'(B, F)$ is equal to 0.

5   Parallel Reducts Based on Matrix of Attribute Significance

We will get parallel reducts through matrix of attribute significance in this section. The idea of the following algorithm is, we get the core of all decision tables in F through the matrix of attribute significance at first, then we get the rest of attributes in parallel reduct through the modified matrix of attribute significance. The algorithm is showed in the following:

Algorithm. Parallel reducts based on matrix of attribute significance.
Input: a series of subsystems $F \subseteq P(DS)$.
Output: a parallel reduct of F.

1. Construct initial matrix of attribute significance $M(A, F)$
2. $B = \bigcup_{j=1}^{m} \{a_j : \exists \sigma_{kj}(\sigma_{kj} \in M(A, F) \wedge \sigma_{kj} \neq 0)\}$ // B is the set of all of attribute cores of decision tables in F
3. Count $M'(B, F)$
4. Do the following steps until $M'(B, F) = 0$
   (a) For $j = 1$ to m do $t_j = 0$
   (b) For $j = 1$ to m do
      For $k = 1$ to n do
         If $\sigma'_{kj} \neq 0$ then $t_j = t_j + 1$;//count the number of $\sigma'_{kj} \neq 0$ in a column.
   (c) $B = B \cup \{a_j : \exists t_j (t_j \neq 0 \wedge \forall t_p (t_j \geq t_p))\}$// add the attribute, of which the number of attribute significance is the most and not equal to zero, to the set of reduct B.
   (d) Count $M'(B, F)$
5. Output the reduct B.

We will estimate the time complex of the above algorithm. Because the time of the above algorithm is mainly consumed in constructing the matrix of attribute significance and improving it. Assume that we use the algorithm in literature [18] to count the attribute significance , its time complex is $O(|B||U|log|U|)$, where $|U|$ denotes the number of instances in one decision table . The time complex of constructing a matrix of attribute significance is $O(mn|B||U'|log|U'|)$, where $|U'|$ could be assigned to the maximal number of instances of the decision tables in F. The matrix of attribute significance must be improved in the algorithm. In the worst condition the number of improving matrix of attribute significance is $|A|$. Therefore, the time complex of improving attribute significance is $O(mn|B||A||U'|log|U'|)$. $|B|$ may be changed in the process of getting attribute reduct, We could assign it the maximal value $|A|$(It's equal to m). Consequently, the time complex of the above algorithm in the worst condition is $O(nm^3|U'|log|U'|)$.

Theorem 1. *The reduct B got by the above algorithm is a parallel reduct. If the dynamic reducts of the series subsystem F exist, then B is also a dynamic reduct.*

Proof. We prove the first proposition at first.

According to the properties of attribute significance, the attribute set B could keep all of positive regions in F. If we reduce an attribute from the set B, there is at least one decision table in F, which positive region has been reduced. This is to say, the attribute set B is a parallel reduct of F.

We prove the second proposition now.

Assume that the attribute set B is not a dynamic reduct. Because the dynamic reduct of the series subsystem F exist. The dynamic reducts are the intersection of all of reducts of decision tables in F. This is to say, any element in the dynamic reduct belongs to one of reducts of per decision table in F and its attribute significance corresponding to the dynamic reduct must be bigger than zero. In our algorithm we select the attribute, the number of whose attribute significance in the matrix, which is bigger than zero, is the most. If the attribute set B is not the dynamic reduct, then the number of attribute significance which is bigger than zero is not the most. It is contradiction with the algorithm. Therefore, the attribute set B is a dynamic reduct.

The above algorithm and theorem provide an efficient method to get the dynamic reduct and parallel reduct, and its time complex is less than that of other algorithms before.

6   Experiments

The data in our experiments come from UCI repository of machince learning databases. We use RIDAS system(developed by Chongqing University of Posts and Telecommunications) to count dynamic reducts and to normalize the data. In our algorithm we create 10 sub-tables from each original data. The first sub-table has 10% data, the second 20% data, and so on.

The indexes of our computer are as follows: Computer Model: DELL OPTI-PLEX GX260, CPU: Intel Pentium IV 1.8Ghz, Memory: 768MB RAM, Hard disk: 40G, OS: Windows XP Professional (5.1 2600). The results of our experiments are in the following table:

Table 1. Comparison of dynamic reducts and parallel reducts

Data	Condition	Decision	Instances	Dynamic	Parallel	Consistent
Abalone	8	1	4177	N/A: > 600	Y:90.125	N/A
Acute Inflammations	6	2(1)	120	Y : 0.925	Y: 0.109	Y
Breast cancer wisconsin	10(9)	1	699	N/A: > 600	Y: 1.594	N/A
Car Evaluation	6	1	1728	Y :112.828	Y: 3.922	Y
Contraceptive Method Choice	9	1	1473	Y : 65.141	Y: 2.672	Y
Ecoli	8(7)	1	336	Y : 7.265	Y: 0.782	Y
Hayes-Roth	5	1	132	Y : 1.109	Y: 0.110	Y

In the above table, The symbol '10(9)' means that there is 10 condition attributes in the original database, but there is an useless attribute for the reduction, '8(7)' is the same meaning. The symbol '2(1)' means that we have to only consider one in the two decision attributes because the RIDAS system is not the multi-attribute decision-making system.

In symbol $x : y$, x denotes whether our experiment is computable, y denotes time of processing. The unit of time is second. The right column 'Consistent' denotes whether the parallel reduct belongs to dynamic reducts.

Limited to the RIDAS system and the performance of compute hardware, if calculation lasts more than ten minutes (600s), we give it up.

From the above table we could find that our method overmatchs the method of dynamic reducts in time complex, and that all of parallel reduct are the dynamic reduct. However, the method of dynimic reducts is not complete. The examples are given in literatures [15, 16].

The experiments show that our algorithm could not only get parallel reduct but also dynamic reduct. The time complex of our method is less than that of dynamic reducts.

7   Conclusion

In this paper, we continue to investigate the properties of parallel reducts, and propose a new method based on matrix of attribute significance to get parallel reduct. If the series of decision tables F exist dynamic reducts, then the parallel reduct got by our algorithm must be the dynamic reduct. This result shows that the theory of parallel reducts is complete, however, the time complex of our algorithm is less than that of dynamic reducts.

In the next future we will investigate the properties of parallel reducts and more efficient algorithms.

References

1. Pawlak, Z.: Rough Sets-Theoretical Aspect of Reasoning about Data. Kluwer Academic Publishers, Dordrecht (1991)
2. Bazan, G.J.: A Comparison of Dynamic Non-dynamic Rough Set Methods for Extracting Laws from Decision Tables. In: Polkowski, L., Skowron, A. (eds.) Rough Sets in Knowledge Discovery 1: Methodology and Applications, pp. 321–365. Physica-Verlag, Heidelberg (1998)
3. Bazan, G.J., Nguyen, H.S., Nguyen, S.H., Synak, P., Wroblewski, J.: Rough Set Algorithms in Classification Problem. In: Polkowski, L., Tsumoto, S., Lin, T.Y. (eds.) Rough Set Methods and Applications, pp. 49–88. Physica-Verlag, Heidelberg (2000)
4. Liu, Q.: Rough Sets and Rough Reasoning. Science Press (2001) (in Chinese)
5. Wang, G.: Calculation Methods for Core Attributes of Decision Table. Chinese Journal of Computers 26(5), 611–615 (2003) (in Chinese)
6. Liu, Z.: An Incremental Arithmetic for the Smallest Reduction of Attributes. Acta Electronica Sinica 27(11), 96–98 (1999) (in Chinese)
7. Wang, J., Wang, J.: Reduction algorithms based on discernibility matrix: The order attributes method. Journal of Computer Science and Technology 16(6), 489–504 (2001)
8. Zheng, Z., Wang, G., Wu, Y.: A Rough Set and Rule Tree Based Incremental Knowledge Acquisition Algorithm. In: Proceedings of 9th International Conference of Rough Sets, Fuzzy Sets, Data Mining, and Granular Computing, pp. 122–129 (2003)
9. Deng, D., Huang, H.: A New Discernibility Matrix and Function. In: Wang, G.-Y., Peters, J.F., Skowron, A., Yao, Y. (eds.) RSKT 2006. LNCS (LNAI), vol. 4062, pp. 114–121. Springer, Heidelberg (2006)
10. Kryszkiewicz, M., Rybinski, H.: Finding Reducts in Composed Information Systems. In: Proceedings of International Workshop on Rough Sets and Knowledge Discovery (RSKD 1993), pp. 259–268 (1993)
11. Deng, D.: Research on Data Reduction Based on Rough Sets and Extension of Rough Set Models (Doctor Dissertation). Beijing Jiaotong University (2007)
12. Deng, D., Huang, H., Li, X.: Comparison of Various Types of Reductions in Inconsistent Decision Systems. Acta Electronica Sinica 35(2), 252–255 (2007)
13. Deng, D.: Attribute Reduction among Decision Tables by Voting. In: Proceedings of 2008 IEEE International Conference of Granular Computing, pp. 183–187 (2008)
14. Deng, D., Wang, J., Li, X.: Parallel Reducts in a Series of Decision Subsystems. In: Proceedings of the Second International Joint Conference on Computational Sciences and Optimization (CSO 2009), Sanya, Hainan, China, pp. 377–380 (2009)
15. Deng, D.: Comparison of Parallel Reducts and Dynamic Reducts in Theory. Computer Science 36(8A), 176–178 (2009) (in Chinese)
16. Deng, D.: Parallel Reducts and Its Properties. In: Proceedings of 2009 IEEE International Conference on Granular Computing, pp. 121–125 (2009)
17. Deng, D.: (F,ε)-Parallel Reducts in a Series of Decision Subsystems (accepted by IIP 2010)
18. Liu, S., Sheng, Q., Shi, Z.: A New Method for Fast Computing Positive Region. Journal of Computer Research and Development 40(5), 637–642 (2003) (in Chinese)

A Rough Sets Approach to User Preference Modeling

Siyuan Jing and Kun She

Department of Computer Science and Engineering, UESTC, Chengdu 610054, China
{jingsiyuan_628,kunshe}@126.com

Abstract. User preference modeling is one of the challenging issues in intelligent information system. Extensive researches have been performed to automatically analyze user preference and utilize it. But one problem still remains: All of them could not deal with semantic preference representation and uncertain data at the same time. To overcome this problem, this paper proposes a rough set approach to user preference modeling. A family of atomic preference granules supporting semantic in knowledge space and two operators, called vertical aggregation operator $\odot$ and horizon combination operator $\oplus$ respectively, are defined to represent user preference. Based on this, a rough granular computing framework is also constructed for user preference analyzing. Experimental results show that the proposed model plays well in recommendation tests.

Keywords: User Preference, Rough Sets, Preference Granules, Semantic, Uncertainty, Entropy.

1 Introduction

User preference model is one of the important infrastructures in intelligent information system, and it is the precondition for personalized recommendation. According to the point proposed in literature [1], user preference modeling should consider two aspects problems, explicit user preference presentation and implicit user preference mining. That's to say, a user preference model should have abilities to represent the semantic and compute the uncertain data. This point of view corresponds to common sense of preference. But the most of the actual user preference models, which can be classified into one of the following three classes, depending on how they represent preference: vector similarity [2,3,4], probability [5] and association [6], can not process the two aspects at the same time, they must rely on other techniques, such as ontology [11].

To overcome this problem, this paper proposes a rough set approach to user preference modeling. A family of atomic preference granules supporting semantic in knowledge space and two operators, called vertical aggregation operator $\odot$ and horizon combination operator $\oplus$ respectively, are defined to represent user preference. Based on this, a rough granular computing framework is also constructed for user preference analyzing. Experimental results show that the proposed model plays well in recommendation tests.

J. Yu et al. (Eds.): RSKT 2010, LNAI 6401, pp. 344–352, 2010.

This paper is organized as follows: Section 2 introduces foundation of rough set theory. Section 3 to section 5 describes the preference model in detail. In section 6, we report on the experimental evaluation of the proposed model. Finally, we draw a conclusion.

2   Rough Sets: Foundation

Rough set theory [7,8] is based on the concept of approximation spaces and models of sets and concepts. The data in rough set theory is collected in a table called a *information table*. Rows of the decision table correspond to objects, and columns correspond to features. In the data set, we also assume that a set of examples with a class label to indicate the class to which each example belongs are given. We call the class label a decision feature, the rest of the features are conditional. Let U , F denote a set of sample objects and a set of functions representing object features, respectively. Assume that $B \subseteq F$, $x \in U$. Further, let $[x]_B$ denote:

$$[x]_B = \{y : x \sim_B y\}$$

Rough set theory defines three regions based on the equivalent classes induced by the feature values: lower approximation $\underline{B}(X)$, upper approximation $\overline{B}(X)$ and boundary $BND(X)$. A lower approximation of a set X contains all equivalence classes $[x]_B$ that are subsets of X, and upper approximation $\overline{B}(X)$ contains all equivalence classes $[x]_B$ that have objects in common with X, while the boundary $BND(X)$ is the set $\overline{B}(X) \backslash \underline{B}(X)$, i.e., the set of all objects in $\overline{B}(X)$ that are not contained in $\underline{B}(X)$. So, we can define a rough set as any set with a non-empty boundary.

The indiscernibility relation $\sim_B$(or byIND_B) is a fundamental principle of rough set theory. In formally, $\sim_B$ is a set of all objects that have matching descriptions. Based on the selection of B, $\sim_B$ is an equivalence relation that partitions a set of objects U into equivalence classes. The set of all classes in a partition is denoted by $U/\sim_B$(or byU/IND_B). The set U/IND_B is called the quotient set. Affinities between objects of interest in the set $X \subseteq U$ and classes in a partition can be discovered by identifying those classes that have objects in common with X. Approximation of the set X begins by determining which elementary sets $[x]_B \in U/IND_B$ are subsets of X.

3   User Preference Modeling

User preference is defined by a function which represents how much a user likes or dislikes a given item. The user has his/her behavior history V , which is provided by user actions such as purchasing, voting, or viewing. Then, preference for an item x, $pref(x)$, can be estimated by a function of item$x$and user history$V$, which can be approximated by using user profile G :

$$pref(x) = f(x, V) \approx f(x, G) \tag{1}$$

User history V is represented by a decision table, just as table 1. Rows of the decision table correspond to a set of selected items ($U = \{x_1, x_2, x_3, x_4, x_5,\}$), and columns correspond to features of item ($C \cup d = \{c_1, c_2, c_3, c_4\} \cup d$).The condition attributes C correspond to the features of item. The decision attribute d corresponds to the user's evaluation, like-1 and dislike-0. Item x is represented by a set of preference granules $x = \{\Theta_1, \Theta_2, \cdots, \Theta_k\}$. The preference granule, denoted by Θ, will be introduced in section 3. User profile is then defined by a set of preference values for each preference granules, $G = \{pref(\Theta_1), pref(\Theta_2), \cdots, pref(\Theta_k)\}$. The $pref(\Theta_1)$ represents how much a user likes a given preference granule and is computed from the user history.

So, the formula (1) can be denoted as the follow:

$$pref(x) = \sum_{\Theta_i \in x} pref(\Theta_i) \tag{2}$$

Then, a novel way to represent semantic preference by a family of preference granules will be proposed in next section.

Table 1. An example of user history

objects	c_1	c_2	c_3	c_4	d
x_1	1	1	2	0	1
x_2	0	1	1	0	1
x_3	1	1	0	1	1
x_4	1	0	1	1	0
x_5	0	0	1	2	0

4 Preference Granules

Take the case of choosing cars; users usually express their preference as: not Japanese, not over 100000 yuan, red or blue. The three conditions in this preference naturally correspond to three logic operations ($\neg, \leq, \vee$). Rough logic [10,12,13,14,15] is an important extension of rough set theory and it also defines a family of operators like the classic logic. But the granules in classic rough sets don't have the ability of semantic representation. Chen proposes an "Entity-Attribute-Value" model which can represent the semantic information in granules [9]. Based on their work, this paper defines a family of atomic preference granules supporting semantic in knowledge space and two operators, called vertical aggregation operator $\odot$ and horizon combination operator $\oplus$ respectively, to represent user preference.

Definition 1. *(Atomic Preference Granule, APG): APG is a semantic cell which contains the minimal preference description in information system. Formally, it can be denoted as ξ or $\xi(u, c, v)$ where u correspond to preference identifier, c correspond to feature and v correspond to the value of feature. APG represents a truth that preference u corresponds to feature c values v.*

Definition 2. *(Vertical Aggregation): Vertical aggregation is an operation, denoted as $\odot$, that constitutes several APGs in the same feature to a complex one called compound preference granule, CPG for short. Formally, the CPG can be denoted as Θ or $\Theta\left(\xi_1 \odot \xi_2 \odot \cdots \odot \xi_n\right)$.*

By defining APG and Vertical aggregation operator $\odot$, other logic operations can be calculated from them.

$$\neg \xi\left(u_i, c_k, v_i\right) = \Theta\left(\odot_{j \neq i}\left(u_j, c_k, v_j\right)\right)$$

$$\xi\left(u_i, c_k, v_i\right) \vee \xi\left(u_j, c_k, v_j\right) = \Theta\left(u_i \odot u_j, c_k, v_i \odot v_j\right)$$

For APG is the minimal preference description in information system, conjunction operation is not necessary.

Definition 3. *(Horizon Aggregation): Horizon aggregation is an operation, denoted as $\oplus$, that constitutes several APGs or CPGs in the different features to a preference sentence used to express integrated user preference.*

Thus, we have proposed a new approach to express user preference by preference granules without semantic loss. For example, the example give above can be expressed as: $\Theta\left(USA \odot Europe \odot China\right) \oplus \Theta\left(5W \odot 10W\right) \oplus \Theta\left(R \odot B\right)$

Next, this paper will propose a framework for user preference measuring based on preference granules.

5   User Preference Measure

The user preference model proposed in section 3 shows that the preference of an item mainly relies on the preference granules including in the given item. So, calculating preference granules from user history V is the key to user preference measure.

The information how much user like the external information contained in preference granules can be derived from the user's evaluation to the items which contain the preference granules. The relation between preference granules and user's evaluation can be classified into three classes depending on the inclusion degree (Fig. 1). In Fig. 1, D_0 and D_1 represent the dislike and like item set respectively; Θ_1, Θ_2 and Θ_3 represent the item set containing these preference granules. We can easily find that the proposed method is accordant to rough set theory (Fig. 2).

Next, a method for preference granules measuring is proposed.

Definition 4. *Given a user history V and a preference granule Θ, the degree of preference can be calculated as follows:*

$$pref\left(\Theta\right) = \frac{\left|\Theta \cap D_1\right|}{\left|\Theta\right|} \tag{3}$$

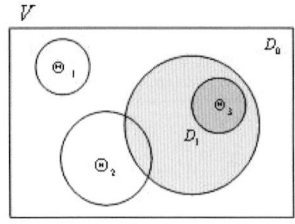

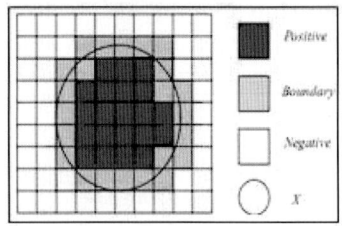

Fig. 1. relation between preference granules and user's evaluation

Fig. 2. Rough Sets Model

In (3), D_1 represents the item set that user likes; Θ represents the item set that contains preference Θ.

Obviously, $0 \leq pref(\Theta) \leq 1$. If $pref(\Theta) = 1$, we say the given user likes this preference granule (in fact, the external information contained in preference granule); if $0 < pref(\Theta) < 1$, we say the given user finitely likes this preference granule; if $pref(\Theta) = 0$, we say the given user doesn't like this preference granule.

The results calculated from table 1. show in table 2. In the result, we use a pair (feature value, preference value) to substitute the form $pref(\Theta)$.

Table 2. User profile G

c_1	c_2	c_3	c_4
(1, 0.67)	(1, 1)	(1, 0.33)	(0, 1)
(0, 0.5)	(0, 0)	(0, 1)	(1, 0.5)
(0 ⊙ 1, 0.6)	(0 ⊙ 1, 0.6)	(2, 1)	(2, 0)
-	-	(0 ⊙ 1, 0.5)	(0 ⊙ 1, 0.75)
-	-	(0 ⊙ 2, 1)	(0 ⊙ 2, 0.67)
-	-	(1 ⊙ 2, 0.5)	(1 ⊙ 2, 0.33)
-	-	(0 ⊙ 1 ⊙ 2, 0.6)	(0 ⊙ 1 ⊙ 2, 0.6)

The method mentioned above gives an effective way to measure preference granules. But it doesn't consider feature preference which is very important actually. Take the case of choosing cars again, somebody cares for price, but somebody cares for brand.

This paper will propose an entropy way to evaluate the uncertainty of feature preference based on APG, the bigger the entropy value is, the less important for user preference analyzing the feature will be. Based on the proposed entropy, a unitary method for feature weight calculating is also introduced.

Definition 5. *Given a user history V, the uncertainty of feature preference c can be calculated as follows:*

$$H(c|D_1) = \sum_{\Theta_i \in C} \frac{|\xi_i|}{|U|} \times pref(\xi_i) \times (1 - pref(\xi_i)) \qquad (4)$$

Obviously, $0 \leq H(c|D_1) \leq \frac{1}{4}$.

Definition 6. *Given a user history V , the weight of feature c_i can be calculated as follows:*

$$W_{c_i} = \frac{L - H\left(c_i\,|D_1\right)}{\sum\limits_{c_i \in C}\left(L - H\left(c_j\,|D_1\right)\right)} \tag{5}$$

In (5), L is a constant and $L \geq \frac{1}{4}$.

If let $L = 0.5$, we can calculate the features weight that shows in table 1.

$$\begin{cases} H\left(c_1\,|d\right) = 0.233 \\ H\left(c_2\,|d\right) = 0 \\ H\left(c_3\,|d\right) = 0.133 \\ H\left(c_4\,|d\right) = 0.1 \end{cases} \Rightarrow \begin{cases} W_{c_1} = 0.17 \\ W_{c_1} = 0.33 \\ W_{c_1} = 0.24 \\ W_{c_1} = 0.26 \end{cases}$$

The user profile G must be rectified by the feature weight as follows:

$$pref^{\sim}\left(\Theta\right) = pref\left(\Theta\right) \times W \tag{6}$$

In (6), W is weight of the feature which preference granule Θ belongs to. The rectified user profile in table 2 is shown in table 3.

Table 3. Rectified user profile G

c_1	c_2	c_3	c_4
(1, 0.11)	(1, 0.33)	(1, 0.08)	(0, 0.26)
(0, 0.09)	(0, 0)	(0, 0.24)	(1, 0.13)
(0 ⊙ 1, 0.1)	(0 ⊙ 1, 0.2)	(2, 0.24)	(2, 0)
-	-	(0 ⊙ 1, 0.12)	(0 ⊙ 1, 0.2)
-	-	(0 ⊙ 2, 0.24)	(0 ⊙ 2, 0.17)
-	-	(1 ⊙ 2, 0.12)	(1 ⊙ 2, 0.09)
-	-	(0 ⊙ 1 ⊙ 2, 0.14)	(0 ⊙ 1 ⊙ 2, 0.16)

Thus, an integrated model of user preference has been introduced. But we should do some experiments to validate the proposed model.

6   Experimental Results

To validate the validity of the proposed method, we establish a recommendation system and do some tests. 37 students participate in our test and their purchase histories on Internet are gathered for an initialized V. Moreover, 2228 goods' information is input into system for recommendation. The students input their explicit preference and the system makes a recommendation. This experiment persists three days and each student performs 100 times tests per day. The test results will be added to history V for new test.

The results of tests in the first day show in Fig. 3. The maximal recommendation accuracy is 0.63. The minimal recommendation accuracy is 0.25. The average recommendation accuracy is 0.41. The distribution of experiment result

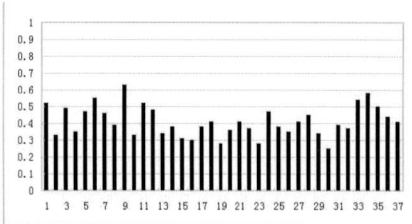

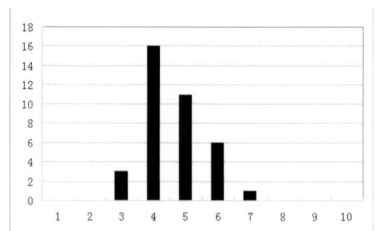

Fig. 3. The results of experiment in the first day

Fig. 4. The distribution of experiment results in Fig.3

shows in Fig. 4. The proportion of the test results between 0.3 and 0.5 is 73%, between 0.5 and 0.6 is 16%, the other is 11%.

The results of tests in the second day show in Fig. 5. The maximal recommendation accuracy is 0.76. The minimal recommendation accuracy is 0.46. The average recommendation accuracy is 0.63. The distribution of experiment result shows in Fig. 6. The proportion of the test results between 0.5 and 0.8 comes to 100%.

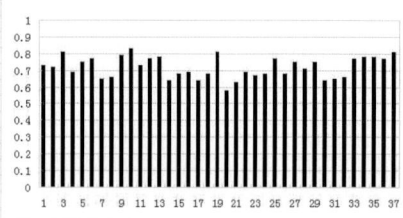

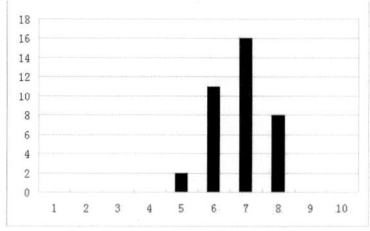

Fig. 5. The results of experiment in the second day

Fig. 6. The distribution of experiment results in Fig.5

The results of tests in the last day show in Fig. 7. The maximal recommendation accuracy is 0.83. The minimal recommendation accuracy is 0.58. The average recommendation accuracy reaches to 0.72. The distribution of experiment result shows in Fig. 8. We can see that the most of the recommendation accuracy is in the interval 0.6 and 0.8.

By analyzing the results, we find the reason leading to recommendation accuracy increase is the user history data. The more sufficient user history data is, the higher recommendation accuracy will be, but the rate of increase will fall. In general, the proposed model plays well in recommendation tests.

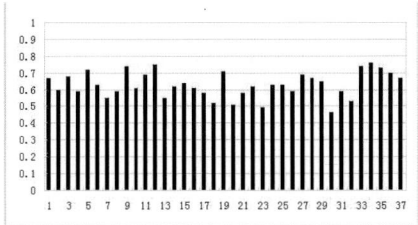
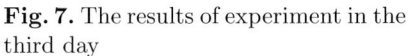
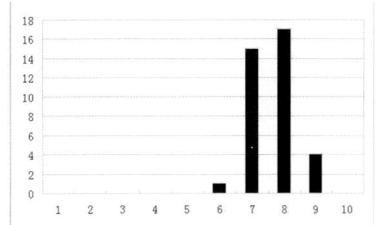

Fig. 7. The results of experiment in the third day

Fig. 8. The distribution of experiment results in Fig.7

7 Conclusion

This paper proposes a rough set approach to user preference modeling to solve the problem mentioned in section 1. The model defines a family of atomic preference granules and two operators $\odot$ and $\oplus$ to represent user preference and also constructs a framework for user preference measuring. The proposed method plays well in the performed experiment in general. Next, we plan to do some research on cold start and fuzzy preference.

Acknowledgement

This work was supported by the National High-tech Research and Development Program of China (Grant No. 2008AA04A107), the key fields Guangdong-Hongkong Technology Cooperation Funding (Grant No. 2009498B21).

References

1. Zeng, C., Xing, C.X., Zhou, L.Z.: A survey of personalization technology. Journal of Software 13(10), 1952–1961 (2002) (in Chinese)
2. Joachims, T.: Text Categorization with Support Vector Machines: Learning with Many Relevant Features. In: Nédellec, C., Rouveirol, C. (eds.) ECML 1998. LNCS, vol. 1398, pp. 137–142. Springer, Heidelberg (1998)
3. Zhang, T.: Text Categorization Based on Regularized Linear Classification Methods. Information Retrieval 4(1), 5–31 (2001)
4. Breeseetal, J.S.: Empirical Analysis of Predictive Algorithms for Collaborative Filtering. In: Proc. Fourth Conf. Uncertainty in Artificial Intelligence (1998)
5. Jung, S.Y., Hong, J.-H., Kim, T.-S.: A Statistic Model for User Preference. IEEE Trans. on Knowledge and data engineering 17(6) (2005)
6. Agrawal, R., et al.: Mining Association Rules between Sets of Items in Large Databases. In: Proc. ACM SIGMOD Int'l. Conf. Management of Data, pp. 207–216 (1993)
7. Pawlak, Z.: Rough sets. International Journal of Computer and Information Science 11(5), 341–356 (1982)
8. Pawlak, Z.: Rough sets – Theoretical Aspects of Reasoning about Data. Kluwer, Dordrecht (1991)

9. Bo, C., Ming-Tian, Z.: Granular Rough Theory Research. Journal of Software 19(3), 565–583 (2008) (in Chinese)
10. Qing, L., Hui, S., Hong-fa, W.: The Present Studying of Granular Computing and Studying of Granular Computing Based on Semantics of Rough Logic. Journal of Computer 31(4) (2008) (in Chinese)
11. Yan, Y., Jian-Zhong, L., Hong, G.: Ontology-based Preference Model in Digital Library. Journal of software 16(12) (2005) (in Chinese)
12. Qing, L.: Rough Set and Rough Reasoning. 3rd edn. Science Press (2005) (in Chinese)
13. Qing, L., Qun, L.: Granules and application of granular computing in logic reasoning. Journal of Computer Research and Development 41(2), 546–551 (2004) (in Chinese)
14. Qing, L., Shao-hui, L., Fei, Z.: Rough logic and its application in data mining. Journal of Software 12(3), 415–419 (2001) (in Chinese)
15. Qing, L., Zhao-Hua, H.: G-logic and resolution reasoning. Chinese Journal of Computer 27(7), 865–874 (in Chinese)

A Tool for Study of Optimal Decision Trees[*]

Abdulaziz Alkhalid, Igor Chikalov, and Mikhail Moshkov

Mathematical and Computer Sciences & Engineering Division
King Abdullah University of Science and Technology
Thuwal 23955-6900, Saudi Arabia
{abdulaziz.alkhalid,igor.chikalov,mikhail.moshkov}@kaust.edu.sa

Abstract. The paper describes a tool which allows us for relatively small decision tables to make consecutive optimization of decision trees relative to various complexity measures such as number of nodes, average depth, and depth, and to find parameters and the number of optimal decision trees.

Keywords: Decision tree, optimization, dynamic programming.

1 Introduction

Decision trees are widely used as algorithms, as predictors, and as a way for knowledge representation in rough set theory [6,7], test theory [3], machine learning and data mining [2]. Unfortunately, the most part of problems connected with decision tree optimization are NP-hard.

In this paper, we describe a tool which allows us to study optimal decision trees for relatively small decision tables with discrete attributes only (decision table mushroom considered in this paper contains 22 conditional attributes and 8124 rows). For such tables it is possible to make consecutive optimization of decision trees relative to various complexity measures (such as number of nodes, number of nonterminal nodes, average depth and depth) and to find parameters and the number of optimal decision trees. This tool is based on an extension of dynamic programming approach [4,5] which allows us not only find one optimal decision tree but describe the whole set of optimal decision trees. As subproblems we consider sub-tables of the initial decision table which are given by systems of equations of the kind "attribute = value".

Our previous program [5] solves a specific problem of classification points in the plane. We modified it in order to be capable of processing an arbitrary decision table. The current program organizes computations in multiple threads and shows good performance on SMP architecture.

The rest of the paper is organized as follows. Section 2 contains basic notions. Section 3 gives a way of representing a set of decision trees in a form of labeled

---

[*] The research has been partially supported by KAUST-Stanford AEA project "Predicting the stability of hydrogen bonds in protein conformations using decision-tree learning methods".

J. Yu et al. (Eds.): RSKT 2010, LNAI 6401, pp. 353–360, 2010.

directed acyclic graph. Section 4 describes a procedure of finding optimal decision trees. In Sect. 5 possibilities of consecutive optimization relative to different complexity measures are studied. Section 6 provides the results of experiments. Section 7 contains short conclusions.

2 Basic Notions

Consider *a decision table* T depicted in Figure 1. Here $f_1, \ldots, f_m$ are names of columns (*conditional* attributes); $w_1, \ldots, w_m$ are positive numbers (weights of columns) each of which could be interpreted as time of computation of corresponding attribute value; $c_1, \ldots, c_N$ are nonnegative integers which can be interpreted as decisions (values of the decision attribute d); $\pi_1, \ldots, \pi_N$ are positive numbers which are interpreted as probabilities of rows (in general case we do not require that the sum of these numbers will be equal to 1); δ_{ij} are nonnegative integers which are interpreted as values of conditional attributes (we assume that the rows $(\delta_{11}, \ldots, \delta_{1m}), \ldots, (\delta_{N1}, \ldots, \delta_{Nm})$ are pairwise different).

w_1	$\ldots$	w_m	
f_1	$\ldots$	f_m	d
δ_{11}	$\ldots$	δ_{1m}	c_1 π_1
	$\ldots$		$\ldots \ldots$
δ_{N1}	$\ldots$	δ_{Nm}	c_N π_N

Fig. 1. Decision table

Let $f_{i_1}, \ldots, f_{i_t} \in \{f_1, \ldots, f_m\}$ and $a_1, \ldots, a_t$ be nonnegative integers. Denote by $T(f_{i_1}, a_1) \ldots (f_{i_t}, a_t)$ the sub-table of the table T, which consists of such and only such rows of T that at the intersection with columns $f_{i_1}, \ldots, f_{i_t}$ have numbers $a_1, \ldots, a_t$ respectively. Such nonempty tables (including the table T) are called *separable* sub-tables of the table T.

A *decision tree* for the table T is a finite directed tree with the root in which each terminal node is labeled with a decision. Each nonterminal node is labeled with a conditional attribute, and for each nonterminal node, edges issuing from this node are labeled with pairwise different nonnegative integers. Let v be an arbitrary node of the considered decision tree. Let us define a sub-table $T(v)$ of the table T. If v is the root then $T(v) = T$. Let v be not the root, and in the path from the root to v, nodes be labeled with attributes $f_{i_1}, \ldots, f_{i_t}$, and edges be labeled with numbers $a_1, \ldots, a_t$ respectively. Then $T(v) = T(f_{i_1}, a_1), \ldots, (f_{i_t}, a_t)$. It is required that for each row r of the table T there exists a terminal node v of the tree such that r belongs to $T(v)$, and v is labeled with the decision attached to the row r.

Denote by $E(T)$ the set of attributes (columns of the table T), each of which contains different numbers. For $f_i \in E(T)$ let $E(T, f_i)$ be the set of numbers from the column f_i.

Among decision trees for the table T we distinguish *irredundant* decision trees. Let v be a node of a decision tree Γ. Let all rows of the sub-table $T(v)$ be labeled with the same decision b. Then, v is a terminal node labeled with b. Let $T(v)$ contain rows labeled with different decisions. Then the node v is labeled with an attribute $f_i \in E(T(v))$. If $E(T(v), f_i) = \{a_1, \ldots, a_t\}$ then t edges issue from the node v, and these edges are labeled with $a_1, \ldots, a_t$ respectively.

3 Representation of the Set of Irredundant Decision Trees

Consider an algorithm for construction of a graph $\Delta(T)$, which represents in some sense the set $D(T)$. Nodes of this graph are some separable sub-tables of the table T. During each step we process one node and mark it with the symbol *. We start with the graph that consists of one node T and finish when all nodes of the graph are processed.

Let the algorithm have already performed p steps. Let us describe step $(p+1)$. If all nodes are processed then the work of the algorithm is finished, and the resulted graph is $\Delta(T)$. Otherwise, choose a node (table) Θ that has not been processed yet. Let all rows of Θ be labeled with the same decision b. Then we label the considered node with b, mark it with the symbol $*$ and proceed to the step $(p+2)$. Let Θ contain rows with different decisions. For each $f_i \in E(\Theta)$ draw a *bundle* of edges from the node Θ. Let $E(\Theta, f_i) = \{a_1, \ldots, a_t\}$. Then draw t edges from Θ and label these edges with pairs $(f_i, a_1), \ldots, (f_i, a_t)$ respectively. These edges enter to nodes $\Theta(f_i, a_1), \ldots, \Theta(f_i, a_t)$. If some of nodes $\Theta(f_i, a_1), \ldots, \Theta(f_i, a_t)$ do not present in the graph then add these nodes to the graph. Mark the node Θ with symbol $*$ and proceed to the step $(p+2)$.

Now for each node of the graph $\Delta(T)$ we describe the set of decision trees corresponding to it. It is clear that $\Delta(T)$ is a directed acyclic graph. A node of such graph will be called *terminal* if there are no edges leaving this node. We will move from terminal nodes, which are labeled with numbers (decisions), to the

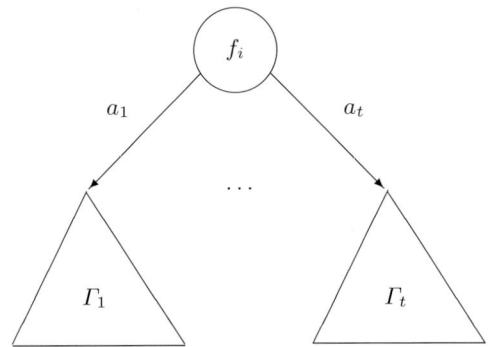

Fig. 2. Trivial decision tree

Fig. 3. Aggregated decision tree

node T. Let Θ be a node, which is labeled with a number b. Then the only trivial decision tree depicted in Figure 2 corresponds to the considered node. Let Θ be a node (table) which contains rows with different decisions. Let $f_i \in E(\Theta)$ and $E(\Theta, f_i) = \{a_1, \ldots, a_t\}$. Let $\Gamma_1, \ldots, \Gamma_t$ be decision trees from sets corresponding to the nodes $\Theta(f_i, a_1), \ldots, \Theta(f_i, a_t)$. Then the decision tree depicted in Figure 3 belongs to the set of decision trees, which corresponds to the node Θ. All such decision trees belong to the considered set. This set does not contain any other decision trees. We denote by $D(\Theta)$ the set of decision trees corresponding to the node Θ. The following proposition shows that the graph $\Delta(T)$ represents all irredundant decision trees for the table T.

Proposition 1. [4] *Let T be a decision table and Θ a node in the graph $\Delta(T)$. Then the set $D(\Theta)$ coincides with the set of all irredundant decision trees for the table Θ.*

4   Selecting Optimal Decision Trees

In this section, we introduce some notions and give the procedure of finding the set of optimal decision trees.

4.1   Proper Sub-Graphs of Graph $\Delta(T)$

Let us introduce the notion of proper sub-graph of the graph $\Delta(T)$. For each node of the graph $\Delta(T)$, which is not terminal, we can remove any but not all bundles that leave the node. Denote the obtained sub-graph by G. Such sub-graphs will be called proper sub-graphs of the graph $\Delta(T)$. It is clear that all terminal nodes of G are terminal nodes of the graph $\Delta(T)$. As it was described earlier, we can associate a set of decision trees to each node Θ of G. It is clear that all these decision trees belong to the set $D(\Theta)$. We denote this set of decision trees by $D_G(\Theta)$.

4.2   Complexity Measures

We will consider complexity measures which are given in the following way: values of considered complexity measure ψ, which are nonnegative numbers, are defined by induction on pairs (T, Γ), where T is a decision table and Γ is a decision tree for T. Let Γ be a decision tree represented in Figure 2. Then $\psi(T, \Gamma) = \psi^0$ where ψ^0 is a nonnegative number. Let Γ be a decision tree depicted in Figure 3. Then $\psi(T, \Gamma) = F(\pi(T), w_i, \psi(T(f_i, a_1), \Gamma_1), \ldots, \psi(T(f_i, a_t), \Gamma_t))$. Here $\pi(T)$ is the sum of probabilities attached to rows of the table T, w_i is the weight of the column f_i and $F(\pi, w, \psi_1, \psi_2, \ldots)$ is an operator which transforms the considered tuple of nonnegative numbers into a nonnegative number. Note that the number of variables $\psi_1, \psi_2, \ldots$ is not bounded from above. So the complexity measure ψ is defined by the pair (ψ^0, F).

The considered complexity measure will be called *monotone* if for any natural $i, t, 1 \leq i \leq t - 1$, and any nonnegative numbers $a, b, c_1, \ldots, c_t, d_1, \ldots, d_t$ the inequality $F(a, b, c_1, \ldots, c_t) \geq \max\{c_1, \ldots, c_t\}$ holds, the equality $F(a, b, c_1, \ldots, c_i,$

$c_{i+1}, \ldots, c_t) = F(a, b, c_1, \ldots, c_{i+1}, c_i, \ldots, c_t)$ holds, the inequality $F(a, b, c_1, \ldots,$
$c_{t-1}) \leq F(a, b, c_1, \ldots, c_t)$ holds if $t \geq 2$, and from inequalities $c_1 \leq d_1, \ldots, c_t \leq d_t$
the inequality $F(a, b, c_1, \ldots, c_t) \leq F(a, b, d_1, \ldots, d_t)$ follows.

The considered complexity measure will be called *strongly monotone* if it is
monotone and for any natural t and any nonnegative numbers $a, b, c_1, \ldots, c_t, d_1,$
$\ldots, d_t$ from inequalities $a > 0$, $b > 0$, $c_1 \leq d_1, \ldots, c_t \leq d_t$ and inequality
$c_i < d_i$, which is true for some $i \in \{1, \ldots, t\}$, the inequality $F(a, b, c_1, \ldots, c_t) <$
$F(a, b, d_1, \ldots, d_t)$ follows.

Now we take a closer view of some complexity measures.

Number of nodes: $\psi(T, \Gamma)$ is the number of nodes in decision tree Γ. For this
complexity measure $\psi^0 = 1$ and $F(\pi, w, \psi_1, \ldots, \psi_t) = 1 + \sum_{i=1}^{t} \psi_i$. This measure
is strongly monotone.

Number of nonterminal nodes: $\psi(T, \Gamma)$ is the number of nonterminal nodes in
the decision tree Γ. For this complexity measure $\psi^0 = 0$ and $F(n, \psi_1, \psi_2, \ldots) =$
$1 + \sum_{i=1}^{t} \psi_i$. This measure is strongly monotone.

Weighted depth: we attach a weight to each path from the root to a terminal node
of tree, which is equal to the sum of weights of attributes attached to nodes of the
path. Then $\psi(T, \Gamma)$ is the maximal weight of a path from the root to a terminal
node of Γ. For this complexity measure $\psi^0 = 0$ and $F(\pi, w, \psi_1, \ldots, \psi_t) = w +$
$\max\{\psi_1, \ldots, \psi_t\}$. This measure is monotone.

Average weighted depth: for an arbitrary row $\bar{\delta}$ of the table T we denote by $\pi(\bar{\delta})$
its probability and by $w(\bar{\delta})$ we denote the weight of the path from the root to
a terminal node of Γ which accepts $\bar{\delta}$. Then $\psi(T, \Gamma) = \sum_{\bar{\delta}} w(\bar{\delta})\pi(\bar{\delta})$, where we
take the sum on all rows $\bar{\delta}$ of the table T. For this complexity measure $\psi^0 = 0$
and $F(\pi, w, \psi_1, \ldots, \psi_t) = w\pi + \sum_{i=1}^{t} \psi_i$. This measure is strongly monotone.

Depth is the weighted depth in the case when all weights of columns are equal
to 1.

Average depth (in this paper) is the weighted average depth in the case when
all weights of columns are equal to 1 and all probabilities π_i are equal to $\frac{1}{N}$
where N is the number of rows in the table.

Proposition 2. [4] *Let T be a decision table and ψ a monotone complexity measure. Then there exists irredundant decision tree for T that is optimal relatively
to complexity measure ψ among all decision trees for T.*

4.3   Procedure of Optimization

Let G be a proper sub-graph of the graph $\Delta(T)$, and ψ be a complexity measure.
Describe a procedure, which transforms the graph G into a proper sub-graph
G_ψ of G. We begin from terminal nodes and move to the node T. We attach a
number to each node, and possible remove some bundles of edges, which start in
the considered node. We attach the number ψ^0 to each terminal node. Consider
a node Θ, which is not terminal, and a bundle of edges, which starts in this
node. Let edges be labeled with pairs $(f_i, a_1), \ldots, (f_i, a_t)$, and edges enter to

nodes $\Theta(f_i, a_1), \ldots, \Theta(f_i, a_t)$, to which numbers $\psi_1, \ldots, \psi_t$ are attached already. Then we attach to the considered bundle the number $F(\pi(\Theta), w_i, \psi_1, \ldots, \psi_t)$.

Among numbers attached to bundles starting in Θ we choose the minimum number p and attach it to the node Θ. We remove all bundles starting in Θ to which numbers are attached that are greater than p. When all nodes will be treated we obtain a graph. Denote this graph by G_ψ. As it was done previously for any node Θ of G_ψ we denote by $D_{G_\psi}(\Theta)$ the set of decision trees corresponding to Θ in the graph G_ψ.

Note that using the graph G_ψ it is easy to find the number of decision trees in the set $D_{G_\psi}(\Theta)$: $|D_{G_\psi}(\Theta)| = 1$ if Θ is terminal node. Consider a node Θ, which is not terminal, and a bundle of edges, which start in this node and enter to nodes $\Theta_1 = \Theta(f_i, a_1), \ldots, \Theta_t = \Theta(f_i, a_t)$. We correspond to this bundle the number $|D_{G_\psi}(\Theta_1)| \cdot \ldots \cdot |D_{G_\psi}(\Theta_t)|$. Then $|D_{G_\psi}(\Theta)|$ is equal to the sum of numbers corresponding to bundles start in Θ.

Let T be a decision table and ψ a monotone complexity measure. Let G be a proper subgraph of $\Delta(T)$ and Θ an arbitrary node in G. We will denote by $D_{\psi,G}^{opt}(\Theta)$ the subset of $D_G(\Theta)$ containing all decision trees having minimum complexity relatively ψ, i.e., $D_{\psi,G}^{opt}(\Theta) = \{\hat{\Gamma} \in D_G(\Theta), \psi(\Theta, \hat{\Gamma}) = \min_{\Gamma \in D_G(\Theta)} \psi(\Theta, \Gamma)\}$.

The following theorems describe important properties of the set $D_{G_\psi}(\Theta)$.

Theorem 1. [4] *Let T be a decision table and ψ a monotone complexity measure. Let G be a proper subgraph of $\Delta(T)$ and Θ an arbitrary node in the graph G. Then $D_{G_\psi}(\Theta) \subseteq D_{\psi,G}^{opt}(\Theta)$.*

Theorem 2. [4] *Let T be a decision table and ψ a strongly monotone complexity measure. Let G be a proper subgraph of $\Delta(T)$ and Θ be an arbitrary node in the graph G. Then $D_{G_\psi}(\Theta) = D_{\psi,G}^{opt}(\Theta)$.*

5   Possibilities of Consecutive Optimization

Let the graph $\Delta(T)$ be constructed for a decision table T, and ψ_1 and ψ_2 be strongly monotone complexity measures. Let us apply the procedure of optimization relative to ψ_1 to the graph $\Delta(T)$. As a result, we obtain the proper subgraph $(\Delta(T))_{\psi_1}$ of the graph $\Delta(T)$. Denote this subgraph by G_1. According to Proposition 1 and Theorem 2, the set of decision trees corresponding to the node T of this graph coincides with the set of all irredundant decision trees for the table T which have minimum complexity relative to ψ_1. Denote this set by D_1. Using Proposition 2 we conclude that decision trees from D_1 are optimal relative to ψ_1 not only among irreducible decision trees but also among all decision trees for the table T.

Now we apply the procedure of optimization relative to ψ_2 to the graph G_1. As a result, we obtain the proper subgraph $(G_1)_{\psi_2}$ of the graph $\Delta(T)$. Denote this subgraph by G_2. The set of decision trees corresponding to the node T of this graph coincides with the set of all decision trees from D_1, which have minimum complexity relative to ψ_2. It is possible to continue this process of consecutive optimization relative to various criteria.

If ψ_2 is a monotone complexity measure, then according to Theorem 1 the set of decision trees, corresponding to the node T of the graph G_2, is a subset of the set of all decision trees from D_1, which have minimum complexity relative to ψ_2.

6   Results of Experiments

We consider in this section the results of experiments with decision table mushroom from UCI ML Repository [1] which contains 22 conditional attributes and 8124 rows. We denote this table by T.

First, we construct the graph $\Delta(T)$ which contains 58800 nonterminal nodes (sub-tables). Applying the procedure of optimization relative to four complexity measures, we found that:

- The minimum number of nodes in a decision tree for T is equal to 21;
- The minimum number of nonterminal nodes in a decision tree for T is equal to 5;
- The minimum average depth of a decision tree for T is equal to 1.52388;
- The minimum depth of a decision tree for T is equal to 3.

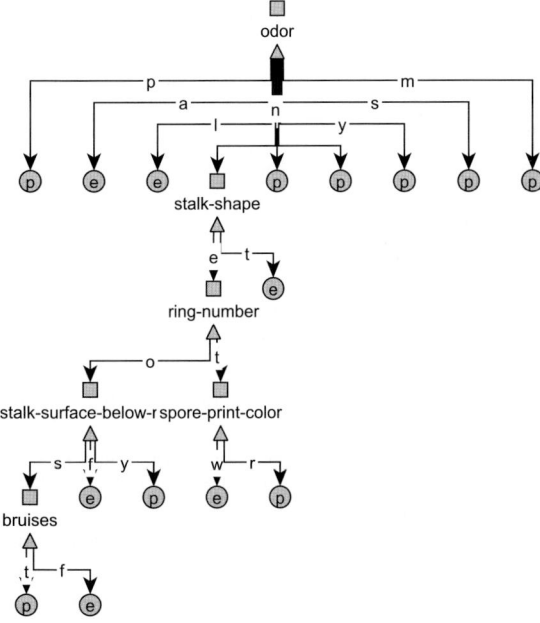

Fig. 4. Decision tree example for decision table mushroom (the picture is rendered using yEd Graph Editor [8])

We make a sequential optimization of decision trees for T regarding the following complexity measures: number of nodes, number of nonterminal nodes, average depth, and depth.

After the first step, we obtain a proper sub-graph of the graph $\Delta(T)$ which contains only 18 nodes, after the second step we have the same number of nodes, after the third step we have 6 nodes, and after the last step we have also 6 nodes.

After the first step, the number of optimal decision trees is equal to 38, after the second step we have the same number of optimal decision trees, and after the third step we have only one optimal decision tree.

After the last step, we obtain only one decision tree (see Figure 4) with the following parameters: the number of nodes is 21 (the minimum is 21), the number of nonterminal nodes is 6 (the minimum is 5), the average depth is 1.72772 (the minimum is 1.52388), and the depth is 5 (the minimum is 3). It means that the results of optimization will depend on the order of complexity measures during sequential optimization.

7   Conclusions

The paper is devoted to consideration of a new tool for study of optimal decision trees. The features of this tool are illustrated by applying it to the decision table mushroom from UCI ML Repository. Further studies will be connected with extension of this tool to decision tables which contain continuous attributes.

References

1. Asuncion, A., Newman, D.J.: UCI Machine Learning Repository. University of California, School of Information and Computer Science, Irvine, CA (2007), http://www.ics.uci.edu/~mlearn/MLRepository.html
2. Breiman, L., Friedman, J.H., Olshen, R.A., Stone, C.J.: Classification and Regression Trees. Wadsworth & Brooks (1984)
3. Chegis, I.A., Yablonskii, S.V.: Logical methods of electric circuit control. Trudy MIAN SSSR 51, 270–360 (1958) (in Russian)
4. Moshkov, M., Chikalov, I.: Consecutive optimization of decision trees concerning various complexity measures. Fundamenta Informaticae 61(2), 87–96 (2004)
5. Chikalov, I., Moshkov, M., Zelentsova, M.: On optimization of decision trees. In: Peters, J.F., Skowron, A. (eds.) Transactions on Rough Sets IV. LNCS, vol. 3700, pp. 18–36. Springer, Heidelberg (2005)
6. Nguyen, H.S.: On efficient construction of decision trees from large datasets. In: Ziarko, W.P., Yao, Y. (eds.) RSCTC 2000. LNCS (LNAI), vol. 2005, pp. 354–361. Springer, Heidelberg (2001)
7. Pawlak, Z.: Rough Sets – Theoretical Aspects of Reasoning about Data. Kluwer Academic Publishers, Dordrecht (1991)
8. yEd Graph Editor, http://www.yworks.com/en/index.html

Automatic Part of Speech Tagging for Arabic: An Experiment Using Bigram Hidden Markov Model

Mohammed Albared, Nazlia Omar, Mohd. Juzaiddin Ab Aziz,
and Mohd Zakree Ahmad Nazri

University Kebangsaan Malaysia, Faculty of Information Science and Technology,
Department of Computer Science
mohammed_albared@yahoo.com, {no,din}@ftsm.ukm.my
http://www.ukm.my

Abstract. Part Of Speech (POS) tagging is the ability to computationally determine which POS of a word is activated by its use in a particular context. POS tagger is a useful preprocessing tool in many natural languages processing (NLP) applications such as information extraction and information retrieval. In this paper, we present the preliminary achievement of Bigram Hidden Markov Model (HMM) to tackle the POS tagging problem of Arabic language. In addition, we have used different smoothing algorithms with HMM model to overcome the data sparseness problem. The Viterbi algorithm is used to assign the most probable tag to each word in the text. Furthermore, several lexical models have been defined and implemented to handle unknown word POS guessing based on word substring i.e. prefix probability, suffix probability or the linear interpolation of both of them. The average overall accuracy for this tagger is 95.8

Keywords: Arabic languages, Hidden Markov model, Smoothing methods, Part of speech tagging.

1   Introduction

Part of speech disambiguation is the ability to computationally determine which POS of a word is activated by its use in a particular context. POS tagging is an important pre-processing activity in many NLP applications like speech processing and machine translation. Being one of the first processing steps in any such application, the performance of the POS tagger directly impacts the performance of any subsequent text processing steps. The main challenges in POS tagging involve resolving the MorphoSyntactic ambiguity where a word can occur with different POS tags in different contexts, and handling unknown words i.e. words that appear in the test data but they do not appear in the training data.

Arabic language is a Semitic language spoken by over 250 million people and the language is used by 1.2 billion Muslims. Arabic language has many varieties that diverge widely from one another. Arabic is a strongly structured, highly

J. Yu et al. (Eds.): RSKT 2010, LNAI 6401, pp. 361–370, 2010.

derivational and grammatically ambiguous language. In spite of the widely use of the Arabic language in the world and the huge amount of electronic available Arabic data online, Arabic is still a language, for which very few computational linguistic resources have been developed. Up to date, there are a few annotated corpora for Arabic language, which are not freely available for research activity. Moreover, the domain of the text in these corpora is limited to newswire data. They are only suitable for some domains. As a matter of fact, taggers trained with tagged corpora from a domain perform quite poorly in others. So our motivation for the development of this work is to provide a tool that can help to accelerate the development of such resources, which in their turn can be used in further research to train more accurate NLP tools.

In this paper, we have developed a bigram HMM POS tagger for Arabic text. The HMM parameters are estimated from our small training data. We have studied the behavior of the HMM applied to Arabic POS tagging using small amount of data. For dealing with sparseness problem, we use several smoothing algorithms. Furthermore, we use word string information (suffix and prefix) features, beside the context information (previous tag) to guess the POS tag of an unknown word. The tagger has an accuracy of about 95% on the test data provided.

The remainder of this paper is organized as follows: First, some previous work in the field is described in Section 2. A brief overview of the Arabic language is introduced in Section 3, and then the used tagset and data is described in Section 4. Section 5 introduces the used smoothing methods. Unknown word handling is described in Section 6. The results of experiments are introduced in Section 7 together with a short discussion. Finally, Section 7 concludes the paper and points to ways in which we believe that the results can be improved in the future.

2   Related Work

Different techniques have been used for POS tagging in the literature. Some of these techniques focused on rule based using linguistic rules. On the other side, POS tagging is viewed as a classification problem: the tagset is the classes and a machine learning algorithm is used to assign each occurrence of a word to one class based on the evidence from the context. This machine learning methods range from methods with full supervision, to fully unsupervised methods. The supervised POS disambiguation approaches use machine learning techniques to learn a classifier from labeled training sets such as Maximum Entropy Model [1], Hidden Markov Model [2] and Conditional Random field [3].

Unsupervised POS disambiguation methods are based on unlabeled corpora, and do not exploit any manually POS tagged corpus such as Contrastive Estimation [4], Fully Bayesian approach [5] and Prototype Driven Learning [6].

Different Arabic taggers have been recently proposed. Some of them use supervised machine learning approaches such as [7] and [8], while others are using hybrid approaches, combining rule based approach and statistical approach such as [9] (for more details about works on Arabic taggers, see [10])

Almost all of these taggers are generally developed for Modern Standard Arabic (MSA) and few works are interested in Classical Arabic [7]. Moreover, most of these taggers are assumed closed vocabulary assumption in which the test data only contain words that appear in the training data.

For these reasons, we developed our tagger which is trained and tested on a hybrid annotated data, which come from both MSA and CA. Moreover, in order to overcome the limitation of the small training data, we propose several lexical models based on word substring information to handle unknown word POS problem.

3 Overview of Arabic Language

Arabic language is a Semitic language spoken by over 250 million people and the language is used by 1.2 billion Muslims. It is one of the seven official languages of the United Nations. Arabic is a strongly structured and highly derivational language. The Arabic language has many varieties that diverge widely from one another. Arabic language consists of Modern standard Arabic and set of spoken dialects.

Arabic words are quite different from English words, and the word formation process for Arabic words is quite complex. The main formation of English word is concatenative i.e. simply attach prefixes and suffixes to derive other words . In contrast, the main word formation process in Arabic language is inherently non-concatenative. Words are derived by interdigitating roots into patterns. Arabic words is derived by inserting root's radical characters into pattern's generic characters. The first radical is inserted into the first generic character, the second radical fills the second generic and the third fills the last generic as shown in Table 1.

Arabic language is grammatically ambiguous. There are many sources of ambiguity in Arabic language such as:

- Arabic is highly inflective language: Arabic words are often formed by concatenating smaller parts. A word may be segmented into different candidate sequences of segments each of which may be ambiguous with respect to the choice of a POS tag. Moreover, Arabic words are segmentation ambiguous;

Table 1. Arabic Word deviation process

Pattern	Pattern in Arabic	Root	The resulting word
XوXX	فعول	علم	علوم
اXتXIX	افتعال	قصد	اقتصاد
يستXXX	يستفعل	عمل	يستعمل
تXXX	تفعل	طور	تطور
يXXXون	يفعلون	عمل	يعملون

there is different alternative segmentation of the words. The result is a relatively high degree of ambiguity.

- The lack of the short vowels in the modern Arabic text. Short vowels are special symbols that smooth the processes of reading and understanding the Arabic text especially for foreign readers. An Arabic word without short vowels can have different meanings and also different POS .However, in almost all modern Arabic text, short vowels are no longer used.

4   The Used Data and Tagset

Arabic linguists and grammarians classify Arabic words into nominals (أَسْمَاء),

verbs (أَفْعَال) and particles (حروف). The nominals include nouns, pronouns, adjectives and adverbs. The verbs include perfect verbs, imperfect verbs and imperative verbs. The particles include prepositions, conjunctions and interrogative particles, as well as many others. Our tagset have been inspired by Arabic TreeBank (ATB) POS Guidelines [11].The used tagset consists of 23 tags (see Table 2). Our training corpus consists of 26631 manually annotated word forms from two types of Arabic texts: the Classical Arabic text and from Modern Standard Arabic. We use this corpus to train and test our tagger. We split the corpus into training set with size 23146 words and test set with size 3485 words.

Table 2. Arabic Part-of-speech tagset

POS Tag	Label	POS Tag	Label
Conjunction	**CC**	Possessive Pronoun	**POSS_PRON**
Number	**CD**	Imperfective Verb	**VBP**
Adverb	**ADV**	Non Inflected Verb	**NIV**
Particle	**PART**	Relative Pronoun	**REL_PRON**
Imperative Verb	**IV**	Interjection	**INTERJ**
Foreign Word	**FOREIGN**	Interrogative Particle	**INTER_PART**
Perfect Verb	**PV**	Interrogative Adverb	**INTER_ADV**
Passive Verb	**PSSV**	Demonstrative Pronoun	**DEM_ PROP**
Preposition	**PREP**	Punctuation	**PUNC**
Adjective	**ADJ**	Proper Noun	**NOUN_PROP**
Singular Noun	**SN**	Personal Pronoun	**PRON**
Plural Noun	**SPN**		

5   Our Approach: The Bigram HMM Tagger

Hidden Markov Model (HMM) is a well-known probabilistic model, which can predict the tag of the current word given the tags of one previous word (bi-gram) or two previous words (trigram). The HMM tagger assign a probability value to each pair$\langle w_1^n, t_1^n \rangle$, where $w_1^n = w_1.....w_n$ is the input sentence and $t_1^n = t_1....t_n$ is the POS tag sequence. In HMM, the POS problem can be defined as the finding the best tag sequence t_1^n given the word sequence w_1^n. This is can be formally expressed as:

$$t_1^n = \arg\max_{t_1^n} \prod_{i=1}^n p\left(t_i|t_{i-1}\ldots.t_1\right).p\left(w_i|t_i\ldots.t_1\right). \tag{1}$$

Since it is actually difficult or even impossible to calculate the probability of the form $p\left(t_i|t_{i-1}\ldots.t_1\right)$ for large i because of the sparse problem, bi-gram or tri-gram model is often used to alleviate data sparseness problem. Here we assumed that the word probability is constrained only by its tag, and that the tag probability is constrained only by its preceding tag, the bigram model:

$$t_1^n = \arg\max_{t_1^n} \prod_{i=1}^n p\left(t_i|t_{i-1}\right).p\left(w_i|t_i\right). \tag{2}$$

The probabilities are estimated with relative frequencies from the training data. The tag-word pair probabilities and tag transition probabilities are commonly called as lexical probabilities, and transition probabilities, respectively. Having computed the lexical and transition probabilities and assigning all possible tag sequences to all words in a sentence, now we need an algorithm that can search the tagging sequences and find the most likely sequence of tags. For this purpose, we use Viterbi algorithm to find the most probable sequence of POS tags (t_1^n) for a given sentence. The Viterbi algorithm is the most common decoding algorithm for HMMs that gives the most likely tag sequence given a set of tags, transition probabilities and lexical probabilities.

6   Smoothing

The simplest way to compute the parameters for our HMM is to use relative frequency estimation, which is to count the frequencies of word/tag and tag/tag. This way is called Maximum Likelihood Estimation (MLE). MLE is a bad estimator for statistical inference, because data tends to be sparse. This is true even for corpus with large number of words. Sparseness means that various events, here bigrams transition, are either infrequent or unseen. This leads to zero probabilities being assigned to unseen events, causing the probability of the whole sequence to be set to zero when multiplying probabilities. There are many different smoothing algorithms in the literature to handle sparseness problem [12], all of them consisting of decreasing the probability assigned to the known event and distributing the remaining mass among the unknown events. Smoothing algorithms are primarily used to better estimate probabilities when there is insufficient data to estimate probabilities accurately. In the following subsections, we give a brief description of the used smoothing techniques.

6.1 Laplace Estimation

Laplace estimation is the simple smoothing method for data sparseness [13]. In this method, we add one for each bigram frequency count either seen or unseen in the training data. Since there is T tags in the tagset, and each one got incremented, we also need to adjust the denominator. We add the number of tags in the tagset to the unigram count of the tag t_{i-1}. The main problem of Laplace method is that it assigns overvalued probabilities to unseen bigram, which is principally not true.

$$P_{Laplace}(t_i|t_{i-1}) = \frac{C(t_{i-1}t_i) + 1}{C(t_{i-1}) + T} . \tag{3}$$

6.2 The Kneser Ney Smoothing

Chen et al. [12] showed that Kneser-Ney smoothing and its variations outperform all other smoothing techniques. In Kneser-Ney smoothing, the transition probabilities is formulated as:

$$P_{KN}(t_i|t_{i-1}) = \frac{\max\{f(t_{i-1}t_i) - D, 0\}}{\sum_{t_i} f(t_{i-1}t_i)} + \frac{D \cdot N_{1+}(t_{i-1}, \bullet)}{\sum_{t_i} f(t_{i-1}t_i)} \cdot \frac{N_{1+}(\bullet, t_i)}{N_{1+}(\bullet\bullet)} . \tag{4}$$

where $N_{1+}(\bullet, \bullet) = |\{(t_{i-1}t_i) : f(t_{i-1}t_i)1\}|$, $D = \frac{n_1}{n_1 + 2n_2}$ and n1 and n2 are the total number of bi-grams with exactly one or two counts, respectively, in the training corpus.

6.3 The Modified Kneser Ney Smoothing

Modified Kneser-Ney smoothing is an interpolated variation of Kneser Ney smoothing with an augmented version of absolute discounting. Modified Kneser-Ney smoothing is proposed by [12].They found that it outperforms all other methods. However, in this smoothing algorithm,the transition propability $p(t_i|t_{i-1})$ is calculated as in the following equation:

$$P_{MKN}(t_i|t_{i-1}) = \frac{f(t_{i-1}t_i) - D(f(t_{i-1}t_i))}{\sum_{t_i} f(t_{i-1}t_i)} + \gamma(t_{i-1}) \cdot \frac{N_{1+}(\bullet, t_i)}{N_{1+}(\bullet\bullet)} . \tag{5}$$

where $D(f) = \begin{cases} 0 & \text{iff} = 0 \\ D_1 & \text{iff} = 1 \\ D_2 & \text{iff} = 2 \\ D_{3+} & \text{iff} \geq 3 \end{cases}$ and $\gamma(t_{i-1}) = \frac{D_1 N_1(t_{i-1}, \bullet) + D_2 N_2(t_{i-1}, \bullet) + D_{3+} N_{3+}(t_{i-1}, \bullet)}{\sum_{t_i} f(t_{i-1}t_i)}$.

The D values for the model are calculated using the following equations:
$Y = \frac{n_1}{n_1 + 2n_2}$,
$D_1 = 1 - 2Y\frac{n_2}{n_1}, D_2 = 2 - 3Y\frac{n_3}{n_2}$ and $D_3 = 3 - 4Y\frac{n_4}{n_3}$.

7   Unknown Words Handling

Applying a POS tagger to new data can result in two types of errors: misclassification of the ambiguous words and failure to find the right class label for an unknown word that appears in the test data but not in the training data. The number of words in the corpus is always finite. It is not possible to cover all words of the language. In addition, new words are being created everyday, so it is very difficult to keep the dictionary up-to-date. Handling of unknown words is an important issue in POS tagging. For words which have not been seen in the training set, is estimated based on features of the unknown words, such as whether the word contains a particular suffix. We use a successive abstraction scheme. The probability distribution for an unknown word suffix is generated from all words in the training set that have the same suffix up to some predefined maximum length. This method was proposed by Samuelsson [14] and implemented for English and German [2]. The method is expressed in the following equation:

$$P\left(t|c_{n-i+1}, ..., c_n\right) = \frac{P(t, c_{n-i+1}, ..., c_n) + \theta P(t, c_{n-i+2}, ..., c_n)}{1 + \theta} . \qquad (6)$$

$$\theta = \frac{1}{S-1} \sum_{J=1}^{S} (P\left(t_j\right) - \overline{P})^2, \overline{P} = \frac{1}{s} \sum_{J=0}^{S} P\left(t_j\right)$$

where $c_{n-i+1}, ..., c_n$ represent the last n characters of the word. In addition to word suffix, the experiments utilize the following features: the presence of non-alphabetic characters and the existence of foreign characters. In addition to the suffix guessing model, we define another lexical model to handle unknown words POS guessing based on word prefix information. The new lexical model is expressed as in the following equation:

$$P(t|w) = P\left(t|c_1, ..., c_m\right) . \qquad (7)$$

where $c_1, ..., c_m$ represent the first m characters of the word w . The probability distribution in Equation 7 is estimated as in Equation 6. Furthermore, we have defined another lexical model based on combining both word suffix information and word prefix information. The main linguistic motivation behind combining affixes information is that in Arabic word sometimes an affix requires or forbids the existence of another affix [15]. Prefix and suffix indicate substrings that come at the beginning and end of a word respectively, and are not necessarily morphologically meaningful. In this model, the lexical probabilities are estimated as follows:

1. Given an unknown word w, the lexical probabilities P(suffix(w)|t) are estimated using the suffix tries as in Equation 6.
2. Then, the lexical probabilities P(prefix(w)|t) are estimated using the prefix tries as in Equation 6. Here, the probability distribution for a unknown word prefix is generated from all words in the training set that have the same prefix up to some predefined maximum length.

3. Finally, we use the linear interpolation of both the lexical probabilities obtained from both word suffix and prefix to calculate the lexical probability of the word w as in the following equation :

$$P\,(w|t) = \lambda P\,(\text{suffix}(w)|t) + (1 - \lambda)P(\text{prefix}(w)|t)\ . \tag{8}$$

where λ is an interpolation factor, experimentally set to 0.7. prefix(w) and suffix(w) are the first m and the last n characters, respectively.

8   Experiments and Results

For our experiments, we evaluated the performance of the bigram HMM with different smoothing techniques. We divide our corpus into training set with size 23146 words and test set with 3485 words. The test set contains 263 unknown words. We define the tagging accuracy as the ratio of the correctly tagged words to the total number of words. Table 3 summarizes the results of Bigram Hidden Markov Model with smoothing methods.Among smoothing algorithms, the modified Kneser-Ney approach yields the highest results, with the Kneser-Ney algorithm close behind, at 95.4. Finally, The Laplace algorithm achieved an accuracy of 95.1. From Table 3, we notice that the different smoothing techniques exhibit the same behaviors due to the reasons that Arabic is free ordering which reduced the number of the missing bigrams transition probabilities and also due to the small size of the used tagset. However, the results of unknown words POS guessing in Table 3 is achieved using suffix guessing technique with successive abstraction. These results also show that this techniques which proved to be effective for some languages does not work well for Arabic language. The suffix guessing algorithem proved to be a good indicator for unknown word POS guessing in English and German[2].

Furthermore, in Table 4, we summarize and compare the results of the three unknown word POS guessing algorithms. These algorithms are the suffix guessing algorithm, the prefix guessing algorithm and the linear interpolation of both prefix guessing algorithm and suffix guessing algorithm for unknown words POS guessing. As we can see, the third algorithm which combine information from both suffix and prefix, gives a considerable rise in accuracy compared to the suffix guessing method. This result of the combined algorithm is achieved when the value of in Equation 8 experimentally set to 0.7. We can see from these expriments that Arabic word suffix is a better indicator than its prefix in guessing

Table 3. Tagging accuracy obtained with bigram HMM coupled with smoothing techniques

Smoothing	Overall acc.	Unknown words acc
Laplace	95.1	65.8
Kneser-Ney	95.4	70.3
modified Kneser-Ney	95.5	70.7

its POS. But, the combination of both word suffix and prefix does better than using one of them alone.In general,unknown word accuracy for Arabic is not so high due to many reasons such as Arabic has complex morphology than English i.e. the non concatenative nature of Arabic words and the Arabic words affixes are ambiguous, short and sparse. However, we have obtained an overall accuracy of 95.8% which is quite encouraging especially with the small size of our training corpus. The main remark which can be extracted from these results is that a competitive Arabic HMM taggers may be built using relatively small train sets, which is interesting especially, with the lack of language resources.

Table 4. Unknown word POS tagging tagger accuracies with the three lexical models

Model	% of unknown word	Unknown acc.	Overall acc.
Suffix guessing algorithm	7.5	70.7	95.5
Prefix guessing algorithm	7.5	60.8	94.7
The combined guessing algorithm	7.5	73.5	95.8

9   Conclusion

Part of speech tagging is an important tool in many NLP applications. In addition, it accelerates the development of large manually annotated corpora. In this paper, we described preliminary experiments in designing HMM-based part-of speech tagger for Arabic language. We have investigated the best configuration of HMM when small amount of data is available. We implemented bigram HMM-based disambiguation procedure. HMM parameters were automatically estimated from our training corpus of size 23146 words. We conducted a series of experiments using different smoothing techniques, such as Laplace and Kneser-Ney and the modified Kneser-Ney. Furthermore, three lexical models have been used to handle unknown words POS tagging. These algorithms are suffix tries and successive abstraction and prefix guessing algorithm and the linear interpolation of both suffix guessing algorithm and prefix guessing algorithm. Our Arabic part-of speech tagger achieves near 95.8 % tagging accuracy. This result is very encouraging especially with the small size of our training corpus. In the future, we plan to improve the tagging accuracy of unknown words by increasing the size of our training corpus as well as by using specific features of Arabic words, which can be better predictor for Arabic words POS than words suffixes and prefixes.

References

1. Ratnaparkhi, A.: A maximum entropy part of speech tagger. In: Brill, E., Church, K. (eds.) Conference on Empirical Methods in Natural Language Processing, University of Pennsylvania (1996)
2. Brants, T.: TnT: A statistical part-of-speech tagger. In: Proceedings of the 6th Conference on Applied Natural Language Processing, Seattle, WA, USA (2000)

3. Lafferty, J., McCallum, A., Pereira, F.: Conditional random fields: Probabilistic models for segmenting and labeling sequence data. In: Proceedings of the International Conference on Machine Learning, MA, USA (2001)

4. Smith, N., Eisner, J.: Contrastive estimation: Training log-linear models on unlabeled data. In: Proceedings of ACL 2005, pp. 354–362 (2005)

5. Goldwater, S., Griffiths, T.: A fully Bayesian approach to unsupervised part-of-speech tagging. In: Proceedings of the 45th Annual Meeting of the Association for Computational Linguistics (2007)

6. Haghighi, A., Klein, D.: Prototype-driven learning for sequence models. In: Proceedings of the Main Conference on Human Language Technology Conference of the North American Chapter of the ACL, pp. 320–327 (2006)

7. El Hadj, Y., Al-Sughayeir, I., Al-Ansari, A.: Arabic Part-Of-Speech Tagging using the Sentence Structure. In: Proceedings of the Second International Conference on Arabic Language Resources and Tools, Cairo, Egypt (2009)

8. Al Shamsi, F., Guessoum, A.: A hidden Markov model-based POS tagger for Arabic. In: Proceeding of the 8th International Conference on the Statistical Analysis of Textual Data, France, pp. 31–42 (2006)

9. Zribi, C., Torjmen, A.: An Efficient Multi-agent System Combining POS-Taggers for Arabic Texts. In: Gelbukh, A. (ed.) CICLing 2006. LNCS, vol. 3878, pp. 121–131. Springer, Heidelberg (2006)

10. Albared, M., Omar, N., Ab Aziz, M.J.: Arabic Part of Speech Disambiguation: A Survey. International Review on Computers and Software, 517–532 (2009)

11. Maamouri, M., Bies, A., Krouna, S., Gaddeche, F., Bouziri, B.: Arabic Tree Banking Morphological Analysis And POS Annotation, Ver. 3.8, Linguistic Data Consortium, Univ. Pennsylvania (2009),
   http://projects.ldc.upenn.edu/ArabicTreebank/
   ATB-POSGuidelines-v3.7.pdf

12. Chen, S.F., Goodman, J.: An empirical study of smoothing techniques for language modeling. Technical Report TR-10-98, Computer Science Group, Harvard University (1998)

13. Jurafsky, D., Martin, J.H.: SPEECH and LANGUAGE PROCESSING: An Introduction to Natural Language Processing, Computational Linguistics and Speech Recognition. Prentice-Hall, Englewood Cliffs (2000)

14. Samuelsson, C.: Handling sparse data by successive abstraction. In: COLING 1996, Copenhagen, Denmark (1996)

15. Attia, M.: Handling Arabic Morphological and Syntactic Ambiguity within the LFG Framework with a View to Machine Translation. PhD thesis, School of Languages, Linguistics and Cultures, Univ. of Manchester, UK (2008)

Application of Rough Sets Theory in Air Quality Assessment

Pavel Jirava, Jiri Krupka, and Miloslava Kasparova

Institute of System Engineering and Informatics, Faculty of Economics and Administration, University of Pardubice, Studentska 84, Pardubice, Czech Republic
{Pavel.Jirava,Jiri.Krupka,Miloslava.Kasparova}@upce.cz

Abstract. This paper analyses rough sets approaches to air quality assessment in given locality of the Czech Republic (CR). Original data for modeling we obtained from the daily observation of air polluting substances concentrations in town Pardubice. Two applications were used for decision rules computation and achieved results are compared and analysed. Output of this paper is the proposal how to assign an air quality index (AQI) to the selected locality, on the basis of various attributes.

Keywords: Rough sets, decision trees, region, air quality assessment.

1 Introduction

In this paper the rough sets theory (RST) in processing of air pollution data is applied. Air pollution, its measurement and control is one of rising environmental threats. There are a number of different types of air pollutants, including suspended particulate matter, lead, sulphur dioxide, carbon monoxide and nitrogen oxide. Sources of suspended particulate matter include incomplete fuel combustion and vehicle exhaust gases, particularly from diesel engines. Influence of the polluted environment on human health should be dual, some negative environmental risks arise because of natural fluctuations; others are due to human activity (anthropogenic). Most known examples of anthropogenic influence on the environment are [1]: pesticides use, stratospheric ozone depletion, intercontinental transport of wind-borne dust and air pollutants, drinking water disinfection and increasing levels of anthropogenic pollutions into air, soil and groundwater. Health effects include increased incidence of respiratory diseases, such asthma, bronchitis and emphysema, and increased mortality among children and elderly people [2]. Due to these negative effects, it is important to have appropriate tools for air quality assessment and analysis. Finally, the air quality protection is part of the State environmental policy of the CR.

This paper deals with approaches to air pollution data classification, which could be helpful for air quality assessment. It is focused on observations of air polluting substances concentrations in Pardubice - county seat of the Pardubice region. This town is also the largest economical center with a high density of industry and commercial and public services. Industrial production has a mixed

J. Yu et al. (Eds.): RSKT 2010, LNAI 6401, pp. 371–378, 2010.

structure. Besides the chemical industry there is also strong engineering, energy, and optical devices production. Pardubice belongs to air pollution areas with regard to an industrial enterprises existence, heavy traffic and other factors. This article aims to apply the RST to classification of air pollution data with a focus on data from a particular region.

2   Data Description and Data Pre-processing

Original data was obtained from the daily observation of air polluting substances concentrations in 2008 in town Pardubice. Measurements were realized by the mobile monitoring system HORIBA. In Pardubice town 8 of these mobile devices are used. They are equipped with automatic analyzer from the Horiba Company, meteorology devices from the company Thies-Clima, and with dust sampler FAG. We obtained 522 measurements (objects, observations) from 7 localities of Pardubice town. These localities are the following: Paramo, Polabiny II, the traffic junction Palachova - Pichlova, Lazne Bohdanec, Namesti Republiky, Rosice, and Rybitvi. Achieved objects were described by 17 attributes.

In step of data pre-processing, that is an important parts of modeling process [3], 3 new attributes were derived (number of locality (attribute l_2); day of the week (attribute c_1); and part of day (attribute c_3) from attribute time of measurement c_5). In the next step AQI y_k was specified. Values of y_k are determined by the Czech Hydrometeorological Institute [4]. This approach to definition of AQI y_k is based on the highest value of measured concentrations of pollutants. In Table 1 we can see different levels of air pollution in $[\mu g/m^3]$. AQI values are from 1 "very good air quality" to 6 "very bad air quality". For more information about specification of AQI y_k in our case (frequencies of linguistic values of AQI y_k are 5 for "Very good", 183 for "Good", 212 for "Favorable", 55 for "Satisfactory", 49 for "Bad", and 18 for "Very bad") and detailed analysis of used attributes see in [5]. Final dimension of data matrix $\mathbf{M}$ is (522×21). Every observation o_i for $i=1, 2, ..., 522$ can be described by the following vector:

$$\mathbf{o}_i = (c_{0i}, c_{1i}, c_{2i}, c_{3i}, c_{5i}, x_{1i}, x_{2i}, ..., x_{6i}, x_{8i}, m_{1i}, m_{2i}, ..., m_{6i}, l_{1i}, l_{2i}, y_k). \quad (1)$$

Description of attributes is following (Name of attribute, Type): c_0 is Number of measurement, integer; c_1 is Day of week (number), integer; c_2 is Month, integer;

Table 1. Air Quality Index

Air quality	Index	SO$_2$ per 1h	NO$_2$ per 1h	CO per 8h	O$_3$ per 1h	PM$_{10}$ per 1h
Very good	1	0-25	0-25	0-1000	0-33	0-15
Good	2	25-50	25-50	1000-2000	33-65	15-30
Favorable	3	50-120	50-100	2000-4000	65-120	30-50
Satisfactory	4	120-250	100-200	4000-10000	120-180	50-70
Bad	5	250-500	200-400	10000-30000	180-240	70-150
Very bad	6	500-	400-	30000-	240-	150-

Table 2. Descriptive Characteristics of Selected Attributes

Attribute	Min	Max	Mean	St.dev	Valid
c_1	1	4	-	-	522
c_2	1	9	-	-	522
c_3	1	5	-	-	522
x_1	2.600	67.000	10.807	6.809	522
x_2	2.800	94.600	27.470	18.596	522
x_3	0.100	2.200	0.464	0.347	522
x_4	0.200	142.200	54.429	29.152	522
x_5	1	391	39.738	47.196	522
x_6	2.400	278.700	25.974	35.296	522
x_8	6.000	502.100	66.710	69.303	522
m_1	0.000	4.200	0.636	0.681	522
m_2	0	359	194.205	105.256	522
m_3	0.300	37.300	14.876	9.028	522
m_4	28.600	95.900	55.581	16.252	522
m_5	876.200	1006.000	988.204	8.718	522
m_6	0	852	303.887	235.557	522
l_2	1	7	-	-	522
y_k	1	6	-	-	522

c_3 is Part of day (number), integer; c_5 is Time of measurement, time; x_1 is Sulfur dioxide [SO_2], real; x_2 is Nitrogen dioxide [NO_2], real; x_3 is Carbon monoxide [CO], real; x_4 is Ozone [O_3], real; x_5 is Suspended particles [PM_{10}], real; x_6 is Nitrogen monoxine [NO], real; x_8 is Nitrogen oxides [No_x], real; m_1 is Wind power [m/s], real; m_2 is Wind direction [grad], real; m_3 is Temperature tree meters above the surface [°C], real; m_4 is Relative air humidity [%], real; m_5 is Atmospheric pressure [hPa], real; m_6 is Solar radiation [W/m²], real; l_1 is Locality (word), string; l_2 is Locality (number), integer; y_k is AQI (number), integer. For units measurement is used International system of units and derived units. Unit [$\mu g/m^3$] is used for attribute $x_1, x_2, x_3, x_4, x_5, x_6, x_8$. Attributes c_0, c_1, c_2, c_3, c_5 and the rest of attributes are numbers without units or values expressed by words.

Basic descriptive characteristics of attributes that are used for modeling are in the Table 2. Mean values and standard deviation (St.dev) are computed for "range type attributes". Any missing values were not found in this data matrix **M**.

3   Rough-Sets Approach to Air Quality Modeling

RST is well known part of soft computing (SC). The RST [6],[7],[8],[9],[10] is based on the research of information system logical properties, and uncertainty in it is expressed by the boundary region. Every investigated object is connected to a specific piece of information, to specific data. The objects which are characterized by the same pieces of information are mutually undistinguishable from the point of view of the accessible pieces of information. This is expressed in this theory by the indiscernibity relations. Many scientific works were focused on generating rules from analyzed data and there are many various methods and processes using SC [11],[12],[13],[14],[15]. Some of them are also based on other components of SC - fuzzy sets, neural networks [16],[17],[18],[19],[20],[21]. Other processes based on areas of SC and Machine Learning are quoted in [22],[23],[24].

In this work was used mathematical apparatus of the RST to derive the rules and subsequent classification of data. For the computations we used Rough Sets Toolbox (RSTx) [25] and well known software RSES [26]. The RSTx is intended for the generation of If-Then rules. The problem of classification in our model is divided in three phases: the first is data preprocessing, the second is classification divided into RSTx rules generation and RSES rules generation and evaluation and the third is air quality assessment. The whole process is displayed in the Fig. 1.

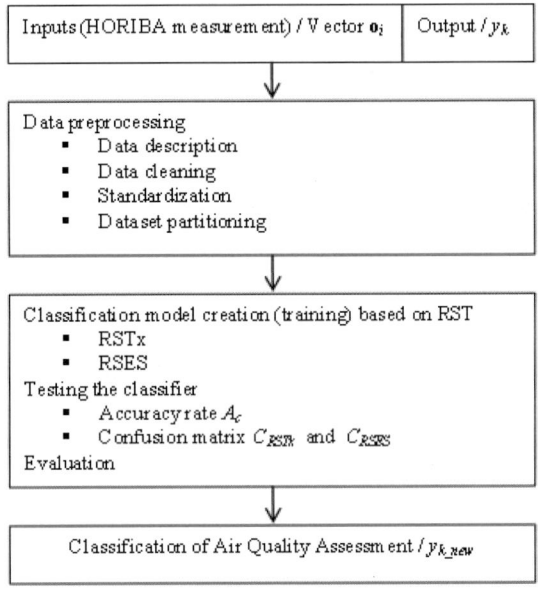

Fig. 1. The classification model design

The holdout method was used in this process. Input data matrix containing 522 objects was divided into training data set (with 417 objects) and testing data set (with 105 objects). Training data were used for decision rules calculation. RSTx calculated set of 326 decision rules. With RSES were calculated set of 50 decision rules (LEM2 algorithm were used [27]). The acquired rules can bring many information about what attributes are important in determining air quality and what attributes are not important. When comparing the obtained rules by their importance, these are the most important: If x_5 is 4 then class is 4; if x_5 is

5 then class is 5; if x_4 is 3 and x_5 is 3 then class is 3; if x_4 is 3 and x_5 is 1 then class is 3; if x_5 is 2 and x_4 is 3 then class is 3. It is clear, that the attributes of x_4 and x_5 which are contained in these rules have the significant effect on the determination of air quality. The total number of calculated rules characterizes the following Table 3.

Table 3. Number of Calculated Rules

Class y_{knew}	No. of rules with RSTx	No. of rules with RSES
1	4	2
2	106	32
3	137	13
4	35	1
5	31	1
6	13	1

To verify and compare results we use the generated rules for testing data matrix classification. Resulting confusion matrix for RSTx and RSES, containing information about actual and predicted (final) classifications is in Table 4. Best results were achieved with a set of rules generated by LEM2 algorithm with RSES. Total accuracy for C_{RSTx} respective C_{RSES} is 0.38 (respective 0.924) and total coverage is 0.45 (1). Outputs (accuracy, coverage) achieved by RSTx are weak and generated rules do not cover the input data. It is clear that the classifier could not assign 58 inputs to appropriate rule. Therefore, the accuracy is low.

Table 4. Confusion matrix C_{RSTx} and C_{RSES}

		Final values of classifier C_{RSTx} (C_{RSES})					
	Class	1	2	3	4	5	6
	1	0 (1)	0 (0)	0 (0)	0 (0)	0 (0)	0 (0)
Actual	2	3 (0)	18 (29)	2 (8)	0 (0)	0 (0)	0 (0)
observed	3	1 (0)	1 (0)	16 (42)	0 (0)	0 (0)	0 (0)
values	4	0 (0)	0 (0)	0 (0)	1 (13)	0 (0)	0 (0)
C_{RSTx} (C_{RSES})	5	0 (0)	0 (0)	0 (0)	0 (0)	4 (9)	0 (0)
	6	0 (0)	0 (0)	0 (0)	0 (0)	0 (0)	1 (3)

RSES by contrast, achieved good results. How we can see, with RSES was achieved classification accuracy 92.4% (best result achieved in our experiments), and this is result comparable with decision trees algorithms [28], which are already used for air quality data classification and analysis. In the article [28] was carried out air quality modeling with the same data as in this work. Various algorithms from the area of decision trees were used in that article. Achieved accuracy rate A_c for all used algorithms [28] was greater than 90% ($A_{cC5.0}$ is

Table 5. Classification Results

Method	Accuracy rate A_c [%]
RSES	92.40
C5.0 boost	94.43
CHAID	91.41

94.20%; $A_{cC5.0boost}$ is 94.43%; and A_{cCHAID} is 91.41%). Therefore now we can compare the results of classification accuracy in Table 5.

4 Conclusion

In the introduced paper was proposed the application of RST in the area of air quality assessment. Air pollution belongs to basic environmental problems in the world and also in the CR air quality belongs also to very important and actual questions. Improvement of air quality and its modeling is a part of a sustainable development of regions and countries. For example in [2],[29] we can found approaches to solving of these problems. In our work we focused on the air quality modeling with data collected in town Pardubice.

In this paper we first defined and described the basic terms related to the topic. Considering the fact, that experimental data on air from the Pardubice region was used, the region is also characterized. For decision rules computation were used two systems based on RST mathematical apparatus. The first is RSTx [25] developed in the MATLAB environment, the second is RSES 2.2.2 [26] developed by the Logic Group, Institute of Mathematics, Warsaw University. The resulting rules were used to classify data on air quality.

Outputs achieved by RSTx are not satisfactory, classification accuracy was only 38%. RSES seems to be a more appropriate tool, because with RSES we achieved classification accuracy 92.4%, which is comparable result as accuracy achieved with decision trees [28] algorithms. Different was also the number of generated rules - 326 decision rules with RSTx and 50 rules with RSES. Based on the results we can say that the RST is applicable to air quality assessment in the selected locality. However, it is necessary to use appropriate tools.

The areas where our future research could be directed can be divided into two groups. Firstly, it is the investigation of theoretical context and the possibilities to use other methods for hybridization - Kohonen maps, or fuzzy sets is in our opinion possible. Secondly, it is important to gain data sets based on daily air pollution observations from other CR regions and realize their analysis and comparison.

Acknowledgments. This work was supported by the project SP/4i2/60/07 "Indicators for Valuation and Modeling of Interactions among Environment, Economics and Social Relations" and 402/08/0849 "Model of Sustainable Regional Development Management".

References

1. National Research Council of The National Academies.: The Environment: Challenges for the Chemical Sciences in the 21st Century. National Academies Press, Washington DC (2003)
2. Hersh, M.: Mathematical Modelling for Sustainable Development. Springer, Heidelberg (2006)
3. Witten, I.H., Frank, E.: Data Mining: Practical Machine Learning Tools and Techniques. Morgan Kaufman Publishers, San Francisco (2005)
4. Czech Hydrometeorological Institute (2009), http://www.chmi.cz
5. Kasparova, M., Krupka, J., Jirava, P.: Approaches to Air Quality Assessment in Locality of the Pardubice Region. In: 5th Int. Conf. Environmental Accounting Sustainable Development Indicators (EMAN 2009), Prague, Czech Repulbic, pp. 1–12 (2009)
6. Komorowski, J., Pawlak, Z., Polkowski, L., Skowron, A.: Rough sets: A tutorial. In: Pal, S.K., Skowron, A. (eds.) Rough-Fuzzy Hybridization: A New Trend in Decision-Making, pp. 3–98. Springer, Singapur (1998)
7. Pawlak, Z.: Rough sets. Int. J. of Information and Computer Sciences 11, 341–356 (1982)
8. Pawlak, Z.: A Primer on Rough Sets: A New Approach to Drawing Conclusions from Data. Cardozo Law Review 22, 1407–1415 (2001)
9. Aviso, K.B., Tan, R.R., Culaba, A.B.: Application of Rough Sets for Environmental Decision Support in Industry. Clean Technologies and Environmental Policy 10, 53–66 (2007)
10. Pal, S.K., Skowron, A. (eds.): Rough-Fuzzy Hybridization: A New Trend in Decision Making. Springer, Singapore (1999)
11. Olej, V., Krupka, J.: Analysis of Decision Processes of Automation Control Systems with Uncertainty. University Press Elfa, Kosice (1996)
12. Sakai, H., Nakata, M.: On Possible Rules and Apriori Algorithm in Nondeterministic Information Systems. In: Greco, S., Hata, Y., Hirano, S., Inuiguchi, M., Miyamoto, S., Nguyen, H.S., Słowiński, R. (eds.) RSCTC 2006. LNCS (LNAI), vol. 4259, pp. 264–273. Springer, Heidelberg (2006)
13. Stanczyk, U.: On Construction of Optimised Rough Set-based Classifier. Int. J. of Mathematical Models and Methods in Applied Sciences 2, 533–542 (2008)
14. Nguyen, S.H., Nguyen, H.S.: Hierarchical Rough Classifiers. In: Kryszkiewicz, M., Peters, J.F., Rybiński, H., Skowron, A. (eds.) RSEISP 2007. LNCS (LNAI), vol. 4585, pp. 40–50. Springer, Heidelberg (2007)
15. Zhou, Z.H., Jiang, Y., Chen, S.F.: Extracting Symbolic Rules from Trained Neural Network Ensembles. AI Communications 16, 3–15 (2003)
16. Kudo, Y., Murai, T.: A method of Generating Decision Rules in Object Oriented Rough Set Models. In: Greco, S., Hata, Y., Hirano, S., Inuiguchi, M., Miyamoto, S., Nguyen, H.S., Słowiński, R. (eds.) RSCTC 2006. LNCS (LNAI), vol. 4259, pp. 338–347. Springer, Heidelberg (2006)
17. Odajima, K., Hayashi, Y., Tianxia, G., et al.: Greedy Rule Generation from Discrete Data and Its Use in Neural Network Rule Extraction. Neur. Networks. 21, 1020–1028 (2008)
18. Vascak, J., Rutrich, M.: Path Planning in Dynamic Environment Using Fuzzy Cognitive Maps. In: 6th International Symposium on Applied Machine Inteligence and Informatics (SAMII), pp. 5–9. University Press Elfa, Kosice (2008)

19. Cyran, K.A.: Comparison of Neural Network and Rule-based Classifiers Used as Selection Determinants in Evolution of Feature Space. WSEAS Transactions on Systems 6, 549–555 (2007)
20. Andoga, R., Fozo, L., Madarasz, L.: Intelligent Approaches in Modeling and Control of a Small Turbojet Engines. In: 11th International Conference on Intelligent Engineering Systems (INES 2007), p. 59. IEEE, Los Alamitos (2007)
21. Tkac, J., Chovanec, A.: The Application of Neural Networks for Detection and Identification of Fault Conditions. Metalurgija/Metallurgy 49, 566–569 (2010)
22. Simonova, S., Panus, J.: Genetic Algorithms for Optimization of Thematic Regional Clusters. In: Conf. on Computer as a Tool (EUROCON 2007), Warsaw, Poland, pp. 798–803 (2007)
23. Athanasiadis, I.N., Kaburlasos, V.G., Mitkas, P.A., Petridis, V.: Applying Machine Learning Techniques on Air Quality Data for Real-Time Decision Support. In: 1st Int. Symposium on Information Technologies in Environmental Engineering (ITEE), Gdansk, Poland, p. 51 (2003)
24. Rutkowski, L., Siekmann, J., Tadeusiewicz, R., Zadeh, L.A.: Artificial Intelligence and Soft Computing. In: Rutkowski, L., Siekmann, J.H., Tadeusiewicz, R., Zadeh, L.A. (eds.) ICAISC 2004. LNCS (LNAI), vol. 3070. Springer, Heidelberg (2004)
25. Jirava, P., Krupka, J.: Classification Model based on Rough and Fuzzy Sets Theory. In: 6th WSEAS Int. Conf. on Computational Intelligence, Man-machine Systems and Cybernetics, Puerto De La Cruz, Tenerife, Spain, pp. 199–203 (2007)
26. Skowron, A., Bazan, J., Szczuka, M.S., Wroblewski, J.: Rough Set Exploration System, version 2.2.2 (2009), http://logic.mimuw.edu.pl/~rses/
27. Grzymala-Busse, J.W., Wang, A.Z.: Modified algorithms LEM1 and LEM2 for Rule Induction from Data with Missing Attribute Values. In: 5th Int. Workshop on Rough Sets and Soft Computing (RSSC 1997) at the Third Joint Conference on Information Sciences (JCIS 1997), pp. 69–72. Research Triangle Park, NC (1997)
28. Kasparova, M., Krupka, J.: Air Quality Modelling by Decision Trees in the Czech Republic Locality. In: 8th WSEAS Int. Conf. on Applied Informatics and Communications (AIC 2008), pp. 196–201. WSEAS Press, Greece (2008)
29. Graf, H.J., Musters, C.J.M., Keurs, W.J.: Regional Opportunities for Sustainable Development: Theory, Methods and Applications. Kluwer Academic Publisher, Dordrecht (1999)

An Interactive Approach to Outlier Detection

R.M. Konijn and W. Kowalczyk

Department of Computer Science, Vrije Universiteit Amsterdam
{rmkonijn,wojtek}@few.vu.nl

Abstract. In this paper we describe an interactive approach for finding outliers in big sets of records, such as collected by banks, insurance companies, web shops. The key idea behind our approach is the usage of an easy-to-compute and easy-to-interpret *outlier score* function. This function is used to identify a set of potential outliers. The outliers, organized in clusters, are then presented to a domain expert, together with some context information, such as characteristics of clusters and distribution of scores. Consequently, they are analyzed, labelled as *non-explainable* or *explainable*, and removed from the data. The whole process is iterated several times, until no more interesting outliers can be found.

1 Introduction

Outlier detection has many applications like fraud detection, medical analysis, etc. A successful approach to detecting anomalies must involve a deep interaction between domain experts and a powerful system that can quickly analyze huge volumes of data, present preliminary findings to experts and process their feedback. In this paper we propose such an interactive system that supports the process of finding unusual records in huge databases.

The paper is organized as follows. In the next section we provide a brief overview of the relevant research on outlier detection. Then we outline our approach to the problem, which is followed by the definition of an *outlier score function*, which is the key element of our system. This function is used to measure the "degree of deviation" of every record. Outlier scores can be calculated very quickly: it takes only two passes through the data to score all records. Next, we elaborate on the interactive aspects of the outlier detection process and provide two more mechanisms to support domain experts: a graphical representation of interesting clusters and a *local outlier score function* which refines the search process even more. Finally, we illustrate our approach on two data sets: the well-known *Abalone* set and a non-public set of 35 million of records from an insurance company.

2 Related Work

The method described in this paper can be categorised as unsupervised outlier detection, with respect to a target variable. Existing methods for unsupervised outlier detection (a missing label problem) can be split into two main categories: statistical methods and distance-based methods, which can be further split into depth-based and density-based methods. Additionally, any clustering algorithm can be used for detecting outliers.

J. Yu et al. (Eds.): RSKT 2010, LNAI 6401, pp. 379–385, 2010.

Most statistical methods use as a starting point a model of probability density function which provides a reference for measuring the degree of deviation: outliers are instances with very low probabilities. The models of pdf's might be *parametric* or *semi-parametric*, like *mixture models*, [1], [5].

The model of a pdf can be used for detecting outliers in a yet another way: the bigger the impact of an instance on the model the more unusual the instance is. This approach is used in [13], where the authors describe an algorithm called SmartSifter that uses this idea to calculate outlier scores for each instance. The outlier score of an instance is defined as the Hellinger distance between two probability distributions of available data: one built for the whole data set and the other one built for the whole data set without the observed instance. In [12], the authors create labelled examples by giving positive labels to a number of higher scored data and negative labels to a number of lower scored data. Then, with the use of supervised learning an outlier filtering rule is constructed. The data is filtered using the constructed rule, and SmartSifter is run again.

Other statistical methods include the regression and time series models. For example, when a linear regression model is fitted to data then the residual can be used to determine if an instance is an outlier or not. If there are outliers in the explanatory variables as well as the target variable, a technique called *robust regression*, [10], can be used for identifying them. There is a lot of literature about outlier diagnostics in regression models, e.g., [6] and [3].

Distance based methods require a distance measure to determine the distance between two instances. The main idea is that the distance between outlying instances and their neighbours is bigger than the distance between normal instances and their neighbours, [7]. To handle differences in densities within a dataset, the Local Outlier Factor (LOF), [4], or the Connectivity-based Outlier Factor (COF), [11], or the Multi-granularity Deviation Factor (MDEF), [8], can be used.

Another approach to outlier detection, which is described in [14], assumes that some outliers are identified before starting the search process. First, the remaining instances are ranked by a score which is based on the Multi-granularity Deviation Factor (MDEF), [8], where a high score indicates a high probability of being an outlier. Next, a training set is formed from records which have low scores (i.e., they are not likely to be outliers) and the known outliers. Finally, a Support Vector Machine model is trained on this set and applied to the remaining records to identify new outliers. The whole process is repeated several times until no new outliers can be found.

3 Our Approach

Let us consider a collection D of n records with $k + 1$ fields, and let us identify one field that for some reasons is important for us. Let Y denote this selected field, and let $X_1, X_2, \ldots, X_k$ denote the remaining fields. For example, in the context of credit card transactions the selected field might represent the amount of money involved in the transaction, while the remaining fields could contain other characteristics of the transaction such as the age and the gender of the cardholder, the time and the location of the transaction, type of the terminal, etcetera. Our data set can be viewed as a sample of n observations drawn from an unknown probability distribution $P(Y, X_1, \ldots, X_k)$, where we slightly misuse the notation using the same symbols to denote random variables and the corresponding fields of our database.

Finally, we will assume that all variables $X_1 \ldots X_k$ are discrete and take a few values; otherwise we discretize them with help of any suitable discretization method. Possible values of variable X_i will be denoted by v_{ij}, where j ranges between 1 and n_i. We will use C_{ij} to denote the set of all records for which $X_i = v_{ij}$. Sometimes we will call the sets C_{ij} *clusters*. Let us notice that for any i, clusters $C_{ij}, j = 1, \ldots, n_i$, form a partitioning of D into n_i disjoint sets.

The key idea behind our approach is based on the concept of an *outlier score function* which, for a given record $(y, x_1, \ldots, x_k)$, provides a heuristic estimate of the probability $P(Y = y | X_1 = x_1, \ldots, X_k = x_k)$, in case Y is discrete, or $P(Y > y | X_1 = x_1, \ldots, X_k = x_k)$, in case Y is continuous.

In the next section we will propose several outlier score functions that are easy to compute and have simple interpretation.

When a score function is fixed, the outlier detection process is organized in the following loop, which is repeated until no new outliers can be identified:

- Calculate outlier score for each record.
- Present to a domain expert the scored data as a collection of sets C_{ij} together with some additional statistics, so (s)he could quickly identify groups of "explainable outliers" and remove them from the data.
- Calculate for each remaining record a *local outlier score* and present to the expert all records with scores smaller than a pre-specified threshold.

In the following two sections we will describe all steps in more detail.

4   Outlier Score

Let us assume for a moment that the variable Y is discrete and let us consider a record $(y, x_1, \ldots, x_k)$ from our data set D. We would like to estimate the probability of observing y in combination with values $x_1, \ldots, x_k$, i.e.,

$$P(Y = y | X_1 = x_1, \ldots, X_k = x_k).$$

We are interested in an estimate that could be quickly calculated and conceptually simple, so it could be explained to domain experts with little or no statistical knowledge. This is achieved by estimating, for every i, $p_i = P(Y = y | X_i = x_i)$ by the ratio:

$$p_i = |(Y = y)\&(X_i = x_i)| \, / \, |X_i = x_i|.$$

where $|F|$ denotes the number of records from D that satisfy formula F, and then defining the outlier score as the product of these "one dimensional" estimates:

$$OutlierScore((y, x_1, \ldots, x_k)) = \prod_{i=1}^{k} p_i.$$

Let us notice that our score function can be computed very quickly: it takes only two scans of the data set D to score every record. The first scan is used to determine the frequencies that are needed to estimate all p_i's, and in the second scan scores of all instances are computed with help of these estimates.

In case the target value Y is continuous it is no longer possible to estimate $P(Y = y | X_i = x_i)$ by counting. However, because we are interested in extreme values of Y, we can determine an upper bound for the probability $P(Y > y | X_i = x_i)$, with help of Chebyshev's inequality, [9], which bounds, for any data sample, the ratio of observations that deviate from the sample mean by more than d sample standard deviations by $1/d^2$. In case of one-tailed estimate, this bound is $1/(1 + d^2)$.

This leads to the following definition of the outlier score function:

$$OutlierScore((y, x_1, \ldots, x_k)) = \prod_{i=1}^{k} 1/(1 + d_i^2),$$

where d_i denotes the standardized score of y with respect to all records from D that satisfy $X_i = x_i$.

More precisely, d_i is calculated in the following way. First, we find all the records in D with $X_i = x_i$ and consider the set Y_i of all y's that occur in these records. Next, we calculate the mean and the standard deviation of Y_i, $mean(Y_i)$ and $std(Y_i)$. Finally, we let: $d_i = (y - mean(Y_i))/std(Y_i)$.

Obviously, when we are interested not only in extremely large values of Y, but also in extremely small values, we should use an alternative formula for the outlier score:

$$OutlierScore((y, x_1, \ldots, x_k)) = \prod_{i=1}^{k} min(1, 1/d_i^2).$$

Values of the $OutlierScore$ function are usually very small, especially for large values of k. Therefore, we will usually work with a logarithm of this function.

5   Interactive Outlier Detection

When all records are scored a domain expert should analyze the results, identifying two groups of outliers: *explainable outliers*–outliers which are not interesting, and *unexplainable outliers*–outliers that require further investigation. In order to support the expert in the process of evaluating outliers, we will proceed in 2 steps.

Step1: Elimination of explainable outliers
First, we analyze the distribution of outlier scores and decide how to set the threshold: records with scores below the threshold are considered to be outliers.

Next, for every cluster C_{ij} we find two numbers: the cluster size (the number of records in the cluster) and the percentage of outliers in this cluster. We use these numbers as the x and y coordinates to visualize properties of all clusters in a scatter plot. We are particularly interested in "big clusters" with a "high percentage" of outliers, i.e., those that are located in the upper right corner of the plot.

Finally, the domain expert can start exploring individual clusters, starting with those that are located on the Pareto front (they are most interesting). For every selected cluster a sorted list of outliers is presented to the user, so (s)he could quickly decide whether they are interesting or not. In practice, big clusters with a high percentage of outliers are not interesting–they usually contain outliers that can be easily "explained" by the domain expert, and such clusters could be excluded from further analysis.

Step2: Identification of non-explainable outliers

At this stage the most interesting (non-explainable) outliers from those that survived Step 1 are identified. The selection is guided by a *local outlier score* which is defined as follows. For every cluster C_{ij} we calculate standardized outlier score for all the remaining outliers. In other words, we calculate the mean and the standard deviation of all scores of records that belong to the cluster (therefore the term "local") and then standardize scores of records we are interested in (outliers that survived Step 1). We call these standardized scores "local outlier scores". Finally, all outliers are sorted by the local outlier score and the top N are presented to the expert for further evaluation.

After completing both steps, i.e., the identification and removal from the data some of the explainable and non-explainable outliers, the whole process (Step 1 and Step 2) is iteratively repeated until no more interesting outliers can be found. Obviously, after each iteration the set of remaining records is changed and consequently the outlier scores have to be recalculated.

6   Results and Analysis

We implemented our method for different sets of insurance data. Because these datasets are not public, we can not show detailed results. Therefore we demonstrate the working of our algorithm with the use of the Abalone dataset that is available at the UCI data repository. After that a summary of the results obtained with the insurance data will be given.

6.1   Abalone Data

The Abalone dataset, [2], contains information about characteristics and measurements of individual abalones. The dataset consists of 4177 records with the following fields: *Sex, Length, Diameter, Height, Whole weight, Shucked weight, Viscera weight, Shell weight*, and *Rings*. In our experiments we used the variable *Whole weight* as target and discretized all remaining continuous variables into 20 categories with equal frequencies.

After applying Step 1 of our algorithm, it turned out that there were no big clusters of outliers we could eliminate. Next, Step 2 of our algorithm is applied. For the top m (we take $m = 50$) outliers, those with the smallest values for the global outlier score, we calculate the local outlier score. We examine the top outliers found. In the Abalone dataset, the top 3 records that have the highest local outlier score, coincide with the top 3 outliers with respect to the global outlier score. An example of one of these outliers is presented on Figure 1 which contains two scatter plots of the most relevant attributes.

6.2   Experimental Results on Health Insurance Data

To test the algorithm on a big real-life data set, we use a health insurance data set. The data contains 35 million pharmacy claims. A single record contains characteristics of the medication, patient, doctor who prescribed the medication, and the pharmacy itself. We are interested in outliers with respect to the variable *costs* (the claim amount), which will be our target variable Y.

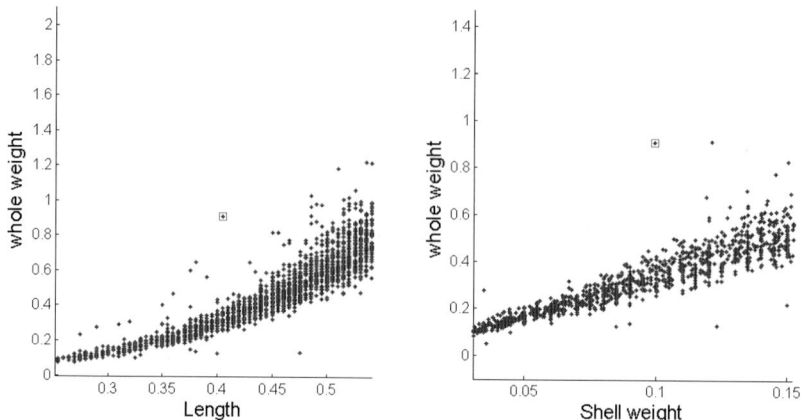

Fig. 1. Outlier 1, record number 2729 of the Abalone data set. On the left figure, the target variable *Whole weight* is plotted against *length* while the figure on the right contains the plot of *Whole weight* against *Shell weight*. The two scatter plots are zoomed-in on the outlier. Note that the scales of the y-axes on both graphs are not the same. The outlier is marked with a square around it. It can be seen that this point is an outlier.

First, we apply Step 1 of the evaluation process. For each cluster we calculate the number of outliers and the percentage of outliers. We plot them in a figure. After removing the explainable outliers, the top 50 outliers were determined with help of the local outlier score function.

In order to compare our approach to the regression-based approach, we constructed a regression tree for the data and determined some outliers by analysing standardized (per cluster) residuals. It turned out that many of the outliers found by both methods were the same. However, the proposed method was able to find small clusters or single points for which the outlying value of the target variable was "caused by" an outlying value of one predictor. For example, several records with an extremely high number of delivered pills and the corresponding extremely high costs were not detected by the regression method (the costs were proportional to the number of pills), while they were detected by our method.

7   Conclusion and Recommendations

The described method works well in practice. It is fast and therefore it can be applied in an interactive way to big data sets. It can be used while literally sitting next to a domain expert, anlyzing and removing outliers on the fly. Furthermore, it turned out to be easy to explain the method to non-statisticians and to implement it in a general purpose data analysis tools like Matlab and SAS. A possible extension of the method, especially if there is an (almost) linear relationship between an explanatory variable X and the target variable Y, is to use the two variables at the same time as target variables. Further research can focus more on feature selection or feature elimination. The method could also be extended to operate on aggregated records instead of single records only (e.g., finding outlying patients instead of single claims). In that case feature extraction and feature selection becomes even more important. Another extension would be to provide a set of rules to the domain expert, instead of single clusters.

References

1. Agarwal, D.: Detecting anomalies in cross-classified streams: a bayesian approach. Knowl. Inf. Syst. 11(1), 29–44 (2006)
2. Asuncion, A., Newman, D.: UCI machine learning repository (2007)
3. Belsley, D.A., Kuh, E., Welsch, R.E.: Regression Diagnostics: Identifying Influential Data and Sources of Collinearity. Wiley and sons, Chichester (2004)
4. Breunig, M.M., Kriegel, H.-P., Ng, R.T., Sander, J.: Lof: identifying density-based local outliers. SIGMOD Rec. 29(2), 93–104 (2000)
5. Eskin, E.: Anomaly detection over noisy data using learned probability distributions. In: Proceedings of the International Conference on Machine Learning, pp. 255–262. Morgan Kaufmann, San Francisco (2000)
6. Jensen, D.R., Ramirez, D.E., Jensenandd, D.R., Ramirez, E.: Bringing order to outlier diagnostics in regression models. In: Proceedings of Recent advances in outlier detection, summer research conference in statistics/ASA (2001)
7. Knorr, E.M., Ng, R.T.: Algorithms for mining distance-based outliers in large datasets. In: VLDB 1998: Proceedings of the 24rd International Conference on Very Large Data Bases, pp. 392–403. Morgan Kaufmann Publishers Inc., San Francisco (1998)
8. Papadimitriou, S., Kitagawa, H., Gibbons, P.B., Faloutsos, C.: Loci: Fast outlier detection using the local correlation integral. In: International Conference on Data Engineering, p. 315. IEEE Computer Society, Los Alamitos (2003)
9. Ross, S.M.: Introduction to Probability and Statistics for Engineers and Scientists, 4th edn. Elsevier Academic Press, Amsterdam (2009)
10. Rousseeuw, P., Leroy, A.: Robust regression and outlier detection. Wiley and Sons, Chichester (2003)
11. Tang, J., Chen, Z., chee Fu, A.W., Cheung, D.: A robust outlier detection scheme for large data sets. In: Proceedings 6th Pacific-Asia Conf. on Knowledge Discovery and Data Mining, pp. 6–8 (2001)
12. Yamanishi, K., Takeuchi, J.-i.: Discovering outlier filtering rules from unlabeled data: combining a supervised learner with an unsupervised learner. In: KDD 2001: Proceedings of the seventh ACM SIGKDD international conference on Knowledge discovery and data mining, pp. 389–394. ACM, New York (2001)
13. Yamanishi, K., Takeuchi, J.-I., Williams, G., Milne, P.: On-line unsupervised outlier detection using finite mixtures with discounting learning algorithms, pp. 320–324 (2000)
14. Zhu, K., Papadimitriou, F.: Example-based outlier detection with relevance feedback. DBSJ Letters 3(2) (2004)

Rules for Ontology Population from Text of Malaysia Medicinal Herbs Domain

Zaharudin Ibrahim[1,2], Shahrul Azman Noah[2], and Mahanem Mat Noor[3]

[1] Department of Information System Management,
Faculty of Information Management, Universiti Teknologi MARA,
40450 Shah Alam, Selangor, Malaysia
[2] Department of Information Science,
Faculty of Information Science and Technology, Universiti Kebangsaan Malaysia,
43600 Selangor, Malaysia
[3] School of Biosciences and Biotechnology,
Faculty Science and Technology, Universiti Kebangsaan Malaysia,
43600 Selangor, Malaysia
zahar347@salam.uitm.edu.my, samn@ftsm.ukm.my, mahanem@ukm.my

Abstract. The primary goal of ontology development is to share and reuse domain knowledge among people or machines. This study focuses on the approach of extracting semantic relationships from unstructured textual documents related to medicinal herb from websites and proposes a lexical pattern technique to acquire semantic relationships such as synonym, hyponym, and part-of relationships. The results show of nine object properties (or relations) and 105 lexico-syntactic patterns have been identified manually, including one from the Hearst hyponym rules. The lexical patterns have linked 7252 terms that have the potential as ontological terms. Based on this study, it is believed that determining the lexical pattern at an early stage is helpful in selecting relevant term from a wide collection of terms in the corpus. However, the relations and lexico-syntactic patterns or rules have to be verified by domain expert before employing the rules to the wider collection in an attempt to find more possible rules.

Keywords: Knowledge management and extraction, medicinal herb, semantic web, Natural Language Processing, knowledge engineering.

1 Introduction

Ontology is a kind of knowledge which was historically introduced by Aristotle and has recently become a topic of interest in computer science. Ontology provides a shared understanding of the domain of interest to support communication among human and computer agents. It is typically represented in a machine processable representation language [1] and is also an explicit formal specification of terms, which represents the intended meaning of concepts, in the domain and relations among them, and considered as a crucial factor for the success of many

J. Yu et al. (Eds.): RSKT 2010, LNAI 6401, pp. 386–394, 2010.

knowledge-based applications [2]. With the overwhelming increase of biomedical literature in digital forms there is a need to extract knowledge from the literature [3]. Ontology may also helpful in fulfilling the need to uncover information present in large and unstructured bodies of text that commonly referred to non-interactive literatures [4].

Ontology is considered as the backbone of many current applications, such as knowledge-based systems, knowledge management systems and semantic web applications. One of the important tasks in the development of such systems is knowledge acquisition. Conventional approaches to knowledge acquisition are mainly from interviewing domain experts and subsequently modeling and transforming the acquired knowledge into some form of knowledge representation technique. However, a huge amount of knowledge is currently embedded in various academic literatures and has the potential of being exploited for knowledge construction. The main inherent issue is that such knowledge is highly unstructured and difficult to transform into meaningful model. Although a number of automated approach in acquiring such knowledge has been proposed by [5] and [6] their success have yet to be seen. Such approaches have only been tested on general domain whereas scientific domains such as the medicinal herbs domain have yet to be explored. While automated approach seems to offer promising solutions, human still play an important role in validating the correctness of the acquired knowledge, particularly in scientific domain. Human experts are still required to construct the TBox of ontology while the ABox can be supported by some form of automated approaches. The creation of the ABox can be considered as an ontology population whereby the TBox is populated with relevant instances.

This study, therefore, proposed a set of rules for populating medicinal herbs domain ontology from unstructured text. The TBox ontology for this domain was constructed from a series of interviews with domain experts as well as analysis of available reputable literatures. Our proposed approach is based on pattern matching and Named Entity Recognition (NER), whereby semantic relations are identified by analyzing given sentences and identified entities are subsequently asserted as instance of concepts of the TBox ontology.

2   Related Research

Due to limited space, we briefly reviewed representative approaches for building herbs ontology domain. There are several researches on ontology building from unstructured text. The Hearst's technique [7] has been employed to extract concept terms from the literature and to discover new patterns through corpus exploration. The technique acquires hyponym relations automatically by identifying a set of frequently used unambiguous lexico-syntactic patterns in the form of regular expressions. Moreover [8] used these techniques to extract Hyponym, Meronym and Synonym relations from agricultural domain corpus. Two processes involved in this corpus-based ontology extraction: (1) the process for finding lexico-syntactic patterns and (2) the process for extract corpus-based ontology. The first process for finding lexico-syntactic patterns was divided into three steps. In the first step, 100 pairs of IS-A concepts, 50 pairs of Part-of concepts and 50 pairs of synonym concepts from AGROVOC [9] were selected. Second step involved extraction of sentences with selected concepts pairs previously obtained from the first step and subsequently deduce to lexico-syntactic pattern. In the third step, the patterns from the second step have been accumulated to the lexico-syntactic patterns database. Meanwhile the second process involved three steps (1) Morphological Analysis Preprocessing, (2) Term extraction, and (3) Semantic Relation Identification. There are also studies by [10] that proposed methods which are very close to the pattern-based approach of extracting the ontology from item lists, especially in technical documents, and by [11] which applied a semi-automatic approach in collecting relevant terms extracted from lexico-patterns from medicinal herb documents to be inserted as ontological term.

The study of lexico-syntactic pattern also was carried out by [12] which discovered domain-specific concepts and relationships in an attempt to extend the WordNet. The pattern-based approach was also applied by [13] which proposed a bootstrapping algorithm to detect new patterns in each iteration. [14] proposed an algorithm based on statistical techniques and association rules of data mining technology for detecting relevant relationships between ontological concepts. Meanwhile [15] presented an unsupervised hybrid text-mining approach for automatic acquisition of domain relevant terms and their relations by deploying TFIDF-based term classification method to acquire domain relevant single-word terms. An existing ontology was used as initial knowledge for learning lexico-syntactic patterns. [16] used patterns and machine learning techniques to learn semantic constraints for the automatic discovery of part-whole relation (meronymy). For this paper, we proposed the methods using the pattern-based approach, but since there are many problems due to the ambiguity of cue words of patterns and the candidate terms selection, this research present an additional method for solving these problems. We suggest extracting the ontology using seed word. In order to extract more complete information concerning concepts and relations the domain expert will determine the seed word to be used as anchor in the creation of semantic rules. This approach involves two-fold activities. First is the identification of semantic relations among instances of available concepts and, second is the construction of rules for detecting named entities corresponding to the semantic relations

3   Rules for Populating Malaysia Medicinal Herb Ontology

The main aim of this research is to automatically populate medicinal herb domain ontology by means of rules created from analyzing a set of textual documents. Our approach mainly concerns with identifying entities related to the domain such as 'herb', 'effect' and etc. These entities are related to the TBox ontology which was constructed from interviewing domain experts from the

Reproductive and Developmental Biology Research Group, Universiti Kebangsaan and analysis of reputable literature. Such ontology is illustrated in Fig. 1.

As can be seen from Fig. 1 there are ten concepts and nine object properties (or relations) in the constructed ontology. The rules designed for name entities recognition are mainly set of patterns which matched with the given object properties and subsequently inferred to be instances of the corresponding concepts. The framework is dynamic which the number of concept and object property can be added if there is any recommendation by experts in the future. These nine semantic relations are constructed according to stages illustrated in Fig. 2.

For the initial stage, several documents related to Malaysia medicinal herbs have been collected from websites and other resources. We then apply Hearst's

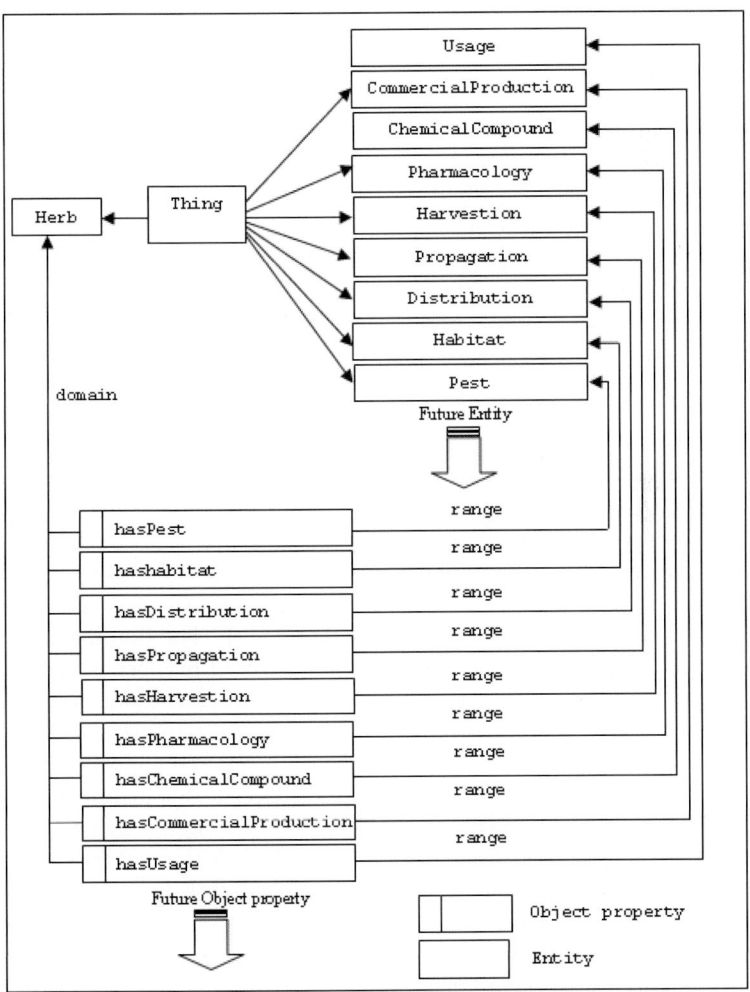

Fig. 1. The TBox ontology for the Malaysia medicinal herb domain

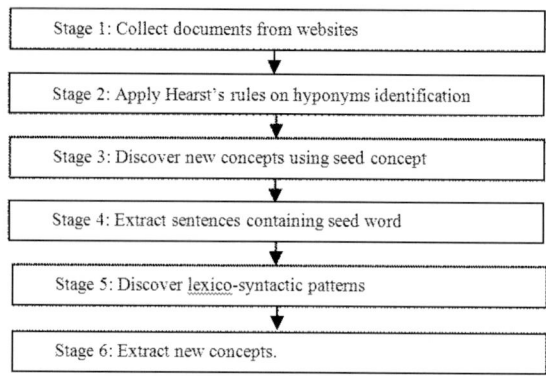

Fig. 2. The stages of the semantic relation identification

rules to identify hyponyms. In this case the collection or corpus was analyzed to extract the semantic relationships enhancement based on the work of [7]. The Hearst rule for detecting hyponym from text includes:

1. $NPo \ldots such \ as\{NP1, NP2 \ldots (and\backslash or)\}NP,,$
2. $such \ NP \ as\{NP,\} * \{(or \ [and)\}NP$
3. $NP\{,NP\} * \{,\} \ or \ other \ NP$
4. $NP\{,NP\} * \{,\} \ and \ other \ NP$
5. $NP\{,\} \ including \ \{NP * \{or\backslash and\} \ NP$
6. $NP\{,\} \ especially \ \{NP,\} * \{or] \ and\}NP$

In order to find other semantic relation rules or lexico-syntactic patterns, the technique from [12] was modified. The identification of the herbs is very important in order to avoid inconsistency of the ontology. As such a list of herbal names has been constructed which acts as seeds for further concept discovery. The documents retrieved were further processed so that only sentences containing the seed word were retained. Each sentence in this corpus was part-of-speech (POS) tagged and then parsed using Genia Tagger [17]. For example parsing the sentence "Sweet flag is also known for its biopesticidal properties and has been used in destroying fleas, lice and white ants" produces the output [NP Sweet flag] [VP is also known] [PP for] [NP its biopesticidal properties] and [VP has been used] [PP in] [VP destroying] [NP fleas] , [NP lice] and [NP white ants]. Each of the parser's output is analyzed to discover lexico-syntactic patterns.

After discovering the lexico-syntactic patterns, the new concepts which are directly related to the seed word were extracted from the sentence. In the example above, the seed word (Sweet flag) was linked with new concepts such as biopesticidal properties, fleas, lice and white ants.

As can be seen, the results of the activities are two-fold. First is the identification of semantic relations among instances of available concepts and, second is the construction of rules for detecting named entities corresponding to the semantic relations. Rules are constructed based on the lexical pattern between

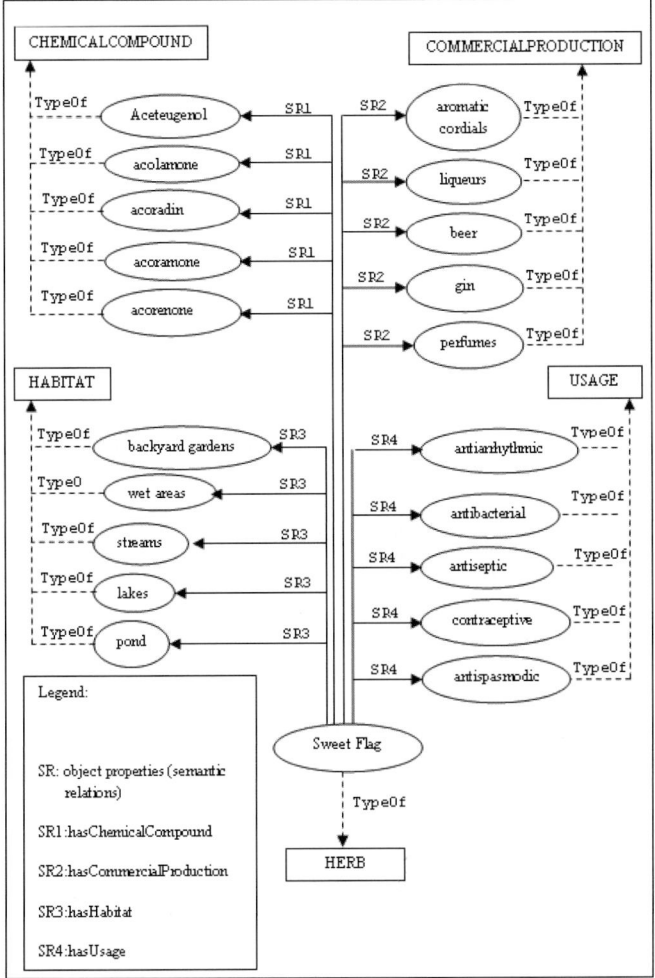

Fig. 3. Malaysia Medicinal Herbs Ontology Diagram

noun phrases in a sentence. As to date, apart from the hyponym-kind of lexico-patterns there are 105 numbers of rules relating to semantic relation have been discovered from nine entities as can be seen from Table 1.

The rules listed below are examples for populating ontology related to the concepts. Only selected concept is discussed in this research such as 'Commer-cialProduction', 'Habitat', 'ChemicalCompound' and 'Usage'.

Type of concept: CommercialProduction

Rule:

IF ... NP *products in the market, ranging from* ... NP_1,
$[NP_2]$, ...$[or|and]$ $[NP_n]$ AND NP typeOf Herb

Table 1. Number of rules for semantic relations

Types of Semantic Relation	Number of rules
hasCommercialProduction	31
hasPharmacology	21
hasDistribution	4
hasChemicalCompound	3
hasHabitat	14
hasPest	3
hasHarvestion	8
hasPropagation	6
HerbUsage	15
Total	105

THEN has CommercialProduction (NP, NP_i) AND NP_i typeOf CommercialProduction;
where $i = 1, 2, \ldots, n$

Type of concept: Habitat
Rule:
IF ... NP *inhabits* NP_1, $[NP_2]$, ...$[or|and]$ $[NP_n]$ AND NP typeOf Herb
THEN hasHabitat (NP, NPi) AND NPi typeOf Habitat;
where $i = 1, 2, \ldots, n$

Type of concept: ChemicalCompound
Rule:
IF ... *The chemical constituents of* NP *are* NP_1, $[NP_2]$, ...$[or|and]$ $[NP_n]$
AND NP typeOf Herb
THEN hasChemicalCompound (NP, NP_i) AND NP_i typeOf ChemicalCompound;
where $i = 1, 2, \ldots, n$

Type of concept: Usage
Rule:
IF ... NP ... *used for treatment of*... NP_1, $[NP_2]$, ...$[or|and]$ $[NP_n]$
AND NP typeOf Herb
THEN hasUsage (NP, NP_i) AND NP_i typeOf Usage;
where $i = 1, 2, ..., n$

Running the aforementioned rules resulted in the identification terms (or instances) of the ten entities as shown in Fig 3. As can be seen from Table 1 there are 105 rules have been discovered from 130 articles and will be tested for rules confidence level using technique proposed by [18]. We anticipate more rules will be discovered when the number of articles increase.

4 Discussion and Conclusion

This study presented an approach of extracting semantic relation from a specific domain. Nine types of semantic relations and 105 new rules were derived from the semantic relationships. At the preliminary stage, all rules or lexical patterns have not been tested for their rules confidence level while the learning process will be conducted in the next study. According to [19] the learning process should involve the application of meta-rules and a control structure to invoke as a framework for processing ill-formed input. By using the approach proposed in this study all terms that linked by the lexical patterns or rules have been considered as ontological terms for constructing an ontology diagram as shown in Fig 3. This study emphasized on finding the lexical patterns at earlier stage after selecting the seed concepts. Although there was an effort of automated seed word selection [20] but in this study the seed concepts have been selected manually from 100 names of Malaysia herbs species. By using seed concept user is able to acquire new domain knowledge around seed concepts that user considers important [12]. This study shows that it is helpful to have background knowledge about the domain prior to finding the lexical pattern. By mapping all the nine semantic relation and the rules to other documents in the corpus the effort to extract relevant knowledge which was done manually will be reduced. This approach therefore will minimize the role of domain expert in the development of the specific domain ontology.

References

1. Haase, P., Sure, Y.: State-of-the-Art on Ontology Evolution. Institute AIFB, University of Karlsruhe (2004),
   http://www.aifb.uni-karlsruhe.de/WBS/ysu/publications/
   SEKT-D3.1.1.1.b.pdf
2. Staab, S., Schnurr, H.-P., Studer, R., Sure, Y.: Knowledge processes and ontologies. IEEE Intelligent Systems, Special Issue on Knowledge Management 16(1), 26–34 (2001)
3. Fuller, S., Revere, D., Bugni, P.F., Martin, G.M.: A knowledgebase system to enhance scientific discovery: Telemakus. Biomedical Digital Libraries 1, 2 (2004)
4. Swanson, D.R., Smalheiser, N.R.: An interactive system for finding complementary literatures: A stimulus to scientific discovery. Artificial Intelligence 91(2), 183–203 (1997)
5. Alani, H., Kim, S., Millard, D.E., Weal, M.J., Hall, W., Lewis, P.H., Shadbolt, N.R.: Automatic Ontology-Based Knowledge Extraction from Web Documents. IEEE Intelligent Systems 18(1), 14–21 (2003)
6. Cimiano, P., Hotho, A., Staab, S.: Learning Concept Hierarchies from Text Corpora using Formal Concept Analysis. Journal of Artificial Intelligence Research 24, 305–339 (2005)
7. Hearst, M.: Automatic acquisition of hyponyms from large text corpora. In: Proceedings of the Fourteenth International Conference on Computational Linguistics, pp. 539–545 (1992)

8. Kawtrakul, A., Suktarachan, M., Imsombut, A.: Automatic Thai Ontology Construction and Maintenance System. In: Workshop on Papillon, Grenoble, France (2004)

9. AGROVOC Thesaurus,
http://jodi.ecs.soton.ac.uk/incoming/Soergel/JoDI_FAO_Soergl_revC.html

10. Imsombut, A., Kawtrakul, A.: Automatic building of an ontology on the basis of text corpora in Thai. Journal of Language Resources and Evaluation 42(2), 137–149 (2007)

11. Zaharudin, I., Noah, S.A., Noor, M.M.: Knowledge Acquisition from Textual Documents for the Construction of Medicinal herb Ontology Domain. J. Applied Science 9(4), 794–798 (2009)

12. Moldovan, D., Girju, R., Rus, V.: Domain-specific knowledge acquisition from text. In: Proceedings of the sixth conference on Applied natural language processing, Seattle, Washington, pp. 268–275 (2000)

13. Pantel, P., Pennacchiotti, M.: Espresso: Leveraging Generic Patterns for Automatically Harvesting Semantic Relations. In: Proceedings of the 21st International Conference on Computational Linguistics and the 44th annual meeting of the Association for Computational Linguistics, Sydney, Australia, pp. 113–120 (2006)

14. Maedche, A., Staab, S.: Ontology Learning for the Semantic Web. IEEE Intelligent Systems 16(2) (2001)

15. Xu., F., Kurz., D., Piskorski., J., Schmeier, S.: A Domain Adaptive Approach to Automatic Acquisition of Domain Relevant Terms and their Relations with Bootstrapping. In: Proceedings of the 3rd International Conference on Language Resources an Evaluation (LREC 2002), Las Palmas, Canary Islands, Spain, May 29-31 (2002)

16. Girju, R., Badulescu, A., Moldovan, D.: Learning Semantic Constraints for the Automatic Discovery of Part-Whole Relations. In: The Proceedings of the Human Language Technology Conference, Edmonton, Canada (2003)

17. Genia Tagger, http://www-tsujii.is.s.u-tokyo.ac.jp/GENIA/tagger/

18. Celjuska, D., Vargas-Vera, M.: Ontosophie: A semi-automatic system for ontology population from text. In: Proceedings of the 3rd International Conference on Natural Language Processing, ICON (2004)

19. Ralph, M.W., Norman, K.S.: Meta-rules as a basis for processing ill-formed input. Computational Linguistics 9(3-4) (1983)

20. Zagibalov., T., Carroll, J.: Automatic seed word selection for unsupervised sentiment classification of Chinese text. In: Proceedings of the 22nd International Conference on Computational Linguistics, vol. 1 (2008)

Gait Recognition Based on Outermost Contour

Lili Liu, Yilong Yin*, and Wei Qin

School of Computer Science and Technology,
Shandong University,
Jinan, Shandong, China, 250101
Tel.: 86-531-88391367; Fax: 86-531-88391367
ylyin@sdu.edu.cn
http://mla.sdu.edu.cn

Abstract. Gait recognition aims to identify people by the way they walk. In this paper, a simple but effective gait recognition method based on Outermost Contour is proposed. For each gait image sequence, an adaptive silhouette extraction algorithm is firstly used to segment the images and a series of postprocessing is applied to the silhouette images to obtain the normalized silhouettes with less noise. Then a novel feature extraction method based on Outermost Contour is proposed. Principal Component Analysis (PCA) and Multiple Discriminant Analysis (MDA) are adopted to reduce the dimensionality of the feature vectors and to optimize the class separability of different gait image sequences simultaneously. Two simple pattern classification methods are used on the low-dimensional eigenspace for recognition. Experimental results on a gait database of 100 people show that the accuracy of our algorithm achieves 97.67%.

Keywords: gait recognition, Outermost Contour, Principal Component Analysis, Multiple Discriminant Analysis.

1 Introduction

Gait recognition is a relatively new research direction in biometrics aiming to identify individuals by the way they walk. In comparison with the first generation biometrics such as fingerprint and iris, gait has many advantages. It does not require users' interaction and it is non-invasive. Also it is difficult to be concealed or disguised. Furthermore, gait can be effective for recognition at low resolution where face information is not available. Gait may be the only perceivable biometric from a great distance. Therefore, it receives increasing interest from researchers and various methods have been proposed on this domain recently[1][2] [3][4][5].

These gait recognition methods can be divided into model-based and model-free methods. Model-based methods construct human model and use the parameters of the model for recognition. An early such attempt [1] models the lower

---

* Corresponding author.

J. Yu et al. (Eds.): RSKT 2010, LNAI 6401, pp. 395–402, 2010.

limbs as two inter-connected pendulum. Lee and Grimson [2] use 7 ellipses to model the human body. In this paper we focus on model-free methods which do not model the structure of human motion. Wang et al. [4] extract features from the outer contour of each silhouette. Han et al. [5] represent gait image sequence by gait energy image and synthetic templates, and use fused feature for recognition.

All these methods promote the development of gait recognition domain. However, there are still many challenges in gait recognition, such as imperfect segmentation of the walking subject, different walking direction of the subject, changes in clothes, and changes of gait as a result of mood or as a result of carrying objects. The gait feature extraction method proposed in this paper can tolerate some imperfection in segmentation and clothing changes.

The gait recognition method in this paper is related to that of Wang [4]. However, instead of extracting gait feature using outer contour of silhouette, we extract gait feature from the Outermost Contour, which is more simple, effective and has better tolerance for imperfect silhouette. And we use not only PCA training but also MDA training for better recognition performance. The overview of our gait recognition system is shown in Fig. 1.

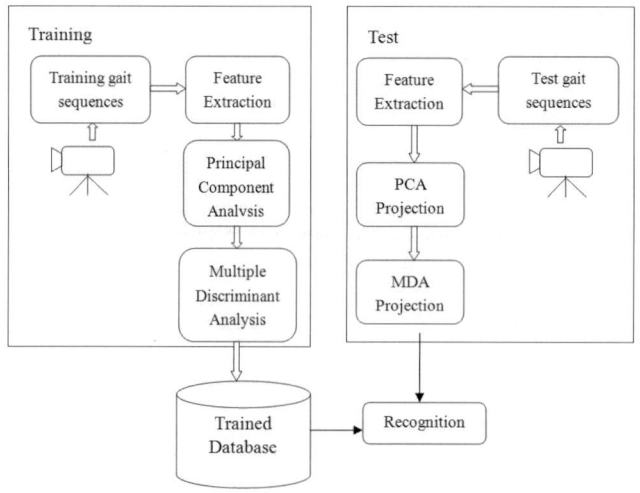

Fig. 1. Overview of our gait recognition system

The remainder of this paper is organized as follows: Section 2 describes the proposed feature extraction method in detail, including gait image preprocessing and silhouette feature representation. In section 3, we present the training and projection process based on PCA and MDA, and show the gait recognition method we used. Experimental results are presented in Section 4, and Section 5 concludes the paper.

2   Gait Representation

In this paper, we only consider gait recognition by regular human walking, which is used in most current approaches of gait recognition.

2.1   Preprocessing

We apply an existing adaptive gait silhouette extraction algorithm using Gauss model developed by Fu [6] to extract the walking figure for better segmentation performance. Then for each binary silhouette image we use morphological operators such as erosion and dilation to filter some small noises on the background area and to fill the small holes inside the silhouette. A binary connected component analysis is applied to extract the connected region with the largest size for ignoring all the remaining noises. In order to make it easier to carry out the following feature extraction process and reduce time consumption, we normalize the silhouette images to the same size (Proportionally resize each silhouette image to make all the silhouettes have the same height. Every image was resized to 128×100 pixels in this paper, and 100 is the normalized height of silhouette) and align the normalized silhouettes to the horizontal center. An example of silhouette extraction is shown in Fig. 2, from which we can see that the silhouette segmentation procedure performs well as a whole.

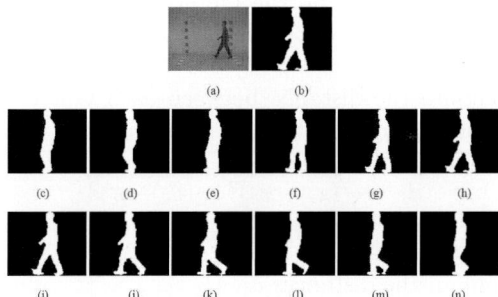

Fig. 2. An example of silhouette extraction: (a) an original image in gait database, (b) the normalized and aligned silhouette of (a), (c)-(n) temporal changes of 12 successive frames in a gait silhouette sequence

2.2   Silhouette Feature Representation

In order to represent the silhouette in a simple and effective way, we propose a new feature extraction method which only uses some of the pixels on the contour. For the sake of description, we make a definition as follows:

Outermost Contour: In each row of a silhouette image, the most right pixel and the most left pixel on the contour belong to Outermost Contour. Fig. 3(a) shows the schematic of Outermost Contour. The bold boundaries in Fig. 3(a)

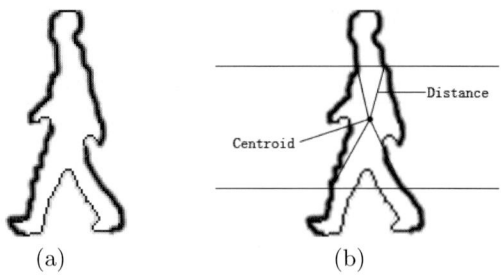

Fig. 3. (a)the schematic of Outermost Contour, (b)illustration of distance signal extraction

is belonged to the Outermost Contour, but the thin boundary between the two legs does not belong to the Outmost Contour. Because all the silhouettes are normalized, the number of pixels on each Outermost Contour is definite, i.e. $2H$ (H is the height of the silhouette).

Firstly, we compute the centroid of the Outermost Contour (x_c, y_c).

$$x_c = \frac{1}{n}\sum_{i=1}^{n} x_i \qquad y_c = \frac{1}{n}\sum_{i=1}^{n} y_i \qquad (1)$$

where n is the number of pixels on the Outermost Contour, (x_i, y_i) is the coordinate of pixel on the Outermost Contour. Actually, $n = 2H$ as mentioned above.

Secondly, we compute the distance between each Outermost Contour pixel (x_i, y_i) to the centroid (x_c, y_c) row by row, as is shown in Fig. 3(b).

$$d_i = \sqrt{(x_i - x_c)^2 + (y_i - y_c)^2} \qquad (2)$$

Thus, for each silhouette image, we obtain a distance signal $D = [d_1, d_2, \cdots, d_{2H}]$ which is composed of all the distances d_i.

Compared with the feature extraction method using outer contour in [4], our feature extraction method using the Outermost Contour ignores the region between two legs where imperfect segmentation often exists as a result of the shadow of legs.

3   Gait Recognition

Although we have enormously reduced the dimensionality of the silhouette image in the gait representation procedure, the dimensionality of the distance signal is still very large. There are two classical linear approaches to find transformations for dimensionality reduction: PCA and MDA which have been effectively used in face recognition[7]. PCA seeks a projection that best represents the data in the least-square sense, while MDA seeks a projection that best separates the data in the least-square sense [5]. Thus, The combination of PCA and MDA can reduce the data dimensionality and optimize the class separability simultaneously. In [3][5] PCA and MDA are used as combination in gait recognition to achieve the best data representation and the best class separability simultaneously. In this paper, we use the same learning method as [3][5] for good performance.

3.1   PCA and MDA Training and Projection

The purpose of PCA training is to reduce the dimensionality of data. We use the same PCA training and projection as [4] in this paper, so we do not discribe it in detail. After the PCA training and projection process, we can obtain the projection centrid C_i of the ith gait image sequence, and the unit vector of the centroid U_i can be easily computed by $\frac{C_i}{\|C_i\|}$.

As we have obtained the n k-dimensional vectors $\{U_1, U_2, \cdots, U_n\}$ belonged to c classes, the within-class scatter matrix S_W and the between-class scatter matrix S_B can be computed as [5] describes. Using the same MDA training and projection method as [5] describes, we can obtain the transformation matrix $[v_1, v_2, \cdots, v_{c-1}]$. And the final feature vector F_i for the ith gait image sequence is obtained from the k-dimensional unit vector U_i:

$$F_i = [v_1, v_2, \cdots, v_{c-1}]^T U_i \tag{3}$$

3.2   Recognition

There are two simple classification methods-the nearest neighbor classifier (NN) and the nearest neighbor classifier with respect to class exemplars (ENN) in pattern recognition. For NN test, each sequence is classified to the same class with its nearest neighbor; for ENN test, each sequence is classified to the same class with its nearest exemplar which is defined as the mean of feature vectors for one given person in training set. In our experiments, we use both of them.

Let G represent a test sequence, we can compute the feature vector F_G according to section 2, 3.1 and 3.2. G is classified to ω_k when

$$d(F_G, F_k) = min_{i=1}^{c} d(F_G, F_i) \tag{4}$$

4   Experiments

4.1   Gait Database

In our experiment, we use the CASIA Gait Database (Dataset B) [8],one of the largest gait databases in gait-research community currently. The database consists of 124 persons. Every subject has 6 natural walk sequences of each view angle (ranging from $0°$ to $180°$, with view angle interval of $18°$). The frame size is 320×240 pixel, and the frame rate is 25 fps.

We use gait videos numbered from 001 to 100 (person id, that is 100 persons) of view angle $90°$ in Dataset B to carry out our experiment. As every person has 6 normal sequences, we assign 3 sequences to training set and the rest 3 sequences to testing set. Fig. 4 shows three images in this gait database.

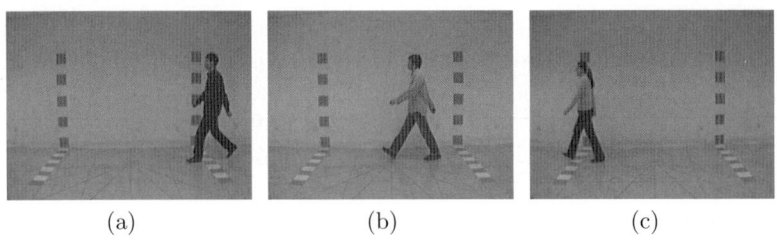

(a)   (b)   (c)

Fig. 4. Three images in CASIA Gait Database (Dataset B) with view angle 90°

4.2   Experimental Results

In our experiment, each gait image sequence is firstly preprocessed and converted into a sequence of distance signals as described in Section 2. For training set, 30 continuous distance signals of each class are chosen for PCA training and the threshold value is chosen as 0.99. As a result, 47 eigenvectors are kept to form the transformation matrix. Then each training sequence is projected to a unit centroid vector. MDA is used on all the unit centroid vectors to form another transformation matrix and to project the unit centroid vectors to a new eigenspace as described in Section 3.1. That is, we obtain the final feature vectors of each training sequence. For the testing sequences, we can easily compute the feature vectors by directly using the two transformation matrices obtained from PAC and MDA training process. Fig. 5 shows the projection of 15 training gait feature vectors belonged to 5 people respectively. For visualization, only first three-dimensional eigenspace is used. The points with the same shape are belonged to the same person. From Fig. 5, we can see that these points can be separated easily.

For comparison, we use four different combinations for recognition which is described in Table 1. And Table 1 also show the CCR (Correct Classification Rate) of the four combinations. From Table 1 we can see that our method can achieve high accuracy at a relative large gait database. Fig. 6(a) shows the cumulative match scores for rank from 1 to 50. It is noted that the cumulative match score of Rank=1 is equivalent to the CCR as showed in Table 1.

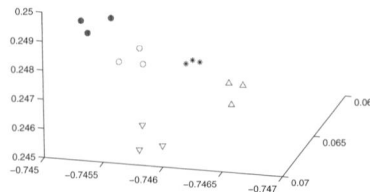

Fig. 5. The projection of 15 training gait feature vectors belonging to 5 people (only the first three-dimensional eigenspace is used for visualization)

Table 1. Correct classification rates using four combinations

Method		
Training	Recognition	CCR(%)
PCA	NN	72.33
PCA+MDA	NN	96.67
PCA	ENN	69.00
PCA+MDA	ENN	97.67

For completeness, we also estimate FAR (False Acceptance Rate) and FRR (False Reject Rate) in verification mode. The ROC (Receiver Operating Characteristic) curve is showed in Fig. 6(b), from which we can see that the EERs (Equal Error Rate) are approximately 16%, 11%, 8% and 5% for PCA+NN, PCA+ENN, PCA+MDA+NN and PCA+MDA+ENN, respectively.

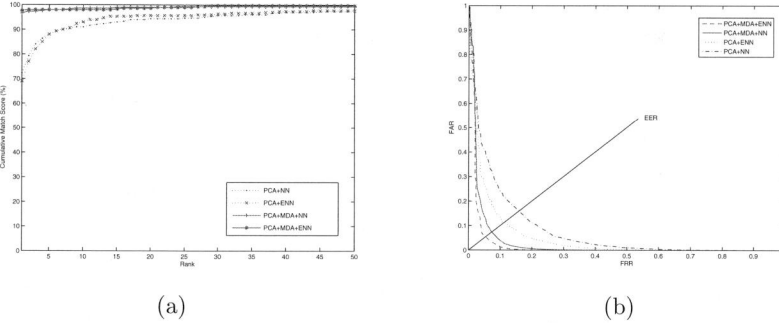

(a)   (b)

Fig. 6. (a)Cumulative match score of the four combinations,(b)ROC curves of the four combinations

From Table 1 and Fig. 6, we can conclude that:(1) Our feature extraction method is an effective method. We only extract feature from the Outermost Contour which includes most of the useful information and ignores much noise on the contour. (2) On a large database, the results of using NN classifier and using ENN classifier are almost the same. (3) The two combinations using PCA+MDA achieve higher performance than the other combinations. This is because MDA can seek a projection that best separates the data, which is beneficial for recognition.

5   Conclusions

In this paper, we propose a novel and simple gait recognition algorithm based on Outermost Contour. An adaptive silhouette extraction algorithm and a series of postprocessing are applied to segment and normalize each image. Simple feature extraction method and PCA, MDA training significantly reduce the data dimensionality and optimize the class separability of different gait sequences simultaneously. Experimental results have demonstrated that the proposed algorithm is encouraging for human gait identification.

Acknowledgments. We would like to express our thanks to the Institute of Automation, Chinese Academy of Sciences for CASIA Gait Database. This work is supported by Shandong University Independent Innovation Foundation under Grant No.2009TS034 and 2009TS035.

References

1. Nixon, M.S., Carte, J.N., Nash, J.M., Huang, P.S., Cunado, D., Stevenage, S.V.: Automatic gait recognition. In: Proceedings of Biometrics Personal Identification in Networked Society, pp. 231–219 (1999)
2. Lee, L., Grimson, W.E.L.: Gait Analysis for Recognition and Classification. In: 5th IEEE International Conference on Automatic Face and Gesture Recognition, Washington, DC, USA, pp. 148–155 (2002)
3. Huang, P.S., Harris, C.J., Nixon, M.S.: Recognizing Humans by Gait via Parameteric Canonical Space. Artificial Intelligence in Eng. 13, 359–366 (1999)
4. Wang, L., Tan, T., Ning, H.S., Hu, W.M.: Silhouette Analysis-Based Gait Recognition for Human Identification. IEEE Trans. Pattern Analysis and Machine Intelligent 25(12) (2003)
5. Han, J., Bhanu, B.: Individual Recognition Using Gait Energy Image. IEEE Trans. Pattern Analysis and Machine Intelligence 28(2) (2006)
6. Fu, C.Y., Li, P., Wen, Y.M., Yuan, H.J., Ye, B.: Gait Sillhouette Extraction Algorithm Using Gauss Model. Chinese Journal of Sensors and Actuators 21(7) (2008)
7. Belhumeur, P.N., Hespanha, J.P., Kriegman, D.J.: Eigenfaces versus Fisherfaces: Recognition Using Class Specific Linear Projection. IEEE Trans. Pattern Analysis and Machine Intelligence 19(7), 711–720 (1997)
8. Yu, S., Tan, D., Tan, T.: A framework for evaluating the effect of view angle clothing and carrying condition on gait recognition. In: International Conference on Pattern Recognition (ICPR), pp. 441–444 (2006)

Pseudofractal 2D Shape Recognition

Krzysztof Gdawiec

Institute of Computer Science, University of Silesia, Poland
kgdawiec@ux2.math.us.edu.pl

Abstract. From the beginning of fractal discovery they found a great number of applications. One of those applications is fractal recognition. In this paper we present some of the weaknesses of the fractal recognition methods and how to eliminate them using the pseudofractal approach. Moreover we introduce a new recognition method of 2D shapes which uses fractal dependence graph introduced by Domaszewicz and Vaishampayan in 1995. The effectiveness of our approach is shown on two test databases.

1 Introduction

The shape of an object is very important in object recognition. Using the shape of an object for object recognition is a growing trend in computer vision. Good shape descriptors and matching measures are the central issue in these applications. Based on the silhouette of objects a variety of shape descriptors and matching methods have been proposed in the literature.

One of approaches in object recognition is the use of fractal geometry. Fractal geometry breaks the way we see everything, it gives us tools to describe many of the natural objects which we cannot describe with the help of classical Euclidean geometry. The idea of fractals was first presented by Mandelbrot in the 1970s [9] and since then have found wide applications not only in object recognition but also in image compression [5], generating terrains [10] or in medicine [7]. Fractal recognition methods are used in face recognition [2], character recognition [11] or as general recognition method [12].

In this paper we present a new approach to the recognition of 2D shapes. In our approach we modified fractal image coding scheme which is base for features in many fractal recognition methods. Moreover we present a new method which uses the so-called fractal dependence graph [3] and our modified coding scheme.

2 Fractals

Because we will modify fractal coding scheme first we must introduce the definition of a fractal. The notion of fractal has several non-equivalent definition, e.g. as attractor or as an invariant measure [1]. The most used definition is fractal as attractor.

J. Yu et al. (Eds.): RSKT 2010, LNAI 6401, pp. 403–410, 2010.

Let us take any complete metric space (X, d) and denote as $\mathcal{H}(X)$ the space of nonempty, compact subsets of X. In this space we take the Haussdorf metric:

$$h(R, S) = \max\{\max_{x \in R} \min_{y \in S} d(x, y), \max_{y \in S} \min_{x \in R} d(y, x)\}, \tag{1}$$

where $R, S \in \mathcal{H}(X)$.

Definition 1 ([1]). *We say that a set* $W = \{w_1, \ldots, w_N\}$, *where* w_n *is a contraction mapping for* $n = 1, \ldots, N$ *is an* iterated function system *(IFS).*

Any IFS $W = \{w_1, \ldots, w_N\}$ determines the so-called Hutchinson operator which is defined as follows:

$$\forall_{A \in \mathcal{H}(X)} W(A) = \bigcup_{n=1}^{N} w_n(A) = \bigcup_{n=1}^{N} \{w_n(a) : a \in A\}. \tag{2}$$

The Hutchinson operator is a contraction mapping in space $(\mathcal{H}(X), h)$ [1].

Definition 2. *We say that the limit* $\lim_{k \to \infty} W^k(A)$, *where* $A \in \mathcal{H}(X)$ *is called* an attractor *of the IFS* $W = \{w_1, \ldots, w_N\}$.

2.1   Fractal Image Coding

Most of the fractal recognition methods use fractal image coding to obtain fractal description of the objects and next extract features from this description. Below we introduce basic algorithm of the fractal coding. Let us start with the notion of partitioned iterated function system (PIFS) [5].

Definition 3. *We say that a set* $P = \{(F_1, D_1), \ldots, (F_N, D_N)\}$ *is a* partitioned iterated function system, *where* F_n *is a contraction mapping,* D_n *is an area of an image which we transform with the help of* F_n *for* $n = 1, \ldots, N$.

Because the definition of PIFS is very general we must restrict to some fixed form for the transformations. In practice we take affine mappings $F : \mathbb{R}^3 \to \mathbb{R}^3$ of the following form:

$$F\left(\begin{bmatrix} x \\ y \\ z \end{bmatrix}\right) = \begin{bmatrix} a_1 & a_2 & 0 \\ a_3 & a_4 & 0 \\ 0 & 0 & a_7 \end{bmatrix} \begin{bmatrix} x \\ y \\ z \end{bmatrix} + \begin{bmatrix} a_5 \\ a_6 \\ a_8 \end{bmatrix}, \tag{3}$$

where coefficients $a_1, \ldots, a_6 \in \mathbb{R}$ describe a geometric transformation, coefficients $a_7, a_8 \in \mathbb{R}$ are responsible for the contrast and brightness and x, y are the co-ordinates in image, z is pixel intensity.

The coding algorithm can be described as follows. We divide an image into a fixed number of non-overlapping areas of the image called range blocks. Next we create a list of domain blocks. The list consists of overlapping areas of the image, larger than the range blocks (usually two times larger) and transformed using four mappings: identity, rotation through $180°$, two symmetries of a rectangle. Now for every range block R we search for a domain block D such the value $d(R, F(D))$ is the smallest, where d is a metric and F is mapping of the form (3) determined by the position of R and D, the size of these in relation to itself, one of the four mappings used to transform D and the coefficients a_7, a_8 are calculated with the help of following equations:

$$a_7 = \frac{k \sum_{i=1}^{k} g_i h_i - \sum_{i=1}^{k} g_i \sum_{i=1}^{k} h_i}{k \sum_{i=1}^{k} g_i^2 - (\sum_{i=1}^{k} g_i)^2}, \tag{4}$$

$$a_8 = \frac{1}{k} \left[\sum_{i=1}^{k} h_i - a_7 \sum_{i=1}^{k} g_i \right], \tag{5}$$

where k is the number of pixels in the range block, $g_1, \ldots, g_k$ are the pixel intensities of the transformed and resized domain block, $h_1, \ldots, h_k$ are the pixel intensities of the range block. If $k \sum_{i=1}^{k} g_i^2 - \left(\sum_{i=1}^{k} g_i \right)^2 = 0$, then $a_7 = 0$ and $a_8 = \frac{1}{k} \sum_{i=1}^{k} h_i$.

The search process in the most time-consuming step of the algorithm [5]. Figure 1 presents the idea of the fractal image coding.

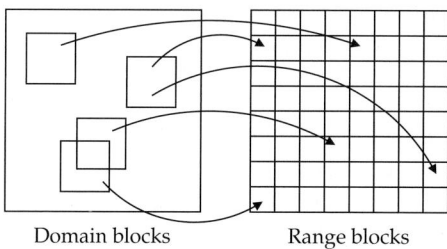

Domain blocks   Range blocks

Fig. 1. The idea of fractal image coding

3   Pseudofractal Approach to 2D Shape Recognition

Fractal methods like method using PIFS Code (PIFSC) [2] or Multiple Mapping Vector Accumulator (MMVA) [11] have one weakness. To show this weakness let us take a look at fractal coding from the point of view of some domain block D. This block can fit to several range blocks. Each of these fits corresponds to one mapping in PIFS. Now let us suppose that block D was changed into block D' (e.g. the shape in this area was cut or it was deformed). In this case domain D' can fit to the same range blocks as D (to all or only to some), it can also fit to some other range blocks. This change of fitting causes a change of the mappings in PIFS. In the worst case all mappings can be changed. If the change of fitting is big then it can cause that object will be wrongly classified [6].

To eliminate the problem we can use pseudofractal approach. Simply in the fractal coding step for the creation of the domain list we take any fixed image called *domain image*. The rest of the coding algorithm remains the same. Using the domain image the change of fitting can appear only for the ranges where the image was changed and the rest of fitting remains the same because the change of the image has no influence on the domain blocks.

Definition 4. *Fractal image coding using fixed domain image is called a* pseudofractal image coding *and the PIFS obtained with the help of this algorithm is called* pseudofractal PIFS *(PPIFS).*

To obtain better recognition results in the methods mentioned earlier (PIFSC, MMVA) instead of the fractal image coding we can use the pseudofractal image coding. Moreover, before the pseudofractal image coding we can resize the image with the object to fixed dimensions (e.g. 128×128). This step is used to shorten the time needed to coding the image. The PIFSC method modified in the presented way is called pseudofractal PIFSC (PPIFSC) and the MMVA method is called pseudofractal MMVA (PMMVA).

3.1   Pseudofractal Dependence Graph Method

In 1995 Domaszewicz and Vaishampayan introduced the notion of a dependence graph [3]. The graph reflects how domain blocks are assigned to range blocks. They used this graph for three different purposes. The first one is an analysis of the convergence of the fractal compression. The second is a reduction of the decoding time and the last is improvement upon collage coding. The definition of a dependence graph is as follows [3]:

Definition 5. *Let W be the PIFS with a set of range blocks $\mathcal{R}$. The dependence graph of W is a directed graph $G = (V, E)$ whose $V = \mathcal{R}$ and for all $R_i, R_j \in \mathcal{R}$ we have $(R_i, R_j) \in E$ if and only if the domain block corresponding to R_j overlaps the R_i.*

Example of an image and a dependence graph corresponding to the PIFS which codes the image are presented in Fig.2.

Definition 6. *The dependence graph for a PPIFS W is called a* pseudofractal dependence graph.

The Pseudofractal Dependence Graph (PDG) method is following:

1. binarise the image and extract the object,
2. find a set of correct orientations Γ,
3. choose any correct orientation $\gamma \in \Gamma$ and rotate the object through γ,
4. resize the image to 128×128 pixels,
5. find a normalised PPIFS W, i.e. for which the space in $[0, 1]^2$,
6. determine the adjacency matrix $\mathbf{G}$ of the pseudofractal dependency graph of W,

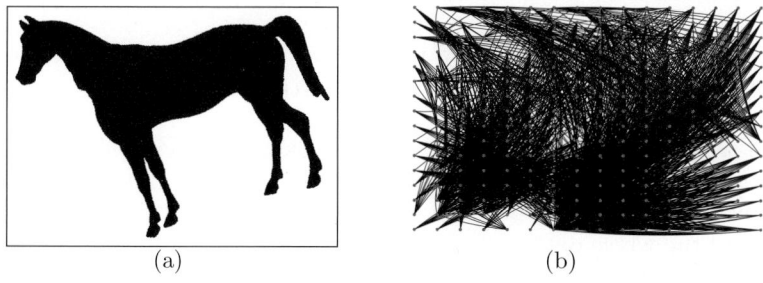

Fig. 2. Example image (a) and its fractal dependence graph (b)

7. in the base find adjacency matrix $\boldsymbol{H}$ which minimizes expression

$$d_{\boldsymbol{H}} = \|\boldsymbol{G} - \boldsymbol{H}\| = \sqrt{\sum_i \sum_j |g_{ij} - h_{ij}|^2}, \qquad (6)$$

8. choose an image from the base which corresponds to the $d_{\boldsymbol{H}}$.

The notion of a correct orientation used in the PDG method is defined as follows.

Definition 7. *A correct orientation is an angle by which we need to rotate an object so that it fulfills following conditions: area of the bounding box is the smallest, height of the bounding box is smaller than the width and left half of the object has at least as many pixels as the right.*

Examples of objects and their correct orientations are presented in Fig.3. In the case of triangle (Fig.3(b)) we see three different orientations. If we want to add such an object to the base for each of the correct orientations, we find the corresponding pseudofractal dependence graph of the PPIFS representing the rotated object and add it to the base.

Fig. 3. Examples of objects and their correct orientations

Giving the object a correct orientation is needed to make the method rotation invariant. Resizing image to 128×128 is used to speed up the fractal coding process and the normalisation is needed to make the method translation and scale invariant.

4   Experiments

Experiments were performed on two databases. The first database was created by the authors and the second one was MPEG7 CE-Shape-1 Part B database [8].

Our base consists of three datasets. For creation of the datasets we have used 5 different base images of different classes of objects. In each of the datasets we have 5 classes of objects, 20 images per class. So each dataset consists of 100 images and the whole database of 300 images.

In the first dataset we have base objects changed by elementary transformations (rotation, translation, change of scale). In the second dataset we have base objects changed by elementary transformations and we add small changes to the shape locally, e.g. shapes are cut and/or they have something added to it. In the last, the third dataset, similar to the other two datasets the objects were modified by elementary transformations and moreover we add to the shape large changes locally. Example images from the authors base are presented in Fig.4(a).

The MPEG7 CE-Shape-1 Part B database consists of 1400 silhouette images from 70 classes. Each class has 20 different shapes. Sample images from the MPEG7 database are presented in Fig.4(b).

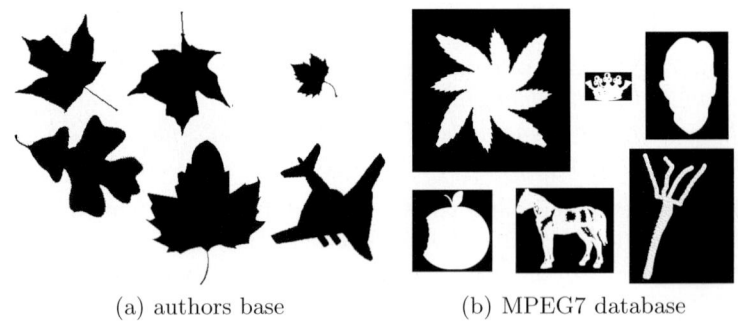

(a) authors base   (b) MPEG7 database

Fig. 4. Example images form the bases used in the tests

In the tests we used one domain image (Fig.5). The number of transformations used in fractal coding was fixed to 256 transformation (division into 16×16 range blocks).

To estimate the error rate of the methods we used leave-one-out method for the three datasets created by the authors and for the MPEG7 CE-Shape-1 Part B database we used stratified 10-fold cross validation. All the algorithms were implemented in Matlab and the tests were performed on computer with AMD Athlon64 X2 6400+, 4GB DDR2 RAM memory and WindowsXP operating system.

The results of the test for the authors database are presented in Tabs 1(a)-1(c). We can clearly see that the proposed pseudofractal approach led to the

Fig. 5. Domain image used in the tests

decrease of the recognition error for the methods known from the literature. The PDG method has obtained the lowest value of the error on the first and the third dataset. Only on the second dataset the PMMVA method was better, but the difference was small. Moreover we see that the proposed resizing of the image before the fractal coding has shorten the time of the recognition.

Table 1. Results of the test for the authors base

(a) elementary

Method	Error [%]	Time [s]
PIFSC	4.00	7915
PPIFSC	2.00	1348
MMVA	10.00	7964
PMMVA	2.00	1411
PDG	2.00	1020

(b) locally small

Method	Error [%]	Time [s]
PIFSC	11.00	7123
PPIFSC	4.00	1304
MMVA	18.00	7054
PMMVA	2.00	1364
PDG	3.00	991

(c) locally large

Method	Error [%]	Time [s]
PIFSC	37.00	6772
PPIFSC	4.00	1313
MMVA	32.00	6734
PMMVA	4.00	1368
PDG	3.00	998

The results of the test for the MPEG7 CE-Shape-1 Part B database are presented in Tab.2. Similarly as in the case of the authors base we can see that proposed pseudofractal approach led to the decrease of the recognition error for the methods known from the literature and the PDG method obtained the lowest value of the recognition error.

Table 2. Results of the tests for the MPEG7 CE-Shape-1 Part B base

Method	Error [%]
PIFSC	52.74
PPIFSC	18.15
MMVA	49.03
PMMVA	19.22
PDG	14.72

5   Conclusions

In this paper we have presented the weaknesses of the fractal recognition methods and proposed a pseudofractal approach to eliminate the weaknesses. Moreover we have presented a new recognition method called Pseudofractal Dependence Graph Method. The proposed approach led to a significant decrease of the recognition error and PDG method obtained the lowest values of errors. Moreover proposed resizing of the image before fractal coding has shorten the time needed for the recognition. We can shorten the time more by implementing the algorithms in C++ and using the General-Purpose Computation on Graphics Processing Units [4].

Because in the test we used only one domain image in our further research we will use different images as the domain image. With this kind of tests we will check if the results are dependent on the choice of the domain image and which properties the domain image must have. Moreover we will try other classification methods. The correct orientation used to align the object is very simple so there is research under way to find a better method for aligning the object.

References

1. Barnsley, M.: Fractals Everywhere. Academic Press, Boston (1988)
2. Chandran, S., Kar, S.: Retrieving Faces by the PIFS Fractal Code. In: Proceedings 6th IEEE Workshop on Applications of Computer Vision, pp. 8–12 (December 2002)
3. Domaszewicz, J., Vaishampayan, V.A.: Graph-theoretical Analysis of the Fractal Transform. In: Proceedings 1995 International Conference on Acoustics, Speech, and Signal Processing, vol. 4, pp. 2559–2562 (May 1995)
4. Erra, U.: Toward Real Time Fractal Image Compression Using Graphics Hardware. In: Bebis, G., Boyle, R., Koracin, D., Parvin, B. (eds.) ISVC 2005. LNCS, vol. 3804, pp. 723–728. Springer, Heidelberg (2005)
5. Fisher, Y.: Fractal Image Compression: Theory and Application. Springer, New York (1995)
6. Gdawiec, K.: Shape Recognition using Partitioned Iterated Function Systems. Advances in Intelligent and Soft Computing 59, 451–458 (2009)
7. Iftekharuddin, K.M., Jia, W., Marsh, R.: Fractal Analysis of Tumor in Brain MR Images. Machine Vision and Applications 13, 352–362 (2003)
8. Latecki, L.J., Lakamper, R., Eckhardt, T.: Shape Descriptors for Non-rigid Shapes with a Single Closed Contour. In: Proceedings IEEE Conference on Computer Vision and Pattern Recognition, vol. 1, pp. 424–429 (June 2000)
9. Mandelbrot, B.: The Fractal Geometry of Nature. W.H. Freeman and Company, New York (1983)
10. Meng, D., Cai, X., Su, Z., Li, J.: Photorealistic Terrain Generation Method Based on Fractal Geometry Theory and Procedural Texture. In: 2nd IEEE International Conference on Computer Science and Information Technology, pp. 341–344 (2009)
11. Mozaffari, S., Faez, K., Faradji, F.: One Dimensional Fractal Coder for Online Signature Recognition. In: Proceedings of the 18th International Conference on Pattern Recognition, vol. 2, pp. 857–860 (August 2006)
12. Neil, G., Curtis, K.M.: Shape Recognition Using Fractal Geometry. Pattern Recognition 30(12), 1957–1969 (1997)

Classification of MMPI Profiles of Patients with Mental Disorders – Experiments with Attribute Reduction and Extension

Jerzy Gomuła[1,2], Krzysztof Pancerz[3], and Jarosław Szkoła[3]

[1] The Andropause Institute, Medan Foundation, Warsaw, Poland
jerzy.gomula@wp.pl
[2] Cardinal Stefan Wyszyński University in Warsaw, Poland
[3] Institute of Biomedical Informatics
University of Information Technology and Management in Rzeszów, Poland
kpancerz@wsiz.rzeszow.pl,
jszkola@wsiz.rzeszow.pl

Abstract. In the research presented in the paper we try to find efficient methods for classification of MMPI profiles of patients with mental disorders. Each profile is described by a set of values of thirteen attributes (scales). Patients can be classified into twenty categories concerning nosological types. It is possible to improve classification accuracy by reduction or extension of the number of attributes with relation to the original data table. We test several techniques of reduction and extension. Experiments show that the proposed attribute extension approach improves classification accuracy, especially in the case of discretized data.

Keywords: classification, attribute reduction, attribute extension, rough sets, MMPI profiles.

1 Introduction

One of the main tasks of building decision support systems is to search for efficient methods of classification of new cases. Our research concerns psychometric data coming from the Minnesota Multiphasic Personality Inventory (MMPI) test [1]. MMPI is used to count the personality-psychometric dimensions which help in diagnosis of mental diseases. In this paper, we describe different experiments with attribute reduction and extension by comparison with the original data table. Endeavors should lead to find the best approaches to rule generation which may be used in a created computer tool supporting diagnosis of mental diseases [2]. Attribute reduction and extension techniques have been applied for several rule generation algorithms based on rough set approaches available in the RSES system - a software system for data exploration, classification support and knowledge discovery. Moreover, we have tested classification accuracy for the data without and with preliminary discretization.

In years 1998-1999 a team of researchers consisting of W. Duch, T. Kucharski, J. Gomuła, R. Adamczak created two independent rule systems devised for the

J. Yu et al. (Eds.): RSKT 2010, LNAI 6401, pp. 411–418, 2010.

nosological diagnosis of persons that may be screened with the MMPI-WISKAD test [3]. In [4], the results of testing different algorithms of rules generation for MMPI data were presented. Our research is focused on creating a new computer tool for multicriteria diagnosis of mental diseases. The first version of this tool is presented in [2].

The rest of the paper is organized as follows. Section 2 gives basic medical background of a considered problem. In section 3, available data, examined techniques as well as results of experiments are presented. Section 4 consists of some conclusions and directions for the further work.

2   Medical Background

The Minnesota Multiphasic Personality Inventory (MMPI) test [1] delivers psychometric data on patients with selected mental disorders. Originally, the MMPI test was developed and published in 1943 by a psychologist S.R. McKinley and a neuropsychiatrist J.Ch. Hathaway from the University of Minnesota. Later the inventory was adapted in above fifty countries. The MMPI-WISKAD personality inventory is the Polish adaptation of the American inventory. It has been used, among other modern tools, for carrying out nosological differential diagnosis in psychiatric wards, counseling services, prison ward services as well as in the public hospitals, for collective screenings and examinations (e.g., professional soldiers for the public/military services such as the army as well as people employed in fire brigades, police and airline pilots). Interestingly enough, effectiveness of a prescribed therapy can be valuated by means of such an inventory. MMPI is also commonly used in scientific research. The test is based upon the empirical approach and originally was translated by M. Chojnowski (as WIO) [5] and elaborated by Z. Płużek (as WISKAD) in 1950 [6]. American norms were accepted there. Based upon received responses ("Yes", "Cannot Say", "No") to selected questions we may make up a total count for the subject being examined measuring obtained raw results against the reference and clinical scales as being directly related to specific questions (items) and recalculate the outcome into T-scores results. The T-scores scale, which is traditionally attributed to the MMPI, represents the following parameters: offset ranging from 0 to 100 T-scores, average equal to 50 T-scores, standard deviation equal to 10 T-scores. The profile (graph) that is built for such a case always has a fixed and invariable order of its constituents as distributed on the scales:

- The validity part consists of three scales: L - the scale of lying which is sensitive to all false statements aiming at representing ourselves in a better light, F - the scale which detects atypical and deviational answers to all items in the test, K - it examines self defensive mechanisms and it detects subtler attempts of the subject being screened at falsifying and aggravation.
- The clinical part consists of ten scales: 1. Hypochondriasis (Hp), 2. Depression (D), 3. Hysteria (Hy), 4. Psychopathic Deviate (Ps), 5. Masculinity/Femininity (Mk), 6. Paranoia (Pa), 7. Psychastenia (Pt), 8. Schizophrenia (Sc), 9. Hypomania (Ma), 0. Social introversion (It).

3   Experiments

Patients were examined using the Minnesota Multiphasic Personality Inventory (MMPI) from the psychic disturbances perspective. Examination results are presented in the form of a profile. Patient's profile is a data vector consisting of fourteen attributes (thirteen descriptive attributes and one decision attribute - a class to which a patient is classified). A data set consists of over 1300 patients (women) classified by a clinic psychologist. Each case is assigned to one of 20 classes. Each class corresponds to one of nosological types: norm (*norm*), neurosis (*neur*), psychopathy (*psych*), organic (*org*), schizophrenia (*schiz*), delusion syndrome (*del.s*), reactive psychosis (*re.psy*), paranoia (*paran*), manic state (*man.st*), criminality (*crim*), alcoholism (*alcoh*), drag induction (*drag*), simulation (*simu*), dissimulation (*dissimu*), and six deviational answering styles (*dev1*, *dev2*, *dev3*, *dev4*, *dev5*, *dev6*). The data set was collected by T. Kucharski and J. Gomuła from the Psychological Outpatient Clinic. The data for the analysis (i.e., profiles of patients) were selected using the competent judge method (the majority of two-thirds of votes of three experts).

The proposed approaches, methods and procedures can also be applied for multidimensional data coming from other inventories (e.g., MMPI-2, EPQ) or collections of inventories if they will improve classification and diagnosis accuracy, what is a goal of research.

In our experiments, we used the following software tools:

- The Rough Set Exploration System (RSES) - a software tool featuring a library of methods and a graphical user interface supporting a variety of rough set based computations [7], [8].
- WEKA - a collection of machine learning algorithms for data mining tasks [9], [10] (the supervised attribute selection methods: ChiSquareAttributeEval, InfoGainAttributeEval, GainRatioAttributeEval and Ranker have been used).
- STATISTICA - a statistics and analytics software package developed by StatSoft (the discriminant analysis module has been used).
- COPERNICUS - a computer tool created by authors for analysis of MMPI profiles of patients with mental disorders [2].

In our experiments, three rule generation algorithms available in the RSES system have been used, namely: genetic algorithm, covering algorithm, and LEM2 algorithm. The first algorithm originates in an order-based genetic algorithm coupled with heuristic (see [11], [12]). It uses reducts for rule generation. Another two algorithms are based on a covering approach. A covering algorithm is described in [11] and [13]. The LEM2 algorithm was proposed by J. Grzymala-Busse in [14]. Covering-based algorithms produce less rules than algorithms based on an explicit reduct calculation. They are also (on average) slightly faster. It seems to be important if we extend a number of attributes in a decision table. For each algorithm, experiments have been performed for two kinds of data: the original data (without discretization) and the data after discretization using the local method available in the RSES system [11].

To determine the accuracy of classification of new cases a cross-validation method has been used. The cross-validation is frequently used as a method for evaluating classification models. It comprises of several training and testing runs. First, the data set is split into several, possibly equal in size, disjoint parts. Then, one of the parts is taken as a training set for the rule generation and the remainder (a sum of all other parts) becomes the test set for rule validation. In our experiments a standard 10 cross-validation (CV-10) test was used.

We tested six different situations according to the number of attributes:

1. Data table with attribute reduction using the standard rough set approach based on indiscernibility relation.
2. Data table with attribute reduction (selection) using three methods implemented in WEKA (AttributeEvaluators, supervised):
   - InfoGainAttributeEval - evaluates attributes by measuring the information gain with respect to the class,
   - GainRatioAttributeEval - evaluates attributes by measuring the gain ratio with respect to the class,
   - ChiSquaredAttributeEval - evaluates attributes by computing the value of the chi-squared statistics with respect to the class.
3. Original data table.
4. Data table with attribute extension by classification functions.

For WEKA, we have assumed the following attribute selection model: Attribute Evaluator + Search Metod (always Ranker) + Attribute Selection Mode (Cross-Validation: Folds 10 and Seed 1).

Formally, we have an original decision system (decision table) $S = (U, A \cup \{d\})$, where $U = \{x_1, x_2, ..., x_{1711}\}$, $A = \{a_1, a_2, ..., a_{13}\}$, and d determines a nosological type (category). a_1 corresponds to the scale of lying, a_2 - the scale which detects atypical and deviational answers, a_3 - the scale which examines self defensive mechanisms, a_4 - the scale of Hypochondriasis, a_5 - the scale of Depression, a_6 - the scale of Hysteria, a_7 - the scale of Psychopathic Deviate, a_8 - the scale of Masculinity/Femininity, a_9 - the scale of Paranoia, a_{10} - the scale of Psychastenia, a_{11} - the scale of Schizophrenia, a_{12} - the scale of Hypomania, a_{13} - he scale of Social Introversion.

A reduct is one of the most important notions in the rough set theory. Reducts preserve indiscernibility among objects in a decision table. Using an exhaustive algorithm of reduct computation for our data table preliminary discretized we obtain 31 reducts of the length from seven to nine attributes. For experiments, we have selected one of them, namely $\{a_1, a_3, a_4, a_6, a_7, a_8, a_9, a_{10}, a_{12}\}$.

In WEKA, three attribute reduction (selection) methods have been chosen: Chi Squared based on a statistics measure, and two based on information theory: Info Gain and Gain Ratio. Info Gain favors selection of multivalue attributes [15]. It is not a desirable feature, especially in strongly disparate attribute value sets constituting the profiles of patients. Therefore, another method, Gain Ratio, has also been chosen. The results of selection using WEKA are collected in Tables 1 and 2. The asterisk symbol (∗) denotes attributes for removing.

Table 1. Attribute selection using (a) Info Gain, (b) Gain Ratio

a)

Average merit	Average rank	Attribute
1.398 ± 0.02	1 ± 0	a_{11}
1.297 ± 0.02	2 ± 0	a_9
1.211 ± 0.04	3 ± 0	a_2
1.145 ± 0.02	4 ± 0	a_{10}
1.063 ± 0.02	5.5 ± 0.5	a_{12}
1.058 ± 0.04	5.6 ± 0.7	a_7
0.99 ± 0.03	7.4 ± 0.9	a_4
0.964 ± 0.04	8.2 ± 0.9	a_5
0.963 ± 0.01	8.3 ± 0.5	a_6
0.711 ± 0.0	10.4 ± 0.5	a_{13}^*
0.706 ± 0.02	10.6 ± 0.5	a_1^*
0.622 ± 0.03	12.3 ± 0.5	a_3^*
0.603 ± 0.03	12.7 ± 0.6	a_8^*

b)

Average merit	Average rank	Attribute
0.508 ± 0.01	1 ± 0	a_{11}
0.443 ± 0.01	2.2 ± 0.4	a_9
0.422 ± 0.02	2.9 ± 0.54	a_7
0.392 ± 0.01	4.4 ± 0.66	a_{12}
0.388 ± 0.01	4.9 ± 0.94	a_{10}
0.383 ± 0.01	5.7 ± 0.64	a_2
0.363 ± 0.01	7.3 ± 0.46	a_6
0.357 ± 0.02	8.3 ± 1.35	a_5
0.347 ± 0.01	8.9 ± 0.3	a_{13}
0.338 ± 0.01	9.4 ± 0.92	a_4
0.313 ± 0.01	11 ± 0	a_1^*
0.259 ± 0.01	12.3 ± 0.46	a_8^*
0.255 ± 0.01	12.7 ± 0.46	a_3^*

Table 2. Attribute selection using Chi Squared

Average merit	Average rank	Attribute
4944 ± 367.2	1 ± 0	a_9
4218 ± 141.9	2.3 ± 0.5	a_{11}
4061 ± 102.7	3.4 ± 0.8	a_{12}
4066 ± 159.5	3.5 ± 0.9	a_8
3793 ± 172.1	4.8 ± 0.4	a_2
3330 ± 50.4	6 ± 0	a_{10}
3120 ± 109.7	7.1 ± 0.3	a_3
3023 ± 76.2	8 ± 0.5	a_1
2925 ± 112.8	8.9 ± 0.3	a_7
2629 ± 160.9	10.2 ± 0.4	a_5
2504 ± 103.0	10.9 ± 0.5	a_4
2343 ± 37.7	11.9 ± 0.3	a_6
1677 ± 36.8	13 ± 0	a_{13}^*

The reduction of the attribute a_{13} only (ChiSquaredAttributeEval) or the attributes a_3, a_8, a_1 (InfoGainAttributeEval and GainRatioAttributeEval) seems to be the most valid approach. It is safe for nosological diagnosis of patients examined to use the MMPI-WISKAD test. In general, analysis of these three methods of attribute reduction (selection) clearly indicates that removing $a_{13}, a_3,$ a_8, and alternatively a_1 should improve classification accuracy and it is consistent with the clinician-diagnostician practice.

Classification functions (linear) should not be confused with canonical and discriminant functions. Classification functions can be used for adjudicating to which classes the particular cases belong. We have as much classification functions as categories. Each function enables us to calculate classification values for each case x in each category (class) i according to the formula $cf_i(x) = w_{i1}a_1(x) + w_{i2}a_2(x) + \ldots + w_{im}a_m(x) + c_i$.

The classification function technique is one of the most classical forms of classifier design. After calculating classification values for a given case, we assign to it a category for which a classification function has the greatest value. In the presented approach, classification functions are not directly used for classification. They define new attributes added to the original data table. Each value of each classification function has been normalized with respect to the greatest value for a given object (case).

We have defined (using the discriminant analysis module from STATISTICA) twenty classification functions ($cf1\_$, $cf2\_$, ..., $cf20\_$). There are the abbreviations of nosological categories after underlines) - each assigned to one decision category:

$cf1\_norm(x) = 1.165a_1(x) + 0.179a_2(x) + 1.566a_3(x) + 0.095a_4(x) - 0.004a_5(x) + 1.248a_6(x) - 0.279a_7(x) + 1.602a_8(x) + 1.101a_9(x) - 0.336a_{10}(x) - 0.876a_{11}(x) + 1.612a_{12}(x) + 5.179a_{13}(x) - 342.449$

$cf2\_neur(x) = 1.186a_1(x) + 0.102a_2(x) + 1.586a_3(x) + 0.407a_4(x) + 0.108a_5(x) + 1.625a_6(x) - 0.392a_7(x) + 1.454a_8(x) + 1.004a_9(x) - 0.212a_{10}(x) - 0.834a_{11}(x) + 1.425a_{12}(x) + 5.438a_{13}(x) - 387.068$

$cf3\_psych(x) = 1.306a_1(x) - 0.001a_2(x) + 1.708a_3(x) - 0.198a_4(x) + 0.035a_5(x) + 0.96a_6(x) + 0.503a_7(x) + 1.573a_8(x) + 1.323a_9(x) - 0.578a_{10}(x) - 0.673a_{11}(x) + 1.527a_{12}(x) + 4.81a_{13}(x) - 356.525$

$cf4\_org(x) = 1.288a_1(x) + 0.105a_2(x) + 1.578a_3(x) + 0.372a_4(x) + 0.105a_5(x) + 1.515a_6(x) - 0.38a_7(x) + 1.567a_8(x) + 1.226a_9(x) - 0.768a_{10}(x) - 0.464a_{11}(x) + 1.692a_{12}(x) + 5.403a_{13}(x) - 407.032$

$cf5\_schiz(x) = 1.271a_1(x) + 0.19a_2(x) + 1.678a_3(x) - 0.015a_4(x) + 0.035a_5(x) + 1.418a_6(x) - 0.211a_7(x) + 1.45a_8(x) + 1.223a_9(x) - 0.491a_{10}(x) - 0.589a_{11}(x) + 1.67a_{12}(x) + 5.459a_{13}(x) - 394.667$

$cf6\_del.s(x) = 1.381a_1(x) - 0.107a_2(x) + 1.803a_3(x) - 0.348a_4(x) + 0.27a_5(x) + 1.466a_6(x) - 0.11a_7(x) + 1.741a_8(x) + 1.356a_9(x) - 0.486a_{10}(x) - 0.613a_{11}(x) + 1.558a_{12}(x) + 5.398a_{13}(x) - 414.517$

$cf7\_re.psy(x) = 1.37a_1(x) + 0.132a_2(x) + 1.772a_3(x) + 0.106a_4(x) - 0.021a_5(x) + 1.244a_6(x) - 0.253a_7(x) + 1.379a_8(x) + 1.176a_9(x) - 0.005a_{10}(x) - 0.636a_{11}(x) + 1.408a_{12}(x) + 5.521a_{13}(x) - 407.826$

$cf8\_paran(x) = 1.486a_1(x) + 0.147a_2(x) + 1.802a_3(x) - 0.232a_4(x) + 0.055a_5(x) + 0.99a_6(x) + 0.05a_7(x) + 1.511a_8(x) + 1.577a_9(x) - 0.596a_{10}(x) - 0.433a_{11}(x) + 1.609a_{12}(x) + 4.455a_{13}(x) - 367.485$

$cf9\_man.st(x) = 1.293a_1(x) + 0.257a_2(x) + 1.55a_3(x) - 0.118a_4(x) - 0.044a_5(x) + 1.237a_6(x) - 0.015a_7(x) + 1.809a_8(x) + 1.406a_9(x) - 0.615a_{10}(x) - 0.809a_{11}(x) + 1.83a_{12}(x) + 4.461a_{13}(x) - 350.636$

$cf10\_crim(x) = 1.355a_1(x) + 0.226a_2(x) + 1.587a_3(x) - 0.018a_4(x) - 0.025a_5(x) + 1.267a_6(x) + 0.229a_7(x) + 1.489a_8(x) + 1.476a_9(x) - 0.499a_{10}(x) - 0.847a_{11}(x) + 1.621a_{12}(x) + 4.913a_{13}(x) - 382.284$

$cf11\_alcoh(x) = 1.023a_1(x) + 0.173a_2(x) + 1.458a_3(x) + 0.438a_4(x) - 0.061a_5(x) + 1.283a_6(x) - 0.19a_7(x) + 1.528a_8(x) + 1.068a_9(x) - 0.442a_{10}(x) - 0.862a_{11}(x) + 1.584a_{12}(x) + 4.65a_{13}(x) - 313.411$

$cf12\_drag(x) = 0.895a_1(x) + 0.179a_2(x) + 1.82a_3(x) + 0.375a_4(x) - 0.012a_5(x) + 1.187a_6(x) + 0.067a_7(x) + 1.445a\_8(x) + 1.253a_9(x) - 0.32a_{10}(x) - 0.829a_{11}(x) + 2.06a_{12}(x) + 4.895a_{13}(x) - 395.528$

$cf13\_simul(x) = 1.12a_1(x) + 0.388a_2(x) + 1.552a_3(x) + 0.1a_4(x) - 0.206a_5(x) + 1.601a_6(x) - 0.195a_7(x) + 1.559a_8(x) + 1.51a_9(x) - 0.393a_{10}(x) - 0.606a_{11}(x) + 1.815a_{12}(x) + 5.441a_{13}(x) - 441.559$

$cf14\_dissimu(x) = 1.506a_1(x) + 0.215a_2(x) + 1.876a_3(x) - 0.054a_4(x) - 0.088a_5(x) + 1.328a_6(x) - 0.135a_7(x) + 1.424a_8(x) + 0.997a_9(x) - 0.347a_{10}(x) - 0.746a_{11}(x) + 1.598a_{12}(x) + 4.78a_{13}(x) - 353.926$

$cf15\_dev1(x) = 1.524a_1(x) + 0.232a_2(x) + 2.058a_3(x) + 0.341a_4(x) + 0.083a_5(x) + 1.47a_6(x) + 0.402a_7(x) + 1.176a_8(x) + 0.806a_9(x) - 0.517a_{10}(x) - 0.557a_{11}(x) + 1.421a_{12}(x) + 4.831a_{13}(x) - 431.444$

$cf16\_dev2(x) = 0.843a_1(x) + 0.644a_2(x) + 1.431a_3(x) + 0.493a_4(x) - 0.618a_5(x) + 1.077a_6(x) - 1.328a_7(x) + 2.444a_8(x) + 2.377a_9(x) - 0.723a_{10}(x) - 0.179a_{11}(x), +.339a_{12}(x) + 5.579a_{13}(x) - 553.103$

$cf17\_dev3(x) = 1.114a_1(x) + 0.778a_2(x) + 1.847a_3(x) + 0.295a_4(x) + 0.099a_5(x) + 1.26a_6(x) - 0.36a_7(x) + 1.89a\_8(x) + 1.9a_9(x) - 0.909a_{10}(x) - 0.901a_{11}(x) + 2.216a_{12}(x) + 4.715a_{13}(x) - 473.711$

$cf18\_dev4(x) = 1.364a_1(x) + 0.73a_2(x) + 1.662a_3(x) + 0.404a_4(x) - 0.715a_5(x) + 1.51a_6(x) - 0.705a_7(x) + 1.614a_8(x) + 1.431a_9(x) - 0.524a_{10}(x) - 0.086a_{11}(x) + 1.6a_{12}(x) + 5.476a_{13}(x) - 464.539$

$cf19\_dev5(x) = 1.867a_1(x) + 0.486a_2(x) + 1.341a_3(x) + 0.354a_4(x) - 0.487a_5(x) + 1.414a_6(x) + 0.44a_7(x) + 2.492a_8(x) + 1.046a_9(x) - 1.27a_{10}(x) - 0.285a_{11}(x) + 2.168a_{12}(x) + 4.844a_{13}(x) - 531.166$

$cf20\_dev6(x) = 0.703a_1(x) + 0.757a_2(x) + 2.146a_3(x) + 0.338a_4(x) - 0.095a_5(x) + 1.163a_6(x) - 1.124a_7(x) + 0.954a_8(x) + 2.218a_9(x) - 0.13a_{10}(x) - 0.586a_{11}(x) + 1.696a_{12}(x) + 5.153a_{13}(x) - 454.695$

All experiment results have been collected in Tables 3, 4, and 5 - for the LEM2 algorithm, the covering algorithm and the genetic algorithm, respectively.

Experiments show that attribute extensions by classification functions for discretized data always improve accuracy of classification. In other cases, attribute reductions and extensions have a different influence on classification accuracy. Such observations are very important for further research, especially in the context of a created computer tool for multicriteria diagnosis of mental diseases.

Table 3. Results of experiments using the LEM2 algorithm

System	Without discretization		With discretization	
	Accuracy	Covering	Accuracy	Covering
After rough set reduction	0.870	0.504	0.934	0.657
After Info Gain reduction	**0.898**	0.527	0.934	0.668
After Gain Ratio reduction	**0.906**	0.522	0.949	0.694
After Chi Squared reduction	**0.906**	0.489	0.951	0.694
Original	0.885	0.498	0.956	0.701
After extension by classification functions	**0.966**	0.708	**0.968**	0.787

Table 4. Results of experiments using the covering algorithm

System	Without discretization		With discretization	
	Accuracy	Covering	Accuracy	Covering
After rough set reduction	**0.794**	0.115	**0.781**	0.518
After Info Gain reduction	0.761	0.137	**0.832**	0.525
After Gain Ratio reduction	**0.772**	0.14	0.749	0.485
After Chi Squared reduction	0.742	0.144	**0.818**	0.574
Original	0.761	0.146	0.764	0.428
After extension by classification functions	0.369	0.997	**0.854**	0.426

Table 5. Results of experiments using the genetic algorithm

System	Without discretization		With discretization	
	Accuracy	Covering	Accuracy	Covering
After rough set reduction	0.741	0.999	0.834	1.000
After Info Gain reduction	0.774	0.999	0.857	1.000
After Gain Ratio reduction	0.775	1.000	**0.878**	1.000
After Chi Squared reduction	0.780	1.000	0.873	1.000
Original	0.785	1.000	0.876	1.000
After extension by classification functions	0.761	1.000	**0.922**	1.000

4   Conclusions

In this paper, we have examined several approaches to attribute reduction and extension in the context of rule generation for data on the mental disorders coming from the MMPI test. Especially, results for extension by classification functions seem to be promising. In the future work a computer tool supporting diagnosis of mental diseases will be developed. Therefore, presented results shall be helpful for selection of suitable algorithms for this tool.

Acknowledgments

This paper has been partially supported by the grant from the University of Information Technology and Management in Rzeszów, Poland.

References

1. Lachar, D.: The MMPI: Clinical assessment and automated interpretations. Western Psychological Services, Fate Angeles (1974)
2. Gomuła, J., Pancerz, K., Szkoła, J.: Analysis of MMPI profiles of patients with mental disorders - the first unveil af a new computer tool. In: Proceedings of the XVII International Conference on Systems Science (ICSS 2010), Wroclaw, Poland (2010)
3. Duch, W., Kucharski, T., Gomuła, J., Adamczak, R.: Machine learning methods in analysis of psychometric data. Application to Multiphasic Personality Inventory MMPI-WISKAD, Toruń (1999) (in polish)
4. Gomuła, J., Paja, W., Pancerz, K., Szkoła, J.: A preliminary attempt to rules generation for mental disorders. In: Proceedings of the International Conference on Human System Interaction (HSI 2010), Rzeszów, Poland (2010)
5. Choynowski, M.: Multiphasic Personality Inventory. Psychometry Laboratory. Polish Academy of Sciences, Warsaw (1964) (in polish)
6. Płużek, Z.: Value of the WISKAD-MMPI test for nosological differential diagnosis. The Catholic University of Lublin (1971) (in polish)
7. Bazan, J.G., Szczuka, M.S.: The Rough Set Exploration System. In: Peters, J.F., Skowron, A. (eds.) Transactions on Rough Sets III. LNCS, vol. 3400, pp. 37–56. Springer, Heidelberg (2005)
8. The Rough Set Exploration System, http://logic.mimuw.edu.pl/~rses/
9. Witten, I.H., Frank, E.: Data Mining: Practical Machine Learning Tools and Techniques. Morgan Kaufmann, San Francisco (2005)
10. WEKA, http://www.cs.waikato.ac.nz/ml/weka/
11. Bazan, J.G., Nguyen, H.S., Nguyen, S.H., Synak, P., Wroblewski, J.: Rough set algorithms in classification problem. In: Polkowski, L., Tsumoto, S., Lin, T.Y. (eds.) Rough Set Methods and Applications. Studies in Fuzziness and Soft Computing, pp. 49–88. Physica-Verlag, Heidelberg (2000)
12. Wróblewski, J.: Genetic algorithms in decomposition and classification problem. In: Rough Sets in Knowledge Discovery, vol. 2, pp. 471–487
13. Wróblewski, J.: Covering with reducts - a fast algorithm for rule generation. In: Polkowski, L., Skowron, A. (eds.) RSCTC 1998. LNCS (LNAI), vol. 1424, pp. 402–407. Springer, Heidelberg (1998)
14. Grzymala-Busse, J.: A new version of the rule induction system LERS. Fundamenta Informaticae 31, 27–39 (1997)
15. Quinlan, J.: Induction of decision trees. Machine Learning 1(1), 81–106 (1986)

Automatic 3D Face Correspondence Based on Feature Extraction in 2D Space

Xun Gong[1], Shuai Cao[1], Xinxin Li[2], Ping Yuan[1], Hemin Zou[1], Chao Ye[1], Junyu Guo[1], and Chunyao Wang[1]

[1] School of Information Science and Technology,
South west Jiaotong University,
Chengdu 600031, P.R. China
[2] Jincheng College of Sichuan University,
Chengdu 611731, P.R. China
xgong@swjtu.cn

Abstract. This present study proposes to solve 3D faces correspondence in 2D space. Texture maps of 3D models are generated at first by unwrapping 3D faces to 2D space. Following this, we build planar templates based on the 2D average shape computed by a group of annotated texture map. In this paper, landmarks on the unwrapped texture images are located automatically and the final correspondence is built according to the templates. Experimental results show that the presented method is stable and can achieve good matching accuracy.

Keywords: Dense correspondence; 3D face; re-sampling.

1 Introduction

Recent advances in 3D digitizing techniques make the acquisition of 3D human face data much easier and cheaper. Constructing alignment of 3D objects is a crucial element of data representations in computer vision.

The raw 3D face data captured by a 3D scanner are stored in a loosening mode, and the topological structure of the face varies from one to another. In order to formulate the 3D data by mathematic tools, the prototypes should be standardized in a uniform format before being used. This problem can also be described as feature points marked in one face—such as the tip of the nose, the eye corner—must be located precisely in other faces. To illustrate it, consider Fig. 1. If we are given two 3D face f_1, f_2, and we would like to produce a third model which is "in between" these two, then we might first attempt to simply compute a superposition $f_1 + f_2$. However, as the figure illustrates, this does not lead to satisfactory results. It produces ghost contours caused by the fact that, for instance, the eyes of the two faces were not aligned before computing the superposition. If, on the other hand, we manage to align all relevant facial features in f_1 with the corresponding ones in f_2, then the two faces are said to be in correspondence. Obviously, such an alignment will typically not be possible in, say, a face dataset that captured by a scanner.

J. Yu et al. (Eds.): RSKT 2010, LNAI 6401, pp. 419–426, 2010.

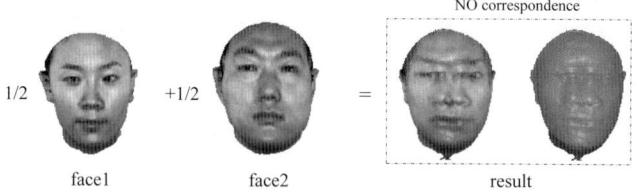

Fig. 1. The models of the two 3D faces are linearly superposed (see the last two), leading to a face which no longer is an admissible human head

Automatic algorithms for computing warps between surfaces have been presented previously. For parameterized surfaces of human faces that were captured with a laser scanner, Blanz and Vetter proposed a modified optical flow algorithm [1]. Yin *et al.*[2] proposed a uniform mesh re-sampling based method to make correspondence between the vertices of the prototypes. To decrease the number of vertices, in their following work, they proposed a uniform mesh re-sampling algorithm combined with mesh simplification and a non-uniform re-sampling method [3]. Inspired by Yin's work, Gong et al. [4,5] proposed a solution based on planar templates (PTs) extracted from feature regions that could run automatically without user interventions. However, the feature regions extracted by their automatic method is not reliable enough and always failed in ear regions.

Based on the basic idea of PTs, this paper proposes a more robust matching method to compute correspondence between 3D faces. In this work, 34 key feature points is used to present 2D facial shape on texture maps, which could be extracted accurately by the feature extraction methods. Texture maps of 3D models are generated at first by unwrapping 3D faces to 2D space. Following this, we build planar templates based on the mean shape computed by a group of annotated texture map. Landmarks on the unwrapped texture images are located automatically by active appearance models (AAMs). Once all texture maps of the prototypes are warped to the PTs, pixel-wise correspondences between the prototypical 3D faces were then constructed. The main idea of our method is illustrated in the framework of Fig. 2.

In the following section, the unwrapping process is described to create texture maps for prototypes. Section 3 presents our PTs based re-sampling method. The experiment results is outlined in Section 4.

2   Unwrapping 3D Faces

The BJUT-3D Face Database [6] is taken for illustration, which contains 500 laser-scanned 3D heads of Chinese people (250 females and 250 males). A prototypic 3D face can be denoted by the following vectors:

$$s = (x_1, y_1, z_1, ..., x_i, y_i, z_i, ..., x_n, y_n, z_n)^{\mathrm{T}} \in \Re^{3n}, 1 \leq i \leq n \tag{1}$$

$$t = (r_1, g_1, b_1, ..., r_i, g_i, b_i, ..., r_n, g_n, b_n)^{\mathrm{T}} \in \Re^{3n}, 1 \leq i \leq n . \tag{2}$$

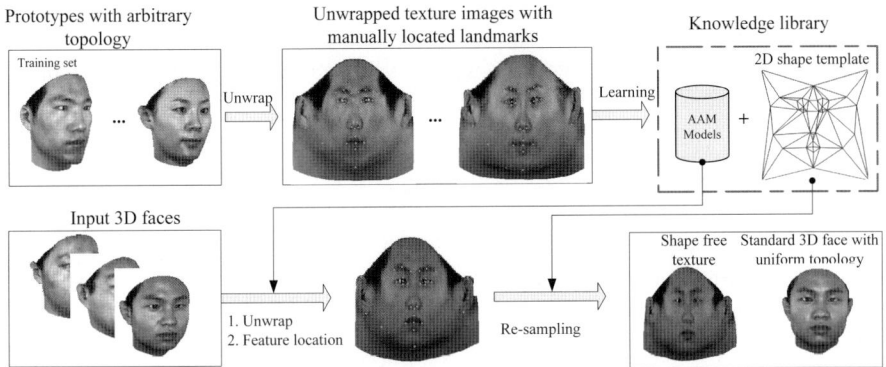

Prototypes with arbitrary topology

Unwrapped texture images with manually located landmarks

Knowledge library

Training set

Unwrap

2D shape template

Learning

AAM Models +

Input 3D faces

1. Unwrap
2. Feature location

Re-sampling

Shape free texture   Standard 3D face with uniform topology

Fig. 2. Framework of our system, which consists of 2 main parts: One is the training step, including building a 2D shape template and the AAMs. The other is the re-sampling step which is responsible for creating a standard 3D face for each input.

where, s is the shape vector, t is the texture vector. n is the number of vertices. (r_i, g_i, b_i) is the 24-bit color values corresponding to a spatial vertex (x_i, y_i, z_i).

Now, we create the 2D texture image from a 3D model by converting the 3D vertices to 2D. At first, we calculate the cylindrical coordinates for each vertex. Let $c_i = (h_i, \theta_i, r_i)$ be the cylindrical coordinates, where h_i, θ_i, r_i denote height, angle and the radius of the i -th vertex, respectively. 3D spatial coordinates and 3D cylindrical coordinates are related to each other:

$$h_i = y_i, r_i = \sqrt{x_i^2 + z_i^2} \tag{3}$$

$$\theta_i = \begin{cases} \arctan(x_i/z_i) & z_i > 0 \\ \arccos(z/r) & z_i < 0, x_i > 0 \\ -\arccos(z_i/r_i) & z_i < 0, x_i < 0 \end{cases} \tag{4}$$

where, $r_i \geq 0$, $-\pi \leq \theta_i \leq \pi$.

Then, we can easily generate the texture map $f(u, v)$ by sampling the continuous values θ. The experimental results show that a visually comfortable texture image can be created while $Height = 273, Width = 280$. Texture map can be seen in Fig. 2. Two mapping tables are obtained simultaneously, which maintain the correspondence relationship between 3D vertices and 2D pixels.

3   Dense Pixel-Wise Alignment by Planar Templates

3.1   2D Shape Extraction Based on AAMs

It's feasible to locate the key points by machine learning algorithms automatically as the 3D models have been captured under controlled environment. AAMs [7] is a popular method for face modeling [8] and feature localization. The *2D shape* of an AAM is defined by a 2D triangulated mesh and in particular the vertex locations of the mesh. Mathematically, we define the shape s^{2D} of an AAM as the 2D coordinates of the n vertices that make up the mesh [9].

In AAMs, the shape matrix s^{2D} can be expressed as a base shape s_0^{2D} plus a linear combination of m shape matrices s_i^{2D}:

$$s^{2D} = s_0^{2D} + \sum_{i=1}^{m} \alpha_i s_i^{2D} \qquad (5)$$

where the coefficients α_i are the shape parameters. The base shape s_0^{2D} is the mean shape and the matrices s_i^{2D} are the (reshaped) eigenvectors corresponding to the m largest eigenvalues.

Likewise, the appearance $A(q)$ can be expressed as a base appearance $A_0(q)$ plus a linear combination of l appearance images $A_i(q)$:

$$A(q) = A_0(q) + \sum_{i=1}^{l} \beta_i A_i(q) \qquad (6)$$

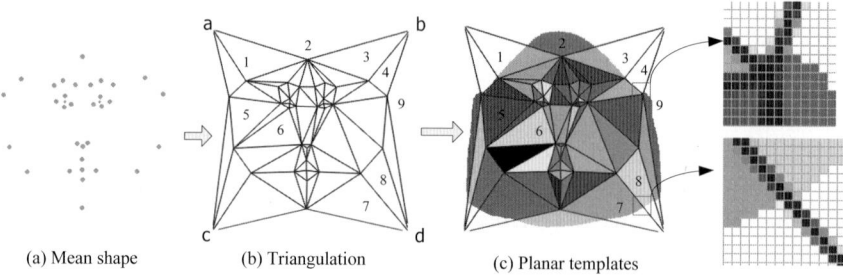

(a) Mean shape   (b) Triangulation   (c) Planar templates

Fig. 3. Training and fitting results of AAMs on texture maps. (b) shows the fitting process of a testing texture map.

Fitting an AAM to an image consists of minimizing the error between the input image and the closest model instance; i.e. solving a nonlinear optimization problem to get proper α and β. However, AAMs usually use a combined set of parameters $c = (c_1, c_2, ..., c_l)^{\mathrm{T}}$ to parameterize shape α and appearance β. In this paper, we choose to use an efficient fitting algorithm for AAMs based on the inverse compositional image alignment algorithm, see [10] for more details.

As shown in Fig. 3, 34 key landmarks on face image plus 4 supplementary points (corners) are chosen to form a planar facial shape. The average texture & shape and the fitting results are demonstrated in Fig. 3. As we expected, after mere 15 iterations, the landmarks on a novel image could be extracted which is accurate enough in our application.

3.2 Planar Templates

After manually annotate the landmarks on training textures, as preprocess, we first align the shapes s_i^{2D} by Generalized Procrustes Analysis [11], including scaling, rotating and translation operators. Then, the mean shape s_0^{2D} can be computed by:

$$s_0^{2D} = \sum_{i=1}^{m} s_i^{2D} \qquad (7)$$

The mean shape of 30 faces is shown in Fig. 4 (a). Since the 34 key facial points with semantic meanings have been located manually in the texture map, a sparse correspondence is already established. Other points within the face region can be interpolated and warped to the PTs, which are used to present the standard resolution and topology of texture map. Once each texture map of a 3D face set is re-sampled according to the PTs, the correspondence is built between texture maps. Due that each point in the texture map corresponds to a vertex in the shape model (see section 2), in this sense, the correspondence between two vertices from different 3D faces are also established. The planar templates can be generated by following steps:

1. Triangulation based on the mean shape and 4 supplementary points at 4 corners, see Fig. 4 (b);
2. Design points distribution in triangles. There are two types, outer triangles around boundaries (see triangles annotated by 1, 2, 3 in Fig. 4 (b, c)) and inner triangles like ones numbered as 5 and 6 in Fig. 4 (b, c). For the inner templates, we use the resolution on mean texture directly even though we can alter it easily for multi-resolution applications. For the outer triangles, as shown in Fig. 4 (c), a pixel is taken as a valid one only if its occurrence rate higher than 0.5 (the majority rule) at the same position across all the textures maps in the training set.

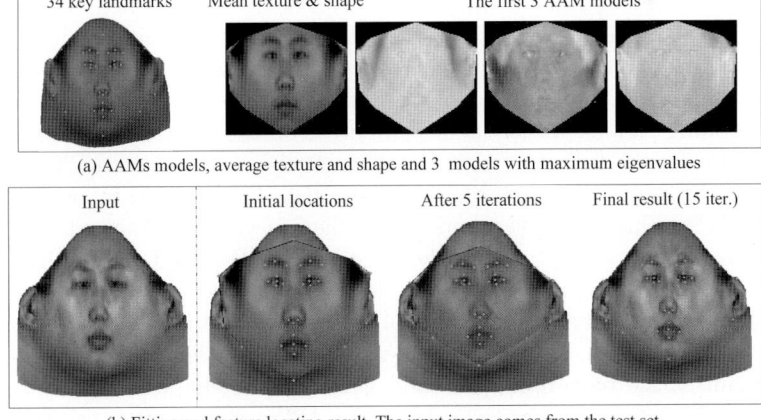

(a) AAMs models, average texture and shape and 3 models with maximum eigenvalues

(b) Fitting and feature locating result. The input image comes from the test set

Fig. 4. Triangulation based on mean shape and the planar templates. (a) mean shape generated by 30 faces; Points a-d in (b) are 4 supplementary points; (c) visualization of planar templates, in which 2 square are zoomed in to display its structures in detail.

4   Experiments and Discussions

To create a standardized 3D face database, we have solved the correspondence problem of all 500 faces in BJUT-3D Face Database [6] by using the proposed PTs based method. After correspondence, all the faces have a uniform topological structure with 59266 vertices and 117530 triangles for each. As shown in Table 1, our system can finish correspondence automatically in mere about 0.1s, while it takes about 2s in our previous work [5] and 2 hours in Bernhard's work [12].

Table 1. Average time costs of standardize one 3D face (CPU: 2.93GHz, RAM:4G)

3D face	Feature location	Re-sampling	Total
Vertices: 57657	26 ms	101 ms	127 ms
Triangles: 117530			

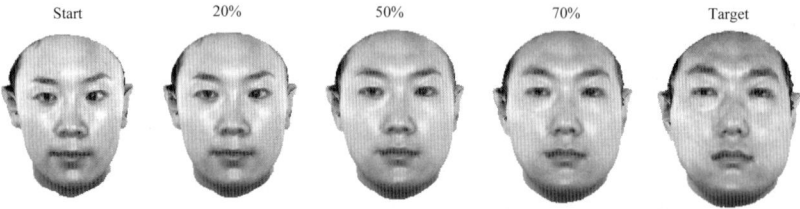

Fig. 5. Average faces of standardized faces generated by two methods

Once the point-to-point correspondence of 3D faces is established, the direct application of the alignment is to create 3D face morphing and animation. In Fig. 5, a series of smooth transitions from a female face to a male face is illustrated. Compared to Fig. 1, we can see that these transitions have achieved very good morphing results, where the feature areas on the intermediate synthesized 3D faces are rather distinct. Fig. 6 (a) shows 2 views of the average face, which is synthesized from 50 females and 50 males that standardized by our proposed method. We see that the correspondence problem is well solved since their feature areas are as distinct as captured by the laser scanner. For comparison, the average face generated by Gong et al. [5] is also demonstrated in Fig. 6 (b). It's not hard to find that the second average face is more obscure due to the losses of accuracy in correspondence process, especially in the key features like eyes and ears.

To quantify the re-sampling method we used, the shape similarity and appearance similarity error functions D_s and e_t [13] are adopt for evaluation:

$$D_s = \left(\sum_{i=1}^{n} S_i\right) \Big/ \left(\sum_{i=1}^{m} SS_i\right), e_t = \frac{1}{m^2}\sum_{i=1}^{m} \|T_1(i) - T_2(i)\|^2 \qquad (8)$$

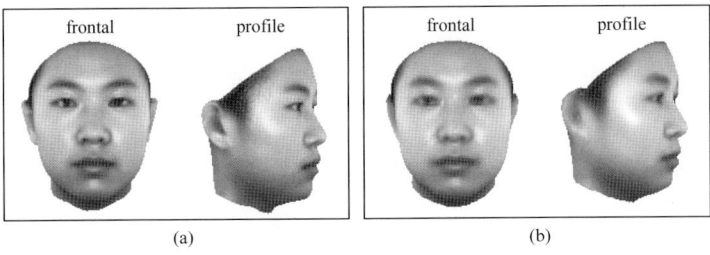

Fig. 6. Average faces of standardized faces generated by two methods

Table 2. The results of surface approximation and appearance approximation error

Methods	Number of vertices	D_s	e_t
Yin's [13]	26058	98%	$0.1 \sim 0.3$
Wang's	132546	99%	$0.0 \sim 0.2$
Gong's [5]	57657	99%	0.0
Our method	59266	99%	0.0

After standardizing 100 3D faces, which are randomly selected from the database, by different algorithms, the comparison results are listed in Table 2. It's clear to see that, the 3D faces create by our method could maintain high approximation compared to the original models in both shape and appearance standpoints. Furthermore, the new created models keep moderate amount of data, which will improve the efficiency of the following applications, such as face modeling and animation.

5 Conclusions

To overcome the limitation of conventional mesh re-sampling algorithm, the work represented in this paper provides a fully automatic resolution a new dense point-to-point alignment method and apply it on scanned 3D faces. The experiment results show that without complex manual interventions, our system is able to run quickly and obtain standardized 3D faces with good quality.

Acknowledgment

This work is supported by the Young Teachers Start Research Project of SWJTU (No. 2009Q086), Fundamental Research Funds for the Central Universities (No. 2009QK17), National Undergraduate Innovation Experiment Project (No. 091061322). Portions of the research in this paper use the BJUT-3D Face Database collected under the joint sponsor of National Natural Science Foundation of China.

References

1. Blanz, V., Vetter, T.: A morphable model for the synthesis of 3D faces. ACM Press/Addison-Wesley Publishing Co., Los Angeles (1999)
2. Gu, C., Yin, B., Hu, Y., et al.: Resampling Based Method for Pixel-wise Correspondence between 3D Faces, Las Vegas, Nevada, USA (2004)
3. Yin, B., He, Y., Sun, Y., et al.: Nonuniform Resampling Based on Method for Pixel-wise Correspondence Between 3D Faces. Journal of Beijing University of Technology 33(2), 213–218 (2007)
4. Gong, X., Wang, G.: 3D Face Deformable Model Based on Feature Points. Journal of Software 20(3), 724–733 (2009)
5. Gong, X., Wang, G.: An automatic approach for pixel-wise correspondence between 3D faces. IEEE Computer Science, Cheju Island (2006)
6. Hu, Y., Yin, B., Cheng, S., et al.: Research on Key Technology in Construction of a Chinese 3D Face Database. Journal of Computer Research and Development 42(4), 622–628 (2005)
7. Cootes, T.F., Edwards, G.J., Taylor, C.J.: Active appearance models. IEEE Trans. Pattern Analysis and Machine Intelligence 23(6), 681–685 (2001)
8. Lanitis, A., Taylor, C., Cootes, T.: Automatic interpretation and coding of face images using flexible models. IEEE Transactions on Pattern Analysis & Machine Intelligence 19(7), 743–756 (1997)
9. Xiao, J., Baker, S., Matthews, I., et al.: Real-Time Combined 2D+3D Active Appearance Models (2004)
10. Matthews, I., Baker, S.: Active Appearance Models Revisited. Robotics Institute. Carnegie Mellon University, Pittsburgh (2003)
11. Gower, J.: Generalizaed Procrustes analysis. Psychometrika 40(1), 33–55 (1975)
12. Scholkopf, B., Steinke, F., Blanz, V.: Object correspondence as a machine learning problem, Bonn, Germany (2005)
13. Yin, B., He, Y., Sun, Y., et al.: Nonuniform Resampling Based on Method for Pixel-wise Correspondence Between 3D Faces. Journal of Beijing University of Technology 33(2), 213–218 (2007)

The Neuropathological Diagnosis of the Alzheimer's Disease under the Consideration of Verbal Decision Analysis Methods

Isabelle Tamanini, Plácido Rogério Pinheiro, and Mirian Calíope D. Pinheiro

University of Fortaleza (UNIFOR) - Graduate Program in Applied Computer Sciences
Av. Washington Soares, 1321 - Bl J Sl 30 - 60.811-905 - Fortaleza - Brazil

Abstract. The Alzheimer's disease incidence and prevalence double every five years with a prevalence of 40% among people with more than 85 years of age, and there is a great challenge in identifying the pathology in its earliest stages. The main focus of the work is to aid in the decision making on the diagnosis of the Alzheimer's disease, which will be made applying the method ORCLASS and the Aranaú Tool. The modeling and evaluation processes were conducted based on bibliographic sources and on the information given by a medical expert.

1 Introduction

The Alzheimer's disease is the most frequent cause of dementia, being responsible (alone or in association with other diseases) for 50% of the cases in western countries [11]. There's a major importance in identifying the cases in which the risks of developing a dementia are higher, considering the few alternative therapies and the greater effectiveness of treatments when an early diagnosis is possible [6]. Besides, according to studies conducted by the Alzheimer's Association [7], the treatments have significant resulting costs, and it is known that it is one of the costliest diseases, second only to cancer and cardiovascular diseases.

A battery of tests from the Consortium to Establish a Registry for Alzheimer's Disease (CERAD) [5] was used in this work, because it encompasses all the steps of the diagnosis and it is used all over the world. The CERAD was founded with the aim of establishing a standard process for evaluation of patients with possible Alzheimer's disease who enrolled in NIA-sponsored Alzheimer's Disease Centers or in other dementia research programs [10].

The aim of this paper is to determine which of these tests are relevant and would detect faster if the patient is developing the disease. This will be made in two stages: at first, a classification method will be applied to determine which questionnaires from a set would, themselves, detect the Alzheimer's disease; then, the questionnaires classified as the ones that could give this diagnosis will be ordered, from the most likely to lead to the diagnosis faster, to the least. To do so, considering the experience of the decision maker (the medical expert), the models were structured based on the methods ORCLASS and ZAPROS. This work was based on several other works [1, 2, 3, 4, 13, 14].

J. Yu et al. (Eds.): RSKT 2010, LNAI 6401, pp. 427–432, 2010.

2   The Ordinal Classification Method - ORCLASS

The ORCLASS (ORdinal CLASSification) method [8] belongs to the Verbal Decision Analysis framework, and aims at classifying the alternatives of a given set. It enables the verbal formulation of the classification rule obtained for the explanation of decisions, such that any alternatives defined by the criteria and criteria values previously structured can be classified.

In the preferences elicitation process, classification boards will be structured such that each cell is composed by a combination of determined values of all the criteria defined to the problem, which represents a possible alternative to the problem. It will be presented to the decision maker classify only the combinations that, when classified, will reflect this information in a subset of the possible alternatives, so only the alternatives having the most informative index will be presented to decision maker. To determine this alternative, one should calculate, for each cell of the board, the number of alternatives that may be also classified based on the decision maker's response (considering all decision classes). A detailed description of the method's operation is available in [8].

3   An Approach Methodology Based on ZAPROS

The ZAPROS Method [9] is a qualitative method that aims at rank ordering any set of alternatives based on the preferences defined by the decision maker upon a set of criteria and criteria values given. A great advantage of the method is that all the questionings made in the preferences elicitation process are presented in the decision maker's natural language. Moreover, verbal descriptions are used to measure the preferences levels. This procedure is psychologically valid, respecting the limitations of the human information processing system.

This characteristic, however, brings about some limitations: the incomparability cases become unavoidable when the scale of preferences is purely verbal, since there is not an accurate measure of the values. Therefore, the method may not be capable of achieving satisfactory results in some situations, presenting, then, an incomplete result to the problem.

Considering the limitation exposed, it is presented in [12] some modifications to the ZAPROS method for aiding at the decision making process on Verbal Decision Analysis, aiming at the presentation of a complete and satisfactory result to the decision maker by reducing the incomparability cases between the alternatives. The modifications were applied mainly to the alternatives comparison process and did not modify the method's computational complexity.

In order to facilitate the decision process and perform it consistently, observing its complexity and with the aim of making it accessible, a tool implemented in Java was used [12]. At the end of the process, it provides a graph to the user, representing the dominating relations of the alternatives.

4   Multicriteria Models for the Diagnosis of the Alzheimer's Disease

In order to establish which tests would be capable of leading to a diagnosis of the Alzheimer's disease, a classification multicriterion model was structured based on the tests of the CERAD's neuropathological battery. We can state the classification problem into two groups of questionnaires:

   I) The ones that would lead to the diagnosis of Alzheimer's disease; and
   II) The ones that would require other data to be able to get the diagnosis.

The criteria were defined considering parameters of great importance to the diagnosis, based on the analysis of each questionnaire data by the decision maker. This analysis was carried out following some patterns, such that if a determined data is directly related to the diagnosis of the disease, and this fact can be found based on the number of its occurrences in the battery data, this fact will be selected as a possible value of criteria. For being relevant to the diagnosis, one can notice that questionnaires that are able to detect this occurrence are more likely to give a diagnosis. Regarding the CERAD data, only the results of the tests of patients that had already died and on which the necropsy has been done were selected (122 cases), because it is known that necropsy is essential for validating the clinical diagnosis of dementing diseases. The criteria defined to the classification model are exposed in Table 1.

Table 1. Criteria for evaluation of the questionnaires capable of leading to the diagnosis

Criteria	Values of Criteria
A: Type of examination performed to answer the questionnaire	A1. Neuroimaging tests were applied
	A2. Tests concerning the gross characteristics of the brain or abnormalities of meninges
	A3. Other tests concerning the patient's health
B: The type of the diagnosis based on the questionnaire's answers	B1. An early diagnosis may be possible
	B2. A diagnosis can be established
	B3. It can only be achieved based on post-mortem tests
C: Type of evaluated patient's clinical history	C1. Complete study of the patient's clinical history
	C2. Questionings concerning the patient's clinical history
	C3. No clinical history is evaluated

The classification was performed based on the analysis by medical expert of the questionnaire's answers and the patients' final diagnosis. The facts that were more constant and extremely connected to the patient's final diagnosis were set as preferable over others. This way, it was possible to establish a order of the facts that had a direct relation with the patients' final diagnosis, and so, which of the questionnaires had the questions that were more likely to lead to it. The classification board obtained is presented in Fig. 1, and the questionnaires defined as alternatives to this problem are shown in Table 2.

	B_1	B_2	B_3
A_1	I	I	II
A_2	I	I	II
A_3	I	I	II

C_1

	B_1	B_2	B_3
A_1	I	I	II
A_2	I	I	II
A_3	I	II	II

C_2

	B_1	B_2	B_3
A_1	I	I	II
A_2	II	II	II
A_3	II	II	II

C_3

Fig. 1. Classification board obtained to the model

Table 2. The questionnaires represented as criteria values and their respective classes

Alternatives	Criteria Evaluations	Class
Clinical History	A3B2C1	I
Gross Examination of the Brain	A2B1C3	II
Cerebral Vascular Disease Gross Findings	A1B1C3	I
Microscopic Vascular Findings	A1B2C3	I
Assessment of Neurohistological Findings	A2B3C3	II
Neuropathological Diagnosis	A1B2C2	I

Then, the questionnaires classified as group I were submitted to an ordination method, such that these would be ordered from the best one to be applied to get to a diagnosis faster, to the least one. The criteria definition followed the same procedure of the previous model. This way, the ones considered relevant to evaluate the CERAD's questionnaires afore mentioned and the values identified for them are exposed in Table 3.

Table 3. Criteria for evaluation of the analyzed questionnaires of CERAD

Criteria	Values of Criteria
A: Verification of Other Brain Problems' Existence	A1. There are questions about the severity of other problems A2. There are questions about its existence A3. There are no questions considering its existence
B: Neuroimaging tests: MRI or CT	B1. One or more cerebral lacunes can be detected B2. May detect extensive periventricular white matter changes B3. Focal or generalized atrophy can be detected
C: Clinical Studies	C1. Electroencephalogram C2. Analysis of Cerebrospinal Fluid C3. Electrocardiogram

The preferences elicitation was based on the same information sources considered on the previous model, and the preferences scale obtained is given as follows: $b_1 \equiv c_1 \prec a_1 \prec c_2 \prec b_2 \prec b_3 \prec c_3 \prec a_2 \prec a_3$.

Then, the alternatives were formulated identifying which facts (the values of criteria) would be identified by each questionnaire, considering the tests it would require to be filled. So, we'll have the questionnaires described as criterion values

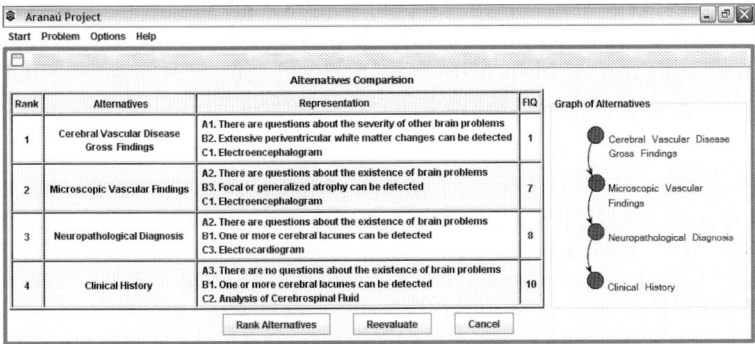

Fig. 2. Results presentation screen of the Aranú Tool

according to the facts verified by each one of them. The model was applied to the Aranaú Tool [12] and the results presentation screen is shown in Fig. 2.

As a conclusion to the modeling, it was identified that the questionnaire Cerebral Vascular Disease Gross Findings is more likely to detect a possible case of Alzheimer's disease than the others, so, it should be applied first.

5   Conclusions

The main purpose of this work is to set up an order of questionnaires from the neuropathological battery of CERAD that were more likely to give the diagnosis. To do so, it was presented a model to classify the questionnaires into two groups: the ones that would lead to a diagnosis of the Alzheimer's disease, and the ones that would require further analysis. Then, a model was formulated based on the questionnaires classified in the first group and submitted to Aranaú tool [12], which is structured on the ZAPROS method but with some modifications to improve its comparison power, and the questionnaires order was obtained.

The criteria were defined considering the characteristics of each questionnaire. The preferences were given through the analysis by a medical expert of the questionnaires results and the postmortem diagnosis of each patient, in a way to establish a relation between them. Then, the alternatives were structured according to the facts that could be detected on answering each of these questionnaires. As the contribution of this paper, we had a ranking of the questionnaires, from the one that had the greatest importance on the diagnosis, to the one that had the least. This would enable a faster detection of patients that would develop the disease, increasing their chances of treatment.

As future works, we intend establish a model to rank the questionnaires considering all the characteristics of each one, aiming at a more detailed description of these questionnaires as criteria values; and to verify which stage of the Alzheimer's disease would be detected based on the order established, in order to detect the disease in its earliest stages.

Acknowledgment. The authors are thankful to the National Counsel of Technological and Scientific Development (CNPq) for all the support received and to the Consortium to Establish a Registry for Alzheimer's Disease (CERAD) for making available the data used in this case study.

References

[1] De Castro, A.K.A., Pinheiro, P.R., Pinheiro, M.C.D.: Applying a Decision Making Model in the Early Diagnosis of Alzheimer's Disease. In: Yao, J., Lingras, P., Wu, W.-Z., Szczuka, M.S., Cercone, N.J., Ślęzak, D. (eds.) RSKT 2007. LNCS (LNAI), vol. 4481, pp. 149–156. Springer, Heidelberg (2007)

[2] De Castro, A.K.A., Pinheiro, P.R., Pinheiro, M.C.D.: A Multicriteria Model Applied in the Diagnosis of Alzheimer's Disease. In: Wang, G., Li, T., Grzymala-Busse, J.W., Miao, D., Skowron, A., Yao, Y. (eds.) RSKT 2008. LNCS (LNAI), vol. 5009, pp. 612–619. Springer, Heidelberg (2008)

[3] de Castro, A.K.A., Pinheiro, P.R., Pinheiro, M.C.D.: An Approach for the Neuropsychological Diagnosis of Alzheimer's Disease. In: Wang, G., Li, T., Grzymsls-Busse, J.W., Miao, D., Skowron, A., Yao, Y.Y. (eds.) RSKT 2009. LNCS, vol. 5589, pp. 216–223. Springer, Heidelberg (2009)

[4] de Castro, A.K.A., Pinheiro, P.R., Pinheiro, M.C.D., Tamanini, I.: Towards the Applied Hybrid Model in Decision Making: A Neuropsychological Diagnosis of Alzheimer's Disease Study Case. International Journal of Computational Intelligence Systems (accepted to publication, 2010)

[5] Fillenbaum, G.G., van Belle, G., Morris, J.C., et al.: Consortium to Establish a Registry for Alzheimers Disease (CERAD): The first twenty years. Alzheimer's & Dementia 4(2), 96–109 (2008)

[6] Hughes, C.P., Berg, L., Danzinger, W.L., et al.: A New Clinical Scale for the Staging of Dementia. British J. of Psychiatry 140(6), 566–572 (1982)

[7] Koppel, R.: Alzheimer's Disease: The Costs to U.S. Businesses in 2002. Alzheimers Association - Report (2002)

[8] Larichev, O., Moshkovich, H.M.: Verbal decision analysis for unstructured problems. Kluwer Academic Publishers, Boston (1997)

[9] Larichev, O.: Ranking Multicriteria Alternatives: The Method ZAPROS III. European Journal of Operational Research 131(3), 550–558 (2001)

[10] Morris, J.C., Heyman, A., Mohs, R.C., et al.: The Consortium to Establish a Registry for Alzheimer's Disease (CERAD): Part 1. Clinical and Neuropsychological Assessment of Alzheimer's Disease. Neurology 39(9), 1159–1165 (1989)

[11] Prince, M.J.: Predicting the Onset of Alzheimer's Disease Using Bayes' Theorem. American Journal of Epidemiology 143(3), 301–308 (1996)

[12] Tamanini, I., Pinheiro, P.R.: Challenging the Incomparability Problem: An Approach Methodology Based on ZAPROS. Modeling, Computation and Optimization in Information Systems and Management Sciences, Communications in Computer and Information Science 14(1), 344–353 (2008)

[13] Tamanini, I., Castro, A.K.A., Pinheiro, P.R., Pinheiro, M.C.D.: Towards the Early Diagnosis of Alzheimer's Disease: A Multicriteria Model Structured on Neuroimaging. Int. J. of Social and Humanistic Computing 1, 203–217 (2009)

[14] Tamanini, I., Castro, A.K.A., Pinheiro, P.R., Pinheiro, M.C.D.: Verbal Decision Analysis Applied on the Optimization of Alzheimer's Disease Diagnosis: A Study Case Based on Neuroimaging. Special Issue: Software Tools and Algorithms for Biological Systems, Advances in Experimental Medicine and Biology (to appear 2010)

Autonomous Adaptive Data Mining
for u-Healthcare

Andrea Zanda, Santiago Eibe, and Ernestina Menasalvas*

Universidad Politecnica Madrid, Facultad de Informatica, Spain
andrea.zanda@alumnos.upm.es, {seibe,emenasalvas}@fi.upm.es

Abstract. Ubiquitous healthcare requires intelligence in order to be able to react to different patients needs. The context and resources constraints of the ubiquitous devices demand a mechanism able to estimate the cost of the data mining algorithm providing the intelligence. The performance of the algorithm is independent of the semantics, this is to say, knowing the input of an algorithm the performance can be calculated. Under this assumption we present formalization of a mechanism able to estimate the cost of an algorithm in terms of efficacy and efficiency. Further, an instantiation of the mechanism for an application predicting glucose level for diabetic patients is presented.

1 Introduction

Applications in several domains require some kind of intelligence. In the domain of intelligent vehicles the devices have to take decisions in real time [2]. The huge number of apps nowadays available for mobile phones are able to collect considerable amount of data that can be exploited into intelligence to provide recommendations to the users. Intelligence is broad in sense, here by intelligence we understand the capability of the machines to learn, therefore a possibility to materialize such intelligence can be by means of data mining models. Nevertheless the data mining models have to represent the domain over time, and we know that the change is the only constant of life. If a data mining model downgrades, it needs to be updated to adapt to the changing world. We can divide the mechanisms providing such intelligence into two types, on one hand the mechanisms extracting intelligence in a central server and later install the intelligence on the device using it, and on the other hand mechanisms that locally compute the intelligence in the device. Efficiency is the main reason for the first mechanism because there will not be costs associated with a local execution, while privacy as well as personalization lay at the roots for the second possibility. Despite the recent advances, the automatization of the mining algorithm and its adaption to external factors is still an open problem. Further, to calculate the cost associated to the execution of the mining algorithm is fundamental in ubiquitous devices; this in order to know if the execution is possible or whether the resources have

---

* This work has been partially funded by Spanish Ministry of Science and Innovation, Project TIN2008-05924.

J. Yu et al. (Eds.): RSKT 2010, LNAI 6401, pp. 433–438, 2010.

to be rationed. In [3] a mechanism able to choose the best conguration of a data mining algorithm according to context and available resources is presented. As part of the mechanism, a cost model (Efficiency Efficacy Model) that is able to estimate efficiency and efficacy of a classication algorithm given information on the inputs of the algorithm is also presented. The mechanism is able to estimate the cost associated to the algorithm execution because of such cost model. In this paper we formalize the steps in order to build the mechanism - as it is not previously done - and we also underline the importance of autonomous adaptive data mining for healthcare sector by presenting an application scenario with the aim of predicting glucose level for patients with diabetes. In fact, the application exploits an instantiation of the mechanism that is also presented and evaluated here. The rest of the paper has been organized as follows: section 2 presents the focus and the requirements of the problem. In section 3 the decision mechanism and the cost model are presented, as well as experimental results. Besides in section 4 we present a mobile application for glucose prediction and the section 5 presents the conclusionts and the future research.

2   Preliminaries

The adaptation of the mining algorithm execution to external factors can be decomposed into two subproblems [3]:

1. to know the best configuration of an algorithm given the external factors;
2. to calculate the cost in terms of efficacy and efficiency associated to an algorithm with given inputs.

By efficiency they refer to the performance of the mining algorithm execution, while by efficacy they refer to the quality of the mining models obtained. The first subproblem needs an analysis for: i) which are the external factors the algorithm has to adapt to, ii) how such factors affect efficacy, efficiency and semantics. Information regarding resources, external policies and priorities, costs of resources can affect the decision on the best input of the algorithm to fulfill a certain goal. We call the factors affecting the inputs that are not related to the algorithm itself external factors and concentrate in this paper in two main groups:

- Factors describing resources: memory, battery, CPU;
- Factors describing Context information: location, temperature, time, etc.

The second sub-problem requires a deep analysis of the algorithm behavior. It is required to know: i) which are the inputs that can affect efficacy, efficiency and semantics of the model obtained, ii) how efficiency, efficacy and semantics change when these inputs are altered.

We tackle these two problems separately, first we plan to build a cost model predicting the algorithm behavior given the algorithm inputs, then we exploit such information for finding the best algorithm configuration according to context and resources.

3   The Decision Mechanism

In [3] a mechanism based on the above division in order to execute data mining algorithms autonomously according to context and resources is proposed. The mechanism that converges to the best algorithm configuration, is based on a cost model (EE-Model) which estimates efficacy and efficiency of the algorithm given dataset metadata and an algorithm configuration. The figure 1 shows how the various phases of the mining process, the mechanism has a central role, it can access dataset information, configuration metadata and external factors, and it gives as result the best configuration for the mining algorithm.

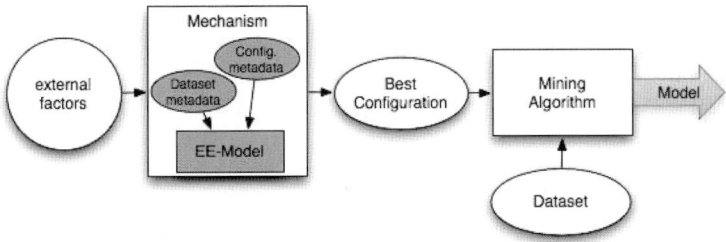

Fig. 1. The process flow

3.1   The EE-Model

In [3] we proposed a machine learning based approach to analyze historical data of past executions of an algorithm to extract, if any, the relationship among used resources, quality of the model and the input dataset features, algorithm configuration. The eficiency-efficacy model (EE-Model) acts as a predictor associated to a particular mining algorithm that for a certain algorithm input information, estimates the efficiency (resource consumption as CPU, Battery, Memory, etc) of the algorithm execution and efficacy (accuracy, and in general any measurement indicating how good the mining model is) of the mining model obtained. In this paper we formalize how to build the EE-Model for a classification algorithm:

1. Decide the set of variables to describe the algorithm executions. *Condition variables:* describe algorithm inputs as algorithm parameters and dataset; *Decision variables:* describe the results of the algorithm execution (i.e. CPU).
2. Obtain a dataset of historical data of algorithm executions. Here it is important to collect data with a representative number of executions, covering the possibles algorithm configurations.
3. Apply a data mining algorithm in order to discover relations between condition variables and decision variables, here we suggest to use mining techniques as linear regression or regression trees that seem to be suitable for numeric decision variables.

3.2   EE-Model for C4.5 Algorithm

Following the guidelines of the previous section we have built an EE-Model for C4.5 algorithm which we will need to calculate the cost of the algorithm for the application we present in 4. We calculated the EE-Model for battery consumption, memory, CPU (efficiency) and and accuracy (efficacy). We obtained a dataset of historical data (20.000 executions) of the algorithm in a system with 2.16 GHz Core 2 processor and 2.5GB 667 MHz DDR2 SDRAM memory, and then we applied linear regression algorithm in order to discover relations between condition variables and decision variables.

3.3   EE-Model Experimentation

Here we evaluate the EE-Model obtained in 3.2, comparing its estimations with the real values calculated ex-post. The EE-Model takes as input an algorithm configuration and dataset metadata, so to test it we define a set of algorithm configurations (table 1), and information metadata regarding the glucose level of diabetic patients dataset: $NCategorical = 3$, $NInteger = 3$, $NReal = 6$, $NAttributes = 12$, $NInstances = 4163$, $Size = 240$, are respectively the number of categorical, integer, real, total attributes, number of instances and size of the dataset. Then we calculated the estimations obtained from the EE-Model and we compare them with the real values. As the table 3 shows, the estimates are close to the real values.

Table 1. Set of algorithm configurations

Config.	Prun	Bin	Laplace	CF	Sub	Min#Obj	#Folds	Seed
Config. 1	1	Yes	Yes	0.25	Yes	3	–	–
Config. 2	0	Yes	No	–	0	2	–	–
Config. 3	2	No	No	–	Yes	5	5	5

Table 2. EE-Model estimations and real resources consumption

Configuration	Memory (real)	CPU (real)	Battery (real)	Accuracy
Config. 1	2045782.4 (2145007)	1344.7 (1781)	388.0 (430)	0.12
Config. 2	3355878.5 (3510288)	1873.4 (2001)	490.3 (498)	0.8
Config. 3	2678120.1 (3212679)	1796.5 (1946)	471.3 (426)	-0.6

3.4   Choosing the Best Algorithm Configuration

Knowing the external factors and knowing the efficiency and efficacy of the algorithm for each possible configuration, the mechanism has to be able to obtain the best configuration for a set of external factors values. We assume we have a set of algorithm configurations among which the mechanism have to select the best one. The problem now can be seen as a MCDM problem, a search for a Pareto-optimal solution in a 2-dimensional space of efficiency and efficacy. We propose a solution with hybrid heuristic, where the situation defines the mining objectives (semantic) and the resource state defines the constraints for the quality of the model. For each situation defined by context is associated a mining schema which goes hand in hand with the project objectives, while the algorithm configuration is tuned according to available resources. The pseudocode of the proposed mechanism is given in Algorithm 1. Given a current situation inferred by context values, the mining schema is obtained. Here is important to underline that the context defines the mining semantic with the mining schema and it is done by mean of static rules the data mining expert decided when setting the project goals. Once the resource state R_i is read, the weights for the resource components (CPU, memory, battery) are calculated. To low resource availability will correspond higher weights. These weights are needed because there could be situations in which is more important to take care of certain resources. Then the resource consumption of the algorithm execution and the accuracy of the model for a set of predefined configurations $Conf[n]$ are estimated by the EE-Model. The configuration chosen to launch the mining algorithm will be the one minimizing the weighed cost of the resources divided by a measure for the accuracy (P is a weight to balance resource cost values and accuracy values).

Algorithm 1. Decision mechanism

Require: Current situation S_i, resources state R_i, set of algorithm configurations $Conf[n]$, dataset D, dataset information DI.
1: Get mining schema for S_i;
2: Calculate weighs W for resources state R_i;
3: **for all** i such that $0 \leq i \leq n$ **do**
4:     Calculate resource cost $ResCost[i] \leftarrow$ EE-Model(GetCost, W, DI, $Conf[i]$);
5:     Calculate accuracy $Acc[i] \leftarrow$ EE-Model(GetAcc, DI, $Conf[i]$);
6: **end for**
7: **return** $Config[x]$ where x MIN $\frac{RestCost[x]}{P*Acc[x]}$

---

4   Diabetes Management by Glucose Prediction

The main problem in the management of diabetes disease is related to insulin therapy to avoid occurrences of hypoglycemia and hyperglycemia. The dosage of the insulin is based on the glucose level a patient has in a certain moment, but many factors impact glucose fluctations in diabetes patients, such as insulin dosage, nutritional intake, daily activities and lifestyle (e.g. sleep-wake cycles and exercise), and emotional states (e.g., stress). In [1] the authors provide a method predicting glucose concentration by mean of neural networks techniques. We believe that the prediction model has to be also based on personal information as we suppose that each patient reacts differently to the factors and further the prediction model can be calculated in the patients device. In this way the patient does not have to share sensitive information and can have always updated models that might better predict in case of changes (disease evolution, lifestyle, age, etc). Here we present an application scenario for glucose prediction that could exploit data mining described in the previous sections. The patient interacts with with a mobile GUI, which records all the information as hypertension level, glucose level, sport activity, intake food, etc. Such information is the dataset to train with the classification algorithm, we identify as target variable the glucose level. We assume to have a mining request, a dataset, a mining schema associated, a number of algorithm configurations from which to select the best one and resource and context information. So the when a mining request comes, the EE-Model presented in 3.2 estimates the efficiency and efficacy values for all the configurations. The mechanism selects the best one for a particular situation of resources and context, so that after the algorithm execution the patient can get recommendations on the insulin dosage.

5   Conclusion

We have presented an instantiation of a mechanism for providing autonomous adaptive data mining on mobile devices for a glucose prediction application. The mechanism is able to choose the optimum configuration of a C4.5 algorithm according to context information and resources. The main advantage of the approach is the possibility to know an estimation of the algorithm behavior (efficiency and efficacy) before the execution, so to avoid possible out of memory problems, CPU bottlenecks, etc. In future research we will focus on the EE-Model accuracy for efficacy and efficiency and on the implementation of a real mobile application.

References

1. Pappada, S.M., Cameron, B.D., Rosman, P.M.: Development of a neural network for prediction of glucose concentration in type 1 diabetes patients
2. Siewiorek, D., Smailagic, A., Furukawa, J., Krause, A., Moraveji, N., Reiger, K., Shaffer, J., Wong, F.L.: Sensay: A context-aware mobile phone. In: ISWC, Washington, DC, USA. IEEE Computer Society, Los Alamitos (2003)
3. Zanda, A., Eibe, S., Menasalvas, E.: Adapting batch learning algorithms execution in ubiquitous devices. In: MDM, Kansas city, USA. IEEE Computer Society, Los Alamitos (2010)

Fast Iris Localization Based on Improved Hough Transform

Lu Wang, Gongping Yang*, and Yilong Yin

School of Computer Science and Technology,
Shandong University
Jinan, Shandong, China, 250101
gpyang@sdu.edu.cn
http://mla.sdu.edu.cn

Abstract. Iris is a new biometric emerging in recent years. Iris identification is gradually applied to a number of important areas because of its simplicity, fast identification and low error recognition rate. Typically, an iris recognition system includes four parts: iris localization, feature extraction, coding and recognition. Among it, iris localization is a critical step. In the paper, a fast iris localization algorithm based on improved Hough transform was proposed. First, the algorithm builds gray histogram of iris image to analyze the gray threshold of the iris boundary. Then takes the pupil image binarization, using corrosion and expansion or region growing to remove noise. As a result, it obtains the radius of the inner edge. Then, we conduct iris location based on Hough transform according to the geometrical feature and gray feature of the human eye image. By narrowing searching scope, localization speed and iris localization accuracy are improved. Besides, it has better robustness for real-time system. Experimental results show that the proposed method is effective and encouraging.

Keywords: iris recognition, iris localization, region growing, gray projection, Hough transform.

1 Introduction

According to the problem of traditional method of identification, a new and more secure identification based on the biometric identity authentication is becoming more and more popular in recent years' research and has achieved rapid development and achievement in three short decades. The so-called biometric identification technology refers to using the inherent physical characteristics of the human body or behavioral characteristics for personal identification through the computer. Current identification technology based on biometrics mainly includes fingerprints, palm prints, iris, face, gait, voice, handwriting recognition, etc, in which iris recognition is very active for its uniqueness, stability, non-intrusion and anti-falsification, and has achieved a good recognition effect. According to some statistics, iris recognition has the lowest error rate in a variety of biometric identification [1]. Typically, iris recognition is composed of iris localization, feature

---

* Corresponding author.

J. Yu et al. (Eds.): RSKT 2010, LNAI 6401, pp. 439–446, 2010.

extraction, coding and recognition. In this paper, we mainly discuss the process of iris localization. Accurately and effectively locating iris is a prerequisite for iris recognition.

There are many iris localization methods, such as Flexible template detection method proposed by Daugman [2][3], two-step approach proposed by Wildes [4], Gradient optimization method proposed by Wildes [8], Least squares fitting method [9], Iris localization method based on texture segmentation [10], An iris Localization method based on FCM [11]. In recent years, there are many new iris localization methods based on Hough transform, which are improvements of the two-step method, such as [5][6][7].

The two classical algorithms for iris localization are as follows.

(1) Flexible template detection method [2][3]proposed by Daugman. Its idea is searching for local maxima throughout the iris image in the region using a circular edge detector as shown in equation (1).

$$\max_{(r,x_c,y_c)} |G(r) * \frac{\partial}{\partial r} \oint_{(r,x_c,y_c)} \frac{I(x,y)}{2\pi r}| \tag{1}$$

$$\text{where} \quad G(r) = \frac{1}{\sqrt{2\pi}\sigma} e^{-\frac{(r-r_0)^2}{2\sigma^2}} \tag{2}$$

$I(x,y)$ is the gray value of the pixel (x,y) of input image, and ds is the arc of a curve element. The symbol "*" represents convolution operation. In equation (2), $G(r)$ represents a Gaussian function with r as its mean and σ as its standard deviation. This approach is seeking the operator's largest group of vectors (x,y,r) in the x,y,r space. By increasing the radius r, it takes the parameters (x,y) and r in the vector (x,y,r) which is largest for the operator as the outer edge of the iris center point and radius respectively, then get the internal radius of the iris by increasing the correction factor. This locating method has the advantage of high locating precision. The disadvantage is that it needs to search in the whole image so it is time consuming computation. Besides, it is vulnerable to local interference (especially light) which can cause localization failure. In addition, the method searches the circle edge along the radius from the center, which may lead to local extreme situation.

(2) Two-step approach [4] proposed by Wildes in 1997. The main idea is to detect edges first, then conduct the Hough transform. It's a more generic way. Firstly, it converts iris image into a boundary map. Secondly, it defines a circle with its center and radius changes in the image. Thirdly, it enumerates all the possible circles in the image after an appropriate discretization of the parameter. Finally, it votes in the parameter space using Hough transform and statistics for the numbers of the points of the boundary included by each round, in which the boundary is the round which has the most points and so it gets the inter and outer parameters of the iris. On one hand, the method is subject to intermittent noise. The boundary has little effect, and it has high precision. On the other hand, it is susceptible to the impact of factors such as uneven illumination, and we have to search the entire area to get the iris radius, which improves complexity and time consuming. In addition, threshold selection is critical to binarization, but there are no good methods for it currently.

Recognizing the deficiency of the iris localization algorithm mentioned above, a fast localization algorithm based on improved Hough transform was proposed. The proposed method is an improvement of Wildes's method. It can effectively avoid the shortcomings above and greatly reduce the computation complexity of improved Hough transform.

The proposed algorithm has the following characteristics:

(1) Find a initial point in the pupil quickly. We use region growing to separate pupil region, which reduces the time for iris locating significantly.
(2) Physical characteristics of the eye of position relation between the iris and pupil are adopted when obtaining the pupil center of the circle, determining the scope of the iris center of a circle.
(3) The searching scope of Hough transform is reduced because of the use of physiological characteristics of human iris.

The remainder parts of this paper are organized as follows. Section 2 describes the proposed method. Experimental results are reported in Section 3. And conclusions are drawn in Section 4.

2   Iris Localization Based on Improved Hough Transform

Iris is the layer of the circular area between pupil and sclera. Its inner and outer edges can be approximately regarded as circles. The localization process is to identify the centers and radiuses of internal and external circles. In the proposed algorithm, we first get the center point and radius of the pupil. Then we get the center point and radius of the iris making use of pupil's center and radius information, which reduces the searching scope. The flow chart of the proposed algorithm is shown in Fig. 1.

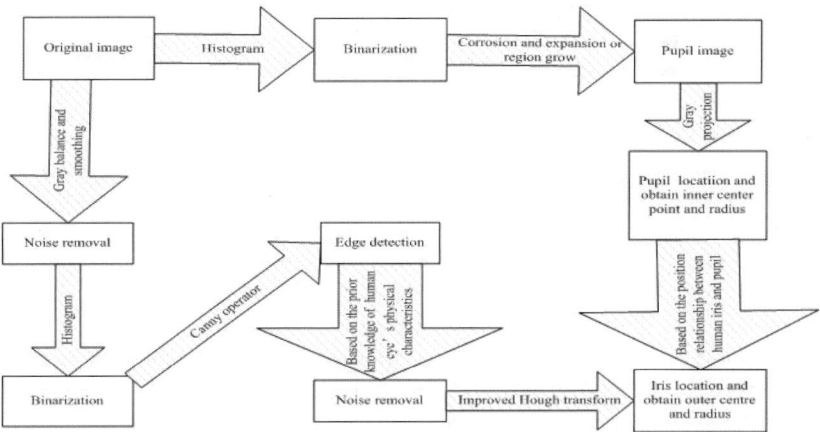

Fig. 1. The flow chart of the proposed method

The process of iris localization is shown in Fig. reffig:b, in which the 001_1_1.bmp in CASIA1.0 iris database is taken as an example.

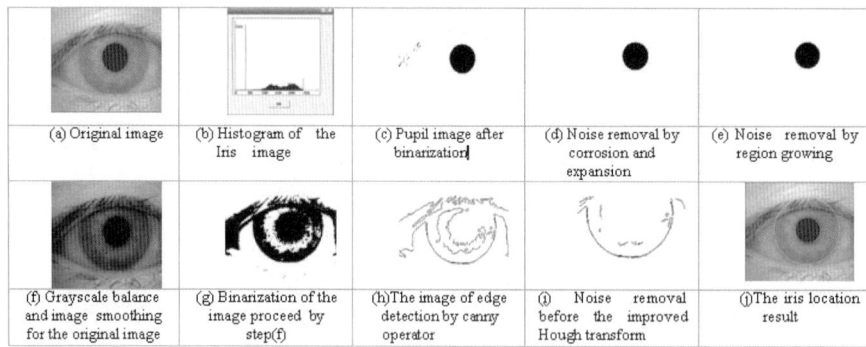

Fig. 2. The flow chart of the proposed method

2.1   Extraction of Inner Boundary

Iris images have certain characteristics in distribution of gray values in different regions. Generally speaking, gray value from pupil is less than that of iris, and gray value of iris is less than that of sclera, so the pupil has the smallest gray value in the gray histogram. We can separate pupil from iris and sclera based on the reason above [12] by taking binarization. The gray histogram of iris image has clear multi-peak characteristics. The peak valley behind the lowest peak gray value of the histogram is the segmentation threshold between pupil and other parts. We use an iterative method to find the first local minimum value on the left of the histogram's first peak value to be the threshold for binarization. The binary image is shown in Fig.2(c). As is shown in Fig.2(c), the binary pupil image using our threshold contains only a small amount of noise points. These noise points can be removed by many corrosion and expansion operations or region growing. As is shown in Fig.2(d) and 2(e), both of these methods can remove noise effectively. The region growing method is described as following.

In the region growing operation, the selection of initial seeds is a crucial step. Generally speaking, when we get an image via sensor, the yaw angle is not big, and the pupil region is near the center of the captured image. So it is unnecessary to search all the pixels in the image, and we can only search much less points, as shown in Fig 3. This ensures low time complexity. It's assumed that the point is in the pupil region if the pixel values within the 5*5 template of the point are all 0 (Namely, black). Firstly, We search in the center of the image(as is shown is Fig. reffig:c marked black), if the point meets the above conditions, it's considered in the pupil region and the search operation is over; otherwise, we search the white pixels which are marked in Fig 3 to determine whether they can meet the above conditions. Once a certain point meets the above conditions, the searching process is terminated. And the point is taken as initial point of region growing. Then we can carry out the remaining operations.

Compared with the expansion and corrosion, the region growing is more accurate and the time consuming is less, so it's more suitable for real-time application.

Compared with the expansion and corrosion, the region growing is more accurate and the time consuming is less, so it's more suitable for real-time application.

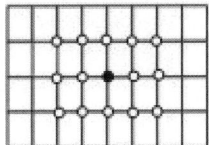

Fig. 3. The pixels that we search when finding a point in the pupil region

2.2   Extraction of Outer Boundary(Iris Boundary)

In this paper, we adopt an improved Hough transform [4] to locate the boundary of the iris. Firstly, we carry out gray balance and smoothing operation on the iris image in order to reduce non-uniform illumination effects on iris recognition, in which gray balance is histogram equalization and smoothing is to wipe off noise by Gaussian filtration. Secondly, we binarize the image using the threshold obtained before. Thirdly, we carry on edge extraction operation on the binary image. Finally, we locate the boundary of iris using improved Hough transform. The result of the iris localization is shown in Fig.2(j).

Assuming all the boundary points of the image after edge detection can be described as $(x_j, y_j) j = 1, \cdots, n$.

Hough transform is defined as

$$H(x_c, y_c, r) = \sum_{j=1}^{n} h(x_j, y_j, x_c, y_c, r) \tag{3}$$

$$\text{where} \quad h(x_j, y_j, x_c, y_c, r) = \begin{cases} 1 & if \quad g(x_j, y_j, x_c, y_c, r) \\ 0 & otherwise \end{cases} \tag{4}$$

$$\text{and} \quad g(x_j, y_j, x_c, y_c, r) = (x_j - x_c)^2 + (y_j - y_c)^2 - r^2 \tag{5}$$

For each boundary point (x_j, y_j), if there is $g(x_j, y_j, x_c, y_c, r) = 0$, it represents the boundary of circumference determined by the parameters (x_c, y_c, r) getting through the boundary point (x_j, y_j). So the maximum value of H corresponds to the circle getting through the largest number of border points.

In this paper, an improved Hough transform is adopted to extract the cylindrical boundary. The radius of iris is about 1.5 to 2 times larger than that of pupil and the difference between the two centers is no more than 5 pixels. Based on these prior knowledge, we can reduce the searching space. By experience, the range of the pupil radius of the CASIA1.0 iris database provided by Chinese Academy of Science is roughly from 30 pixels to 60 pixels and the range of the iris radius is roughly from 80 to 130 pixels.

Improved Hough transform consists of the following steps. (1) From Figure 2(h), it can be seen that the extracted image after using canny operator contains the entire eye contour, but we only need the outer edge of iris for localization. It's necessary to remove the noise points [13]. Provided with the prior knowledge that the difference between the two centers is no more than 5 pixels, and the center point and radius of the pupil, taking the pupil center as the center point, we define a rectangle template that is 90 pixels to the top, 65 pixels to the bottom, 70 pixels to the left and 70 pixels to the right. Then the pixels covered by the rectangular region are set to white.

Improved Hough transform consists of the following steps.

(1) From Fig.2(h), it can be seen that the extracted image after using canny operator contains the entire eye contour, but we only need the outer edge of iris for localization. It's necessary to remove the noise points [13]. Provided with the prior knowledge that the difference between the two centers is no more than 5 pixels, and the center point and radius of the pupil, taking the pupil center as the center point, we define a rectangle template that is 90 pixels to the top, 65 pixels to the bottom, 70 pixels to the left and 70 pixels to the right. Then the pixels covered by the rectangular region are set to white.

The next step is to eliminate noise outside the iris region. According to the iris radius and pupil radius and the relationship between their centers points described above, we take the pupil center as the center, taking 120 pixels to the left, 120 pixels to the right, 60 pixels to the top and 120 pixels to the bottom, all the pixels outside this scope are set to white. As is shown in 2 (i), it can be seen that a lot of noises are removed.

(2) Taking the pupil center as the centre point, we define a 9×9 matrix. It's assumed that the iris is located within the matrix.

(3) Taking each point in the matrix as the center point respectively, we compute the distance from every black point (i.e. outer boundary point) to the center point. Then, take these distances as the radius and we record the maximum value corresponding to each radius.

(4) Taking the radius that has the maximum value as the iris radius, and the corresponding point is the iris center.

(5) Taking the pupil center and iris center as the centers and draw circle with the inner circle radius and outer circle radius respectively. The result is shown as Fig.2(j).

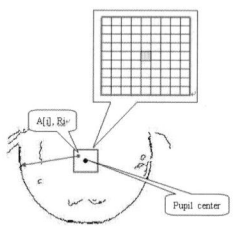

Fig. 4. Improved Hough Transform

3   Experiment Result and Analysis

CASIA1.0 version of the iris image database was adopted in the experiment. The proposed algorithm ran under VC++6.0 Window platform with 2GB physical memory. We take the all 756 iris images of CAISA1.0 version (80 subjects, 108 eyes and each seven in two sessions with a month gap acquisition) [14] for localization operation. If the corrosion and expansion method is used to remove noise (the proposed method 1), it'll require the average time of 0.15 seconds to locate an iris image and the localization accuracy is 99.2%. If the region growing method is used to remove noise (the proposed method 2), it'll require the average time of 0.065 seconds to locate an iris image and the localization accuracy is 99.3%. We have implemented methods in the references [2][4] and run them under the same environment, and the comparative results are shown in Table 1. It can be seen that compared with these classical iris localization algorithms, the proposed method achieves great reduction in positioning time, and at the same time, increasing a certain in accuracy.

Table 1. Comparison among different iris localization method of localization time and accuracy

Methods of localization	Average time(s)	Accuracy(%)
Reference[2]	4.24	98.7
Reference[4]	4.38	99.6
Proposed method 1	0.15	99.2
Proposed method 2	0.065	99.3

Compared with the algorithm proposed by Daugman [2], the proposed method can avoid the possibility of falling into local extreme point, so it can correctly locate these images which have low contrast between iris and sclera. Compared with the algorithm proposed by Wildes [4], the proposed method obtains binarization threshold using an effective method to ensure the smooth operation of the back, estimates the scope of the iris center according to the pupil center, and remove most of the noise according to the iris radius range of the physiological characteristics of the human eye, which can reduce the searching scope of Hough transform and the time complexity largely. To sum up, the proposed method has a fast iris localization speed and a high localization accuracy, so it's suitable for real-time system.

4   Conclusion and Discussion

Based on the two-step method proposed by Wildes, we proposed a fast iris localization algorithm based on improved Hough transform. Compared with the classical algorithms, the proposed method achieves a certain degree of increase in both speed and accuracy. However, there is still much space for improvement. For example, the pupil and the iris may fit elliptic curves better instead of standard circle, so the iris and pupil can be defined as ellipses in future work. Moreover, the proposed method did not eliminate eyelash, eyelid and other noise in view of algorithm complexity and real-time requirements. Regarding this, we can use a curve fitting method to obtain iris images without noise point in the future.

Acknowledgments. The authors thank Laboratory of Pattern Recognition Institute of Automation Chinese Academy of Sciences for providing the CASIA1.0 image database. This work is supported by Shandong University Independent Innovation Foundation under Grant No.2009TS034 and 2009TS035.

References

1. Lu, C.S., Liao, H.M.: Multipurpose watermarking for image authentication and protection. J. IEEE Transaction on Image Processing 10(10), 1579–1592 (2001)
2. Daugman, J.G.: How iris recognition works. IEEE Transaction on circuits and systems for video technology 14(1), 21–30 (2004)
3. Daugman, J.G.: High confidence visual identification of person by a test of statistical independence. IEEE Trans. Pattern Anal. Machine Intell. 15(11), 1148–1161 (1993)
4. Wildes, R.P.: Iris recognition: An emerging biometric technology. Proceedings of the IEEE 85(9), 1348–1363 (1997)
5. Liu, X., Bowyer, K.W., Flynn, P.J.: Experiments with an Improved Iris Segmentation Algorithm. In: Proceedings of the Fourth IEEE Workshop on Automatic Identification Advanced Technologies, Autoid, October 17 - 18, pp. 118–123. IEEE Computer Society, Washington (2005)
6. Toennies, K.D., Behrens, F., Aurnhammer, M.: Feasibility of Hough-Transform-based Iris Localisation for Real-Time-Application. In: Internat. Conf. Pattern Recognition, pp. 1053–1056 (2002)
7. Cui, J., Wang, Y., Tan, T., Ma, L., Sun, Z.: A fast and robust iris localization method based on texture segmentation, For Biometric Authentication and Testing, National Laboratory of Pattern Recognition. Chinese Academy of Sciences, 401-408 (2004)
8. Camus, T., Wildes, R.P.: Reliable and Fast Eye Finding in Close-up Images. In: Proceedings of the IEEE International Conference on Pattern Recognition, pp. 389–394 (2002)
9. Wang, Y., Zhu, Y., Tan, T.: Identification based on iris recognition. Acta Automatica Sinica 28(1), 1–10 (2002)
10. Cui, J., Ma, L., Wang, Y., Tan, T., Sun, Z.: A Fast and Robust Iris Localization Method based on Texture Segmentation. In: SPIE, vol. 5404, pp. 401–408 (2004)
11. Proenca, H., Alexandre, L.A.: Iris segmentation methodology for non-cooperative recognition. IEEE Proc. Vis. Image Signal Process. 153(2) (2006)
12. Wildes, R.P., Asmuth, J.C., Green, G.L., et al.: A system for automated iris recognition. In: Proceedings of the IEEE Workshop on Applications of Computer Vision, Sarasota, FL, USA, pp. 121–128 (1994)
13. Sun, Y.: A Fast Iris Localization Method Based on Mathematical Morphology. Computer Applications 28(4) (2007)
14. C.A. of Sciences-Institute of Automation, Database of 756 grey scale eye images (2003), http://www.sinobiometrics.com

Face Recognition Using Consistency Method and Its Variants

Kai Li, Nan Yang, and Xiuchen Ye

School of Mathematics and Computer, Hebei
University, Baoding 071002, China
likai_njtu@163.com

Abstract. Semi-supervised learning has become an active area of recent research in machine learning. To date,many approaches to semi-supervised learning are presented. In this paper,Consistency method and its some variants are deeply studied. The proof about the important condition for convergence of consistency method is given in detail. Moreover,we further study the validity of some variants of consistency method. Finally we conduct the experimental study on the parameters involved in consistency method to face recognition. Meanwhile, the performance of Consistency method and its some variants are compared with that of support vector machine supervised learning methods.

Keywords: Semi-supervised learning; Consistency; Convergence; Face recognition.

1 Introduction

The machine learning is one of the important research domain of artificial intelligence, which mainly includes supervised learning, unsupervised learning, and semi-supervised learning. Given a dataset(x_1,y_1), (x_2,y_2),..., (x_N,y_N),where $x_i=(x_{i1},x_{i2},...,x_{iN})$ is the properties vector of the i-th sample, element x_{ij} is the j-th properties value of the i-th sample, and this value can be a discrete or continuous one. The primary differences between the three kinds of learning methods above are whether the system has prior knowledge or not and how much prior knowledge there is in the system. It is supervised learning or unsupervised learning when all values of y_i(i=1,2,...,N)in N sample points are known or unknown. However, labeled instances in real life are often difficult, expensive, or time consuming to obtain, meanwhile unlabeled data may be relatively easy to collect, which is rarely used, it is called semi-supervised classification learning [1]. In other words, when the values of very few sample points in N sample points are known, then it is semi-supervised learning. The main goal of semi-supervised classification learning is that how to use few labeled data and a large amount of unlabeled data to enhance the generalization performance of system. Recently, researchers addressed many semi-supervised learning algorithms [2-6]. In this paper we mainly carries out the study on CM (Consistency Method) algorithm and give the proof about the important condition of CM algorithm's convergence in detail and apply CM algorithm to face recognition.

J. Yu et al. (Eds.): RSKT 2010, LNAI 6401, pp. 447–452, 2010.

2   Proof of Convergence's Condition for Consistency Method and Its Variants

Given a data set $X = \{(x_1, y_1), \ldots, (x_l, y_l), x_{l+1}, \ldots, x_n\}$, $x_i \in R^m$ and $y_i \in \{1, 2, \ldots, c\}$, note that points for $x_{l+1}, \ldots, x_n$ don't have label in data set X and the goal of learning is to predict the labels of the unlabeled data $(x_{l+1}, \ldots, x_n)$. In general, l is far less than n $(l<<n)$. Let F denote the set of matrices with non-negative entries. A matrix $F = [F_1^T, \ldots, F_n^T]$ corresponds to a classification on the dataset X by labeling each point x_i as a label. We can get the last label of each point by taking the maximum value from each row in matrix F, namely $y_i = arg\ max_j\{F_{ij}\}$. In addition, define matrix $Y = (Y_{ij})$, where If $y_i = j$ then $Y_{ij} = 1$ else $Y_{ij} = 0$ and $W = (W_{ij})$, where if $i \neq j$ then $W_{ij} = \exp(-\|x_i - x_j\|^2/2\sigma^2))$ else $W_{ij} = 0$. Aim to this, Zhou et al.[3] presented the CM algorithm and gave a proof framework for the convergence of this algorithm. However, they didn't give the proof why the absolute value of eigenvalues for matrix S is less than 1. In the following, we further analyze CM algorithm and its variants. Especially, the detailed derivative procedures are given in this paper.

Proposition 1: The eigenvalues of matrix $S = D^{-1/2}WD^{-1/2}$ is in [-1, 1], where D is a diagonal matrix with its (i,i)-element equal to the sum of the i-th row of W.

Proof: As $D^{-1}W = D^{-1/2}SD^{1/2}$, denote $D^{-1}W$ by P, so the matrix P is similar with S. As similarity matrix has the same eigenvalues, the eigenvalues of matrix S is equal to that of matrix P. Thus, we just need to prove that the eigenvalues of matrix P is in [-1,1]. As P is equal to $D^{-1}W$, we get the following expression:

$$P = \begin{pmatrix} P_{11} & 0 & \ldots & 0 \\ \ldots & \ldots & \ldots & \ldots \\ 0 & 0 & \ldots & P_{nn} \end{pmatrix} \begin{pmatrix} 0 & W_{12} & \ldots & W_{1n} \\ \ldots & \ldots & \ldots & \ldots \\ W_{n1} & W_{n2} & \ldots & 0 \end{pmatrix},$$

where $P_{ii} = (\sum_{j=1, j\neq i}^{n} W_{ij})^{-1}$. Let P_i denotes the sum of the i-th row elements in matrix P, we have the following expression

$$P_i = \sum_{j=1, j\neq i}^{n} (W_{ij} / \sum_{j=1, j\neq i}^{n} W_{ij}) = 1.$$

It may be known by the Gerschgorin theorem[7] that the eigenvalues of matrix P is in [-1, 1]. That is to say that the eigenvalues of matrix S is also in [-1, 1].

Proposition 2[3]: Let F^* be the limit of the sequence $\{F(t)\}$, we have $F^* = (I - \alpha S)^{-1}Y$ for the classification.

Proof: From the iteration equation $F(t + 1) = \alpha SF(t) + (1 - \alpha)Y$, we have $F(t) = (\alpha S)^{t-1}Y + (1 - \alpha)\sum_{i=0}^{t-1}(\alpha S)^i Y$. As $0 < \alpha < 1$ and the eigenvalues of matrix S is in [-1, 1] (known by Proposition 1), then $\lim_{t\to\infty}(\alpha S)^{t-1} = 0, \lim_{t\to\infty}\sum_{i=0}^{t-1}(\alpha S)^i = \frac{1-(\alpha S)^t}{1-\alpha S} = (I-\alpha S)^{-1}$. So we can get $F^* = \lim_{t\to\infty} F(t) =$

$(1 - \alpha)(I - \alpha S)^{-1}Y$.In reality,equation above is equivalent to the following expression for classification $F^* = (I - \alpha S)^{-1}Y$.

Proposition 3: As P is a stochastic matrix of S and $P = D^{-1}W = D^{-1/2}SD^{1/2}$, $F^* = (I - \alpha S)^{-1}Y$ can be approximately expressed as $F^* = (I - \alpha P)^{-1}Y$.

Proof: For $P = D^{-1}W = D^{-1/2}SD^{1/2}$, multiply $D^{1/2}$ in the left and $D^{-1/2}$ in the right at two sides of the equation, we can get $S = D^{1/2}PD-1/2$, substitute S into the objective function $F^* = (I - \alpha S)^{-1}Y$, then $F^* = (I - \alpha D^{1/2}PD^{-1/2})^{-1}Y$. We substitute I with $D^{1/2}D^{-1/2}$ in F^*, then above equation become $F^* = (D^{1/2}D^{-1/2} - \alpha D^{1/2}PD-1/2)^{-1}Y$. Further it will transform into $F^* = D^{1/2}(I - \alpha P)^{-1}D^{-1/2}Y$. Since $D^{1/2}(I - \alpha P)^{-1}D^{-1/2}$ is similar with $(I - \alpha P)^{-1}$, equation F^* can be approximately expressed as $F^* = (I - \alpha P)^{-1}Y$.

Proposition 4: As P is a similar stochastic matrix of S and $P = D^{-1}W = D^{-1/2}SD^{1/2}$,$(I - \alpha S)^{-1}Y$ can be approximately expressed as $F^* = (I - \alpha P^T)^{-1}Y$.

Proof: Since $P = D^{-1/2}SD^{1/2}$,by transposing both sides of this equation, we can get $P^T = (D^{-1/2}SD^{1/2})^T$ and further have $P^T = ((D^{1/2})^T S^T (D^{-1/2})^T$. As D is a diagonal matrix, $(D^{-1/2})^T = D^{-1/2}$and $(D^{1/2})^T = D^{1/2}$. Thus,above equation can be transformed into $P^T = D^{1/2}S^T D^{-1/2}$. By $S = D^{-1/2}WD^{-1/2}$ and D is a diagonal matrix, we obtain $S^T = D^{-1/2}W^T D^{-1/2}$.And as W is a symmetric matrix, so $S^T = D^{-1/2}WD^{-1/2}$ and $S^T = S$. Then above equation can be expressed as $S = D^{-1/2}P^T D^{1/2}$. Substituting S into $F^* = (I - \alpha S)^{-1}Y$, then function is transformed into the following equation $F^* = (I - \alpha\alpha D^{-1/2}P^T D^{1/2})^{-1}Y$. We write identity matrix I as $D^{-1/2}D^{1/2}$, then the above equation can be expressed as $F^* = (D^{-1/2}D^{1/2} - \alpha D^{-1/2}P^T D^{1/2})^{-1}Y$,namely $F^* = D^{-1/2}(I - \alpha P^T)^{-1}D^{1/2}Y$. Since $D^{-1/2}(I - \alpha P^T)^{-1}D^{1/2}$ is similar with $(I - \alpha P^T)^{-1}$, we can get the $F^* = (I - \alpha P^T)^{-1}Y$.

3   Experimental Results

In this section, we mainly study performance with the CM algorithm and its variants for face recognition.In addition,we also experiment with the parameters involved in them. For this, we choose Orl face database as experimental data,which contains 10 different images of each of 40 distinct subjects and the resolution of each image is 112*92 pixels. In all the images of Orl, the subjects are in an upright, frontal position with some variation in lighting, facial expression, facial details etc.In the experiment on scale factor σ of exponential function, we choose 5 values for scale factor σ,which is S1=0.15, S2=0.35, S3=0.55, S4=0.75 and S5=0.95,respectively. In the semi-supervised learning method,for face database, each chosen number of samples with class label is the value of integer between [1,10].The results reported here are the arithmetic mean of 10 test runs under same scale factor. The experimental results are seen in the Fig. 1–3,which are the experimental results of the standard CM algorithm (denoted by CMS)and two kinds of variants of CM algorithm (denoted by CMP and CMPT),respectively.

From Fig. 1–3, we can see that there has not significantly change with performance when replace S matrix with the P^T, while when replace S matrix with P, the test accurate rate has a large fluctuations. When S5 is equal to 0.95, there has a significantly change. In addition, for different values of the scale factor σ, CM algorithm and its variants's test accurate rates are quite different.Especially when there is the small number of known labeled samples,performance for above three kinds of algorithms is more sensitive to the value of scale factor.

In addition, for comparing the performance of CM algorithm and its variants with SVM, we select LIBSVM as our experimental environment.For Orl face

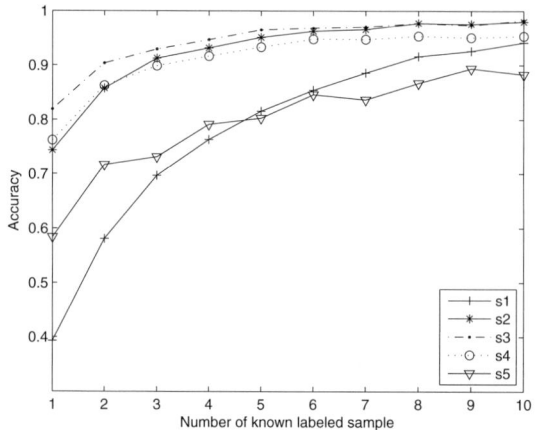

Fig. 1. Relationship between the number of known labeled sample and accuracy for CMS

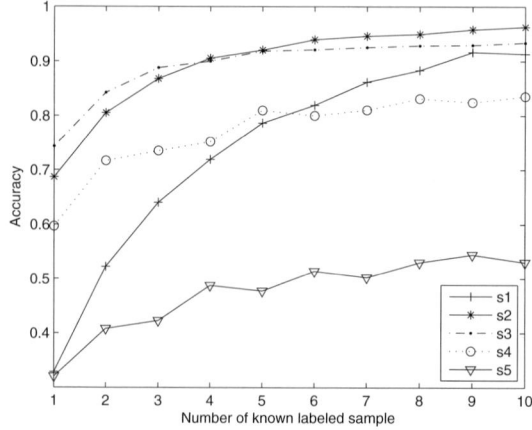

Fig. 2. Relationship between the number of known labeled sample and accuracy for CMP

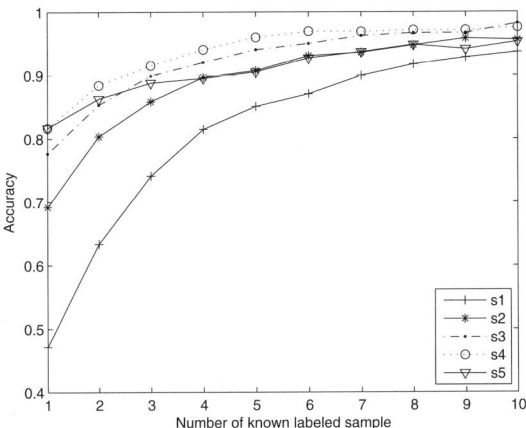

Fig. 3. Relationship between the number of known labeled sample and accuracy for CMPT

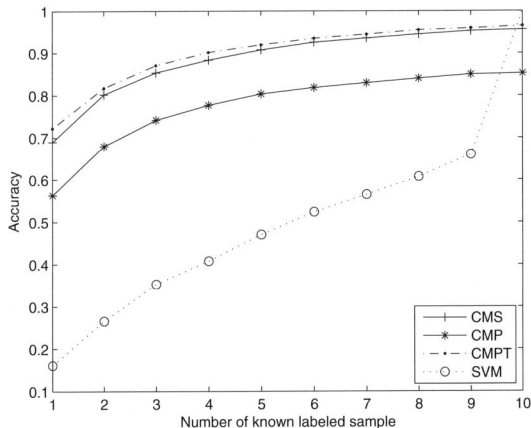

Fig. 4. Comparison of experimental results for CMS, CMP, CMPT and SVM between the number of known labeled sample and accuracy

database, chosen number of training samples is 40, 80, 120,160, 200, 240,280,320 and 400,respectively and the remaining face samples are used for testing. When the number of training samples is 400, the test data set is the same as the train data set. In SVM, the selected kernel is radial basis function, other parameters use default values for LIBSVM. We conduct 10 times for SVM classifier. The results reported here is the arithmetic mean of 10 test runs. The experimental results are seen in Fig. 4.

From the experimental results, we can know that performance for CM algorithm outperforms one of SVM algorithm, the performance of CMS is equivalent to the one of CMPT method and the results show a large deviation between CMS and CMP method. We think that the reason of the results showing a large deviation between CMS and CMP algorithm consist in P matrix instead of symmetrical matrix.

4   Conclusions

Consistency method and its variants are deeply analyzed, the proof about the important condition for consistency method's convergence is given in detail. In addition, this paper carries out deep research on the convergence of some other transformations of this algorithm, and proved that using the transformation matrix in CM algorithm is reasonable. Finally we carry out the study on the parameters and the transformation matrix involved in this algorithm for face recognition. Meanwhile, performance of consistency method and its variants are compared with that of support vector machine.

Acknowledgments. The authors would like to thank Hebei Natural Science Foundation for its financial support (No: F2009000236) and National Natural Science Foundation for its financial support (No: 60773062).

References

1. Zhu, X.J.: Semi-supervised Learning Literature Survey. Computer Sciences TR 1530, University of Wisconsin - Madison (2008)
2. Chapelle, O., Zien, A.: Semi-supervised Classification by Low Density Separation. In: The 10th international workshop on Artificial Intelligence and Statistics, Barbados (2005)
3. Zhou, D.Y., Bousquet, O., Thomas, N.L., et al.: Learning with Local and Global Consistency. In: Thrun, S., Saul, L., Scholkopf, B. (eds.) Advances in neural information processing systems, vol. 16, pp. 321–328. MIT press, Cambridge (2004)
4. Zhou, Z.H., Zhan, D.C., Yang, Q.: Semi-supervised Learning with Very Few Labeled Training Examples. In: The 22nd AAAI Conference on Artificial Intelligence, Vancouver, pp. 675–680 (2007)
5. Yin, X.S., Chen, S.C., Hu, E.L., Zhang, D.Q.: Semi-supervised Clustering with Metric Learning: an Adaptive Kernel Method. Pattern Recognition 43, 1320–1333 (2010)
6. Mahdieh, S.B., Saeed, B.S.: Kernel-based Metric Learning for Semi-supervised Clustering. Neurocomputing 73, 1352–1361 (2010)
7. Su, Y.C., Jiang, C.B., Zhang, Y.H.: Theory of Matrix. Science Press, Beijing (2007)

Clonal Selection Algorithm for Learning Concept Hierarchy from Malay Text

Mohd Zakree Ahmad Nazri[1,2], Siti Mariyam Shamsuddin[1], and Azuraliza Abu Bakar[2]

[1] Soft Computing Research Group, Faculty of Computer Science & Information System, Universiti Teknologi Malaysia, 81100, Skudai, Johor, Malaysia
mzan@ftsm.ukm.my;mariyam@utm.my
[2] Data Mining and Optimization Research Group (DMO), Center for Artificial Intelligence Technology, Universiti Kebangsaan Malaysia, 43600 Bangi, Selangor, Malaysia
aab@ftsm.ukm.my

Abstract. Concept hierarchy is an integral part of ontology which is the backbone of the Semantic Web. This paper describes a new hierarchical clustering algorithm for learning concept hierarchy named Clonal Selection Algorithm for Learning Concept Hierarchy, or CLONACH. The proposed algorithm resembles the CLONALG. CLONACH's effectiveness is evaluated on three data sets. The results show that the concept hierarchy produced by CLONACH is better than the agglomerative clustering technique in terms of taxonomic overlaps. Thus, the CLONALG based algorithm has been regarded as a promising technique in learning from texts, in particular small collection of texts.

Keywords: Artificial Immune System, Ontology Learning, Hierarchical Clustering, Machine Learning, Semantic Web.

1 Introduction

Concept hierarchy (also called taxonomy) is an essential element of ontology for a hierarchical classification in a machine-processable form. However, the traditional and labour-intensive ontology modelling, as described by Gulla et al. [1] is a process that is tedious, complex, time consuming and expensive. Thus, there has been a flurry of papers reporting the attempts in semi or fully automatic concept hierarchy construction. In this paper, we proposed an immune-inspired algorithm for learning concept hierarchies.

Natural immune system has several important abilities that can be very useful to ontology engineering community. In this paper, we will review the clonal selection concept and demonstrate that these biological principles can lead to the development of (semi-)automatic ontology learning tools. We will describe an idea for hierarchical data clustering using techniques based on the CLONALG algorithm [2]. However, the proposed algorithm is not developed to mimic exactly the natural adaptive immune system but to show that some basic immune principles can be used in computation.

J. Yu et al. (Eds.): RSKT 2010, LNAI 6401, pp. 453–461, 2010.
© Springer-Verlag Berlin Heidelberg 2010

The basic notion of this paper is that there are distinct advantages of using CLONACH in learning concept hierarchy from text.

2   Background and Related Work

The intended concept hierarchy is concerned with the organization of terms with hyponym/hypernym relation, which is also known as "is-a" and "a-type-of" relations. In practical terms, Han and Fu [3] defined concept hierarchy as one or a set of an attribute domains. Formally, concept hierarchy is expressed in Eq. 1:

$$H_l : D_i \times \ldots D_k \longrightarrow H_{l-1} \longrightarrow \ldots \longrightarrow H_0 \tag{1}$$

Where a hierarchy H is defined on a set of domains $D_i \ldots D_k$, in which different level of concepts are organized into hierarchies. The concepts are partially ordered according to a general-to-specific ordering. H_l represents a set of concepts at the primitive level, H_{l-1} represents the concepts at one level higher than those at H_l. H_0 is the highest level hierarchy that may contain the most general concept called Root. A survey on ontology learning procedures reported in [4] has found that the techniques that are based on statistical and linguistic approaches are often used. The Harris' distributional hypothesis [5] has been accepted by many as the underlying assumption in statistical approaches for ontology learning from text. This hypothesis states that terms with similar distributional patterns tend to have the same meaning. Harris's hypothesis has created a group of methods for extracting "is-a" relations such as in [6] and [7] who used hierarchical agglomerative clustering methods to learn ontologies, and more recently by Reinberger and Daelemans [8].

In this paper, terms extracted from texts are organized into a hierarchy structure that can be transformed directly into a concept hierarchy. Terms are extracted from texts using a pre-determined syntactic rule and then grouped according to the semantic distance (affinity) between them. However, ontology learning approaches that are based on unsupervised learning paradigms shared two major problems. First problem is the inability to appropriately label abstract concepts and second the data sparsity.

2.1   Clonal Selection

In this section, basic clonal selection theory is discussed at a level to aid understanding of the proposed algorithms presented in this paper. For a detail explanation, the reader is directed towards the literature such as [9], which much of this section is derived. In brief, clonal selection theory explains how the adaptive immune system works against pathogenic microorganisms. In the human immune system, B cells are a vital component in the humoral immune response. It main functions are to produce antibodies (Ab) against substances that the immune system perceives as foreign or nonself, the antigens (Ag). Thus, an Ag is any molecule that stimulates an adaptive immune response (especially the production of Ab). Produced Ab is attached primarily to the surface of B cell whose aim is to recognize and bind to Ag.

When an Ab as a B-cell receptor recognizes an Ag with a certain affinity, the B cell is selected to proliferate and respond by producing the Abs in high volumes. The stimulation caused the B-cell to proliferate (divide) and mature into terminal (non-dividing) Ab secreting cells, called plasma cells. These are short lived cells. The process of cell division (mitosis) generates a clone, i.e., a cell or a set of cells that are the progenies of a single cell. During reproduction, the clones undergo a mutation process together with a selection mechanism which results in Abs with higher affinities for a specific Ag. The process of mutation and selection is called the affinity maturation while the process of the proliferation and secretion of antibodies is called clonal expansion. In addition to plasma cells, B cell can differentiate into long-lived B memory cells. Memory B-cells are formed from a single type of B-cells, which is relatively aimed to respond quickly following a second exposure to the same specific Ag. De Castro and von Zuben [10] studied the clonal selection and expansion process and affinity maturation principles within the immune system and proposed an unsupervised learning technique known as CLONALG. Noteworthy researches for hierarchical clustering inspired by CLONALG can be seen applied in [11] and [12].

The main features of the clonal selection theory from the viewpoint of ontology engineering are: 1) based on affinity where the antigen selects several immune cells to proliferate and differentiate. The proliferation rate is proportional to its affinity, the higher the affinity the higher the number of clones generated; 2) affinity maturation process creates clones which are expressed as diverse Ab patterns. The mutation rate is inversely proportional to the affinity of the cell receptor with the antigen, which means the higher the affinity then the smaller rate of mutation.

3   Proposed Algorithm: CLONACH

The proposed algorithm is called Clonal Selection Algorithm for Learning Concept Hierarchy, or CLONACH. CLONACH processes are divided into two stages; the pre-processing stage and the concept hierarchy induction.

3.1   Stage 1: Pre-processing

The aim of this stage is to build the feature vector. The features are extracted between the verbs appearing in a sentence and the heads of the subject, object, and prepositional phrase-complements. For each noun (i.e. object/term) that appears as the head of the argument positions, the corresponding verbs are used as attributes that results in a pair of object-attribute (Cimiano, 2006). In order to acquire the features and its corresponding vectors, we used Malay natural language processing (NLP) tools developed by [13]. However, as reported in [14], the available Malay NLP tools produce erroneous outputs as the tools were developed for particular domain and text genre. Therefore, in this work, the NLP outputs are subject to human inspection to ensure that the extracted features are accordingly to the predefined linguistic patterns.

We represent the extracted attributes for each feature as vector in high dimensional space $\mathbb{R}^n$, where the dimensions corresponding to verbs found in the context of the word in question. Cimiano in [15] stated that the vector-based context representation constitutes the core of the vector space model used in information retrieval in [16]. Each noun is a feature of which each verb is a binary representing whether or not the corresponding verb appears along the noun in a sentence. Thus, 1 represents presence and 0 represent absence of the feature. In this stage, the similarity between the terms is also calculated on the basis of texts. The measurements will be utilised by the bisecting k-means divisive clustering algorithm in the stage on the basis of the computed similarities by using the cosine metric. On the basis of these feature vectors, the similarity between two terms t_1 and t_2 as the cosine between their corresponding vectors:

$$cos(t_1, t_2) = \frac{\sum v \in C(t_1) \cap C(t_2)^{P(t_1|v) \times P(t_1|v)}}{\sqrt{\sum v \in C(t_1)^{P(t_1|v)^2} \times \sum v \in C(t_1)^{P(t_2|v)^2}}} \tag{2}$$

3.2   Stage 2: Concept Hierarchy Induction Using CLONACH

In this section, we describe the proposed algorithm which is based on the Clonal Selection Algorithm (CLONALG), applied in this work with slight modifications. A detailed and thorough discussion on CLONALG algorithm can be found in [2]. The input for this stage is list T of n objects (terms) to be ordered hierarchically. Each object has corresponding vectors. The proposed algorithm can be summarized as follows:

1. Initialization: Create an initial random population of antibodies as a cluster initiator called cluster seed.
2. Antigenic presentation: for each antigenic pattern in the clonal memory, DO:
   (a) Clonal selection: for each cluster seed (Ab), determine its affinity with the antigen presented. Select n highest affinity Ab.
   (b) Cell binding and stimulation: The n highest affinity Abs binds the presented Ag in clonal memory. The Ag will stimulate its Ab to proliferate.
   (c) Clonal expansion: For each of these highest affinity Abs, if an Ab affinity is larger than threshold σ_d , clone the Ab. The clone size is proportional to the affinities between an Ab and the antigen.
   (d) Affinity maturation: mutate each clone inversely proportional to affinity. Diversity: If none of the elements in the clonal memory can 'bind' the presented antigen, classified the antigens as the new cluster seed.
3. Clonal interactions: For each cluster in the memory, determine the affinity of its elements to the cluster seed.
4. Clonal suppression: For each cluster, DO:
   (a) If an element's affinity to the cluster seed is smaller than threshold σ_s , then remove the element from the cluster.
   (b) Remove the clones by replacing the clones with its progenitor.

5. Concept hierarchy construction: For each cluster, DO:
   (a) Build a hierarchical tree using Bisecting K-means algorithm, σ_g
   (b) Label the abstract concepts of the hierarchy using Caraballo's method.
6. Cycle: Repeat steps 2 to 5 until a pre-specified number of iterations are reached.

In the clonal selection and expansion step, the number of clones Nc is determined proportionally to its affinity as Eq. (3) and (4):

$$Nc = round(\eta \times Ab.affinity) \tag{3}$$

$$M = round(\frac{\omega}{Ab.affinity}) \tag{4}$$

Eq. 3 is used to determine the number of genes randomly selected for mutation while Eq. 4 is used to calculate the mutation rate. The η and ω are the coefficients selected by users that can significantly affect the number of clone and the mutation rate respectively. However, the values of η and ω coefficients depends on the length of the cell. Equation 5 depicts the Hamming distance measure used to calculate the affinity.

$$D_{HAMMING} = \sum_{i=1}^{k} \delta_i, where \ \delta_i = \begin{cases} 1 \ \text{if} \ Ab_i \neq Ab_i \\ 0 \ \text{otherwise} \end{cases} \tag{5}$$

The divisive bisecting k-means clustering procedure is used to partition any given number of clusters in a cluster resulting the clusters so-obtained are structured as a hierarchical concept tree. This is the reason why the bisecting divisive approach is used in CLONACH. Moreover, the total number of clusters is not required prior to running the algorithm. However, the bisecting divisive approach is not able to appropriately label the produced cluster [15]. Therefore, we implemented the simple Caraballo's method described in [17]. Following Caraballo's method, the produced hierarchical tree is compressed by eliminating internal nodes without a label. In depth-first order, examine the children of each internal node, if the child is itself an internal node and not labeled, delete this child and make its children into children of the parent instead.

4   Experimentations and Results

We test the effectiveness of CLONACH on a relatively small collection of texts obtained from the Center for Artificial Intelligence Technology (CAIT) of Universiti Kebangsaan Malaysia (UKM). The collection of texts contains documents from three different domains i.e. Information Technology, Biochemistry and Fiqh.

4.1   Evaluation of Concept Hierarchy Quality

In this experiment, the performance of CLONACH is evaluated by means of comparing the automatically produced concept hierarchy against a handcrafted reference concept hierarchy [18]. The quality of the induced concept hierarchy is measured according to the F-Measure of Taxonomic Overlap or FTO [15]. FTO is the main indicator of quality in this work because we want to maximize both recall and precision. In this section, we briefly present measures to compare the lexical and taxonomic overlap (TO) (Eq. 12 and 13) between two concept hierarchies. For a comprehensive discussion of the measures presented in this paper, the reader is directed towards [15]. The lexical recall (Eq. 6) and lexical precision (Eq. 7) of two concept hierarchies C_1 and C_2 are measured as follows:

$$LR(C_1, C_2) = |C_1 \cap C_2| \div |C_2| \tag{6}$$
$$LP(C_1, C_2) = |C_1 \cap C_2| \div |C_1| \tag{7}$$

In order to compare two concept hierarchies, we use the common semantic cotopy (SC) introduced by Cimiano [15] as Eq.(8):

$$SC''(c, O_1, O_2) := c_j \in C_1 \cap C_2 | c_j < c_j^{c_i \vee c_1} c_j \tag{8}$$

We calculate the precision of taxonomic overlap (PTO) as Eq. (9) and recall of taxonomic overlap (RTO) as Eq. (10) between two concept hierarchies as follows [15]:

$$P_{TO}(O_1, O_2) = \overline{TO'}(O_1, O_2) \tag{9}$$
$$P_{TO}(O_1, O_2) = \overline{TO'}(O_2, O_1) \tag{10}$$

We also calculate the harmonic mean of both PTO and RTO as Eq. (11):

$$F_{TO}(O_1, O_2) = \frac{(2.P_{TO}(O_1, O_2).R_{TO}(O_1, O_2))}{P_{TO}(O_1, O_2) + R_{TO}(O_1, O_2)} \tag{11}$$

where

$$\overline{TO'}(O_1, O_2) = \frac{1}{|C_2, C_1|} \sum_{c \in C_1 / C_2} max_{c' \in C_2 \cup \{root\}} TO''(c, c', O_1, O_2) \tag{12}$$

$$TO(c, c', O_1, O_2) = \frac{|SC''(c, O_1, O_2) \cap SC''(c', O_2, O_1)|}{|SC''(c, O_1, O_2) \cup SC''(c', O_2, O_1)|} \tag{13}$$

A group of parameter used in CLONACH is determined by trial and error. During parameter setting, a check for redundancy is performed before measuring its taxonomy overlaps. The parameters are listed in Table 1.

In this paper, the iteration n is set to 1. The decision is made based on our observations that if the iteration is more than 1, the suppression steps will fail to remove redundant concepts, which are not allowed in a concept hierarchy. Gomez et al. [19] stated that redundancies of Subclass-Of relations occur between classes that have more than one SubClass-Of . The existing suppression step is not effective to make the right decision if two or more hypernym have the same affinity to the term in question. A term would be classified to more than one hypernym because of the clonal expansion and affinity maturation process. The

Table 1. Results comparison

Threshold for divisive clustering	σ_g	0.15
Threshold for clonal selection	σ_d	0.02860
Threshold for cloning suppression	σ_s	0.9104
The mutation coefficient	ω	9
The percentage number of bases will be copied to improve the performance.	ζ	4
The clone coefficient	μ	1

generated clone's vectors can be so diverse and bind more than one irrelevant antigen. The resulting FTO and other measurement will significantly decreased which indicates that there are multi-copies of an antigen in the produced concept hierarchy. If there are many copies of the same hyponym/hypernym in the clonal memory, a solution has to be found on how to select the right one and remove the false relation.

5   Results

We compare our CLONALG inspired hierarchical clustering algorithm with three different hierarchical agglomerative clustering (HAC) algorithms which differ in the similarity measures they employ: single-link, average-link and complete-link.

The LF and F_{TO} value is slightly superior to that of HACs. It's noteworthy to remind the readers that the divisive bisecting k-mean is used to build the concept hierarchy from the clusters produced by the CLONALG-inspired algorithm. The capabilities to diversify the vectors and to detect patterns are the striking features of CLONALG inherited by CLONACH. The length of vectors in representation space for IT dataset is 137, Biochemistry 80 and the Fiqh is 162. Compared to other datasets, Biochemistry datasets has the lowest LF and F_{TO}. This is due to the lesser amount of features and its attributes. However, the precision taxonomic overlap, i.e. F_{TO} for Biochemistry dataset is 100% for all methods which shows that even it has lesser amount of features but the terms are placed at the precise place in a hierarchy.

Results in Table 2 and 3 have shown that our CLONALG-based approach is a reasonable alternative to similarity-based agglomerative clustering approaches, even yielding better results on CAIT's datasets with respect to the LF and F_{TO} measures defined in section 4.1. The main reason for this is that the concept

Table 2. Comparison with other methods on IT texts

	LR	LP	LF	P_{TO}	R_{TO}	F_{TO}
CLONACH	56.49	59.31	57.87	46.92	16.48	24.39
HAC Single	45.15	56.74	50.29	11.38	20.03	14.51
HAC Average	45.15	56.74	50.29	11.38	20.03	14.51
HAC Complete	45.21	56.07	50.06	22.01	10.96	14.63

Table 3. Comparison with other methods on Biochemistry Texts and Fiqh Texts

	Biochemistry Texts						Fiqh Texts					
	LR	LP	LF	P_{TO}	R_{TO}	F_{TO}	LR	LP	LF	P_{TO}	R_{TO}	F_{TO}
CLONACH	13.58	51.16	21.46	100	11.26	20.24	43.57	58.49	49.94	60.67	11.98	20.01
HAC Single	11.11	56.25	18.56	100	6.68	12.58	33.73	56.48	42.24	11.07	22.12	14.75
HAC Average	11.11	54.55	18.46	100	6.68	12.52	33.73	56.86	42.34	10.97	21.22	14.47
HAC Complete	9.88	53.33	16.67	100	7.62	14.15	33.73	57.05	42.39	11.21	23.1	15.09

hierarchies produced by CLONACH yield a higher lexical recall due to higher number of concepts. The P_{TO} is better than other methods on two datasets except for Biochemistry which the P_{TO} for all methods are 100%. This is due to the low LF experienced by each methods on Biochemistry texts which resulting a lower R_{TO} compared to other datasets. The reason for the better result achieved by CLONACH is that the feature vectors representing any terms has been diversified into many forms caused by the cloning expansion and affinity maturation (mutations) process.

6   Conclusions and Future Work

This paper proposes the use of hierarchical approach for a hybrid algorithm to the learning of concept hierarchy named CLONACH (Clonal Selection Algorithm for Learing Concept Hierarchy). The hybrid algorithm is a combination of an artificial immune system, named CLONALG and bisecting k-means. An ontology learning system was built based on CLONACH and was tested on texts from three different domains which are IT, Biochemistry and Fiqh (i.e, Islamic Jurisprudence). Results have shown that CLONACH is better than GAHC in two domains and produced lower result on the Biochemistry texts. The biochemistry text is an academic text and very technical compared to the other two texts. The length of bit strings which representing vectors from Biochemistry texts are significantly less than the other texts. However further studies and larger trials on huge corpuses are needed in order to further examine and confirm the efficacy of CLONACH in the field of ontology learning from text. Our future work will incorporate other concept hierarchy induction techniques in order to compare the performance of CLONACH against different learning paradigms.

References

1. Gulla, J.A., Brasethvik, T.: A Hybrid Approach to Ontology Relationship Learning. In: Kapetanios, E., Sugumaran, V., Spiliopoulou, M. (eds.) NLDB 2008. LNCS, vol. 5039, pp. 79–90. Springer, Heidelberg (2008)
2. de Castro, L., Von Zuben, F.J.: Learning and Optimization Using the Clonal Selection Principle. IEEE Transactions on Evolutionary Computation 6(3) (2002)
3. Han, J., Fu, Y.: Dynamic Generation and Refinement of Concept Hierarchies for Knowledge Discovery in Databases. In: Workshop on Knowledge Discovery in Databases (AAAI 1994). AAAI, Burnaby (1994)

 4. Drumond, L., Girardi, R.: A Survey of Ontology Learning Procedures. In: Freitas, F.L.G.d., et al. (eds.) The 3rd Workshop on Ontologies and their Applications, Salvador, Bahia, Brazil, October 26. CEUR-WS.org (2008)
 5. Harris, Z.: Mathematical Structure of Language. Wiley, Chichester (1968)
 6. Cimiano, P., et al.: Comparing conceptual, divisive and agglomerative clustering for learning taxonomies from text. In: Proceedings of the European Conference on Artificial Intelligence, ECAI (2004)
 7. Schickel-Zuber, V., Faltings, B.: Using hierarchical clustering for learning the ontologies used in recommendation systems. In: International Conference on Knowledge Discovery and Data Mining 2007, pp. 599–608. ACM, San Jose (2007)
 8. Reinberger, M.-L., Daelemans, W.: Is Shallow Parsing Useful for Unsupervised Learning of Semantic Clusters? In: Computational Linguistics and Intelligent Text Processing, pp. 24–34. Springer, Heidelberg (2010)
 9. de Castro, L.N., Timmis, J.: Artificial Immune Systems: A New Computational Intelligence Approach. Springer, Heidelberg (2002)
10. de Castro, L.N., Zuben., F.v.: The clonal selection algorithm with engineering applications. In: Proceedings of Genetic and Evolutionary Computation, Las Vegas, USA (2000)
11. Secker, A., et al.: An Artificial Immune System for Clustering Amino Acids in the Context of Protein Function Classification. Journal of Mathematical Modelling and Algorithms 8(2), 103–123 (2009)
12. Zhang, C., Yi, Z.: Tree structured artificial immune network with self-organizing reaction operator. Neurocomputing 73(1-3), 336–349 (2009)
13. Hamzah, M.P., dan Hubungan, F., Dalam, S., Pengetahuan, P.: kesan Terhadap Keberkesanan Capaian Dokumen Melayu. In: Information Science, p. 214. Universiti Kebangsaan Malaysia, Bangi (2006)
14. Nazri, M.Z.A., et al.: An Exploratory Study on Malay Processing Tool for Acquisition of Taxonomy Using FCA. In: Eighth International Conference on Intelligent Systems Design and Applications, ISDA 2008 (2008)
15. Cimiano, P.: Ontology Learning and Population from Text. Springer, Berlin (2006)
16. Salton, G., McGill, M.J.: Introduction to Modern Information Retrieval. McGraw Hill, New York (1983)
17. Caraballo, S.A.: Automatic construction of a hypernym-labeled noun hierarchy from text. In: 7th Annual Meeting of the Association for Computational Linguistics, College Park, Maryland. ACM, New York (1999)
18. Cimiano, P., Staab, S.: Learning Concept Hierarchies from Text with a Guided Agglomerative Clustering Algorithm. In: International Conference on Machine Learning, ICML 2005, Bonn Germany (2005)
19. Gomez-Perez, A., et al.: Ontological Engineering, 4th edn., p. 416. Springer, New York (2005)

Action Potential Classification Based on LVQ Neural Network

Jian-Hua Dai[1,2], Qing Xu[1], Mianrui Chai[1], and Qida Hu[3]

[1] School of Computer Science and Technology,
Zhejiang University, Hangzhou 310027, P.R. China
[2] Center for the Study of Language and Cognition,
Zhejiang University, Hangzhou 310028, P.R. China
[3] College of Medicine,
Zhejiang University, Hangzhou 310027, P.R. China
jhdai@zju.edu.cn

Abstract. Neural action potential classification is the prerequisite condition of further neural information process, but there are difficulties in accurate action potential classification due to the existence of great amount of noise. In this paper, we propose a method of action potential classification based on Learning Vector Quantization (LVQ) network. In the experimental stage, the performance of the presented system was tested at various signal-to-noise ratio levels based on synthetic data. The results show that the proposed action potential classification method is effective. The proposed method supplies a new thought for action potential classification problem.

Keywords: LVQ Network, Action potential classification, SNR.

1 Introduction

Action potential, also called spike potential, is the principal manifestation of neural network activities. For any network built with neurons, spike pulses form signals of neurons, which indicate that spike potentials are the information carriers. Multi-channel extracellular recording permits neurophysiologist to study information transmission within the nervous system by measuring neuronal activities. A major challenge here is that one microelectrode may pick up spikes originating from several adjacent neurons. In such case, separation of spikes generated by adjacent neurons is necessary. This separation task is called action potential classification. Therefore, neural action potential classification has become the basis and premise of research on neuroscience [1-4]. In the past several years, some action potential classification algorithms have been proposed [5-7].

Action potential classification is actually a clustering problem. Classification methods based on usual neural networks can get good performance [6,7], but the neural networks used are completely supervised.

In this paper, we introduce learning vector quantization network into action potential classification problem and construct a corresponding method.

J. Yu et al. (Eds.): RSKT 2010, LNAI 6401, pp. 462–467, 2010.

The performance of the presented system was tested at various signal-to-noise ratio levels based on synthetic data. The results show that the proposed action potential classification method is effective. The proposed method supplies a new thought for action potential classification problem.

2   Action Potential Classification Based on LVQ Network

LVQ neural network, derived from Kohonen's SOFM (self organizing feature map) neural network, is a kind of self-organizing network. LVQ network is a mixed network combined with supervised learning and competitive learning (unsupervised learning). Figure 1 shows the structure of LVQ network. An LVQ network has a first competitive layer and a second linear layer. In terms of neurons, an LVQ has 3 layers by defining a hidden layer based on the self-organizing competitive network: input layer, hidden layer and output layer. The connection between the input layers and the hidden layer is full join and each neuron in the input layer connects to each neuron in the hidden layer (Kohonen neuron); the connection between the hidden layer and the output layer is linear join and each neuron in the hidden layer connects to one neuron in the output layer.

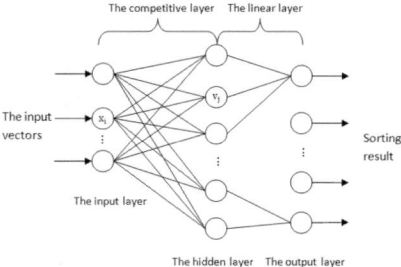

Fig. 1. The structure of LVQ network

LVQ network is an evolutional algorithm of SOFM network. The competitive layer can classify input vectors based on their distances, that is, two input vectors will be sorted into the same class if the distance is quite short. Finally every neuron in the input layer will be sorted into the subclass represented by a certain neuron in the hidden layer. The linear layer can convert the classification information of input vectors in the competitive layer into defined classes and then optimize the network weights of the neurons in the hidden layer to obtain the final sorting result. Sorting with supervised learning will be carried out in the linear layer with fixed weights. After getting the output results, the LVQ network will properly modify the weights of the neurons in the hidden layer to get expected output by deciding whether the results are correct or not.

Steps to create a LVQ network: Since the LVQ network is a mixed network combined with supervised learning and competitive learning, a set of examples of correct network input and output are needed to train the network when creating an LVQ network. Suppose X is the input vector set in the input layer, Y the sorting result in the output layer, T the maximum iteration times, E the maximum permissible error, the learning rate ((0, 1), the default value 0= 0.01) [5-6].

Step 1 Initialize the hidden layer as v1 = v1, v2, ...,vn (n is the number of neurons in the competitive layer).

Step 2 Competitive learning: Calculate the Euclidean distances from each input vector Xi to all hidden layer neurons and find out the vector's neuron corresponding to the minimum distance. Suppose Vj is the neuron in the hidden layer with minimum distance to Xi, m=d(xi, vj) the distance between Xi and Vj.

Step 3 Supervised learning: For any Xi, learn and optimize the weight of vj comparing the sorting result in step 2 and Y. Let Vj$'$ be the subclass after optimization

(1)If the classification result corresponds with the input sorting result, then $v_j' = v_j + \alpha(x_i - v_j)$

(2)If the classification result doesn't correspond with the input sorting result, then $v_j' = v_j - \alpha(x_i - v_j)$

Step 4 Calculate errors $\varepsilon = \sum_1^n |v_j' - v_j|$. If $\varepsilon < E$, stop calculation; otherwise adjust the learning rate to be ', return to step 2 to continue running, until the error condition is satisfied or the iteration number is reached. The LVQ network is a progress with continuous learning and optimization. When an input vector is assigned to the neuron with closest distance by competition, the weight of the neuron will be optimized based on the input vector and therefore the neuron can represents the features of its input vector more precisely. It sorts the similar input vectors into the same class and instruct the class by output [4].

3   Build the LVQ Network Model

3.1   Build the LVQ Network Model

To build the LVQ Network, there are several steps to be done, such as problem description, data preparation, network building, network training and network testing. The LVQ network could sort any kind of input, no matter whether they are linearly separable. It could also solve complex sorting problems easily. In this paper, the LVQ network was built on the MATLAB 2007b platform. The matrices with 33 variables were selected as the input, and the output was one of neurons, as shown in figure 2.

1.Build the LVQ network.

Use the matlab function newlvq() to builds the LVQ network, and use the function train()to train the LVQ network. The format of the function is: net= newlvq (PR,SI,PC,LR,LF). PR is an R-by-2 matrix of minimum and maximum values for R input elements; SI is the number of first-layer hidden neurons; LR is the learning rate (default 0.01); LF is the learning function (default is learnlv1).[5-6]

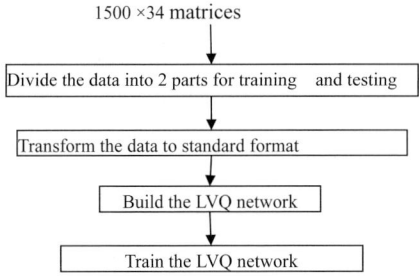

Fig. 2. Flow process diagram of building an LVQ network

2. Train the LVQ network net = train(net,P,T). net is the LVQ network build by using function, P is training matricesT is the sorting result of P. To use function train() the net, it requires to set the max training time EPOCHS previously (default 100).

3.2   Parameter Selection

During the process of building the LVQ network, different learning rate LR and the max training time EPOCHS could lead to different results. The bigger LR is, the faster the net converges. As a result, it needs less time to reach the max cycle time, but with stronger network shocks. At the same time EPOCHS relates to the quality of the net. If EPOCHS is too small, it would not be able to study all of the inputs to build the net. On the contrary, when EPOCHS is too big, the training time would be greatly lengthened, which leads to over learning and makes the net inefficient [4]. In this experiment, since the number of training data was 100034, EPOCHS was set to 150. In consideration of the stability of learning, LR was set to 0.01. In LVQ network, since the number of final output class was 3, and the linear parts combine many hidden layer neurons into 1, the number of hidden layer neurons SI was set to 6.

4   Experiments

The signals in this paper were made by electrical signal simulative generator. The frequency of sampling is 30kHz and it includes 3 neuron modes. To ensure the feasibility and accuracy of the experiment result, 150014 neural spike signals made by the signal simulative generator were selected and 33 moments successive neural spike signals were selected as the input data. Another variable was selected to sign the 3 neuron modes. To train and test the network, the data was divided equally into several parts: 1500 34 matrices were firstly sequenced based on their sorting result, and then divided the data in the proportion of 2:1,to have 2/3 of the neural spike signals (1000 totally) selected randomly and 500 sets of data left as the testing parts.

In order to show the performance of the LVQ network, some noise was added to the testing data in the experiment. At the same time two other networks-the SOM network and the BP network were used as well, to make a comparison with the LVQ network. Due to a large number of differences between the networks when they are built, some special steps were taken to make it accurate and representative. In the experiment, groups of noised action potentials are acquired with SNRs from 0 dB to 10 dB, and 14 sets of data were selected in these SNR. In each SNR, the LVQ network, the SOM network and the BP network were used, their results recorded and accuracy calculated. Finally, the average values of each network in different SNRs were calculated. The result is shown in figure 3.

In this experiment, the other variables in the BP network were set properly to make it comparable. The hidden layer was set to 1 and the hidden layer neuron number was set to 6, which is the same as the LVQ network. However, this would affect the accuracy and efficiency of the BP network.

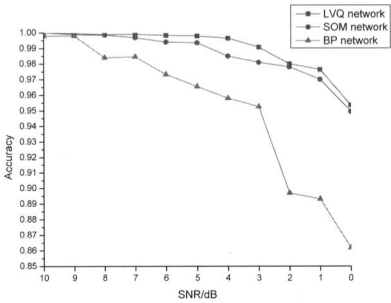

Fig. 3. Results of the performances at different level of SNB

5   Conclusion

LVQ network theory combines theoretical modeling technology with people's feeling and judgment to the real world, which increases the accuracy of judgment. The network has low standard for data and strong fault-tolerant ability. The experiment shows that the accuracy of the LVQ network is more than 99% when SNR is high (SNR=10, 9, 8). The accuracy decreases slightly as SNR decreases, but it's still satisfying (¿95%). In comparison with SOM and BP network, LVQ network can give us higher output accuracy in different SNRs, which also applies to action potential classification under the same condition.

There are several neural network applied and they all have advantages respectively. SOM network can use the input data and the final number of classes to establish a neural network without the known sorting results. And creating and training a SOM network is easier and more convenient than LVQ network. Establishing an LVQ network needs input data and the sorting results, but LVQ network is more complex and has higher accuracy of sorting. The result shows that sorting based on those two networks under different SNR has considerable accuracy, which makes both of them suitable to be applied to action potential classification.

In the experiment, as the noise grew louder, the time of iteration time to train the LVQ network increased as well. For example, when the SNR is 10dB, only 9 iteration could reach the training requirements; when the SNR is 6dB, this time would increase to 34; when the SNR is 5dB, this time required would increase to 119; when the SNR is 4dB, in the most cases, the LVQ network would be forced to stop training as it could not meet the training requirement in presented 119 iteration time. So under high SNR, the training progress of the LVQ network could meet the training requirement quickly. But when the SNR becomes lower, the iteration time would increase obviously, the network should be chosen carefully under this condition.

Acknowledgements

The work is supported by the National Natural Science Foundation of China (No. 60703038) ,the Excellent Young Teachers Program of Zhejiang University and the Research Foundation of Center for the Study of Language and Cognition of Zhejiang University.

References

1. Hulata, E., Segev, R., Ben-Jacob, E.: A method for action potential classification and detection based on wavelet packets and Shannon's mutual information. Journal of Neuroscience Methods 117(1), 1–12 (2002)
2. Letelier, J.C., Weber, P.P.: Action potential classification based on discrete wavelet transform coefficients. Journal of Neuroscience Methods 101(2), 93–106 (2000)
3. Pazienti, A., Grun, S.: Robustness of the significance of spike synchrony with respect to sorting errors. Journal of Computational Neuroscience 21(3), 329–342 (2006)
4. Zumsteg, Z.S., Kemere, C., O'Driscoll, S., Santhanam, G., Ahmed, R.E., Shenoy, K.V., Meng, T.H.: Power feasibility of implantable digital action potential classification circuits for neural prosthetic systems. IEEE Transactions on Neural Systems and Rehabilitation Engineering 13(3), 272–279 (2005)
5. Shi, N.-y., Chen, K., Li, C.-h.: The application of fuzzy clustering in teacher-evaluating model. In: 2009 IEEE International Symposium on IT in Medicine and Education (ITME 2009), pp. 872–875 (2009)
6. Ni, S., Qingming, L., Xiangning, L.: Action potential classification in networks of cultured neurons on multi-electrode arrays. In: Proc. SPIE - Int. Soc. Opt. Eng., vol. 5254(1), pp. 513–520 (2003)
7. Ding, Y., Yuan, J.-q.: The neural action potential classification under low signal-to-noise ratio based on artificial neural networks. J. Shanghai Jiaotong Univ. 40(5), 852–855 (2006)

Back Propagation Approach for Semi-supervised Learning in Granular Computing

Hong Hu[1], Weimin Liu[1,2], and Zhongzhi Shi[1]

[1] The Key Laboratory of Intelligent Information Processing, Institute of Computing Technology,
Chinese Academy of Sciences, Beijing, 100190, China
[2] Graduate University of Chinese Academy of Sciences, Beijing, 100190, China
huhong@ict.ac.cn, {liuwm,shizz}@ics.ict.ac.cn

Abstract. Zadeh proposes that there are three basic concepts underlying human cognition: granulation, organization and causation and that a granule is a clump of points (objects) drawn together by indistinguishability, similarity, proximity or functionality. Tolerance relation can describe the concept of Granular systems. In this paper, a novel definition of Granular System(GS), which is described by metric function under the framework of tolerance relation, is presented, concepts are created upon GS, and we introduce semi-supervised learning into the Granular computing for concepts creating. For this purpose, a novel back propagation approach is developed for concepts learning. The experiment shows that the new BP is better than traditional EM algorithm when samples do not come from a random source, which has the density we want to estimate.

Keywords: granular computing, semi-supervised learning, back propagation learning.

1 Introduction

As we know the concepts of our knowledge come from a finite and discrete set, and so-called learning in our cognition tries to create or recognize concepts from continuous information of our world, such kind process is defined as "Granular Computing GC" by Zadeh[1]. For GC, Zadeh [1] proposed that granulation of an object A leads to a collection of granules of A, with a granule being a clump of points (objects) drawn together by indistinguishability, similarity, proximity or functionality.

The Granular Computing (GrC) is characterized by a set of granules(defined by similarity of information), which create a leveled granular system based on tolerance relation of distance function (Definition 4).

In this paper, we introduce semi-supervised learning into the Granular computing, and a novel back-propagation approach (denoted as "*DBP*") is developed for concepts learning in our Granular Computing. The experiment shows that *DBP* is better than traditional EM algorithm when samples do not come from a random source, which has the density we want to estimate.

J. Yu et al. (Eds.): RSKT 2010, LNAI 6401, pp. 468–474, 2010.

2   Granular System Based on Tolerance Relation

2.1   The Definition of a Granular System

If all elements of a class are distributed in a region, which can be described by some kind distance function (denoted as a *granule*), similarity between two objects can be intuitively described by distance functions. If $dis(x, y)$ is a distance function in the n-dimensional space R^n and c is a point in R^n, the formula $dis(c, y) < r$ described a convex region D in R^n which takes c as its center. Every point y in this region is equal to $y = c + e$, here e is some kind noise, if D is just a ball, e can be viewed as white noise which has an amplitude less than r.

Definition 1. (Weighted distance function) If $d(\alpha, \beta)$ is a distance function,then $d(\alpha, \beta)$ satisfies the three axioms of a distance function. Now we define $dis(\alpha, \beta|\omega) = d(\alpha_\omega, \beta_\omega)$, where α_ω and β_ω are the projections of α and β on ω separately, i.e. $dis(\alpha, \beta|\omega) = d(\omega \otimes \alpha, \omega \otimes \beta)$ here $x = (x_1, \cdots, x_n) \otimes y = (y_1, \cdots, y_n) = (x_1 \cdot y_1, \cdots, x_n \cdot y_n)$

Definition 2. (A simple fuzzy logical formula based on distance function SFL or a simple concept). A simple fuzzy logical formula based on distance function $sp(a, c|dis, d, \omega)$ is denoted as:

$$sp(a, c|dis, \omega) = m(dis(a, c|\omega)) \tag{1}$$

where $dis()$ is a point-wise distance function,m() can be viewed as a membership function of a fuzzy set,usually m() should be a continuous function. In this paper m() is defined as:

$$\phi(x|\omega) = a \cdot exp\{-1/2 \cdot dis(x, c|\omega)\} \tag{2}$$

If $dis(x, c|\omega) = (x - c)^T \Sigma^{-1}(x - c)$, and Σ is a diagonal matrix, then above equation is just a Gaussian formula.

if P is a fuzzy logical formula, we denote the negative P as: $\neg P = 1 - P$

We define the r-cut set of $sp(a, c|dis, d, \omega)$ as:

$sp^r(a, c|dis, d, \omega) = sp(a, c|dis, d, \omega) > r$(strong r-cut set)or $sp^r(a, c|dis, d, \omega) = sp(a, c|dis, d, \omega)r$(r-cut set), where a and β are two n-dimensional vectors in R_n,and $\omega = (\omega_0, \omega_1, \cdots, \omega_{n-1})$, $\omega_i \geqslant 0$,is a weight vector. We can select suitable membership function $m()$ to make $spr(x, c|dis, d, \omega)$ as an open convex region and denoted as a *granule*. We denote the set described by $sp(a, c|dis, d, c, \omega)$ as a fuzzy convex set.

Every region R_T in R^n can be described by a compound fuzzy logical proposition (Definition 3) with arbitrary precision.

Definition 3. (A compound fuzzy logical proposition based on distance function CSFL or a complex concept). A CSFL is a formula of SFLs combined by fuzzy logical operators.

Definition 4. (A leveled granular system Gsys) A granular system is an open sets' system and sets are described by CSFLs. For the detailed description please refer to [3].

2.2   The Fuzzy Operator Used in CSFLs or Complex Concepts

After the theory of fuzzy logic was conceived by Zadeh in 1965, many fuzzy logical systems have been presented. In this paper, we use Bounded operator.

Definition 5. (Bounded operator $F(\oplus_f, \otimes_f)$) Bounded product: $x \otimes_f y = max(0, x+y-1)$, and Bounded sum: $x \oplus_f y = min(1, x + y)$, where $0 \leqslant x, y \leqslant 1$.

In order to describe different granules, it is necessary to extend the Bounded Operator to Weighted Bounded Operator. The fuzzy formulas defined by q-value weighted bounded operators is denoted as q-value weighted fuzzy logical functions.

Definition 6. (q-value weighted bounded operator)
Weighted Bounded product:

$$p_1 \otimes p_2 = F_{\otimes_f}(p_1, p_2, w_1, w_2) = max(0, w_1 p_1 + w_2 p_2 - (w_1 + w_2 - 1)q) \qquad (3)$$

Weighted Bounded sum:

$$p_1 \oplus_f p_2 = F_{\oplus_f}(p_1, q_2, w_1, w_2) = min(q, w_1 \cdot p_1 + w_2 \cdot p_2) \qquad (4)$$

N operator: $N(x) = q - x$

For association and distribution rules, we define

$(p_1 \Delta_f p_2) \Theta_f p_3 = F_{\Theta_f}(F_{\Delta_f}(p_1, p_2, w_1, w_2), p_3, 1, w_3)$
$p_1 \Delta_f (p_2 \Theta_f p_3) = F_{\Delta_f}(p_1, F \Theta_f(p_2, p_3, w_2, w_3), w_1, 1)$
Here $\Delta_f, \Theta_f = \otimes_f \, or \oplus_f$.

The value q changes the fuzzy value from $[0, 1]$ to $[0, q]$. For the sake of simplicity, in this paper, a leveled granular system Gsys only tries to find a covering system in the samples' space, so we don't use N operator in a CSFL, and an enough large q value is selected and

$$\sum_{1 \leqslant i \leqslant n} w_i = 1, \qquad (5)$$

here $w_i \geqslant 0$ for all $i = 1, \cdots, n$; so the *max* or *min* operator can be omitted in the definition of the $q - value$ Weighted Bounded operator. In this case, the compound fuzzy logical proposition based on distance function CSFL using $f_{cs}(x|\theta)$ q-value fuzzy logic is simplified to Eq.6:

$$f_{CS}(x|\theta) = \sum_{i=1}^{m} a_j \varphi_j(x|\theta_j) + c \qquad (6)$$

Where every $\varphi_j(x|\theta_j)$ is a simple fuzzy logical formula with a parameter θ, $\sum_{i=1}^{m} a_i = 1, a_i \geqslant 0$, and c is a constant.

3   Granular Computing in Concept Computing

3.1   Semi-supervised Learning in Granular Computing

Semi-supervised learning deals with methods for automatically exploiting unlabeled data in addition to labeled data to improve learning performance. That is, the exploitation of unlabeled data does not need human intervene. Here the key is to use the unlabeled data to help estimate the data distribution (Z.-H. Zhou, 2006[2]). The Granular Computing in concept computing takes place in following new semi-supervised learning scenario: The concepts are divided into two kind sets $C = c_1, c_2, \cdots, c_p$ and $C_h = h_{c1}, h_{c2}, \cdots, h_{ck}$. The concepts in the set C are just the normal concepts in our knowledge, whereas, the meaning of concepts in C_h is hidden or unknown by our knowledge. Concepts' set and hidden concepts' set are created from on a leveled granular system based on tolerance relation of distance function created by granular computing.

To find a new complex concept based on already known C and C_h is the aim of a granular computing. Suppose the n samples $\chi = c_1, c_2, \cdots, c_p$, which mutually independent come from a random k-dimensional source in the real vector space R^k, and these samples belong to p fuzzy concepts $C = c_1, c_2, \cdots, c_p$ and q hidden fuzzy concepts $C_h = h_{c1}, h_{c2}, \cdots, h_{cq}$. Fuzzy concepts must have concrete meaning, for example, in image classification, a concept may have meaning such as "car", "tree" or "house" etc. Whereas, a hidden fuzzy concept has no concrete meaning, and only represents a definite feature set which is useful in machine learning. Every concept $\xi_l = c_l$ or h_{cl} denotes a fuzzy set with membership function $f(\xi_l|x)$. The sample set χ is divided into two sets L and U, here L is a labeled set and U is an unlabeled set. Every sample x_i in L belongs to concept set C and is assigned a set of fuzzy labels $l(c_l|x_i) = f(c_l|x_i), l = 1, \cdots, p$. When $f(c_l|x_i) \in 0, 1$, a fuzzy label is degenerated to a usual binary label. Every sample in U belongs to C or C_h, and every sample x_l in U are created by a random source with probability density $f(x_l)$(here we consider the meaning of fuzzy has a probability meaning). In this case, we can estimate the fuzzy label $f(x_l)$ by Eq.7

$$f(x_l) = p_e(x_l|\chi = \{x_1, \cdots, x_n\}) = \frac{|S_l(r)|}{V_r \cdot n} \tag{7}$$

where $S_l(r)$ is the set $S_l(r) = \{x_l - |x_t - x_l| < r \, x_t \in \{x_1, x_2, \cdots, x_n\}\}$ and all samples in $S_l(r)$ are created by a random source with probability density $f(\zeta_l|x_i)$, and V_r is the volume of a ball with a radius r or a cube with edges length r.

In the learning process of our Granular computing, we try to use the risk formula Eq.9 to learn the compound fuzzy logical proposition based on distance function CSFL (see Eq.8).

$$f_{CS}(x|\theta) = \sum_{i=1}^{m} a_i f(c_i|x) + \sum_{i=1}^{n} b_i f(h_{c_i}|x) + \rho \tag{8}$$

Where $f(t_i|x) = \sum_{j=1}^{j=m_i} \alpha_j^i \cdot \varphi_j(x|\theta_j^i)$ for $t_i = c_i$ or h_{c_i}, $j = 1, \cdots, m_i, i = 1, \cdots, m$, and every $\varphi_j(x|\theta_j^i)$ is a simple fuzzy logical formula, $\sum_{i=1}^{m} a_i + \sum_{j=1}^{n} b_j^i = 1, a_i \geqslant 0, b_j \geqslant 0$, and θ is the model's coefficient.

$$R_{pe}(\theta, \chi) = \sum_{x \in U} (f_{CS}(x \mid \theta) - p_e(x \mid \chi))^2 + \sum_{l=1}^{p} \sum_{x \in L} (l(c_l|x_i) - f(c_l|x_i))^2 \tag{9}$$

Where $p_e(x|\chi)$ is the estimate of $f_{CS}(x|\theta)$ based on the samples set χ, and $l(c_l|x_i)$ is a fuzzy label.

3.2   Dual Back Propagation Algorithm of EM (DBP)

For the sake of pages, the experiment is only show the ability of DBP, not the semi-supervised learning. Eq.(8) has similar form as mixture Gaussian, DBP can compare with the traditional Expectation Maximization (EM) algorithm.

Suppose we already have an reasonable estimate $p_e(x_i|\chi)$ of the unknown compound fuzzy logical proposition based on distance function CSFL $f_{CS}(x|\theta)$ on n samples $X = x_1, x_2, \cdots, x_n$, which mutually independently come from a random k dimensional source with the density $f_{CS}(x|\theta)$, here $x_i = (x_{i,1}, x_{i,2}, \cdots, x_{i,m})$, $i = 1, 2, \cdots, n$, then we can use the least square risk functional Eq.10 to estimate the unknown parameter θ.

The coefficient θ_k of Eq.8 can be computed by gradient descending approach of back propagation. Now we try to compute the partial differential of Eq.10 on μ_h, Σ_h and α_h respectively.

$$R_{p_e}(\theta, \chi) = \sum_{x \in \chi} (f_{CS}(x|\theta) - p_e(x|\chi))^2 \tag{10}$$

Based on above analysis, the iteration steps of the means, mixing coefficients, and the covariances can be defined as Eq.(11), Eq.(13) and Eq.(17or18), sp_μ, sp_α and sp_Σ are dynamic parameters which control the iteration speed. In the part1 of DBPEM, $sp_\mu = 0.0371$, and $sp_\alpha = sp_\Sigma = 0.00056$; In the part2 of DBPEM, $sp_\mu = 0.0027$, $sp_\Sigma = 0.0069$ and $sp_\alpha = 0.000005$.

In the following pages:
$delta = -\sum_{i=1}^{n}(p_e(x_i|X) - \sum_{h=1}^{m} a_h\phi(x_i|\Sigma_h, \mu_h)); delta1 = delta \cdot a_h \cdot \phi(x_i|\Sigma_h, \mu_h); M(\Sigma_h) = 1/2 \cdot diag(\Sigma_h) - \Sigma_h; M(x_i, \mu_h) = M(T)$, here $T = (x_i - \mu_h) \cdot (x_i - \mu_h)^T$.

$$\mu_h(t+1) = \mu(t) - sp_\mu \cdot delta_1 \cdot \Sigma_h^{-1}(t) \cdot (x_t - \mu_h(t)) \tag{11}$$

The mixing coefficients α_h should satisfy the constraint:

$$\alpha_h \geqslant 0, h = 1, 2, \cdots, m \text{ and } \sum_{1 \leq h \leq m} \alpha_h = 1 \tag{12}$$

In order to satisfy the constraints of Eq.(12), the iteration of the mixing coefficients should be as Eq.(13).

The experiments show that the mixing coefficients a_h and the means μ_h are easily learned by back-propagation, but the correct covariance Σ_h are not easy to achieve, in order to find the correct, a new kind back-propagation-dual form back propagation should be proposed here. The dual form back-propagation uses two kind risk functions.

$$a_h(t+1) = \begin{cases} \dfrac{a_h(t) - \Delta a_h(t)}{\sum_{1 \leq l \leq n} a_l(t+1)}, & a_h(t) + \Delta a_h(t) > 0, \tag{13} \\ 0, & a_h(t) + \Delta a_h(t) < 0. \qquad (13') \\ 1, & a_h(t) + \Delta a_h(t) > 1. \qquad (13'') \end{cases}$$

Where $\Delta a_h(t) = sp_\alpha \cdot delta \cdot \varphi(x_i, \theta_h)$.

The 1st kind risk is just Eq.10, the 2nd risk is showed as Eq.14. In Eq.14, the estimated density $p_e(x_i|\chi)$ is used as a weight in the risk function, so Eq.14 is just the mean value of the 1st risk function. The 1st part of DBPEM uses the empirical risk Eq.10 whereas the 2nd part uses Eq.14.

$$E(R_{p_e}(\theta, \chi)) = \sum_{i=1}^{n} p_e(x_i|\theta) \cdot (p_e(x_i|\theta) - p_e(x_i|\chi))^2, \theta \in \Lambda \tag{14}$$

Eq.10 accords with Eq.15, but Eq.14 is the discrete case of Eq.16.

$$R_{p_e}(\theta) = \int (p(x|\theta) - p(x|\theta_0))^2 p(x|\theta_0) \cdot dx \tag{15}$$

$$R_{p_e}(\theta) = \int (p(x|\theta) - p(x|\theta_0))^2 p^2(x|\theta_0) \cdot dx \tag{16}$$

Eq.14 pays much more attention to the samples which have high $p_e(x_i|\chi)$. In order to leap out the local minimum of covariance, two approaches (roughly approach Eq.17 and fine approach Eq.18) of iteration of Σ_h are used in DBPEM. The 1st part of DBPEM uses rough approach, in this stage, the covariances Σ_h are adapted by Eq.17, but in the 2nd part, the covariances Σ_h are alternatively modified by Eq.17 and Eq.18. The strategy in this fine stage is that if the risk estimate Eq.14 increases continuously in T_{cov} steps, then the modification should be changed from rough to fine or vice versa.

The rough approach of covariance modification:

$$\Sigma_h(t + 1) = \Sigma_h(t) - sp_\Sigma \cdot delta_1 \cdot M(x_i, \mu_h(t)). \tag{17}$$

The fine approach of covariance modification:

$$\Sigma_h(t + 1) = \Sigma_h(t) - sp_\Sigma \cdot delta_1 \cdot [M(x_i, \mu_h) - M(\Sigma_h)]. \tag{18}$$

Because the covariances Σ_h should be positive definite matrixes, when Σ_h loses the positive definite character in any stages, our back-propagation algorithm will enter a dead corner. In this case, we should set Σ_h randomly, then a new cycle of the iteration is started based on this new covariances Σ_h .

3.3   Experiment Results

The size of samples' set is from 1300 to 2600, and are created by a m-mixture Gaussian in the region A , but in the region B some samples are deleted. In the region B, every sample x is assigned a fuzzy label which is just the density . Experiment result shows that the average error rate for EM is 7.745% and 4.70% for DBP.

Acknowledgement

This work is supported by the National Nature Science Foundation of China (No.60903141, 60933004) and National Nature Basic Research Priorities Programme (No. 2007CB311004).

References

1. Zadeh, L.A.: Toward a theory of fuzzy information granulation and its centrality in human reasoning and fuzzy logic. Fuzzy Sets Syst. 19, 111–127 (1997)
2. Zhou, Z.-H.: Learning with unlabeled data and its application to image retrieval. In: Yang, Q., Webb, G. (eds.) PRICAI 2006. LNCS (LNAI), vol. 4099, pp. 5–10. Springer, Heidelberg (2006)
3. Hu, H., Shi, Z.: Perception Learning as Granular Computing. In: Proceedings of the 2008 Fourth International Conference on Natural Computation, vol. 03, pp. 272–276 (2008)

WebRank: A Hybrid Page Scoring Approach Based on Social Network Analysis[*,**]

Shaojie Qiao[1,***], Jing Peng[2,***], Hong Li[1]
Tianrui Li[1], Liangxu Liu[3], and Hongjun Li[4]

[1] School of Information Science and Technology, Southwest Jiaotong University,
Chengdu 610031, China
sjqiao@swjtu.edu.cn
[2] Department of Science and Technology, Chengdu Municipal Public Security
Bureau, Chengdu 610017, China
pj@tfol.com
[3] School of Electronics and Information Engineering, Ningbo University of
Technology, Ningbo 315016, China
[4] School of Computer Science, Sichuan University, Chengdu 610065, China

Abstract. Applying the centrality measures from social network analysis to score web pages may well represent the essential role of pages and distribute their authorities in a web social network with complex link structures. To effectively score the pages, we propose a hybrid page scoring algorithm, called WebRank, based on the PageRank algorithm and three centrality measures including degree, betweenness, and closeness. The basis idea of WebRank is that: (1) use PageRank to accurately rank pages, and (2) apply centrality measures to compute the importance of pages in web social networks. In order to evaluate the performance of WebRank, we develop a web social network analysis system which can partition web pages into distinct groups and score them in an effective fashion. Experiments conducted on real data show that WebRank is effective at scoring web pages with less time deficiency than centrality measures based social network analysis algorithm and PageRank.

Keywords: social network analysis, web social network, PageRank, centrality measures.

---

[*] This work is partially supported by the National Science Foundation for Post-doctoral Scientists of China under Grant No. 20090461346; the Fundamental Research Funds for the Central Universities under Grant No. SWJTU09CX035; the Sichuan Youth Science and Technology Foundation of China under Grant No. 08ZQ026-016; the Sichuan Science and Technology Support Projects under Grant No. 2010GZ0123; the Innovative Application Projects of the Ministry of Public Security under Grant No. 2009YYCXSCST083; the Education Ministry Youth Fund of Humanities and Social Science of China under Grant No. 09YJCZH101.
[**] Scientific and Technological Major Special Projects-Significant Creation of New Drugs under Grant No.2009ZX09313-024.
[***] Corresponding author.

J. Yu et al. (Eds.): RSKT 2010, LNAI 6401, pp. 475–482, 2010.

1 Introduction

As the communication throughout the Internet becomes ubiquitous, the social (or communication) networks in the Web are becoming more and more complex as well as their structures become huge and change dynamically. Simultaneously, there exists a large amount of useful information in a web social network (WSN for short) which raises the urgent need to analyze complex network structures. Herein, an important and challenging problem is to score the importance of members and discover the essential players from complex networks.

PageRank is originally designed to compute the importance of web pages based on the link structure and is used in the search engine Google to score the search results. Particularly, it can also be used to determine the importance of social actors in a proper social network where links imply similar "recommendation" relationships as the hyperlinks in Web graph [1]. PageRank adapts the voting strategy as well as analyzes the content or motif of adjacent web pages to effectively rank pages. It is of great practical value to develop an effective (i.e., the accuracy of scoring) and efficient (i.e., the time performance) page scoring approach for WSNs that consist of a huge volume of web pages.

In this paper, we make the following contributions:

1. We propose an improved PageRank algorithm, called WebRank, which is a hybrid approach based on PageRank and three centrality measures in order to evaluate the importance of distinct pages.
2. We visualize web social networks in order to help understand the structure of the networks, and develop a web social network analysis system.
3. We conduct experiments to estimate the effectiveness and efficiency of the WebPank algorithm.

2 Related Work

Recently, there emerge several social or community networks in the Web, for example, Facebook, Flickr, TraveBlog are all famous web social networks that possess a large number of users. Users are more interested in web pages that are authoritative and influential, and resort to the search engine to find pages that rank in the top sequences. Google is the most popular commercial search engine by the Internet users. Its most important merit is the correctness of the query results due to the famous PageRank algorithm.

Zhang et al. [2] proposed an improved PageRank algorithm to accelerate the phase of scoring in order to distribute the authority in a fast manner. The key idea of this algorithm is to calculate the rank value of one URL published later by using time series based PageRank method. Simultaneously, due to the dated web pages, the efficiency of page scoring will fall down quickly. Another commonly used PageRank algorithm is proposed by Haveliwala [3], which is a topic-sensitive PageRank algorithm. This approach can handle the case that some pages may not be treated to be important in other fields though they obtain a high score by the traditional PageRank algorithm in some field.

Several centrality measures in relational analysis can be used to identify key members who play important roles in a network, that is, degree, betweenness, and closeness. These centrality measures are particularly appropriate to analyze the importance of members and can be applied to WSNs for scoring pages as well. Qiao et al. [4] proposed a centrality measure based approach and used it to identify the central players in criminal networks.

3   Problem Statement and Preliminaries

In this section, we will introduce the important concepts used by the proposed WebRank algorithm in Section 5. Then, we will formalize three important centrality measures in SNA.

Definition 1 (Web social network). *Let $\aleph$ be a web social network, $\aleph$ is defined as a graph $\aleph = (V, E)$, where V is a set of web pages and E is a set of directed edges (p, q, w), where $p, q \in V$, and w is a weight between p and q. The pages in $\aleph$ should satisfy that there is at least one ingoing (outgoing) link from one page to another, i.e., the nodes in $\aleph$ directly (indirectly) connected to others.*

Based on the concept of web social networks, the page scoring problem is defined as ranking web pages in WSNs by their importance.

Definition 2 (Geodesic). *Geodesic is the shortest path between two web pages and is computed by the shortest path algorithm [5].*

Geodesic is represented by Equation 1 and obtained by iteratively computing the elements in an adjacent matrix which indicates the length of shortest path from one page to another one.

$$L(x_i, x_j) = \min_{0 \le k < n} L(x_i, x_k) + L(x_k, x_j) \tag{1}$$

In this study, we use the following PageRank scoring formula to compute the rank value of each page.

$$PageRank(p) = 1 - e + e \times \sum_{all\_q\_link\_to\_p} \frac{PageRank(q)}{N(q)} \tag{2}$$

where $PageRank(\cdot)$ is the scoring function, p represents a web page, e is a dampening factor between 0 to 1 and is set to 0.85 which is proved to be a reasonable value by Qin in [1], q is any page linking to p, and $N(q)$ is the number of out-going links in q.

In this study, we use three popular centrality measures: degree, betweenness, and closeness, which is proposed by Freeman [6,5] as follows.

Degree measures how active a particular web page is. It is defined as the number of direct links a page k has:

$$C_D(k) = \sum_{i=1}^{n} a(i, k) \tag{3}$$

where n is the total number of pages in a network, and $a(i, k)$ is a binary variable indicating whether a link exists between pages i and k. A page with a high degree could be the "hub" of a WSN.

Betweenness measures the extent to which a particular page lies between other pages in a network. The betweenness of a node k is defined as the number of geodesics passing through it:

$$C_B(k) = \sum_i^n \sum_j^n g_{ij}(k), i \neq j \tag{4}$$

where $g_{ij}(k)$ indicates whether the shortest path between two other pages i and j passes through the page k. A page with high betweenness may act as a gatekeeper or "broker" of a WSN.

Closeness is the sum of the length of geodesics between a particular page k and all the other pages in a network. It actually measures how far away one page is from other pages.

$$C_C(k) = \sum_{i=1}^n l(i, k) \tag{5}$$

where $l(i, k)$ is the length of the shortest path connecting pages i and k.

4   System Description

In order to evaluate the effectiveness of WebRank algorithm, we design a web social network analysis system called WebExplorer as shown in Fig. 1.

Fig. 1 is a snapshot of the WebRank analysis module which consists of distinct parts. The user can specify the weigh vector in the text fields following the labels "PageRank", "Degree", "Betweenness", and "Closeness". When clicking the tab labeled "rank", the scored URLs are listed in the canvas labeled as "Graph". In addition, the detail of these four measures are given in the right

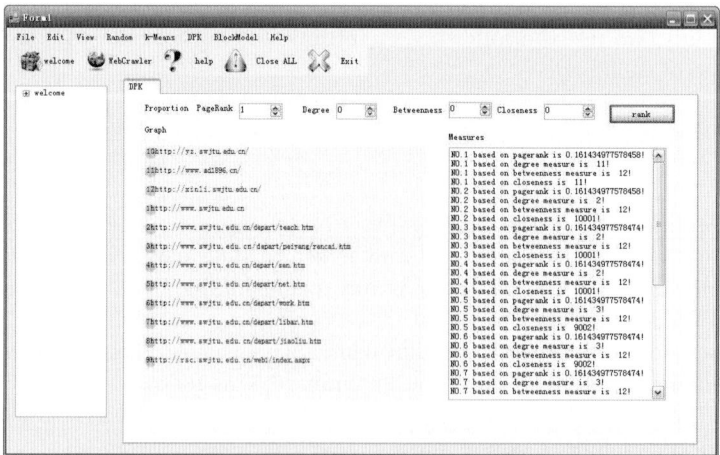

Fig. 1. Scenario of the WebExplorer system

text area labeled "Measures", which can help understand the sequence of URLs. Particularly, WebExplorer can be used to partition WSNs into several clusters.

5 WebRank Algorithm

In this section, we will introduce a hybrid page scoring algorithm and the detail of this approach is shown in Algorithm 1. This algorithm contains three essential phases: (a) use Equations 2~5 to compute the rank value, degree, betweenness, and closeness of a page, respectively (Steps 2-6); (b) standardize these four measures into the values between 0 and 1, respectively (Steps 7-11); (c) compute the scoring value by using the weighted average formula (Step 12).

Algorithm 1. WebRank: A Hybrid Page Scoring Approach Based on Social Network Analysis

---

Input: A webpage dataset $\mathbb{D}$, a weight vector $w = <w_1, w_2, w_3, w_4>$, where the elements in w represent the weights of rank value, degree, betweenness, and closeness, respectively.

Output: A scoring list of pages $score[n]$, where n is the number of pages.

1. n=GetNumber($\mathbb{D}$);
2. **for** (each page $p \in \mathbb{D}$) **do**
3. $\quad$ $rank[p]$ = PageRank(p);
4. $\quad$ $degree[p] = C_D(p)$;
5. $\quad$ $betweenness[p] = C_B(p))$;
6. $\quad$ $closeness[p] = C_C(p)$;
7. **for** ($i = 0$; $i < n$; i++) **do**
8. $\quad$ $rank[i]'$ = Standardize($rank[i]$);
9. $\quad$ $degree[i]'$ = Standardize($degree[i]$);
10. $\quad$ $betweenness[i]'$ = Standardize($betweenness[i]$);
11. $\quad$ $closeness[i]'$ = Standardize($closeness[i]$);
12. $\quad$ $score[i] = w_1 * rank[i]' + w_2 * degree[i]' + w_3 * betweenness[i]' + w_4 * closeness[i]'$;
13. $\quad$ **output** $score[i]$;

---

6 Experiments

We report the experimental studies by comparing WebRank with centrality measures based SNA algorithm (called CentralityRank), and PageRank.

6.1 Weight Vector Tuning

In this section, we perform four sets of experiments by using distinct proportion of the weights corresponding to the rank value, degree, betweenness and closeness. The results are given in Table 1.

Table 1. Page rank under distinct proportion of weights among four measures

Page Sequence No.	Proportion of four measures			
	1:1:1:1	1:0:0:0	0:3:2:1	3:3:2:1
1	1	4	1	4
2	5	5	2	5
3	6	6	3	6
4	7	7	4	7
5	8	8	5	8
6	9	9	6	9
7	10	10	7	10
8	11	11	8	11
9	12	12	9	12
10	2	1	10	1
11	3	2	11	2
12	4	3	12	3

In Table 1, the first column is the sequence number of each page, the numbers from the second to the fifth columns represent the page rank that is scored by the WebRank algorithm under distinct weight combinations. As we can see from the above table, the order of pages in the second, the fourth, and the fifth columns are nearly the same and consistent with the real-world situation. Whereas, the results in the third column deviate from the truth. We conclude that we cannot only use the page rank measure without taking into account the SNA measures. After a careful estimation by the experts from the web mining committee, the best proportion of these four measures is $(1 : 1 : 1 : 1)$. In the following section, we use a weight vector $< 1, 1, 1, 1 >$ to evaluate the performance among PageRank, CentralityRank and WebRank.

6.2   Accuracy and Efficiency Comparison of Page Ranking

In this section, we compare the scoring accuracy and execution time of WebRank with the PageRank and CentralityRank algorithms. Firstly, we observe the scoring accuracy as the number of pages grows from 10 to 500, and the results are given in Table 2.

By Table 2, we can see that: (1) the scoring accuracy of CentralityRank increases when the cardinality of pages is small (less than 50 pages), but when the number of pages grow larger than 50, the accuracy decreases. This is because the prediction accuracy of centrality measures will fall down when the WSN becomes very complex; (2) the scoring accuracy of PageRank increases with the number of pages. Because PageRank adapts the authority transition mechanism to score pages. If the authoritative pages increase, the authority will transfer to other pages as well as improve the scoring accuracy; (3) as for WebRank, the accuracy keeps at a similar stage, which demonstrates its stability. This is because WebRank combines the advantages of the above two methods, i.e., CentralityRank

Table 2. Accuracy of page scoring among three algorithms

Number of pages	CentralityRank	PageRank	WebRank
10	78%	62%	85%
20	80%	65%	84%
30	81%	70%	86%
50	82%	71%	86%
100	80%	72%	87%
200	78%	74%	88%
300	77%	75%	86%
400	76%	78%	87%
500	75%	78%	83%

is especially suitable for social network analysis and PageRank is a good page scoring strategy; (4) WebRank outperforms PageRank and CentralityRank in all cases with an average gap of 14.1% and 7.2%, respectively. Whereas, the performance of PageRank is worse than CentralityRank and WebRank. Because the PageRank algorithm is mainly used to compute the importance of pages in the Web, instead of a web social network.

We have to further compare the execution time of WebRank to the CentralityRank and PageRank algorithms. The experimental results are given in Fig. 2, where the x-axis represents the number of pages and the y-axis is the execution time corresponding to each algorithm.

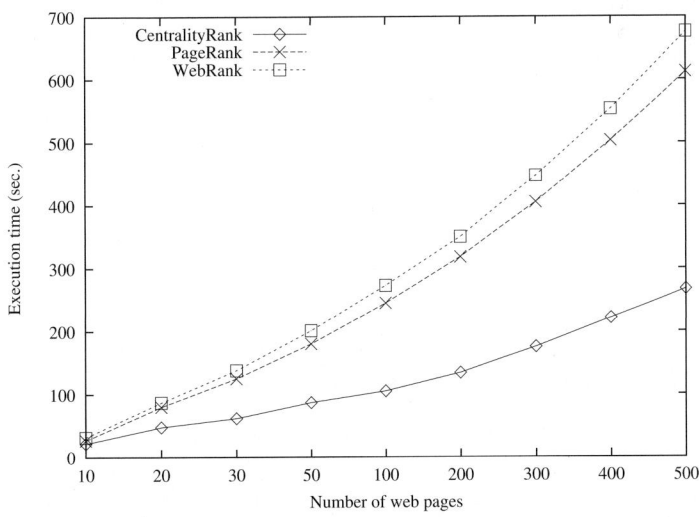

Fig. 2. Execution time comparison among three algorithms

As we can see from Fig. 2, WebRank has comparable time efficiency to Page Rank, while CentralityRank takes the most economical time cost. In addition, as the number of pages increases, the time gap between WebRank (or PageRank) and CentralityRank is big. This is because WebRank and PageRank need to recursively compute the score of each page, which is time-intensive. Particularly, we find that the time complexity of WebRank increases in an approximately linear fasion as the number of pages grows. This conclusion is consistent with the time complexity of Algorithm 1.

7   Conclusions and Future Work

In this paper, we provide a solution called WebRank to score pages in a web social network by applying the centrality measures (i.e., degree, betweenness, and closeness) to the PageRank algorithm. By empirical studies, we prove that WebRank is an effective page scoring approach with good time performance by comparing it with CentralityRank and PageRank algorithms. Furthermore, we develop an easy-to-use web social network analysis system, called WebExplorer, which can group and score web pages in an effective fashion.

In terms of future research directions, we will extend WebRank to find the leader or key members in a dark web (terrorism social network) by taking into account terrorists' intricate surfing behavior. Moreover, we will further improve the WebExplorer system by integrating other web mining algorithms.

References

1. Qin, J., Xu, J.J., Hu, D., Sageman, M., Chen, H.: Analyzing terrorist networks: a case study of the global salafi jihad network. In: ISI 2005: IEEE International Conference on Intelligence and Security Informatics, Atlanta, Georgia, pp. 287–304. IEEE, Los Alamitos (2005)
2. Zhang, L., Ma, F.: Accelerated ranking: a new method to improve web structure mining quality. Journal of Computer Research and Development 41(1), 98–103 (2004)
3. Haveliwala, T.H.: Topic-sensitive pagerank: a context-sensitive ranking algorithm for web search. IEEE Transactions on Knowledge and Data Engineering 15(4), 784–796 (2003)
4. Qiao, S., Tang, C., Peng, J., Liu, W., Wen, F., Jiangtao, Q.: Mining key members of crime networks based on personality trait simulation e-mail analyzing system. Chinese Journal of computers 31(10), 1795–1803 (2008)
5. Xu, J.J., Chen, H.: Crimenet explorer: a framework for criminal network knowledge discovery. ACM Transactions on Information Systems 23(2), 201–226 (2005)
6. Freeman, L.C.: Centrality in social networks: Conceptual clarification. Social Networks 1(10), 215–239 (1979)

Superficial Method for Extracting Social Network for Academics Using Web Snippets

Mahyuddin K.M. Nasution and Shahrul Azman Noah

Knowledge Technology Research Group, Faculty of Information Science & Technology
Universiti Kebangsaan Malaysia, Bangi 43600 UKM Selangor Malaysia
mahyunst@yahoo.com, samn@ftsm.ukm.my

Abstract. Social network analysis (SNA) has become one of the main
themes in the Semantic Web agenda. The use of web is steadily gaining
ground in the study of social networks. Few researchers have shown the
possibility of extracting social network from the Web via search engine.
However to get a rich and trusted social network from such an approach
proved to be difficult. In this paper we proposed an Information Retrieval
(IR) driven method for dealing with the heterogeneity of features in the
Web. We demontrate the possibility of exploiting features in Web snippets returned by search engines for disambiguating entities and building
relations among entities during the process of extracting social networks.
Our approach has shown the capacity to extract underlying strength relations which are beyond recognition using the standard co-occurrence
analysis employed by many research.

Keywords: Social networks analysis, Semantic Web, TFIDF, Jaccard
coefficient.

1 Introduction

Social networks is an approach for representing relations between individuals,
groups or organizations. In the semantic Web area, two research streams for
extracting social networks from various sources of informations exists i.e. the
superficial method and the supervised method. Methods proposed by Mika [1],
Matsuo et al. [2] and Mori [3] for instance belong to the superficial research
stream. In this case the method exploit search engines for obtaining documents
related to the participating entities, and subsequently calculate the co-occurrence
measures. Mika [1] developed Flink which extracted, aggregate and visualize social network of a Semantic Web community by utilizing the friend-of-a-friend
(FOAF) semantic profile. E-mail address available in the FOAF is used disambiguating among entities. Matsuo et al. [2] on the other hand created procedures
to expand superficial method by providing keywords in query submitted to a
search engine and to implement is to system called Polyphonet. Polyphonet is
developed to identify the relations in the Japan AI conference. Works done Mika
[1] and Matsuo et al. [2] mainly concern with extraction of network but do not
address the issue of naming the relations or giving clues to the nature of the

J. Yu et al. (Eds.): RSKT 2010, LNAI 6401, pp. 483–490, 2010.

relations. Mori et al. [3] , therefore, developed an approach that not only assigns strength of its relation, but also assigns underlying relation exists behind the link.

The supervised research stream involves corpus analysis for identifying the description of a given relations in a social network. In this case, each entity will be assigned with multiple labels extracted from e-mails, publications or Web pages related to that entity. Suitable labels will be assigned to relations by using the Information Retrieval technique, mainly the generative probabilistic model (GPM) [4]. The parameters of GPM are used as modalities to get the knowledge from the corpus. For example, the label of relation between entities based on participants in the same conference or workshop [5].

	1	2	3	4	5	6	7
1. Abdul Razak Hamdan		0.04425	0.00112	0.00221	0.00009	0.00011	0.00010
2. Abdullah Mohd Zin	0.04425		0.00043	0.00046	0.00001	0.00001	0.00005
3. Shahrul Azman Mohd Noah	0.00112	0.00043		0.07865	0.02978	0.00321	0.00000
4. Tengku Mohd Tengku Sembok	0.00221	0.00046	0.07865		0.00761	0.00204	0.00000
5. Opim Salim Sitompul	0.00009	0.00001	0.02978	0.00761		0.01379	0.00000
6. Erna Budhiarti Nababan	0.00011	0.00001	0.00321	0.00204	0.01379		0.00000
7. Michael D. Williams	0.00010	0.00005	0.00000	0.00000	0.00000	0.00000	

Fig. 1. The strength relations among entities based on Jaccard coefficient

Works which belong to the superficial research stream depend heavily on the co-occurrence measure which therefore may result in the omission of potential and important relations among entities. Our goal is to enhance research belong to the superficial research stream by extracting social network from Web document by not only relying on the co-occurrence but to consider other information such as keywords and URL analysis available in Web snippets provided by search engines. We also assign suitable description of relations of the participating entities.

2   Problem Definition

Assume a set of Web pages indexed by a search engines is defined as Ω, whereby its cardinalit is given as $|\Omega|$. Suppose that each term x represents *singleton event* $\mathbf{x} \subseteq \Omega$ on web pages stating occurrence of $x \in S$, S is a singleton term universal. We define $P : \Omega \to [0, 1]$ to be an uniform mass probability function. Probability of event $\mathbf{x}$ is $P(\mathbf{x}) = |\mathbf{x}|/|\Omega|$. Similarly, for "$x$ AND y", representing *doubleton event* $\mathbf{x} \cap \mathbf{y} \subseteq \Omega$ is a set of Web pages that contain occurrence of both term $x \in S$ and $y \in S$, where its probability is $P(\mathbf{x} \cap \mathbf{y}) = |\mathbf{x} \cap \mathbf{y}|/|\Omega|$, and the probability function through $\{\{x, y\}, x, y \in S\}$ as searching terms by a search engine based on probability of event. There is $|S|$ singleton terms, and 2-combination of $|S|$ doubleton consisting of terms which are not identical, $x \neq y$, $\{x, y\} \subseteq S$. Suppose $z \in \mathbf{x} \cap \mathbf{y}$, $\mathbf{x} = \mathbf{x} \cap \mathbf{x}$ and $\mathbf{y} = \mathbf{y} \cap \mathbf{y}$, then $z \in \mathbf{x} \cap \mathbf{x}$ and $z \in \mathbf{y} \cap \mathbf{y}$. Consequently, there is collection of web page with the cardinality, $|\psi| \geq \Omega$, containing bias for each search term. Therefore, we define a similarity measure between two terms, as Jaccard coefficient: $|\mathbf{x} \cap \mathbf{y}|/|\mathbf{x} \cup \mathbf{y}| = |\mathbf{x} \cap \mathbf{y}|/(|\mathbf{x}| + |\mathbf{y}|)$, $\mathbf{x}$ and $\mathbf{y}$ is a singleton event with respect to query "x" and "y", and $\mathbf{x} \cap \mathbf{y}$ is the doubleton the query of "x AND y". For example, the adjacency matrix shows on Fig. 1 represents a social network between some person extracted from Web by involving singleton event and doubleton event in superficial method. At the time of doing this experiment, a Yahoo! Search for "Abdul Razak Hamdan", returned 114000 hits. The number of hits for the search "Abdullah Mohd Zin" was 28600. Searching the pages where both "Abdul Razak Hamdan" and "Abdullah Mohd Zin" gave 17700 hits. Using these numbers in Jaccard coefficient, yield a strength relation of ≈ 0.04425 between "Abdullah Razak Hamdan" dan "Abdullah Mohd Zin". Moreover, this matrix of strength relation also shows bias in relation, due to the limitations of search engine which amt due ambiguity of results. For example, it is known that person 1 and 3 is the supervisor of person 6 and 5, respectively. Furthermore it is also known that person 7 is also the supervisor of person 3, and person 6 is the spouse of person 5. However, the strength relations between these known entity pairs are weaker than the remaining. The hit count provided by a search engine is the main feature used to form the social network. Other features of search engines hit list such as the selection and ranking web pages in accordance with a given query and the provision of brief summaries can be exploited to further reveal more relations among entities. For example, each lines of selected Web snippet S also contain URL address which can be used to provide underlying strength relations (USRs) as will be shown in this paper. The scope of this research is on academics. Therefore, the research also explores other features such as academic organizations and citations information available for example in DBLP and Springer.

3   The Proposed Aproach

Studies for extracting academic social networks can be found in [1], [2], [4], [5], [6], whereby the emphasis has been on identifying the co-occurrence among entities using co-authors information, co-members in a project and co-participation in confereces. Such co-occurrence information is not really sufficient for identifying the strength relation. Relations among academics in a social network require further information about entities (i.e., persons, organization, and citation). Such information is usually readily available from the web but is highly unstructured. The proposed approach in this paper makes uses of the information retrieval techniques for extracting the academics sosial network.

Our proposed approach involves a main two-stage of task. The first task is to extract cues for the purpose of disambiguating entities. The next task is to provide an USR from URL based on the layers structure. The next section discusses our proposed approach.

3.1   Name Entities Disambiguation

The approach extracts keywords that are useful for representing features of entities (i.e. person names) by analyzing lines of Web snippet S. We employ a windows of size 50 for analyzing the context words surrounding the ambiguous names. In this case, the 50 words to the left and right of the ambiguous name is used both for feature selection and context representation.

The words are ranked using a modified TFIDF (term frequency−inverse document frequency) which is a method widely used in many information retrieval research [3], [7], [8]. In our case, the method is meant for giving score of individual words within the text Web snippets. The intuition is that words which appeared frequently in certain Web snippets but rarely in the remaining of snippets are strongly associated with that entity. The modified TFIDF equation is as follows:

$$\text{TFIDF} = tf(w, da) \cdot idf(w) \tag{1}$$

where $idf(w) = \log(N/df(w))$; $tf(w, da) = \sum_{i=1}^{m} \frac{1}{n}$; w = word in S; n total number of words in a Web snippet; N = number of web snippets S; da = Web snippets containing the word w; and $fd(w)$ = the number of Web snippets containing the word w.

In this case of term frequency, $tf(w, da)$, is measured as the probability of occurrence of word w in the Web snippet, $P(w|da)$. The TFIDF can be considered as a bag of words model to represent the context surround an entity.

There are a few potential keywords for identifying characteristic of entities. However, irrelevant keywords also exist in the same list. To select relevant keywords among all candidates, we deleloped a method based on graph theory.

Consider the Web as a large corpus of relationship between words. We define the labeled undirected graph $G = (V, E)$ to describe the relations between words, where $V = \{v_1, v_2, \ldots, v_m\}$ is a set of unique words and m is a total number of unique words, and each edge $(v_i, v_j) \in E$ represents co-occurrence of words between v_i and v_j in the same Web page. In this case the co-occurrence is based on Jaccard coefficient similarity measure. Therefore, G for any given name is not always the connected graph because of the possible relationship between the words based on co-occurrence analysis do not meet the specified threshold α. We assume that the graph of words G denotes the relationships of initial co-occurrence between words for all entities with same name. As a result we need to extract graph G' which represents the strong co-occurrence of words based on the threshold α, based on the following definition.

Definition 1. *Assume a sub-graph G', $G' \subset G$, G' is a* micro-cluster, *i.e. maximal clique sub-graph of entity name where the node represents word that meet the highhest score in document and the weighted relations between words are above the threshold α.*

In constructing the graph G', we use the TFIDF scores. In this case, we obtain a set W of words whose TFIDF values above the threshold. To ensure that a collection of words refer to the same entity, and not the surname, we build a tree of words T by eliminating the weakest relations one by one so that the tree has no single cycle on the graph. A cycle is a sequence of two or more edges $(v_i, v_j), (v_j, v_k), \ldots, (v_{k+1}, v_j) \in E$ such that there is an optimal edge $(v_i, v_j) \in E$ connects both ends of sequence. In graph theory, a tree is an optimal representation of connected networks. Therefore, we assume that the words tree is an optimal representation of words for a single entity. After eliminations of weak relations among words and no single cycle exists in graph G, we expect to see more than one words-tree if there exist more than one entity for the same entity names. The definition of such trees is provided in the following.

Definition 2. *A tree T is an* optimal micro-cluster *for an entity if and onl if T is a sub-graph of graph G' where there are at least one word $v_i \in T$ have stronger relation with an intrusive word of entity.*

In social networks, the entities which have stronger relations with each other, socially located in the same community. This means that words which refer to the entities are also assumed to describe the communities which the entities belong to, which results in the following definitions.

Definition 3. *Assume a sub-graph G'', $G'' \subset G$, G'' is a* macro-cluster*, i.e. the maximal clique sub-graph of entities names queries in a community, where the node represents word that meet the highest score in document and the weighted relations between words are above the threshold α such that micro-clusters interconnected by the strong connectivity.*

Definition 4. *A set of word W is the* current context *of an entity if and only if W is optimal micro-cluster, and also a sub-graph of macro-cluster as the learning.*

3.2   Identifying Underlying Strength Relations (USRs)

The next stage in social network analysis is identifying strength relations among entities. Our proposed approach is to exploit URL addresses and its organization since URL address is always available in Web snippets. URL address also indicates the layered structure of a Web site which can be logically shown as a hierarchy. As such the URL of web pages which provides and indicator of its logical position in the hierarchy structure that can be considered as the underlying strength of the relationship. The hyperlink has the form: [www].$d_m \cdots .d_2.d_1/p_1/p_2/ \cdots /p_{n-1}$ and its layer is counted as n. URLs are separated by a slash into several parts, each part can be considered as a domain name, directory or file. For example, URL `http://www.ftsm.ukm.my/amz` has 2 layers. we have $d_1 = $ my, $d_2 = $ ukm, $d_3 = $ ftsm, and $p_1 = $ amz, i.e. DNS layer and amz layer. Therefore, for each URL address with n layers will be given n generated values: $1, ..., n$. Web site editors usually tend to put similar or related web pages as close as possible underlying relations among entities in the case that co-occurrence measures unable to provide such relations. Such a case is provided by the URL addresses `paper.ijcsns.org/07_book/200902/20090258.pdf` and `paper.ijcsns.org/07_book/200932/20090341.pdf` involving entity names "Shahrul Azman Mohd Noah" and "Tengku Mohd Tengku Sembok" respectively. Both URLs indicate that both entities located at the same site.

For any Web snippets produced as a results of the entity name query, there exists a set of k URL addresses. Therefore, there will be $\sum_{i=1}^{k} n_i$ URL addresses whereby n_i is the number of layers for each i. For these generated URLs, there is a possibility of redundant URLs. Let u is the number of same URL address. For each entity $\mathbf{a}$ we can derive a vector space $\mathbf{a} = [a_1, \ldots, a_K]$ where $a_j = un_i$. We, therefore, can measure the distance between the two entities based on the list of URL addresses containing in Web snippets.

Suppose dissimilarity d can be defined as $1 - sim()$, Vitanyi [11] has defined the normalized distance based on Kolmorov complexity involving singleton and doubleton, i.e. $N_d(x, y) = (|\mathbf{x} \cap \mathbf{y}| - \min(|\mathbf{x}|, |\mathbf{y}|))/(\max(|\mathbf{x}|, |\mathbf{y}|))$, where $\mathbf{x}$ and $\mathbf{y}$ are set of terms and $x \in \mathbf{x}$ and $y \in \mathbf{y}$. For $sim() = 0$, we have $1 = (|\mathbf{x} \cap \mathbf{y}| - \min(|\mathbf{x}|, |\mathbf{y}|))/(\max(|\mathbf{x}|, |\mathbf{y}|))$ and we know that the prevailing $|\mathbf{x}| + |\mathbf{y}| \geq |\mathbf{x} \cap \mathbf{y}|$. Therefore, $1 = 2(|\mathbf{x} \cap \mathbf{y}|)/(|\mathbf{x}| + |\mathbf{y}|)$. As a result, we have $sim() = 2(|\mathbf{x} \cap \mathbf{y}|)/(|\mathbf{x}| + |\mathbf{y}|) + 1$. We then normalize the $sim()$ measure by assigning $|\mathbf{x}| + |\mathbf{y}| = \log(|\mathbf{a}| + |\mathbf{b}|)$ and $2(|\mathbf{x} \cap \mathbf{y}|) = \log(2|\mathbf{ab}|)$, which results in the following equation:

$$sim(\mathbf{a}, \mathbf{b}) = \frac{\log(2|\mathbf{ab}|)}{\log(|\mathbf{a}| + |\mathbf{b}|)}), \tag{2}$$

where $|\mathbf{a}| = \sum_{k=1}^{K} a_k^2$; $|\mathbf{b}| = \sum_{l=1}^{L} b_l^2$; and $|\mathbf{ab}| = \sum_{p=1}^{P} ab_p$; $ab = u_a \cdot n \cdot u_b \cdot n$.

Suppose X and Y are similarity measure between pair of entities, and there exists $f : X \to Y$. For instance, X and Y represent the similarity measure using equation (2) and Jaccard coefficient respectively. We shall call f mapping if, and only if, for each Z subset of X, there is an image $f(Z)$ in Y. A real sequence $x = (x_k)_1^{\infty}$ is said to statistically converge to the number L if for each $\epsilon > 0$, $\lim_n 1/n|\{k \geq n : |x_k - L| \geq \epsilon\}| = 0$ and if the set $K = K(\epsilon) = \{k \in K : |x_k - L| \geq \epsilon\}$, where $x_k \in X$ and $k \in K$ be the set of positive integers [10]. The condition $\lim_n 1/n|\{k \geq n : |z_k - L| \geq \alpha\}| = 0$ hold on $z_k \in Z$, where $\alpha > 0$ and $z_k \in X$. Similarly, we define the statistical convergent for y and $f(Z)$ as follows.

Definition 5. *A function $g : X \to Y$ is the mapping relations that converge if and only if there is a measurement $|z_k - L| - |f(z_k) - L| = |z_k - f(z_k)|$ for each $z_k \in Z$ such that $g = |x_k - y_k|$, $y_k = f(z_k)$ for all $z_k, x_k \in X$, $y_k \in Y$.*

4   Experiment

Based from the aforementioned approaches, we build a social network according to the following steps: (a) collect the occurrence information and context of an entity via query to search engine; (b) collect co-occurrence informaton of an entity pair; (c) generate a current context model of each entity; (d) extract a social network composing of entity pairs using a standard method; (e) disambiguation occurrence information of an entity with the current context; (f) extract an USR based on URL structure similarity of each entity pair; (g) map each USR to the basic social network with labels of relation. Due to limited space it is beyond the scope of this paper to discuss the above steps in greater details.

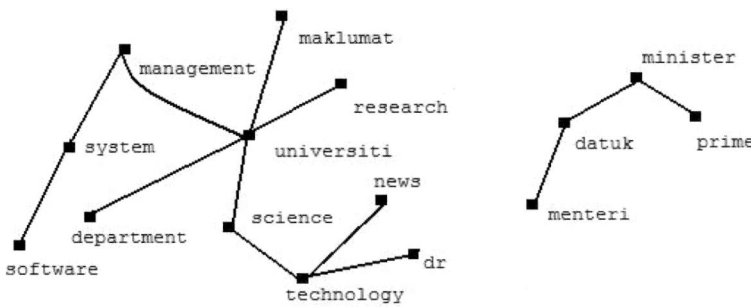

Fig. 2. Example of keywords obtained from a current context of an entity

Abdul Razak Hamdan		Abdullah Mohd Zin		Shahrul Azman Mohd Noah	
UKM	0.0223	UKM	0.0270	UKM	0.0882
Informatik	0.0117	Informatik	0.0044	DBLP	0.0540
Bibsonomy	0.0004	Eurojournals	0.0002	Informatik	0.0524
Eurojournals	0.0002			Bibsonomy	0.0029
				Springer	0.0597
Abdullah Mohd Zin	0.8362	T Mohd T Sembok	0.6758	Abdul R Hamdan	0.7006
T Mohd T Sembok	0.7001	Shahrul A M Noah	0.6894	T Mohd T Sembok	0.8300
Opim S Sitompul	0.5969	Opim S Sitompul	0.3970	Opim S Sitompul	0.7352
Erna B Nababan	0.3489	Erna B Nababan	0.1966	Erna B Nababan	0.4851
Michael D Williams	0.2945	Michael D Williams	0.3600	Michael D Williams	0.6084

Fig. 3. Example of USRs between entities and the possible labels

Our method requires a list of entity names as an input to form a social network. The output does not only contain relation among nodes but also the labels of relations. Using our proposed method, we extract the current context of each entity from Web snippets. We remove stop words, symbols, and highly frequent name words. For each entity, we download all or a maximum of 1,000 Web snippets. We only consider words where the TFIDF value $> \delta$, where $\delta = 0.3 \times highest\ value\ of$ TFIDF. Fig. 2 shows an example words-tree for the entity name "Abdullah Mohd Zin". It clearly shows two different dominant trees of which the left hand-side tree belongs to 'academic' and the other tree belongs to 'politic'. These cue words are then used to further build the social network involving the entity.

We test for 150 entity names and there are 11,110 potential relations, but at the time of doing this experiment there are only 5,778 (52%) relations which satisfy $\alpha = 0.0001$. However, our proposed method able to identify 6,549 (59%) relations and 8,103 (73%) for $\alpha = 0.001$ and $\alpha = 0.0001$, respectively. For example in Fig. 3 the relations between "Shahrul Azman Mohd Noah" and "Michael D. Williams" is identified as potential relations which was unresolved earlier. The label for each relations can also be derived from the weights of USRs as shown in Fig. 3. For example, the USR weight for "Abdullah Mohd Zin" and "Tengku

Mohd Tengku Sembok" ≈ 0.6758 and the possible label is UKM, Informatik or Eurojournals.

5   Conclusions and Future Work

The proposed method has the potential to be incorporated into existing extracting method of social network. It shows how to uncover underlying strength relations by exploiting Web snippets and URL structure. Our near future work is to further experiment the proposed method and look into the possibility of enhancing IR performance by using social networks.

References

1. Mika, P.: Semantic web technology for the extraction and analysis of social networks. Journal of Web Semantics 3, 211–223 (2005)
2. Matsuo, Y., Mori, Y., Hamasaki, M., Nishimura, T., Takeda, T., Hasida, K., Ishizuka, M.: POLYPHONET: An advanced social network extraction system from the Web. Journal of Web Semantics 5, 262–278 (2007)
3. Mori, J., Tsujishita, T., Matsuo, Y., Ishizuka, M.: Extracting relations in social networks from the Web using similarity between collective contexts. In: Cruz, I., Decker, S., Allemang, D., Preist, C., Schwabe, D., Mika, P., Uschold, M., Aroyo, L.M. (eds.) ISWC 2006. LNCS, vol. 4273, pp. 487–500. Springer, Heidelberg (2006)
4. McCallum, A., Wang, X., Corrada-Emmanuel, A.: Topic and role discovery in social networks with experiments on Enron and Academic Email. Journal of Artificial Intelligence Research 30, 249–272 (2007)
5. Tang, J., Zhang, D., Yao, L., Li, J., Zhang, L., Su, Z.: ArnetMiner: Extraction and mining of academic social networks. In: KDD 2008, Las Vegas, Nevada, USA, pp. 990–998 (2008)
6. Adamic, L.A., Adar, E.: Friends and neighbors on the Web. Social Network 25, 211–230 (2003)
7. Chen, C., Junfeng, H., Houfeng, W.: Clustering technique in multi-document personal name disambiguation. In: Proceedings of the ACL-IJNCLP 2009 Student Research Workshop, Suntex, Singaore, pp. 88–95 (2009)
8. Balog, K., Azzopardi, L., de Rijke, M.: Resolving person names in Web people search. In: King, I., Baeza-Yates, R. (eds.) Weaving services and people on the World Wide Web, pp. 301–323. Springer, Heidelberg (2009)
9. Vitanyi, P.: Universal similarity. In: Dinneen, M.J. (ed.) Proc. of IEEE ISOC ITW 2005 on Coding and Complexity, pp. 238–243 (2005)
10. Friday, J.A.: On statistical convergence. Analysis 5, 301–313 (1985)

Research of Spatio-temporal Similarity Measure on Network Constrained Trajectory Data

Ying Xia[1,2], Guo-Yin Wang[2], Xu Zhang[2],
Gyoung-Bae Kim[3], and Hae-Young Bae[4]

[1] School of Information Science and Technology,
Southwest Jiaotong University, China
xiaying@cqupt.edu.cn
[2] College of Computer Science and Technology,
Chongqing University of Posts and Telecommunications, China
wanggy@cqupt.edu.cn, zhangxu.jn@gmail.com
[3] Department of Computer Education, Seowon University, Korea
gbkim@seowon.ac.kr
[4] Department of Computer Science & Engineering, Inha University, Korea
hybae@inha.ac.kr

Abstract. Similarity measure between trajectories is considered as a pre-processing procedure of trajectory data mining. A lot of shaped-based and time-based methods on trajectory similarity measure have been proposed by researchers recently. However, these methods can not perform very well on constrained trajectories in road network because of the inappropriateness of Euclidean distance. In this paper, we study spatio-temporal similarity measure for trajectories in road network. We partition constrained trajectories on road network into segments by considering both the temporal and spatial properties firstly, then propose a spatio-temporal similarity measure method for trajectory similarity analysis. Experimental results exhibit the performance of the proposed methods and its availability used for trajectory clustering.

Keywords: constrained trajectory, road network, spatio-temporal similarity measure, moving objects.

1 Introduction

With the development of the wireless communication and positioning technologies, how to analyze large size of trajectory data generated by moving objects has been a great challenge. Generally, we assume that a moving object moves in the X-Y plane and the traversed path is a set of line segments in (x, y, t) space. We define these paths as its trajectory. According to the motion environment of the moving objects, there are two kinds of trajectories [1]: First is generated by free moving objects, which assume that objects can move freely without any restrictions. Another is the constrained trajectories moving on predefined spatial networks such as road segments.

J. Yu et al. (Eds.): RSKT 2010, LNAI 6401, pp. 491–498, 2010.

Trajectory similarity measure is originated from the similarity analysis of time series data and has attracted a lot of research works. By identifying similar trajectories, various data mining techniques like clustering can be applied to discover useful patterns. Trajectory similarity is useful for many applications in intelligent transport system, for example, traffic navigation which need analyzing of trajectory congestion, traffic prediction which need mining of historical trajectories pattern, etc. In this paper, we focus on measuring spatio-temporal similarity of trajectories constrained by road network effectively and efficiently.

The rest of the work is organized as follows: Section 2 discusses the related works and motivation. In Section 3, we propose to reconstruct and partition trajectories into sub-trajectories according to road network characteristics. A new spatio-temporal similarity measure are covered in Section 4. Section 5 presents the results of experimental evaluation. Finally, Section 6 concludes our study.

2   Related Work and Motivation

Recently, a lot of researches have been done on trajectory similarity measure, and several methods have been proposed such as Euclidean, DTW, LCSS, ERP and EDP [2-5]. However, these methods are inappropriate when dealing with road network constrained trajectories. First, the Euclidean distance is no longer applicable in road network space. Although, the Euclidean distance is short between two trajectories, the trajectories in road network may not applicable or there is a great cost for them to reach each other. Second, the early research did not fully explore the spatio-temporal characteristics between trajectories.

Hwang et al. [2] were the first to propose a similarity measure based on the spatio-temporal distance between two trajectories using the network distance. They proposed to search similar trajectories with two steps, filtering phase with the spatial similarity on the road network and refinement phase for discovering similar trajectories with temporal distance. Many researches [3-5] proposed to measure similar trajectories by adjusting weight between spatial and temporal constraint, however the performance need to be improved. Other researches [6, 9] focused on considering the spatial feature to get shape-based similar trajectories. However, ignoring the time features have a great weak point in discovering spatio-temporal evolution trends of trajectories and their hidden knowledge.

3   Reconstruction and Partition of Trajectories

Raw data acquired from positioning devices such as GPS are discrete position data of moving objects, which can be expressed by the point P(x, y, t). It is easy to acquire the spatio-temporal trajectories of moving objects by approaching linear or curve interpolation method on these sampling points. These line segments connect from end to end to form a polyline in (x, y, t) space and its projections in the X-Y plane are the routine of moving objects, which exactly overlap with road segments.

The raw trajectory covers the entire semantic and spatio-temporal character-istic of moving objects, however, this makes the storage and query processing difficult. The purpose of trajectory reconstruction and partition is to get rid of useless position data, and possess representative location data as less as possible to reduce the storage usage. On the other hand, more *characteristic points* [6] of trajectories should be preserved to improve the accuracy of similarity mea-surement. Therefore, we are aiming to find the points where the behavior of a trajectory changes rapidly to fully explore the spatio-temporal properties.

Unlike the free moving trajectories, partition of road network constrained trajectories suffers more from road network rather than the speed and direc-tion changes of moving objects. It is obvious that the road network constrained trajectories always have an overlap with the linear road segments in spatial, on which the linear interpolation method is easy to be performed. We propose to record location data at the crossing and get rid of other raw sampling data with an observation that the crossing always has a trajectory joining or distributing. As a result this, we propose to choose the crossings as the *characteristic points* to partition the whole trajectory into sub-trajectories. Obviously, our partitioned sub-trajectories have a exact overlap with road segments.

4   Spatio-temporal Similarity Measure Method

With the observation of different metric of spatial and temporal similarity, we aim to normalize them to [0-1], where 0 represents two trajectories are irrelevant and 1 means they have the most similarity. We notice the effective use of *Jaccard coefficient* in binary variable [7] when dealing with their similarity. Our proposed method is mainly from this idea and we describe it in this section.

We first define 2 trajectories with their sub-trajectories as following:

$TR_1 = (tr_1, tr_5, tr_7, tr_9)$, $TR_2 = (tr_1, tr_2, tr_5, tr_8, tr_9)$

According to the knowledge from Jaccard coefficient, we define the similarity between trajectories as the ratio of the common part to the summation of the common and uncommon parts. We could calculate the similarity as:

$$Sim\left(TR_1, TR_2\right) = \frac{L\left(tr_1\right) + L\left(tr_5\right) + L\left(tr_9\right)}{L\left(tr_1\right) + L\left(tr_5\right) + L\left(tr_7\right) + L\left(tr_9\right) + L\left(tr_2\right) + L\left(tr_8\right)} \quad (1)$$

Then, we have our spatial and temporal similarity measure definition:

Definition 1. The spatial and temporal similarity measure of trajectories can be calculated in one form as the ratio of the common part to the summation of the common and uncommon parts.

$$Sim\left(TR_i, TR_j\right) = \frac{L_c\left(TR_i, TR_j\right)}{L\left(TR_i\right) + L\left(TR_j\right) - L_c\left(TR_i, TR_j\right)} \quad (2)$$

Where, $L_c(TR_i, TR_j)$ means the total length of the common part between TR_i and TR_j. $L(TR_i)$ indicate the total length of trajectory with $SL(TR_i)$ in spatial and $TL(TR_i)$ in temporal.

4.1   Spatial and Temporal Similarity

The spatial features of trajectories mainly represent the location and relationship between trajectories and it plays an important role in knowledge discovery. According to definition 1, our spatial similarity (SSim) is mainly affect by the common part of trajectories. Obviously, when it covers the whole length of the trajectory, then SSim = 1, which represent that they have the most similarity in spatial. On the contrary, if there is not a common part between TR_i and TR_j, SSim = 0, we assume that they are irrelevant.

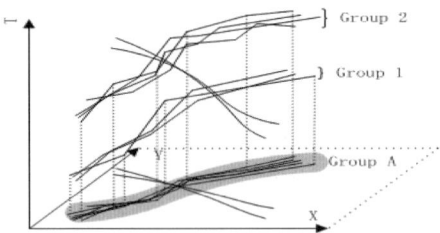

Fig. 1. Temporal feature of trajectories

The temporal features of a trajectory represent the moving trend as the time varies. Most of the trajectories on road network are heavy overlapping with each other from the spatial viewpoint. We proposed to find similar trajectories have both high spatial similarity and temporal similarity, which can be defined as spatio-temporal neighbors. As it shown in Fig. 1, trajectories in Group A possess spatial similarity. However, we can refine Group A into Group 1 and Group 2 according to the temporal similarity. Obviously, trajectories in Group 1 are spatio-temporal similar, while irrelevant with trajectories in Group 2.

It is easy to find that our temporal similarity (TSim) value also ranges from 0 to 1. Obviously, when the common part covers the whole length of the trajectory, then TSim = 1, which represent that they have the most similarity in temporal. On the contrary, if there is not a time overlap between TR_i and TR_j, TSim = 0, we assume that they are irrelevant or dissimilar.

4.2   Spatio-temporal Similarity

Other researchers in [10, 11] propose that both the spatial and temporal features of trajectories should be analyzed, but due to the separate calculation, it split the spatio-temporal correlation of trajectories and mix single-feature similarity together with spatio-temporal similarity. The research in [10] proposed to use the coefficient method in Formula (2).

$$D_{total}\left(TR_i, TR_j\right) = W_{net} \times D_{net}\left(TR_i, TR_j\right) + W_{time} \times D_{time}\left(TR_i, TR_j\right) \quad (3)$$

Where, $D_{net}(TR_i, TR_j)$ is the road network costs, $D_{time}\left(TR_i, TR_j\right)$ is the time costs for two trajectories to reach each other. W_{net} and W_{time} correspond to weight of spatial and temporal features. The great shortcoming of this method is that it is a must to predefine the coefficient of spatial and temporal features.

In this paper, we aim to find the trajectories have the most spatio-temporal similarity, the spatial and temporal features have the same impact to trajectories, as a result of this, the coefficient of them should be considered to be equivalent.

We analyze the above formula with mathematic knowledge. For the function $f(x,y) = x + y$ and $g(x,y) = xy$, when x and y are in [0-1], $g(x,y) <= f(x,y)$ .So, the similarity measure method which satisfy $g(x,y)$ must satisfy $f(x,y)$. On the other hand, the character of $g(x,y)$ is that when one of the value approaches zero, the other can not influence the final value, and it meet the definition of spatio-temporal proximity of trajectory. It minimizes the impact of any single-feature. Only when the spatial and temporal similarity both gets maximum value, trajectories are regarded as spatio-temporal neighbors. We define our spatio-temporal similarity as follow:

Definition 2. (Spatio-Temporal Similarity Measure) Similarity measure between two trajectories is the value measure of similarity degree of these two objects, which can be calculated as the product of spatial similarity and temporal similarity.

$$STSim\,(TR_i, TR_j) = SSim\,(TR_i, TR_j) \times TSim\,(TR_i, TR_j) \qquad (4)$$

The value of spatio-temporal similarity calculated with Formula (3) make the result still between [0-1], where 0 indicates that there is no spatio-temporal relationship between two trajectories, 1 indicates that two trajectories are completely overlap, which means they are spatio-temporal closing to each other. Our proposed method is based on two observations. First, trajectories are restricted by both spatial and temporal features, the influence of them are equal to each other. Second, any single-feature similarity could not be considered as spatio-temporal similarity. For example, when the spatial similarity of two trajectories is 0, the two trajectories are far from each other in spatial, even if their possess the max value in temporal similarity, the spatio-temporal similarity measure is still 0 to indicate that they are far away from each other in spatio-temporal, vice versa. As it shown in Fig. 1, this method can distinguish Group 1 and Group 2 very well, trajectories in each group are spatio-temporal similar to each other, while trajectories between groups have a spatio-temporal dissimilarity.

According to definition 1, we do not need to measure all the spatial and temporal similarity of the trajectories. For example, we first calculate spatial similarity of all trajectories, when dealing with temporal similarity, it is not necessary to calculate its temporal similarity if its spatial similarity value is 0.

Dissimilarity is used to indicate the degree of measure value differences between two objects, the more similar the objects, the lower in their dissimilarity. Generally, we regard the term of distance as the synonyms of dissimilarity [12]. In reality, similarity and dissimilarity are often interchangeable. In this paper, we proposed to converse spatio-temporal similarity measure to its dissimilarity to indicate the spatio-temporal distance between trajectories.

Definition 3. (Spatio-temporal Distance) Spatio-temporal distance between two trajectories can be expressed by its spatio-temporal dissimilarity:

$$STDist\,(TR_i, TR_j) = 1 - STSim\,(TR_i, TR_j) \qquad (5)$$

When STDist = 0, it means trajectories possess a full overlap in both spatial and temporal, they are overlap in spatio-temporal. When STDist = 1, it means trajectories are far away from each other. Our purpose of retrieving similarity trajectory is to find spatio-temporal neighbors, the real distance value or cost of reaching is not necessary for the calculation in the proposed method.

5   Experiments and Evaluation

We conduct experiments on a Intel(R) Pentium(R) Dual CPU T2330 @1.80GHz PC with 3GBytes of main memory, running on Windows XP Professional. We implement our algorithm and simulation with Java 1.6 and Matlab 7.0. We generate trajectory data with Road Network Data Generator developed by Brinkhoff [13] on the map of Oldenburg, and conduct three experiments on it to evaluate our proposed methods.

Experiment I. Memory use of the trajectories

We test the memory use with 5 sets consist of 100, 1000, 2000, 5000 and 10000 trajectories. As it shown in Table 1, reconstructed and partitioned trajectories have sharply decreased the memory use. Because of most of the raw location points on road segments have been erased, we just possess the crossing points as characteristic points and partition the trajectories into sub-trajectories.

Table 1. Memory use of trajectories

	100	1K	2K	5K	10K
Raw Trajectory	71KB	759KB	1529KB	3941KB	8126KB
Partitioned Trajectory	40KB	395KB	1102KB	1892KB	4431KB

Experiment II. Time cost of querying spatio-temporal similar trajectories

We compare our spatio-temporal similarity measure method with the algorithms MA and AA proposed in [5]. The performance result is shown in Fig. 2, our proposed method is shown to be more efficient. This is because the MA and AA algorithm searching temporal similarity with different time period as hours, days and weeks. There is a weight selection and calculation time cost before temporal similarity measure. Furthermore, a good spatio-temporal similarity measuring result is needed through adjusting the parameters because of the different time and spatio-temporal measuring units. However, our proposed method can significant reduce the calculation time cost because we only calculate the similarity where its spatial similarity is not 0. The result shows our proposed method is adaptive and effective.

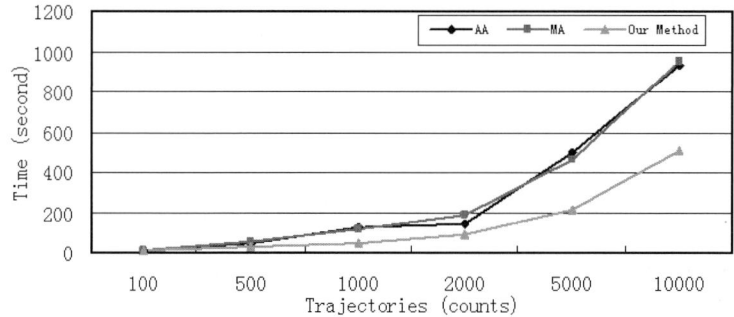

Fig. 2. Trajectory clustering result

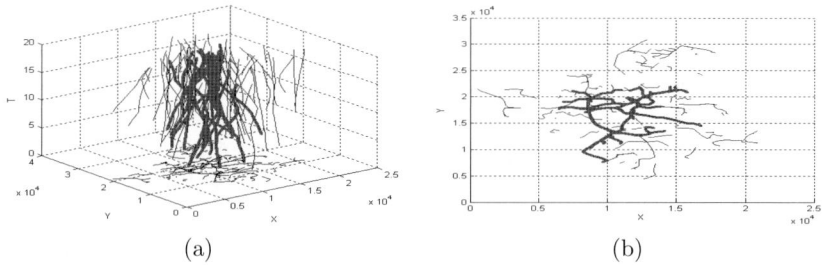

(a)   (b)

Fig. 3. Trajectory clustering result

Experiment III. Application of spatio-temporal similarity measure

We proposed to use a cluster method to find spatio-temporally patterns of trajectories and predict the moving trends of moving objects. We conduct this experiment to prove that our spatio-temporal similarity measure is useful and effective in data mining area as a pre-processing procedure. First, we generate 100 trajectories on Oldenburg, and calculate the spatio-temporal distance between trajectories with our proposed method. Then, we use the spatio-temporal distance as an Eps parameter to cluster trajectories. As it shown in Fig. 3, we can easily distinguish two spatio-temporally closing trajectory groups, which have obvious similar motion patterns, from others. All of the correspondent routes of the trajectories are projected on X-Y plane in (a) and (b).

6   Experiments and Evaluation

In this paper, we proposed to reconstruct trajectories and partition them into sub-trajectories with characteristics points, and then proposed a new spatio-temporal similarity measure. Theoretical analysis and experimental evaluation show that our method is reasonable, effective and efficient. It reduces the memory use effectively, and efficiently improves the time cost of retrieving spatial-temporal similar trajectories. The definition of spatial-temporal distance helps us to apply our similarity method for trajectory clustering and knowledge discovery. The future research should be paid more attention to extracting of spatial-temporal features for trajectory partition and reduce the time cost of similarity measure.

Acknowledgments. This paper is supported by National Natural Science Foundation of China (No. 60773113) and the Open Foundation Project of Chongqing Key Lab of Computer Network and Communication Technology, China (No. CY-CNCL-2009-01).

References

1. Won, J.-I., Kim, S.-W., Baek, J.-H., Lee, J.: Trajectory clustering in road network environment. Computational Intelligence and Data Mining, 299–305 (2009)
2. Meng, X.-F., Ding, Z.-M.: Management of Mobility Data: Concepts and Technology, p. 150. Tsinghua (2009)
3. Hwang, J.-R., Kang, H.-Y., Li, K.-J.: Spatio-temporal Analysis Between Trajectories on Road Networks, pp. 280–289 (2005)
4. Tiakas, E., Papadopoulos, A.N., Nanopoulos, A., Manolopoulos, Y.: Trajectory Similarity Search in Spatial Networks, pp. 185–192. IDEAS (2006)
5. Chang, J.-W., Bista, R., Kim, Y.-C., Kim, Y.-K.: Spatio-temporal Similarity Measure Algorithm for Moving Objects on Spatial Networks. In: Gervasi, O., Gavrilova, M.L. (eds.) ICCSA 2007, Part III. LNCS, vol. 4707, pp. 1165–1178. Springer, Heidelberg (2007)
6. Lee, J.-G., Han, J., Whang, K.-Y.: Trajectory Clustering: A Partition-and-Group Framework. In: Proceedings of the 2007 ACM SIGMOD international conference on management of data, pp. 593–604 (2007)
7. Han, J., Kamber, M.: Data mining: Concepts and Technology, 2nd edn (2007)
8. Yanagisawa, Y., Akahani, J., Satoch, T.: Shape-Based Similarity Query for Trajectory of Mobile Objects. In: Chen, M.-S., Chrysanthis, P.K., Sloman, M., Zaslavsky, A. (eds.) MDM 2003. LNCS, vol. 2574, pp. 63–77. Springer, Heidelberg (2003)
9. Vlachos, M., Kollios, G., Gunopulos, D.: Discovering similar multidimensional trajectories. In: Proceedings of the 18th International Conference on Data Engineering (ICDE 2002), pp. 673–684. IEEE, San Jose (2002)
10. Tiakas, E., Papadopoulos, A.N., Nanopoulos, A., Manolopoulos, Y., Stojanovic, D., Djordjevic-Kajan, S.: Searching for Similar Trajectories in Spatial Networks. Journal of Systems and Software, 772–788 (2009)
11. Hwang, J.-R., Kang, H.-Y., Li, K.-J.: Searching for Similar Trajectories on Road Networks Using Spatio-temporal Similarity. In: Manolopoulos, Y., Pokorný, J., Sellis, T.K. (eds.) ADBIS 2006. LNCS, vol. 4152, pp. 282–295. Springer, Heidelberg (2006)
12. Tan, P.-N., Steinbach, M.: Introduction to Data Mining. Addison Wesley, Reading (2006)
13. Brinkhoff, T.: A Framework for Generating Network-Based Moving Objects. GeoInformatica 6(2), 153–180 (2002)

Dampster-Shafer Evidence Theory Based Multi-Characteristics Fusion for Clustering Evaluation

Shihong Yue[1], Teresa Wu[2], Yamin Wang[1], Kai Zhang[1], and Weixia Liu[1]

[1] School of Electric Engineering and Automation, Tianjin University, Tianjin 300072, China
[2] School of Computing, Informatics, Decision Systems Engineering, Arizona State University Tempe, AZ, 85287, USA
shyue1999@tju.edu.cn

Abstract. Clustering is a widely used unsupervised learning method to group data with similar characteristics. The performance of the clustering method can be in general evaluated through some validity indices. However, most validity indices are designed for the specific algorithms along with specific structure of data space. Moreover, these indices consist of a few within- and between- clustering distance functions. The applicability of these indices heavily relies on the correctness of combining these functions. In this research, we first summarize three common characteristics of any clustering evaluation: (1) the clustering outcome can be evaluated by a group of validity indices if some efficient validity indices are available, (2) the clustering outcome can be measured by an independent intra-cluster distance function and (3) the clustering outcome can be measured by the neighborhood based functions. Considering the complementary and unstable natures among the clustering evaluation, we then apply Dampster-Shafter (D-S) Evidence Theory to fuse the three characteristics to generate a new index, termed fused Multiple Characteristic Indices (fMCI). The fMCI generally is capable to evaluate clustering outcomes of arbitrary clustering methods associated with more complex structures of data space. We conduct a number of experiments to demonstrate that the fMCI is applicable to evaluate different clustering algorithms on different datasets and the fMCI can achieve more accurate and robust clustering evaluation comparing to existing indices.

Keywords: Validity index, data structure, clustering algorithm, Dampster-Shafer evidence theory.

1 Introduction

Clustering analysis is an unsupervised technique used to assign the objects with similar characteristics to the same group [1], [2]. One critical issue in clustering analysis is to validate the clustering results quantitatively, e.g., determining the optimal number of clusters and their optimal clustering configurations. The performance of the clustering method can be in general evaluated through various

J. Yu et al. (Eds.): RSKT 2010, LNAI 6401, pp. 499–519, 2010.

validity indices. The most commonly accepted clustering partition is to maximize the intra-cluster distance and to minimize inter-cluster distance [3],[4],[5]. Consequently, most validity indices consist of several specified intra- and inter-cluster distance functions. The applicability of these indices heavily relies on the correctness of combining these functions. To seek more rational combination of several intra- and inter- cluster distance functions has ever increasing needs in the past decades. For example, some validity indices, such as the statistic [6], the Davies-Bouldin's DB measure [7], and the Xie-Beni's separation measure [8] have greatly been reformulated to the normalized Γ statistic [9], the Kim's DB** index [10], and the Pakhira and Bandyopadhyay's PB index [11], respectively. Consequently, the structure of the existing validity indices can greatly affect the efficiency of any validity index. Although extensive research has explored various new validity indices, existing indices are highly dependent on the characteristics of the clustering algorithms associated with specific structure. For example, Saha et al proposed the symmetry distance-based validity index based on a point symmetry distance [12], sym-index, which can evaluate the clustering outcomes of dataset of symmetric clusters. Tantrum et al proposed model-based Fractionation and Refractionation method [13] which mainly is used to identify the clusters of Gaussian distribution and validate the corresponding clustering results. More general, most clustering validity indices originate from partitional clustering algorithms: C-Means (CM) [14] and fuzzy C-Means (FCM) [15] such as Huberts Γ statistic [6], Dunn's separation measure [16], Davies-Bouldin's measure [7], Tibshirani's Gap statistic measure [17], and etc. These indices mainly are effective to evaluate the clustering outcomes of spherical clusters and may not be applicable to evaluate hierarchical clustering algorithms such as CLIQUE [18], DBSCAN [19], SHIFT [20]. One can find those problems that most clustering algorithms are designed to cluster specific structure of data space. In fact, the validity indices also involve these problems. Consequently, the existing validity indices are dependent to specific algorithms along with specific datasets. These methods suffer inconsistency and instability for different structures of data space. Literature also indicates that existing indices are constrained to handle the dataset with large variability in geometric shapes and densities [21],[22],[23]. So far these problems are solved with little satisfactory.

Instead of improving single index, another research focus is to integrate multiple indices for better performance. Bezdek [2] attempted to use voting mechanism to determine the appropriate number of clusters collectively from different validity indices. Literature claimed that fusing multiple classifiers can obtain higher accuracy and better accepted classification of the data than each classifier in itself [24],[25]. Fred and Jain [26] proposed a heuristic framework to combine several clustering algorithms with validity evaluations. Sheng et al [27] developed a weighted sum function (WSVF) on a group of validity indices for clustering evaluation. Their experimental results indicate the success of the WSVF to achieve robust and accurate clustering outcomes in some specific datasets, but the challenge remains how to determine the appropriate weights for each index. Although these methods are instructive to reformulate the efficiency of clustering evaluation, there are none general and feasible way to solve clustering evaluation of arbitrary algorithms and arbitrary structures. In addition, these methods cannot solve the problem on how to combine a group of intra- and inter- cluster distance functions for better performances [23],[28].

In this research, we first investigate the inherit problems of the clustering evaluations of the existing validity indices, termed uncertainty, instability and incompleteness, and conclude that the consecutive numbers of clusters should be dependent and associated with each other. Then we study the use of the D-S evidence theory [29],[30] to fuse multiple complementary characteristics in clustering evaluation. We introduce three measures as the information resources in the D-S evidence theory tool: the weighted average of a group of existing validity indices; the new two-order difference of an independent intra-cluster distance function; and the neighborhood function which approximate the center and boundary of all clusters. The fused validity index, termed fused Multiple Characteristic Index (fMCI) is generalized index that not only improves the confidence of clustering solutions but also is applicable to evaluate various clustering outcomes. To what follows, the overview of the clustering algorithms and their validity measures are presented in Section 2. Section 3 includes the three major components - basic characteristics of cluster evaluation, characteristic extraction of clustering evaluation, and fusion of these characteristic for an estimation of the number of clusters. Computer simulations on the identification of different data configurations for several examples are verified in Section 4. Section 5 is the conclusions.

2   Related Work

In this section, we first review four commonly used validity indices followed by a combined index, termed WSVF. The D-S evidence theory is then explained.

2.1   Four Representative Validity Indices and the WSVF Method

Let $X = x_1, x_2, \ldots, x_n$ be a dataset in a d-dimensional data space, containing n data points distributed in c clusters, $C_1, C_2, \ldots, C_c$ Hereafter, we denote $NC_{optimal}$ as the optimal number of clusters by any validity index.

Tibshirani's Gap Statistic (GS) Index. The GS index [17] is based on a pooled within- cluster sum of squares around the cluster mean $\bar{X}$ such that

$$W_c = \sum_{i=1}^{c} D_i/2|C_i|) s.t. D_i = \sum_{l \in C_i} \sum_{s \in C_i} ||x_s - x_l||^2 = 2|C_i| \sum_{j \in C_i} ||x_j - \bar{x}|| \quad (1)$$

where $\bar{X} = \sum_{i=1}^{|C|} x_i/|C_i|$. Originally, the optimal number of clusters is encountered in the elbow point where W_c is a faster varying ratio as the value of c

varies. To represent the "elbow" point, the GS index investigates the relationship between the $log(W_c)$ for different values of c and the expectation of $log(W_c)$ for a suitable null reference distribution by the gap statistics:

$$Gap_c = E^*[log(W_c)] - log(W_c) \tag{2}$$

Here, E^* denotes the expectation under a null reference distribution, and the expected value is estimated by drawing B samples from the null distribution. For $i = 1$ to c, the value of $log(W_c)$ needs to be estimated after the standard deviations sd_c are obtained, where $sd_c = [(1/B)\sum_b log(W_{cb} - \bar{l})^2]^{1/2}$, $\bar{l} = (1/B)\sum_b logW_{cb}$ and $S_c = sd_c(1 + 1/B)^{1/2}$. Finally, the optimal number of clusters c^* is determined by

$$c^* = \mathbf{min}c s.t., Gap(c)(c+1) - S_{c+1} \tag{3}$$

Davies-Bouldin (DB) Index. The DB index [7] is an equation based on both intra-cluster distance and inter-cluster distance functions. Let $\Delta_{i,q}$ be the i-th intra-cluster distance function, both t and q are user-specified exponents. $\Delta_{i,q} = \sum_{x_i} ||x - z_i||^q/|C_i|^{1/q}$, where, z_i is the i-th cluster center; $\delta_{ij,t}$ is the inter-cluster distance function of clusters C_i and C_j, $\delta_{ij,t} = ||z_i - z_j|| = \{\sum_{s=1}^{d} |z_{is} - z_{js}|^t\}^{1/t}$. The DB index is then defined as

$$DB = \sum_{i=1}^{c} R_{i,qt}/c \tag{4}$$

where $R_{i,qt}/c = \mathbf{max}_{j,j\neq i}\{(\Delta_{i,q} + \Delta_{j,q})/\delta_{ij,t}\}$. The optimal number of clusters can be obtained by minimizing Eq.(4) for all possible numbers of clusters.

Saha and Bandyopadhyay's (SB) Index. Consider a partition of the data set X into c clusters. The center of each cluster C_i is computed by using $c_i = \sum_{j=1}^{n_i} x_{ij}/n_i$, where $n_i(i = 1, 2, \ldots, c)$ is the number of points in cluster i, and x_{ij} is the j th point of the ith cluster. The SB's validity function $sym[12]$ is defined as

$$Sym(c) = D_c/(c\xi_c) \tag{5}$$

where $\xi_c = \sum_{i=1}^{c} E_i$, such that $E_i = \sum_{j=1}^{n} d_{ps}(x_j, c_i)$, and $D_c = \mathbf{max}_{i,j=1}^{n}||c_i - c_j||$. D_c is the maximum Euclidean distance between two cluster centers among all centers. $d_{ps}(x_j, c_i)$ is the so-called PS distance associated with point x_j with respect to a center c_i . Let the symmetrical (reflected) point of x_j with respect to c_i is $2 \times c_i - x_j$ and denote this by x_j^*. Let $knear$ be unique nearest neighbors of x_j^* at Euclidean distances of d_k , $k = 1, 2, \cdots, knear$. Then

$$d_{ps}(x_j, c_i) = (\sum_{k=1}^{n} d_i/knear) \times d(x_j, c_i) \tag{6}$$

where $d(x_j, c_i)$ is the Euclidean distance between the point x_j and c_i. The objective is to maximize this index in order to obtain the optimal number of clusters.

Kim's DB** Index:Kim et al. proposed a series of validity indices [10] for both crisp and fuzzy clustering. The DB^{**} index is characterized as

$$DB^{**}(c) = \frac{1}{c} \sum_{i=1}^{c} \left(\frac{max_{j=1,c,\ldots,c,j\neq i}\{S_j + S_i\} + \mathbf{max}Diff_i(c)}{min_{l=1,\ldots,c,l\neq i}} \right) \qquad (7)$$

where $\mathbf{max}Diff_i(c) = \mathbf{max}_{c_{max},\ldots,c}diff_i(c)$, $c = 1, 2, \cdots, c_{max}$, c_{max} is the maximal number of clusters, $diff_i(c) = \mathbf{max}_{j=1,\ldots,c,j\neq i}\{S_i(c) + S_j(c)\} - max_{j=1,\ldots,c+1,j\neq i}\{S_i(c+1) + S_j(c+1)\}$, $S_i(c) = \sum_{x\in C_i} d(x, c_i)/|C_i|$. The optimal number of clusters can be obtained by finding the number of clusters for minimizing Eq. (7).

Sheng's WSVF Index: Sheng et al. proposed an objective function called the Weighted Sum Validity Function (WSVF) [27], which is a weighted sum of several cluster validity functions

$$\mathbf{max}(WSVF) = \sum_{i=1}^{m} w_i f_i(x) \qquad (8)$$

where m is the number of component functions, w_i is the non-negative weighting coefficients representing the relative importance of f_i such that $\sum_{i=1}^{m} w_i = 1$, and f_i are component functions corresponding to a group of validity indices, for $i = 1, 2, \ldots, m$. Since there is no priori information about the relative importance of any individual function, WSVF method initializes the weighting coefficients as $w_1 = w_2 = \ldots = w_m = 1/m$. A hybrid niching genetic algorithm is then applied to search for the optimal setting until the proper number of clusters as well as appropriate partitioning of the data set is reached. Inspired by the idea of combining indices, we explore the use of the D-S evidence theory to fuse various characteristics in the process of evaluating cluster outcomes which is explained in the following section.

2.2   The D-S Evidence Theory

The D-S evidence theory is a mathematical tool for handling uncertain, imprecise and incomplete information [29], [30], [31]. First, by representing the uncertainty and the imprecision of a body of knowledge via the notion of evidence, belief can be committed to single hypothesis (singleton) or a composite hypothesis (union of hypotheses). A useful operator following the evidence combination rule is introduced to integrate information from different sources. Finally, the decision on the optimal hypotheses choice can be made in a flexible and rational manner. Let us assume (i) a frame of discernment X consisting of the exhaustive and exclusive hypothesis and (ii) the reference set 2^X of all the subsets of the elements of X. In the D-S evidence theory, a basic probability assignment (BPA) is an elementary mass function: $m : 2^X \rightarrow [0, 1]$ satisfying: $m(\phi) = 1$ and $\sum_{A\subseteq 2^X} m(A) = 1$. The element of 2^A that has a non-zero mass value are called focal element, and the union of all the focal elements is called the core of the mass function. In the evidence theory tool $m(A)$ is called as a focal elements if $m(A) \neq 0$, and expresses the certainty degree of A. A BPA is characterized by two functions: the belief function Bel and the plausibility function Pl. The belief in a subset $A \in 2^X$ is the sum of all pieces of evidence that support A and the plausibility of A the sum of pieces of evidence not supporting $\bar{A}$:

$$Bel(A) = \sum_{B} m(B) \, and \, pl(A) = \sum_{B \wedge A \neq \phi} m(B) = 1 - bel(\bar{A}) \qquad (9)$$

Based on these definitions, combination operators can be characterized as

$$m(A) = ((1 - \sum_{A_i \cap B_j = C} m(A_i)m(B_j))^{-1} \sum_{A_i \cap B_j = C} m(A_i)m(B_j) \qquad (10)$$

It is possible to build a unique elementary mass function m from n elementary mass functions $m_1, m_2, \ldots, m_n$ arising from n distinct and independent sources but characterized on the same set. The combination rule in the DS evidence theory consists of calculating:

$$bel(A) = (((bel_1 \oplus bel_2) \oplus bel_3) \oplus \ldots) \oplus bel_n \qquad (11)$$

where $\oplus$ denotes the combination operator. A similar equation for the plausibility function is

$$pl(A) = (((pl_1 \oplus pl_2) \oplus pl_3) \oplus \ldots) \oplus pl_n \qquad (12)$$

The combination is commutative and associative. So, the combination rule can be easily extended to combine several belief functions by repeating the rule for new belief functions. The basic determinations of the BPA of each source in X are different shapes or densities of pattern clusters formed during unsupervised analysis of the pattern data. Finally, the last step is the decision making process which is supported by the results provided by the combination rules. After the decision principles are determined, the evidence theory can efficiently fuse a group of information sources in the given frame of discernment. The evidence theory tool has the ability to effectively deal with uncertain, imprecise, and incomplete information [13], and to increase the confidence degree of the final decision result while the uncertain information decreasing rapidly as the information quantity increases. These abilities are desirable for identifying the real number of clusters. However, Eq.(10) may cause a most unreasonable case called "rejected by one vote" for a given hypothesis [32], [33]. To solve the problem the mass function must be reasonably determined such that all possible numbers of clusters are assured to have chance to be included into the final fused system.

3   Fused Multiple Characteristic Indices Using the D-S Evidence Theory

We first investigate the problems of existing validity indices, termed uncertainty, instability, and incompleteness, then extract two natural characteristics in the clustering evaluation for arbitrary clustering algorithms along arbitrary structures of data space, and finally fuse these characteristics using the D-S evidence theory to a new index, termed fMCI, as explained below.

3.1 Uncertainty, Instability and Incompleteness of Existing Validity Indices

Any clustering algorithm has its own applicable range, which is designed to cluster some specific structures of data space. The structure of a data space can be very complex. It can represented by six basic features including (1) Size: it contains clusters with different sizes (2) Shape: it has arbitrary-shape clusters (3) Overlap: it has overlapped clusters (4) Noise: it contains noises in the data (5) Density: it has clusters with different densities (6) Subspace: it has clusters of different subspaces in high-dimensional space. For illustration purpose, we study the performance of three most used cluster algorithms, CM, FCM, DBSCAN, the variants, and the validity indices based on the six features, for different data structures. The results are summarized in Table 1. As shown in Table 1, the original CM performs well on spherical clusters, fails on partitioning the irregular dataset that contains clusters of large variations. To overcome these limitations, some variants of the CM have been proposed to partition the shape-diverse clusters (e.g., k-modes algorithm [34]) or attribute-diverse clusters (e.g., FWKA algorithm [35]) in high-dimensional data space. Different from the CM, the FCM partitions the dataset in a "soft" or fuzzy way [1] ,[15]. The FCM can correctly partition the overlapped clusters. As two variants of the FCM, the conditional FCM (c-FCM) [36] can further handle density-diverse clusters, and the robust FCM (r-FCM) [37] can effectively partition the noisy data. The DBSCAN algorithm can handle arbitrary-shape clusters but cannot be used in a high-dimensional data space [1]. As an extension on DBSCAN, the OPTICS [38] was developed to hierarchically partition density-diverse clusters. The most important property in evaluating a clustering algorithm is the capability of the algorithm to locate the correct number of clusters. Another important property is the applicability of the validity index on various datasets. Apparently, the above validity indices are dependent on specific clustering algorithms along specific structures of data space. Consequently, any validity index must be incomplete to their applicable ranges when facing arbitrary clustering algorithms along arbitrary structure of data space. However, the above clustering algorithms and their validity indices have shown strong complementary natures among their

Table 1. The performances of the three most used clustering algorithms and their validity indices

Characters	(1) Size	(2) Shape	(3) Overlap	(4) Noise	(5) Density	(6) Subspace
CM/k-modes/ FKWA	×/×/√||×/√/√	×/×/√||×/×/×	×/×/√||×/×/×	×/×/√||×/×/√	×/×/√||×/×/×	×/√/×||×/√/√
FCM/c-FCM/r-FCM	×/×/√||×/×/×	×/×/√||×/×/×	√/√/√||√/√/√	×/√/√||×/√/√	×/√/×||×/√/×	×/√/√||×/×/×
DBSCAN/OPTICS/DBSCAN*	√/√/√||√/√/√	√/√/√||√/√/√	√/×/√||√/×/√	×/×/×||×/×/×	√/×/√||√/×/√	×/×/×||×/×/×

Note: the sign "√" means that the algorithm/validity index is effective to the corresponding feature and the sign "×" means inefficient. In the sign "*||*" the items in left stand for three cluster algorithms, and the items in right are their corresponding validity indices.

applicable ranges. Consequently, one promising way to generalize the validity index may be to integrate several validity indices for a more generalized accuracy measures. Let "NC" denote the number of clusters for any dataset. Most exiting validity indices suggest the optimal NC by their optimums (maximum or minimum) along a group of consecutive values of NC. Usually, a common acceptable principle is, the closer the optimal value of a validity index is, the higher probability the calculated NC is the true NC for a given dataset. If any value of a validity index at a NC is dominantly larger than other values, such NC is highly believable to be suggested as the true NC. To examine the impacts of different structures of data space for any validity index, we constructed a number of variations on the Satimage dataset provided by the UCI Machine Learning Repository [39]. Satimage dataset contains 6435 data samples of 36 attributes and it is known the data samples belong to six different clusters. We generated four datasets with 30Consequently, the clustering evaluation needs to consider the stability issue. That is, similar problems should always have similar solutions, and trial change of different parameter settings in a validity index must not lead to great distinct clustering evaluations. This stability concept has been defined as robustness measure in our previous work [23]. With a close look of clustering evaluation of the above four datasets, the performances of the GS

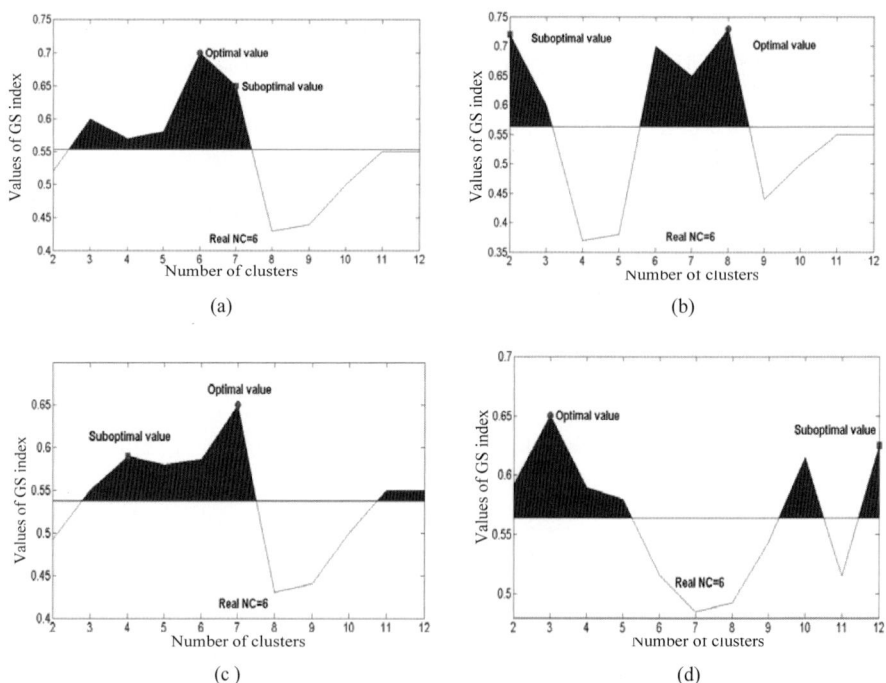

Fig. 1. The clustering evaluation of four variations on the original Satimage dataset by the GS index. Curves of validity indices when adding 30% noises to (a) smallest, (b) second, (c)third, fourth small clusters in the Satimage dataset.

index is very instable and inconsistent when facing the similar problems. Consequently, the existing validity indices suffer from these above three problems, termed uncertainty, incompleteness, and unstableness. To find way to overcome these problems are used to guide the development of the generalized validity index for clustering evaluation.

3.2 Characteristic Extraction to General Clustering Evaluation

To effectively develop a general validity index, it is necessary to extract the characteristics that are independent to arbitrary clustering algorithm along arbitrary structure of data space. Moreover, these characteristics should be hardly involved in the problems how to determine the optimal structure of intra-cluster and inter-cluster distance functions, have robust solution when facing similar problems (datasets). The first characteristic we derive is an intra-cluster distance function. Most existing validity indices consist of the intra- and/or inter-cluster distance functions. A combined structure of both intra- and inter-cluster distance functions will impact the effectiveness of these validity indices. Usually the intra-cluster distance function is monotonically decreasing as the NC increases. If the calculated NC is far from the true NC, the declining tendency tends to be flat. The optimal NC can be found in the position where the within-cluster distance appears with the changes of the magnitude, which is called "elbow" [17] (see Fig.2). Locating the "elbow" point is highly subjective and dependent upon the user's perception. To overcome the problem, Tibshirani applied a reference distribution curve to construct GS index, to objectively locate the "elbow" point by finding a maximum of the GS index. The approach may suffer from the three problems: 1) a global "elbow" point may not be found. Fig.2 (a) shows that the "elbow" position cannot be seen by the GS index in the Satimage dataset, though there is clear "elbow" point observed at $NC=6$. 2) A convex function of the intra-cluster distance function is difficult to be represented by the GS index. 3) The necessary parameter B in the GS index can affect its clustering

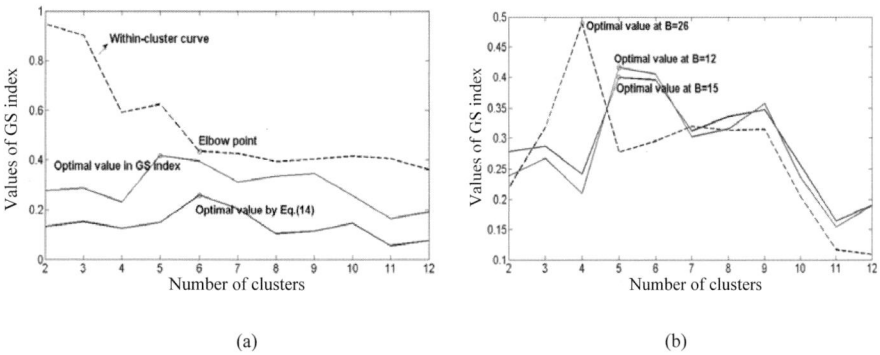

(a)   (b)

Fig. 2. Comparison of the GS index and Eq.(13). (a) The GS index cannot find the elbow point at NC=6 but Eq.(13) can. (b) Different values of B in the GS index lead to very different clus-tering evaluations.

evaluation. Fig.2 (b) shows that the GS index can cause large differences at the different values of B. Different from the GS index, we previously have demonstrated [3],[4] that the two-order difference of within-cluster distance function can help locate the elbow position as the reference curves in the GS index. The two-order difference is defined as

$$T(NC) = max_{NC,NC',NC'' \in A NC'<NC<NC''}(W_{NC'} + W_{NC''} - 2W_{NC})$$

$$s.t., NC = NC'' + |A|/2, and 5 \leq |A| \leq NC_{max} \tag{13}$$

Where W is GS index defined in Eq. (1), A is a set of consecutive values of NC, $|A|$ is larger than 3 which is the smallest number of consisting of any two-order difference. When $NC' = NC - 1, NC'' = NC + 1, T(NC)$ reduces to the standard two-order difference [23]. In light of Eq.(13) the "elbow" point can be found when $|A| = 4$. After applying Eq.(13) to the above four datasets of *Satimage*, Fig.3 shows that each of their curves of $T(NC)$ is divided into two parts located above and below the average of optimums respectively. The real NC is close to or contained in the parts above the average horizontal line. This demonstrates an appropriate intra-cluster distance function, such as $T(NC)$, may better suggest the true NC since it only involve an independent intra-cluster distance function and thus can ignore the problem of how to combine a group of

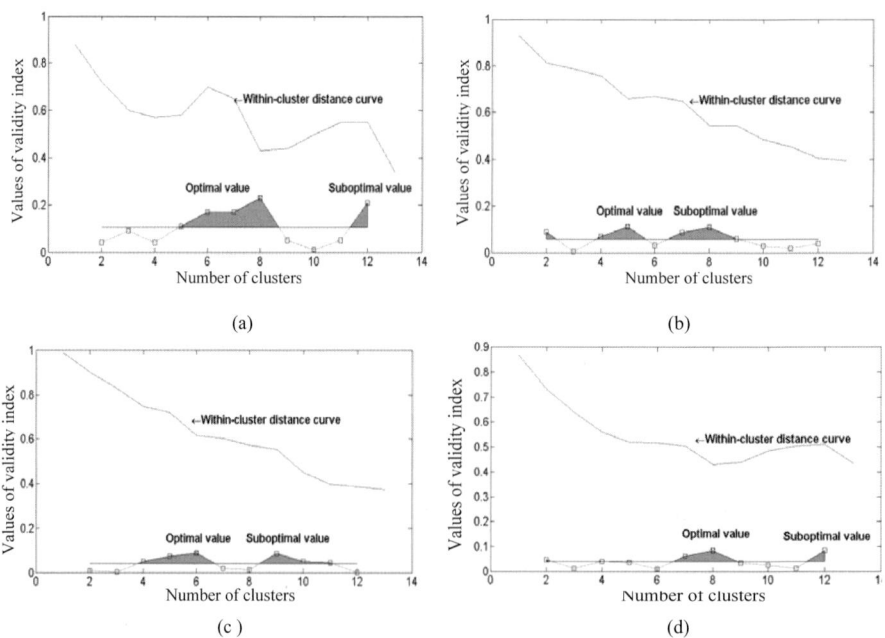

Fig. 3. Four curves of two-order difference in the four datasets of the Satimage dataset. Fig.3 (a)-(d) shows the different characters of these variants based on Eq.(13).

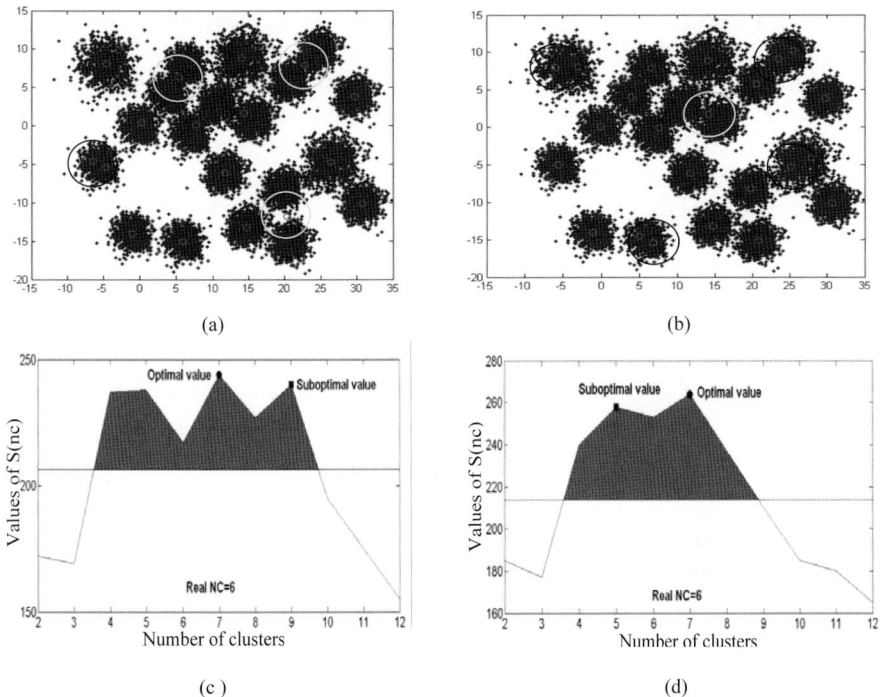

Fig. 4. The effect of Eq.(14) in clustering evaluation. (a) and (b) show the r-neighborhoods when the considered NC is less than the real NC respectively. (c) and (d) is the curves of Eq.(14) associated with the two situations of S(NC). The yellow and blue circles in figures stand for the r-neighborhoods of clustering prototypes and the middle point of any pair of clustering prototypes.

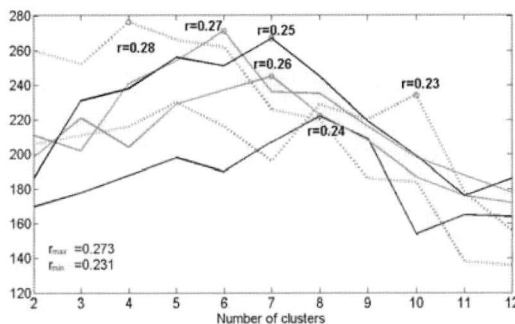

Fig. 5. Different curves of S(NC) from different values of r in the Satimage dataset

specified intra- and inter -cluster distance functions in considerations. The second characteristic we consider in this research is a measure termed $S(NC)$ that consists of ratio between average center density $CD(NC)$ and average boundary density $BD(NC)$ of all clusters under the calculated NC. For example, $CD(NC)$ and $BD(NC)$ in DBSCAN algorithm may be calculated by the average data number across all r-neighborhoods of core data and across all r-neighborhoods of boundary data [12], [13], respectively, where the r-neighborhood is a set of data that is centered in the related data objective with radius of r; $CD(NC)$ in CM algorithm is the average number of data in the r-neighborhood of any clustering prototype, while $BD(NC)$ is the average number of data that fall into the r-neighborhood of the middle point between any pair of clustering prototypes. To locate the true NC, $S(NC)$ should satisfy

$$S(NC^*) = \mathbf{max}_{NC}CD(NC)/BD(NC) \tag{14}$$

The effect of $S(NC)$ can be examined in a dataset of 20 clusters in which some clusters are partially overlapped (see Fig.4(a)). If NC is less than 20 when using the CM algorithm, some r-neighborhoods that contain clustering prototypes are located in the overlaps of the in-between clusters that contain few of data points due to the effect to minimize the objective function of the CM algorithm. Hence, $CD(NC)$is less than the value at $NC = NC_{optimal}$ (see Fig.4(b)). If $NC$is larger than 20, in $S(NC)$ increases so $S(NC)$ naturally decreases as NC increases. We conclude $S(NC)$ can attain its maximum in the vicinity of $NC_{optimal}$. Consequently, $S(NC)$ may better predict the true NC (see Figs.4 (c)(d)). However, the prediction is conditional on the values of r (see Fig.4). Let order all clusters in a dataset decreasingly by their magnitudes and let r_{max} and r_{min} be the maximal and minimal sizes of cluster respectively. If the value of r is larger than r_{max} , $S(NC)$ tends to predict smaller NC than the true NC; if r is smaller than r_{min} , $S(NC)$ tends to predict larger NC than the true NC. When r satisfies $r_{max} > r >_{r_{min}}$ the results of $S(NC)$ are desirable (see Fig.5). Thus a rough but correct estimation of r is key to assure the correctness of $S(NC)$. In summary, we propose two general characteristics in clustering evaluation, one is based on an independent intra-cluster distance function and the other is based on a neighborhood based function. Integrating these characteristics and available existing validity indices is a challenging task since the clustering evaluation is uncertain, incomplete and unstable in nature. Different from the limitation of locating single point NC in the existing indices, we suggest finding a few of sets of consecutive values of NC with high confidence to identify the true NC. Then we propose the application of the D-S evidence theory for a generalized index by fuse the multiple characteristics since the advantage of the D-S evidence theory in fusing uncertain, incomplete and unstable information sources for an optimal decision, as discussed in the following section.

3.3   Proposed Index: Fused Multiple Characteristic Index (fMCI)

For any given dataset, if a group of validity indices $f_{11}, f12, \ldots, f_{1m}$ are applicable along a used algorithm, then we first combine the group of validity indices to construct a characteristic, termed V_1 by a weighting sum of values of the group of indices

$$V_1(A) = w_1 f_{11}(A) + W_2 f_{12}(A) + \ldots + w_m f_{1m}(A), for A \in 2^X \quad (15)$$

where w_j is the weighting value to represent the importance degree of $f_j(A)$ for all $j, j = 1, 2, \ldots, m$. The weighting value is characterized as

$$w_j = m_j(A)/(\sum_{j=1}^{m} \sum_{A \in 2^X} m_j(A))/|A|, j = 1, 2, \ldots, m \quad (16)$$

where $\sum_{j=1}^{m} w_j = 1$. If w_j is greater (less) than 1, the impact of f_{1j} increases (decreases). Specially, if there are two or more sets of consecutive numbers of clusters above the average line, the calculated weighting value of Eq.(15) for any A must decease since these sets have to share the value 1 according to the definition of the mass function. Inversely, if there is a dominant set of number of clusters above average of line, the value of Eq.(15) must be 1. Thus Eq.(15) can automatically increase and decrease the importance of each index in the group of indices. Eq.(15) indicates even though the real NC is incorrectly rejected by one index, it still can be suggested as the optimal NC if it has larger values in the other indices. This can prevent the unreasonable situation of "rejected by one vote" that often occurs in the D-S evidence theory. Let $F_1(NC)$ be , $F_2(NC)$ is $T(NC)$ and $F_3(NC)$ is $S(NC)$, X be a frame of discernment consisting of a set of values of NC. Then for any $A \in 2^X$, $F_i, i = 1, 2, 3$,the focal element in the D-S evidence theory is computed as

$$f_i(A) = \begin{cases} \sum_{NC \in A}(F_i(NC) - Aver F_i(NC)) & F_i(NC) \geq Aver F_i(NC) \text{for all } NC \in A \\ 0 & otherwise \end{cases} \quad (17)$$

where $Aver F(NC)$ is the average of all values of $F_i(NC)$ for all values of NC, Eq. (17) explains the basic idea that the true NC should be within a set of consecutive values of NC in which the value of the validity index at each NC is larger than the average of all values at all NCs in a determined range. In addition, since the values of different validity indices in Eq.(17) may have different orders of magnitude, using a weighting sum approach to combine a group of validity indices should be normalized in order to make the values in units of approximately the same numerical value. The mass functions in the D-S evidence theory are normalized as

$$m_i(A) = f_i(A/)\bar{f}_i, for all A \in 2^A, i = 1, 2, 3 \quad (18)$$

where $\bar{f}_i$ is the average of all values of $F_i(A)$. After fusing the above three characteristics by the D-S evidence theory, the decision-making rule is that $A_1, A_2, \in 2^A$, which satisfies

$$m(A_1) = \max\{m(A_i), A_i \subset U\} m(A_2) = \max\{m(A_i), A_i \subset U and A_i \neq A_1\} (19)$$

If there exists

$$\{ m(A_1) - m(A_2) > \varepsilon_1 m(X) < \varepsilon_2 m(A_1) > m(X) \quad (20)$$

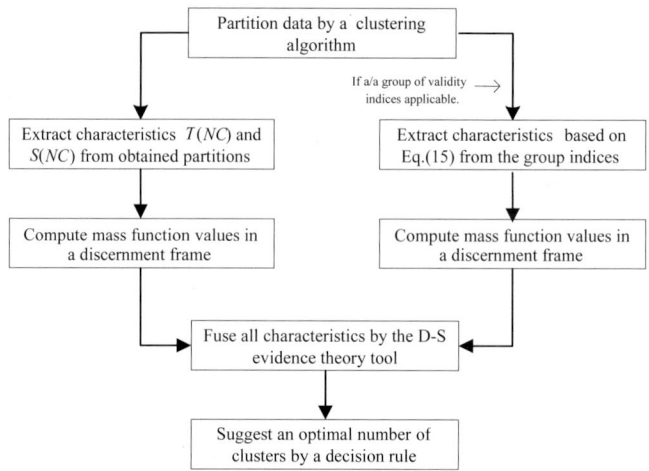

Fig. 6. Flowchart of fusing a group of information sources of validity indices

then, A_1 is the decision-making result, where $varepsilon_1$ and $varepsilon_2$ are pre-established threshold. In this paper we have taken $varepsilon_1 = varepsilon_2 = 0.05$.

4   Experiments

In our computer simulations, all the results are calculated by a desktop computer with Matlab 6.5, 3.2 GHz CPU, 512M of RAM and Windows XP professional version 2002. Our proposed approach is applied to evaluating the cluster outcomes on six synthetic datasets and four real world datasets. The three clustering algorithms, CM, FCM, and DBSCAN are adopted from the data mining workshop: AlphaMiner 2.0 [14]. To compare the performances of various validly indices, we define three notations, accuracy, robustness and algorithm-reachable property. For a given dataset the robustness measure of a validity index under a cluster algorithm is defined as

$$Robustness = |NC_{optimal} - NC_{second}| \qquad (20)$$

where NCoptimal and NCsecond stand for the optimal and the second optimal values of NC of any validity index respectively. Based on Eq.(21), we conclude that the smaller the values of robustness measure is, the easier it is to search for the optimal number of clusters. Please note the above notation of robustness measure in fact generalizes the concept of stability in literature [29], [33]. On the other hand, If a given algorithm can correctly partition half of total data of each cluster in any dataset, the dataset is called algorithm-reachable. The algorithm-reachable notation shows whether a clustering algorithm has ability to correctly partition a given dataset. If a dataset is not algorithm-reachable, it is difficult to suggest a correct NC by an existing validity index under the clustering algorithm. All validity indices are evaluated by two indices: Accuracy and robustness measure after classifying these clustering evaluations by the algorithm-reachable properties.

4.1   Experiments on Synthetic Data

Six Synthetic Datasets. We generated six synthetic datasets using the functions in the Matlab? toolbox. The correct clustering label of the data is known as a prior. These labels are used to evaluate the correctness of the clustering results. The six synthetic datasets are created representing different structure of the data space (shown in Fig. 7). Dataset 1 (Fig. 7a) contains 6,000 data points from three clusters. One cluster is high-density with 4,000 data points and the other two are low-density clusters with 1,000 data points for each. In Dataset 2, there are 20,000 data points in 20 sphere-shaped, density-diverse and partially overlapping clusters (see Fig. 7(b)). Dataset 3 consists of thirteen clusters with 2000 three-dimensional data points. There are six pairs of strongly overlapped clusters (see Fig. 7(c)). Dataset 4 contains 366 data points in five low-density ellipsoidal clusters and one high-density line-like cluster (see Fig. 7(d)). The data in the six clusters are all evenly distributed. Dataset 5 contains 6000 data in three ellipse-shaped clusters (see Fig. 7(e)). Dataset 6 is a very particular example of arbitrary-shaped clusters, where some clusters are perfectly contained in other clusters. There are twelve identifiable clusters with some noisy data in this dataset (see Fig. 7(f)).

Clustering Algorithms. We clustered the above six datasets using CM, FCM, and DBSCAN algorithm. Datasets 1, 2 and 3 are algorithm-reachable by all the three clustering algorithms since these clusters are approximately sphere-shaped, while datasets 4, 5 and 6 are algorithm-reachable only when using the DBSCAN since these clusters in the three datasets are irregular. Thus we can examine the clustering evaluation under different algorithm-reachable conditions. We assigned a data point to the corresponding clusters using the maximum membership degree criterion if the clustering result is achieved by the FCM algorithm. Since the CM and FCM algorithms can be greatly impacted by the different initializations, we take the optimal results when the objective functions obtain the minimal values under different initial conditions. In addition, to run the validity indices sym and WSVF, we apply the genetic algorithm operator $ga()$ in the Matlab toolbox instead of the optimization steps in the two validity indices. This may affect the runtime of the two validity indices but has no impact on their accuracy. We combined the three validity indices sym, GS and DB^{**} to the F_1 by Eq.(15), then fused F_1, F_2, and F_3 to fMCI, and finally compared fMCI the fused index with the three validity indices themselves as well as the WSVF, where the values of t and q in DB_{**} both are set as 1. Specifically, the following functions $1/(1 + sym)$, GS, $1/(1 + DB_{**})$ consist of the component functions in the WSVF to suggest the optimal NC. The continuous values ε in the DBSCAN must be partitioned into discrete values in the range of their

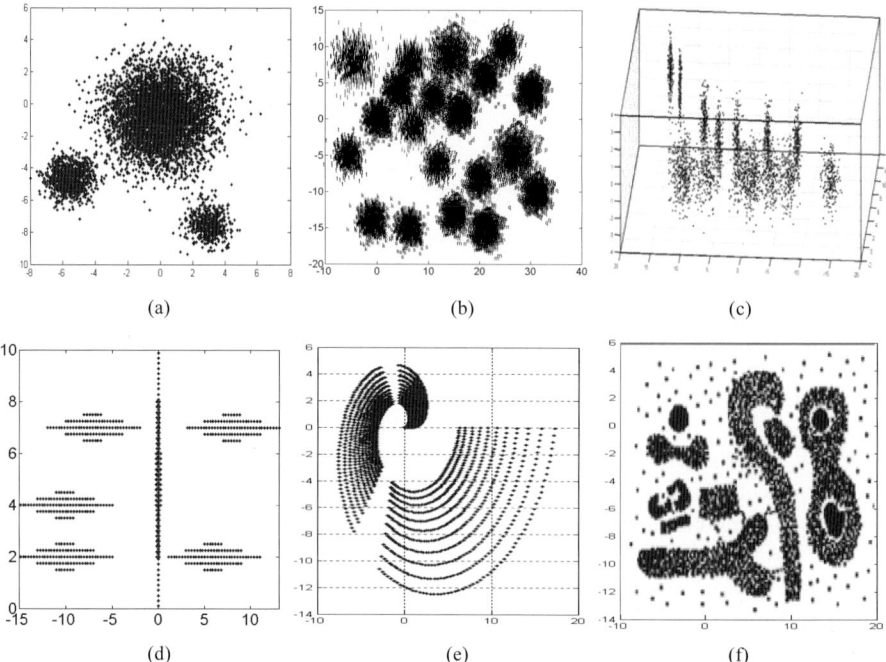

Fig. 7. Six synthetic datasets with diverse characters. (a) Density-diverse clusters. (b) Large size dataset. (c) Overlapped clusters. (d) Even density and arbitrary shape of clusters. (e) Three non-convex -shaped clusters. (f) Twelve clusters in which some clusters are contained in other clusters.

possible values. We increasingly take their discrete values from their minimum at every 10^{-2} and label these corresponding numbers of clusters in the axis of abscissas. Consequently, the performances of the above validity indices can be pictorially evaluated in the same coordinate system.

Clustering Evaluation of Validity Indices. The clustering evaluations of these validity indices are shown in Table 3. As the most representative clustering results, we have the results from the FCM algorithm shown in Fig. 8, where the values of m are the best partitions for these datasets. Moreover, we have uniformly added or subtracted a constant respectively for each validity index to make their values have the similar order of magnitude, and the values of the corresponding mass functions in the fMCI are shown in Table 2. Table 2 indicates that the focal elements of the determined mass functions in the fMCI mostly contain the true NC except dataset 6. This ascribes that at least one character function show larger values when the true NC attains. In addition, since the nonzero values of mass function can complement each other when identifying the real NC, the problem of "rejected by one vote" in the D-S evidence theory is resolved. After fusing the mass functions of F_1, F_2, and F_3, the final focal elements become a smaller set of consecutive values of NC with larger mass values.

Table 2. Values of the mass functions in the six synthetic datasets

dataset 1/f_1	[4]/0.0485; [7,8]/0.295; [6,7,8] /0.322; [3,4,5,6]/0.333.
dataset 1/f_2	[2]/0.509; [4]/0.467; [8]/0.0249.
dataset 1/f_3	[2]/0.321; [4,5,6,7]/0.678.
fMCI	[4]/1.000.
dataset 2/f_1	[18,19]/0.193; [22]/0.140; [20,21,22]/0.416; [17,18]/0.251.
dataset 2/f_2	[17]/0.004; [19,20,21]/0.458; [23]/0.537.
dataset 2/f_3	[17,18,19,20,21,22]/1.00.
fMCI	[17]/0.002; [19]/0.188; [20]/0.405; [21]/0.405.
dataset 3/f_1	[14]/0.189; [16,17,18]/0.166; [15]/0.056; [17,18]/0.558; [12]/0.030.
dataset 3/f_2	[12]/0.355; [16]/0.249; [18]/0.396.
dataset 3/f_3	[15,16,17]/0.736; [13]/0.264.
fMCI	[16]/1.000.
dataset 4/f_1	[6,7,8,9]/0.667; [5,6,7,8,9]/ 0.333.
dataset 4/f_2	[3]/0.207; [6]/0.637; [8,9]/0.156.
dataset 4/f_3	[3]/0.133; [5,6,7,8]/0.867.
fMCI	[6]/0.804; [8]/0.196.
dataset 5/f_1	[2]/0.023; [6,7,8]/0.310; [7,8]/0.333; [4,5]/0.208; [8]/0.125.
dataset 5/f_2	[2,3]/1.000.
dataset 5/f_3	[2,3,4,5]/0.917; [7]/0.083.
fMCI	[2]/1.00.
dataset 6/f_1	[11,12,13,14]/0.333; [11,12,13,14,15]/0.333; [9]/0.175; [12,13]/0.158.
dataset 6/f_2	[9]/0.047198; [12]/0.5083; [14,15]/ 0.4445.
dataset 6/f_3	[9,10,11,12]/1.000.
fMCI	[9]/0.019; [12]/0.981.

Note: The "A" in sign " A/B" from the second to fifth columns stands for a subset of numbers of clusters and "B" mass values of the subset respectively. These mass values of each set are decreasingly ordered in each row from the left to the right.

Therefore these sets that contain the true NC are tightened while those sets that cannot contain the real NC are reduced. The advantages of D-S evidence theory assure that the FMCI can effectively approximate the real NC. Table 3 shows that the fMCI can accurately determine the NC and have the smaller values of robustness measure over all the six data sets. The other validity indices fail at least one of the above six data sets. Compared the fMCI with the WSVF, Fig. 8 shows that the WSVF tends to agree the NC that most validity indices support. This is an inherit defect of the class of independent weighting coefficient function when some numbers of clusters are irrelative to each others. Considering algorithm-reachable notation, the accuracy and robustness measure of the above six datasets are further concluded as follows: 1) *Accuracy*: Both the CM and the FCM are not algorithm-reachable in dataset 4, 5 and 6. The incorrectly partitioned numbers of data are far more than that from the DBSCAN which is algorithm-reachable for all six datasets. Following the order of CM, FCM, and DBSCAN, the incorrectly partitioned numbers of data decreases. Consequently, their corresponding validity indices present the same order of accuracies. Hence, the algorithm-reachable properties can greatly affect the accuracy of clustering evaluation of these validity indices. Nevertheless, the accuracies of the fMCI are the highest compared with the other indices since the fMCI can take advantage of the complementary characteristics of the three extracted characteristics. 2) *Robustness*: Table 3 shows that the robustness measures of any validity index are slightly affected by the algorithm-reachable property. For a determined data structure these validity indices give very similar values of robustness measure

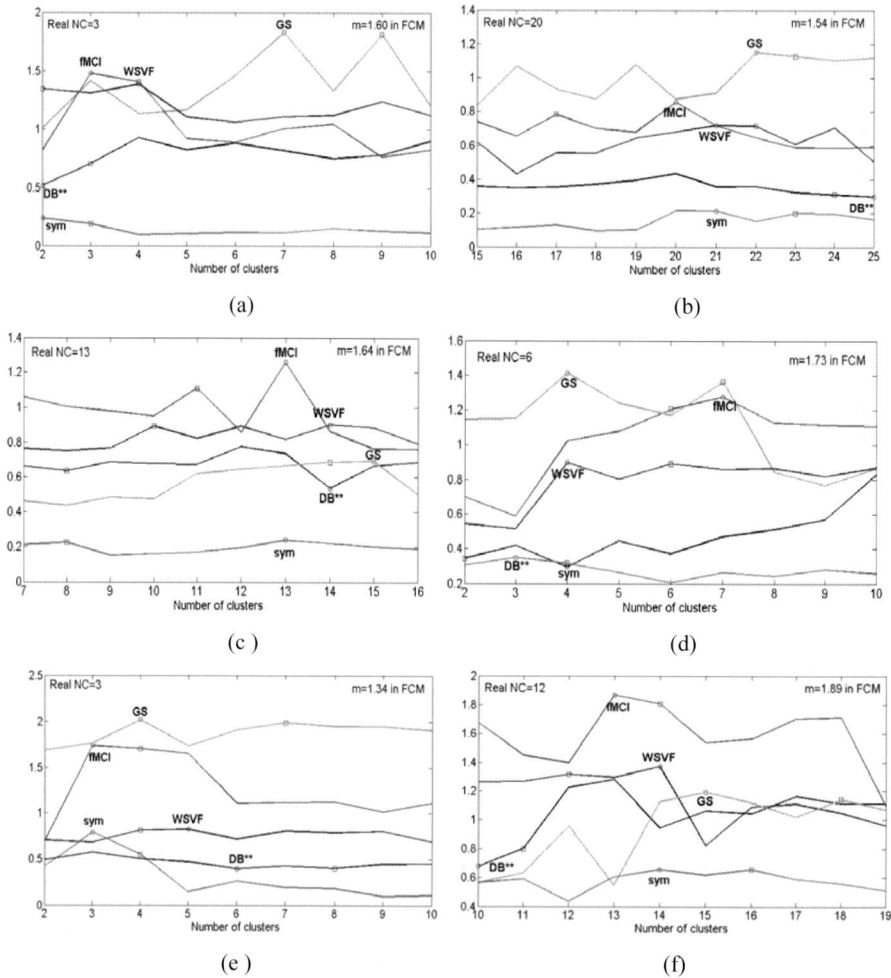

Fig. 8. Performances of the fMCI when using the FCM algorithm under different numbers of clusters. The figures (a)-(f) respond to the clustering evaluations in Set 1-Set 6, respectively.

under different clustering algorithms. However, if fixing each clustering algorithm for different data structures, their values of robustness measure are very different. Considering the six features of the data space (Table 1), noisy data has more impact on the robustness while the size of clusters has the lowest impact. The density-diverse and overlapped clusters have same level of impact on the result. As for shaped-diverse clusters, it slightly affects the values of robustness measure in the case of strong separation among clusters. However, the impact dramatically increases when no strong separation exists in the dataset. This is one of the major reasons some existing validity indices suggest incorrect number of clusters. 3) If the structure of data space is both unclear and not

Table 3. Performances of the validity indices in six synthetic datasets

	sym			GS			DB**			WSVF			FMCI			NC
	CM	FCM	DBS	CM	FCM	DBS	CM	FCM	DBS	CM	FCM	DBS	CM	FCM	DBS	
Set 1	+0/1	- 1/1	+0/2	+0/1	+4/2	+0/1	+0/1	-1/1	+0/1	+0/1	+0/1	+0/1	+0/1	+0/1	+0/1	3
Set 2	- 1/3	+1/2	+0/2	- 1/2	+1/2	-1/2	- 1/2	+5/1	- 1/2	- 1/2	+2/1	+0/2	- 1/2	+0/2	+0/2	20
Set 3	- 2/3	+0/5	- 2/4	- 1/3	+2/1	-2/2	- 2/4	+1/1	- 2/3	- 1/3	+1/4	- 2/3	+0/2	+0/2	+0/1	13
Set 4	- 3/1	- 2/1	- 3/1	- 3/1	+1/3	-3/1	- 2/1	- 3/1	- 3/1	+1/1	- 2/2	- 3/1	- 1/1	+1/1	+0/1	6
Set 5	- 1/1	+0/1	- 1/1	- 1/1	+1/3	-1/1	- 1/1	+3/1	- 1/1	- 1/1	+1/1	- 1/1	+0/1	+0/1	- 1/1	3
Set 6	+3/1	+2/3	+3/4	+3/1	+3/3	+3/4	+3/3	+1/1	+3/5	+3/2	+2/2	+3/4	+1/3	+1/1	- 1/3	12
	m=1.71 in FCM, ε=2.12			m=1.71, ε=2.12			m=1.71 in FCM, ε=2.12			m=1.71 in FCM, ε=2.12			m=1.71 in FCM, ε=2.12			

Note: The "A" in sign " A/B" stands for accuracy and "B" robustness measure respectively. The "+" in "A" shows that the suggested number of clusters exceeds the real one, and "-" shows shorter than the real one. The "NC" refers to the actual number of clusters.

algorithm-reachable (e.g., dataset 6), all validity indices can hardly give a good accuracy. The first characteristic in the fMCI may be inefficient but the other two characteristics still are effective.

5   Conclusion

In general, the existing indices suffer from the problems of uncertainty, instability and incompleteness in clustering evaluations. To address these issues, we propose a novel approach by using the D-S evidence theory to introduce a generalized index by combining multiple aspects from the clustering process. To accomplish this, we first extract three efficient characteristics of clustering evaluation. The first characteristic can make use of the complementary natures among the existing validity indices, and provide a group of candidates of possible numbers of clusters. The other two are independent from the used clustering algorithms and structure of data space. A fused multiple characteristics index (fMCI) is introduced by fusing the three major characteristics in clustering evaluation based on the D-S evidence theory. Different from the use of a single optimum to predict the true number of clusters in the existing validity indices, the fMCI applies a group of set functions of consecutive numbers of clusters to approximate the true number of clusters. Experimental results indicate that the fMCI can generally handle uncertainty, imprecision, and instability issues in clustering evaluation and hardly is affected by any validity index structure. While promising, there are several directions in which the work may be extended further. For example, how to effectively define the optimal mass functions for the fMCI to further improve the results remains unanswered well. Secondly, how to choose the combination rules in the DS-evidence theory for better estimation of the true number of clusters shall be explored. These are our research in the future.

References

1. Xu, R., Wunsch, D.: Survey of clustering algorithm. IEEE Trans. Neural Network 16(3), 645–678 (2005)
2. Bezdek, J.C., Pal, N.R.: Some new indexes of cluster validity. IEEE Trans. SMC-B 28(3), 301–315 (1998)

3. Maulik, U., Bandyop, S.: Performance evaluation of some clustering algorithms and validity indices. IEEE Trans. Pattern Anal. Mach. Intel. 24(12), 1650–1654 (2002)

4. Pakhira, M.K., Bandyopadhyay, S., Maulik, U.: A study of some fuzzy cluster validity indices, genetic clustering and application to pixel classification. Fuzzy Sets Syst. 155(3), 191–214 (2005)

5. Wang, J., Chiang, J.: A Cluster Validity Measure with Outlier Detection for Support Vector Clustering. IEEE Trans. SMC-B 38(1), 78–89 (2008)

6. Hubert, L.J., Arabie, P.: Comparing partitions. J. Classification 2, 193–218 (1985)

7. Davies, D.L., Bouldin, D.W.: A cluster separation measure. IEEE Trans. Pattern Anal. Machine Intell. 1(4), 224–227 (1979)

8. Xie, X.L., Beni, G.: A validity measure for fuzzy clustering. IEEE Trans. Pattern Anal. Mach. Intell. 13(8), 841–847 (1991)

9. Bezdek, J.C.: Pattern Recognition with fuzzy objective function algorithms. Plenum Press, New York (1981)

10. Kim, M., Ramakrishna, R.S.: New indices for cluster validity assessment. Patt. Recog. Lett. 26, 2353–2363 (2005)

11. Pakhira, M.K., Bandyopadhyay, S., Maulik, U.: Validity index for crisp and fuzzy clusters. Pattern Recognition 37(3), 487–501 (2004)

12. Saha, S., Bandyopadhyay, S.: Application of a new symmetry based cluster validity index for satellite image anghamitra. IEEE Geos. Remote Sensing Letter 5(2), 166–170 (2008)

13. Tantrum, J., Murua, A., Stuetzle, W.: Hierarchical model-based clustering of large datasets through fractionation and refractionation. Information Systems 29, 315–326 (2004)

14. MacQueen, J.B.: Some methods for classification and analysis of multivariate observations. In: The 5th Berkeley Symposium on Mathematical and Probability, Berkeley, vol. 1, pp. 281–297 (1967)

15. Bezdek, J.C., Pal, S.K.: Fuzzy models for Pattern recognition. Plenum Press, New York (1992)

16. Dunn, J.C.: A fuzzy relative of the ISODATA process and its use in detecting compact well-separated clusters. J. Cybern. 3(3), 32–57 (1973)

17. Tibshirani, R., Walther, G., Hastie, T.: Estimation the number of clusters in a dataset via the gap statistic. J. Royal Society-B 63(2), 411–423 (2000)

18. Agrawal, R., Gehrke, J., Gunopulos, D., et al.: Automatic subspace clustering of high dimensional data. Data Mining. Knowl. Disc. 11(1), 5–33 (2005)

19. Ester, M., Kriegel, H.P., et al.: A density-based algorithm for discovering clusters in large spatial datasets with noise. In: Proc. 2nd Int. Conf. KDDD 1996, Portland, Oregon, pp. 226–239 (1996)

20. Ma, E.W.M., Chow, T.W.S.: A new shifting grid clustering algorithm. Pattern Recognition 37, 503–514 (2004)

21. Wang, J., Chiang, J.: A cluster validity measure with a hybrid parameter search method for support vector clustering algorithm. Pattern Recognition 41(2), 506–520 (2008)

22. Kim, D.J., Lee, K.H., Lee, D.: On cluster validity index for estimation of the optimal number of fuzzy clusters. Pattern Recognition 37(10), 2009–2025 (2004)

23. Yue, S., Li, P., Song, Z.: On the index of cluster validity. J. Chinese Electronic 14(3), 535–539 (2005)

24. Kittler, J., Hatef, M., Duin, R.P., Matas, J.: On Combining Classifiers. IEEE Trans. Patt. Anal. Mach. Intell. 20(3), 226–239 (1998)

25. Kaftandjian, V., Zhu, Y., Dupuis, O., Lyon, I.: The combined use of the evidence theory and fuzzy logic for improving multimodal nondestructive testing system. IEEE Trans. Instr. Mea. 54(4), 1968–1977 (2005)
26. Fred, A.L.N., Jain, A.K.: Combining multiple Clusterings using evidence accumulation. IEEE Trans. Patt. Anal. Mach. Intell. 27(6), 835–851 (2005)
27. Sheng, W., Swift, S., Zhang, L., Liu, X.: A Weighted Sum Validity Function for Clustering With a Hybrid Niching Genetic Algorithm. IEEE Trans. SMC-B 35(6), 1156–1167 (2005)
28. Wu, S., Chow, W.S.: Clustering of the self-organizing map using a clustering validity index based on inter-cluster and intra-cluster density. Pattern Recognition 37(2), 175–188 (2004)
29. Zhang, W., Lee, Y.: The uncertainty of reasoning principles. Xi'an Jiaotong University Press, Xi'an (1999)
30. Cuzzolin, F.: A geometric approach to the theory of evidence. IEEE Trans. SMC-C 38(4), 522–534 (2008)
31. Regis, M., Doncescu, A., Desachy, J.: Use of Evidence theory for the fusion and the estimation of relevance of data sources: application to an alcoholic bioprocess. Traitements Signal 24(2), 115–132 (2007)
32. Boudraa, A., Bentabet, A., Salzensten, F., Guillon, L.: Dempster-Shafer's probability assignment based on fuzzy membership functions. Elec. Lett. Comp. Vison. Image Anal. 4(1), 1–9 (2004)
33. Salzenstein, F., Boudraa, A.: Iterative estimation of Dempster-Shafer's basic probability assignment: application tomultisensor image segment. Opt. Eng. 43(6), 1–7 (2004)
34. Huang, Z., Ng, M.: A Fuzzy k-Modes Algorithm for Clustering Categorical Data. IEEE Trans. Fuzzy Systems 7(4), 446–452 (1999)
35. Huang, Z., Ng, M.K., Rong, H.: Automated variable weighting in k-means type clustering. IEEE Trans. Patt. Anal. Mach. Intell. 27(3), 657–668 (2005)
36. Pedrycz, W.: Conditional fuzzy clustering. Patt. Recog. Lett. 18(7), 791–807 (2005)
37. http://morden.csee.usf.edu/brfcm/brfcm-src/brfcm.c
38. Ankerst, M., Breunig, M., Kriegel, H.P.: Ordering points to identify the clustering structure. SIGMOD Record 28(2), 49–60 (1999)
39. UCI Machine Learning Repository, ftp://ftp.cs.cornell.edu/pub/smart/
40. AlphaMiner2.0., http://bi.hitsz.edu.cn/alphaminer/index.htm
41. Yue, S., Wei, M., Wang, J., Wang, H.: A general grid-clustering approach. Patt. Recog. Lett. 29(9), 1372–1384 (2008)
42. Lange, T., Roth, V., BrauM, L., Buhmann, J.M.: Stability-Based Validation of Clustering Solutions. Neural Comput. 16(6), 1299–1323 (2004)

Recognition of Internet Portal Users on the Basis of Their Behaviour

Wojciech Jaworski

Institute of Informatics
University of Warsaw, Banacha 2, 02-097 Warsaw, Poland
wjaworski@mimuw.edu.pl

Abstract. Our aim is to develop methodology for recognition of Internet portal users on the basis of their behaviours. This is a classification task in which we have thousands values of decision attribute and objects are described by means of sequences of symbols. We develop feature selectors which make our task tractable. Since the behaviour usually does not distinguish users, we introduce user profiles which are clusters of indiscernible users and we construct classifiers which assign descriptions of user behaviour with user profiles. We also derive specific for our task quality measures.

Keywords: Web traffic analysis, sequential data classification, quality measures.

1   Introduction

Web traffic analysis is an area of still growing importance. The ability of determining which pages were displayed by the same user is crucial for this domain. Usually users are identified by the cookies technology. However, this method fails in many cases: user may delete cookie, use more than one computer etc. The only trace that allow us to identify user in such case is his behaviour - the similarity between sequences of Internet pages visited by user which we try to recognize with behaviour of known users.

   In this paper we analyze it in the context of the following case study: we are given an Internet portal that provides web pages. Each of them belongs to one of 20 thematic categories denoted by $A, B, \ldots, T$ [1]. User activity is reported as a sequence of category ids of pages displayed by user. In order to compress the data, subsequent occurrences of the same category were represented as a pair composed of category id and number of occurrences. Formally we denote such sequence as $\{c_i, m_i\}_{i=1}^{n}$, where each $c_i \in C$ is a category id and m_i is a number of its occurrences; let S be the set of all sequences.

   We are given two tables. Each of them consists of two fields: user id and user activity. First table describe users' activity during first three weeks and the

---

[1] In the original dataset categories were denoted by means numbers from 1 to 20. However for the sake of clarity we mapped numeric ids into letters.

J. Yu et al. (Eds.): RSKT 2010, LNAI 6401, pp. 520–527, 2010.

second describe the activity during fourth week. Our goal is to recognize user id on the basis of his activity in fourth week, having given activities of all users during the first three weeks labeled by user ids.

This is a classification task in which we have 4882 values of decision attribute (there are 4882 users) and objects are described by means of sequences.

However, the behaviour does not distinguish users, rather split them into clusters of indiscernible objects. As a consequence the classifier should indicate a set of users indiscernible with the one whose description is given as a classifier input, rather that point a single user.

Formally, describe our task as follows: we are given a set of users U, a function $g : U \rightarrow S$ that describe user behaviour during the first three weeks. We are also given a multiset V of sequences generated by users during the fourth week. Our goal is to generate function $f : V \rightarrow P(U)$ that will be a classifier which indicates users indiscernible with the one who generated classified sequence.

For the classifier validation purposes we use also a function $h : U \rightarrow S$ which describe user behaviour during the fourth week. h is the function approximated by classifier.

2   Data Description and Preprocessing

We analyze web traffic data from Polish web sites employing *gemiusTraffic* study. Data was acquired by Gemius company through the use of scripts, placed in code of the monitored web page. The scripts reported to the *gemiusTraffic* platform each Page View - event of displaying the monitored web page. Page Views were grouped into Visits - uninterrupted series of Page Views on a given web site executed by the same user.

This dataset was used as subject of one tasks in ECML/PKDD'2007 Discovery Challenge [3]. However the of that task was different that the one we pursue here. The problem objective was to predict user behaviour, while we are recognizing users on the basis of their behaviour.

The original dataset was composed out of two tables. First of them consisted of information about users: their country, operating system, browser, etc. This information is irrelevant for our task, so we did not use the first table.

Second table described users' activity in terms of Visits. It consisted of the following fields: user id, timestamp and sequences of categories of pages displayed during the Visit. For each user we have merged subsequent Visits into one sequence of Page Views. This operation significantly simplified the structure of data. According to our tests, the information about how Page Views are grouped into Visits is not helpful in our task.

3   Quality Measures

Before we start to construct a classifier we must define measures of its quality. Our task is different from a typical classification problem, because of the large number of decision attribute values. That is why traditional quality measures [2] developed for classification tasks (such as accuracy) does not suit to our task.

Since the classifier indicates a set of users indiscernible with the one whose description is given as its input, the quality measures should compare the set of users indicated by the classifier with the set of users indiscernible with a given one (the latter set we denote also as a set of relevant users).

For a single set of relevant users our task is similar to the problem of information retrieval: searching for documents relevant for a given query. In this case *precision* and *recall* are adequate quality measures. They can be seen as a measures of classifier exactness and completeness.

We define *precision* as the number of relevant users indicated by a classifier divided by the total number of users indicated by a classifier, and we define *recall* as the number of relevant users indicated by a classifier divided by the total number of relevant users.

Formally, let $[u]$ denote a set of users indiscernible with $u \in U$ (equivalence class of the indiscernibility relation). For a given classifier f and user $u \in U$ we define precision and recall as follows:

$$\mathbf{Precision}(f, u) = \frac{|[u] \cap f(h(u))|}{|f(h(u))|}, \mathbf{Recall}(f, u) = \frac{|[u] \cap f(h(u))|}{|[u]|},$$

where $h(u)$ is a sequence generated by user u and $f(h(u))$ is a value calculated by classifier for this sequence.

Now, we extend these measures on the sets of users. Since precision and recall are identical for indiscernible objects it is natural to state that

$$\mathbf{Precision}(f, [u]) = \mathbf{Precision}(f, u), \mathbf{Recall}(f, [u]) = \mathbf{Recall}(f, u)$$

and precision and recall of sets that are sums of indiscernibility classes are weighted means of their values calculated separately for each indiscernibility class.

On the other hand, values of precision and recall for sets of users should be an arithmetic mean of their values calculated for each user:

$$\mathbf{Precision}(f, X) = \frac{1}{|X|} \sum_{u \in X} \mathbf{Precision}(f, u), \mathbf{Recall}(f, X) = \frac{1}{|X|} \sum_{u \in X} \mathbf{Recall}(f, u).$$

Now, we show that the above definitions are equivalent. Let $X \subset U$ be such that $X = [u_1] \cup [u_2] \cup \ldots \cup [u_n]$.

$$\mathbf{Recall}(f, X) = \frac{1}{|X|} \sum_{u \in X} \mathbf{Recall}(f, u) = \frac{\sum_{k=1}^{n} |[u_k]| \mathbf{Recall}(f, u_k)}{\sum_{k=1}^{n} |[u_k]|} =$$

$$= \frac{\sum_{k=1}^{n} |[u_k]| \mathbf{Recall}(f, [u_k])}{\sum_{k=1}^{n} |[u_k]|} = \frac{\sum_{k=1}^{n} |[u_k] \cap f(h(u_k))|}{\sum_{k=1}^{n} |[u_k]|}.$$

Analogically

$$\mathbf{Precision}(f, X) = \frac{\sum_{k=1}^{n} |[u_k]| \mathbf{Precision}(f, [u_k])}{\sum_{k=1}^{n} |[u_k]|} = \frac{\sum_{k=1}^{n} |[u_k]| \frac{|[u_k] \cap f(h(u_k))|}{|f(h(u_k))|}}{\sum_{k=1}^{n} |[u_k]|}.$$

Precision and recall may be combined into a single measure *F-score* which is their harmonic mean:

$$F(f, X) = 2 \cdot \frac{\mathbf{Precision}(f, X) \cdot \mathbf{Recall}(f, X)}{\mathbf{Precision}(f, X) + \mathbf{Recall}(f, X)}.$$

Apart from the above measures we will consider also *average cluster size* defined as

$$\mathbf{ClusterSize}(f, X) = \frac{1}{|X|} \sum_{u \in X} |f(h(u))| = \frac{\sum_{k=1}^{n} |[u_k]| \cdot |f(h(u_k))|}{\sum_{k=1}^{n} |[u_k]|},$$

where $X \subset U$ is such that $X = [u_1] \cup [u_2] \cup \ldots \cup [u_n]$.

We assume that recall is the most important quality measure: high recall guarantee that the relevant user will be among users indicated by classifier. The second important measure is the cluster size: it tells us how many users will be on average indicated by classifier.

4 Feature Selection

Users cannot be classified by classifier that simply receive sequences of Page Views as input data and tries to find dependence between user id and sequence. The reason is that such classifier would have an infinite Vapnik–Chervonenkis dimension [4], so it would not be learnable [1].

In order to construct learnable classifier, we had to select relevant features from the original data. We accomplished this goal adaptively guessing how relevant information is encoded in data: we prepared and tested several features and, on the basis of tests results, we developed feature selectors described in this and the following sections.

We assume that user preferences did not change during 4 weeks covered by data. The reason for this assumptions is the fact that 4 weeks is a very short period so there is a small probability that user will change his preferences, on the other hand users behave pretty randomly so it is very hard to distinguish preference change from random noise.

As a consequence, categories that appear in sequence generated by a given user during first three weeks should be identical to those which he generated during fourth week. However, some rare categories may appear only in one sequence as a statistical noise. In order to eliminate such noise we introduce thresholds: we eliminate from sequence all categories that were viewed less then a given number of times.

For a given threshold t function $a_t : \mathcal{S} \to P(\mathcal{C})$ selects features as follows:

$$a_t \big(\{c_i, m_i\}_{i=1}^n \big) = \{c : \sum_{i:c_i=c} m_i > t\} \cup \{\arg\max_c \sum_{i:c_i=c} m_i\}.$$

It generates a set of categories, such that user displayed more than t pages belonging to each of them. In order to avoid users described by empty feature set, the most frequent category is always added to the set.

Function a_t generates indiscernibility classes, which may be interpreted as user profiles. Here are presented the most common user profiles, generated with threshold $t = 30$:

1176 Q	80 GNQ	27 F	15 CG	11 ILQ
834 GQ	73 IQ	27 CGQ	14 NQ	11 HPQ
676 G	69 I	25 FQ	14 N	11 $GLNQ$
186 LQ	61 GP	23 CQ	14 GH	11 FG
157 GPQ	58 P	22 $GLPQ$	14 C	10 LPQ
116 GN	52 GIQ	19 $GNPQ$	13 GNP	10 FGQ
108 L	45 GHQ	17 GOQ	13 $CGPQ$	10 EHQ
99 PQ	40 H	17 EQ	12 OQ	10 EGQ
89 GLQ	36 GL	15 MQ	12 $GHPQ$	9 KQ
87 HQ	29 GI	15 GKQ	11 IPQ	9 GK

Size of profiles decreases exponentially, we have a few large clusters and many small ones: there are 245 profiles total. The number of categories in profiles is small, typical user visits pages belonging to 1, 2 or 3 categories.

We tested also alternative feature selection function $b_t : \mathcal{S} \to P(\mathcal{C})$

$$b_t\big(\{c_i, m_i\}_{i=1}^n\big) = \{c : \sum_{i : c_i = c} m_i > t \cdot \sum_{i=1}^n m_i\}.$$

It selects categories which are present in at least t percent of pages displayed by user.

Our feature selectors seek for information concerning the thematic categories in which user is interested. That is why, the order of visited pages is irrelevant for them and the number of pages belonging to a given category is important only in terms of being satisfying required threshold.

5   Classifier $f_{\mathrm{simple} \subseteq}$

Now, we construct a simple classifier which will test our features selectors.

The classifier $f_{\mathrm{simple} \subseteq} : V \to P(U)$ selects a set $a_t(s)$ of categories which are present in at least t pages in input sequence. Then it returns set of users for whom set $a_t(s)$ is a subset of a set of categories of pages visited by this user during first three weeks. It is defined as:

$$f_{\mathrm{simple} \subseteq}(s) = \{u \in U : a_t(s) \subseteq a_0(g(u))\}.$$

Classifier has the following performance when used with features selected by a_t:

t	Recall	Precision	F	ClusterSize
0	0.4318	0.0119	0.0232	1211.6
1	0.6059	0.0226	0.0436	1647.3
2	0.6956	0.0312	0.0597	1910.3
3	0.7522	0.0385	0.0733	2099.7
5	0.8191	0.0504	0.0949	2356.7
7	0.8558	0.0597	0.1116	2543.1
10	0.8947	0.0721	0.1335	2747.9
15	0.9295	0.0892	0.1629	3000.6
20	0.9480	0.1044	0.1881	3177.1
25	0.9592	0.1156	0.2064	3301.4
30	0.9664	0.1268	0.2242	3405.6
35	0.9746	0.1397	0.2443	3497.0
40	0.9801	0.1512	0.2620	3567.9
50	0.9846	0.1721	0.2931	3674.5

It offers arbitrarily high recall, however it has large cluster size and small precision.

Performance of an analogical classifier that selected features according to b_t was inferior to $f_{\text{simple}\subseteq}$

Performance of classifier $f_{\text{simple}\subseteq}$ is significantly dependent from number of categories present in cluster (below, the dependence $f_{\text{simple}\subseteq}$ classifier performance from number of categories present in cluster, with threshold $t = 30$ is presented):

No cats	No users	Recall	Precision	F	ClusterSize
1	2224	0.9977	0.1797	0.3046	4276.1
2	1752	0.9657	0.1126	0.2018	3194.1
3	692	0.9031	0.0291	0.0565	1909.7
4	185	0.8648	0.0083	0.0165	1011.3
5	25	0.7600	0.0025	0.0050	420.9
6	3	0.6666	0.0059	0.0118	214.0
7	1	1.0000	0.0045	0.0091	218.0

Observe that large cluster size is a consequence of poor distinctive ability of classifier for users with small number of categories. On the other hand, for clusters specified by more than 3 categories, we observe decrease of recall. That is why, now, we will develop classifiers dedicated to clusters with a certain number of categories, then we combine them into one classifier which will work with recall 0.9.

6 Classification of Sequences Specified by One Category

In this case we know that examined user during the fourth week displayed pages belonging to only one category. The only information that may help us to indicate such user is the percent of Page Views that belong to that category during the first three weeks of user activity.

Let s be a sequence such that $a_{30}(s)$ is a singleton. Let $\mathbf{t} = (t_A, t_B, \ldots, t_T)$. The classifier $f_{\mathbf{t}} : V \to P(U)$ is defined as:

$$f_{\mathbf{t}}(s) = \{u \in U : a_{30}(s) = \{c\} \wedge c \in b_{t_c}(g(u))\}.$$

In the definition above $a_{30}(s) = \{c\}$ means that c is the only category that appears in sequence s more than 30 times and $c \in b_{t_c}(g(u))$ means that more than t_c percent of pages displayed by user u is labeled with category c.

The following table presents optimal $\mathbf{t}$ and quality of $f_{\mathbf{t}}$ classifier for single categorial clusters (Last verse describes classifier on all single categorial clusters together):

Cluster	No users	t	Recall	Precision	F	ClusterSize
G	676	0.50	0.9038	0.3790	0.5341	1612.0
H	40	0.19	0.9000	0.2000	0.3273	180.0
K	8	0.43	1.0000	0.2857	0.4444	28.0
P	58	0.03	0.9310	0.0432	0.0826	1250.0
Q	1176	0.46	0.9022	0.5247	0.6635	2022.0
joint	2224		0.9069	0.4356	0.5885	1624.1
$f_{\text{simple}\subseteq}$	2224		0.9977	0.1797	0.3046	4276.1

As we can see each cluster has its own specific character: clusters differs in size, threshold value and precision. Some clusters are simple to classify while others (see P-cluster for example) are pretty hard. In general f_t reduced average cluster size from 4276.1 to 1624.1 maintaining recall at the level 0.9.

7   Classification of Sequences Specified by Two Categories

Consider the following sequences:

$$CQCGQGQGQGQGQGQGQGQC,$$

$$FGFGFGFGFGFGFGFGQFQF,$$

$$FGFIFIFIFGFGFGFGFIF.$$

They consists of long periods of alternating occurrences of two categories. We observed that mosts of sequences have such structure.

We recognize such periods when they have at least four occurrences and we replace recognized periods with new categories. Each occurrence of such new category is assigned with the number of Page Views equal to the sum of Page Views that appear in recognized period.

For example the above sequences are processed into:

$$CQC\,\underline{GQ}\,C, \quad \underline{FG}\,FQ, \quad \underline{FG}\,\underline{FI}\,\underline{FG}\,IF.$$

While classifying sequence specified by two categories (we will denote them as c_1 and c_2) we considered following features:

- the quantity of occurrence of c_1 category in sequence,
- the quantity of occurrence of c_2 category in sequence,
- the quantity of occurrence of c_1c_2 category created from alternating occurrences of c_1 and c_2.

We analyzed performance of 4 classifiers:

$$f_{t,t'}(s) = \{u \in U : a_{30}(s) = \{c_1, c_2\} \wedge c_1 \in b_{t_{c_1c_2}}(g(u)) \wedge c_2 \in b_{t'_{c_1c_2}}(g(u))\},$$
$$f_{t''}(s) = \{u \in U : a_{30}(s) = \{c_1, c_2\} \wedge c_1c_2 \in d_{t''_{c_1c_2}}(g(u))\},$$
$$f_{t,t'\wedge t''}(s) = f_{t,t'}(s) \cap f_{t''}(s),$$
$$f_{t,t'\wedge t''}(s) = f_{t,t'}(s) \cup f_{t''}(s).$$

Performance of $f_{t,t'}$ and $f_{t''}$ that both quantity of single categories occurrences and quantity of alternating occurrences of categories carry important information. $f_{t,t'\wedge t''}$ and $f_{t,t'\wedge t''}$ tested possible combinations of this information. Test showed that $f_{t,t'\vee t''}$ classifier gives best results.

A possible reason of such behaviour is that quantity of single categories occurrences is relevant feature for short sequences and quantity of alternating category occurrences is relevant for long sequences. So, these features operate on disjoint sets of sequences and should be connected by disjunction.

The following table presents optimal parameters and performance of $f_{t,t'\vee t''}$ classifier for clusters that have more then 10 users:

Cluster	No user	t	t'	t"	Recall	Precision	F	ClusterSize
CG	15	0.01	0.01	0.88	0.9333	0.0648	0.1212	216.0
CQ	23	0.22	0.02	0.33	0.8261	0.0662	0.1226	287.0
GN	116	0.95	0.48	0.05	0.9052	0.4217	0.5753	249.0
GP	61	0.01	0.66	0.00	0.9016	0.0469	0.0891	1173.0
GQ	834	0.09	0.11	0.37	0.9005	0.2950	0.4444	2546.0
HQ	87	0.45	0.01	0.15	0.9080	0.0607	0.1138	1302.0
MQ	15	0.37	0.23	0.11	0.8000	0.4286	0.5581	28.0
PQ	99	0.79	0.00	0.04	0.9091	0.0317	0.0612	2841.0
joint	1667				0.9010	0.2431	0.3829	1714.0
$f_{\text{simple}\subseteq}$	1752				0.9657	0.1126	0.2018	3194.1

Last verse describes performance of classifier on all two categorial clusters together. Cluster have their own specific character similarly as single categorial clusters. For some of them, it was impossible to find recall greater then 0.9. Classifier reduced average cluster size from 3194.1 to 1714.0 maintaining recall at the level 0.9.

In order to avoid overfitting we used $f_{\text{simple}\subseteq}$ for all clusters that have less then 11 users.

8 Results

We defined our ultimate classifier f_{combined} as follows: for a given sequence, we select relevant categories using feature selector a_{30}, then we check the number of categories in obtained set:

- if there is only one category, we classify sequence using $f_{\mathbf{t}}$ classifier with parameters specified in section 6;
- if there are two categories and this pair of categories is present in table in section 7, we classify sequence using $f_{\mathbf{t},\mathbf{t}'\vee\mathbf{t}''}$ classifier with parameters specified in that table;
- otherwise we classify sequence by means of $f_{\text{simple}\subseteq}$ classifier.

Classifier f_{combined} has the following performance for our set of users U:
Recall$(f_{\text{combined}}, U) = 0.9015$, **Precision**$(f_{\text{combined}}, U) = 0.2860$,
$F(f_{\text{combined}}, U) = 0.4342$, **ClusterSize**$(f_{\text{combined}}, U) = 1651.9$.

Acknowledgment. The research has been partially supported by grants N N516 368334 and N N516 077837 from Ministry of Science and Higher Education of the Republic of Poland.

References

1. Angluin, D.: Inductive inference of formal languages from positive data. Information and Control 45, 117–135 (1980)
2. Guillet, F., Hamilton, H.J. (eds.): Quality Measures in Data Mining. Studies in Computational Intelligence, vol. 43. Springer, Heidelberg (2007)
3. Jaworska, J., Nguyen, H.S.: Users Behaviour Prediction Challenge. In: Nguyen, H.S. (ed.) Proc. ECML/PKDD 2007 Discovery Challenge (2007)
4. Vapnik, V.N.: Statistical Learning Theory. Wiley, New York (1998)

Hierarchical Information System and Its Properties

Qinrong Feng

School of mathematics and computer science,
Shanxi Normal University, Linfen, 041004, Shanxi, P.R. China

Abstract. Starting point of rough set based data analysis is a data set, called an information system, whose columns are labeled by attributes, rows are labeled by objects of interest and entries of the table are attribute values. In fact, hierarchical attribute values exists impliedly in many real-world applications, but it has seldom been taken into consideration in traditional rough set theory and its extensions. In this paper, each attribute in an information system is generalized to a concept hierarchy tree by considering hierarchical attribute values. A hierarchical information system is obtained, it is induced by generalizing a given flat table to multiple data tables with different degrees of abstraction, which can be organized as a lattice. Moreover, we can choose any level for any attribute according to the need of problem solving, thus we can discovery knowledge from different levels of abstraction. Hierarchical information system can process data from multilevel and multiview authentically.

Keywords: rough sets, information system, concept hierarchy, hierarchical information system.

1 Introduction

Granular computing concerns problem solving at multiple levels of abstraction, which is a powerful strategy widely used to cope with complex problems. Rough set theory, as a one of main models of granular computing, was introduced by Pawlak in 1982 [1]. In the past ten years, some extensions of rough set model have been proposed, such as the variable precision rough set model [2], the Bayesian rough set model [3], the fuzzy rough set model and the rough fuzzy set model [4], the probabilistic rough set model [5,6] etc. However, the traditional rough set theory and all of existing extensions are all process data at single level of abstraction. In essence, they are still process data from multiview (each attribute can be regarded as a view of problem solving) and single level for each view.

Some researchers have noticed that hierarchical attribute values exist impliedly and applied it to rules extraction [7,8,9]. But they were still process data at single level of abstraction, and have not considered the case of multiple levels of abstraction for each attribute. Yao [10,11] examined two special classes of granulations structures induced by a set of equivalence relations. One is based on a nested sequence of equivalence relations, and the other is based on a hierarchy

J. Yu et al. (Eds.): RSKT 2010, LNAI 6401, pp. 528–536, 2010.

on the universe. But he has not pointed out how to construct a nested sequence of equivalence relations from the given data. Wang [12] introduced concept hierarchy into rough set theory and only redefined the basic concepts of rough sets in the context of concept hierarchy, but they didn't analyze what will be after concept hierarchy was introduced in rough sets.

Hierarchical attribute values has seldom been taken into consideration in an information system, namely, most existing data set in rough data analysis are all flat data tables. Allowing for hierarchical attribute values exists impliedly in some real-world applications, In this paper, each attribute in an information system will be extended to a concept hierarchy tree. A hierarchical information system is induced, it can be organized as a lattice, in which each node represents an information system with different degree of abstraction. Moreover, in hierarchical information system, we can choose any information system with different degrees of abstraction according to the need of problem solving, thus we can discovery knowledge from different levels of abstraction [13]. Hierarchical information system really provides a framework to analyze knowledge from multiple levels of abstraction.

2   Concept Hierarchy

Concept hierarchies are used to express knowledge in concise and high-level terms, which is usually in the form of tree. In every concept hierarchy tree, the root is the name of an attribute, leaves nodes are attribute values appearing in the given data table, and internal nodes represent generalized attribute values of their child nodes. In a concept hierarchy, each level can be labelled by a number k. We stipulate that the level label of leaves nodes be assigned the number zero, and the level label for each internal node is one plus its child's level label. In this case, the generalized attribute values at level k can be called an attribute value at level k.

In [14], the authors pointed out that the granulation process transforms the semantics of the granulated entities. At the lowest level in a concept hierarchy, basic concepts are feature values available from a data table. At a higher level, a more complex concept is synthesized from lower level concepts (layered learning for concept synthesis).

As we all know, there is usually more than one possible way to building concept hierarchies for each attribute because of different users may have different preferences. So concept hierarchies are usually built by combining the given data with some relevant domain knowledge, or it can be given by domain experts directly.

3   Multilevel Information System

In this section, concept hierarchy is introduced into an information system, and each attribute is generalized to a concept hierarchy tree. When concept hierarchy trees of all attributes are given, we can choose any levels of abstraction for any attribute to satisfy the need of problem solving. Each chosen combination of different levels of abstraction for all attributes will determine a decision table uniquely. Thus, we will obtain some information systems with different degrees of abstraction, and these information systems can be organized as a lattice.

3.1 Basic Concepts

An information system can be represented by a quadruple $IS = (U, A, V, f)$, where U is a non-empty finite set of objects, called the universe, and A is a non-empty finite set of attributes, $V = \bigcup\limits_{a \in A} V_a$, where V_a is the value set of a, called the domain of a, f is an information function from U to V, which assigns particular values from domains of attributes to objects such that $f_a(x_i) \in V_a$ for all $x_i \in U$ and $a \in A$.

Definition 1. [12] Let $IS = (U, A, V, f)$ be an information system, then denote $IS_H = (U, A, V_H, H_A, f)$ as a hierarchical information system induced by IS, where $H_A = \{H_a | a \in A\}$, $H_a(a \in A)$ denote the concept hierarchy tree of attribute a. The root of H_a represents the name of attribute a and the leave nodes are measurable value of attribute a or the domain of a in the primitive information system. The internal nodes are values of a with different degrees of abstraction. $V_H = \bigcup\limits_{a \in A} V_a^{range}$, V_a^{range} denote all values of attribute a in all levels of its concept hierarchy tree.

If $A = C \cup D$ and $C \cap D = \emptyset$, then $DT_H = (U, C \cup D, V_H, H_{C \cup D}, f)$ is called hierarchical decision table.

When concept hierarchy trees of all attributes in a decision table are given, the domains of these attributes can be changed when one or more attributes climb up along their concept hierarchy trees. Correspondingly, the universe of the information system and the information function will also be changed. Thus, the information system will be changed, and each combination of the domains with different levels of abstraction for all attributes will uniquely determine an information system.

As mentioned above, we can denote a generalized information system by its attributes levels of abstraction. For example, we can denote the information system $IS_{k_1 k_2 \cdots k_m} = (U_{k_1 k_2 \cdots k_m}, A, V^{k_1 k_2 \cdots k_m}, f_{k_1 k_2 \cdots k_m})$ simply as $(k_1, k_2, \cdots, k_m)$, where $A = \{A_1, A_2, \cdots, A_m\}$ is the set of attributes, and the domain of A_1 is at k_1-th level of its concept hierarchy, the domain of A_2 is at k_2-th level of its concept hierarchy, $\cdots\cdots$, the domain of A_m is at k_m-th level of its concept hierarchy. $U_{k_1 k_2 \cdots k_m}$ is the universe, $V^{k_1 k_2 \cdots k_m}$ is the domain, and $f_{k_1 k_2 \cdots k_m}$ is the information function of the $(k_1, k_2, \cdots, k_m)$ information system.

Remark. In $U_{k_1 k_2 \cdots k_m}$, each element can be regarded as a granule represented by a set. We will omit the brace if there has only one element in a granule.

Definition 2. [15] Let (G, M, I) be a formal context, where the elements of G and M are called objects and attributes, I is a relation between them.

A concept of the context (G, M, I) is a pair (A, B) with $A \subseteq G$, $B \subseteq M$, and $A' = B$, $B' = A$, where

$$A' = \{m \in M | \, gIm \text{ for all } g \in A\} \, , B' = \{g \in G | \, gIm \text{ for all } m \in B\}$$

The set A is called the extent of the concept, the set B the intent. We call (A_1, B_1) is a subconcept of (A_2, B_2) if and only if $A_1 \subseteq A_2$ (or $B_1 \supseteq B_2$) and denotes $(A_1, B_1) \leq (A_2, B_2)$.

Definition 3. Let $IS = (U, A, V, f)$ be an information system, denote the i-th level domain of $A_t \in A$ as V_t^i, we call V_t^i is finer (more specific) than V_t^j, or V_t^j is coarser (more abstract) than V_t^i if and only if for any $a \in V_t^i$, there always exists $b \in V_t^j$, such that (U_a, a) is a subconcept of (U_b, b) (where U_a and U_b are sets of object determined by attribute value a and b respectively), and denote it by $V_t^i \leq V_t^j$.

Obviously, if $i \leq j$, then we have $V_t^i \leq V_t^j$. That is, when the domain of A_t is rolled up along its concept hierarchy from a lower level to a higher level, the values of A_t is generalized.

Definition 4. Let $IS = (U, A, V, f)$ be an information system, where $A = \{A_1, A_2, \cdots, A_m\}$ is the set of attributes, $V^{k_1 k_2 \cdots k_m}$ is the domain of A in the $(k_1, k_2, \cdots, k_m)$ information system, we call the domain $V^{i_1 i_2 \cdots i_m}$ in the $(i_1, i_2, \cdots, i_m)$ information system is finer than the domain $V^{j_1 j_2 \cdots j_m}$ in the $(j_1, j_2, \cdots, j_m)$ information system if and only if for any $k \in \{1, 2, \cdots, m\}$, $V_k^{i_k} \leq V_k^{j_k}$, denote as $V^{i_1 i_2 \cdots i_m} \leq V^{j_1 j_2 \cdots j_m}$.

If there exists at least one $k \in \{1, 2, \cdots, m\}$, such that $V_k^{i_k} < V_k^{j_k}$, then we will call $V^{i_1 i_2 \cdots i_m}$ is strictly finer than $V^{j_1 j_2 \cdots j_m}$, denote as $V^{i_1 i_2 \cdots i_m} < V^{j_1 j_2 \cdots j_m}$.

Definition 5. Denote $IS_{i_1 i_2 \cdots i_m} = (U_{i_1 i_2 \cdots i_m}, A, V^{i_1 i_2 \cdots i_m}, f_{i_1 i_2 \cdots i_m})$ as the $(i_1, i_2, \cdots, i_m)$ information system, $IS_{j_1 j_2 \cdots j_m} = (U_{j_1 j_2 \cdots j_m}, A, V^{j_1 j_2 \cdots j_m}, f_{j_1 j_2 \cdots j_m})$ as the $(j_1, j_2, \cdots, j_m)$ information system, then we will call $U_{i_1 i_2 \cdots i_m}$ is finer than $U_{j_1 j_2 \cdots j_m}$ if and only if for any $x \in U_{i_1 i_2 \cdots i_m}$, there always exists $y \in U_{j_1 j_2 \cdots j_m}$, such that $x \subseteq y$, denote as $U_{i_1 i_2 \cdots i_m} \leq U_{j_1 j_2 \cdots j_m}$.

Obviously, if $V^{i_1 i_2 \cdots i_m} \leq V^{j_1 j_2 \cdots j_m}$, then we have $U_{i_1 i_2 \cdots i_m} \leq U_{j_1 j_2 \cdots j_m}$.

Definition 6. Denote $IS_{i_1 i_2 \cdots i_m} = (U_{i_1 i_2 \cdots i_m}, A, V^{i_1 i_2 \cdots i_m}, f_{i_1 i_2 \cdots i_m})$ be the $(i_1, i_2, \cdots, i_m)$ information system, $IS_{j_1 j_2 \cdots j_m} = (U_{j_1 j_2 \cdots j_m}, A, V^{j_1 j_2 \cdots j_m}, f_{j_1 j_2 \cdots j_m})$ be the $(j_1, j_2, \cdots, j_m)$ information system, if $V^{i_1 i_2 \cdots i_m} \leq V^{j_1 j_2 \cdots j_m}$, then we call information system $IS_{i_1 i_2 \cdots i_m}$ is finer than $IS_{j_1 j_2 \cdots j_m}$, denote as $IS_{i_1 i_2 \cdots i_m} \leq IS_{j_1 j_2 \cdots j_m}$. If there exists a $A_k \in A$ such that $V_k^{i_k} < V_k^{j_k}$, then we call $IS_{i_1 i_2 \cdots i_m}$ is strictly finer than $IS_{j_1 j_2 \cdots j_m}$, denote as $IS_{i_1 i_2 \cdots i_m} < IS_{j_1 j_2 \cdots j_m}$.

3.2   Hierarchical Properties

In this subsection, some hierarchical properties existed in hierarchical information system are revealed.

For the sake of convenient description in the following, we will denote $l(A_t)(t = 1, 2, \cdots, m)$ as the level number of concept hierarchy tree of A_t.

Property 1. Let $IS_H = (U, A, V_H, H_A, f)$ be a given hierarchical information system, for any $A_t \in A = \{A_1, A_2, \cdots, A_m\}$, if $i \leq j$, then we have $V_t^i \leq V_t^j$. Especially, we have $V_t^0 \leq V_t^1 \leq \cdots \leq V_t^{l(A_t)-1}$.

This property indicates that the domain of an attribute at lower level of abstraction is finer than that at higher level of abstraction. All of these domains with different levels of abstraction formed a nested series.

Property 2. Denote $IS_{i_1 i_2 \cdots i_m} = (U_{i_1 i_2 \cdots i_m}, A, V^{i_1 i_2 \cdots i_m}, f_{i_1 i_2 \cdots i_m})$ as the $(i_1, i_2, \cdots, i_m)$ information system, $IS_{j_1 j_2 \cdots j_m} = (U_{j_1 j_2 \cdots j_m}, A, V^{j_1 j_2 \cdots j_m}, f_{j_1 j_2 \cdots j_m})$ as the $(j_1, j_2, \cdots, j_m)$ information system, if $i_1 \leq j_1, i_2 \leq j_2, \cdots, i_m \leq j_m$, then we have $V^{i_1 i_2 \cdots i_m} \leq V^{j_1 j_2 \cdots j_m}$.

Property 3. Denote $IS_{i_1 i_2 \cdots i_m} = (U_{i_1 i_2 \cdots i_m}, A, V^{i_1 i_2 \cdots i_m}, f_{i_1 i_2 \cdots i_m})$ as the $(i_1, i_2, \cdots, i_m)$ information system, $IS_{j_1 j_2 \cdots j_m} = (U_{j_1 j_2 \cdots j_m}, A, V^{j_1 j_2 \cdots j_m}, f_{j_1 j_2 \cdots j_m})$ as the $(j_1, j_2, \cdots, j_m)$ information system, if $i_1 \leq j_1, i_2 \leq j_2, \cdots, i_m \leq j_m$, then we have $U_{i_1 i_2 \cdots i_m} \leq U_{j_1 j_2 \cdots j_m}$, which means that for any $x \in U_{i_1 i_2 \cdots i_m}$, there always exists $y \in U_{j_1 j_2 \cdots j_m}$, such that $x \subseteq y$.

This property indicates that the universe at lower levels of abstraction is finer than that at higher levels of abstraction.

Property 4. Denote $IS_{i_1 i_2 \cdots i_m} = (U_{i_1 i_2 \cdots i_m}, A, V^{i_1 i_2 \cdots i_m}, f_{i_1 i_2 \cdots i_m})$ as the $(i_1, i_2, \cdots, i_m)$ information system, $IS_{j_1 j_2 \cdots j_m} = (U_{j_1 j_2 \cdots j_m}, A, V^{j_1 j_2 \cdots j_m}, f_{j_1 j_2 \cdots j_m})$ as the $(j_1, j_2, \cdots, j_m)$ information system, if $i_1 \leq j_1, i_2 \leq j_2, \cdots, i_m \leq j_m$, then we have $IS_{i_1 i_2 \cdots i_m} \leq IS_{j_1 j_2 \cdots j_m}$.

This property indicates that the information system at lower level of abstraction is finer than that at higher level of abstraction.

Property 5. Let $IS_H = (U, A, V_H, H_A, f)$ be a given hierarchical information system, where $A = \{A_1, A_2, \cdots, A_m\}$ is the set of attributes, denote

$$ISS = \{IS_{k_1 k_2 \cdots k_m} | 0 \leq k_1 \leq l(A_1) - 1, 0 \leq k_2 \leq l(A_2) - 1,$$

$$\cdots, 0 \leq k_m \leq l(A_m) - 1\}$$

as the set of information system in IS_H with different degrees of abstraction, then $(ISS, \leq)$ is a lattice, where $l(A_t)$ denote the depth of concept hierarchy of attribute A_t and the relation of $\leq$ can refer to definition 6.

Proof. Obviously, the relation $\leq$ is a partial order relation over ISS, so $(ISS, \leq)$ is a poset. In order to prove $(ISS, \leq)$ can be organized as a lattice, we only need to prove that any two elements of ISS have the supremum and infimum.

For any two information system $IS_{i_1 i_2 \cdots i_m}$, $IS_{j_1 j_2 \cdots j_m} \in ISS$, denote $IS_{i_1 i_2 \cdots i_m} \vee IS_{j_1 j_2 \cdots j_m}$ and $IS_{i_1 i_2 \cdots i_m} \wedge IS_{j_1 j_2 \cdots j_m}$ as the supremum and the infimum of $IS_{i_1 i_2 \cdots i_m}$ and $IS_{j_1 j_2 \cdots j_m}$, respectively. If we set

$$(k_1, k_2, \cdots, k_m) = (\max\{i_1, j_1\}, \max\{i_2, j_2\}, \cdots, \max\{i_m, j_m\})$$

$$(l_1, l_2, \cdots, l_m) = (\min\{i_1, j_1\}, \min\{i_2, j_2\}, \cdots, \min\{i_m, j_m\}).$$

Then we have

$$IS_{i_1 i_2 \cdots i_m} \vee IS_{j_1 j_2 \cdots j_m} = IS_{k_1 k_2 \cdots k_m}$$

and

$$IS_{i_1 i_2 \cdots i_m} \wedge IS_{j_1 j_2 \cdots j_m} = IS_{l_1 l_2 \cdots l_m}$$

that is, $IS_{k_1 k_2 \cdots k_m}$ is the supremum of $IS_{i_1 i_2 \cdots i_m}$ and $IS_{j_1 j_2 \cdots j_m}$, $IS_{l_1 l_2 \cdots l_m}$ is the infimum of $IS_{i_1 i_2 \cdots i_m}$ and $IS_{j_1 j_2 \cdots j_m}$. Thus $(ISS, \leq)$ is a lattice.

4   Example

In this section, we will show how to extend an information system to a hierarchical information system.

Example 1. Extending the information system illustrated as table 1 [16] to a hierarchical information system, and analyzing nested relations hid in it.

Obviously, table 1 is a flat table. We assume that concept hierarchy trees of attributes in table 1 are given and illustrated as figure 1,2,3. We can choose any levels of abstraction for any attributes, and each chosen combination of abstract levels for all attributes will determine an information system uniquely. All of these information systems can be organized as a lattice as figure 4.

Table 1. An information system

U	Age	Education	Occupation	U	Age	Education	Occupation
1	39	Bachelor	Adm-clerical	10	37	Some-college	Exec-managerial
2	38	Hs-grad	Handlers-cleaners	11	23	Bachelor	Adm-clerical
3	53	11th	Handlers-cleaners	12	32	Associate	Sales
4	28	Bachelor	Prof-specialty	13	40	Associate	Craft-repair
5	37	Master	Exec-managerial	14	34	7th-8th	Transport-moving
6	49	9th	Other-service	15	43	Master	Exec-managerial
7	52	Hs-grad	Exec-managerial	16	40	Doctorate	Prof-specialty
8	31	Master	Prof-specialty	17	54	Hs-grad	Other-service
9	42	Bachelor	Exec-managerial	18	35	9th	Farming-fishing

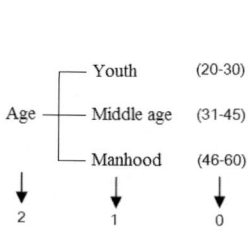

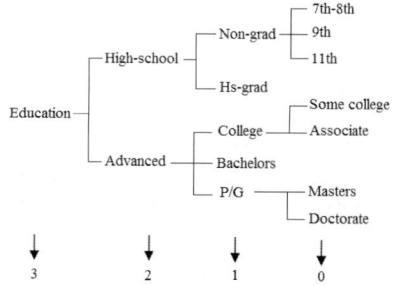

Fig. 1. Concept hierarchy of Age   **Fig. 2.** Concept hierarchy of Education

Table 2. Information system IS_{111}

U	Age	Education	Occupation
1	middle	bachelors	intellectual
2	middle	hs-grad	manual
3	manhood	nongrad	manual
$\{4, 11\}$	youth	bachelors	intellectual
5	middle	P/G	intellectual
6	manhood	nongrad	mixed
7	manhood	hs-grad	intellectual
$\{8, 15, 16\}$	middle	P/G	intellectual
9	middle	bachelors	intellectual
10	middle	college	intellectual
12	middle	college	intellectual
13	middle	college	manual
$\{14, 18\}$	middle	nongrad	manual
17	manhood	hs-grad	mixed

Table 3. Information system IS_{121}

U	Age	Education	Occupation
$\{1, 5, 12\}$	middle	advanced	intellectual
$\{2, 14, 18\}$	middle	high-school	manual
3	manhood	high-school	manual
$\{4, 11\}$	youth	advanced	intellectual
$\{6, 17\}$	manhood	high-school	mixed
7	manhood	high-school	intellectual
$\{8, 9, 10, 15, 16\}$	middle	advanced	intellectual
13	middle	advanced	manual

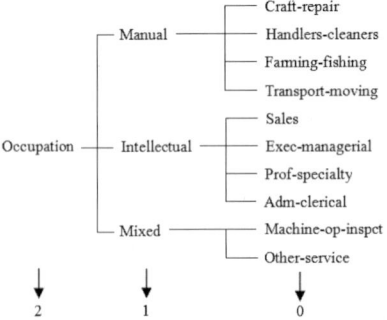

Fig. 3. Concept hierarchy of Occupation

Fig. 4. Lattice of information systems

We choose three information systems IS_{000}, IS_{111} and IS_{121} from the lattice illustrated as figure 4 and show some nested relations among these information systems. For any chosen object '8' from IS_{000}, there obviously exists partial relation among equivalence classes includes '8' in these three tables, e.g. $[8]_{000} \subseteq [8]_{111} \subseteq [8]_{121}$. Moreover, we have $V^{000} \leq V^{111} \leq V^{121}$, $U_{000} \leq U_{111} \leq U_{121}$ and $IS_{000} \leq IS_{111} \leq IS_{121}$.

5   Conclusion

In this paper, hierarchical attribute values are taken into consideration, and each attribute is generalized to a concept hierarchy tree. Thus, a flat information system is extended to a hierarchical information system, which can process data from multiple levels of abstraction. This can improve the ability of data processing for rough set theory greatly.

Acknowledgments. This work was supported by Youth Technology Research Foundation of Shanxi Province(2008021007).

References

1. Pawlak, Z.: Rough sets. International Journal of Computer and Information Sciences 11, 341–356 (1982)
2. Ziarko, W.: Variable precision rough set model. Journal of computer and system sciences 46(1), 39–59 (1993)
3. Slezak, D., Ziarko, W.: Variable precision Bayesian rough set model. In: Wang, G., Liu, Q., Yao, Y., Skowron, A. (eds.) RSFDGrC 2003. LNCS (LNAI), vol. 2639, pp. 312–315. Springer, Heidelberg (2003)
4. Dubois, D., Prade, H.: Rough fuzzy sets and fuzzy rough sets. International journal of general systems 17, 191–209 (1990)
5. Ziarko, W.: Probabilistic rough sets. In: Ślęzak, D., Wang, G., Szczuka, M.S., Düntsch, I., Yao, Y. (eds.) RSFDGrC 2005. LNCS (LNAI), vol. 3641, pp. 283–293. Springer, Heidelberg (2005)
6. Yao, Y.Y.: Probabilistic rough set approximations. International Journal of Approximate Reasoning 49(2), 255–271 (2008)
7. Hu, X., Cercone, N.: Discovery maximal generalized decision rules through horizontal and vertical data reduction. Computational Intelligence 17(4), 685–702 (2001)
8. Shan, N., Hamilton, H.J., Cercone, N.: GRG:Knowledge Discovery Using Information Generalization, Information Reduction, and Rule Generation. In: Proceedings of the seventh international conference on tools with artificial intelligence, Washington, DC, USA, pp. 372–379. IEEE Computer Society, Los Alamitos (1995)
9. Hong, T.-P., Lin, C.-E., Lin, J.-H., Wang, s.-L.: Learning cross-level certain and possible rules by rough sets. Expert Systems with Applications 34(3), 1698–1706 (2008)
10. Yao, Y.Y.: Stratified rough sets and granular computing. In: Proceedings of the 18th International Conference of the North American on Fuzzy Information Processing Society, pp. 800–804 (1999)
11. Yao, Y.Y.: Information Granulation and Rough Set Approximation. International Journal of Intelligent Systems 16(1), 87–104 (2001)
12. Wang, Q., Wang, X., Zhao, M., Wang, D.: Conceptual hierarcy based rough set model. In: Proceedings of the second of International Conference on Machine Learning and Cybernetics, pp. 402–406 (2003)
13. Feng, Q., Miao, D., Cheng, Y.: Hierarchical decision rules mining. Expert systems with applications 37(3), 2081–2091 (2010)

14. Bargiela, A., Pedrycz, W.: The roots of granular computing. In: Proceedings of the 2006 IEEE international conference on granular computing, pp. 806–809 (2006)
15. Wille, R.: Concept lattices and conceptual knowledge systems. Computers & Mathematics with Applications 23(6-9), 493–515 (1992)
16. Yi, H.Y., de la Iglesia, B., Rayward-Smith, V.J.: Using Concept Taxonomies for Effective Tree Induction. In: Hao, Y., Liu, J., Wang, Y.-P., Cheung, Y.-m., Yin, H., Jiao, L., Ma, J., Jiao, Y.-C. (eds.) CIS 2005, Part II. LNCS (LNAI), vol. 3802, pp. 1011–1016. Springer, Heidelberg (2005)

Feature-Weighted Mountain Method with Its Application to Color Image Segmentation

Wen-Liang Hung[1], Miin-Shen Yang[2,*], Jian Yu[3], and Chao-Ming Hwang[4]

[1] Graduate Institute of Computer Science, National Hsinchu University of
Education, Hsin-Chu, Taiwan
[2] Department of Applied Mathematics, Chung Yung Christian University,
Chung-Li 32023, Taiwan
[3] Department of Computer Science, Beijing Jiaotong University, Beijing, China
[4] Department of Applied Mathematics, Chinese Culture University Yangminshan,
Taipei, Taiwan
msyang@math.cycu.edu.tw

Abstract. In this paper, we propose a feature-weighted mountain clustering method. The proposed method can work well when there are noisy feature variables and could be useful for obtaining initially estimated cluster centers for other clustering algorithms. Results from color image segmentation illustrate the proposed method actually produces better segmentation than previous methods.

Keywords: Mountain method, Feature weight, Color image segmentation.

1   Introduction

Cluster analysis is a method of clustering a data set into groups. It is an approach to unsupervised learning and one of major techniques used in pattern recognition. Yager and Filev [4] proposed a simple and effective algorithm, called the mountain method, as an approximate clustering technique. Chiu [6] modified the original mountain method by considering the mountain function on the data points instead of the grid nodes. The approach is based on the density estimation in feature space with the highest potential value chosen as a cluster center and then new density estimation is created for the extraction of the next cluster center. The process is repeated until a stopping condition is satisfied. This method can be used to obtain initial guesses of cluster centers for other clustering algorithms. It can also be used for an approximate estimate of cluster centers as a clustering method.

Yang and Wu [3] created another modified mountain clustering algorithm. The proposed algorithm can automatically estimate the parameters in the modified mountain function in accordance with the structure of the data set based on the correlation of self-comparison method. However, the modified mountain function

---

[*] Corresponding author.

J. Yu et al. (Eds.): RSKT 2010, LNAI 6401, pp. 537–544, 2010.

treats all the features of equal importance. In practice, there may be some noisy variables in the data set in which these variables may influence the performance of clustering results. To solve this problem, we propose a modified algorithm, called feature-weighted mountain method. This method can work well for noisy feature variables.

The remainder of this paper is organized as follows. In Section 2, we first describe Yang and Wu's mountain clustering algorithm. We then present the feature-weighted mountain method. For estimating feature weights, we introduce an attribute weight method using a variation approach. Image segmentation is an important step for many image processing and computer vision. We use the proposed algorithm to obtain approximate cluster centers and apply it to color image segmentation. The segmentation results with comparisons are given in Section 3. Finally, we make our conclusions in Section 4.

2 The Proposed Feature-Weighted Mountain Method

Let $X = \{X_1, \cdots, X_n\}$ be a data set where $X_i = (x_{i1}, x_{i2}, \cdots, x_{ip})$, $i = 1, \cdots, n$ are feature vectors in p-dimensional Euclidean space R^p. Yang and Wu [3] modified the mountain method (cf. Yager and Filev [4]) and proposed the modified mountain function for each data vector X_i on all data points as :

$$M_1(X_i) = \sum_{j=1}^{n} \exp\left(-m\|X_i - X_j\|^2 / \sigma^2\right), i = 1, \cdots, n \tag{1}$$

where $\|X_i - X_j\|^2 = \sum_{l=1}^{p} (x_{il} - x_{jl})^2$ is the Euclidean distance between the ith data point X_i and the jth data point X_j, $\sigma^2 = \sum_{i=1}^{n} \|X_i - \overline{X}\|^2 / n$, with $\overline{X} = \sum_{i=1}^{n} X_i / n$. The parameter m in Eq.(1) is to determine the approximate density shape of the data set. Thus, the role of m is similar to the bandwidth in a kernel density estimate defined on the data set X. The kernel density estimate with the kernel K and the bandwidth h is defined by

$$\hat{f}(x) = \frac{1}{n\sigma^p h^p} \sum_{j=1}^{n} K\left(\frac{x - X_j}{\sigma h}\right)$$

In this section, we consider the standard multivariant normal density function

$$K(x) = (2\pi)^{-p/2} \exp\left(\frac{-1}{2}\|x\|^2\right)$$

Thus,

$$\hat{f}(x) = \frac{1}{n\sigma^p h^p (2\pi)^{p/2}} \sum_{j=1}^{n} \exp\left(\frac{-1}{2h\sigma^2}\|x - X_j\|^2\right) \tag{2}$$

If the underlying density is the multivariant normal, then the optimal bandwidth is given by $h_{opt} = A \cdot n^{-1/(p+4)}$ (cf. [2]) where $A = 0.96$ if $p = 2$ and $A =$

$(4/(2p+1))^{1/(p+4)}$ if $p > 2$. By comparing the estimated density function (2) with the mountain function, we obtain $m = n^{2/(p+4)}/2A^2$. We then choose $m_0 = n^{2/(p+4)}/2A^2$ as the initial value in Yang and Wu's correlation self-comparison algorithm. To implement this algorithm, the modified mountain method can be rewritten as $M_1^{m_0}(X_i) = \sum_{j=1}^{n} \exp(-m_0\|X_i - X_j\|^2/\sigma^2)$ and $M_1^{m_t}(X_i) = \sum_{j=1}^{n} \exp(-m_t\|Xi - Xj\|^2/\sigma^2)$ where $m_t = m_0 + t$, $t = 1, 2, 3 \cdots$. The correlation self-comparison procedure is summarized as follows:

S1: Set $t = 1$ and $\rho = 0.99$.
S2: Calculate the correlation between $\{M_1^{m_{(t-1)}}(X_i) \,|\, i = 1, \cdots, n\}$ and $\{M_1^{m_t}(X_i) \,|\, i = 1, \cdots, n\}$.
S3: IF the correlation is greater than or equal to the specified ρ, Then choose $M_1^{m_{(t-1)}}(X_i)$ as the modified mountain function; Else $t = t + 1$ and goto S2.

After the parameter m is estimated by the correlation self-comparison algorithm, the modified mountain function is obtained. Next, the search for the kth cluster center is used by the following modified revised mountain function

$$M_k(X_i) = M_{k-1}(X_i) - M_{k-1}(X_i) \cdot \exp(-\|Xi - X_{k-1}^*\|^2/\sigma^2), \; k = 2, 3, \cdots \quad (3)$$

where X_{k-1}^* is the $(k-1)$th cluster center which satisfies

$$M_{k-1}(X_{k-1}^*) = \max_i \{M_{k-1}(X_i)\}, \; k = 2, 3 \cdots \quad (4)$$

Thus, Yang and Wu's modified mountain clustering algorithm is summarized as follows:

S1: Obtain the modified mountain function using the correlation self-comparison algorithm.
S2: Fix the kth cluster center X_k^*, using the modified revised mountain function (3) and condition (4).

From Eq.(1), the modified mountain function treats all features of equal importance. In practice, there may be some noisy variables in the data set in which these variables may influence the performance of clustering results. To overcome this problem, we propose a feature-weighted mountain function as follows. Let $W = (w_1, \cdots, w_p)$ be the weights for p variables. We define the feature-weighted mountain function as

$$M_1^w(X_i) = \sum_{j=1}^{n} \exp(-\gamma \|X_i - X_j\|_w^2/\sigma^2), \; i = 1, \cdots, n \quad (5)$$

where $\gamma > 0$, $\sum_{l=1}^{p} w_l = 1$, $0 < w_l \leq 1$ and $\|X_i - X_j\|_w^2 = \sum_{l=1}^{p} w_l^b (x_{il} - x_{jl})^2$ with $b < 0$ or $b > 1$. In this paper, we take $b = -2$. The weighted revised mountain function can be described as follows:

$$M_k^w(X_i) = M_{k-1}^w(X_i) - M_{k-1}^w(X_i) \cdot \exp\left(\left\|Xi - X_{k-1}^*\right\|_w^2 / \sigma^2\right), \ k = 2, 3, \cdots \quad (6)$$

where X_{k-1}^* is the $(k-1)$th cluster center which satisfies

$$M_{k-1}^w(X_{k-1}^*) = \max_i \{M_{k-1}^w(X_i)\}, \ k = 2, 3, \cdots \quad (7)$$

It is known that variation plays an important role in statistics. Let us start from scratching and devising a measure of variability that uses a random sample of size n with $\{x_1, \cdots, x_n\} \subset R$, where R is the one dimensional Euclidean space. It would logically indicate what we construct should measure how the data vary from average. The sample standard deviation, s, is a measure of variability, defined as $s = \sqrt{\frac{\sum_{i=1}^n (x_i - \bar{x})^2}{n-1}}$, where $\bar{x} = \frac{1}{n} \sum_{i=1}^n x_i$. In practice, Karl Pearson's coefficient of variation (CV) has been used extensively, defined by $cv = s/\bar{x}$. On the other hand, if we have a random sample $X = \{X_1, \cdots, X_n\} \subset R^p$ and $X_i = (x_{i1}, \cdots, x_{ip})$ represents the ith sample, then the CV of the lth attribute is defined as

$$cv_l = \frac{\sqrt{\sum_{i=1}^n (x_{il} - \bar{x}_l)^2 / (n-1)}}{\bar{x}_l}, \text{ where } \bar{x}_l = \frac{1}{n} \sum_{i=1}^n x_{il}, \ l = 1, \cdots, p.$$

We know that attributes with small variations can provide more reliable information than those with large variations. Therefore, the attribute weight should be inversely related to its variation. It means that an attribute that has a large variation receives less weight than the attribute that has a smaller variation. Since attribute weights are considered to be non-negative with summation to one, they could be defined as the inverse of absolute CV values. However, we consider those applications to color image segmentation in which all data points are non-negative. Thus, the lth attribute weight w_l is proportional to $1/cv_l$ and defined as

$$w_l = \frac{1/cv_l}{\sum_{t=1}^p 1/cv_t}, l = 1, \cdots, p \quad (8)$$

3   Applications to Color Image Segmentation

For most image processing and computer vision algorithms, image segmentation is an important step. Thus, in this section, we compare the proposed method with Yang and Wu [3] with random generated initial cluster centers on color image segmentation. We use two color images shown in Figs.1~2: butterfly with the size 127×96 and clown with the size 128×128 from Kim et al. [7]. The parameters in the proposed algorithm, we set as follows: (i) the termination criteria $\varepsilon = 0.0001$; (ii) the number of clusters $k = 4$ in butterfly image and $k = 8$ in clown image; (iii) $\beta = 8$ (cf. [1]) ; (iv) the weight $w = 1/cv$. For simplicity, we choose the raw color data in the RGB color space. Thus, we run the proposed algorithm to the RGB space of these images with 10 sets of random

Fig. 1   the original butterfly image

Fig.1-1 (weighted mountain, F(I)= **6.7999**)

Fig. 1-2 (mountain, F(I)= 6.8072)

Fig. 1-3 (random, F(I)= 20.9213)

Fig. 1-4 (random, F(I)= **6.7999**)

Fig. 1-5 (random, F(I)= 20.9213)

Fig. 1-6 (random, F(I)= 6.8044)

Fig. 1-7 (random, F(I)= 6.8072)

Fig. 1-8 (random, F(I)= 20.9213)

Fig. 1-9 (random, F(I)= 20.9161)

Fig. 1-10 (random, F(I)= 20.9161)

Fig. 1-11 (random, F(I)=6.8072)

Fig. 1-12 (random, F(I)= 20.9213)

Fig. 2   the original clown image

Fig. 2-1 (weighted mountain, F(I)= **14.9012**)

Fig. 2-2

(mountain, F(I)= 20.2506)

Fig. 2-3

(random, F(I)= 18.5847)

Fig. 2-4

(random, F(I)= 14.9018)

Fig. 2-5

(random, F(I)= 43.1246)

Fig. 2-6

(random, F(I)= 41.9133)

Fig. 2-7

(random, F(I)= 23.8368)

Fig. 2-8

(random, F(I)= 25.6799)

Fig. 2-9

(random, F(I)= 58.6490)

Fig. 2-10

(random, F(I)= 49.6979)

Fig. 2-11

(random, F(I)= 26.4106)

Fig. 2-12

(random, F(I)= 41.5850)

generated initial cluster centers, the proposed method and Yang and Wu's [3] with the same initial attribute weights. The segmentation results of these images are shown in Figs.1-1~1-12 and 2-1~2-12. Figures 1-1 and 2-1 are segmentation results with the initial cluster centers using the proposed approach. Figures 1-2 and 2-2 are segmentation results with initial cluster centers using the Yang and Wu's [3]. Figures 1-3~1-12 and 2-3~2-12 are segmentation results by the 10 sets of random generated initial cluster centers. To evaluate the results of color image segmentation, it is necessary for us to make a quantitative comparison of segmented images by different initial cluster centers in the proposed algorithm.

The following evaluation function F(I) given by Liu and Yang [5] is used for our comparisons

$$F(I) = \sqrt{R} \times \sum_{i=1}^{R} \frac{(e_i/256)^2}{\sqrt{A_i}}$$

where I is the segmented image, R, the number of regions in the segmented image, A_i, the area, or the number of pixels of the ith region, and ei, the color error of region i. The ei is defined as the sum of the Euclidean distance of the color vectors between the original image and the segmented image of each pixel in the region. In this paper, R is equal to k. Note that the smaller the value of $F(I)$ is, the better the segmentation result should be. Figures 1~2 also show the values of $F(I)$ corresponding to segmented images. From Figs.1~2, we find that for butterfly and clown images, the proposed approach (Fig.1-1, Fig.2-1) has better segmentation results than Yang and Wu's [3] and these randomly generated initial cluster centers. According to the evaluation function $F(I)$, we can also see that the performance of the proposed approach is better than Yang and Wu's [3] and these randomly generated initial cluster centers.

4   Conclusions

We proposed a feature-weighted mountain clustering method so that it can work well for noisy feature variables. The proposed method can be also for obtaining initial estimate cluster centers for other clustering algorithms. Results from color image segmentation with evaluation function illustrate the proposed method actually produces better segmentation than previous methods.

References

1. Huang, J.Z., Ng, M.K., Rong, H., Li, Z.: Automated Variable Weighting in K-Means Type Clustering. IEEE Trans. Pattern Anal. Machine Intelligence 27, 657–668 (2005)
2. Silverman, B.W.: Density Estimation for Statistics and Data Analysis. Chapman & Hall, New York (1986)
3. Yang, M.S., Wu, W.L.: A Modified Mountain Clustering Algorithm. Pattern Anal. Applic. 8, 125–138 (2005)
4. Yager, R.R., Filev, D.P.: Approximate Clustering Via the Mountain Method. IEEE Trans. Syst. Man Cybern. 24, 1279–1284 (1994)

5. Liu, J., Yang, Y.H.: Multiresolution Color Image Segmentation Technique. IEEE Trans. on Pattern Analysis and Machine Intelligence 16, 689–700 (1994)
6. Chiu, S.L.: Fuzzy Model Identification Based on Cluster Estimation. J. Intel. Fuzzy Syst. 2, 267–278 (1994)
7. Kim, D.W., Lee, K.H., Lee, D.: A Novel Initialization for the Fuzzy C-Means Algorithm for Color Clustering. Pattern Recognition Letters 25, 227–237 (2004)

An Improved FCM Clustering Method for Interval Data

Shen-Ming Gu[1], Jian-Wen Zhao[1,2], and Ling He[3]

[1] School of Mathematics, Physics and Information Science,
Zhejiang Ocean University, Zhoushan, Zhejiang 316000, China
[2] Xiaoshan School, Zhejiang Ocean University, Hangzhou, 311200, China
[3] School of Foreign Languages, Zhejiang Ocean University, Zhoushan 316000, China
gsm@zjou.edu.cn, hxzjw@zjou.edu.cn, heling9707@163.com

Abstract. In fuzzy c-means (FCM) clustering algorithm, each data point belongs to a cluster with a degree specified by a membership grade. Furthermore, FCM partitions a collection of vectors in c fuzzy groups and finds a cluster center in each group so that the objective function is minimized. This paper introduces a clustering method for objects described by interval data. It extends the FCM clustering algorithm by using combined distances. Moreover, simulated experiments with interval data sets have been performed in order to show the usefulness of this method.

Keywords: Clustering; combined distances; FCM; interval data.

1 Introduction

Clustering is to partition an unlabelled sample set into subsets according to some certain criteria [16]. In clustering, homogeneous samples are aggregated into same cluster. So it can be used to determine the closeness quantificationally among objects under study. Thus, valid classification and analysis can be achieved [11].

In recent years the synthesis between clustering and fuzzy theory has led to the development of fuzzy clustering [10]. Fuzzy c-means (FCM) algorithm is one of the most important fuzzy clustering algorithm [14]. The FCM algorithm was originally developed by Dunn [9], and later generalised by Bezdek [1]. Compared with other clustering algorithms, the FCM has a higher efficiency [12]. Therefore, FCM has been widely used in various application fields [5].

In classical FCM clustering methods, data is represented by a $n \times s$ matrix where n individuals take exactly one value for each variable [15]. However, this model is too restrictive to represent data with more complex information. Clustering analysis has extended the classical tabular model by allowing interval data for each variable. Interval data may occur in many different situations. We may have native interval data, describing ranges of variable values, for instance, daily stock prices or monthly temperature ranges. Interval data may also arise from the aggregation of data bases, when real values are generalized by intervals.

Clustering interval data is an interesting issue in data mining. For example, De Carvalho et al. [7] have proposed a partitioning clustering method following the dynamic clustering approach and using an L_2 distance. Chavent and

J. Yu et al. (Eds.): RSKT 2010, LNAI 6401, pp. 545–550, 2010.

Lechevallier [4] proposed a dynamic clustering algorithm for interval data where the class representatives are defined by an optimality criterion based on a modified Hausdorff distance.

This paper mainly focuses on some techniques for clustering interval data. The rest of the paper is organized as follows. In Section 2, we give some basic notions related to interval data. In Section 3, we review the classical definition of the distance, and give a new definition of combined distances for interval data. In Section 4, an improved FCM clustering method is proposed. In order to validate this new method, in Section 5 we provide the experiments with simulated interval data sets. We then conclude the paper with a summary in Section 6.

2   Interval Data

There are numerous situations where the information gathered for n individuals $\{x_1, x_2, \ldots, x_n\}$, each individual x_i is described by s interval-type variables $\{a_1, a_2, \ldots, a_s\}$, such that each (x_i, a_j) is an interval $[l_{ij}, u_{ij}]$ and the i-th row $([l_{i1}, u_{i1}], [l_{i2}, u_{i2}], \ldots, [l_{is}, u_{is}])$ describes the individual x_i (see Table 1).

Table 1. Interval data

X	a_1	a_2	$\ldots$	a_s
x_1	$[l_{11}, u_{11}]$	$[l_{12}, u_{12}]$	$\ldots$	$[l_{1s}, u_{1s}]$
x_2	$[l_{21}, u_{21}]$	$[l_{22}, u_{22}]$	$\ldots$	$[l_{2s}, u_{2s}]$
$\ldots$	$\ldots$	$\ldots$		$\ldots$
x_n	$[l_{n1}, u_{n1}]$	$[l_{n2}, u_{n2}]$	$\ldots$	$[l_{ns}, u_{ns}]$

This is equivalent to considering the s-dimensional interval $x_i = [l_{i1}, u_{i1}] \times [l_{i2}, u_{i2}] \times \ldots \times [l_{is}, u_{is}]$ in the space $\Re^s$. Note that if $l_{ij} = u_{ij}$ for some (x_i, a_j), this means that the underlying cell is single-valued.

For a cell (x_i, a_j) contains an interval $[l_{ij}, u_{ij}]$, let

$$c_{ij} = \frac{l_{ij} + u_{ij}}{2}, r_{ij} = \frac{u_{ij} - l_{ij}}{2}, (i = 1, \ldots, n, j = 1, \ldots, s). \tag{1}$$

c_{ij} denotes the midpoint of the interval $[l_{ij}, u_{ij}]$, and r_{ij} denotes the radius. The lower and upper bounds of the interval can be denoted as follows:

$$l_{ij} = c_{ij} - r_{ij}, u_{ij} = c_{ij} + r_{ij}, (i = 1, \ldots, n, j = 1, \ldots, s). \tag{2}$$

c_i denotes the vector of midpoint c_{ij}, and r_i denotes the vector of radius r_{ij}. They can be denoted as follows:

$$c_i = <c_{i1}, c_{i2}, \ldots, c_{is}>, r_i = <r_{i1}, r_{i2}, \ldots, r_{is}>, (i = 1, \ldots, n). \tag{3}$$

3   Distance Measure

3.1   Classical Distance

An important component of a clustering algorithm is the distance measure between data points. Given two vectors $x_i, x_j \in R^s$, where $x_i = <x_{i1}, x_{i2}, \ldots, x_{is}>$ and $x_j = <x_{j1}, x_{j2}, \ldots, x_{js}>$, there are some most frequently used distances between x_i and x_j.

- Minkowski distance (L_p): $d(x_i, x_j) = \left(\sum_{k=1}^{s} |x_{ik} - x_{jk}|^p \right)^{1/p}$.

- Manhattan distance (L_1): $d(x_i, x_j) = \sum_{k=1}^{s} |x_{ik} - x_{jk}|$.

- Euclidean distance (L_2): $d(x_i, x_j) = \sqrt{\sum_{k=1}^{s} (x_{ik} - x_{jk})^2}$.

- Chebychev distance (L_∞): $d(x_i, x_j) = \max_{k=1,\ldots,s} (|x_{ik} - x_{jk}|)$.

3.2   Distance for Interval Data

However, there are no general theoretical guidelines for selecting a measure for any given application. Some researchers use the L_2 distance to measure interval data. In [7] a distance defined as $d(x_i, x_j) = \sqrt{\sum_{k=1}^{s}[(l_{ik} - l_{jk})^2 + (u_{ik} - u_{jk})^2]}$.

Some researchers use the L_p distance to measure interval data. In [13] a distance defined as $d(x_i, x_j) = \sum_{k=1}^{s} (|l_{ik} - l_{jk}|^p + |u_{ik} - u_{jk}|^p)^{1/p}$.

Some researchers use the L_1 distance to measure interval data. In [3] a distance defined as $d(x_i, x_j) = \sum_{k=1}^{s} (|c_{ik} - c_{jk}| + |r_{ik} - r_{jk}|)$.

Some researchers use the L_∞ distance to measure interval data. In [8] a weighted L_∞ distance defined as $d(x_i, x_j) = \sum_{k=1}^{s} \lambda_k \max(|l_{ik} - l_{jk}|, |u_{ik} - u_{jk}|)$, where $\lambda_k > 0 (k = 1, \ldots, s)$ and $\Pi_{k=1}^{s} \lambda_k = 1$.

In this paper, we use a distance defined as

$$D(x_i, x_j) = \omega_c d_c(x_i, x_j) + \omega_r d_r(x_i, x_j), \tag{4}$$

where $d_c(x_i, x_j) = \sqrt{\sum_{k=1}^{s}(c_{ik} - c_{jk})^2}, d_r(x_i, x_j) = \sum_{k=1}^{s} |r_{ik} - r_{jk}|$, and $\omega_c, \omega_r \in [0, 1]$, $\omega_c + \omega_r = 1$. ω_c is the weight for the midpoints distance, ω_r is the weight for the radius distance, respectively. Obviously, $d_c(x_i, x_j)$ is a L_2 distance, $d_r(x_i, x_j)$ is a L_1 distance. And $D(x_i, x_j)$ also is a kind of distance, which combines L_1 distance and L_2 distance.

4   Fuzzy Clustering

4.1   FCM Clustering Method

The fuzzy c-means algorithm has successful applications in a wide variety of clustering problems. To describe the algorithm, we set some notations. The set of all points considered is $X = \{x_1, x_2, \ldots, x_n\}$. We write $u_q : X \to [0, 1]$ for the qth cluster, $(q = 1, \ldots, c)$, c is a fixed number of clusters. We use u_{qk} to denote $u_q(x_k)$, i.e. the grade of membership of x_k in the qth cluster. And we use $U = \langle u_{qk} \rangle$ for the fuzzy partition matrix of all membership values. The centroid of the qth cluster is v_q, and is computed by

$$v_q = \frac{\sum\limits_{k=1}^{n} (u_{qk})^m x_k}{\sum\limits_{k=1}^{n} (u_{qk})^m}, (q = 1, \ldots, c). \tag{5}$$

A parameter m, $1 \le m < \infty$, will be used as a weighted exponent, and the particular choice of value m is application dependent [2]. For $m = 1$, it coincides with the HCM algorithm, and if $m \to \infty$, then all u_{qk} values tend to $1/c$ [6]. Membership in fuzzy clusters must fulfil the following conditions

$$\sum_{q=1}^{c} u_{qk} = 1, (k = 1, \ldots, n), 0 < \sum_{k=1}^{n} u_{qk} < n, (q = 1, \ldots, c). \tag{6}$$

And the fuzzy partition matrix is computed by

$$u_{qk} = \left(\sum_{j=1}^{c} \left(\frac{\|x_k - v_q\|^2}{\|x_k - v_j\|^2} \right)^{\frac{1}{m-1}} \right)^{-1}, \tag{7}$$

where the measure $\| \bullet \|$ is the Euclidean distance. The objective of the clustering algorithm is to select u_q so as to minimize the objective function

$$J_m = \sum_{q=1}^{c} \sum_{k=1}^{n} (u_{qk})^m \|x_k - v_q\|^2. \tag{8}$$

The following is the algorithm for FCM clustering:
  Step 1: Set the values for c, m and ε.
  Step 2: Initialize fuzzy partition matrix $U^{(1)}$.
  Step 3: Calculate the objective function $J_m^{(1)}$.
  Step 4: Set the iteration number $b = 1$.
  Step 5: Calculate the centroid $v_q^{(b)}$ with $U^{(b)}$ for each cluster.
  Step 6: Calculate the fuzzy partition matrix $U^{(b+1)}$.
  Step 7: Calculate the objective function $J_m^{(b+1)}$.
  Step 8: If $|J_m^{(b+1)} - J_m^{(b)}| < \varepsilon$, stop; otherwise, set $b = b + 1$ and go to Step 5.

4.2   Improved FCM for Interval Data

In order to cluster interval data, we modify the objective function as follow:

$$J_m = \sum_{q=1}^{c} \sum_{k=1}^{n} (u_{qk})^m (D(x_k - v_q))^2. \tag{9}$$

The objective function J_m can be optimized with respect to the midpoint centroid, radius centroid and the fuzzy partition matrix in a similar manner to the FCM model. In particular, the fuzzy partition matrix is computed by

$$u_{qk} = \left(\sum_{j=1}^{c} \left(\frac{(D(x_k - v_q))^2}{(D(x_k - v_j))^2} \right)^{\frac{1}{m-1}} \right)^{-1}. \tag{10}$$

The midpoints centroid and radius centroid are, respectively:

$$vc_q = \frac{\sum_{k=1}^{n} (u_{qk})^m c_k}{\sum_{k=1}^{n} (u_{qk})^m}, vr_q = \frac{\sum_{k=1}^{n} (u_{qk})^m r_k}{\sum_{k=1}^{n} (u_{qk})^m}, (q = 1, \ldots, c). \tag{11}$$

5 Experiments

In this section we apply the previous method to experiments with simulated interval data. The data points were divided into three groups.

Data set 1 is generated according to the following parameters:

- Class 1: $u_1 = 15$, $u_2 = 35$, $\sigma_1^2 = 25$, $\sigma_2^2 = 45$, and $r_1, r_2 \in [4, 9]$;
- Class 2: $u_1 = 42$, $u_2 = 72$, $\sigma_1^2 = 20$, $\sigma_2^2 = 40$, and $r_1, r_2 \in [3, 8]$;
- Class 3: $u_1 = 65$, $u_2 = 35$, $\sigma_1^2 = 15$, $\sigma_2^2 = 20$, and $r_1, r_2 \in [3, 8]$.

Data set 2 is generated according to the following parameters:

- Class 1: $c_1 = c_2 = 50$, $u_1 = 2$, $u_2 = 3$, $\sigma_1^2 = 1.2$, $\sigma_2^2 = 2.2$;
- Class 2: $c_1 = c_2 = 50$, $u_1 = 7$, $u_2 = 8$, $\sigma_1^2 = 1.2$, $\sigma_2^2 = 2.2$;
- Class 3: $c_1 = c_2 = 50$, $u_1 = 13$, $u_2 = 14$, $\sigma_1^2 = 1.2$, $\sigma_2^2 = 2.2$.

Data set 3 is generated according to the following parameters:

- Class 1: $u_{11} = 25$, $u_{12} = 40$, $\sigma_{11}^2 = 35$, $\sigma_{12}^2 = 35$, $u_{21} = 8$, $u_{22} = 9$, $\sigma_{21}^2 = 1.2$, $\sigma_{22}^2 = 2.2$;
- Class 2: $u_{11} = 46$, $u_{12} = 60$, $\sigma_{11}^2 = 35$, $\sigma_{12}^2 = 40$, $u_{21} = 6$, $u_{22} = 7$, $\sigma_{21}^2 = 1.2$, $\sigma_{22}^2 = 2.2$;
- Class 3: $u_{11} = 60$, $u_{12} = 40$, $\sigma_{11}^2 = 25$, $\sigma_{12}^2 = 30$, $u_{21} = 4$, $u_{22} = 5$, $\sigma_{21}^2 = 1.2$, $\sigma_{22}^2 = 2.2$.

The simulated data sets have been produced several times. And the proposed method was applied to those data sets. Experimental results have shown that the improved FCM algorithm perform well in clustering interval data.

6 Conclusions

In this paper, an improved FCM clustering method for clustering interval data with combined distances was presented. Experiments with simulated interval data sets showed the usefulness of this clustering method. Our future work will focus on the following issues: to conduct more experiments to validate the scalability of our approach, and meanwhile to develop other methods by combining various types of real world data.

Acknowledgments. This work is supported by grants from the National Natural Science Foundation of China (No. 60673096), and Scientific Research Project of Science and Technology Department of Zhejiang, China(No. 2008C13068), and the Scientific Research Project of Science and Technology Bureau of Zhoushan, Zhejiang, China (No. 091057).

References

1. Bezdek, J.C.: Cluster validity with fuzzy sets. Journal of Cybernetics 4, 58–73 (1974)
2. Bezdek, J.C.: Pattern Recognition with Fuzzy Objective Function Algorithms. Plenum, New York (1981)
3. Chavent, M., de Carvalho, F.A.T., Lechevallier, Y., et al.: New clustering methods for interval data. Computational Statistics 21, 211–229 (2006)
4. Chavent, M., Lechevallier, Y.: Dynamical clustering algorithm of interval data: optimization of an adequacy criterion based on hausdorff distance. In: Classification, Clustering and Data Analysis, pp. 53–59. Springer, Heidelberg (2002)
5. Chen, S., Zhang, D.Q.: Robust image segmentation using FCM with spatial constraints based on new kernel-induced distance measure. IEEE Transactions on Systems, Man and Cybernetics-B 34, 1907–1916 (2004)
6. Choe, H., Jordan, J.B.: On the optimal choice of parameters in a fuzzy c-means algorithm. In: Proc. IEEE International Conference on Fuzzy Systems, pp. 349–354. IEEE Press, San Diego (1992)
7. De Carvalho, F.A.T., Brito, P., Bock, H.H.: Dynamic clustering for interval data based on L_2 distance. Computational Statistics 21, 231–250 (2006)
8. De Carvalho, F.A.T., de Souza, R.M.C.R., Silva, F.C.D.: A clustering method for symbolic interval-type data using adaptive chebyshev distances. In: Bazzan, A.L.C., Labidi, S. (eds.) SBIA 2004. LNCS (LNAI), vol. 3171, pp. 266–275. Springer, Heidelberg (2004)
9. Dunn, J.C.: A fuzzy relative of ISODATA process and its use in detecting compact well-separated clusters. Journal of Cybernetics 3, 32–57 (1973)
10. Jian, J.Q., Ma, C., Jia, H.B.: Improved-FCM-based readout segmentation and PRML detection for photochromic optical disks. In: Wang, L., Jin, Y. (eds.) FSKD 2005. LNCS (LNAI), vol. 3613, pp. 514–522. Springer, Heidelberg (2005)
11. Li, J., Gao, X.B., Tian, C.N.: FCM-based clustering algorithm ensemble for large data sets. In: Wang, L., Jiao, L., Shi, G., Li, X., Liu, J. (eds.) FSKD 2006. LNCS (LNAI), vol. 4223, pp. 559–567. Springer, Heidelberg (2006)
12. Pal, N.R., Bezdek, J.C.: On cluster validity for the fuzzy c-mean model. IEEE Transactions on Fuzzy Systems 3, 370–379 (1995)
13. Souza, R.M.C.R., Salazar, D.R.S.: A non-linear classifier for symbolic interval data based on a region oriented approach. In: Köppen, M., Kasabov, N., Coghill, G. (eds.) ICONIP 2008, Part II. LNCS, vol. 5507, pp. 11–18. Springer, Heidelberg (2009)
14. Yu, J., Cheng, Q.S., Huang, H.K.: Analysis of the weighting exponent in the FCM. IEEE Transactions on Systems, Man and Cybernetics-B 34, 634–639 (2004)
15. Yu, J., Yang, M.S.: Optimality test for generalized FCM and its application to parameter selection. IEEE Transactions on Fuzzy Systems 13, 164–176 (2005)
16. Zhang, J.S., Leung, Y., Xu, Z.B.: Clustering methods by simulating visual systems. Chinese Journal of Computers 24, 496–501 (2001)

An Improved FCM Algorithm for Image Segmentation[*]

Kunlun Li[1], Zheng Cao[1], Liping Cao[2], and Ming Liu[1]

[1] College of Electronic and Information Engineering, Hebei University,
Baoding 071002, China
[2] Department of Electrical and Mechanical Engineering,
Baoding Vocational and Technical College, Baoding 071002, China
likunlun@hbu.edu.cn

Abstract. A novel image segmentation method based on modified fuzzy c-means (FCM) is proposed in this paper. By using the neighborhood pixels as spatial information, a spatial constraint penalty term is added in the objective function of standard FCM to improve the robustness. Meanwhile, the neighbor pixels information is used selectively to reduce computing time and iterations. Experiments on real images segmentation proved the availability of the improvement.

Keywords: Fuzzy C-Means; Image Segmentation; Spatial Constraint.

1 Introduction

In research and application of images, only some regions of an image can attract people's interest. These regions are called targets or foreground, corresponding to specific regions that have unique nature. In order to identify and analyze the targets, these regions need to be partitioned out and analysis. Image segmentation is the technology and process that an image is divided into a number of non-overlapping regions which have consistency in properties and extract the interesting target.

Fuzzy c-means (FCM) clustering is one of the most widely used algorithms in real application. FCM was first proposed by Dunn[1], and modified by Bezdek[2] who provided the iterative optimization method based on least squares principle and proved the convergence of the algorithm[3]. FCM is widely and successfully used in magnetic resonance images[4], [5], remote sensing satellite images[6].

In order to overcome the drawbacks of traditional clustering algorithms, many methods have been proposed by incorporating local spatial information[5], [6], [7], [8]. Among them, Ahmed et al. proposed a bias-corrected FCM[8] (BCFCM) by regularizing the objective function with a spatial neighborhood regularization term. Chen and Zhang[8] proposed FCM_S which improved the BCFCM

---

[*] This work is supported by the National Natural Science Foundation of China No.60773062, the Natural Science Foundation of Hebei Province of China No.F2009000215 and Scientific Research Plan Projects of Hebei Educational Bureau No.2008312.

J. Yu et al. (Eds.): RSKT 2010, LNAI 6401, pp. 551–556, 2010.

to lower the computational complexity by replacing the local neighborhood of their mean or median value. However, the choice of mean or median value needs prior knowledge. In this paper,we choose the local spatial neighbors selectively. It shows more effective than FCM and FCM_S.

2   FCM Clustering

Based on the concept of fuzzy c-partition defined by Ruspini, Dunn1 extended the crisp c-means to fuzzy c-means. In order to make the extension make sense, Dunn extended the objective function of within group sum of square error to within group sum of square weighted error version. It is given as follows:

$$J(U, V) = \sum_{i=1}^{n} \sum_{j=1}^{c} u_{ij}^2 \cdot d^2 (x_i, \nu_j) . \tag{1}$$

3   Modified FCM

If images affected by noise are segmented by standard FCM, the segmentation performance of anti-noise would be bad. But images are often affected by noise when imaging, storage or transmission. So the noise must be considered in image segmentation to suppress effect of noise.

By adding a spatial constraint term in objective function of FCM, Ahmed et al proposed BCFCM (Bias-Corrected FCM) [7]. The new objective function is:

$$J_m = \sum_{i=1}^{c} \sum_{k=1}^{n} u_{ik}^p \|x_k - \nu_i\|^2 + \frac{\alpha}{N_R} \sum_{i=1}^{c} \sum_{k=1}^{n} u_{ik}^p \left(\sum_{x_\gamma \in N_k} \|x_\gamma - \nu_i\|^2 \right) . \tag{2}$$

In (2), $[u_{ik}] = U$ is membership degree matrix, N_k stands for the set of neighbors falling into a window around x_k and N_R is its cardinality. The parameter α in the second term controls the effect of the penalty.

A shortcoming of (2) is that computing the neighborhood terms will take much more time than FCM. When the candidate image is large in size, the computing time is intolerable. Based on (2), Chen et al proposed a simplification approach with low-complexity called FCM_S[8], their new objective function is:

$$J_m = \sum_{i=1}^{c} \sum_{k=1}^{N} u_{ik}^m \|x_k - \nu_i\|^2 + \alpha \sum_{i=1}^{c} \sum_{k=1}^{N} u_{ik}^m \|\overline{x_k} - \nu_i\|^2 . \tag{3}$$

Where $\overline{x_k}$ is a mean or median of neighboring pixels lying within a window around x_k. Unlike BCFCM computing the distances among pixels in N_k and prototype, FCM_S use $\overline{x_k}$ instead the set of neighbors, and $\overline{x_k}$ can be computed in advance, therefore, computing complexity and computing time can be reduced.

4 An Improved FCM Algorithm

A shortcoming of FCM_S is how to choose of $\overline{x_k}$. When the noise in images is Gaussian noises, take $\overline{x_k}$ be the mean of neighbors set, and when the noise is salt and pepper noises, take $\overline{x_k}$ be the median. But in real application, the type of noise is unpredictable. To deal with this problem, we propose an improved method - spatial constraint FCM (SCFCM).

We use two kind of information in images, the gray value and space distributed structure. Taking a window for pixel x_k in an image, neighbors in the window constitute a set. Based on the relevance of nearby pixels, the neighbors in the set should be similar in feature value. When the image is affected by noises, the gray values of pixels suffer interference. But the relevance of the neighbors in the set still exists. So we realign the neighbors in the set and compute the average of several pixels in such case. Take the average as spatial constraint of and add it to the objective function of FCM. So, our new objective function is:

$$J_m = \sum_{i=1}^{c}\sum_{k=1}^{N} u_{ik}^{m}\|x_k - \nu_i\|^2 + \frac{\alpha}{2n+1}\sum_{i=1}^{c}\sum_{k=1}^{N} u_{ik}^{m}\left(\sum_{\gamma=k-n}^{k+n}\|x_\gamma - \nu_i\|^2\right) . \quad (4)$$

Where, x_γ is the average, $2n+1$ is the length of side of rectangular window used. Using the new computing method of x_γ, the type of noise is no longer considered. Adopting uniform computing fashion for two types of noises, the algorithm can be played better in real application.

Subject to:

$$U \in \left\{ u_{ik} \in [0,1] \,|\, \sum_{i=1}^{c} u_{ik} = 1 \ \forall k \ and \ 0 < \sum_{k=1}^{N} u_{ik} < N \ \forall i \right\} . \quad (5)$$

Solve this optimization problem by Lagrange multiplier method as follows:

$$F_m = \sum_{j=1}^{c}\sum_{x=1}^{n}\left(u_{ij}^{m} \cdot d_{ij}^2 + \frac{\alpha}{2n+1} u_{ij}^{m} \cdot d_{\gamma j}^2 \right) + \lambda\left(\sum_{j=1}^{c} u_{ij} - 1\right) . \quad (6)$$

The first order requirement of optimization is:

$$\begin{cases} \frac{\partial F_m}{\partial \lambda} = \left(\sum_{j=1}^{c} u_{ij} - 1\right) = 0 \\ \frac{\partial F_m}{\partial u} = \left[m(u_{ij})^{m-1} \cdot d_{ij}^2 + \frac{\alpha}{2n+1} m(u_{ij})^{m-1} \cdot d_{\gamma j}^2 - \lambda \right] = 0 \end{cases} . \quad (7)$$

The last iteration formula is:

$$\begin{cases} u_{ik} = \dfrac{\left(\|x_k - \nu_i\|^2 + \frac{\alpha}{2n+1}\sum_{\gamma=k-n}^{k+n}\|x_k - \nu_i\|^2\right)^{-1/(m-1)}}{\sum_{j=1}^{c}\left(\|x_k - \nu_i\|^2 + \frac{\alpha}{2n+1}\sum_{\gamma=k-n}^{k+n}\|x_k - \nu_i\|^2\right)^{-1/(m-1)}} \\ \nu_i = \dfrac{\sum_{k=1}^{n} u_{ik}^{m}\left(x_k + \frac{\alpha}{2n+1}\sum_{\gamma=k-n}^{k+n} x_\gamma\right)}{(1+\alpha)\sum_{k=1}^{n} u_{ik}^{m}} \end{cases} . \quad (8)$$

5   Experiment Results and Analysis

5.1   Experiment Results

To inspect and verify the effectively and availability of our algorithm, we compare the proposed SCFCM with FCM and FCM_S.

Experiment 1: Comparison on segment results

Consider three situations: Gaussian noise (mean is set to 0, variance is set to 0.02), "salt and pepper" noises (density is set to 0.1) and the mixture of the

Fig. 1. Segments result on Gaussian noises (There, a1 is the image affected by Gaussian noises; a2 is the result segmented by FCM; a3 is the result segmented by FCM_S1; a4 is the result segmented by SCFCM)

Fig. 2. Segment results on "salt and pepper" noises (There, a1 is the image affected by salt and pepper noises; a2 is the result segmented by FCM; a3 is the result segmented by FCM_S2; a4, is the result segmented by SCFCM)

Fig. 3. Segment results on mixture noises (There, a1 is the image affected by mixture noises; a2 is the result segmented by FCM; a3 is the result segmented by FCM_S1; a4 is the result segmented by FCM_S2; a5 is the result segmented by SCFCM)

two type noises. Figure 1 gives the result of Gaussian noises. Figure 2 shows the result of "salt and pepper" noises. Figure 3 is the result of mixture noises. There, name the FCM_S to FCM_S1 and FCM_S2 when $\overline{x_k}$ be mean or median.

Experiment 2: Comparison of efficiency

To test the efficiency and rate of convergence, contrast the computing time and iterations of the three algorithms on the images used in experiment 1. All data are average results of 300 trials to exclude random factors and reflect fair. The results are show in the following tables.

Table 1. Compare result on computing time (seconds)

Image	Noises	FCM	FCM_S1	FCM_S2	SCFCM
Cameraman	Gaussian	0.98	4.40	–	3.95
Cameraman	Salt and Pepper	0.71	–	3.66	3.89
Cameraman	Mixture	0.96	4.60	4.61	4.49

Table 2. Compare on iterations

Image	Noises	FCM	FCM_S1	FCM_S2	SCFCM
Cameraman	Gaussian	28.79	25.43	–	23.68
Cameraman	Salt and Pepper	18.99	–	21.95	21.62
Cameraman	Mixture	28.63	27.80	25.74	24.57

5.2   Experiment Analysis

From the results in experiment 1, we can see that the performance of SCFCM is not only better than FCM, but also excels FCM_S.

From figure 3, the images are affected by Gaussian noises and "salt and pepper" noises simultaneously, the results segmented by FCM_S1 and FCM_S2 are both undesirable. The results of SCFCM are much better than FCM_S, and the type of noise needs not to be considered. So little priori knowledge is request in SCFCM.

From results of experiment 2, the computing time of SCFCM is obvious less than FCM_S when the noise type is Gaussian and mixture, slightly more than FCM_S when salt and pepper. The iterations used in SCFCM are less than FCM_S in all situations. It proves that SCFCM has faster rate of convergence and better performance.

6   Conclusion

A new image segmentation method is proposed based on improved FCM in this paper. The gray value and spatial information are used in our work simultaneously. Spatial constraint is added in the objective function of FCM as a penalize term to improve the performance of anti-noise. The neighbor pixels information is used selectively to reduce computing time and iterations. The experimental results show that our method is effective and available.

References

1. Dunn, J.C.: A fuzzy relative of the ISODATA process and its use in detecting compact well separated cluster. J. Cybernet. 3, 32–57 (1973)
2. Bezdek, J.C.: Pattern Recognition with Fuzzy Objective Function Algorithms. Plenum Press, New York (1981)
3. Bezdek, J.C.: A Convergence Theorem for the Fuzzy ISODATA Clustering Algorithm. J. IEEE PAMI 1, 1–8 (1980)
4. Liao, L., Li, B.: MRI brain image segmentation and bias field correction based on fast spatially constrained kernel clustering approach. Pattern Recognition Letters 29, 1580–1588 (2008)
5. Sikka, K., Sinha, N.: A fully automated algorithm under modified FCM framework for improved brain MR image segmentation. Magnetic Resonance Imaging 27, 994–1004 (2009)
6. Das, S., Kanor, A.: Automatic image pixel clustering with an improved differential evolution. Applied Soft Computing 9, 226–236 (2009)
7. Ahmed, M.N., Yamany, S.M., Mohamed, N., Farag, A.A., Moriarty, T.: A modified fuzzy c-means algorithm for bias field estimation and segmentation of MRI data. IEEE Trans. Med. Imaging 21, 193–199 (2002)
8. Chen, S.C., Zhang, D.Q.: Robust image segmentation using FCM with spatial constraints based on new kernel-induced distance measure. IEEE Transaction on System, Man, and Cybernetics B 34, 1907–1916 (2004)

A Neighborhood Density Estimation Clustering Algorithm Based on Minimum Spanning Tree

Ting Luo and Caiming Zhong

College of Science and Technology, Ningbo University
Ningbo 315211, PR China
luoting@nbu.edu.cn

Abstract. In this paper a clustering algorithm based on the minimum spanning tree (MST) with neighborhood density difference estimation is proposed. Neighborhood are defined by patterns connected with the edges in the MST of a given dataset. In terms of the difference between patterns and their neighbor density, boundary patterns and corresponding boundary edges are detected. Then boundary edges are cut, and as a result the dataset is split into defined number clusters. For reducing time complexity of detecting boundary patterns, an rough and a refined boundary candidates estimation approach are employed, respectively. The experiments are performed on synthetic and real data. The clustering results demonstrate the proposed algorithm can deal with not well separated, shape-diverse clusters.

Keywords: Minimum Spanning Tree, Clustering, Density, Neighbor.

1 Introduction

Cluster analysis is an important component of data analysis, and it is applied in a variety of engineering and scientific disciplines such as biology, psychology, medicine, marketing, computer vision, and remote sensing [1]. There are many types of clustering algorithms, such as center-based clustering algorithm, density-based clustering algorithm [2] and so on. The K-means clustering algorithm [1] is a typical one for center-based, however the performance is based on initial centers. Ester et al. presented a density-based algorithm called DBSCAN (density based spatial clustering of applications with noise) [2], and the important concept of the algorithm is the ε-neighborhood (the number of patterns in ε radius area), but the parameter ε is difficult to be defined properly.

Recently the graph theory based clustering algorithms have been paid more attention to. A graph-based clustering algorithm first constructs a graph and then partitions the graph [2]. An edge-based clustering algorithm is one of them, in which the patterns are regarded as nodes and the similarity between patterns as edges of the graph. K-nearest neighbor is often employed for constructing the graph: Karypis et al. presented chameleon [3] algorithm that constructed a sparse graph representation of patterns based on k-nearest neighbor. It used

J. Yu et al. (Eds.): RSKT 2010, LNAI 6401, pp. 557–565, 2010.

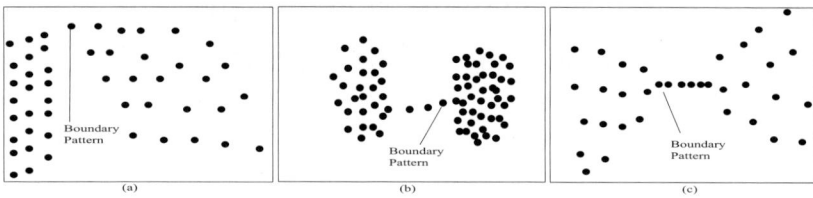

Fig. 1. Boundary Patterns locate in the datasets

the agglomerative hierarchical clustering algorithm to merge the similar subclusters by the relative interconnectivity of clusters. Minimum spanning tree (MST) graph was employed for clustering as well in the past years. Zahn demonstrated how the MST can be used to detect clusters [5]. His algorithm is to identify and remove inconsistent edges in the MST for grouping patterns. However his algorithm works well for disjoint clusters with arbitrary shapes but homogeneous clusters of different point density.

The MST reflects the similarity of patterns to some extent, however it cannot express the density of patterns accurately. The proposed clustering algorithm employs the neighborhood density difference estimation based on MST for finding boundary pattern. The remaining content is organized as follows: section 2 introduces the proposed clustering algorithm; In section 3 experiments are performed on synthetic and real datasets; Section 4 gives the conclusions.

2   The Proposed Clustering Algorithm

In general, patterns in a same cluster have a similar density. Even if the densities of the patterns are different from each other, the densities of corresponding neighborhoods would be similar. The proposed clustering algorithm is to detect the difference between the density of patterns and the corresponding neighborhood density based on the MST.

2.1   Definitions

Suppose a given dataset X contains N patterns, that is $X=\{x_1,x_2,...x_N\}$. The given dataset is constructed by a MST based on Euclidean distances.

Definition 1. *The number of edges directly connected to a given pattern is called the degree of the pattern.*

Definition 2. *The neighbor patterns of x_i are those connected directly to x_i by edges on the MST.*

Definition 3. *If two clusters are connected directly via an edge of MST, then the edge is called boundary edge, and the corresponding patterns connected with the edge are called boundary patterns.*

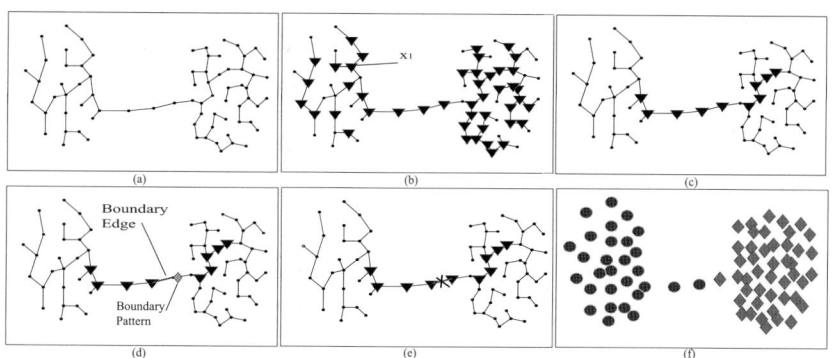

Fig. 2. Part of boundary pattern estimation processes on one dataset: (a) constructs the MST. (b) is the result of rough boundary candidates estimation(Triangles are boundary candidates).(c) is the result of refined boundary candidates estimation.(d) estimates the boundary pattern(Diamond) and boundary edge.(e) cuts the boundary edge. (f) is the clustering result.

Definition 4. *Let $ME(i)$ be the MST density of pattern x_i as in:*

$$ME(i) = \frac{1}{|E_i|} \sum_{e_j \in E_i} \|e_j\| \qquad (1)$$

where E_i is the set of edges connected with x_i, $|E_i|$ is the number of set E_i, and $\|e_j\|$ is the value of edge e_j.

    In fact, $ME(i)$ depicts the pattern density to some extent from the viewpoint of MST. If $ME(i)$ is greater, the pattern x_i locates in the sparse region, and vice versa.

Definition 5. *Let $DN(i)$ be the ε-density of pattern x_i as in:*

$$DN(i) = Card(\{x_j | dist(x_j, x_i) < \varepsilon\}) \qquad (2)$$

where $1 \le i \le C$, $1 \le j \le N$, C is the number of boundary candidates, $Card$ is number of set elements, $dist$ is the function for distance calculation, ε is a parameter produced automatically by the proposed clustering algorithm and will be discussed in subsection 2.3.

2.2 Boundary Candidates Estimation

Intuitively, a boundary pattern can be detected by density difference between the pattern and its neighbor patterns. Obviously, this process is time consuming. For saving time complexity we employ a two-stage approach to select boundary candidates: rough boundary candidates estimation and refined boundary candidates estimation.

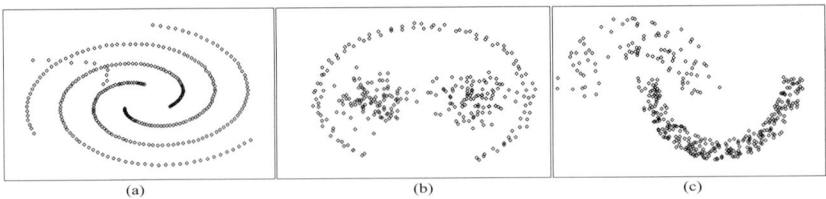

Fig. 3. Three datasets are used in our experiments: (a) is DS1. (b) is DS2. (c) is DS3.

2.2.1   Rough Boundary Candidates Estimation

In general, boundary patterns locate in the sparse regions or high-density regions compared with neighbors density. The boundary patterns may locate in the sparse region, as illustrated in the Fig.1(a)(b) , or in the high-density region, as illustrated in Fig.1(c). Thus we choose patterns locating in the sparse region or high-density region as boundary candidates. Patterns with one degree are definitely not boundary patterns between two clusters. The main process of estimation is as follows:

1. $ME(i)$ is computed for patterns except patterns with one degree. Assume N' is the number of patterns whose degree is greater than one and $1 \leq i \leq N'$.
2. Compute the mean and maximum of $ME(i)$ as MME and $MaxME$.
3. Compute the difference between ME and $MaxME$ as DME, and the difference $ME(i)$ and MME as $DE(i)$.
4. If $DE(i) > \alpha \times DME$, x_i is considered to be in the sparse or high-density region, and will be selected as the boundary candidates as illustrated in Fig. 2(b). The parameter α will be discussed in the section 3.

2.2.2   Refined Boundary Candidates Estimation

This approach is for refining boundary candidates. If the size of a cluster is relative small, it can be considered as a noise cluster. When one edge is cut from one cluster on MST, the cluster is split into two clusters. If the size of one cluster is smaller than $Nmin$, this cut is invalid and should be undone, where $Nmin = \beta \times N/K$, K is the final cluster number and the parameter of β will be discussed in section 3.

Edges connected to a boundary candidate will be judged one by one. If all edges connected to a boundary candidate are cut invalidly, the boundary candidate is removed. Illustrated in Fig.2(b), the pattern x_1 will be removed from boundary candidates.

After this approach the boundary candidates set, say Y, is estimated, where $Y = \{y_1, y_2, .., y_C\}$. Illustrated in the Fig.2(c) some fake boundary candidates are removed from Fig.2(b).

2.3   Boundary Pattern Estimation

In this subsection density differences between boundary candidates and their neighbor patterns are computed. The ε-density $DN(i)$ for y_i and ε-density

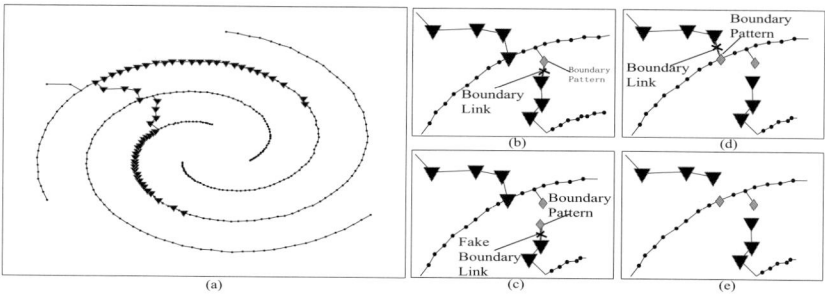

Fig. 4. The part processes of boundary pattern estimation and boundary edge cut on DS1:(a) is the result of rough and further boundary candidates estimation. (b) gets the first boundary pattern and first valid boundary edge, and the boundary edge is cut. (c) gets the next boundary pattern and cuts the fake boundary edge. (d) puts back the fake edge, chooses the next boundary pattern and cuts the boundary edge. (e) is the result of all boundary edges cut.

$NDN(j)$ for z_j are calculated, where $1 \leq i \leq C$, z_j is the neighbor patterns of y_i, $1 \leq j \leq |E_j|$, and E_j is the set of edges connected with y_i.

The density difference is calculated between boundary candidates and their neighbors. When computing the ε-density for y_j and z_j, ε should be same. However, ε-density for every boundary candidate will not be compared, thus ε is not constant for every boundary candidate. The parameter ε is set to be the longest edge connected with y_i.

After computing every density difference between y_i and its neighbor patterns, the greatest difference is chosen. The density difference for y_i as $DDN(i)$ is computed by

$$DDN(i) = Max(|DN(i) - NDN(j)|) \tag{3}$$

The greater $DDN(i)$, the earlier y_i is to be chosen as boundary patterns. The rule of selecting boundary edges is: if the boundary pattern density is lower than the density neighbor pattern, the longest one is chosen as boundary edge, and vice versa. When boundary edge is cut and if both sizes of one clusters are greater than $\beta \times N/K$, one cluster is split into two clusters as illustrated in Fig. 2(d)(e). Fig. 2(f) shows the clustering result after the boundary edge is cut.

If the size of one cluster is less than $\beta \times N/K$, the removed edge should be put back. That kind of edges are called fake boundary edges. The next edges connected with the boundary pattern will be cut until there are no edges connected.

The main steps for boundary patterns estimation and boundary edges cut are illustrated in Fig.4, and described as follows:

1. Get the maximum value $DDN(i)$, and recognize y_i as the boundary pattern.
2. Cut the corresponding edges according to the edge cut rules. Remove $DDN(i)$ from DDN.
3. Go back to the step 1 until K valid boundary edges are cut.

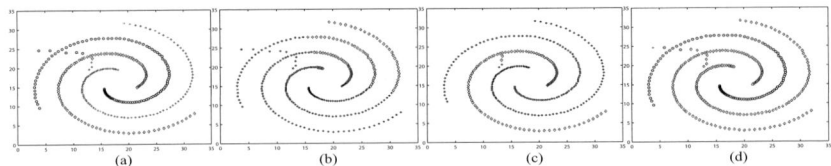

Fig. 5. Clustering results on DS1:(a) is the clustering result of the proposed clustering algorithm. (b) is the clustering result of K-means. (c) is the clustering result of DBSCAN. (d) is the clustering result of Single link

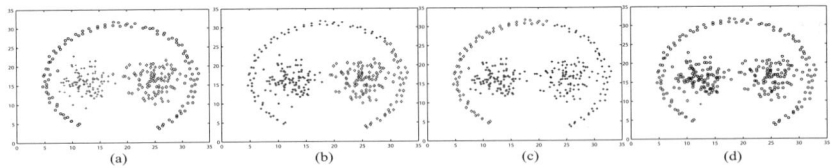

Fig. 6. Clustering results on DS2:(a) is the clustering result of the proposed clustering algorithm. (b) is the clustering result of K-means. (c) is the clustering result of DBSCAN. (d) is the clustering result of Single link

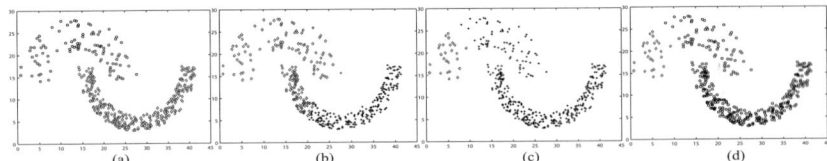

Fig. 7. Clustering results on DS3:(a) is the clustering result of the proposed clustering algorithm. (b) is the clustering result of K-means. (c) is the clustering result of DBSCAN. (d) is the clustering result of Single link

2.4   Main Processes of the Proposed Clustering Algorithm

The proposed clustering algorithm is completely described as follows:
   Input: Dataset X, Parameters:K, α and β.
   Output: Clusters $C^1, C^2, ..C^K$.

1. Construct the MST for datasets.
2. Rough boundary candidates are estimated.
3. Boundary candidates are refined.
4. Boundary patterns are obtained by computing the density differences and valid boundary edges are cut.
5. K clusters are achieved.

3   Experiments

In this section, we perform a series experiments on both synthetic and real datasets to demonstrate the validity and effectiveness.

In out experiments, the parameters α and β are constant for all datasets. α is set to 0.1 and β is to 0.5.

3.1   Synthetic Data

We test the proposed clustering algorithm on 2D datasets of different shapes as illustrated in Fig.3. The clustering results on DS1, DS2, DS3 will be compared with K-means, DBScan and single link clustering algorithms.

The dataset DS1 is composed of three spiral clusters which are not well separated. In Fig.5, compared with K-means and single edge clustering algorihtms, the clustering result of the proposed algorithm is better. Although DBSCAN performs well, the parameter ε for DBSCAN is sensitive to the clustering results. In the proposed algorithm, ε is produced automatically. The other two parameters α and β are robust to the clustering results.

In the experiment on DS1, if the parameter α is from 0 to 0.4, the clustering results are same. If α is smaller, the computational cost is more expensive, and vice versa. For parameter β, it can be from 0.1 to 0.9 for the correct clustering results.

DS2 is composed of three clusters, and the shapes of clusters are different. Moreover, the boundary patterns between clusters are implicit. Compared with K-means, DBSCAN and single link clustering algorihtms, the proposed clustering algorithm performs best, as illustrated in Fig.6.

For DS3, although two clusters are well separated, the shapes are arbitrary and the patterns density distribution in clusters are different. K-means, DBSCAN and single link algorithms fail on DS3 and the proposed clustering algorithm can detect two clusters accurately as illustrated in Fig.7.

3.2   Real Data

Two real datasets from UCI Machine Learning Repository are employed to test the applicability of the proposed clustering algorithm. The first one is Iris. The performances of algorithms on Iris are showed in Table 1. Four frequently used external clustering validity indices are employed to evaluate the clustering results: Rand statistic, Adjusted rand, Jaccard cofficient, and Folkes and Mallows index(FM) [2]. It is evident that clustering result of the proposed clustering algorithm outperform the others.

The second one is Wine, which is 13-dimensional. The performances on Wine are illustrated in the Table2. Although the performance is slightly worse than DBSCAN, it is better than K-means and single link clustering algorithms.

Table 1. Performance Comparison on Iris

Method	Rand	Ajusted Rand	Jaccard	FM
Proposed	0.9341	0.8512	0.8188	0.9004
K-means	0.8797	0.7302	0.6959	0.8208
DBScan	0.8834	0.7388	0.7044	0.8368
Single Link	0.7766	0.5638	0.5891	0.7635

Table 2. Performance Comparison on Wine

Method	Rand	Ajusted Rand	Jaccard	FM
Proposed	0.7311	0.4001	0.4322	0.6036
K-means	0.7183	0.3711	0.4120	0.7302
DBScan	0.7610	0.5291	0.5902	0.7512
Single Link	0.3628	0.0054	0.3325	0.5650

4   Conclusion

In this paper we propose a neighborhood density difference estimation clustering algorithm based on MST. A two-stage approach is designed to detect suitable boundary candidates. Differences between boundary candidates density and their neighbor density are computed for boundary patterns and boundary edges estimating. Experiments on both the synthetic and real data demonstrate that the proposed clustering algorithm outperform the traditional clustering algorithms. However, the proposed clustering algorithm has slightly low ability in dealing with high-dimension datasets, and our future work will focus on it.

Acknowledgment

This work was supported by Zhejiang Provincial Natural Science Foundation of China, Grant No. Y1090851 and Scientific Research Fund of Zhejiang Provincial Education Department, Grant No. Y201016652.

References

1. Jain, A., Dubes, R.: Algorithms for Clustering Data. Prentice-Hall, NJ (1988)
2. Guojun, G., Ma, C., Wu, J.: Data Clustering:Theory, Algorithms, and Applications. ASA-SIAM Series on Statistics and Applied Probability, Philadelphis (2007)
3. Karypis, G., Han, E., Kumar, V.: Chameleon: Hierarchical clustering using dynamic modeling. Computer 32(8), 68–75 (1999)
4. Park, N., Lee, W.: Statistical grid-based clustering over data streams. SIGMOD Record 33(1), 32–37 (2004)
5. Zahn, C.T.: Graph-theoretical methods for detecting and describing Gestalt clusters. IEEE Transactions on Computers C 20, 68–86 (1971)

6. Shi, J., Malik, J.: Normailized cuts and image segmentation. IEEE Trans. Pattern Anal, Mach. Intell. 22(8), 888–905 (2000)
7. Davies, D.L., Bouldin, D.W.: A Cluster Separation Measure. IEEE Trans. Pattern Analysis and Machine Intelligence 1, 224–227 (1979)
8. Bock, H.H.: Clustering Algorithms and Kohone Maps for Symbolic Data. Society of computational statistical, 1–13 (2002)

Hybrid Differential Evolution for Global Numerical Optimization

Liyuan Jia[1], Lei Li[2], Wenyin Gong[3,*], and Li Huang[1]

[1] Department of Computer Science,
Hunan City University, Yiyang Hunan 413000, P.R. China
jia_4211003@126.com
[2] College of Computer & Information Technology,
HeNan XinYang Normal College, XinYang HeNan, 464000, China
[3] School of Computer Science,
China University of Geosciences, Wuhan 430074, P.R. China
cug11100304@yahoo.com.cn

Abstract. Differential evolution (DE) is an efficient and versatile evolutionary algorithm for global numerical optimization over continuous domain. Although DE is good at exploring the search space, it is slow at the exploitation of the solutions. To alleviate this drawback, in this paper, we propose a generalized hybrid generation scheme, which attempts to enhance the exploitation and accelerate the convergence velocity of the original DE algorithm. In the hybrid generation scheme the operator with powerful exploitation is hybridized with the original DE operator. In addition, a self-adaptive exploitation factor is introduced to control the frequency of the exploitation operation. In order to evaluate the performance of our proposed generation scheme, the migration operator of biogeography-based optimization is employed as the exploitation operator. Moreover, 23 benchmark functions (including 10 test functions provided by CEC2005 special session) are chosen from the literature as the test suite. Experimental results confirm that the new hybrid generation scheme is able to enhance the exploitation of the original DE algorithm and speed up its convergence rate.

1 Introduction

Differential Evolution (DE), proposed by Storn and Price [1], is a simple yet powerful population-based, direct search algorithm with the generation-and-test feature for global optimization using real-valued parameters. Among DE's advantages are its simple structure, ease of use, speed and robustness. Due to these advantages, it has many real-world applications, such as data mining [2], [3], neural network training, pattern recognition, digital filter design, etc. [4], [5], [6].

Although the DE algorithm has been widely used in many fields, it has been shown to have certain weaknesses, especially if the global optimum should be located using a limited number of fitness function evaluations (NFFEs). In addition, DE is good at exploring the search space and locating the region of global minimum, but it is slow at the exploitation of the solutions [7]. To remedy this drawback, this paper proposes a

---

* Corresponding author.

J. Yu et al. (Eds.): RSKT 2010, LNAI 6401, pp. 566–573, 2010.

new hybrid generation scheme, which attempts to balance the exploration and exploitation of DE. The proposed hybrid generation scheme, which is based on the binomial recombination operator of DE, is hybridized with the exploitative operation. Thus, the new hybrid generation scheme is able to enhance the exploitation of the original DE algorithm. Additionally, a self-adaptive exploitation factor, η, is introduced to control the frequency of the exploitation operation.

2   DE and BBO

2.1   Differential Evolution

The DE algorithm [1] is a simple evolutionary algorithm (EA) for global numerical optimization. In DE, the mutation operator is based on the distribution of solutions in the current population. New candidates are created by combining the parent solution and the mutant. A candidate replaces the parent only if it has a better fitness value. In [1], Price and Storn gave the working principle of DE with single mutation strategy. Later on, they suggested ten different mutation strategies of DE [8], [4]. More details of the DE algorithm can be found in [4].

2.2   Biogeography-Based Optimization

Since we will use the migration operator of BBO as the exploitation operator in this work, the BBO algorithm will briefly be introduced in this section. BBO [9] is a new population-based, biogeography inspired global optimization algorithm. BBO is a generalization of biogeography to EA. In BBO, it is modeled after the immigration and emigration of species between islands in search of more friendly habitats. The islands represent the solutions and they are ranked by their island suitability index (ISI), where a higher ISI signifies a superior fitness value. The islands are comprised of solution features named suitability index variables, equivalent to GA's genes.

In BBO, the k-th ($k = 1, 2, \cdots, NP$) individual has its own immigration rate λ_k and emigration rate μ_k. A good solution has higher μ_k and lower λ_k, vice versa. The immigration rate and the emigration rate are functions of the number of species in the habitat. They can be calculated as follows:

$$\lambda_k = I\left(1 - \frac{k}{n}\right) \tag{1}$$

$$\mu_k = E\left(\frac{k}{n}\right) \tag{2}$$

where I is the maximum possible immigration rate; E is the maximum possible emigration rate; k is the number of species of the k-th individual; and $n = NP$ is the maximum number of species. Note that Equations 1 and 2 are just one method for calculating λ_k and μ_k. There are other different options to assign them based on different specie models [9].

3   Our Proposed Hybrid Generation Scheme

From the literature reviewed, we can observe that there are many DE variants to remedy some pitfalls of DE mentioned above. In this study, we will propose a novel hybrid generation scheme to enhance the exploitation of DE. Our approach, which is different from the above-mentioned DE variants, is presented in detail as follows.

3.1   Motivations

The DE's behavior is determined by the trade-off between the exploration and exploitation. As stated in [7], DE is good at exploring the search space, but it is slow at the exploitation of the solutions. Recently, hybridization of EAs is getting more popular due to their capabilities in handling several real world problems [10]. Thus, hybridization of DE with other algorithm, which has powerful exploitation, might balance the exploration and exploitation of the DE algorithm. Based on this consideration, we propose a new hybrid generation scheme, where the operator with powerful exploitation is hybridized with the binomial recombination operator of the original DE algorithm.

As we briefly described in Section 2.2, the migration operator of BBO has good exploitation; it can share the useful information among individuals. Since the BBO algorithm is biogeography theoretical support, we will first hybridize the migration operator with DE to verify the performance of our proposed hybrid generation scheme.

3.2   Hybrid Generation Scheme

The new hybrid generation scheme of DE is based on the binomial recombination operator of the original DE algorithm. The trial vector u_i is possibly composed of three parts: the mutant variables generated by the DE mutation, variables generated by the exploitative operation, and the variables inherited from the target vector x_i. The pseudo-code of the hybrid generation scheme is described in Algorithm 1.

From Algorithm 1, we can observe that:

1) The proposed hybrid generation scheme is able to enhance the exploitation of DE because of the exploitative operation.

Algorithm 1. The new hybrid generation scheme

```
 1: for i = 1 to NP do
 2:     Select uniform randomly r₁ ≠ r₂ ≠ r₃ ≠ i
 3:     j_rand = rndint(1, D)
 4:     for j = 1 to D do
 5:         if rndreal_j[0, 1) < CR or j == j_rand then
 6:             u_i,j is generated by the mutation strategy of DE {explorative operation}
 7:         else if rndreal_j[0, 1) < η then
 8:             u_i,j is generated by the operator with powerful exploitation {exploitative operation}
 9:         else
10:             u_i,j = x_i,j
11:         end if
12:     end for
13: end for
```

---

2) It is a generalized scheme. For example, in line 6 of Algorithm 1, it is the explorative operation. The mutation strategy of DE, which is good at exploring the search space, can be used here, such as "DE/rand/1", "DE/rand/2", and so on. In line 8, the operator with powerful exploitation can be employed, such as the migration operator of BBO.
3) Compared with the original DE generation scheme, our proposed scheme is also very simple and easy to implement.
4) The hybrid generation scheme maintains the main property of the original DE generation scheme. The crossover rate CR mainly controls the whole generation scheme. If the CR value is higher, the explorative operation plays a more important role, and the exploitative operation has less influence to the offspring. Especially, when CR is close to 1.0, the generation scheme is more rotationally invariant [4], and hence it can be used to solve the parameter-dependent problems.

3.3 Self-adaptive Exploitation Factor

In our proposed hybrid generation scheme shown in Algorithm 1, we introduced an additional parameter, i.e., exploitation factor η. The parameter is used to control the frequency of the exploitative operation. If η is set higher, it may lead to over-exploitation. The algorithm may trap into the local optimum. Contrarily, if η is lower, it may result in under-exploitation. The algorithm can not exploit the solutions sufficiently.

Self-adaptation [11] is highly beneficial for adjusting control parameters, especially when done without any user interaction [12]. Thus, in this study, we propose a self-adaptive strategy to control the exploitation factor η. This strategy is similar to the self-adaptive parameter control method proposed in [13]. In our proposed method, η is encoded into the chromosome. The i-th individual X_i is represented as follows.

$$X_i = \langle \boldsymbol{x}_i, \eta_i \rangle = \langle x_{i,1}, x_{i,2}, \cdots, x_{i,D}, \eta_i \rangle \tag{3}$$

where η_i is the exploitation factor of the i-th individual; and it is initially randomly generated between 0 and 1. The new parameter is calculated as

$$\eta_i = \begin{cases} \text{rndreal}[0, dynamic\_gen], & \text{rndreal}[0, 1] < \delta \\ \eta_i, & \text{otherwise} \end{cases} \tag{4}$$

where rndreal$[a, b]$ is a uniform random value generated in $[a, b]$. δ indicates the probability to adjust the exploitation factor η_i. It is similar to the parameters τ_1 and τ_2 used in [13]. $dynamic\_gen$ is calculated as:

$$dynamic\_gen = \frac{gen}{Max\_gen} \tag{5}$$

where gen is the current generation, and Max_gen is the maximum generation. The reason of calculating $\eta_i = \text{rndreal}[0, dynamic\_gen]$ is that at the beginning of the evolutionary process the useful information of the population is less, and hence, exploiting more information might degrade the performance of the algorithm, especially for the multi-modal functions. As the evolution progresses, the population contains more and more useful information, thus the probability of the exploitative operation needs to be increased to utilize the useful information.

4   Experimental Results and Analysis

4.1   Test Functions

In this work, we have carried out different experiments using a test suite, consisting of 23 unconstrained single-objective benchmark functions with different characteristics chosen from the literature. All of the functions are minimization and scalable problems. The first 13 functions, f01 - f13, are chosen from [14]. The rest 10 functions, F01 - F10, are the new test functions provided by the CEC2005 special session [15]. Since we do not make any changes to these problems, we do not describe them herein. More details can be found in [14] and [15].

4.2   Experimental Setup

In this work, we will mainly use the DE-BBO algorithm to evaluate the performance of our proposed hybrid generation scheme. For DE-BBO, we have chosen a reasonable set of value and have not made any effort in finding the best parameter settings. For all experiments, we use the following parameters unless a change is mentioned. In addition, the self-adaptive control parameter proposed in [13] for CR and F values is used in DE-BBO, referred to as jDE-BBO.

- Dimension of each function: $D = 30$;
- Population size: $NP = 100$ [14], [13];
- $E = I = 1.0$ [9];
- Probability of adjusting η: $\delta = 0.1$;
- Value to reach: VTR = 10^{-8} [15], [16], except for f07 of VTR = 10^{-2};
- Max_NFFEs: For f01, f06, f10, f12, and f13, Max_NFFEs = $150,000$; for f03 - f05, Max_NFFEs = $500,000$; for f02 and f11, Max_NFFEs = $200,000$; for f07 - f09 and F01 - F10, Max_NFFEs = $300,000$.

4.3   Performance Criteria

Five performance criteria are selected from the literature [15], [16] to evaluate the performance of the algorithms. These criteria are described as follows.

- **Error** [15]: The error of a solution x is defined as $f(x) - f(x^*)$, where x^* is the global minimum of the function. The minimum error is recorded when the Max_NFFEs is reached in 50 runs. The average and standard deviation of the error values are calculated as well.
- **NFFEs** [15]: The number of fitness function evaluations (NFFEs) is also recorded when the VTR is reached. The average and standard deviation of the NFFEs values are calculated.
- **Number of successful runs (SR)** [15]: The number of successful runs is recorded when the VTR is reached before the Max_NFFEs condition terminates the trial.
- **Convergence graphs** [15]: The convergence graphs show the average error performance of the best solution over the total runs, in the respective experiments.
- **Acceleration rate (AR)** [16]: This criterion is used to compare the convergence speeds between our approach and other algorithms. It is defined as follows: $AR = \frac{\text{NFFEs}_{other}}{\text{NFFEs}_{ours}}$, where $AR > 1$ indicates our approach is faster than its competitor.

Table 1. Comparison the Performance Between jDE-BBO and jDE at $D = 30$. "NA" Means Not Available.

Prob	Best Error Values						NFFEs				AR
	jDE			jDE-BBO			jDE		jDE-BBO		
	Mean	Std Dev	SR	Mean	Std Dev	SR	Mean	Std Dev	Mean	Std Dev	
f01	1.75E-27	1.57E-27	50	**0.00E+00**	0.00E+00	50†	6.11E+04	1.12E+03	**5.24E+04**	9.54E+02	1.165
f02	0.00E+00	0.00E+00	50	0.00E+00	0.00E+00	50	8.45E+04	1.40E+03	**7.11E+04**	1.13E+03	1.189
f03	4.03E-13	6.20E-13	50	**1.42E-15**	3.34E-15	50†	3.57E+05	1.84E+04	**3.22E+05**	1.61E+04	1.109
f04	**3.44E-14**	1.83E-13	50	7.81E-07	3.13E-06	44	3.09E+05	4.54E+03	**2.41E+05**	8.04E+03	1.282
f05	9.99E-02	1.23E-01	1	**7.98E-02**	5.64E-01	4	4.81E+05	0.00E+00	4.83E+05	1.31E+04	0.996
f06	0.00E+00	0.00E+00	50	0.00E+00	0.00E+00	50	2.30E+04	7.05E+02	**2.01E+04**	6.68E+02	1.140
f07	3.46E-03	9.00E-04	50	**2.89E-03**	6.88E-04	50†	1.11E+05	2.23E+04	**1.00E+05**	2.38E+04	1.106
f08	0.00E+00	0.00E+00	50	0.00E+00	0.00E+00	50	9.58E+04	2.27E+03	**7.29E+04**	1.87E+03	1.314
f09	0.00E+00	0.00E+00	50	0.00E+00	0.00E+00	50	1.19E+05	4.08E+03	**8.90E+04**	2.05E+03	1.338
f10	1.54E-14	5.48E-15	50	**4.14E-15**	0.00E+00	50†	9.31E+04	1.63E+03	**7.64E+04**	1.32E+03	1.218
f11	0.00E+00	0.00E+00	50	0.00E+00	0.00E+00	50	6.47E+04	3.14E+03	**5.64E+04**	1.76E+03	1.148
f12	1.15E-28	1.15E-28	50	**1.57E-32**	0.00E+00	50†	5.52E+04	1.28E+03	**4.78E+04**	1.09E+03	1.156
f13	3.92E-26	5.22E-26	50	**1.35E-32**	0.00E+00	50†	6.71E+04	1.45E+03	**5.62E+04**	1.27E+03	1.195
F01	0.00E+00	0.00E+00	50	0.00E+00	0.00E+00	50	6.24E+04	1.04E+03	**5.58E+04**	1.03E+03	1.119
F02	8.33E-05	1.28E-04	0	**4.72E-07**	6.96E-07	0†	NA	NA	NA	NA	NA
F03	2.26E+05	1.05E+05	0	**1.77E+05**	1.07E+05	0	NA	NA	NA	NA	NA
F04	1.06E-04	1.50E-04	0	**5.90E-07**	8.89E-07	0†	NA	NA	NA	NA	NA
F05	**9.44E+02**	3.55E+02	0	9.86E+02	4.07E+02	0	NA	NA	NA	NA	NA
F06	3.34E+01	3.79E+01	0	**1.82E+01**	2.36E+01	0	NA	NA	NA	NA	NA
F07	1.32E-02	1.01E-02	6	**1.05E-02**	8.63E-03	10	**1.89E+05**	1.08E+04	1.95E+05	1.52E+04	0.973
F08	2.10E+01	4.61E-02	0	2.10E+01	4.85E-02	0	NA	NA	NA	NA	NA
F09	0.00E+00	0.00E+00	50	0.00E+00	0.00E+00	50	1.12E+05	3.41E+03	**8.75E+04**	2.03E+03	1.284
F10	5.89E+01	8.79E+00	0	**5.84E+01**	1.10E+01	0	NA	NA	NA	NA	NA

† The value of t with 49 degrees of freedom is significant at $\alpha = 0.05$ by two-tailed t-test.

4.4   General Performance of jDE-BBO

In order to show the performance of jDE-BBO, we compare it with the jDE and BBO algorithms. The parameters used for jDE-BBO and jDE are the same as described in Section 4.2. The parameters of BBO are set as in [9], and the mutation operator with $m_{max} = 0.005$ is also used in our experiments. All functions are conducted for 50 independent runs. Table 1 summarizes the results of jDE-BBO and jDE on all test functions[1]. In addition, some representative convergence graphs of jDE-BBO, jDE, and BBO are shown in Figure 1.

With respect to the best error values shown in Table 1, it can be seen that jDE-BBO significantly outperforms jDE on 9 out of 23 functions. For 7 functions (f02, f06, f08, f09, f11, F01, and F09), both jDE-BBO and jDE can obtain the global optimum within the Max_NFFEs. For functions f05, F06, F07, and F10, jDE-BBO is slightly better than jDE. Only for 2 functions, f04 and F05, jDE is slightly better than jDE-BBO.

With respect to the NFFEs to reach the VTR, Table 1 indicates that jDE-BBO requires fewer NFFEs to reach the VTR for 14 functions. For functions f05 and F07, jDE is sightly faster than jDE-BBO. However, for these two functions, jDE-BBO obtains higher SR values than jDE. For the rest 7 functions (F02 - F06, F08, and F10), both jDE-BBO and jDE can not reach the VTR within the Max_NFFEs. In addition, for

---

[1] For the sake of clarity and brevity, in Table 1 the results of BBO are not reported, because BBO is significantly outperformed by both jDE and jDE-BBO for all test functions. The reader may contact the authors for details.

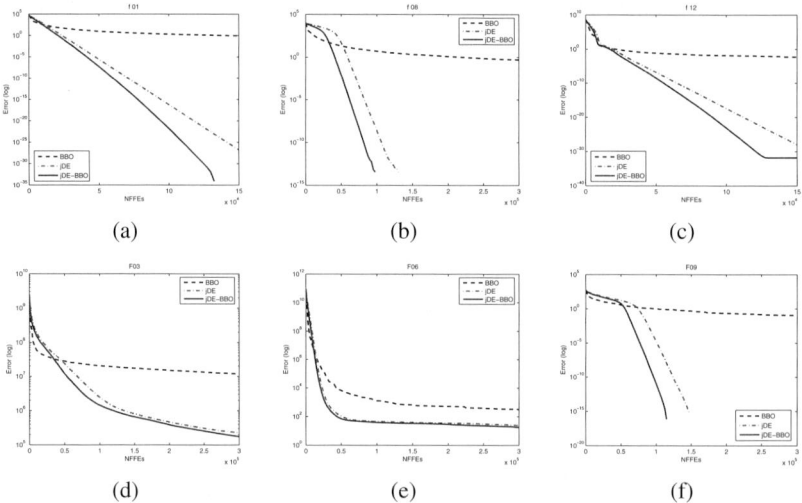

Fig. 1. Mean error curves of BBO, jDE, and jDE-BBO for selected functions. (a) f01. (b) f08. (c) f12. (d) F03. (e) F06. (f) F09.

the successful 16 functions the overall average AR value is 1.171, which indicates that jDE-BBO is on average 17.1% faster than jDE for these functions.

From Fig. 1, it is apparent that in the early stages BBO converges faster than jDE and jDE-BBO, since the migration operator of BBO is good at exploiting the solutions. As the evolution progresses, both jDE and jDE-BBO are faster than BBO. Moreover, jDE-BBO achieves faster convergence rate than jDE.

In general, the performance of jDE-BBO is highly competitive with jDE. Our proposed hybrid generation scheme combined with the migration operator of BBO is able to enhance the exploitation of DE, and hence it can accelerate the original DE algorithm.

5   Conclusion and Future Work

In this paper, we propose a generalized hybrid generation scheme of DE. The hybrid generation scheme is based on the binomial recombination operator of DE, where an additional exploitative operation is incorporated in the original DE generation scheme. Hence, our proposed hybrid generation scheme has the ability to enhance the exploitation and accelerate the convergence rate of DE. Compared with the original DE generation scheme, the hybrid generation scheme is also very simple and easy to implement. In addition, an exploitation factor η is introduced to control the frequency of the exploitative operation. Through a comprehensive set of experimental verifications of our hybrid generations scheme, the results confirm the enhancement of the exploitation by the scheme. The hybrid generation scheme is able to make the original DE algorithm more effective in terms of the quality of the final results and the reduction of the NFFEs. Thus, our proposed hybrid generation scheme can be regarded as an alternative scheme of DE for further research.

Since the proposed generation scheme is a generalized scheme, other exploitation operators (e.g., the SPX operator [17] and the PCX operator [18]) can also be used to enhance the performance of the original DE algorithm. Our future work is required to verify this expectation.

References

1. Storn, R., Price, K.: Differential evolution - A simple and efficient heuristic for global optimization over continuous spaces. Journal of Global Optimization 11(4), 341–359 (1997)
2. Alatas, B., Akin, E., Karci, A.: MODENAR: Multi-objective differential evolution algorithm for mining numeric association rules. Applied Soft Computing 8(1), 646–656 (2008)
3. Das, S., Abraham, A., Konar, A.: Automatic clustering using an improved differential evolution algorithm. IEEE Transaction on Systems Man and Cybernetics: Part A 38(1), 218–237 (2008)
4. Price, K., Storn, R., Lampinen, J.: Differential Evolution: A Practical Approach to Global Optimization. Springer, Berlin (2005)
5. Feoktistov, V.: Differential Evolution: In Search of Solutions. Springer, New York (2006)
6. Chakraborty, U.: Advances in Differential Evolution. Springer, Berlin (2008)
7. Noman, N., Iba, H.: Accelerating differential evolution using an adaptive local search. IEEE Transactions on Evolutionary Computation 12(1), 107–125 (2008)
8. Storn, R., Price, K.: Home page of differential evolution (2008)
9. Simon, D.: Biogeography-based optimization. IEEE Transactions on Evolutionary Computation 12(6), 702–713 (2008)
10. Grosan, C., Abraham, A., Ishibuchi, H.: Hybrid Evolutionary Algorithms. Springer, Berlin (2009)
11. Eiben, Á.E., Hinterding, R., Michalewicz, Z.: Parameter control in evolutionary algorithms. IEEE Transations on Evolutionary Computation 3(2), 124–141 (1999)
12. Brest, J., Bošković, B., Greiner, S., Žumer, V., Maučec, M.S.: Performance comparison of self-adaptive and adaptive differential evolution algorithms. Soft Computing 11(7), 617–629 (2007)
13. Brest, J., Greiner, S., Bošković, B., Mernik, M., Žumer, V.: Self-adapting control parameters in differential evolution: A comparative study on numerical benchmark problems. IEEE Transactions on Evolutionary Computation 10(6), 646–657 (2006)
14. Yao, X., Liu, Y., Lin, G.: Evolutionary programming made faster. IEEE Transations on Evolutionary Computation 3(2), 82–102 (1999)
15. Suganthan, P.N., Hansen, N., Liang, J.J., Deb, K., Chen, Y.P., Auger, A., Tiwari, S.: Problem definitions and evaluation criteria for the CEC 2005 special session on real-parameter optimization (2005)
16. Rahnamayan, S., Tizhoosh, H., Salama, M.: Opposition-based differential evolution. IEEE Transactions on Evolutionary Computation 12(1), 64–79 (2008)
17. Tsutsui, S., Yamamura, M., Higuchi, T.: Multi-parent recombination with simplex crossover in real coded genetic algorithms. In: Proceedings of Genetic and Evolutionary Computation Conference, GECCO 1999, pp. 657–664 (1999)
18. Deb, K., Anand, A., Joshi, D.: A computationally efficient evolutionary algorithm for real-parameter optimization. Evolutionary Computation 10(4), 345–369 (2002)

A Tabu-Based Memetic Approach for Examination Timetabling Problems

Salwani Abdullah[1], Hamza Turabieh[1], Barry McCollum[2], and Paul McMullan[2]

[1] Data Mining and Optimization Research Group (DMO),
Center for Artificial Intelligence Technology,
Universiti Kebangsaan Malaysia, 43600 Bangi, Selangor, Malaysia
{salwani,hamza}@ftsm.ukm.my
[2] Department of Computer Science, Queen's University Belfast,
Belfast BT7 1NN United Kingdom
{b.mccollum,p.p.mcmullan}@qub.ac.uk

Abstract. Constructing examination timetable for higher educational institutions is a very complex task due to the complexity of the issues involved. The objective of examination timetabling problem is to satisfy the hard constraints and minimize the violations of soft constraints. In this work, a tabu-based memetic approach has been applied and evaluated against the latest methodologies in the literature on standard benchmark problems. The approach hybridizes the concepts of tabu search and memetic algorithms. A tabu list is used to penalise neighbourhood structures that are unable to generate better solutions after the crossover and mutation operators have been applied to the selected solutions from the population pool. We demonstrate that our approach is able to enhance the quality of the solutions by carefully selecting the effective neighbourhood structures. Hence, some best known results have been obtained.

Keywords: Examination Timetabling, Memetic Algorithm, Tabu List.

1 Introduction

Timetabling problems have long been a challenging area for researchers across both Operational Research and Artificial Intelligence. Among the wide variety of timetabling problems, educational timetabling (which cover areas such as school, course and exam timetabling) is one of the most widely studied. In the literature, much work has been carried out in relation to examination timetabling problems which can be generally defined as assigning a set of exams into a limited number of timeslots and rooms subject to a set of constraints [2].

In the past, a wide variety of approaches for solving the examination timetable problem have been described and discussed in the literature. For a recent detailed overview readers should consult Qu et al. [6]. Carter et al. [4] categorized the approaches taken into four types: sequential methods, cluster methods, constraint-based methods and generalized search. Petrovic and Burke [11] added the following categories: hybrid evolutionary algorithms, meta-heuristics,

J. Yu et al. (Eds.): RSKT 2010, LNAI 6401, pp. 574–581, 2010.

multi-criteria approaches, case based reasoning techniques, hyper-heuristics and adaptive approaches. These approaches are tested on various examinations timetabling datasets that can be found from http://www.asap.cs.nott.ac.uk/resource/data. These are often termed the 'Carter datsets' and are composed of 13 real-world exam timetabling problems [6]. Due to the existence of multi formulations for university timetabling problem, the 2nd International Timetabling Competition (ITC2007) [13] attempted to standardize the problems found within educational timetabling by introducing three tracks (one on exam and two on course timetabling) where the problems incorporated more real-world constraints that we refer as 'McCollum datasets' [13]. In doing so, the organizers attempted to reduce the acknowledged gap between research and practice which exists in this area. Interested readers can find more details about examination timetabling research in the comprehensive survey paper [6].

The rest of the paper is organized as follows. The next section formally presents the examination timetabling problem and formulation. The solution approach is outlined in Section 3. Our results are presented, discussed and evaluated in Section 4. This is followed by some brief concluding comments in Section 5.

2   Problem Description

The examination timetabling problem is a command problem found in schools and higher educational institutes which are concerned with allocating exams into a limited number of timeslots (periods) subject to a set of constraints (see [6]). The problem description that is employed in this paper is adapted from the description presented in [2]. The input for the examination timetabling problem can be stated as follows:

- E_i is a collection of N examinations $(i = 1, \ldots, N)$.
- T is the number of timeslots.
- $C = (c_{ij})_{N \times N}$ is the conflict matrix where each record, denoted by $c_{ij} (i, j \in \{1, \ldots, N\})$, represents the number of students taking exams i and j.
- M is the number of students.
- $t_k (1 \leq t_k \leq T)$ specifies the assigned timeslots for exam $k(k \in \{1, \ldots, N\})$.

The following hard constraint is considered based on Carter et al. [4]:
no students should be required to sit two examinations simultaneously.

In this problem, we formulate an objective function which tries to spread out students' exams throughout the exam period (Expression (1)).

$$Min \quad \frac{\sum_{i=1}^{N-1} F(i)}{M} \qquad (1)$$

where:

$$F(i) = \sum_{j=i+1}^{N} c_{ij} \cdot proximity(t_i, t_j) \qquad (2)$$

$$proximity(t_i, t_j) = \begin{cases} 2^5/2^{|t_i - t_j|} & \text{if } 1 \leq |t_i - t_j| \leq 5 \\ 0 & \text{otherwise} \end{cases} \qquad (3)$$

Equation (2) presents the cost for an exam i which is given by the proximity value multiplied by the number of students in conflict. Equation (3) represents a proximity value between two exams [6]. Equation (4) represents a clash-free requirement so that no student is asked to sit two exams at the same time. The clash-free requirement is considered to be a hard constraint.

3   The Tabu-Based Memetic Approach

The approach described here consists of a construction stage followed by an improvement stage.

3.1   Construction Heuristic

A construction algorithm which is based on a saturation degree graph coloring heuristic is used to generate large populations of feasible timetables. The approach starts with an empty timetable. The exams with highest number of exams in conflict and more likely difficult to be scheduled will be attempted first without taking into consideration the violation of any soft constraints until the hard constraints are met. This will help the algorithm to easily schedule the rest of the exams with less conflict. More details on graph coloring applications to timetabling can be found in [2].

3.2   Improvement Algorithm

In this paper, a tabu-based memetic approach is proposed as an improvement algorithm that operates on a set of possible solutions (generated from the construction heuristic) to solve examination timetabling problem, where a set of neighbourhood structures (as discussed in subsection 3.3) have been used as a local search mechanism. The aim of using a set of neighbourhood structure inside genetic operators is to produce significant improvements in a solution quality. In this approach, a concept of tabu list is employed to penalize neighbourhood structures that are unable to generate better solutions after the crossover and mutation operators being applied to the selected solutions from the population pool.

The pseudo code of the algorithm is presented in Fig. 1. The algorithm begins by creating initial populations. A best solution from the population is selected, S_{best}. A tabu list with the size $TabuSize$ is created with an aim to hold ineffective neighbourhood structures from being selected in the next iteration and will give more chances for the remaining neighbourhood structures to be explored (see Table 1 for parameter setting). A variable $CanSelect$, (represents a boolean value) that allows the algorithm to control the selection of a neighbourhood structure. For example if a neighbourhood structures, Nbs, shows a good performance (in terms of producing a lower penalty solution), so the algorithm will continue uses Nbs in the next iterations until no good solution can be obtained.

In this case Nbs will be pushed into the FIFO (First In First Out) tabu list, T_{List}. This move is not allowed to be part of any search process for a certain number of iterations. In a do-while loop, two solution are randomly selected, S_1 and S_2. The crossover and mutation operators are applied on S_1 and S_2 to obtain S_1' and S_2'. Randomly select a neighbourhood structure, Nbs, and applied on S_1' and S_2' to obtain S_1'' and S_2''. The best solution among S_1', S_2', S_1'' and S_2'' is chosen an assigned to a current solution S^*. If S^* is better than the best solution in hand, S_{best}, then S^* is accepted. Otherwise Nbs will be added into the T_{List}. The member of the populations will be updated by removing the worst solution and inserting the new solution obtained from the search process while maintaining the size of the population and to be used in the next generation. The process is repeated and stops when the termination criterion is met (in this work the termination criterion is set as a computational times).

---

for $i = 1$ *To* $i \leq Popsize$
    Generate random solutions, S_i;
end for
Set the length of Tabu, TabuSize;
Choose a best solution from the population, S_{best};
Create an empty tabu list with TabuSize, T_{List};
Set CanSelect = True;
do while*(termination condition dose not met)*
    Select two parents from the population using RWS, S_1 *and* S_2;
    Apply crossover and mutation operators on S_1 *and* S_2 *to produce* S_1' *and* S_2' ;
    if *CanSelect == True*
        Randomly select a neighbourhood structure, which is not in T_{List}, *called Nbs;*
    end if
    Apply Nbs on S_1' *and* S_2' *to produce* S_1'' *and* S_2'';
    Get a minimum solution penalty from S_1', S_2', S_1'' *and* S_2'' *called current solution* S^*;
    if $(S^* < S_{best})$
        $S_{best} = S^*$;
        CanSelect = False;
    else
        Push Nbs to T_{List};
        CanSelect = True;
    end if
    Update the sorted population by inserting and removing good and worst solutions,
    respectively. While maintaining the size of population.
end while
Output the best solution, S_{best};

Fig. 1. The pseudo code for tabu-based memetic algorithm

3.3   Neighbourhood Structures

Some techniques in the literature, like simulated annealing and tabu search, generally use a single neighbourhood structure throughout the search and focus more on the parameters that affect the acceptance of the moves rather than the neighbourhood structure. We can say that the success of finding good solutions for these problems is determined by the technique itself and the neighbourhood structure employed during the search. In this paper, a set of different neighbourhood structures have been employed. Their explanation can be outlined as follows (adapted from [8]):

Nbs_1: Select two events at random and swap timeslots.

Nbs_2: Choose a single events at random and move to a new random feasible timeslots.

Nbs_3: Select two timeslots at random and simply swap all the events in one timeslots with all the events in the other timeslots.

Nbs_4: Take two timeslots at random, say t_i and t_j (where j> i) where timeslots are ordered t_1, t_2, t_3, ... t_p. Take all events that in t_i and allocate them to t_j, then allocate those that were in t_{j-1} to t_{j-2} and so on until we allocate those that were t_{j+1} to t_i and terminate the process.

Nbs_5: Move the highest penalty events from a random 10% selection of the events to a random feasible timeslots.

Nbs_6: Carry out the same process as in Nbs_5 but with 20% of the events.

Nbs_7: Move the highest penalty events from a random 10% selection of the events to a new feasible timeslots which can generate the lowest penalty cost.

Nbs_8: Carry out the same process as in Nbs_7 but with 20% of the events.

Nbs_9: Select two timeslots based on the maximum enrolled events, say t_i and t_j. Select the most conflicted event in t_i with t_j and then apply Kempe chain.

3.4   The Memetic Operators: Crossover and Mutation

There are different types of crossover operators available in the literature. For example Cheong et al. [1] applied a crossover operation based on the best days (three periods per day and minimum number of clashes per students). In this paper, we applied a period exchange crossover. This crossover operator allows a number of exams (from one timeslot) to be added to another timeslot and vice versa based on a crossover rate. The feasibility is maintained through a crossover operation between the childs. The crossover operation is illustrated in Fig. 2, represents a period-exchange crossover. The shaded periods represent the selected periods for a crossover operation. These periods are selected based on a crossover rate using roulette wheel selection method, for example timeslots T_3 and T_5 are chosen as parents (a) and (b), respectively. Crossover is performed by inserting all exams from timeslot T_5 in parent (b) to timeslot T_3 in parent (a), which then will produce a child (a). The same process is applied to obtain child (b). This operation leads to an infeasible solution due to a conflict appeared between a number of exams. For example; in child (a), $e10$ is repeated in T_3 (occurs twice), which should be removed; e_{16} and e_{12} that are located in $T1$ and $T2$, respectively, should also be removed to insure that child (a) is feasible. In child (b), e_1 is conflicted with e_{16}, as a result e_1 cannot be inserted in T_5, while e_{10} and e_{18} are occur twice in T_5 and T_2, respectively, so these exams should be removed to obtain feasibility. Removing conflicts and repeating exams in each timeslot is considered as a repair function that changed the infeasibility of each offspring to feasible once. This mechanism of crossover improves the exploration of the search space. Moving a set of exams under feasibility conditions discover

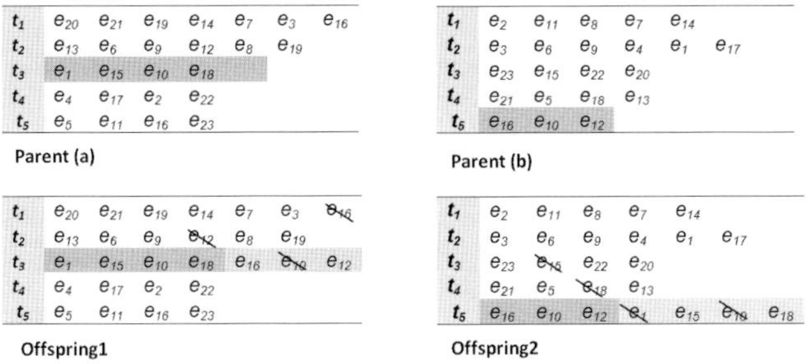

Fig. 2. Chromosome representation after crossover

Table 1. Parameters setting

Parameter	Population size	Crossover Rate	Mutation Rate	Selection mechanism
Value	50	0.8	0.04	Roulette wheel selection

more feasible solutions of the search space. The mutation is used to enhance the performance of crossover operation in allowing a large search space to be explored. Random selection of neighbourhood structures is used in a mutation process based on a mutation rate obtained after preliminary experiments (see Table 1 for a parameter setting).

4 Experimental Results

The proposed algorithm was programmed using Matlab and simulations were performed on the Intel Pentium 4 2.33 GHz computer. The algorithm is tested on standard examination benchmark datasets proposed by Carter et al. [4] and also can be found at [6]. The parameters used in the algorithm are chosen after preliminary experiments as shown in Table 1 (and are comparable similar with the paper [7]). In this approach, we reduced the population size (compared to the one in [7]) with an aim to speed up the searching process and to reduce the time taken to generate the populations. The crossover rate is set to be high with an aim to exchange more exams that results more changes in the solution and then will give more effect the quality of the solution. On the other hand, the mutation rate is set to be low to allow few variations on the solution.

4.1 Comparison Results

We ran the experiments between 5 to 8 hours for each of the datasets. Other techniques reported here run their experiments between 5 to 9 hours. For example Yang and Petrovic [12] took more than 6 hours and Cote et al. [10] about 9 hours. Note that running a system within these periods is not unreasonable in the context of examination timetabling where the timetables are usually produced months before the actual schedule is required. Table 2 provides the comparison

Table 2. Comparison results

Datasets	Our Approach	Best known	Datasets	Our Approach	Best known
Car-f-92	3.82	**3.76** [9]	Sta-f-83	**156.90**	156.94 [9]
Car-s-91	**4.35**	4.46 [9]	Tre-s-92	8.21	**7.86** [9]
Ear-f-83	33.76	**29.3** [3]	Uta-s-92	3.22	**2.99** [9]
Hec-s-92	10.29	**9.2** [3]	Ute-s-92	25.41	**24.4** [3]
Kfu-s-93	12.86	**12.62** [9]	Yor-f-83	36.35	**36.2** [3], [10]
Lse-f-91	10.23	**9.6** [3]			

of our results with the best known results for these benchmark datasets (taken from [6]). The best results out of 5 runs are shown in bold.

It is clearly shown that our tabu-based memetic approach is able to produce quite enough good solutions on the most of the instances, and out performs on two datasets compared to other population based algorithms. We believe that a small decimal improvement (with respect to the objective function value) is considered a good enough contribution, which is not easy to be achieved. We also believe that with the help of a tabu list, the algorithm performs better and able to find a better solution because the non-effective neighbourhood structures will not be employed in the next iterations i.e. the algorithm will only be feed with the currently effective neighborhood structure.

5   Conclusions and Future Work

The overall goal of this paper is to investigate a tabu-based memetic approach for the examination timetabling problem. Preliminary comparisons indicate that our approach outperforms the current state of the art on four benchmark problems. The key feature of our approach is the combination of a multiple neighbourhood structures and a search approach that incorporates the concept of a tabu list. Our future work will be aimed to testing this algorithm on the International Timetabling Competition datasets (ITC2007) introduced by University of Udine. We believe that the performance of a tabu-based memetic algorithm for the examination timetabling problem can be improved by applying advanced memetic operators and by intelligently selects the appropriate neighbourhood structures based on the current solution obtained.

References

1. Cheong, C.Y., Tan, K.C., Veeravalli, B.: Solving the exam timetabling problem via a multi-objective evolutionary algorithm: A more general approach. In: IEEE symposium on computational intelligence in scheduling, Honolulu, HI, USA, pp. 165–172 (2007)
2. Burke, E.K., Bykov, Y., Newall, J.P., Petrovic, S.: A time-predefined local search approach to exam timetabling problem. IIE Transactions 36(6), 509–528 (2004)

3. Caramia, M., Dell'Olmo, P., Italiano, G.F.: New algorithms for examination timetabling. In: Näher, S., Wagner, D. (eds.) WAE 2000. LNCS, vol. 1982, pp. 230–241. Springer, Heidelberg (2001)
4. Carter, M.W., Laporte, G., Lee, S.Y.: Examination Timetabling: Algorithmic Strategies and Applications. Journal of the Operational Research Society 47, 373–383 (1996)
5. Ct, P., Wong, T., Sabourin, R.: A hybrid multi-objective evolutionary algorithm for the uncapacitated exam proximity problem. In: Burke, E.K., Trick, M.A. (eds.) PATAT 2004. LNCS, vol. 3616, pp. 294–312. Springer, Heidelberg (2005)
6. Qu, R., Burke, E.K., McCollum, B., Merlot, L.T.G.: A survey of search methodologies and automated system development for examination timetabling. Journal of Scheduling 12, 55–89 (2009)
7. Abdullah, S., Turabieh, H.: Generating university course timetable using genetic algorithm and local search. In: Proceeding of the 3rd International Conference on Hybrid Information Technology, pp. 254–260 (2008)
8. Abdullah, S., Burke, E.K., McCollum, B.: Using a Randomised Iterative Improvement Algorithm with Composite Neighbourhood Structures for University Course Timetabling. In: Metaheuristics: Progress in complex systems optimization (Operations Research / Computer Science Interfaces Series), ch. 8. Springer, Heidelberg (2007) ISBN:978-0-387-71919-1
9. Abdullah, S., Turabeih, H., McCollum, B.: A hybridization of electromagnetic like mechanism and great deluge for examination timetabling problems. In: Blesa, M.J., Blum, C., Di Gaspero, L., Roli, A., Sampels, M., Schaerf, A. (eds.) HM 2009. LNCS, vol. 5818, pp. 60–72. Springer, Heidelberg (2009)
10. Abdullah, S., Ahmadi, S., Burke, E.K., Dror, M.: Investigating Ahuja-Orlin's large neighbourhood search approach for examination timetabling. OR Spectrum 29(2), 351–372 (2007)
11. Petrovic, S., Burke, E.K.: University timetabling. In: Leung, J. (ed.) Handbook of Scheduling: Algorithms, Models, and Performance Analysis, ch. 45. CRC Press, Boca Raton (April 2004)
12. Yang, Y., Petrovic, S.: A novel similarity measure for heuristic selection in examination timetabling. In: Burke, E.K., Trick, M.A. (eds.) PATAT 2004. LNCS, vol. 3616, pp. 247–269. Springer, Heidelberg (2005)
13. http://www.cs.qub.ac.uk/itc2007/

The Geometric Constraint Solving Based on the Quantum Particle Swarm

Cao Chunhong[1], Wang Limin[2], and Li Wenhui[2]

[1] Collge of Information Science and Engineering, Northeastern University,
Shenyang 110004, P.R. China
caochunhong@ise.neu.edu.cn
[2] College of Computer Science and Technology, Jilin University,
Changchun 130012, P.R. China
Jeffreywlm@sina.com, liwh@public.cc.jl.cn

Abstract. Geometric constraint problem is equivalent to the problem of solving a set of nonlinear equations substantially. The constraint problem can be transformed to an optimization problem. We can solve the problem with quantum particle swarm. The PSO is driven by the simulation of a social psychological metaphor motivated by collective behaviors of bird and other social organisms instead of the survival of the fittest individual. Inspired by the classical PSO method and quantum mechanics theories, this work presents a novel Quantum-behaved PSO (QPSO). The experiment shows that it can improve the algorithmic efficiency.

Keyword: geometric constraint solving, particle swarm optimization, quantum particle swarm optimization.

1 Introduction

Geometric constraint solving approaches are made of three approaches: algebraic-based solving approach, rule-based solving approach and graph-based solving approach [1] One constraint describes a relation that should be satisfied. Once a user defines a series of relations, the system will satisfy the constraints by selecting proper states after the parameters are modified.

The particle swarm optimization (PSO) originally developed by Kennedy and Eberhart in 1995 [2,3] is a population based swarm algorithm. Similarly to genetic algorithms [4], an evolutionary algorithm approach, PSO is an optimization tool based on a population, where each member is seen as a particle, and each particle is a potential solution to the problem under analysis. Each particle in PSO has a randomized velocity associated to it, which moves through the space of the problem. However, unlike genetic algorithms, PSO does not have operators, such as crossover and mutation. PSO does not implement the survival of the fittest individuals; rather, it implements the simulation of social behavior.

At the end of the 19th century, classical mechanics encountered major difficulties in describing motions of microscopic particles with extremely light masses and extremely high velocities, and the physical phenomena related to such motions. These

J. Yu et al. (Eds.): RSKT 2010, LNAI 6401, pp. 582–587, 2010.

aspects forced scientists to rethink the applicability of classical mechanics and lead to fundamental changes in their traditional understanding of the nature of motions of microscopic objects [5]. The studies of Bohr, de Broglie, Schro¨dinger, Heisenberg and Bohn in 1920s inspire the conception of a new area, the quantum mechanics [6]. Recently, the concepts of quantum mechanics and physics have motivated the generation of optimization methods; see [7–9]. Inspired by the PSO and quantum mechanics theories, this work presents a new Quantum-behaved PSO (QPSO) approach. In QPSO design, the main advantage of chaotic sequences application is maintaining the population diversity in the problem of interest.

2 Classical Particle Swarm Optimization

The proposal of PSO algorithm was put forward by several scientists who developed computational simulations of the movement of organisms such as flocks of birds and schools of fish. Such simulations were heavily based on manipulating the distances between individuals, i.e., the synchrony of the behavior of the swarm was seen as an effort to keep an optimal distance between them.

In theory, at least, individuals of a swarm may benefit from the prior discoveries and experiences of all the members of a swarm when foraging. The fundamental point of developing PSO is a hypothesis in which the exchange of information among creatures of the same species offers some sort of evolutionary advantage.

The procedure for implementing the global version of PSO is given by the following steps:

Step 1. *Initialization of swarm positions and velocities*: Initialize a population (array) of particles with random positions and velocities in the n-dimensional problem space using uniform probability distribution function.

Step 2. *Evaluation of particle's fitness*: Evaluate each particle's fitness value.

Step 3. *Comparison to pbest (personal best)*: Compare each particle's fitness with the particle's *pbest*. If the current value is better than *pbest*, then set the pbest value equal to the current value and the *pbest* location equal to the current location in n-dimensional space.

Step4. *Comparison to gbest (global best)*: Compare the fitness with the population's overall previous best. If the current value is better than *gbest*, then reset *gbest* to the current particle's array index and value.

Step 5. *Updating of each particle's velocity and position*: Change the velocity, v_i, and position of the particle, x_i, according to Eqs. (1) and (2):

$$v_i(t+1)=w \bullet v_i(t)+c_1 \bullet ud \bullet [p_i(t)\text{-}x_i(t)]+c_2 \cdot Ud \bullet [p_g(t)\text{-}x_i(t)] \tag{1}$$

$$x_i(t+1)=x_i(t)+\triangle t \bullet v_i(t+1) \tag{2}$$

where w is the inertia weight; $i = 1, 2, \ldots , N$ indicates the number of particles of population (swarm); $t = 1,2,\ldots , t_{max}$ indicates the iterations, w is a parameter called the inertia weight; $v_i = [v_{i1}, v_{i2}, \ldots , v_{in}]^T$ stands for the velocity of the ith particle, $x_i =$

$[x_{i1}, x_{i2}, \ldots, x_{in}]^{T}$ stands for the position of the ith particle of population, and $p_i = [p_{i1}, p_{i2}, \ldots, p_{in}]^{T}$ represents the best previous position of the ith particle. Positive constants c_1 and c_2 are the cognitive and social components, respectively, which are the acceleration constants responsible for varying the particle velocity towards *pbest* and *gbest*, respectively. Index g represents the index of the best particle among all the particles in the swarm. Variables ud and Ud are two random functions in the range [0, 1]. Eq. (1) represents the position update, according to its previous position and its velocity, considering $\triangle t = 1$.

Step 6. *Repeating the evolutionary cycle:* Return to *Step 2* until a stop criterion is met, usually a sufficiently good fitness or a maximum number of iterations (generations).

3   Quantum-Behaved Particle Swarm Optimization

In terms of classical mechanics, a particle is depicted by its position vector xi and velocity vector v_i, which determine the trajectory of the particle. The particle moves along a determined trajectory in Newtonian mechanics, but this is not the case in quantum mechanics. In quantum world, the term trajectory is meaningless, because xi and v_i of a particle cannot be determined simultaneously according to uncertainty principle. Therefore, if individual particles in a PSO system have quantum behavior, the PSO algorithm is bound to work in a different fashion [10,11].

In the quantum model of a PSO called here QPSO, the state of a particle is depicted by wavefunction$\varphi(x, t)$ (Schrodinger equation) [6,13], instead of position and velocity. The dynamic behavior of the particle is widely divergent form that of that the particle in classical PSO systems in that the exact values of x_i and v_i cannot be determined simultaneously.

In this context, the probability of the particle's appearing in position x_i from probability density function $|\varphi(x,t)|^{2}$, the form of which depends on the potential field the particle lies in [12]. Employing the Monte Carlo method, the particles move according to the following iterative equation [12–14]:

$$\begin{cases} x_i(t+1) = p + \beta \bullet |Mbset_i - x_i(t)| \bullet \ln(1/u), if k \geq 0.5 \\ x_i(t+1) = p - \beta \bullet |Mbset_i - x_i(t)| \bullet \ln(1/u), if k \prec 0.5 \end{cases} \tag{3}$$

where β is a design parameter called contraction-expansion coefficient [11]; u and k are values generated according to a uniform probability distribution in range [0, 1]. The global point called Mainstream Thought or Mean Best (*Mbest*) of the population is defined as the mean of the *pbest* positions of all particles and it given by

$$M_{best} = \frac{1}{N} \sum_{d=1}^{N} P_{g,d}(t) \tag{4}$$

where g represents the index of the best particle among all the particles in the swarm. In this case, the local attractor [14] to guarantee convergence of the PSO presents the following coordinates:

$$p = (c_1 p_{i,d} + c_2 p_{g,d}) / (c_1 + c_2) \tag{5}$$

The procedure for implementing the QPSO is given by the following steps:

Step 1. *Initialization of swarm positions*: Initialize a population (array) of particles with random positions in the *n*-dimensional problem space using a uniform probability distribution function.

Step 2. *Evaluation of particle's fitness*: Evaluate the fitness value of each particle.

Step 3. *Comparison to pbest (personal best)*: Compare each particle's fitness with the particle's *pbest*. If the current value is better than *pbest*, then set the *pbest* value equal to the current value and the *pbest* location equal to the current location in *n*-dimensional space.

Step 4. *Comparison to gbest (global best)*: Compare the fitness with the population's overall previous best. If the current value is better than *gbest*, then reset *gbest* to the current particle's array index and value.

Step 5. *Updating of global point*: Calculate the *Mbest* using Eq. (4).

Step 6. *Updating of particles'position*: Change the position of the particles where 1 and c_2 are two random numbers generated using a uniform probability distribution in the range [0, 1].

Step 7. *Repeating the evolutionary cycle*: Loop to *Step 2* until a stop criterion is met, usually a sufficiently good fitness or a maximum number of iterations (generations).

4   QPSO in Geometric Constraint Solving

The constraint problem can be formalized as (E,C) [15], here $E= (e_1,e_2,\ldots\ldots,e_n)$, it can express geometric elements, such as point, line, circle, etc; $C= (c_1,c_2,\ldots\ ,c_m)$, c_i is the constraint set in these geometric elements. Usually one constraint is represented by an algebraic equation, so the constraint can be expressed as follows:

$$\begin{cases} f_1(x_0, x_1, x_2, \ldots, x_n) = 0 \\ \quad \ldots \\ f_m(x_0, x_1, x_2, \ldots, x_n) = 0 \end{cases} \tag{6}$$

$X= (x_0, x_1, \ldots, x_n)$,

X_i are some parameters, for example, planar point can be expressed as (x_1, x_2) . Constraint solving is to get a solution x to satisfy formula (6).

$$F\ (X_j)\ = \sum_1^m |f_i| \tag{7}$$

Apparently if X_j can satisfy $F\ (X_j)$ =0, then X_j can satisfy formula (6). So the constraint problem can be transformed to an optimization problem and we only need to solve min $(F\ (X_j)\ ) < \varepsilon$. ε is a threshold. In order to improve the speed of QPSO, we adopt the absolute value of f_i not the square sum to express constraint equation set. From formula (7) and the solving min $(F(X_j)) < \varepsilon$ by QPSO, we can realize it is not necessary $m=n$, so the algorithm can solve under-constraint and over-constraint problem.

5   Application Instance and Result Analysis

The fig 1(b) is an auto-produced graph after some sizes of the fig 1(a) are modified by QPSO. From the fig 1 we can realize from the above figures that once a user defines a series of relations, the system will satisfy the constraints by selecting proper state after the parameters are modified.

Now we compare the searching capability of PSO, QPSO and GA. Table 1 can indicate the capability comparison of three algorithms.

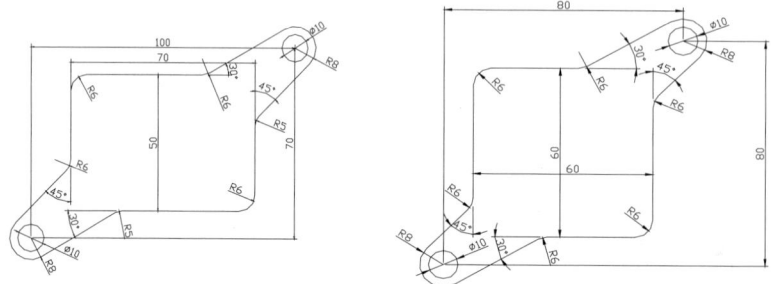

Fig. 1.   (a) A design instance   (b) Solving result

Table 1. The comparison of GA, PSO and IPSO

Algorithm	Generation of evolution	The time of occupying CPU	The generation of appearing the best solution
GA	500	48h	488
PSO	20	10s	15
QPSO	8	30s	5

The experiment results show that the QPSO algorithm has obtained better performance and possesses better convergence property than the compared algorithms.

6   Conclusions

Inspired by the PSO and quantum mechanics theories, this work presents a new QPSO approach for geometric constraint solving. The experiment indicates that QPSO display a satisfactory performance in geometric constraint solving. The aim of future works is to investigate the use of QPSO combined with other computational intelligence methodologies, such as immune system and neural networks for optimization in geometric constraint solving.

References

[1] Bo, Y.: Research and Implementation of Geometric Constraint Solving Technology, doctor dissertation of Tsinghua University, Beijing (1999)

[2] Kennedy, J.F., Eberhart, R.C.: Particle swarm optimization. In: Proceedings of IEEE international conference on neural networks, Perth, Australia, pp. 1942–1948 (1995)

[3] Eberhart, R.C., Kennedy, J.F.: A new optimizer using particle swarm theory. In: Proceedings of international symposium on micromachine and human science, Japan, pp. 39–43 (1995)

[4] Goldberg, D.E.: Genetic algorithms in search, optimization, and machine learning. Addison Wesley, Reading (1989)

[5] Pang, X.F.: Quantum mechanics in nonlinear systems. World Scientific Publishing Company, River Edge (2005)

[6] Schweizer, W.: Numerical quantum dynamics. Hingham, MA, USA (2001)

[7] Hogg, T., Portnov, D.S.: Quantum optimization. Inform. Sci. 128, 181–197 (2000)

[8] Protopescu, V., Barhen, J.: Solving a class of continuous global optimization problems using quantum algorithms. Phys. Lett. A. 296, 9–14 (2002)

[9] Bulger, D., Baritompa, W.P., Wood, G.R.: Implementing pure adaptive search with Grover's quantum algorithm. J. Optim: Theor. Appl. 116(3), 517–529 (2003)

[10] Sun, J., Feng, B., Xu, W.: Particle swarm optimization with particles having quantum behavior. In: Proceedings of congress on evolutionary computation, Portland, OR, USA, pp. 325–331 (2004)

[11] Sun, J., Xu, W., Feng, B.: Adaptive parameter control for quantum-behaved particle swarm optimization on individual level. In: Proceedings of IEEE international conference on systems, man and cybernetics, Big Island, HI, USA, pp. 3049–3054 (2005)

[12] Liu, J., Xu, W., Sun, J.: Quantum-behaved particle swarm optimization with mutation operator. In: Proceedings of 17th international conference on tools with artificial intelligence, Hong Kong, China (2005)

[13] Levin, F.S.: An introduction to quantum theory. Cambridge University Press, Cambridge (2002)

[14] Clerc, M., Kennedy, J.F.: The particle swarm: explosion, stability and convergence in a multi-dimensional complex space. IEEE Trans. Evolution Comput. 6(1), 58–73 (2002)

[15] Shengli, L., Min, T., Jinxiang, D.: Geometric Constraint Satisfaction Using Genetic Simulated Annealing Algorithm. Journal of Image and Graphics 8(8), 938–945 (2003)

Fish Swarm Intelligent Algorithm for the Course Timetabling Problem

Hamza Turabieh[1], Salwani Abdullah[1], Barry McCollum[2], and Paul McMullan[2]

[1] Data Mining and Optimization Research Group (DMO),
Center for Artificial Intelligence Technology,
Universiti Kebangsaan Malaysia, 43600 Bangi, Selangor, Malaysia
{hamza,salwani}@ftsm.ukm.my
[2] Department of Computer Science, Queen's University Belfast,
Belfast BT7 1NN United Kingdom
{b.mccollum,p.p.mcmullan}@qub.ac.uk

Abstract. In this work, a simulation of fish swarm intelligence has been applied on the course timetabling problem. The proposed algorithm simulates the movements of the fish when searching for food inside a body of water (refer as a search space). The search space is classified based on the visual scope of fishes into three categories which are crowded, not crowded and empty areas. Each fish represents a solution in the solution population. The movement direction of solutions is determined based on a Nelder-Mead simplex algorithm. Two types of local search i.e. a multi decay rate great deluge (where the decay rate is intelligently controlled by the movement direction) and a steepest descent algorithm have been applied to enhance the quality of the solution. The performance of the proposed approach has been tested on a standard course timetabling problem. Computational experiments indicate that our approach produces best known results on a number of these benchmark problems.

Keywords: Course Timetabling, Fish Swarm Optimization.

1 Introduction

Timetabling problems can be referred as an assignment of resources for tasks under predefined constraints with an aim to minimize the violation of constraints. This paper concentrates on the university course timetabling problem where events (subjects, courses) have to be assigned into a number of periods and rooms subject to a variety of hard and soft constraints. Hard constraints represent absolute requirements that must be satisfied, while soft constraints (if possible) try to be minimized to reach an optimal or near optimal solution. A solution (timetable in this case) which satisfies the hard constraints is known as a feasible solution. For more details about course timetabling see [11].

In this work, we applied a modified fish swarm optimization algorithm [8,15] that works on selected solutions on the course timetabling problem. It is a

J. Yu et al. (Eds.): RSKT 2010, LNAI 6401, pp. 588–595, 2010.

simulating search algorithm that imitates the fish behavior (such as searching, swarming and chasing behaviors) and hybridizing this biological behavior with local search algorithm to reach optimal or near optimal solution. The main contribution in this work is to control the direction of movements to provide exploration in the search space, and intelligently control the decay rate (in the proposed great deluge algorithm as discussed in Subsection 3.3) to better exploit the search space.

The rest of the paper is organized as follows: next section represents the course timetabling problems. Section 3 represents a fish swarm intelligent algorithm. Section 4 represents the settings and simulation results and finally the conclusion of this work is represented in section 5.

2   Problem Description

The problem description that is employed in this paper is adapted from the description presented in Socha et al. [12] and similar as the description used in the first international competition [16]. The problem has a set of N courses, $e = e_1,\ldots,e_N$, 45 timeslots, a set of R rooms, a set of F room features and a set of M students. The hard constraints are presented as: (i) No student can be assigned to more than one course at the same time; (ii) The room should satisfy the features required by the course; (iii) The number of students attending the course should be less than or equal to the capacity of the room; and (iv) No more than one course is allowed at a timeslot in each room. Soft constraints that are equally penalized are: (i) A student has a course scheduled in the last timeslot of the day; (ii) A student has more than 2 consecutive courses; and (iii) A student has a single course on a day. Minimizing the violation of the soft constraints under feasibility condition is the main objective of this problem.

3   The Fish Swarm Optimization Algorithm

The main idea of fish swarm intelligent algorithm is to simulate the behavior of a fish inside water (search space) while they are searching for food, following with local search of individual fish for reaching the global optimum. We will use the words "fish" and "solution" interchangeably throughout the paper. In this work, we proposed a modification on the original fish swarm intelligent algorithm [8,15] where we works on selected solutions rather than all solutions, with an aim to reduce the computational time. The basic point of a fish swarm intelligent algorithm is the visual scope. We categorized three possible cases for the visual scope for Sol_i as: 1)**Empty area**: no solution is closed to Sol_i. 2)**Crowded area**: many solutions are closed to Sol_i. 3)**Not crowded area**: a few solutions are closed to Sol_i.

A visual scope for a selected solution Sol_i can be represented as the scope of closed solutions. The determination of the number of solutions that are closed to Sol_i is based on the number of solutions inside the visual scope. The visual is used to determine the closeness of two solutions which is based on the distance (solution quality) between two solutions (i.e. $f(x') - f(x)$). Note that in this work, if the distance is less or equal to 10 (with respect to the objective function), then the solution x' is closed to x. The category of the visual scope is determined as((Number of close solution/populating size)$\leq \theta$), where θ is set to 0.5. If the number of the closed solution is more than the size of the visual scope, then the visual scope is considered as a crowded area. The size of visual scope is calculated as: (population size/2)+1. i.e. in other words, the visual scope is considered as a crowded area if the number of the closed solutions inside this visual scope is greater than half of the population size. The simulation behavior of the fish swarm intelligent algorithm can be described in three steps i.e. swarming, chasing and searching behaviors which are based on the category of the visual scope. If the visual scope is a crowded area then a searching behavior is applied; if the visual scope is a not crowded area then a swarming or a chasing behavior is applied; if the visual scope is empty, then again a chasing behavior is applied. The details of the simulation behavior of the fish swarm intelligent algorithm that is employed in this work are discussed as below:

- Swarming/Chasing behavior
  A fish (solution) swarms towards a central point (which is an average of the visual scope) of the search area. This behavior is applied if and only if the central point (in terms of the quality of the solution) is better than the selected solution Sol_i. Otherwise a fish will chase a best solution so far. These behaviors are represented by a great deluge algorithm where a central point is treated as an estimated quality or a best solution is treated as an estimated quality, respectively. However, if the visual scope is empty, a chasing behavior employed is a steepest descent algorithm.
- Searching behavior
  In this behavior, a Nelder-Mead simplex algorithm [10] is used to determine the movement directions of the fish. There are three directions called a $Contraction - External$ (CE), a $Reflection$ (R) and an $Expansion$ (E). These movement directions will be used to control a decay rate in the great deluge algorithm. Note that the details on the decay rate great deluge algorithm are discussed in Subsection 3.3.

3.1   Local Search

The searching behavior within the fish swarm intelligent algorithm is based on two mechanisms. The first mechanism is based on a Nelder-Mead simplex that determines the movement position ($Contraction\_External, Reflection$ and $Expansion$) of the selected solution. These positions are treated as an estimated quality of solution to be achieved later. The second mechanism is based on a great deluge algorithm that acts as a local search to enhance the quality of solution. The following two subsections describe a Nelder-Mead simplex and a great deluge algorithm, respectively. Note that in this work we called this local search as a multi decay rate great deluge algorithm. The pseudo code for the fish swarm intelligent algorithm is represented in Fig. 1.

Input
m: size of population;
Set Iteration =1 ;
Initialization Phase
       $(x_1, x_2, x_3, \ldots, x_n)=$ Construction()
Improvement Phase
While *termination condition is not satisfied do*
    Select a solution using RWS, Sol_i
    Compute the visual scope for Sol_i
    if *visual scope is empty then*
        $Sol'_i=$ *Steepest descent* (Sol_i)
    else
        if *visual scope is crowded then*
            $Sol'_i=$ *Great deluge with Nelder-Mead simplex* (Sol_i)
        else
            Central=compute the central point of visual
            if *Central better than Sol_i then*
                $Sol_{iA}=$ *Swarm-Standard Great Deluge* (Sol_i)
            else
                $Sol_{iA}=$ *Chase-Standard Great Deuce* (Sol_i)
            end if
            if $Sol_{best} < Sol_i$ *then*
                $Sol_{iB}=$ *Chase-Standard Great Deuce* (Sol_i)
            else
                $Sol_{iB}=$ *Steepest descent* (Sol_i)
            end if
            $Sol'_i=$ $min(Sol_{iA}, Sol_{iB})$
        end if
    end if
    $Sol''_i=$ $min(Sol_i, Sol'_i)$
    Iteration ++
end While

Fig. 1. Pseudo code for the fish swarm intelligent algorithm

The pseudo code of the fish swarm intelligent algorithm is divided into the initialization and improvement phases as shown in Fig. 1. At the initialization phase, feasible initial solutions are generated using a constructive heuristic as in Turabieh and Abdullah [14].

In the improvement phase, a visual scope of the fish of the selected solution is identified as discussed in Section 3. If the visual scope is an empty area, a steepest descent algorithm is employed. A great deluge algorithm with a different estimated quality value calculated based on a Nelder-Mead simplex algorithm is employed for the case where the visual scope is a crowded area. Otherwise, a central point of the not crowded area is calculated as an average of the visual scope (with respect to the objective function). It is then will be compared to the current solution Sol_i. If a central point is better than the current solution, then a great deluge with a central point as an estimated quality will be employed. Otherwise, a quality of the best solution is set as an estimated quality.

3.2 Nelder-Mead Simplex Algorithm

Nelder-Mead simplex algorithm has been proposed by Nelder and Mead [10] to find an approximation of a local optimum for a problem with N variables. The method is based on the theory of simplex, which is a special polytope of $N + 1$ vertices in N dimensions. Examples of simplexes include a line segment on a line, a triangle on a plane, a tetrahedron in three-dimensional space etc. In this work, we use a triangle simplex where three points are needed to represent a triangle which they are a selected solution and the nearest two solutions within the same visual scope. These three points (solutions) are used to in the Nelder-Mead simplex algorithm to determine the movement position i.e.:

$$Contraction\_External(CE) = \left(\frac{x_1 + x_2}{2} \right) - \left(\frac{\frac{x_1+x_2}{2} + x_3}{2} \right)$$

Calculate estimated qualities based on Nelder-Mead simplex algorithm (Contraction_External(CE), Reflection(R) and Expansion(E));
Calculate force decay rate, $\beta_{CE} = Contraction\_External(CE)/Iteration$;
Calculate force decay rate, $\beta_R = Reflection(R)/Iteration$;
Calculate force decay rate, $\beta_E = Expansion(E)/Iteration$;
Running decay rate $\beta = \beta_{CE}$;
While *termination condition is not satisfied do*
        Define neighbourhood (Nbs$_1$ and Nbs$_2$) for Sol$_i$ by randomly assigning event
        to a valid timeslots to generate a new solution called Sol$_i^*$
        Calculate $f(Sol_i^*)$;
        if *$f(Sol_i^*) < f(Sol_{best})$ where $f(Sol_{best})$ represent the best solution found so far*
                Sol$_i$ = Sol$_i^*$
                Sol$_{best}$ = Sol$_i^*$
        else
                if *(Sol$_i^* \leq$ level)*
                        Level = Level − β
                end if
        end if
                if *Sol$_{best}$ < Reflection(R)*
                        $\beta = \beta_R$
                else
                        if *Sol$_{best}$ < Expansion(E)*
                                $\beta = \beta_E$
                        end if
                end if
Iteration + +
end while

Fig. 2. The pseudo code for the multi-decay rate great deluge algorithm

$$Reflection(R) = Contraction\_External - \left(\frac{\frac{x_1+x_2}{2} + x_3}{2} \right)$$

$$Expansion(E) = Reflection - \left(\frac{\frac{x_1+x_2}{2} + x_3}{2} \right)$$

3.3 A Multi Decay Rate Great Deluge Algorithm

The Great Deluge (GD) algorithm was introduced by Dueck [6]. Apart from accepting a move that improves the solution quality, the great deluge algorithm also accepts a worse solution if the quality of the solution is less than or equal the level. In this approach the level is set using a Nelder-Mead simplex algorithm (as discussed above). The great deluge algorithm will intelligently control the decay rate / level. By accepting worse solutions, the algorithm is able to escape from local optimum that allows an exploration behavior to take place. Intelligently adjust the decay rate that starts with a high value (with respect to the quality of the solution) and by gradually decreased the level will help the algorithm to better exploit the search space during the searching process. Fig. 2 represents the pseudo code of the proposed multi decay rate great deluge algorithm.

There are two neighborhood moves employed within the local search algorithms i.e.: **Nbs1**: Select a course at random and move to another random feasible timeslot-room. **Nbs2**: Select two courses randomly and swap their timeslots and rooms while ensuring feasibility is maintained.

4   Simulation Results

The proposed algorithm was programmed using Matlab and simulations were performed on the Intel Pentium 4 2.33 GHz computer. The algorithm was tested on a standard benchmark course timetabling problem originally proposed by the Meteheuristic Network [16]. The settings used in this work are shown in Table 1.

Table 2 shows the comparison of our results in terms of penalty cost compared to other recent published results in the literature. The best results out of 11 runs are reported. The best results are presented in bold. We run two set of experiments i.e. (i) for 100000 iterations and (ii) for 500000 iterations (with longer run). In both cases, our algorithm is capable to find feasible solutions on all datasets.

In general, our approach is able to obtain competitive results with other approaches in the literature. We extended our experiments by increasing the number of iterations with the objective to demonstrate that our algorithm is able to produce good results given extra processing time. The emphasis in this paper is

Table 1. Parameters setting

Parameter	Iteration	Population size	GD-iteration	Steepest_Descent_iteration	Visual
Value	100000	50	2000	2000	10

Table 2. Results comparison

Datasets	Our Approach		[4]	[2]	[1]	[9]	[7]	[14]	[13]	[5]	[3]
	100K	500K									
small1	**0**	**0**	2	**0**	**0**	**0**	3	**0**	**0**	**0**	**0**
small2	**0**	**0**	4	**0**	**0**	**0**	4	**0**	**0**	**0**	**0**
small3	**0**	**0**	2	**0**	**0**	**0**	6	**0**	**0**	**0**	**0**
small4	**0**	**0**	0	**0**	**0**	**0**	6	**0**	**0**	**0**	**0**
small5	**0**	**0**	4	**0**	**0**	**0**	0	**0**	**0**	**0**	**0**
medium1	52	**45**	254	317	221	80	140	50	96	124	242
medium2	67	**40**	258	313	147	105	130	70	96	117	161
medium3	65	**61**	251	357	246	139	189	102	135	190	265
medium4	41	35	321	247	165	88	112	**32**	79	132	181
medium5	50	**49**	276	292	130	88	141	61	87	73	151
large	648	**407**	1026	-	529	730	876	653	683	424	-

on generating good quality solutions and the price to pay for this can be taken as being extended amount of computational time. Table 2 shows the comparison of our approach by with other results in the literature. Our approach is better than other methods on all cases except for m4 dataset. The extended experiments are able to improve the solutions between 2% to 40% compared to our previous results. This illustrates the effectiveness of our approach that is able to intelligently tune the decay rate within the great deluge algorithm given extra processing time.

5   Conclusions and Future Work

The proposed algorithm is adapted from the biological behavior of the fish that search for food. Each fish represents a solution in the population. The movement of the solution is based on its location that is categorized into crowded, not crowded and empty area. The movement direction of a selected solution is determined based on Nelder-Mead simplex algorithm which later will be used as an estimated quality to control a decay rate in a great deluge algorithm. A steepest descent algorithm is also been used as an improvement heuristic. This is a new procedure in the timetabling arena and it represents an approach that outperforms the current state of the art on a number of benchmark problems. The key feature of our approach is the combination of a mechanism to identify the visual scope and the search approach that has been controlled by an estimated quality. Our future work will be aimed to testing this algorithm on the International Timetabling Competition datasets (ITC2007) introduced by University of Udine.

References

1. Abdullah, S., Burke, E.K., McCollum, B.: A hybrid evolutionary approach to the university course timetabling problem. IEEE CEC, 1764–1768 (2007) ISBN: 1-4244-1340-0

2. Abdullah, S., Burke, E.K., McCollum, B.: An investigation of variable neighbourhood search for university course timetabling. MISTA, 413–427 (2005)
3. Abdullah, S., Burke, E.K., McCollum, B.: Using a randomised iterative improvement algorithm with composite neighbourhood structures for university course timetabling. In: Metaheuristics: Progress in complex systems optimization (Operations Research / Computer Science Interfaces Series), ch. 8. Springer, Heidelberg (2007)
4. Abdullah, S., Turabieh, H.: Generating university course timetable using genetic algorithms and local search. In: The Third 2008 International Conference on Convergence and Hybrid Information Technology, ICCIT, vol. I, pp. 254–260 (2008)
5. Al-Betar, M.Z., Khader, A. T., Liao, I.Y.: A Harmony Search with Multi-pitch Adjusting Rate for the University Course Timetabling. In: Recent Advances in Harmony Search Algorithm. Studies in Computational Intelligence, pp. 147–161 (2010)
6. Dueck, G.: New Optimization Heuristics. The great deluge algorithm and the record-to-record travel. Journal of Computational Physics 104, 86–92 (1993)
7. Landa-Silva, D., Obit, J.H.: Great deluge with non-linear decay rate for solving course timetabling problem. In: The fourth international IEEE conference on Intelligent Systems, Varna, Bulgaria (2008)
8. Li, X.L., Shao, Z.J., Qian, J.X.: An optimizing method based on autonomous animate: fish swarm algorithm. System Engineering Theory and Practice 11, 32–38 (2002)
9. McMullan, P.: An extended implementation of the great deluge algorithm for course timetabling. In: Shi, Y., van Albada, G.D., Dongarra, J., Sloot, P.M.A. (eds.) ICCS 2007. LNCS, vol. 4487, pp. 538–545. Springer, Heidelberg (2007)
10. Nelder, J.A., Mead, R.: A simplex method for function minimization. Computer Journal 7, 308–313 (1965)
11. Schaerf, A.: A survey of automated timetabling. Artificial Intelligence Review 13(2), 87–127 (1999)
12. Socha, K., Knowles, J., Samples, M.: A max-min ant system for the university course timetabling problem. In: Dorigo, M., Di Caro, G.A., Sampels, M. (eds.) ANTS 2002. LNCS, vol. 2463, pp. 1–3. Springer, Heidelberg (2002)
13. Turabieh, H., Abdullah, S., McCollum, B.: Electromagnetism-like Mechanism with Force Decay Rate Great Deluge for the Course Timetabling Problem. In: Wen, P., Li, Y., Polkowski, L., Yao, Y., Tsumoto, S., Wang, G. (eds.) RSKT 2009. LNCS, vol. 5589, pp. 497–504. Springer, Heidelberg (2009)
14. Turabieh, H., Abdullah, S.: Incorporating Tabu Search into Memetic approach for enrolment-based course timetabling problems. In: 2nd Data Mining and Optimization Conference, DMO 2009, vol. I, pp. 122–126 (2009)
15. Wang, C.R., Zhou, C.L., Ma, J.W.: An improved artificial fish-swarm algorithm and its application in feed forward neural networks. In: Proc. of the Fourth Int. Conf. on Machine Learning and Cybernetics, pp. 2890–2894 (2005)
16. http://www.idsia.ch/Files/ttcomp2002

A Supervised and Multivariate Discretization Algorithm for Rough Sets

Feng Jiang[1,*], Zhixi Zhao[2], and Yan Ge[1]

[1] College of Information Science and Technology, Qingdao University of Science and Technology, Qingdao 266061, P.R. China
[2] Training Department, Jining Institute of Technology, Jining 272013, P.R. China
jiangkong@163.net

Abstract. Rough set theory has become an important mathematical tool to deal with imprecise, incomplete and inconsistent data. As we all know, rough set theory works better on discretized or binarized data. However, most real life data sets consist of not only discrete attributes but also continuous attributes. In this paper, we propose a supervised and multivariate discretization algorithm — SMD for rough sets. SMD uses both class information and relations between attributes to determine the discretization scheme. To evaluate algorithm SMD, we ran the algorithm on real life data sets obtained from the UCI Machine Learning Repository. The experimental results show that our algorithm is effective. And the time complexity of our algorithm is relatively low, compared with the current multivariate discretization algorithms.

Keywords: Rough sets, discretization, supervised, multivariate.

1 Introduction

As an extension of naive set theory, rough set theory is introduced by Pawlak [5, 6], for the study of intelligent systems characterized by insufficient and incomplete information. In recent years, there has been a fast growing interest in rough set theory. Successful applications of the rough set model in a variety of problems have demonstrated its importance and versatility.

Continuous data are very common in real world problems. However, many machine learning approaches such as rough sets and decision trees would work better on discretized or binarized data [7, 8]. Therefore, to utilize such approaches more effectively, we should replace continuous attributes with discrete attributes. This process is called "discretization". Discretization is in fact a process to partition continuous attributes into a finite set of adjacent intervals [15].

Recently, discretization has received much attention in rough sets. Many traditional discretization algorithms have been applied to rough sets. In addition,

* This work is supported by the National Natural Science Foundation of China (grant nos. 60802042, 60674004), the Natural Science Foundation of Shandong Province, China (grant no. ZR2009GQ013).

J. Yu et al. (Eds.): RSKT 2010, LNAI 6401, pp. 596–603, 2010.

some new discretization algorithms have also been proposed from the view of rough sets [1-4, 21]. Usually, discretization algorithms can be divided into two categories: unsupervised and supervised [9]. Unsupervised algorithms, for instance, the equal width (EW) [16] or equal frequency binning (EF) algorithms [4], do not take advantage of class information to increase their performance. Thus, the resulting discretization schemes do not provide much efficiency when used in the classification process, e.g. they contain more intervals than necessary [20]. Therefore, different supervised algorithms have been proposed, such as statistics-based [8, 17], entropy-based [19], naive scaler (naive) [10], and semi-naive scaler (semi-naive) [10], etc.

Although supervised algorithms can use class information to improve their performance. But that is not enough. Since most of the current discretization algorithms used in rough sets focus on univariate, which discretize each continuous attribute independently, without considering interactions with other attributes. Since univariate algorithms do not make use of interdependence among the condition attributes in a decision table, the classification ability of the decision table may decrease after discretization. To avoid that problem, we need some algorithms that use both class information and relations between attributes in their discretization process, i.e., multivariate supervised algorithms. Bay [12, 13] and Monti et al. [14] have done some works in this area. But their methods are not computationally efficient, which limits their applications.

In this paper, we shall propose a supervised and multivariate discretization algorithm based on naive algorithm. As a simple supervised discretization algorithm, naive algorithm discretizes a continuous attribute by sorting the attribute values, and making a cut between attribute values when they belong to different classes [10]. Since naive algorithm does not make use of interdependence among continuous attributes, it generates too many cuts and too many small width intervals. To solve these problems, our algorithm simultaneously considers all continuous attributes in a given decision table $DT = (U, C, D, V, f)$, that is, concerns the interdependence among attributes in C. Through merging adjacent intervals of continuous values according to a given criterion and keeping the cardinality of C-positive region of D unchanged, the cuts obtained are much less than that obtained by naive algorithm, while the original classification ability of DT keeps invariant. In addition, the computational complexity of our algorithm is relatively low, compared with the current multivariate discretization algorithms.

The remainder of this paper is organized as follows. In the next section, we present some preliminaries that are relevant to this paper. In section 3, we present our discretization algorithm. Experimental results are given in Section 4. Finally, section 5 concludes the paper.

2   Preliminaries

In rough set terminology, a data table is also called an information system, which is formally defined as follows [6, 21].

Definition 1 Information System. An information system is a quadruple $IS = (U, A, V, f)$, where:
(1) U is a non-empty finite set of objects;
(2) A is a non-empty finite set of attributes;
(3) V is the union of attribute domains, i.e., $V = \bigcup_{a \in A} V_a$, where V_a denotes the domain of attribute a;
(4) $f : U \times A \rightarrow V$ is an information function which associates a unique value of each attribute with every object belonging to U, such that for any $a \in A$ and $x \in U$, $f(x, a) \in V_a$.

If some of the attributes are interpreted as outcomes of classification, the information system $IS = (U, A, V, f)$ can also be defined as a *decision table* by $DT = (U, C, D, V, f)$, where $C \cup D = A$, $C \cap D = \emptyset$, and C is called the *set of condition attributes*, D the *set of decision attributes*, respectively [6, 21].

Definition 2 Indiscernibility Relation. Given a decision table $DT = (U, C, D, V, f)$, let $B \subseteq C \cup D$, we call binary relation $IND(B)$ an indiscernibility relation, which is defined as follows:

$$IND(B) = \{(x, y) \in U \times U : \forall a \in B \ (f(x, a) = f(y, a))\}. \tag{1}$$

Since the indiscernibility relation $IND(B)$ is also an equivalence relation on set U, it partitions U into disjoint subsets (or equivalence classes), let $U/IND(B)$ denote the family of all equivalence classes of $IND(B)$. For simplicity, U/B will be written instead of $U/IND(B)$. For every object $x \in U$, let $[x]_B$ denote the equivalence class of relation $IND(B)$ that contains element x, called the equivalence class of x under relation $IND(B)$ [6, 21].

Definition 3 Positive Region. Given a decision table $DT = (U, C, D, V, f)$, let $B \subseteq C$ and $X \subseteq U$, the *B-positive region* $POS_B^X(D)$ of D wrt set X is the set of all objects from X which can be classified with certainty to classes of U/D employing attributes from B, i.e.,

$$POS_B^X(D) = \bigcup_{E \in X/B \wedge \forall x, y \in E((x,y) \in IND(D))} E, \tag{2}$$

where X/B denotes the partition of set X induced by relation $IND(B)$.

Since rough set theory is a logically founded approach based on indiscernibility, the discretization of continuous attributes is a key transformation in rough sets [3]. Discretization is usually performed prior to the learning process, which aims to partition continuous attributes into a finite set of adjacent intervals in order to generate attributes with a small number of distinct values. A good discretization algorithm not only can produce a concise summarization of continuous attributes to help the experts and users understand the data more easily, but also make learning more accurate and faster [18].

3   The Discretization Algorithm SMD for Rough Sets

In this section, we give a supervised and multivariate discretization algorithm SMD.

Algorithm 1. SMD

Input: The original decision table $DT = (U, C, D, V, f)$, set $B \subseteq C$ of all condition attributes to be discretized, where $|U| = n, |B| = m, |C| = k$.

Output: The discretized decision table $DT^d = (U^d, C^d, D^d, V^d, f^d)$.

Initialization: Let $P := \emptyset$, where P denotes the set of cuts.
Function Main(DT, B)
(1) By sorting all objects from U, calculate the partition U/C;
(2) Calculate the C-positive region $POS_C^U(D)$ of D wrt U;
(3) For any $a \in B$
(4) {
(5)   By sorting all objects from U, calculate the partition $U/(C - \{a\})$;
(6)   Calculate the $(C - \{a\})$-positive region $POS_{C-\{a\}}^U(D)$ of D wrt U;
(7)   Obtain the significance $SGF(a) = POS_C^U(D) - POS_{C-\{a\}}^U(D)$ of a;
(8) }
(9) Sort all attributes from B according to their significances in descending order. And let $B' = \{a_1, ..., a_m\}$ denote the result of sorting;
(10) For $1 \leq i \leq m$
(11) {
(12)   By sorting all objects from U, calculate the partition $U/\{a_i\} = \{X_1, ..., X_k\}$;
(13)   Combining $(U/\{a_i\}, a_i)$; //Call a function defined below.
(14)   Discretize attribute a_i using the cuts from P;
(15)   Let $P := \emptyset$;
(16) }
(17) Return the discretized decision table DT^d.

Function Combining(T, a) // $T = \{X_1, ..., X_{|T|}\}$ contains some subsets of U, and $a \in B$ is a condition attribute to be discretized.

(1) Let $j := 1, t := |T|$;
(2) While $(j \leq t)$ do
(3) {
(4)   Calculate $(C - \{a\})$-positive regions of D wrt X_j, X_{j+1} and $X_j \cup X_{j+1}$;

(5)   If $(|POS^{X_j}_{C-\{a\}}(D)| + |POS^{X_{j+1}}_{C-\{a\}}(D)| = |POS^{X_j \cup X_{j+1}}_{C-\{a\}}(D)|)$ and $(|X_j \cup X_{j+1}| < |U|/3)$ then

(6)   {

(7)   Merge sets X_j and X_{j+1} in T, that is, let $X_j := X_j \cup X_{j+1}$, and delete X_{j+1} from T;

(8)   Let $j := j + 2$;

(9)   }

(10)   else

(11)   {

(12)   Obtain a cut $p = (f(x,a) + f(y,a))/2$, where $x \in X_j$ is the last object in X_j, $y \in X_{j+1}$ is the first object in X_{j+1};

(13)   Let $P := P \cup \{p\}$;

(14)   Split T into two parts: $T_1 = \{X_1, ..., X_j\}$ and $T_2 = \{X_{j+1}, ..., X_t\}$;

(15)   Combining(T_1, a);

(16)   Combining(T_2, a);

(17)   Return;

(18)   }

(19)   }

(20)   If $|T| \geq 1$, then Combining(T, a).

Usually, the time complexity for calculating the partition U/B of U induced by indiscernibility relation $IND(B)$ is $O(n^2)$, where $n = |U|$. In algorithm 1, we use a method proposed by Nguyen and Nguyen [2] which can calculate the partition U/B in $O(m \times n \log n)$ time, where $m = |B|$, and $n = |U|$.

Algorithm 1 is composed of two parts: function Main and function Combining, where function Combining is called by function Main for m times. In the worst case, the time complexity of function Combining is $O(k \times n \times (\log n)^2)$. Therefore, in the worst case, the time complexity of algorithm 1 is $O(m \times k \times n \times (\log n)^2)$, and its space complexity is $O(n)$, where m, k, n are the cardinalities of B, C and U respectively.

In algorithm 1, the aim of function Combining is to merge adjacent intervals of attribute a according to a given criterion. The criterion requires that the merging operation in function Combining will not change $|POS^U_C(D)|$, that is, the original classification ability of decision table DT should keep invariant after discretization.

4   Experiment

To evaluate our algorithm for discretization, we ran it on five well-known data sets from the UCI Machine Leaning Repository [11]: Iris data set (iris), Glass Identification data set (glass), Ecoli data set (ecoli), Heart Disease data set (heart), and Pima Indians Diabetes data set (pima). Properties of these data sets are listed in Table 1. And algorithm SMD is compared with five state of the art discretization algorithms including one unsupervised algorithm and four supervised algorithms. The unsupervised algorithm is EF [4], and the supervised algorithms are naive [10], semi-naive [10], boolean reasoning (BR) [1] and entropy/MDL (EM) [9], respectively. Algorithm SMD is implemented in Pascal. And the other five algorithms are available in the Rosetta software. Rosetta is a toolkit developed by Alexander Øhrn [10] used for analyzing tabular data within the framework of rough set theory.

Table 1. Properties of data sets used in our experiments

Properties	Data sets				
	iris	glass	ecoli	heart	pima
Number of classes	3	6	8	2	2
Number of objects	150	214	336	270	768
Number of attributes	4	9	7	13	8
Number of continuous attributes	4	9	5	6	8

There are four steps in the experiments: data preparation, data discretization, data reduction as well as rule extraction, and data classification.

(1) The first step is to prepare the training and test sets. For each data set T, through *random sampling without replacement*, the data set T is split randomly into a training set of 50% of the data with the remaining 50% used for the test set.

(2) Next is the discretization step. We first use algorithm SMD to discretize the training set through generating a set of cuts. Then the test set is discretized using these cuts. Analogously, EF, naive, semi-naive, BR and EM algorithms supplied by Rosetta are respectively applied to the training and test sets. In our experiments, for the EF algorithm, the number of bins is set to 3 [10].

(3) The third step is creating reducts for the training set preprocessed by each discretization algorithm, as well as generating decision rules using the reducts. Rosetta provides 8 reduct extraction algorithms [10]. And we choose the "Genetic Algorithm (SAVGeneticReducer)" for our experiments. Options selected for this algorithm are: {DISCERNIBILITY= Object; MODULO.DECISION = True; SELECTION =All; COST=False; APPROXIMATE = True; FRACTION= 0.95; KEEP.LEVELS=1} [10].

Once the reducts have been derived, Rosetta can extract minimal "if-then" rules. These rules can then be used on the correspondingly discretized test set.

(4) The last step is applying the decision rules and a classification method to the discretized test set. We choose "Batch Classifier (BatchClassifier)". Options selected are: {CLASSIFIER= StandardVoter; FALLBACK= False; MULTIPLE=Best; LOG= False; CALIBRATION = False; ROC=False} [10].

The results achieved by Rosetta on five data sets using the six discretization algorithms are summarized in Table 2.

Table 2. Comparison of the classification accuracies achieved by Rosetta on five data sets using the six discretization algorithms

Discretization	Data sets				
Algorithms	iris(%)	glass(%)	ecoli(%)	heart(%)	pima(%)
EF	97.3333	67.2897	75	83.7037	69.5313
naive	90.6667	60.7477	60.119	82.963	67.1875
semi-naive	74.6667	66.3551	70.2381	85.1852	68.2292
BR	97.3333	59.8131	74.4048	85.9259	67.7083
EM	89.3333	57.0093	68.4524	82.2222	58.0729
SMD	**98.6667**	**68.2243**	**76.1905**	**87.4074**	**73.6979**

From Table 2, it is easy to see that SMD algorithm exhibits the highest accuracy for all the five data sets. Therefore, SMD algorithm has the best performance with respect to the other five algorithms.

5   Conclusion

Discretization of continuous attributes is a key issue in rough sets. Since most of the current discretization algorithms used in rough sets focus on univariate, which discretize each continuous attribute independently, without considering interactions with other attributes. Univariate algorithms may change the classification ability of the decision table. To guarantee that the original classification ability of the decision table keeps invariant, we presented a supervised and multivariate discretization algorithm based on naive algorithm. Through merging adjacent intervals according to a given criterion that requires the merging operation does not change the classification ability of decision table, our algorithm can produce less cuts than naive algorithm. Experimental results on real data sets demonstrated the effectiveness of our method for discretization.

References

1. Nguyen, H.S., Skowron, A.: Quantization of real value attributes:rough set and boolean reasoning approach. In: Proceedings of the Second Joint Annual Conference on Information Sciences, pp. 34–37. Society for Information Processing,
2. Nguyen, S.H., Nguyen, H.S.: Some efficient algorithms for rough set methods. In: Proceedings of IPMU 1996, Granada, Spain, pp. 1451–1456 (1996)
3. Nguyen, H.S.: Discretization problem for rough sets methods. In: Polkowski, L., Skowron, A. (eds.) RSCTC 1998. LNCS (LNAI), vol. 1424, pp. 545–555. Springer, Heidelberg (1998)
4. Nguyen, H.S., Nguyen, S.H.: Discretization Methods in Data Mining. In: Rough Sets in Knowledge Discovery, Physica, pp. 451–482 (1998)
5. Pawlak, Z.: Rough sets. International Journal of Computer and Information Sciences 11, 341–356 (1982)

6. Pawlak, Z.: Rough sets: Theoretical Aspects of Reasoning about Data. Kluwer Academic Publishers, Dordrecht (1991)
7. Catlett, J.: On Changing Continuous Attributes into Ordered Discrete Attributes. In: Kodratoff, Y. (ed.) EWSL 1991. LNCS, vol. 482, pp. 164–178. Springer, Heidelberg (1991)
8. Kerber, R.: Chimerge: Discretization of Numeric Attributes. In: Proc. of the Ninth National Conference of Articial Intelligence, pp. 123–128. AAAI Press, Menlo Park (1992)
9. Dougherty, J., Kohavi, R., Sahami, M.: Supervised and Unsupervised Discretization of Continuous Features. In: Proceedings of the 12th International Conference on Machine Learning, pp. 194–202. Morgan Kaufmann Publishers, San Francisco (1995)
10. Øhrn, A.: Rosetta Technical Reference Manual (1999), http://www.idi.ntnu.no/_aleks/rosetta
11. Blake, C.L., Merz, C.J.: UCI Machine Learning Repository, http://archive.ics.uci.edu/ml/
12. Bay, S.D.: Multivariate Discretization of Continuous Variables for Set Mining. In: Proceedings of the Sixth ACM SIGKDD International Conference on Knowledge Discovery and Data Mining, pp. 315–319 (2000)
13. Bay, S.D.: Multivariate Discretization for Set Mining. Knowledge and Information Systems 3(4), 491–512 (2001)
14. Monti, S., Cooper, G.F.: A Multivariate Discretization Method for Learning Bayesian Networks from Mixed Data. In: Proceedings of 14th Conference of Uncertainty in AI, pp. 404–413 (1998)
15. Tsai, C.J., Lee, C.I., Yang, W.P.: A discretization algorithm based on Class-Attribute Contingency Coefficient. Information Sciences 178, 714–731 (2008)
16. Wong, A.K.C., Chiu, D.K.Y.: Synthesizing Statistical Knowledge from Incomplete Mixed- Mode Data. IEEE Trans. Pattern Analysis and Machine Intelligence, 796–805 (1987)
17. Liu, H., Setiono, R.: Chi2: Feature Selection and Discretization of Numeric Attributes, pp. 388–391. IEEE Computer Society, Los Alamitos (1995)
18. Liu, H., Hussain, F., Tan, C.L., Dash, M.: Discretization: an enabling technique. Journal of Data Mining and Knowledge Discovery 6(4), 393–423 (2002)
19. Fayyad, U.M., Irani, K.B.: Multi-interval discretization of continuous-valued attributes for classification learning. In: Proceeding of Thirteenth International Conference on Artificial Intelligence, pp. 1022–1027 (1993)
20. Pongaksorn, P., Rakthanmanon, T., Waiyamai, K.: DCR: Discretization using Class Information to Reduce Number of Intervals. In: QIMIE 2009: Quality issues, measures of interestingness and evaluation of data mining models, pp. 17–28 (2009)
21. Wang, G.Y.: Rough set theory and knowledge acquisition. Xian Jiaotong University Press (2001)

Comparative Study of Type-2 Fuzzy Sets and Cloud Model

Kun Qin[1], Deyi Li[2], Tao Wu[1], Yuchao Liu[3],
Guisheng Chen[2], and Baohua Cao[4]

[1] School of Remote Sensing Information Engineering,
Wuhan University, Wuhan, China
[2] The Institute of Beijing Electronic System Engineering, Beijing, China
[3] Department of Computer Science and Technology,
Tsinghua University, Beijing, China
[4] Ming Hsieh Department of Electrical Engineering,
University of Southern California, Los Angeles, CA 90089, USA
qinkun163@163.com

Abstract. The mathematical representation of a concept with uncertainty is one of foundations of Artificial Intelligence. Type-2 fuzzy sets study fuzziness of the membership grade to a concept. Cloud model, based on probability measure space, automatically produces random membership grades of a concept through a cloud generator. The two methods both concentrate on the essentials of uncertainty and have been applied in many fields for more than ten years. However, their mathematical foundations are quite different. The detailed comparative study will discover the relationship between each other, and provide a fundamental contribution to Artificial Intelligence with uncertainty.

Keywords: type-2 fuzzy sets; cloud model; comparative study.

1 Introduction

The mathematical representation of a concept with uncertainty is one of foundations of Artificial Intelligence. Zadeh proposed fuzzy sets in 1965 [1], which uses precise membership or membership function to measure the fuzziness. Based on type-2 fuzzy sets introduced by Zadeh [2], Mendel proposed type-2 fuzzy logic system in 1998 [3], which studies the fuzziness of membership grade. Deyi Li proposed cloud model in 1995 [4], which automatically produces membership grade based on probability measure space and forms the inference mechanism through cloud rule generator [5].

The research on type-2 fuzzy sets focus on interval type-2 fuzzy sets [6,7]. With respect to the uncertain analysis of qualitative concept, Mendel pointed out that words mean different things to different people, so the uncertain analysis of words needs type-2 fuzzy sets [8]. Mendel proposed the processing methods for transformation between qualitative concept and quantitative data using type-2 fuzzy sets. Type-2 fuzzy sets have been used in many applications, such as forecasting of time-series [6], equalization of nonlinear time-varying channels [9],

J. Yu et al. (Eds.): RSKT 2010, LNAI 6401, pp. 604–611, 2010.

MPEG VBR video traffic modeling and classification [10], and aggregation using the linguistic weighted average [11].

Deyi Li proposed cloud model in 1995 [4,12,13]. Without defining precise membership function, cloud model uses forward cloud generator by computer program to produce disperse cloud drops, and describes qualitative concept using the whole cloud drops. Cloud model is different from statistical methods and fuzzy set methods, it cannot be considered as randomness compensated by fuzziness, fuzziness compensated by randomness, second-order fuzziness or second-order randomness [12,13]. Cloud model is a transformation model between qualitative concepts and quantitative values, which not only considers the fuzziness and randomness but also the association of the two [12,13]. In theory, there are several forms of cloud model, but the normal cloud model is commonly used in practice. The universalities of normal distribution and bell membership function are the theoretical foundation for the universality of normal cloud model [14]. Cloud model has been successfully applied in many fields, such as stabilization of the triple-inverted pendulum system [5], evaluation of electronic products [15], landslide monitoring in data mining and knowledge discovery [16], and image segmentation [17], and so on.

Type-2 fuzzy sets and cloud model both are methods to analyze qualitative concept with uncertainty, but there are some differences. The comparative studies illustrate their advantages and disadvantages, and make fundamental contributions to the development of Artificial Intelligence with uncertainty.

2   Basic Concepts

2.1   Type-2 Fuzzy Sets

Zadeh proposed type-2 fuzzy sets [2], whose membership grades themselves are type-1 fuzzy sets. Mendel provided a clear definition [18]:

A type-2 fuzzy set, denoted $\tilde{A}$, is characterized by a type-2 membership function $\mu_{\tilde{A}}(x, u)$, where x, the primary variable, has domain X, and u, the secondary variable, has domain $u \in J_x \subseteq [0, 1]$ at each $x \in X$, i.e.

$$\tilde{A} = \{((x, u), \mu_{\tilde{A}}(x, u)) | \forall x \in X, \forall u \in J_x \subseteq [0, 1]\} . \tag{1}$$

where $0 \leq \mu_{\tilde{A}}(x, u) \leq 1$. $\tilde{A}$ can also be expressed as

$$\tilde{A} = \int_{x \in X} \int_{u \in J_x} \mu_{\tilde{A}}(x, u)/(x, u), J_x \subseteq [0, 1] . \tag{2}$$

For discrete universe of discourse, $\int$ is replaced by Σ.

The membership grades of type-2 fuzzy sets themselves are type-1 fuzzy sets, which are denoted by a type-1 membership function, called secondary membership function. If the secondary membership functions are type-1 Gaussian membership functions, the type-2 fuzzy set is called a Gaussian type-2 fuzzy set. If the secondary membership functions are type-1 interval sets, the type-2 fuzzy set is called an interval type-2 fuzzy set. Interval type-2 fuzzy set is defined by lower membership function (LMF) and upper membership function (UMF). The region bounded by LMF and UMF is called footprint of uncertainty (FOU). Each value of the primary variable x of interval type-2 fuzzy sets is an interval.

2.2   Cloud Model

Let U be a universe set described by precise numbers, and C be a qualitative concept related to U. If there is a number $x \in U$, which randomly realizes the concept C, and the certainty degree of x for C, i.e., $\mu(x) \in [0, 1]$ is a random value with stabilization tendency

$$\mu : U \to [0, 1], \forall x \in U, x \to \mu(x) . \tag{3}$$

then the distribution of x on U is defined as a cloud, and every x is defined as a cloud drop [12,13].

The overall property of a concept can be represented by three numerical characters, the Expected value Ex, the Entropy En and the Hyper-entropy He. Ex is the mathematical expectation of the cloud drops. En is the uncertainty measurement of the qualitative concept, and determined by both the randomness and the fuzziness of the concept. He is the uncertain measurement of entropy [12,13].

Normal cloud model: Let U be a quantitative universal set described by precise numbers, and C be the qualitative concept related to U, if there is a number $x \in U$, which is a random realization of the concept C and x satisfies $x \sim N(Ex, En'^2)$ where $En' \sim N(En, He^2)$, and the certainty degree of x on C is

$$\mu(x) = e^{-\frac{(x - Ex)^2}{2(En')^2}} . \tag{4}$$

then the distribution of x on U is a normal cloud [12,13].

3   Comparative Study of Type-2 Fuzzy Sets and Cloud Model

Both type-2 fuzzy sets and cloud model analyze the uncertainty of membership grade, but their mathematical methods are different, which leads the differences between their methods of uncertain analysis of qualitative concept, the transformation approaches between qualitative concepts and quantitative values.

3.1   Mathematical Methods of Uncertain Analysis of Qualitative Concept

The mathematical methods of uncertain analysis of qualitative concept between fuzzy sets and cloud model are different, which is illustrated in Fig. 1. Fuzzy sets use mathematical functions to solve problems, while cloud model uses probability and mathematical statistics methods.

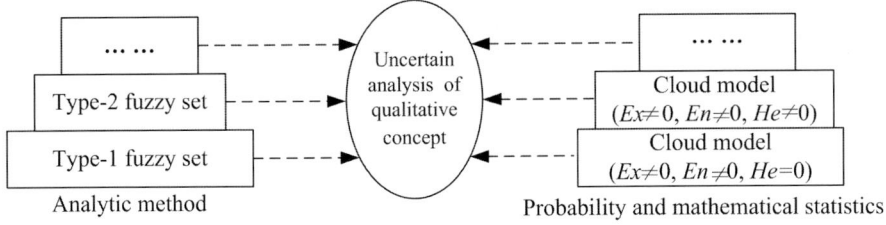

Fig. 1. Comparisons of mathematical methods

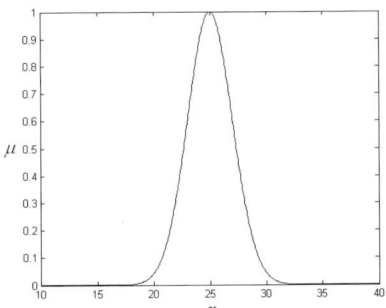

Fig. 2. Membership of "young"

Type-1 fuzzy sets deal with fuzziness of a qualitative concept through a certain membership function. For an example, Fig. 2 illustrated the quality concept of "young" with membership function as $\exp[-\frac{(x-\mu)^2}{2\sigma^2}]$, where $\mu = 25$ and $\sigma = 2$.

The research on type-2 fuzzy sets focus on interval type-2 fuzzy sets, which represent the uncertainty of membership function through FOU determined by the mathematical functions of UMF and LMF. For an example, Fig. 3(a) illustrates quality concept "young" represented by an interval type-2 fuzzy set, whose UMF and LMF are defined as $\exp[-\frac{(x-\mu)^2}{2\sigma_2^2}]$ and $\exp[-\frac{(x-\mu)^2}{2\sigma_1^2}]$ respectively, where $\mu = 25, \sigma_1 = 2$ and $\sigma_2 = 3$. So interval type-2 fuzzy set uses two mathematical functions to solve problems.

Based on probability measure space, cloud model generates the membership grades using forward normal cloud generator. The membership grade is a random value with stabilization tendency, hence, the model has no clear boundary but stable tendency, as illustrated in Fig. 3(b), there is no certain membership function in cloud model. Forward normal cloud generator uses probability and mathematical statistics to solve problems.

3.2 Uncertain Expression and Analysis for Qualitative Concept

Take the interval type-2 fuzzy set with Gaussian primary membership function and normal cloud model for example. Fig. 4 illustrated the qualitative concept

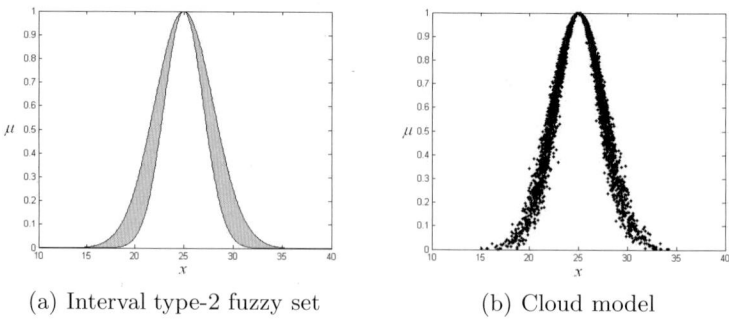

(a) Interval type-2 fuzzy set   (b) Cloud model

Fig. 3. Interval type-2 fuzzy set and cloud model

"young" expressed by the two. In Fig. 4(a), the concept "young" is expressed by an interval type-2 fuzzy set, which is the same as the one in Fig. 3(a). Fig. 4(b) illustrates the expression by cloud model with $Ex = 25, En = 2.5, He = 0.25$.

Interval type-2 fuzzy sets represent the uncertainty of qualitative concept through FOU. Cloud model represents qualitative concept through a model consisting of cloud drops. The overall shapes of both are similar but the expression forms are different. The FOU is consecutive, and has clear boundary. Cloud is composed of discrete points, and has no clear boundary. Interval type-2 fuzzy sets represent the uncertain boundary of FOU through two mathematical functions. While cloud model uses probability and mathematical statistics, and generates cloud drops through two times generation of normal random numbers. The membership grade is a random number with a stable tendency.

Type-2 fuzzy sets study the fuzziness of membership grade, which is considered as second-order fuzziness. Cloud model studies the randomness of membership grade, which is considered as fuzziness, randomness and the connection of them. Interval type-2 fuzzy sets take the membership grade as an interval. For example, as illustrated in Fig. 4(a), the membership grade corresponding to $x = 28$ is an interval [0.3, 0.6], but the uncertainty of the interval is ignored. Interval type-2 fuzzy sets only capture first-order uncertainty. For cloud model, the membership grade is generated by forward normal cloud generator which is realized by program, when $x = 28$, the membership grade is neither an exact value nor an interval, but a series of discrete points, as illustrated in Fig. 4(b). Cloud model capture the randomness of the membership grade.

The difference of uncertain analysis for membership grade between cloud model and interval type-2 fuzzy sets is also resulted from mathematical methods differences. Interval type-2 fuzzy sets represent the uncertainty of membership grade through FOU, which is determined by two functions, and the membership grade is an interval. Interval type-2 fuzzy sets study the fuzziness of membership grade, which is considered as second-order fuzziness. Based on the method of probability and mathematical statistics, cloud model studies the randomness of membership grade, which reflects fuzziness, randomness and their connection.

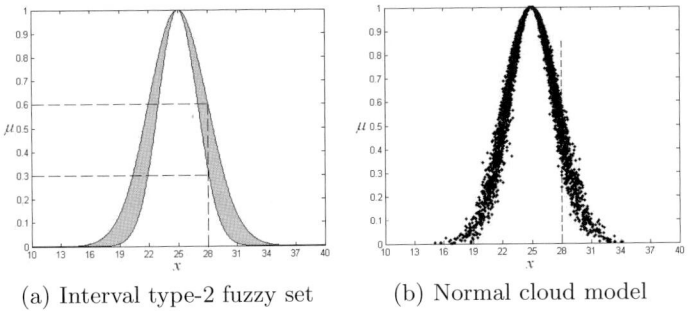

(a) Interval type-2 fuzzy set   (b) Normal cloud model

Fig. 4. Comparison of uncertain expression for qualitative concept "young"

3.3   Qualitative and Quantitative Transformation Approaches

The transformation is the interchange of quantitative values and qualitative concept. For this problem, type-2 fuzzy sets and cloud model provide corresponding methods respectively.

(1) Forward problem processing method and forward cloud generator
The forward problem processing method defines two membership functions (i.e. UMF and LMF) to express the quality concept [19]. Given the shape and parameters of UMF and LMF, FOU of a type-2 fuzzy set can be generated.

Forward normal cloud generator is a mapping from qualitative concept to quantitative values, when three numerical characters and the number of cloud drops are input, it produces the cloud drops to describe a concept. Based on probability measure space, cloud model generates the membership grade using forward normal cloud generator, in which the key point is the second-ordered relationship, i.e., within the two random numbers, one is the input of the other in generation, and the process is a kind of nonlinear mapping from uncertain variable to uncertain membership grade. The process cannot be expressed by a precise function, the membership grade generated by cloud model is a random number with a stable tendency, and the following processes are uncertain.

(2) Inverse problem processing method and backward cloud generator
Type-2 fuzzy sets provide inverse problem processing method, and cloud model provides backward cloud generator. They provide the transformation methods from quantitative numerical values to quality concept, but their processing methods are different.

1) Person membership function and backward cloud generator with membership grades
For person membership function method of type-2 fuzzy sets, the model of quality concept is extracted from the FOU provides by some persons [19,20]. Firstly, person membership data are collected from a group of subjects, who are very knowledgeable about fuzzy sets. And then, a region is filled by all person membership functions. Then, an interval type-2 fuzzy set model for a word is defined

as a specific aggregation of all such person membership functions. Finally, this aggregation is mathematically modeled and approximated.

The data of backward cloud generator with membership grade include samples and their membership grades, and data are collected from a group of subjects. A cloud model with three numerical characters is defined by computing mean and variance of sample data based on probability and mathematical statistics methods.

Person membership data are collected from a group of subjects, who must provide FOU about a word, and may therefore be quite difficult in practice. But backward cloud generator only need provide samples and their membership grades, no need membership functions. Their difference still lies in whether or not using mathematical function to solve problem.

2) Interval end-point approach and backward cloud generator without membership grades

Interval end-point approach collects interval end-point data for a word from a group of subjects [21]. The same results will not be obtained from all of them because words mean different things to different people. The left-hand and the right-hand end-point statistics for the data, such as sample average and standard deviations, are established. And then, these statistics are mapped into a pre-specified parametric FOU. Finally, parameters of FOU can be computed.

For backward cloud generator without membership grades, sample data for a word are collected from a group of subjects. A cloud model with three numerical characters is defined by computing mean, absolute central moment with first order, and variance of sample data based on probability and statistics.

Although both type-2 fuzzy sets and cloud model compute parameters using statistics analysis method, but the former must define the types of UMF and LMF, and the latter no need. The difference of them still lies in whether or not using mathematical function to solve problem.

4   Conclusions

Representation of uncertain knowledge is a critical topic of semantic computing, in which the mathematical expressions of qualitative concept is one of the basic scientific problems. Type-2 fuzzy sets and cloud model all provide the solutions, and they have connections and differences. Type-2 fuzzy set uses two crisp functions to analyze the uncertainty based on analytic mathematical methods; cloud model uses probability and mathematical statistics to analyze the uncertainty. To sum up, they all analyze qualitative concept with uncertainty, but since human think with natural languages, exact membership function and crisp set do not exist in human thought, which is easy to be brought into precise kingdom. The methods of probability and mathematical statistics consider randomness; hence, this kind of methods handles the uncertainties in qualitative concept analysis with uncertainty, qualitative and quantitative transformation. Maybe it is a good method to analyze uncertainty based on probability and mathematical statistics. Cloud model has entered a new stage to study fuzziness through probability and mathematical statistics methods.

Acknowledgments. The research is supported by 973 Program (2006CB701305; 2007CB311003), National Natural Science Foundation of China (60875007).

References

1. Zadeh, L.A.: Fuzzy sets. Information and Control 8(3), 338–353 (1965)
2. Zadeh, L.A.: The concept of a linguistic variable and its application to approximate reasoning-1. Information Science 8, 199–249 (1975)
3. Karnik, N.N., Mendel, J.M.: Introduction to type-2 fuzzy logic systems. In: Proceeding of FUZZ-IEEE 1998, Anchorage, AK, pp. 915–920 (1998)
4. Li, D.Y., Meng, H.J., Shi, X.M.: Membership cloud and membership cloud generator. Journal of Computer Research and Development 32(6), 15–20 (1995)
5. Li, D.Y.: The Cloud Control Method and Balancing Patterns of Triple Link Inverted Pendulum Systems. Engineering Sciences 1(2), 41–46 (1999)
6. Liang, Q., Mendel, J.M.: Interval type-2 fuzzy logic systems: theory and design. IEEE Transactions on Fuzzy Systems 8(5), 535–550 (2000)
7. Mendel, J.M.: Type-2 fuzzy sets and systems: an overview. IEEE Computational Intelligence Magazine 2(1), 20–29 (2007)
8. Mendel, J.M.: The perceptual computer: Architecture for computing with words. In: Proceedings of IEEE Conference on Fuzzy Systems, Melbourne, Australia, pp. 35–38 (2001)
9. Liang, Q., Mendel, J.M.: Equalization of Nonlinear Time-Varying Channels Using Type-2 Fuzzy Adaptive Filters. IEEE Transaction on Fuzzy Systems 8(5), 551–563 (2000)
10. Liang, Q., Mendel, J.M.: MPEG VBR Video Traffic Modeling and Classification Using Fuzzy Technique. IEEE Transactions on Fuzzy Systems 9(1), 183–193 (2001)
11. Liu, F., Mendel, J.M.: Aggregation Using the Fuzzy Weighted Average as Computed by the KarnikCMendel Algorithms. IEEE Transactions on Fuzzy Systems 16(1), 1–12 (2008)
12. Li, D.Y., Du, Y.: Artificial intelligent with uncertainty. National Defence Industry Press, Beijing (2005)
13. Li, D.Y., Du, Y.: Artificial intelligent with uncertainty. Chapman and Hall/CRC, Boca Raton (2007)
14. Li, D.Y., Liu, C.Y.: Study on the Universality of the Normal Cloud Model. Engineering Sciences 6(8), 28–34 (2004)
15. Song, Y.J., Li, D.Y., Yang, X.Z., Cui, D.H.: Reliability evaluation of electronic products based on cloud models. Acta Electronica Sinica 28(12), 74–76 (2000)
16. Wang, S.L.: Spatial data mining and knowledge discovery based on cloud model and data field. Ph.D. Thesis of Wuhan University, Wuhan (2002)
17. Qin, K.: Novel methods for image segmentation with uncertainty. Postdoctoral Thesis of Wuhan University, Wuhan (2007)
18. Mendel, J.M., John, R.I.: Type-2 fuzzy sets made simple. IEEE Transaction on Fuzzy Systems 10(2), 117–127 (2002)
19. Mendel, J.M., Wu, H.W.: Type-2 fuzzistics for symmetric interval type-2 fuzzy sets: part 1, forward problems. IEEE Transactions on Fuzzy Systems 14(6), 781–792 (2006)
20. Mendel, J.M., Wu, H.W.: Type-2 fuzzistics for symmetric interval type-2 fuzzy sets: part 2, inverse problems. IEEE Transactions on Fuzzy Systems 15(2), 301–308 (2007)
21. Mendel, J.M.: Computing with words and its relationships with fuzzistics. Information Sciences 177(4), 988–1006 (2007)

Operations of Fuzzy Numbers via Genuine Set

Jun Han and Baoqing Hu

School of Mathematics and Statistics, Wuhan University
Wuhan, 430072, P.R. of China
guihai-4728@hotmail.com, bqhu@whu.edu.cn

Abstract. In this paper, we discuss the arithmetic operations on fuzzy numbers from the point of view of Genuine set. The paper specially studies the operations on triangular fuzzy numbers and presents an effect algorithm to compute them. We divide one general triangular fuzzy number into two symmetric triangular fuzzy numbers by the properties of extension maximum and minimum operations for the convenience of the calculation.

Keywords: Fuzzy set, Genuine set, triangular fuzzy number.

1   Introduction

For any two real numbers x and y, they must have the relation either $x \leq y$ or $x \geq y$. Since any two fuzzy numbers are not linearly ordered in general, the operations of fuzzy numbers are quite different and much more complicated than those of real numbers. And the operations have been already considered by many researchers in recent years (see [1–9]).

The general references for genuine sets are [10–14]. Let us recall some definitions and denote the crisp sets $[0, 1]$ by I.

A map A is called an nth-order genuine set, if A is a map defined as follows.

$$A : X \times I^n \to I, (x, \varphi_1, \varphi_2, \cdots, \varphi_n) \mapsto \varphi$$

In that case, a fuzzy set can be viewed as a 0th-order genuine set. Although apparently there is no resemblance between the definition of type-n fuzzy sets and the definition of genuine sets, some connections can be established between the type-n fuzzy sets in X and the $(n-1)$st-order genuine sets in X (see [10]).

For an nth-order genuine set A, we can define the fuzzy set $G(A)$ in X whose membership function $G(A) : X \to I$ is given by

$$G(A)(x) = \begin{cases} \bigvee_{(\varphi_1, \varphi_2, \cdots, \varphi_n) \in I^n} \{\varphi \wedge \varphi_1 \wedge \varphi_2 \wedge \cdots \wedge \varphi_n\}, & n > 0 \\ A(x), & n = 0 \end{cases}$$

where $A(x, \varphi_1, \varphi_2, \cdots, \varphi_n) = \varphi$.

J. Yu et al. (Eds.): RSKT 2010, LNAI 6401, pp. 612–617, 2010.

2   Operations of Fuzzy Numbers via Genuine Set

We denote by $\widetilde{\mathbb{R}}$ and $\widetilde{\mathbb{R}}^+$ respectively the set of all fuzzy numbers and all positve fuzzy numbers for convenience's sake.

Let $\{A_i : 1 \leq i \leq n\}$ be n fuzzy sets, then their operation $f(A_1, A_2, \cdots, A_n)$ can be viewed as a function of n variables.

Let $f : \prod_{i=1}^{n} X_i \to Y$, $(x_1, x_2, \cdots, x_n) \mapsto y$ and $\{A_i : 1 \leq i \leq n\} \subseteq \widetilde{\mathbb{R}}$, then an $(n-1)$th-order genuine set A is induced by

$$A : Y \times I^{n-1} \to I, (y, A_1(x_1), A_2(x_2), \cdots, A_{n-1}(x_{n-1})) \mapsto A_n(x_n),$$

where $y = f(x_1, x_2, \cdots, x_n)$.

Since there must be a fuzzy set B such that $f(A_1, A_2, \cdots, A_n) = B$, it can be indicated as $G(A)$, whose membership function is given by

$$G(A)(y) = \bigvee_{\mathbf{Ax} \in I^n} \{\bigwedge_{i=1}^{n} A_i(x_i) : y = f(\mathbf{x})\} = \bigvee_{\mathbf{x} \in \mathbb{R}^n} \{\bigwedge_{i=1}^{n} A_i(x_i) : y = f(\mathbf{x})\}$$

where $\mathbf{Ax} = (A_1(x_1), A_2(x_2), \cdots, A_n(x_n))$, and $\mathbf{x} = (x_1, x_2, \cdots, x_n)$.

For example, $f_1(\widetilde{3}, \widetilde{4}, \widetilde{3}) = \widetilde{3} \cdot (\widetilde{4} + \widetilde{3})$, $f_2(\widetilde{1}) = \widetilde{1}^4$ and $f_3(\widetilde{3}, \widetilde{4}) = \widetilde{3} \cdot (\widetilde{4} + \widetilde{3})$ induce 2nd-, 0th- and 1st-order genuine sets respectively.

Proposition 1. *Let* A, $B, C \in \widetilde{\mathbb{R}}$, $\lambda, \mu \in \mathbb{R}$, $D, E, F \in \widetilde{\mathbb{R}}^+$ *and* $\phi, \varphi \in \mathbb{R}^+$. *Then (1)* $(A + B) + C = A + (B + C)$, $(A \cdot B) \cdot C = A \cdot (B \cdot C)$; *(2)* $A + B = B + A$, $A \cdot B = B \cdot A$; *(3)* $\lambda A + \lambda B = \lambda(A + B)$; *(4)* $A - B = A + (-B)$; *(5)* $A \div D = A \cdot D^{-1}$.

Proof. It follows a straightforward verification.

Proposition 2. *Let* $A \in \widetilde{\mathbb{R}}$ *and* $x_0 \in suppA$. *And let* $f(x_1, x_2, \cdots, x_n)$ *be increasing on* $(suppA)^n$ *and strictly increasing on at least one variable. If there is* $\{x_i' : 1 \leq i \leq n\} \subseteq \mathbb{R}$ *such that* $f(x_1', x_2', \cdots, x_n') = f(x_0, x_0, \cdots, x_0)$, *then* $\bigwedge_{i=1}^{n} A(x_i') \leq A(x_0)$.

Proof. We break the proof into two parts.
(1) If there is a $x_k' \in \{x_i' : 1 \leq i \leq n\}$ such that $x_k' \notin suppA$, then

$$\bigwedge_{i=1}^{n} A(x_i') \leq A(x_k') = 0 \leq A(x_0).$$

(2) Suppose $x_k' \in suppA$ for any $x_k' \in \{x_i' : 1 \leq i \leq n\}$. Let $x_{\min} \overset{\Delta}{=} \min_{1 \leq i \leq n} \{x_i'\}$ and $x_{\max} \overset{\Delta}{=} \max_{1 \leq i \leq n} \{x_i'\}$. And we claim that $x_{\min} \leq x \leq x_{\max}$. In fact, if $x_0 < x_{\min}$, then it follows that

$$f(x_1', x_2', \cdots, x_n') \geq f(x_{\min}, x_{\min}, \cdots, x_{\min}) > f(x_0, x_0, \cdots, x_0),$$

which is impossible. Likewise $x_0 > x_{\max}$ also lead to a contradictory. Since $x_{\min} \leq x \leq x_{\max}$ and fuzzy number A must be a convex fuzzy set, we obtain that

$$\bigwedge_{i=1}^{n} A(x_i') \leq A(x_{\min}) \wedge A(x_{\max}) \leq A(x_0).$$

Proposition 3. *Let $A \in \tilde{\mathbb{R}}$. And let $f(x_1, x_2, \cdots, x_n)$ be increasing on $(suppA)^n$ and strictly increasing on at least one variable. If we write $F(x) = f(x, x, \cdots, x)$, then $F(A) = f(A, A, \cdots, A)$.*

Proof. On one hand, we have that

$$G(f(A, A, \cdots, A))(y) = \sup_{\mathbf{x} \in \mathbb{R}^n} \{\bigwedge_{i=1}^{n} A(x_i) : y = f(\mathbf{x})\}$$

$$\geq \sup_{\mathbf{x} \in \mathbb{R}^n} \{\bigwedge_{i=1}^{n} A(x_i) : y = f(\mathbf{x}) \text{ and } x_1 = x_2 = \cdots = x_n\}$$

$$\geq \sup_{x_1 \in \mathbb{R}} \{A(x_1) : y = f(x_1, x_1, \cdots, x_1)\}$$

$$= \sup_{x_1 \in \mathbb{R}} \{A(x_1) : y = F(x_1)\}$$

$$= G(F(A))(y)$$

On the other hand, it follows from Proposition 2 that

$$G(f(A, A, \cdots, A))(y) = \sup_{\mathbf{x} \in \mathbb{R}^n} \{\bigwedge_{i=1}^{n} A(x_i) : y = f(\mathbf{x})\}$$

$$= \sup_{\mathbf{x} \in (suppA)^n} \{\bigwedge_{i=1}^{n} A(x_i) : y = f(\mathbf{x})\}$$

$$\leq \sup_{x_0 \in suppA} \{A(x_0) : y = f(x_0, x_0, \cdots, x_0)\}$$

$$= \sup_{x_0 \in \mathbb{R}} \{A(x_0) : y = f(x_0, x_0, \cdots, x_0)\}$$

$$= \sup_{x_0 \in \mathbb{R}} \{A(x_0) : y = F(x_0)\}$$

$$= G(F(A))(y)$$

Hence $F(A) = f(A, A, \cdots, A)$.

Proposition 4. *Let $D, E, F, G \in \tilde{\mathbb{R}}^+, A \in \tilde{\mathbb{R}}$ and $\{\lambda_i : 1 \leq i \leq n\} \subseteq \mathbb{R}^+$. Then*

(1) $\sum_{i=1}^{n} (\lambda_i A) = \left(\sum_{i=1}^{n} \lambda_i\right) A;$

(2) $\prod_{i=1}^{n} (\lambda_i D) = \left(\prod_{i=1}^{n} \lambda_i\right) D^n;$

(3) $(D+E)(F+G) = DF + DG + EF + EG;$

Proof. We only proof (1). Let $f(x_1, x_2, \cdots, x_n) = \sum_{i=1}^{n} \lambda_i x_i$ and $F(x) = \left(\sum_{i=1}^{n} \lambda_i \right) x.$ Then it follows from Proposition 3 that $F(A) = f(A, A, \cdots, A).$

3   Triangular Fuzzy Numbers

The triangular fuzzy number is one of the familiar fuzzy numbers, and its membership function is given by

$$A(x) = \begin{cases} \frac{x - a_A}{b_A - a_A}, & x \in [a_A, b_A] \\ \frac{c_A - x}{c_A - b_A}, & x \in (b_A, c_A] \\ 0, & otherwise \end{cases}$$

We denote by $\widetilde{\mathbb{R}}_t$ the set of all triangular fuzzy numbers.

Chiu studied the operations $\vee$ and $\wedge$ for fuzzy numbers. Based on Chiu's idea, we express a common triangular fuzzy number as a linear represetation of a unit triangular fuzzy number for the convenience of the calculation.

Proposition 5 (Chiu, 2002). *For $A, B \in \widetilde{\mathbb{R}}$, with continuous membership function and $A \cap B \neq \emptyset$, and let x_m be the point such that $(A \cap B)(x_m) \geq (A \cap B)(x)$ for all $x \in \mathbb{R}$ and $A(x_m) = B(x_m)$, moreover, x_m is between two mean values of A and B (if the number of x_m is not unique, any one point of those x_m is suitable). Then the operations $\vee$ and $\wedge$ can be implimented as*

(1) $(A \vee B)(z) = \begin{cases} (A \cap B)(z), & z < x_m, \\ (A \cup B)(z), & z \geq x_m, \end{cases}$

(2) $(A \wedge B)(z) = \begin{cases} (A \cup B)(z), & z < x_m, \\ (A \cap B)(z), & z \geq x_m. \end{cases}$

From the above-mentioned propositions, we can have some results as follows.

Proposition 6. *Let $\{A_i : 1 \leq i \leq n\} \subseteq \widetilde{\mathbb{R}}$ and $\{D_i : 1 \leq i \leq n\} \subseteq \widetilde{\mathbb{R}}^+$. Then*

(1) $-(\bigvee_{i=1}^{n} A_i) = \bigwedge_{i=1}^{n} (-A_i), \quad -(\bigwedge_{i=1}^{n} A_i) = \bigvee_{i=1}^{n} (-A_i);$
(2) $(\bigvee_{i=1}^{n} D_i)^{-1} = \bigwedge_{i=1}^{n} (D_i^{-1}), \quad (\bigwedge_{i=1}^{n} D_i)^{-1} = \bigvee_{i=1}^{n} (D_i^{-1}).$

Proposition 7. *For $\{A_i : 1 \leq i \leq n\} \subseteq \widetilde{\mathbb{R}}_t$ with the same kernel, let $A_1 \subseteq A_2 \subseteq \cdots \subseteq A_n$. Then we have that $\bigvee_{i=1}^{n} A_i = A_1 \vee A_n$ and $\bigwedge_{i=1}^{n} A_i = A_1 \wedge A_n.$*

Proposition 8. *For $\{A_i : 1 \leq i \leq n\} \subseteq \widetilde{\mathbb{R}}_t$ with the same kernel, let $A_1 \subseteq A_s \subseteq A_2$ and $A_1 \subseteq A_t \subseteq A_2$. Then $(A_1 \vee A_s) \wedge (A_t \vee A_2) = A_s \wedge A_t$ and $(A_1 \wedge A_s) \vee (A_t \wedge A_2) = A_s \vee A_t.$*

The following Proposition is easiliy checked.

Proposition 9. *For $A_1, A_2, B_1, B_2 \in \tilde{\mathbb{R}}_t$ and $D_1, D_2, E_1, E_2 \in \tilde{\mathbb{R}}_t^+$, let $A_1 \subseteq A_2$ with the same kernel, $B_1 \subseteq B_2$ with the same kernel, $D_1 \subseteq D_2$ with the same kernel and $E_1 \subseteq E_2$ with the same kernel. Then*

(1) $(A_1 \vee A_2) + (B_1 \vee B_2) = (A_1 + B_1) \vee (A_2 + B_2)$;
(2) $(A_1 \vee A_2) + (B_1 \wedge B_2) = (A_1 + B_2) \wedge (A_2 + B_1)$;
(3) $(D_1 \vee D_2) \cdot (E_1 \vee E_2) = (D_1 \cdot E_1) \vee (D_2 \cdot E_2)$;
(4) $(D_1 \vee D_2) \cdot (E_1 \wedge E_2) = (D_1 \cdot E_2) \wedge (D_2 \cdot E_1)$.

Example 1. Calculate $\widetilde{0.5} \div \tilde{1}$, where $\widetilde{0.5}(x) = \begin{cases} x + \frac{1}{2}, & x \in [-\frac{1}{2}, \frac{1}{2}] \\ 2(1 - x), & x \in (\frac{1}{2}, 1] \\ 0, & otherwise \end{cases}$ and

$\tilde{1}(x) = \begin{cases} 2 - x, & x \in [1, 2] \\ 0, & otherwise \end{cases}$

In fact, for the sake of convenience, we can define a fuzzy number θ as follows.

$$\theta(x) = \begin{cases} x + 1, & x \in [-1, 0] \\ 1 - x, & x \in (0, 1] \\ 0, & otherwise \end{cases}$$

And then from the above propositions it follows that

$$\widetilde{0.5} \div \tilde{1} = ((\frac{1}{2}\theta + \frac{1}{2}) \wedge (\theta + \frac{1}{2})) \div (1 \vee (\theta + 1))$$

$$= ((\frac{1}{2}\theta + \frac{1}{2}) \wedge (\theta + \frac{1}{2})) \cdot (1 \wedge \frac{1}{\theta + 1})$$

$$= ((\frac{1}{2}\theta + \frac{1}{2}) \cdot 1) \wedge ((\theta + \frac{1}{2}) \cdot \frac{1}{\theta + 1})$$

$$= (\frac{1}{2}\theta + \frac{1}{2}) \wedge (\frac{2\theta + 1}{2\theta + 2})$$

$$= (\frac{1}{2}\theta + \frac{1}{2}) \wedge (\frac{2\theta + 1}{-2\theta + 2})$$

Hence

$$G(\widetilde{0.5} \div \tilde{1})(y) = \begin{cases} 1 + \frac{2y-1}{2y+2}, & \frac{2y-1}{2y+2} \in [-1, 0] \\ 1 - 2(y - \frac{1}{2}), & 2(y - \frac{1}{2}) \in (0, 1] \\ 0, & otherwise \end{cases} = \begin{cases} \frac{4y+1}{2y+2}, & y \in [-\frac{1}{4}, \frac{1}{2}] \\ 2(1 - y), & y \in (\frac{1}{2}, 1] \\ 0, & otherwise \end{cases}$$

Remark 1. Now let us consider the product of two fuzzy numbers. Let $\tilde{2} = \theta + 2$ $\widetilde{4} = \theta + 4$ $\widetilde{8} = \theta + 8$ and $\widetilde{16} = \theta + 16$. Since they are obviously triangular fuzzy numbers, it follows that

$$\tilde{1} \cdot \widetilde{16} = (\theta + 1)(\theta + 16) = \theta^2 + 17\theta + 16 ,$$

$$\tilde{2} \cdot \widetilde{8} = (\theta + 2)(\theta + 8) = \theta^2 + 10\theta + 16 ,$$

$$\widetilde{4} \cdot \widetilde{4} = (\theta + 4)(\theta + 4) = \theta^2 + 4\theta + 16 .$$

We can obtain such a conclusion that the uncertainty of the operation of fuzzy numbers has something to do with the distance between them.

Acknowledgement

This research was supported by the Natural Scientific Foundation of China (Grand No. 70771081, 60974086) and 973 National Basic Research Program of China (Grand No. 2007CB310804).

References

1. Taheri, S.M.: C-fuzzy numbers and a dual of extension principle. Inform. Sci. 178, 827–835 (2008)
2. Huang, H.L., Shi, F.G.: L-fuzzy numbers and their properties. Inform. Sci. 178, 1141–1151 (2008)
3. Yeh, C.T.: A note on trapezoidal approximations of fuzzy numbers. Fuzzy Sets and Systems 158, 747–754 (2007)
4. Ban, A.: Approximation of fuzzy numbers by trapezoidal fuzzy numbers preserving the expected interval. Fuzzy Sets and Systems 159, 1327–1344 (2008)
5. Allahviranloo, T., Adabitabar, F.M.: Note on Trapezoidal approximation of fuzzy numbers. Fuzzy Sets and Systems 158, 755–756 (2007)
6. Grzegorzewski, P., Mrowka, E.: Trapezoidal approximations of fuzzy numbers—revisited. Fuzzy Sets and Systems 158, 757–768 (2007)
7. Grzegorzewski, P.: Trapezoidal approximations of fuzzy numbers preserving the expected interval—Algorithms and properties. Fuzzy Sets and Systems 159, 1354–1364 (2008)
8. Yeh, C.T.: On improving trapezoidal and triangular approximations of fuzzy numbers. Internat. J. Approx. Reason. 48, 297–313 (2008)
9. Yeh, C.T.: Trapezoidal and triangular approximations preserving the expected interval. Fuzzy Sets and Systems 159, 1345–1353 (2008)
10. Demirci, M., Waterman, M.S.: Genuine sets. Fuzzy Sets and Systems 105, 377–384 (1999)
11. Demirci, M.: Some notices on genuine sets. Fuzzy Sets and Systems 110, 275–278 (2000)
12. Coker, D., Demirci, M.: An invitation to topological structures on first order genuine sets. Fuzzy Sets and Systems 119, 521–527 (2001)
13. Demirci, M.: Genuine sets, various kinds of fuzzy sets and fuzzy rough sets. International Journal of Uncertainty. Fuzziness Knowledge-Based Systems 11(4), 467–494 (2003)
14. Demirci, M.: Generalized set-theoretic operations on genuine sets. Hacet. J. Math. Stat. 34S, 1–6 (2005)
15. Chiu, C.H., Wang, W.J.: A simple computation of MIN and MAX operations for fuzzy numbers. Fuzzy Sets and Systems 126, 273–276 (2002)

An Uncertain Control Framework of Cloud Model

Baohua Cao[1], Deyi Li[2], Kun Qin[3],
Guisheng Chen[2], Yuchao Liu[4], and Peng Han[5]

[1] Ming Hsieh Department of Electrical Engineering,
University of Southern California, Los Angeles, CA 90089, USA
[2] The Institute of Beijing Electronic System Engineering, Beijing, China
[3] School of Remote Sensing Information Engineering, Wuhan University, China
[4] Department of Computer Science and Technology, Tsinghua University, China
[5] Chongqing Academy of Science and Technology, ChongQing, China
baohuacao@yahoo.com, leedeyi@tsinghua.edu.cn, qinkun163@163.com,
cgs00@mails.tsinghua.edu.cn, yuchao_liu@163.com, han.peng@gmail.com

Abstract. The mathematical representation of a concept with uncertainty is one of the foundations of Artificial Intelligence. Uncertain Control has been the core in VSC systems and nonlinear control systems, as the representation of Uncertainty is required. Cloud Model represents the uncertainty with expectation Ex, entropy En and Hyper-entropy He by combining Fuzziness and Randomness together. Randomness and fuzziness make uncertain control be a difficult problem, hence we propose an uncertain control framework of Cloud Model called UCF-CM to solve it. UCF-CM tunes the parameters of Ex, En and He with Cloud, Cloud Controller and Cloud Adapter to generate self-adaptive control in dealing with uncertainties. Finally, an experience of a representative application with UCF-CM is implemented by controlling the growing process of artificial plants to verify the validity and feasibility.

Keywords: Uncertainty, Uncertain Control, Cloud Model, UCF-CM.

1   Introduction

Uncertain control has been a solid challenge in the domain of Artificial Intelligence and also been the core issue in VSC (variable structure control) systems and nonlinear control systems [1]. VSC design trends to use parameters tuning with a variable structure control to improve the control system against disturbances and uncertainties. However, in VSC, the lack of a priori knowledge on the target control signal leads the designer to seek for alternative methods predicting the error on the control [2, 3]. Nonlinear control system alters its parameters to adapt to a changing environment. The changes in environment can represent variations in process dynamics or changes in the characteristics of the disturbances. There are, however, many situations where the changes in process dynamics are so large that a constant linear feedback controller will not work satisfactorily.

J. Yu et al. (Eds.): RSKT 2010, LNAI 6401, pp. 618–625, 2010.

Cloud model is proposed by Deyi Li in 1995 to represent the uncertainty transition between qualitative concept and quantitative description [4]. Cloud Model has three digital characteristics: Expected value (Ex), Entropy (En) and Hyper-Entropy (He), which well integrate the fuzziness and randomness of spatial concepts in unified way. In the discourse universe, Ex is the position corresponding to the center of the cloud gravity, whose elements are fully compatible with the spatial linguistic concept; En is a measure of the concept coverage, i.e. a measure of the spatial fuzziness, which indicates how many elements could be accepted to the spatial linguistic concept; He is a measure of the dispersion on the cloud drops, which can also be considered as the entropy of En.

Cloud Model has the nature for uncertain control, as it uses three mathematical parameters to handle the uncertainty. Cloud Model has already been applied to solve the world puzzle of inverted-pendulum's uncertain control successfully. With Cloud Model, an uncertain control framework called UCF-CM is proposed by us. UCF-CM is a framework to handle specific uncertain control problems with parameters tuning of Ex, En and He defined in Cloud Model. To verify its validity and feasibility, a representative application of UCF-CM is implemented. Controlling the growing process of artificial plants is a challenging topic as it is too difficult to take effective control in artificial plant' generating course for the uncertain randomness and fuzziness during its variation process. Simulating the natural growing process of artificial plants has always been the challenge in Artificial Life, the challengeable domain of Artificial Intelligence. UCF-CM is used to have a try to control the uncertain variation of artificial plants to show its validity and feasibility.

The paper is organized as follows. The next section introduces Cloud Model's definition and its three parameters. Section 3 proposes the UCF-CM framework and section 4 details the application study. The last section summarizes our discovery and presents future research goals.

2   Cloud Model

As a result of random, fuzziness, incompleteness and disagreement of description, uncertainty study has been the core in artificial intelligence [4, 5]. Cloud model theory is proposed to research the uncertainty [4] based on three parameters: Ex, En and He. Membership cloud employs expectation Ex, entropy En and hyper-entropy He to describe a specific concept. Ex is the expectation. En represents a granularity and it reflects the range of domain space. He describes the uncertain measurement of entropy. It can be used to express the relationship between randomness and fuzziness [5].

2.1   Definition

Definition 1: Membership Cloud. Let U denote a quantitative domain composed by precise numerical variables; C is a qualitative concept on U. If the quantitative value $x \in U$ is a random realization of qualitative concept C, then the x 's confirmation on C can be denoted $\mu(x) \in [0, 1]$, which is a random number with stable tendency.

$\mu : U \to [0, 1], x \in U, x \to \mu(x)$

The distribution of x on U is called Cloud, x is called a Cloud Droplet. The cloud is from a series of cloud drops. In the rocess of the formation of clouds, a cloud droplet is a realization of qualitative concept through numeric measurement.

Definition 2 [4,5]: Normal Cloud. Let U denote a quantitative domain composed of precise numerical variables; C is a qualitative concept on U. If the quantitative value $x \in U$ is a random realization of qualitative concept C, $x \sim N(Ex, En'^2)$, $En' \sim N(En, He^2)$, and the certainty degree of x on C is $\mu = exp(-\frac{(x-Ex)^2}{2En'^2})$

The distribution of x on U is called Normal Cloud.

2.2   Algorithm

Given three digital characteristics Ex, En, and He, a set of cloud drops could be generated by the following algorithm [6,7]: Algorithm: Cloud (Ex, En, He, N)

Input: the expectation of cloud Ex, the entropy of cloud En, the hyper entropy of cloud He, the number of drops N.

Output: a normal cloud with digital characteristics Ex, En, and He.

step1: Generate a normal random number En' with the expectation En and stand deviation He, $En' \sim N(En, He^2)$.

step2: Generate a normal random number x_i with the expectation Ex and stand deviation En', $x \sim N(Ex, En'^2)$.

step3: Compute the μ_i of the cloud droplet x_i.

step4: Generate cloud droplet (x_i, μ_i)

step5: Repeat step 1-4, until i = N.

Figure 1 shows some samples of cloud generated by the algorithm.

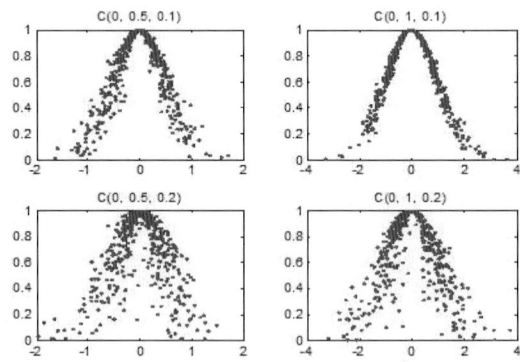

Fig. 1. Clouds

3   UCF-CM

With Cloud Model, an uncertain control framework is designed for handling uncertain control problems. Figure 2 shows the design of UCF-CM.

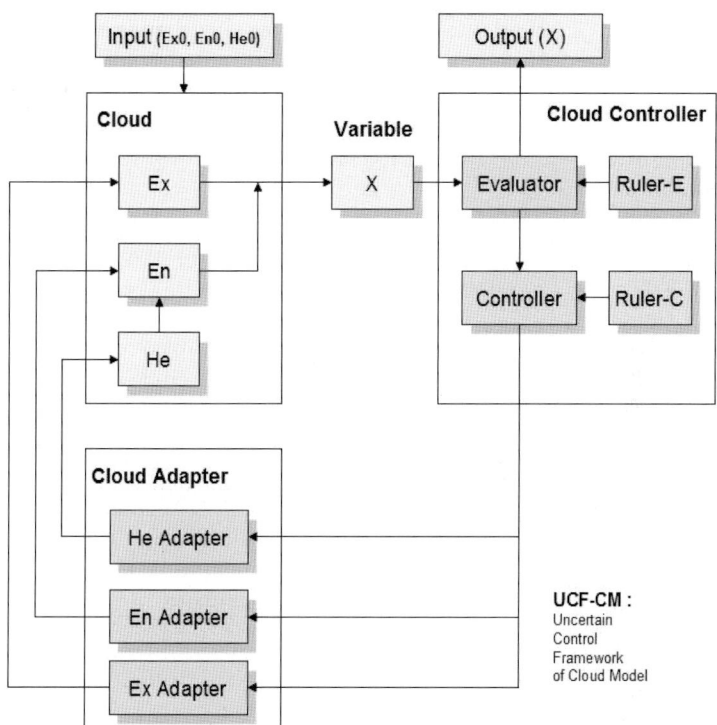

Fig. 2. UCF-CM : An Uncertain Control Framework of Cloud Model

UCF-CM is composed of six parts: input, cloud, variable, cloud controller, cloud adapter, output. Here is the list of description for the six modules.

Input: Initialized values of the three parameters of Cloud Model are required for the input. Besides the parameters, pre-defined rules for uncertain control are also required. They could be taken as part of the input.

Cloud: With the input, Cloud module will calculate a new variable with Ex, En, and He. The algorithm of Cloud Model is used for the generation. In this model, the change of He can affect the value of En, while the change of En can affect the value of variable.

Variable: Variable deposits the variable generated by cloud model and delivers it out as the input of cloud controller.

Cloud Controller: Cloud controller is composed of three components: ruler, evaluator and controller. Ruler provides kinds of pre-defined rules for evaluator and controller. Some rules are used to evaluate the validity of the newly generated variable while others are used to control the adaptive change of parameters of the cloud model. Evaluator executes the evaluation for the variable to judge the situation that whether the variable meets the requirement and provide guidance for the controller. Controller implements the adjustment operations to control the change of the parameters.

Cloud Adapter: Cloud adapter takes the actions to change the three parameters defined in Cloud Model. Three adapters are included for Ex, En, and He separately. Cloud adapter calls the cloud model to execute the next calculation.

Output: Finally, the variable generated by the cloud model, which meets the evaluated conditions and needs no further control actions, is on the output for further usage.

In the overview of UCF-CM, three parts are the core: Cloud, Cloud Controller, and Cloud Adapter. Cloud calculates the value of a variable, Cloud Controller provides decisions for the change of controlled parameters, and Cloud Adapter implements the change.

Meanwhile, the UCF-CM can also support a repeatable control to enable the output of X injects into the input. With this support, circle control can be implemented automatically by the system. Ruler in the Cloud Control decides the breakout. The extended framework can be illustrated by the following experiment.

With more flexible and accurate control rules, UCF-CM can make more strong uncertain control on the variables. Furthermore, self-learning mechanism can be integrated into the UCF-CM for a stronger and unbelievable ability for uncertain control. This topic of research can be pioneered in future.

4   Experimentation

To verify the validity and feasibility of UCF-CM, an experience with a representative application of controlling the growing process of artificial plants is designed and implemented. As much randomness and fuzziness exist in the whole course, although there are series of algorithms in generating artificial plants based on fractal technology at present, such as L-System, iterative function system, interpolates model, etc [8-11], no available solution could be claimed to enable the system to control the growing process of artificial plants.

Generation of artificial plants is always implemented by fractal technology. Fractal is a rough or fragmented geometric shape that can be split into parts, each of which is (at least approximately) a reduced-size copy of the whole[8,9]. This recursive nature is obvious in these examples-a branch from a tree or a frond from a fern is a miniature replica of the whole: not identical, but similar in nature [10,11]. To generate a simple artificial plant, at least four parameters are required: $[\alpha, \beta, sl, sr]$. Here α, β, sl, sr separately stand for left partial angle, right partial angle, left rate of branch length, right rate of branch length. With the four parameters, a normal and simple fractal tree can be generated by computer with the fractal algorithm: With UCF-CM framework for uncertain

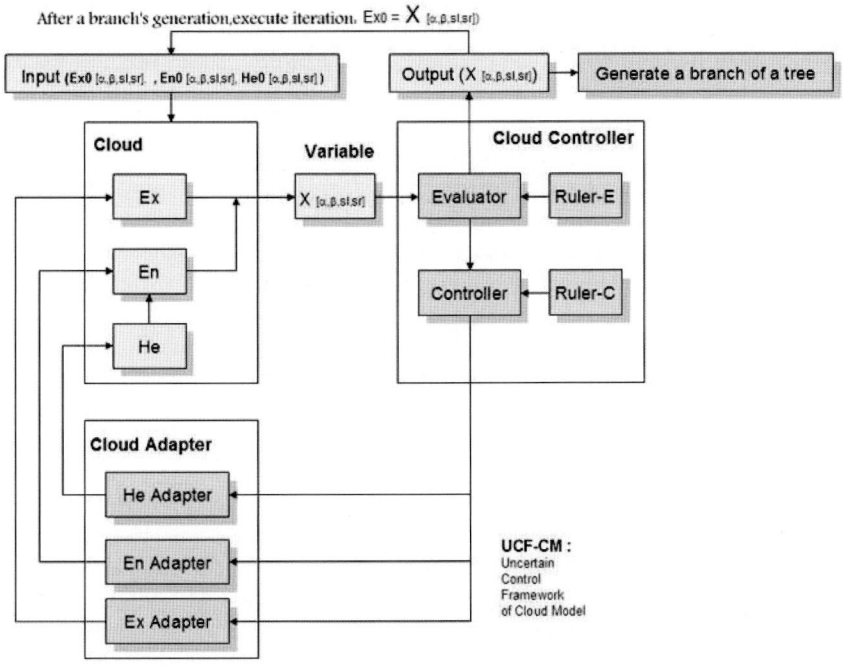

Fig. 3. UCF-CM for controlling the growing process of Artificial Trees

control and the fractal algorithm, an experiment is implemented for controlling the uncertain growing process of an artificial tree. Here is the updated UCF-CM for this experiment.

With the updated UCF-CM for artificial tree, series of detailed steps for this experiment is designed as follows:

1. Locate the starting generation point as P0(0,0), define the original trunk length L, get the coordinate P0(0,L), and draw the trunk.

2. Define the tree depth M and initialize values of $Ex = [\alpha, \beta, sl, sr], En = [En_\alpha, En_\beta, En_s l, En_s r], He = [He_\alpha, He_\beta, He_s l, He_s r]$;

3. Define pre-rules for evaluator and controller. In this experiment, here five attributes of a tree are considered and designed: symmetry(S), tightness (T), hierarchy (H), dense (D) and ornament (O).

$$S = 1 - 0.5(\frac{\alpha - \beta}{\pi/2} + |sl - sr|)$$
$$T = 1 - \frac{\alpha + \beta}{\pi}$$
$$H = 1 - \frac{|sl - 0.618| + |sr - 0.618|}{2}$$
$$D = 0.5(1 - \frac{sl + sr}{2}) + 0.5(\frac{BestDeep - M}{BestDeep}), BestDeep = 12$$
$$O = 0.1S + 0.1(1 - |T - 0.618|) + 0.35H + 0.45D$$

Once generated the parameter $[\alpha, \beta, sl, sr]$, the values of S,T,H,D,O can be calculated according to the pre-defined formulas. The expected values of the five attributes can be defined for the evaluation. Meanwhile, series of rules for evaluation of the branch can be defined. There are always compound rules which include many simple rules.

4. Calculate the values of S, T, H, D, O with the generated variables of $[\alpha, \beta, sl, sr]$.

5. Control the change of variables of $[\alpha, \beta, sl, sr]$ with the evaluator and controller. If all the rules of the evaluation are satisfied, the variables of $[\alpha, \beta, sl, sr]$ will be finalized for generating a branch of the artificial tree. Otherwise, the rules of the controller will generate decisions for the change of the variables with the control of cloud model.

6. Generate a branch of the tree with $[\alpha, \beta, sl, sr]$, or keep tailoring the parameters with the cloud model until the output.

7. Repeat the steps from 4-6 until the depth of the tree satisfies the pre-defined value, or the time is over.

Finally, an artificial tree will be generated when the whole process is completed. Through the relationship between Expected value (Ex), Entropy (En), Hyper-Entropy (He) and the fractal characteristic, the connection and difference are shown while controlling the growing process of the fractal course. An expected fractal tree will be generated under the UCF-CM control. With the UCF-CM control on the fractal growing process, various of artificial trees can be generated. Here are some samples of the result.

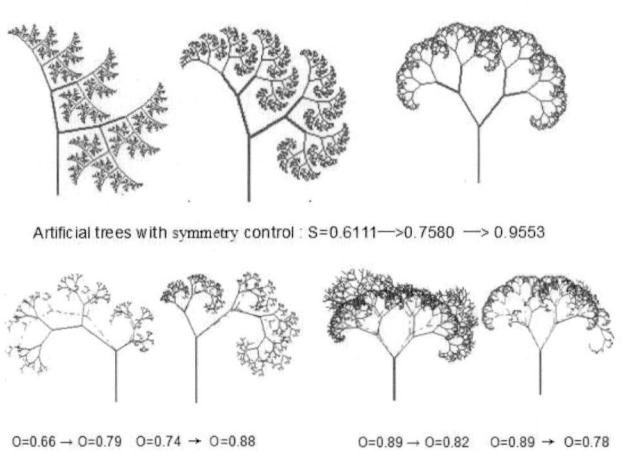

Artificial trees with symmetry control : S=0.6111—>0.7580 —> 0.9553

O=0.66 → O=0.79   O=0.74 → O=0.88   O=0.89 → O=0.82   O=0.89 → O=0.78

Fig. 4. Experiment result of UCF-CM

The trend of tree variation can be controlled by man and also can be mapped to a reference, such as the sun. Now suppose that the sun rises in the left of tree, and sinks in the right of the tree. The trend of the variation can be mapped to the trend of the sun's situation as the figure shows.

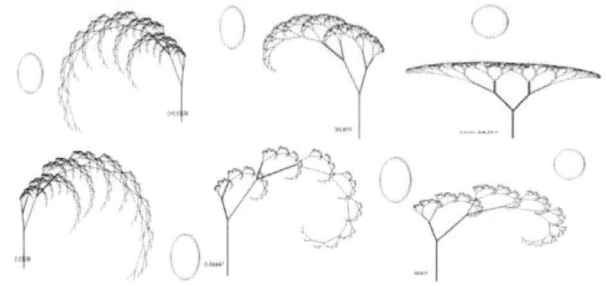

Fig. 5. Uncertain control of UCF-CM on Artificial Plants

5   Conclusion

Randomness and fuzziness make uncertain control be a difficult problem; therefore, we propose an uncertain control framework of Cloud Model called UCF-CM. UCF-CM tunes the parameters of Ex, En and He with Cloud, Cloud Controller and Cloud Adapter to generate self-adaptive control in dealing with uncertainties. Finally, an experience with a representative application of UCF-CM is implemented to verify its validity. Furthermore, integrating self-learning mechanism to the UCF-CM is on the schedule for next steps.

References

1. Hung, J.Y., Gao, W., Hung, J.C.: Variable structure control: A survey. IEEE Transactions on Industrial Electronics 40(1), 2–22 (1993)
2. Parma, G.G., Menezes, B.R., Braga, A.P.: Sliding mode algorithm for training multilayer artificial neural networks. Electronics Letters 34(1), 97–98 (1998)
3. Poznyak, A.S., Yu, W., Sanchez, E.N.: Identication and control of unknown chaotic systems via dynamic neural networks. IEEE Transactions on Circuits and Systems-I: Fundamental Theory and Applications 46(12), 1491–1495 (1999)
4. Li, D.-Y., Du, Y.: Artificial Intelligence with Uncertainty. National Defense Industry Press, Beijing (2005)
5. Li, D.-Y., Liu, C.-Y., Du, Y., Han, X.: Artificial Intelligence with Uncertainty. Journal of software 15, 1583–1594 (2004)
6. Li, D.-Y., Du, Y.: Artificial intelligent with uncertainty. Chapman, Hall/CRC, Boca Raton (2007)
7. Li, D.-Y., Liu, C.Y.: Study on the Universality of the Normal Cloud Model. Engineering Sciences 6(8), 28–34 (2004)
8. Mandelbrot, B.B.: The Fractal Geometry of Nature. W.H. Freeman and Company, New York (1982) ISBN 0-7167-1186-9
9. Falconer, Kenneth: Fractal Geometry: Mathematical Foundations and Applications. vol. xxv. John Wiley, Sons, Ltd., Chichester (2003) ISBN 0-470-84862-6
10. The Hilbert curve map is not a homeomorhpism, so it does not preserve topological dimension. The topological dimension and Hausdorff dimension of the image of the Hilbert map in R2 are both 2. Note, however, that the topological dimension of the graph of the Hilbert map (a set in R3) is 1
11. Hunting the Hidden Dimension. Nova. PBS. WPMB-Maryland (October 28, 2008)

A Comparative Study of Cloud Model and Extended Fuzzy Sets

Changyu Liu[1], Wenyan Gan[1], and Tao Wu[2]

[1] Institute of Command Automation, PLA University of Science and Technology,
Nanjing 210007, China
[2] State Key Laboratory of Software Engineering, Wuhan University,
Wuhan 430072, China
chyu_liu@163.com

Abstract. Since Zadeh introduced fuzzy sets in 1965 to generalize the indicator functions of classical sets to a membership function valued in the interval [0,1], a lot of extensions of original fuzzy sets have been proposed. For example, the interval-valued fuzzy sets, intuitionistic fuzzy sets and vague sets approximate the crisp degree of membership by an interval, while the type-2 fuzzy sets generalize it by a fuzzy set in [0,1]. Other than the above, a recent extension named *cloud model* attracts many attentions, which incorporates probability measure into fuzzy sets, and permits the simultaneous consideration of randomness and fuzziness of linguistic concepts, by extending the degree of membership to random variables defined in the interval [0,1]. Basically, with the three numerical characteristics, cloud model can randomly generate a degree of membership of an element and implement the uncertain transformation between linguistic concepts and its quantitative instantiations. This paper mainly focuses on a comparative study of cloud model and other extensions of fuzzy sets, especially the theoretical significance of cloud model. And the comparative study shows that, compared with other extensions, cloud model suggests a brand new method to handle and represent the inherent uncertainty of linguistic concepts, especially fuzziness, randomness and its relationship.

Keywords: Fuzzy sets; cloud model; random membership degree.

1 Introduction

Uncertainty is the inherent property of nature. As the reflection of the objective world, natural language is uncertain in essence, which combines different types of uncertainty, especially randomness and fuzziness. Inevitably, linguistic concepts, as the basic building blocks of a natural language, also contain uncertainty. However, such uncertainty doesn't affect the human-to-human communication, and even is accepted as a natural part of life. This means that human thought is not a process of mathematical computations, and natural languages occur as the carrier of human thought. Therefore, in order to study and mimic human thought, one of the most important things should be to quantify or formalize

J. Yu et al. (Eds.): RSKT 2010, LNAI 6401, pp. 626–631, 2010.

natural languages as some mathematical models, by which the uncertain transformation between a qualitative concept and its quantitative instantiations can be described.

However, it is very difficult for computers to deal with uncertain information. One reason is that traditional mathematics usually cannot describe the uncertain phenomena of *both-this-and-that*. Considering that human being usually are good at expressing objective and subjective uncertainty in the way of natural languages, people think that maybe it should begin with the uncertainty of linguistic variables. Therefore, Prof. Zadeh proposed the fuzzy set theory [1,2], which has made great contribution for the development of artificial intelligence and automatic control.

During the development of fuzzy set theory and its applications, a lot of advanced concepts and well-known extensions have been proposed, such as interval-valued fuzzy sets proposed by Dubois et.al constitute an extension of fuzzy sets which give an interval approximating the *real* (but unknown) membership degree. The intuitionistic fuzzy sets proposed by Atanassov in 1986 separate the degree of membership and the degree of non-membership of an element in the set, and introduce the hesitation degree to reflect and model the hesitancy of human behavior in real life situation, which is found to be more flexible compared with the original fuzzy sets The vague sets introduced by Gau et al. in 1993 is another generalization of fuzzy set, but Bustince et al. pointed out that the notion of vague set is the same as that of the intuitionistic fuzzy set. The type-2 fuzzy sets proposed by Mendel et al. are another well-known generalization of fuzzy sets which focus on the fuzziness of the membership degree. Although the above extensions generalize the original fuzzy sets in different ways, they all focus on the uncertainty of degree of membership. Let's take a look at the mathematical definitions of the above extensions firstly.

2   Well-Known Extensions of Fuzzy Sets

Definition 1. *Interval-valued fuzzy sets [3].*

Let X be the universe of discourse. An interval-valued fuzzy set in X is a mapping: $X \to Int([0,1])$. Here $Int([0,1])$ is a set of all closed subsets on $[0,1]$.

Definition 2. *Intuitionistic fuzzy sets [4].*

Let X be the universe of discourse. $A = \{(x, \mu(x), \gamma(x)) | x \in X\}$ is an Intuitionistic fuzzy set in X if $\mu(x) : X \to [0,1]$, $\gamma(x) : X \to [0,1]$. And for all $x \in X$, $0 \leq \mu(x) + \gamma(x) \leq 1$ holds.

Definition 3. *Vague sets [5].*

Let X be the universe of discourse. For any $x \in X$, the vague set V is represented by a true membership function t_V and a false function f_V. That is, $V = [t_V, 1 - f_V]$. Here $t_V : X \to [0,1]$, $f_V : X \to [0,1]$, and $t_V + f_V \leq 1$.

Definition 4. *Type-2 fuzzy sets [6].*

Let X be the universe of discourse. A type-2 fuzzy set is defined by

$$\tilde{A} = \{((x,u), \mu_{\tilde{A}}(x,u)) | \forall x \in X, \forall u \in J_x \subseteq [0,1]\} . \tag{1}$$

But the way of calculating membership degree of type-2 fuzzy set is not given.

Although the above extensions generalize the original fuzzy set in different ways, they all focus on the uncertainty of membership degree and share a number of common characteristics which is more expressive in capturing vagueness of data.

3   Super-Measurable Function and Cloud Model

Let (Ω_1, A_1, P) be a measurable space.

$$\Lambda = \{X : (\Omega_1, A_1, P) \to (R, B(R)), \forall B \in B(R), X^{-1}(B) \in A_1\}$$

Distance on Λ is defined as $d(X, Y) = E|X - Y|, X, Y \in \Lambda$. Then:

(1) $d(X, Y) \geq 0$ and $d(X, Y) = 0 \Leftrightarrow X = Y$;

(2) $d(X, Y) = d(Y, X)$;

(3) For $\forall Z \in \Lambda$, $d(X, Z) + d(Z, Y) = E|X - Z| + E|Z - Y| \geq E|X - Z + Z - Y| = d(X, Y)$

Thus d is a distance measurement and (Λ, d) is a measurement space.

Definition 5. *Super-measurable function.*

Let $f : (\Omega_2, A_2, P) \to (\Lambda, B(\Lambda, d))$. If $\forall B \in B(\Lambda, d)$, $f^{-1}(B) \in A_2$ holds, then f is a measurable function on measurable space (Ω_2, A_2, P) valued as variables. We named as a super-measurable function.

Definition 6. *Cloud model.*

Let U be a universe set described by precise numbers, and $\tilde{A}$ be a qualitative concept related to U. The certainty degree $\mu_{\tilde{A}}(x)$ of a random sample x of $\tilde{A}$ in U to the concept $\tilde{A}$ is a random number with a stable tendency

$$\mu : U \to [0, 1], \forall x \in U, x \to \mu(x) . \tag{2}$$

then the distribution of x on U is defined as a cloud, and every x is defined as a cloud drop [7,8].

Obviously, cloud model is a super-measurable function. We can come to this conclusion if we let

$$\Lambda = \{X : (\Omega_1, A_1, P) \to ([0, 1], B([0, 1])), \forall B \in B([0, 1]), X^{-1}(B) \in A_1\}$$

in definition 5.

Considering universality of Gaussian membership function and the normal distribution, the reference [8] gives a method to calculate random membership degree with three numeric characteristics, which can implement the uncertain transformation between linguistic concepts and its quantitative expression.

Cloud model describes numerical characters of a certain qualitative concept through three parameters: Expectation Ex, Entropy En and Hyper Entropy He. By three numeral characters we can design a special algorithm which can also describes the sample distributions in universe of discourse and calculate random certain degrees on the same time. The profound meaning of these three numeric characters is described as following.

Expectation Ex is the point that represents the qualitative concept properly in the universe of discourse. It is the most typical sample of the concept.

Entropy En represents the uncertainty measurement of the qualitative concept. En not only represents fuzziness of the concept but also randomness and their relations. Also En represents the granularity of the concept. Generally, the larger En is, the higher abstraction the concept is. Meantime, En reflects the uncertainty of a qualitative concept, that is, represents the range of values that could be accepted in the universe of discourse.

Hyper Entropy He is the uncertain degree of Entropy En.

For example, *about 20 years old* is a qualitative concept containing uncertain information, which is illustrated in Fig. 1. Obviously, *20* is the most representative value of the qualitative concept. It is the expectation Ex of cloud model. *about* is the extension of the concept and it also has randomness and fuzziness. Let $En = 1$, means the range of *above* on definition field, which can depict the randomness of *above*. Moreover, let hyper-entropy $He = 0.1$ can depict the uncertainty of entropy.

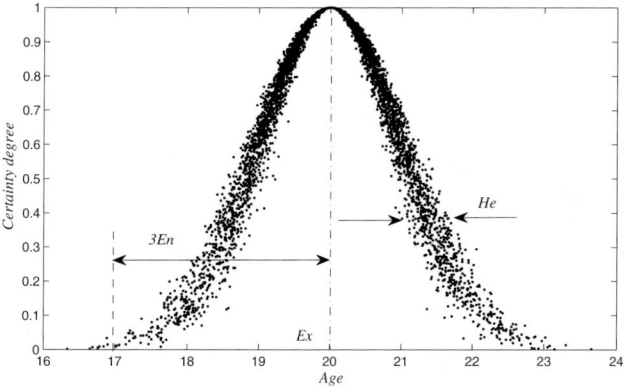

Fig. 1. Numeric characters of a cloud model

Normal cloud model is based on the normal distribution and Gauss membership function.

Given three parameters (Ex, En, He) and the number of cloud drops N, we can present the normal cloud by steps as follows.

(1) Produce a normally distributed random number En' with mean En and standard deviation He;

(2) Produce a normally distributed random number x with mean Ex and standard deviation absolute value of En';

(3) Calculate $y = e^{-\frac{(x-Ex)^2}{2(En')^2}}$;

(4) x is a cloud drop in the universe of discourse and y is certain degree of x belongs to the qualitative concept $\tilde{A}$;

(5) Repeat step (1)-(4) until N cloud drops are generated. Accordingly, $X = CG(Ex, En, He, N)$ for short.

The above algorithm is called a normal cloud model. Membership degree is a degenerate certain degree when $He = 0$.

According to the above discussion, we can see that normal cloud model doesn't emphasize the exact functional expression any more, but try to grasp the uncertainty of concepts at the macro-level. On the basis of normal distribution and Gaussian membership function, cloud models describe the vagueness of membership degree of an element by a random variable defined in the universe.

The essence of cloud model is a measurable function. Being an uncertain transition way between a qualitative concept described by linguistic terms and its numerical representation, cloud has depicted such abundant uncertainties in linguistic terms as randomness, fuzziness, and the relationship between them. Meanwhile, the two uncertainties have been integrated by cloud in a unified way to make mapping between qualitative concepts and quantitative data. It can acquire the range and distributing law of the quantitative data from the qualitative information expressed in linguistic terms.

4   Cloud Model and Extended Fuzzy Sets Theories

In the original fuzzy setsthe degree of membership of an element is defined as a single real value on the interval [0,1]. Instead of using point-based membership, the interval-valued fuzzy sets, the intuitionistic fuzzy sets and vague sets focus on the uncertainty of membership degree and generalize point-based membership to interval-based membership which is more expressive in capturing vagueness of data. Although these extensions describe the uncertainty of membership degree in different ways, they share a number of common characteristics. For example, the vague sets defined by Gau and Buehrer (1993) was proved by Bustine and Burillo (1996) to be the same as the intuitionistic fuzzy sets, because its true-membership function and false-membership function have proved to be equivalent to the degree of membership and non-membership in the intuitionistic fuzzy sets respectively. Even the interval-based membership in an interval-valued fuzzy set also is shown to be equivalent to that of an intuitionistic fuzzy set. Compared with the above extensions, the type-2 fuzzy sets focus on the fuzziness of degree of membership and generalize point-based membership to fuzzy-set-based membership. But the type-2 fuzzy sets theory cannot give the effective methods to calculate the degree of membership. Moreover, the above extensions all exclude the randomness, without regarding the statistical structure of elements in universe of discourse and the randomness of membership degree of an element.

Cloud model is quite different from the above extensions. The theoretical foundation of cloud model is probability measure, i.e. the measure function in the sense of probability. More specifically, cloud model not only focuses on the studies about the distribution of samples in the universe, but also try to generalize the point-based membership to a random variable on the interval [0,1], which can give a brand new method to study the relationship between the randomness of samples and uncertainty of membership degree. More practically speaking, cloud model give the method to calculate the random membership degree with the aid of three numeric characteristics, by which the transformation between linguistic concepts and numeric values will become possible.

In conclusion, all extended fuzzy sets theories are still based on classical set theory. But the theoretical foundation of cloud model is probability measure. Although all above extensions study of the uncertainty of membership degree from different points of view, they are complementary to describe the uncertainty of the objective world in more detail.

References

1. Zadeh, L.A.: Fuzzy sets. Information and Control 8(3), 338–353 (1965)
2. Zadeh, L.A.: The concept of a linguistic variable and its application to approximate reasoning-1. Information Science 8, 199–249 (1975)
3. Lczany, G.: A Method of Inference in Approximate Reasoning Based on Interval Valued Fuzzy Sets. Fuzzy Sets and Systems 21, 1–17 (1987)
4. Atanassov, K.: Intuitionistic Fuzzy Sets. Fuzzy Sets and Systems 20, 87–96 (1986)
5. Bustince, H., Burillo, P.: Vague Sets are Intuitionistic Fuzzy Sets. Fuzzy Sets and Systems 79, 403–405 (1996)
6. Mendel, J.M., John, R.I.: Type-2 fuzzy sets made simple. IEEE Transaction on Fuzzy Systems 10(2), 117–127 (2002)
7. Li, D.Y., Meng, H.J., Shi, X.M.: Membership cloud and membership cloud generator. Journal of Computer Research and Development 32(6), 15–20 (1995)
8. Mendel, J.M.: Computing with words and its relationships with fuzzistics. Information Sciences 177(4), 988–1006 (2007)

A Variable Step-Size LMS Algorithm Based on Cloud Model with Application to Multiuser Interference Cancellation

Wen He[1,2], Deyi Li[3], Guisheng Chen[3], and Songlin Zhang[3,4]

[1] Department of Computer Science and Technology, Tsinghua University,
Beijing 100084, China
[2] Xi'an Communication Institute, Xi'an 710106, China
[3] Institute of China Electronic System Engineering, Beijing 100039, China
[4] PLA University of Science and Technology, Nanjing 210007, China
`he-w09@mails.tsinghua.edu.cn`, `leedeyi@tsinghua.edu.cn`,
`cgs@tsinghua.edu.cn`, `lbzsl@163.com`

Abstract. This paper presents a variable step-size Least Mean Square (LMS) algorithm based on cloud model which is a new cognitive model for uncertain transformation between linguistic concepts and quantitative values. In this algorithm, we use the error differences between two adjacent iteration periods to qualitatively estimate the state of the algorithm, and translate it into a propriety step-size in number according to the linguistic description of basic principle of variable step-size LMS. Simulation results show that the proposed algorithm is able to improve the steady-state performance while keeping a better convergence rate. We also apply this new algorithm to the multiuser interference cancellation, and results are also satisfied.

Keywords: LMS algorithm, cloud model, variable step-size, multiuser interference cancellation.

1   Introduction

The Least Mean Square (LMS) algorithm[1] is one of the adaptive filtering algorithms, which is proposed by Widrow in 1976. Due to its low computational complexity and real-time implementation, LMS has got a widely use in many areas such as signal processing, control and communications.

The conventional LMS algorithm is able to track a system dynamically, but its performance is largely depended on the choice of the step-size parameter. This means a larger step-size may speed the convergence, but results in a larger misadjustment in the steady-state, while a smaller step-size tends to improve steady-state performance, but will need a long time to realize the convergence.

In order to improve such problems, lots of variable step-size LMS algorithms have been proposed. In 1986, Harris et al. proposed a variable step-size strategy by examining the polarity of the successive samples of estimation errors[2]. In 1996, Kwong and Johnston developed a scheme to adjust the step-size based on

J. Yu et al. (Eds.): RSKT 2010, LNAI 6401, pp. 632–639, 2010.

the fluctuation of the prediction squared error[3]. The algorithm proposed by Kwong et al. has also been demonstrated by Lopes et al.[4], that this was probably the best low-complexity variable step-size LMS algorithm in the literature. During recent years, there are still many new variable step-size algorithms[5,6,7] proposed by researchers.

In general, these variable step-size algorithms all follow such principles: use a large step-size at the initial stage to speed the convergence, and progressively reduce the value while the algorithm approaches steady-state. Thus the algorithms may provide both fast convergence and good performances in steady-state.

Though these variable step-size algorithms have greatly improved the performance of basic LMS algorithm, there is a new problem that we need to choose additional parameters to reflect the stage of the algorithm. This means we need to translate the uncertainty linguistic terms on the step-size adjustment into mathematical algorithms, and decide the uncertainty stage of algorithm according to some precise parameters. In this paper, we propose a new variable step-size algorithm based on cloud model which is a new cognitive model to implement the uncertain transformation between linguistic concepts and quantitative values. Thus we can estimate the algorithm stage based on a qualitative analysis and directly change it into a propriety step size.

2   LMS Algorithm

The basic block diagram of the LMS algorithm is shown in Fig. 1 Where the $\mathbf{x}(n)$, $\mathbf{d}(n)$, $\mathbf{y}(n)$, $\mathbf{e}(n)$ and $\mathbf{W}(n)$ are present for input signal, desired signal, output of the adaptive filter, error signal, and the adaptive weight vector separately.

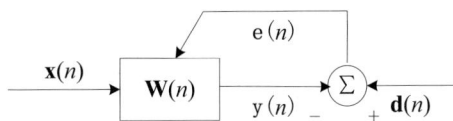

Fig. 1. The block diagram for the LMS System

The steps of the basic LMS algorithm can be given as:

1. Initialize: $\mathbf{w}(0)=0$, $n=1$;
2. Iterate: $\mathbf{e}(n) = \mathbf{d}(n) - \mathbf{w}^T(n-1)\mathbf{x}(n)$
   $\mathbf{w}(n) = \mathbf{w}(n-1) + \mu\mathbf{x}(n)\mathbf{e}(n)$
3. Update: $n = n+1$.

Where μ is the step-size, and it's a constant in the basic LMS algorithm. To make sure the convergence of the algorithm, step-size μ should be defined between the range $(0, 1/\lambda_{max})$, where the λ_{max} is the maximum eigenvalue in the autocorrelation matrix of the input signal.

3   The Variable Step-Size LMS Algorithm Based on the Cloud Model

From the basic LMS algorithm, we may find that the performance of the LMS algorithm is largely depended on the choice of step-sizeμ. If we choose a largerμ, algorithm may convergent with a faster speed, but the stability in the steady stage will be worse. In order to get a better stability, we need to use a smaller step-size, but this will slow down the speed of convergence. To improve such problem, we propose a new algorithm based on cloud model. With the method of cloud model, this algorithm will dynamically estimate the current stage of the algorithm, and then translate it into a propriety step-size in number.

3.1   Cloud Model

The Cloud Model is a new cognitive model proposed by Prof. Li [8], which offers a transformation method between linguistic concepts and quantitative values. This model uses three numerical characteristics (Ex, En, He) to describe the overall quantitative property of a concept. Expectation value Ex is the mathematical expectation of the cloud drops belonging to a concept in the universal. Entropy En represents the uncertainty measurement of a qualitative concept. It is determined by both the randomness and the fuzziness of the concept. Hyperentropy He is the uncertain degree of entropy En.

Cloud model can also be used to realize the qualitative knowledge inference base on cloud rule generator.

If there are sets of rules like:

If $A1$ then $B1$;

If $A2$ then $B2$;

. . . .

If An then Bn.

In which An and Bn are represent a linguistic concepts, such as "high altitude", "low temperature" and so on. Then we can use a multi-rule generator to output a quantitative value in the data field of Bn according to the input of a specific value in the data field of An. Thus we can realize the uncertainty inference between number and concepts. More details of the rule generator based on cloud model can be seen in [9].

3.2   Improved LMS Algorithm

In this paper, we proposed an algorithm where the cloud model is used to adjust the step-size for LMS algorithm dynamically. According to the basic principle for step size adjusting, we first build the rule sets in linguistic description:

If the convergence speed is fast, then the step-size is large;

If the convergence speed is medium, then the step-size is medium;

If the convergence speed is low, then the step-size is small;

We can see that, this is exactly the common idea in adjusting the step-size in LMS algorithm. In order to estimate the convergence speed of the algorithm, we introduce another parameter E_{diff} which be defined as the error difference between two adjacent iteration periods.

$$E_{diff} = (\sum_{i=t}^{t+T} e_i^2 - \sum_{j=t-T}^{t} e_j^2)/T \tag{1}$$

Where e is the error signals and T is the accumulative period. To actually reflect the convergence state, a too small or too large value of T should be avoided. Since a too small value of T will introduce lots of noise and cannot properly reflect the convergence trends, while a too large value will miss the proper time in changing the step-size. In actual, this value should be set experimentally.

Then the general steps for the variable step-size LMS algorithm based on cloud model can be written as:

(1) Initialize $\mathbf{w}(0)=0$, $n=1$;
   (2) Iterate $e(n) = d(n) - \mathbf{w}^T(n-1)\mathbf{x}(n)$
   $\mathbf{w}(n) = \mathbf{w}(n-1) + \mu(n)\mathbf{x}(n)e(n)$
   $\mu(n) = \begin{cases} Cloud\,(E_{diff}(n,T)) \; if(n(\bmod T)) = 0) \\ \mu(n-1) \qquad\qquad otherwise \end{cases}$
   (3) Update $n = n+1$

Where, $\mathbf{x}(n)$, $\mathbf{e}(n)$, $\mathbf{d}(n)$ and $\mathbf{w}(n)$ are the input signal, error signal, desired signal and weight vector. $\mu(n)$ is the time-varying step-size, Cloud(.) represents the cloud generator. And we would like to call this algorithm as Cloud-LMS for short.

The details of Cloud-LMS are described as follows. 1) Calculate the error difference E_{diff}. To avoid the step-size fluctuated too frequently, we only change the step-size at the end of a period of T. Thus we also only need to calculate the error difference between two adjacent periods at each end of the computing period. 2) Calculate certainty degrees μ_i to each qualitative concept. With the input of E_{diff}, we can get the certainty degree of the three qualitative concepts in the preconditions of each of the three cloud rules. 3) Rule activated. If there is only one positive certainty degree of μ_i get in step 2), then the ith rule will be activated with activation strength of μ_i. Else, if there is more than one certainty degree such as μ_i and μ_{i+1} are get in the last step. Then the ith and $i+1$th rules will be activated respectively with each certainty degree. 4) Step-size output. If there is only one rule activated in step 3), the output step-size will be generated directly by the single-rule generator. If there is more than one rule activated, the output will be constructed by means of a virtual cloud[9], which is used for combining more than one cloud, and the output step-size will be the Expected Value Ex of the virtual cloud.

4   Simulation Result

In this section, simulations are performed to verify the method developed in the previous sections. Experimental performance are compared under the background of DS-CDMA system and used for multiuser interference cancellation.

Multiuser interference cancellation detection[10] is a method to joint detects all the users in the same channel simultaneously. Compare to the traditional Match Filter (MF) detector, which simply treats all other users as background noise, multiuser detector can greatly decrease the Multiple Access Inference (MAI) in CDMA system.

The CDMA system under consideration is a chip-synchronous system using Direct-Sequence Spread-Spectrum (DS-SS) employing BPSK. The system model is given in Fig. 2.

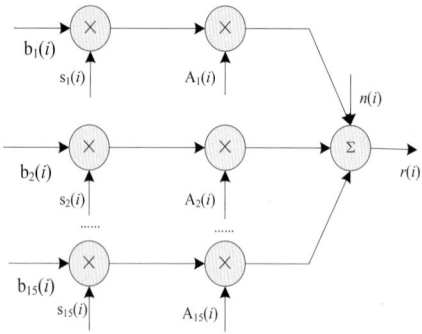

Fig. 2. System model of DS-CDMA

We assume that there are 15 users in the channel, each user k is spread by a Gold sequence s_k and sent at amplitude A_k. Then the received signal $r(i)$ can be written as:

$$r(i) = \sum_{k=1}^{15} A_k(i) b_k(i) s_k(i) + n(i) \tag{2}$$

Where $b_k(i)$ is the data bit of user k, and $n(i)$ is the AWGN vector.

We can also represent this in a more compact matrix form:

$$\mathbf{r} = \mathbf{SAb} + \mathbf{n} \tag{3}$$

where $\mathbf{S}=[\mathbf{s}_1,\ldots,\mathbf{s}_k]$ is the signature matrix whose k_{th} column is the $N\times1$ signature sequence vector for user k, $\mathbf{A}=\mathrm{diag}[A_1,\ldots A_k]$ is the diagonal matrix with $(\mathbf{A})_{k,k} = A_k$ and $\mathbf{b}=[b_1,\ldots b_k]^T$ is the column vector of transmitted bits of each user. $\mathbf{n}$ is the AWGN vector.

We first defined a three-rule cloud generator according to linguistic description and experimental result. The parameters are given in Table 1, and are written in the format of (Ex, En, He). The three precondition generators are denote the concepts of fast, medium and low in convergence, and the three postcondition generators are denote the concepts of large, medium and small in step-size. Fig. 3 and Fig. 4 give the generated cloud graph separately.

In experiment one, we give the comparison between Cloud-LMS and basic LMS algorithm. The step-size of the basic LMS is set at 3×10^{-2} and 5×10^{-3}

Table 1. The parameters of cloud generator

	1st rule generator	2nd rule generator	3rd rule generator
Precondition	$(0.02,0.003,10^{-4})$	$(0.01,0.003,10^{-4)}$	$(0.006,0.001,10^{-4})$
Postcondition	$(0.03,0.006,10^{-4})$	$(0.01,0.004,10^{-4})$	$(0.005,0.0015,10^{-4})$

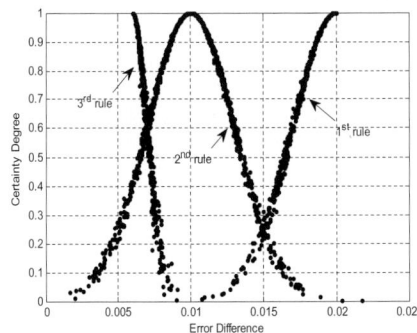
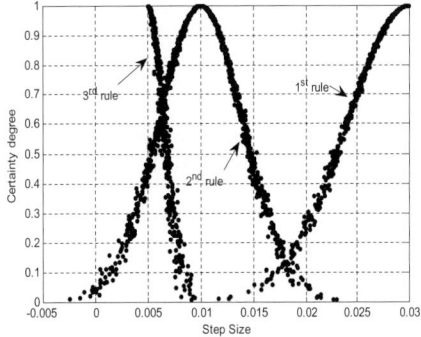

Fig. 3. The cloud graph of the three pre-condition cloud generators

Fig. 4. The cloud graph of the three post-condition cloud generators

separately. The compute period for error difference of Cloud-LMS is 20, and the SNR in CDMA system is 10dB. Fig. 5 shows the result, which is obtained by averaging a number of 100 Monte-Carlo simulations. We can see that Cloud-LMS realize a good combination with fast convergence and low disadjustment in steady stage.

Experiment two gives the performance comparison between Cloud-LMS and two variable step-size algorithms proposed by Kwong et al. and Zhang zhonghua et al. separately in 1992 and 2009, and we call them as Kwong-LMS and ZZ-LMS for short. In experiment, the maximum and minimum step-size for Kwong-LMS are set to be $3\times10^{-2}, 5\times10^{-3}$. The parameters of ZZ-LMS are set to be $\alpha=80$, $\beta=0.03$, $h=1000$. From the result in Fig.6 we can see that, Kwong-LMS need about 350 iterations to let MSE down to -12dB and ZZ-LMS need about 200 iterations, while Cloud-LMS need only 150 iterations. The final MSE of Kwong-LMS is about-12.79dB and ZZ-LMS is about -12.59dB, while Cloud-LMS is about -12.81 dB. So Cloud-LMS obtains the good performance both on the convergence speed and the steady-state error.

Experiment three gives the performance comparison between multiuser detector based on Cloud-LMS and the MF detector. On Fig. 7, we also give the curve of the optimum detector on theory. We can see that with the increase of the SNR in CDMA system, detector based on Cloud-LMS can get a much lower bit error rate (BER) than the traditional MF detector.

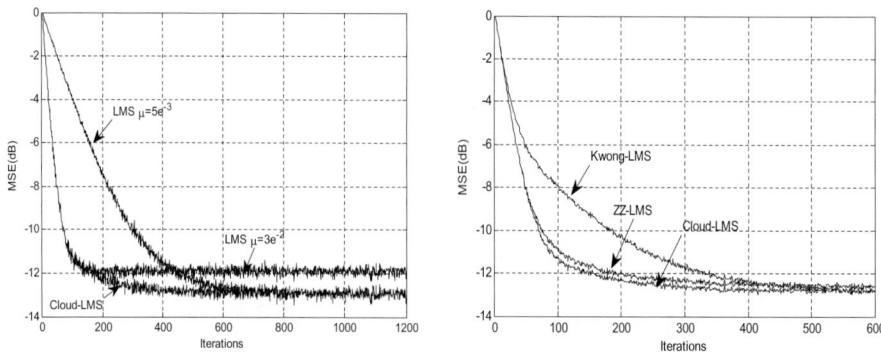

Fig. 5. Performance comparison between Cloud-LMS and basic LMS

Fig. 6. Performance comparison between Cloud-LMS and Kwong-LMS, ZZ-LMS

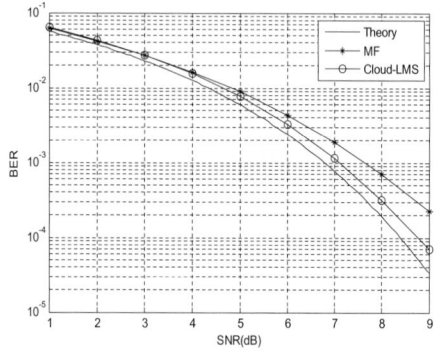

Fig. 7. Performance comparison between Cloud-LMS and MF detectors

5   Conclusion

In this paper, we introduce the method of cloud model into LMS algorithm and propose a new variable step-size LMS algorithm. The step-size of the algorithm is adjusted according to the stage estimation based on the error difference between two adjacent periods. And with the method of cloud model, we realize the direct translation from the stage estimation into a propriety step-size in number. As a result, the algorithm can adjust the step-size in a proper time only according to the basic principle of variable step-size LMS in linguistic term. The performance of the algorithm is compared with the basic LMS as well as other variable step-size LMS algorithms[3,7] through simulations. We also apply the new algorithm to the multiuser interference cancellation in CDMA system, simulation result is also satisfied.

Acknowledgement

This research is supported by the National Natural Science Foundation of China under Grand No. 60974086 and the National Grand Fundamental Research 973 Program of China under Grant No. 2007CB310804.

References

1. Widrow, B., McCool, J.M., Larimore, M., Johnson, C.: Stationary and nonstationary learning characteristics of the LMS adaptive filter. Proceedings of the IEEE 64(8), 1151–1162 (1976)
2. Harris, R., Chabries, D., Bishop, F.: A variable step (VS) adaptive filter algorithm. IEEE Transactions on Acoustics, Speech and Signal Processing 34(2), 309–316 (1986)
3. Kwong, R.H., Johnston, E.W.: A variable step size LMS algorithm. IEEE Transactions on Signal Processing 40(7), 1633–1642 (1992)
4. Guimaraes Lopes, C., Moreira Bermudez, J.C.: Evaluation and design of variable step size adaptive algorithms. In: Proceedings of IEEE International Conference on Acoustics, Speech, and Signal Processing, ICASSP 2001, vol. 6, pp. 3845–3848 (2001)
5. Costa, M.H., Bermudez, J.C.M.: A noise resilient variable step-size LMS algorithm. Signal Process 88(3), 733–748 (2008)
6. Zhao, S., Man, Z., Khoo, S., Wu, H.: Variable step-size LMS algorithm with a quotient form. Signal Process. 89(1), 67–76 (2009)
7. Zhang, Z., Zhang, R.: New variable step size LMS adaptive filtering algorithm and its performance analysis. System Engineering and Electronics 31(9), 2238–2241 (2009)
8. Li, D., Liu, C., Gan, W.: A new cognitive model: Cloud model. International Journal of Intelligent Systems 24(3), 357–375 (2009)
9. Li, D., Di, K., Li, D., Shi, X.: Mining association rules with linguistic cloud models. Journal of Software 11(2), 143–158 (2000)
10. Verdú, S.: Minimum probability of error for asynchronous gaussian multiple-access channels. IEEE Transactions on Information Theory 32(1), 85–96 (1986)

A Qualitative Requirement and Quantitative Data Transform Model

Yuchao Liu[1,2], Junsong Yin[2], Guisheng Chen[2], and Songlin Zhang[2,3]

[1] Department of Computer Science and Technology, Tsinghua University,
Beijing 100084, China
[2] Institute of China Electronic System Engineering, Beijing 100141, China
[3] PLA University of Science and Technology, Nanjing 210007, China
yuchao_liu@163.com, jsyin@nudt.edu.cn, cgs@tsinghua.edu.cn, lbzsl@163.com

Abstract. Development of Internet technology and human computing
has changed the traditional software work pattern which faces the single
computer greatly. Service oriented computing becomes the new tendency
and everything is over services. The appearance of mobile Internet provides a more convenient way for people to join the Internet activity. Thus,
the meeting between the requirements from a number of users and services is the main problem. Cloud model, a qualitative and quantitative
model can be taken to bridge the qualitative requirement on the problem
domain and quantitative data services on the solution domain.

Keywords: cloud computing, cloud model, qualitative requirement,
quantitative data.

1 Introduction

Since 1984, TCP/IP protocol has been used widely on Internet gradually-
"Everything is over IP[1]". The invention of WWW[2] in 1989 push Internet
to the people, Web technology makes network information convenient sharing
and interaction-"Everything is over Web". Seen from the gradual progress of
network evolvement, the communication (especially the photo-communication
and mobile-communication) and network science [3,4] (especially Internet) develop faster than the calculability model and software theory. George Gilder's
Law predicts the total bandwidth of communications systems would triple every
year for the next twenty-five years. The growth speed of bandwidth is even 3
times more than that of computer chip performance[5]. Since 2000, the development of Web Service, semantic network, Web2.0 and cloud computing has
accelerated the formation of network computation environment-"everything is
over Services".

When services description faces anywhere-access and massive demands from
its huge amount of users, described by natural languages with different scales, the
capabilities of interoperation such as semantic and granular computing with uncertainty are essential. That is to say, the meeting between the requirements from
a number of users and services is the essential problem. We propose a method to
bridge qualitative requirement and quantitative services data by Cloud model.

J. Yu et al. (Eds.): RSKT 2010, LNAI 6401, pp. 640–645, 2010.

2   Cloud Model

Cloud model is a method which transforms between quantitative data and qualitative concepts by means of second order normal distribution[6].

2.1   Background of Cloud Model

Approximately speaking, normal distributions with two parameters: expectation(Ex) and entropy (En) are universally existent in natural and social life, such as preferential common requirements, collective interaction behavior and services aggregation process in Internet, etc. If the above phenomena are determined by the sum of a quantity of equal and independent statistical factors, and if every factor has minor influence, then the phenomena obtain normal distribution approximately. However, there exist many occasions in which there are some dependent factors with great influence. The precondition of a normal distribution is no longer existed. To solve this problem, a new parameter of hyper entropy (He) is adopted in order to relieve the precondition. As shown in Fig. 1, different people have different understanding about "Young", and a precise normal distribution cannot reflect the essential relevantly. Normal cloud model can describe this uncertainty, meanwhile, demonstrates the basic certainty of uncertainty.

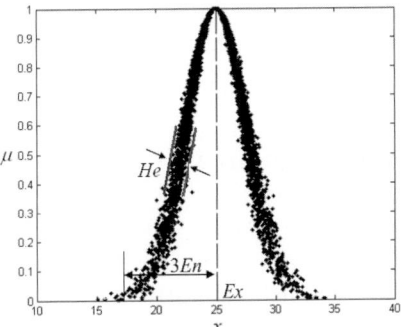

Fig. 1. Describe "Young" by normal cloud model

In cloud model, the overall property of cloud drops is represented by three numerical characters: the expected value Ex, the entropy En, and the hyperentropy He, corresponding to a qualitative concept. Ex is the mathematical expectation of the cloud drops of a concept. In other words, the cloud drop located at the Ex point is the most representative of the qualitative concept. En is the granular measurement of the qualitative concept, which connects both the randomness and the fuzziness of the concept. He is the uncertainty measurement of the entropy, i.e. the entropy of the entropy, showing to what degree a number of cloud drops become a common concept. That is to say, if He is greater than

En, the concept cannot be formed any longer. The uncertainty of the concept can be represented by multiple numerical characters. In general, people get used to perform reasoning with the help of languages, rather than excessive mathematic calculation. So, it is adequate to employ these three numerical characters to reflect common concepts.

2.2   Cloud Model Generator

Forward Cloud Generator (FCG) [7] algorithm transforms a qualitative concept with *Ex*, En and He into a number of cloud drops representing the quantitative description of the concept.

The key point in FCG is the combined relationship, i.e. within the two random numbers, one is the input of the other in generation. If $He = 0$, it will become a normal distribution. If $He = 0$, $En = 0$, the generated x will be a constant *Ex* and $\mu=1$. By this means, we can say that certainty is the special case of the uncertainty.

Reverse Cloud Generator (RCG) algorithm transforms a number of cloud drops (xi) into three numerical characters (*Ex'*, *En'*, *He'*) representing the concept.

According to statistical principles, the more samples there are, the less the error in the backward cloud algorithm is. As a result, constrained by the finity of samples, the error is inevitable no matter which algorithm is used. Because the backward cloud algorithms are based on the statistical result of a large quantity of cloud drops, the performance is determined by the amount of drops. In general, the errors of the 3 parameters will all decrease with the increasing of cloud drop numbers.

3   Transform between Quantitative Data and Qualitative Concepts

Given a group of quantitative data with same property, however, one key challenge is how to form the relevant qualitative concepts, which can reflect the essence and internal relation of the data. According to feature-list theory, concepts are the common properties shared by a class of objects, which can be characterized by the typical and significant features. On the other hand, prototype theory holds that concepts can be modeled as a most exemplary case, as well as the degree of its members deviating from the prototype, i.e. membership degree. Cloud model, a transforming model between qualitative concepts and the quantitative expressions, not only can directly abstract concepts from original data and formally describe the uncertainty of concepts, but also can represent the imprecise background knowledge impermissible for the classical rough set theory.

Generally speaking, a cloud generator is used to form a special concept for a group of data only. Different groups of data in the whole distribution domain may lead to many different concepts by using Gaussian Mixture Model (GMM) and Reverse Cloud Generator (RCG). Any quantitative data will belong to a certain concept. Furthermore, fewer qualitative concepts with high level granular can be further extracted by Cloud Synthesis Method (CSM). These concepts may be the foundation of the ontology construction as semantic interoperation.

3.1   Concept Forming By GMM and RCG

If the quantitative data approximates to normal distribution, Cloud Model can be used to describe the commonsense knowledge. Otherwise, the commonsense knowledge cannot be formed. Let $f(x)$ is the frequency distribution function of the use of data services. By GMM and RCG, $f(x)$ can be divided into n discrete concepts with error threshold . The result might not be unique, however, it is in common sense that the contribution to the qualitative concept by the data value with higher occurrence frequency is more than by that with lower occurrence frequency. The qualitative concepts forming process in small granular is formulated as:

$$f(x) \rightarrow \sum_{i=1}^{n} a_i * C(Ex_i, En_i, He_i)$$

Where ai is the magnitude coefficient and n is the number of discrete concepts after transformation.

From the view of data mining, the process is to extract concepts from data distribution of a certain attribute. It is the transformation from quantitative expression to qualitative description, and is the process of concept induction and learning. It is in common sense that the contribution to the qualitative concept by the data value with higher occurrence frequency is more than by that with lower occurrence frequency.

3.2   Concept Rising By Cloud Synthesis Method

As is well known, to switch among different granularities is a powerful ability of human intelligence in problem solving. That is to say, people not only can solve problems based on information at single granularity, but also can deal with problems involving information at extremely different granularities. Just like the zooming in telescope and microscope, the ability helps to clarify the problems at macro, meso and micro levels respectively. The change from a finer granularity up to a coarser means data reduction and abstraction, which can greatly simplify the amount of data in problem solving. A coarser granularity scheme will help to grasp the common properties of problems and obtain a better global understanding.

By GMM and RCG, the concepts with little granular can be obtained. But in human intelligence, commonsense knowledge, which usually is the knowledge of knowledge, called meta-knowledge can be extracted from the lower level concepts.

The earlier concept rising method [6] is to establish the virtual pan concept tree. Concept set can be defined as composed by the fundamental concepts in the universal set, in another word, the concept set C can be expressed by $C\{C_1(Ex_1, En_1, He_1), C_2(Ex_2, En_2, He_2), \ldots, C_m(Ex_m, En_m, He_m)\}$, where $C_1, C_2, \ldots,$

C_m are the fundamental concepts represented by cloud models. Different concept set can be constructed at different abstract levels, and level by level, the concept tree can be generated gradually. In this method the original distribution data are lost, and the root concept can be obtained by all means. However, these are conflict to the fact that the original data maybe can not be extracted a common concept.

We propose a new method to rise concept based on raw data. When two clouds with common parts or the shortest distance on Ex, Use RCG to their original data to merge them into a new cloud if and only if He of the new cloud is small than En. The process ceases until no two clouds can be merged, then several concepts in different granular will be formed. The process of concept rising can be thought as that of data classifying.

3.3   Matching Qualitative Requirement to Quantitative Service Data

Under Internet environment, the computing orient to a number of users with changing requirements from time to time, at the same time, and take the response of a requirement as an up-to-date best effort rather than a unique precise one. Because the uncertainty of requirements and concepts described by language, the matching between users' qualitative requirement with quantitative service data must be loose coupling, just like hand-shake in actual life. The hand-shake between requirements and services could be closely or loosely.

Like the basic rule of hand-shake is offering right hand, the Interoperation of requirements and services also need a basic standard. The State Key Laboratory of Software Engineering of Wuhan University puts forward Metamodel Framework for Interoperability from four angles (role, goal, process and services) [8] to solve interoperability key technology and serial international standards.(Fig.2)

After meeting requirement to a concept extracted from raw data (such as faster search services), Forward Cloud Generator can be used to provide a quantitative service (Google or Baidu) to user. If the requirement is certain, the matching can be set up directly between users and services.

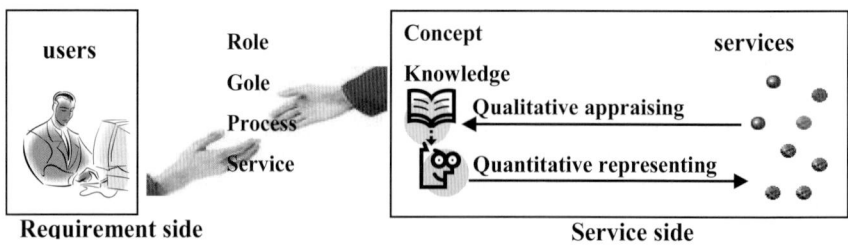

Fig. 2. The framework of matching qualitative requirement to quantitative service

4   Conclusion and Future Work

User-oriented individualized, diversified services are increased rapidly, and Internet becomes the reflection of human society. People's qualitative requirements on problem domain and quantitative service on solution domain is two main aspects in Internet, this paper proposes Cloud model, GMM and CSM can be used to complete the transform between them. Future, a series of experiments will be done to prove it.

Acknowledgement. This research is supported by the National Natural Science Foundation of China (Grand No. 60974086)and the National Grand Fundamental Research 973 Program of China (Grant No. 2007CB310804).

References

1. Haas, R., Marty, R.: Everything over IP, IP over everything. Technical Report Internet Economics Seminar 35-649, ETH Zurich (2000),
   ftp://ftp.tik.ee.ethz.ch/pub/lehre/inteco/WS0001/V1-IP-Everything.PDF
2. Berners-Lee, T., Cailliau, R.: World Wide Web:proposal for a hypertext project (1990), http://www.w3.org/Proposal.html
3. Watts, D., Strogatz, S.: Collective dynamics of 'small-world' networks. Nature 393(6), 440–442 (1998)
4. Barabási, A.L., Albert, R.: Emergence of scaling in random networks. Science 286(5439), 509–512 (1999)
5. Gilder, G.: Wealth and Poverty. ICS Press (1993)
6. Li, D., Du, Y.: Artificial Intelligence with Uncertainty. CRC Press, Taylor and Francis Group (2007)
7. Li, D., Shi, X., Meng, H.: Memebership clouds and membership cloud generators. Journal of Computer Research and Development 32(6), 15–20 (1995)
8. Wang, J., He, K., Gong, P.: RGPS: a unified requirements meta-modeling frame for networked software. In: Proceedings of the 3rd International Workshop on Applications and Advances of Problem Frames in International Conference on Software Engineering, pp. 29–35 (2008)

The High-Activity Parallel Implementation of Data Preprocessing Based on MapReduce

Qing He[1], Qing Tan[1,2], Xudong Ma[1,2], and Zhongzhi Shi[1]

[1] The Key Laboratory of Intelligent Information Processing,
Institute of Computing Technology, Chinese Academy of Sciences,
Beijing, 100190, China
[2] Graduate University of Chinese Academy of Sciences, Beijing, 100190, China
{heq,tanq,maxd,shizz}@ics.ict.ac.cn

Abstract. Data preprocessing is an important and basic technique for data mining and machine learning. Due to the dramatic increasing of information, traditional data preprocessing techniques are time-consuming and not fit for processing mass data. In order to tackle this problem, we present parallel data preprocessing techniques based on MapReduce which is a programming model to implement parallelization easily. This paper gives the implementation details of the techniques including data integration, data cleaning, data normalization and so on. The proposed parallel techniques can deal with large-scale data (up to terabytes) efficiently. Our experimental results show considerable speedup performances with an increasing number of processors.

Keywords: Data preprocessing, Speedup performance, Parallel computing, Hadoop, MapReduce.

1 Introduction

With the rapid development of information technology, huge amounts of data spring up in our everyday life. Due to the wide availability of these data and the need for turning such data into useful information and knowledge, data mining has attracted a great deal of attention in recent years. Data mining [1], or called knowledge discovery from data (KDD), refers to extracting knowledge from large amounts of data. Raw data in real world are always incomplete, noisy and unsatisfied for data mining algorithms. In order to provide high-quality and satisfactory data for mining, data preprocessing techniques are used before extracting and evaluating the data patterns.

There are a number of data preprocessing techniques serving for data mining algorithms. Data summarization is a foundation operator for data preprocessing. Data transformation transforms or consolidates data into forms appropriate for mining. For example, data normalization improves the accuracy and efficiency of mining algorithms involving distance measurements. Data integration merges data from multiple sources into a data warehouse. For all the data preprocessing techniques mentioned above, there are many ways to implement. Survey and

J. Yu et al. (Eds.): RSKT 2010, LNAI 6401, pp. 646–654, 2010.

details could be found in many literatures and books [1-3]. However, the existing preprocessing techniques are developed serially which are time-consuming for large-scale data processing. This paper does not talk about new ways to realize the techniques but focus on the parallel implementation of the processing operations.

Designing and implementing some parallel data preprocessing techniques which could run on a cluster of computers is the main purpose of this paper. Google's MapReduce [4] is a programming model and an associated implementation for processing large-scale datasets. It conceals the parallel details and offers an easy model for programmers by means of a map function and a reduce function.

Based on MapReduce programming mode, we implement a series of high-activity data preprocessing operations. The operations implemented by this way have very high degree of parallelism. Our experimental results show high speedup and sizeup performances. This characteristic makes it a reasonable time to deal with large-scale data using a cluster of computers.

2 Preliminary Knowledge

Hadoop [5] is a framework for writing and running applications on large clusters built of commodity hardware. It mainly includes two parts: Hadoop Distributed File System (HDFS) [6] and MapReduce programming mode. HDFS stores data on the compute nodes and provides very high aggregate bandwidth across the cluster. By MapReduce, the application is divided into many small fragments of work, each of which may be executed on any node in the cluster. MapReduce [7] is a simplified programming model and computation platform for processing distributed large-scale data. Fig.1. shows the flow of a MapReduce operation. Under this model, programs are automatically distributed to a cluster of machines.

As its name shows, map and reduce are two basic operations in the model. Users specify a map function that processes a key-value pair to generate a set of intermediate key-value pairs, and a reduce function that merges all intermediate values associated with the same intermediate key.

More specifically, in the first stage, the map function is called once for each input record. For each call, it may produce any number of intermediate key-value pairs. A map function is used to take a single key-value pair and outputs a list of new key-value pairs. It could be formalized as:

$$map :: (key_1, value_1) \Rightarrow list(key_2, value_2)$$

In the second stage, these intermediate pairs are sorted and grouped by key, and the reduce function is called once for each key. Finally, the reduce function is given all associated values for the key and outputs a new list of values. Mathematically, this could be represented as:

$$reduce :: (key_2, list(value_2)) \Rightarrow (key_2, value_3)$$

The MapReduce model provides sufficient high-level parallelization. Since the map function only takes a single record, all map operations are independent of each other and fully parallelizable. Reduce function can be executed in parallel on each set of intermediate pairs with the same key.

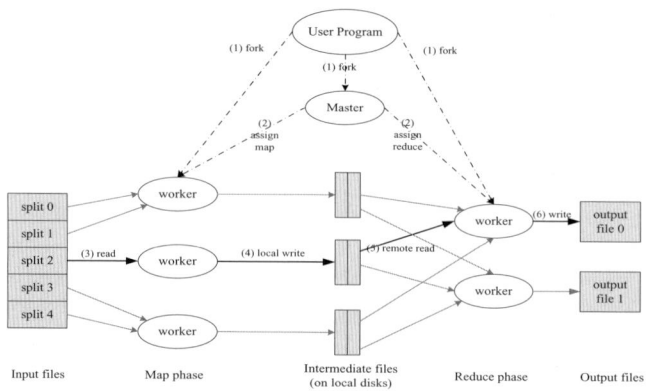

Fig. 1. Execution overview of MapReduce

3   MapReduce Based Parallel Data Preprocessing

In order to bridge the gap between the raw data and the required input data for mining algorithms, we implement a series of parallel data preprocessing operations based on MapReduce. According to the actual user customization, these operations mainly include attribute deleting, attribute exchanging, filling missing values, data summarization, string substitution, adding labels, data integration, data sampling, data interval values, data discretization and data normalization. In this section, very detailed implementation of data integration will be given first. Then, the way to realize other operations will be presented.

3.1   Implementation of Parallel Data Integration

Data integration is a frequently used data preprocessing operation in practical applications. It merges data from different data sets into a single data set. Data integration considered in this paper is table-join, which merges two tables into a single table according to some key attributes. The operations of table-join include natural-join, left-join, right-join and full-join.

Assume two tables denoted as A and B, both of which have the same key attributes to uniquely determine a record. Natural-join refers to the operator that extracts records not only in A but also in B, and the attributes of the new records are composed of the different attributes between A and B adding the key attributes. The result of a left join for table A and B always contains all records of the "left" table (A), even if the join-condition does not find any matching record in the "right" table (B). This means that a left join returns all the values from the left table, plus matched values from the right table (or NULL in case of no matching join predicate). A right join closely resembles a left join, except with the treatment of the tables reversed. A full join combines the results of both left and right outer joins. The joined table will contain all records from both tables, and fill in NULL for missing matches on either side.

In order to implement parallelism based on MapReduce programming model, we naturally code a record as a key-value pair. The *key* is the text formed by the values of key attributes and *value* is the text made up of values of other attributes. And when we get a new key-value pair, we can easily decode it to a record. So the merger operation can be implemented by merging two key-value pairs coming from A and B. Obviously, two records that can be merged must have the same *key*. And these two records could be represented as $(key,\ value_A)$ and $(key,\ value_B)$. In the natural-join, we get the merged key-value pair $(key,\ value_{AB})$ by appending the $value_B$ to the $value_A$ to form $value_{AB}$. And in the left-join, $value_B$ must be a null string, which means that all the values corresponding to attributes from B are null. So the merged key-value pair will be $(key,\ value_A+null\_string_B)$. Similarly, we can get merged key-value pair as $(key,\ null\_string_A+value_B)$ in the right-join. Finally, the full-join can be divided into left-join and right-join to handle.

Map step: In the Map step, input data is a series of key-value pairs, each of which represents a record of a data file. The key is the offset in bytes of this record to the start point of the data file, and the value is a string of the content of this record. The Map-Function is described as pseudo code below.

Algorithm 1. PDIntegrationMapper $(key,\ value)$

Input: $(key$: offset in bytes; $value$: text of a record$)$
Output: $(key'$: key of the record, $value'$: value of the record$)$

1. $key' = key\_attributes(value)$;
2. $value' = value - key'$;
3. output $(key',\ value')$.

After Map step, the MapReduce framework automatically collects all pairs with the same key and groups them together, thus two records that have the same key respectively from A and B are put in a same group.

Reduce step: In the Reduce step, input data is the key-value pairs collected after Map step. The Reduce step realizes the merger operation. The Reduce-Function is described as pseudo code below.

Algorithm 2. PDIntegrationReducer $(key,\ values)$

Input:$(key$: key of the group records, $values$: values of the group records$)$
Output: $(value_{AB}$: value of the merged record$)$

1. If no_value_from_A$(values)$:
2. $value_A = null\_string_A$.
3. Else: $value_A = extrac\_valueA(values)$
4. If no_value_from_B$(values)$:
5. $value_B = null\_string_B$.
6. Else: $value_B = extrac\_valueB(values)$
7. $value_{AB} = value_A + value_B$;
8. Switch$(join\text{-}type)$:
9.   Case natural-join: If $value_A \neq null\_string_A$ and $value_B \neq null\_string_B$:
10.   output$(value_{AB})$.
11.   Case left-join: output$(value_{AB})$.
12.   Case right-join: output$(value_{AB})$.
13.   Case full-join: output$(value_{AB})$.

In the above code, we must distinguish the $value_A$ and $value_B$ from $values$ at the beginning of Reduce procedure. To handle this problem, we attach a table-mark to each key-value pair before passing it to the Map function, so a pretreatment process is applied to both tables A and B before the merger operation. This step is implemented by a MapReduce process described as follows.

Algorithm 3. PDIntegrationPreMapper $(key,\ value)$

Input:$(key$: offset in bytes; $value$: string of a record)
Output: $(key$: meaningless, same to the input, $value'$: adjusted value of the record)

1. $key\_attributes = \text{extract\_key\_attributes}(value)$;
2. $not\_key = \text{not\_key\_attributes}(value)$;
3. $value' = key\_attributes + not\_key + table\_mark$;
4. output $(key,\ value')$.

The Reduce step of Pretreatment is identity reducer: every record from the mapper is sent to the output directly. Fig.2 presents an example of data integration progress.

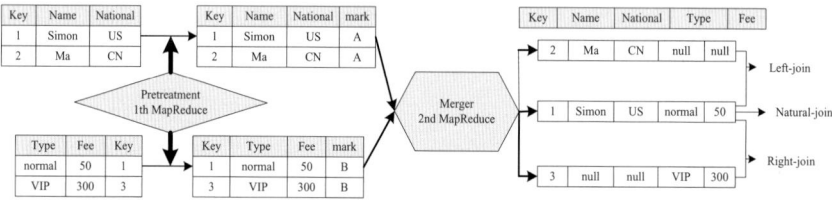

Fig. 2. An illustration of data integration

3.2   Implementation of Other Parallel Preprocessing Operations

For most preprocessing operations such as attribute deleting, attribute exchanging, data normalization and so on, they are simple and similar to each other in programming model. So they can share a uniform frame as follows.

Map step: In this step, the input value is a line of text. The input key is ignored but arbitrarily set to be the line number for the input value. The output key is same to the input, and the output value is the text of record after preprocessing.

Algorithm 4. PDUniformMapper $(key,\ value)$

---

1. $value' = \text{value\_processing}(value)$;
2. output$(key,\ value')$;

The function of value_processing is different for different preprocessing operations and will be discussed later.

Reduce step: The Reduce step of these preprocessing operations is identity reducer, viz. every record from the Mapper is sent to the output directly.

The functions of value_processing for these preprocessing operations are given below. Considering the similarity, we do not list the detailed implementation of all operations, but give several representative ones as examples.

Attribute deleting: Attribute deleting is an operation that deletes some customized attributes from a certain table.

1.   split the text of *value* and store the seprated value in *values[]*;
2.   *value'* = "";
3.   For each column *i*:
4.   If not_deleting_attribute(*i*):
5.   *value'* = *value'* + *values[i]*;
6.   EndFor
7.   return *value'*;

Data normalization: Data normalization is the operation normalizes some numerical attributes with a mean and a standard deviation specified by user.

1.   split the text of *value* and store the seprated value in *values[]*;
2.   *value'* = "";
3.   For each column *i*:
4.   If numerical(i) and normalize(i):
5.   *value'* = *value'* + (*values[i]*-*mean*)/*std*;
6.   Else: *value'* = *value'* + *values[i]*;
7.   EndFor
8.   return *value'*;

4   Experiments

In this section, we evaluate the performances of preprocessing operations with respect to the speedup, sizeup and their efficiencies to deal with large-scale datasets. Performance experiments were run on a cluster of computers, each of which has two 2.8 GHz cores and 4GB of memory. Hadoop version 0.17.0 and Java 1.5.0_14 are used as the MapReduce system for all experiments.

Speedup: Speedup is used to evaluate the effects of parallelism, which is defined as: Speedup(m) = run-time on one computer/run-time on m computers.

In order to measure the speedup, we keep the dataset constant and increase the number of computers in the system. More specifically, we apply our preprocessing operations in a system consisting of one computer. Then we gradually increase the number of computers in the system.

The speedup performances are obtained on datasets of 4GB, 8GB and 16GB respectively. It should be noted that in the operation of data integration, both tables have the size of 4GB, 8GB and 16GB.

For fully parallel operations, it demonstrates an ideal linear speedup: a system with m times the number of computers yields a speedup of m. However, linear speedup is difficult or impossible to achieve because of the communication cost and the skew of the slaves, viz. the problem that the slowest slave determines the total time needed.

Fig. 3 shows the speedup for some representative data preprocessing operations. From the graphs, we can see that: (1) These operations have good speedup performances. (2) Compared with 4GB and 8GB, speedup performances on 16GB dataset are better. The reason is that the time cost of fulfilling a task includes two parts: processing time cost and cost of managing the task. When dealing with larger datasets, processing cost takes up more proportion and this part could be reduced by parallelism. The speedup performances could be further improved by increasing the size of datasets. (3) Among all the operations, data summarization has the best speedup performance. That is because this operation needs not output the filtered dataset, but only a summarization file with very small size.

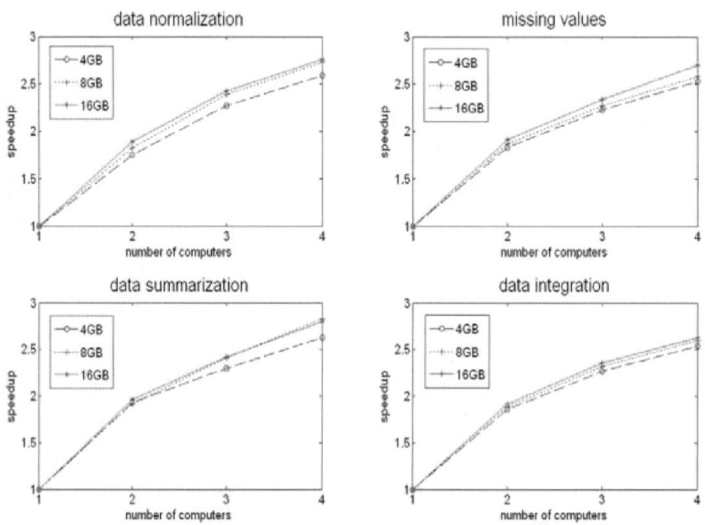

Fig. 3. Speedup results of the preprocessing operations

Sizeup: Sizeup analysis holds the number of computers in the system constant, and grows the size of the datasets by the factor m. Sizeup measures how much longer it takes on a given system, when the dataset size is m-times larger than the original dataset. The sizeup metric is defined as follows: Sizeup($data$, m) = run-time for processing $m*data$/run-time for processing $data$. To measure the performance of sizeup, we have fixed the number of processors to 2, 4, and 8 respectively.

The sizeup results in fig. 4 show that the preprocessing operations have very good sizeup performances. In an 8 processors running environment, a 16 times

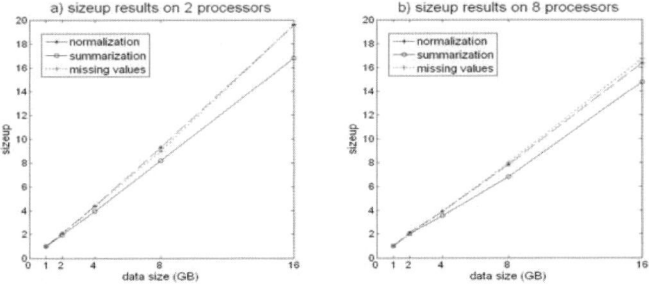

Fig. 4. Sizeup results on 2 and 8 processors

larger preprocessing problem needs about 16 times more time. Even if in a 2 processors running environment, it only needs less than 20 times more time.

Large-scale dataset processing efficiency: A dataset occupied more than 1 terabyte is generated to test the processing efficiency. The task is executed on a cluster of computers with 8 processors totally. The operation of data summarization costs about 63 hours and the corresponding number for filling missing values and data normalization are 99 hours and 103 hours respectively. Although these sound unacceptable, the running time could be greatly reduced to a reasonable time as the number of processors increase. Experiments running in a company's cluster of computers with 512 processors verified that preprocessing tasks for terabyte datasets could be accomplished in less than 2 hours.

5   Conclusions

With the development of information technology, the demand for processing very large-scale datasets is increasing. For the traditional serially implemented data preprocessing techniques, some processing tasks could not be fulfilled in a reasonable time. This work presents the parallel implementation of a series of data preprocessing techniques based on MapReduce. Experimental results show that we achieve a considerable speedup with an increasing number of computers. By means of these parallel implementations, large-scale data preprocessing tasks could be executed on a cluster of computers, which greatly improves the processing efficiency.

Acknowledgement

This work is supported by the National Science Foundation of China (No.60933004, 60975039), 863 National High-Tech Program (No.2007AA01Z132), National Basic Research Priorities Programme (No.2007C-B311004) and National Science and Technology Support Plan (No.2006BAC08B06).

References

1. Han, J.W., Kamber, M.: Data Mining: Concepts and Techniques. Morgan Kaufmann, San Francisco (2000)
2. Dunham, M.H.: Data Mining: Introductory and Advanced Topics. Tsinghua University Press, Beijing (2005)
3. Jian, Z.G., Jin, X.: Research on Data Preprocess in Data Mining and Its Application. Application Research of Computers 7, 117–118, 157 (2004)
4. Dean, J., Ghemawat, S.: MapReduce: Simplified Data Processing on Large Clusters. Communications of the ACM 51, 107–113 (2008)
5. http://hadoop.apache.org/core/
6. Borthakur, D.: The Hadoop Distributed File System: Architecture and Design (2007)
7. Lammel, R.: Google's MapReduce Programming Model – Revisited. Science of Computer Programming 70, 1–30 (2008)

Parallel Implementation of Classification Algorithms Based on MapReduce

Qing He[1], Fuzhen Zhuang[1,2], Jincheng Li[1,2], and Zhongzhi Shi[1]

[1] The Key Laboratory of Intelligent Information Processing, Institute of Computing Technology, Chinese Academy of Sciences, Beijing, 100190, China
[2] Graduate University of Chinese Academy of Sciences, Beijing, 100190, China
{heq,zhuangfz,lijincheng,shizz}@ics.ict.ac.cn

Abstract. Data mining has attracted extensive research for several decades. As an important task of data mining, classification plays an important role in information retrieval, web searching, CRM, etc. Most of the present classification techniques are serial, which become impractical for large dataset. The computing resource is under-utilized and the executing time is not waitable. Provided the program mode of MapReduce, we propose the parallel implementation methods of several classification algorithms, such as k-nearest neighbors, naive bayesian model and decision tree, etc. Preparatory experiments show that the proposed parallel methods can not only process large dataset, but also can be extended to execute on a cluster, which can significantly improve the efficiency.

Keywords: Data Mining, Classification, Parallel Implementation, Large Dataset, MapReduce.

1 Introduction

With the increasing collected data universally, data mining technique is introduced to discover the hidden knowledge which can contribute to decision-making. Classification is one category of data mining which extracts models to describe data classes or to predict future data trends. Classification algorithms predict categorical labels, many of which have been proposed by researchers in the field of machine learning, expert systems, and statistics, such as decision tree [1,2], naïve bayesian model [3,4,5,6], nearest neighbors algorithm [7,8], and so on. While, most of these algorithms are initially memory-resident, assuming a small data size. Recently, the research has focused on developing scalable techniques which are able to process large and disk-resident dataset. These techniques often consider parallel and distributed processing.

However, due to the communication and synchronization between different distributed components, it is difficult to implement the parallel classification algorithms in a distributed environment. The existing models for parallel classifications are mainly implemented through message passing interface (MPI) [9], which is complex and hard to master. For the emergence of cloud computing [10,11], parallel techniques are able to solve more challenging problems, such as

J. Yu et al. (Eds.): RSKT 2010, LNAI 6401, pp. 655–662, 2010.

heterogeneity and frequent failures. The MapReduce model [12,13,14] provides a new parallel implementation mode in the distributed environment. It allows users to benefit from the advanced features of distributed computing without programming to coordinate the tasks in the distributed environment. The large task is partitioned into small pieces which can be executed simultaneously by the workers in the cluster.

In this paper, we introduced the parallel implementation of several classification algorithms based on MapReduce, which make them be applicable to mine large dataset. The key is to design the proper *key/value* pairs.

The rest of the paper is organized as follows. Section 2 introduces MapReduce. Section 3 presents the details of the parallel implementation of the classification algorithms based on MapReduce. Some preparatory experimental results and evaluations are showed in Section 4 with respect to scalability and sizeup. Finally, the conclusions and future work are presented in Section 5.

2   MapReduce Overview

MapReduce [15], as the framework showed in Figure 1, specifies the computation in terms of a *map* and a *reduce* function, and the underlying runtime system automatically parallelizes the computation across large-scale clusters of machines, handles machine failures, and schedules inter-machine communication to make efficient use of the network and disks.

Essentially, the MapReduce model allows users to write Map/Reduce components with functional-style code. These components are then composed as a dataflow graph with fixed dependency relationship to explicitly specify its parallelism. Finally, the MapReduce runtime system can transparently explore the parallelism and schedule these components to distribute resources for execution. All problems formulated in this way can be parallelized automatically.

All data processed by MapReduce are in the form of *key/value* pairs. The execution happens in two phases. In the first phase, a *map* function is invoked once

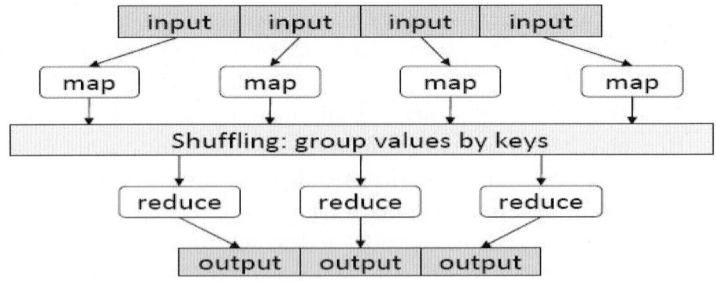

Fig. 1. Illustration of the MapReduce framework: the "map" is applied to all input records, which generates intermediate results that are aggregated by the "reduce"

for each input *key/value* pair and it can generate output *key/value* pairs as intermediate results. In the second one, all the intermediate results are merged and grouped by keys. The *reduce* function is called once for each *key* with associated values and produces output values as final results.

2.1   MapReduce Model

Key/value pairs form the basic data structure in MapReduce (see Figure 1). A *map* function takes a *key/value* pair as input and produces a list of *key/value* pairs as output. The type of output *key* and *value* can be different from input *key* and *value*:

$$map :: (key_1, value_1) \Rightarrow list(key_2, value_2)$$

A reduce function takes a key and an associated value list as input and generates a list of new values as output:

$$reduce :: (key_2, list(value_2)) \Rightarrow (key_2, value_3)$$

2.2   MapReduce Execution

A MapReduce application is executed in a parallel manner through two phases. In the first phase, all *map* functions can be executed independently with each other. In the second phase, each reduce function may depend on the outputs generated by any number of map operations. However, similar to map operations, all reduce operations can be executed independently.

From the perspective of dataflow, MapReduce execution consists of m independent map tasks and r independent reduce tasks. Generally the intermediate results are partitioned by keys with ASCII order, in order words, the values associate with the same key will be placed into the same data block.

The MapReduce runtime system schedules map and reduce tasks to distributed resources, which handle many tough problems, including parallelization, concurrency control, network communication and fault tolerance. Furthermore, it performs several optimizations to decrease overhead involved in the scheduling, network communication and intermediate grouping of results.

As MapReduce can be easily understood, its abstraction considerably improves the productivity of parallel/distributed development, especially for novice programmers.

3   The Parallel Implementation of Classification

3.1   Parallel k-Nearest Neighbors (PkNN)

For k-nearest neighbors (kNN) algorithm [7], the computation of similarity between training and testing samples is independent, thus the algorithm can be easily extended to parallel mode. Specially, the input dataset can be partitioned into n blocks, which can be simultaneously processed on different nodes. Fortunately, the hadoop platform has solved the problem that how to divide the dataset and how to communicate between different nodes, which enables the programmers to program conveniently. A uniform interface is provided to set parameters. For PkNN, the parameters include the number of nearest neighbors k, the paths of the input and output files, the number of map tasks and reduce tasks.

The key to the parallel implementation is the design of *key/value* pairs. In *map* function, the input *key* is the offset and the input *value* is the content of samples. Each sample includes three parts: the identification, the value of conditional attribute and the label[1]. The intermediate *key* is designed as the identification, and the *value* the combination of similarity value and label. For the *reduce* function, the *key* is the identification, and the *value* is the predicted label. Algorithm 1 and Algorithm 2 give the pseudocode of the *map* and *reduce* functions for PkNN.

Algorithm 1. $map(key, value)$

Input: the number of nearest neighbors k, the training dataset and the testing dataset
Output: $< key', value' >$ pair, where key' is id, $value'$ is the combination of the similarity value S and label y' (y' is the label of sample x');

1. for each sample
   - parse the *value* as $< id, x, y >$;
   - compute the similarity $S = sim(x, x')$;
   - output $< key', value' >$
2. end for

Algorithm 2. $reduce(key, value)$

Input: the result of map function
Output: $< key', value' >$ pair, where key' is id, $value'$ is the predicted label

1. Iterator<Double> *Sims*
2. Iterator<String> *labels*
3. for each sample
   - parse v as $< S, y' >$, where S is the similarity value, y' is the label;
   - push the S into *Sims*, Push the y' into *labels*
4. end for
5. sort *Sims* and select the k-nearest neighbors
6. output $< key', value' >$

3.2   Parallel Naive Bayesian Model (PNBM)

Naïve Bayesian classifier has been presented in [3,4,5,6]. The most intensive calculation is to calculate the conditional probabilities, which involves the frequency of the label or the combination of the label, attribute name and attribute value. Firstly the parameters are set, including the paths of the input and output files. When training the model, the *map* function (Algorithm 3) parses the label and

---

[1] if exist, y is the true label of sample x, else $y = 0$.

the attribute value of each attribute. Therefore, the *key* can be designed as the label or the combination of the label, attribute name and attribute value, and the *value* of the *key* is 1. For the *reduce* function (Algorithm 4), we count the frequency of each *key*. So far, the parameters of the naïve Bayesian classifier can be estimated (or calculated), including $P(c_j)$ and $P(A_i|c_j)$, where c_j denotes the j-th category, A_i the i-th attribute. When testing, the *map* function (Algorithm 5) first indexes the *key* in the results produced by the training step, and reads the corresponding probabilities, then calculates the probability of the test sample belonging to each category. So the label can be predicted according to the *maximum posterior* and a pair is output for each sample which takes the "correct" or "wrong" as the *key*, 1 as the value. For the *reduce* function (whcih is the same as Algorithm 4), the number of the correctly or wrongly predicted samples can be stated. Therefore, the correct rate and error rate can be further calculated.

Algorithm 3. *train_map(key, value)*

Input: the training dataset
Output: $< key', value' >$ pair, where key' is the label or the combination of the label, attribute name and attribute value, and $value'$ the frequency

1. for each sample do
   - parse the label and the value of each attribute;
   - take the label as key', and 1 as $value'$;
   - output $< key', value' >$ pair;
   - for each attribute value do
     - construct a string as the combination of the label, attribute name and attribute value;
     - take the string as key', and 1 as $value'$;
     - output $< key', value' >$ pair;
   - end for
2. end for

Algorithm 4. *train_reduce(key, value)*

Input: *key* and *value* (the key' and $value'$ output by map function, respectively)
Output: $< key', value' >$ pair, where key' is the label or the combination of the label, attribute name and attribute value, and $value'$ the result of frequency

1. initialize a counter *sum* as 0 to record the current statistical frequency of the *key*;
2. while(*value*.hasNext())
   - *sum* += *value*.next().get();
3. take *key* as key', and *sum* as $value'$;
4. output $< key', value' >$ pair;

3.3   Parallel Decision Tree (PDT)

The basic technique of constructing decision tree is recursively choosing the best splitting attribute for each node in the tree. The evaluating measure of attribute can be information gain, Gini index and so on. To compute the information gain, we must first get the statistical information, such as the frequency of the combination of the label with each value of each attribute, which is similar to the

---

Algorithm 5. *test_map(key, value)*

Input: the testing dataset and the reduce result of training step
Output: $< key', value' >$ pair, where key' is the label or the combination of the label, attribute name and attribute value, and $value'$ the frequency

1. parse the label and the value of each attribute;
2. initialize an array $prob[]$, the length is set as the size of the labels set;
3. for each label in the labels set do
   - initialize $prob[i]$ as 1.0, i is the index of the label in the labels set;
   - for each attribute do
     - initialize a string as the combination of the label with the attribute name and its value;
     - index the string in the $keys$ of the reduce result, record the corresponding $value$;
     - $prob[i]*= value$;
   - end for
4. end for
5. index the label with the maximum value of $prob$;
6. if the label is the same to that of the instance, take "correct" as key', and 1 as $value'$, output $< key', value' >$; else take "wrong" as key', and 1 as $value'$, output $< key', value' >$.

---

training process of PNBM. Therefore, we skip the details of the map and reduce functions for this process. After selecting an attribute, the training dataset is split into several partitions according to each value of the splitting attribute. Then a new Map/Reduce process is started for each partition of sub-dataset, which recursively computes the best splitting attribute at the current node. Meanwhile, the decision rules are stored when new rules are generated. When testing, the map function is mainly to match each sample with the decision rules and predict the label. Only the map function is needed to complete the testing process. Algorithm 6 has presented the pseudocode of testing process of PDT based on MapReduce.

Algorithm 6. *test_map(key, value)*

Input: the testing dataset and the decision rules
Output: $< key', value' >$ pair, where key' is the sample and $value'$ the predicted label

1. for each sample do
   - match the sample with the decision rules, and take the the matched rule as the predicted label
2. end for
3. take key as key', and the predicted label as $value'$, output $< key', value' >$

---

4   Preparatory Experiments

In this section, we perform some preparatory experiments to test the scalability of parallel classification algorithms proposed in this paper. We build a small cluster with 4 business machines (1 master and 3 slaves) on Linux, and each machine has two cores with 2.8GHz, 4GB memory, and 1TB disk. We use the Hadoop version 0.17.0[2], and Java version 1.5.0_14.

---

[2] http://hadoop.apache.org/core/releases.html.

The datasets *iris* and *car*, from UCI depository, are used in our experiments. The original datasets are replicated to form large datasets. The experiments are designed as follows:

- we use *iris* to test PkNN, and construct 6 test datasets, whose sizes are 5MB, 10MB, 20MB, 40MB, 80MB and 160MB, respectively. The training set is 6 copies (900 samples) of the original dataset and the results are shown in Figure 2(a).
- To test PNBM and PDT, we use *car* data, and the datasets are constructed similarly with PKNN. The sizes of training datasets are 0.25GB, 0.5GB, 1GB, 2GB, 4GB and 8GB. Note that the characteristic of computation time for testing is similar with training, we only list the results (in Figure 2(b)) of training time for PNBM and PDT.

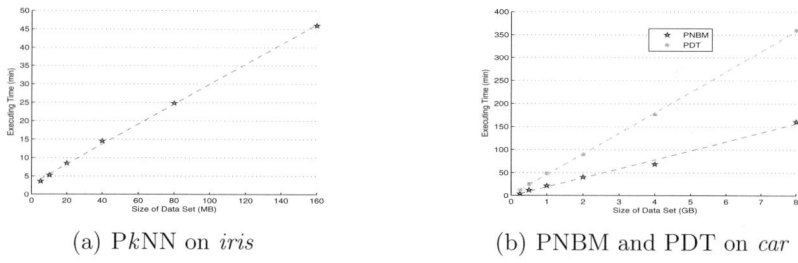

(a) PkNN on *iris*   (b) PNBM and PDT on *car*

Fig. 2. Executing time with different sizes

From Figure 2(a) and 2(b), we can empirically find that 1) the executing time increases approximatively linearly with the size of dataset, which is very important for parallel algorithms; 2) the size of data dealt with in our experiments is much larger than the one serial algorithms can tackle. Additionally, we analyze the executing time of PkNN when dealing with small dataset (eg. 5MB). It is indicated that the time for communication and assigning tasks takes almost the same as executing time of algorithm. So we suggest that the parallel algorithms are more suitable for tackling large scale dataset.

Furthermore, we have tested our algorithms on much larger scale datasets (some are up to terabyte) in our cooperant enterprise's computation environment with more than four hundred processors. The results also reveal promising scalability.

5   Conclusions

This paper mainly presents the parallel implementation methods of classification algorithms based on MapReduce, including parallel k-nearest neighbors algorithm (PkNN), parallel naive bayesian model (PNBM) and parallel decision tree (PDT). Preparatory experiments show that the parallel algorithms can not only process large dataset, but also have the required property of linear scalability. In the future work, we will further conduct the experiments and consummate the parallel algorithms to improve usage efficiency of computing resources.

Acknowledgement

This work is supported by the National Science Foundation of China (No.60933004, 60975039), National Basic Research Priorities Programme (No.2007CB311004) and National Science and Technology Support Plan (No.2006BAC08B06).

References

1. Quinlan, J.R.: Induction of decision trees. Machine Learning 1(1), 81–106 (1986)
2. Quinlan, J.R.: C4.5: Programs for Machine Learning. Morgan Kaufmann, San Mateo (1997)
3. Weiss, S.M., Kulikowski, C.A.: Computer Systems that Learn: Classification and Prediction Methods from Statistics. In: Neural Nets, Machine Learning, and Expert Systems. Morgan Kaufmann, San Francisco (1991)
4. Michie, D., Spiegelhalter, D.J., Taylor, C.C.: Machine learning, neural and statistical classification. Ellis Horwood, New York (1994)
5. Mitchell, T.M.: Machine Learning. McGraw-Hill, New York (1997)
6. Duda, R.O., Hart, P.E., Stork, D.G.: Pattern Classification, 2nd edn. John Wiley and Sons, Chichester (2001)
7. Cover, T., Hart, P.E.: Nearest neighbor pattern classification. IEEE Trans. Information Theory 13, 21–27 (1967)
8. Dasarathy, B.V.: Nearest Neighbor (NN) Norms: NN Pattern Classification Techniques. IEEE Computer Society Press, Los Alamitos (1991)
9. http://www.mcs.anl.gov/research/projects/mpi/
10. Weiss, A.: Computing in the Clouds. netWorker 11(4), 16–25 (2007)
11. Buyya, R., Yeo, C.S., Venugopal, S.: Market-Oriented Cloud Computing: Vision, Hype, and Reality for Delivering IT Services as Computing Utilities. In: Proc. of the 10th IEEE International Conference on High Performance Computing and Communications, China (2008)
12. Dean, J., Ghemawat, S.: MapReduce: Simplified Data Processing on Large Clusters. In: Proc. of the 6th Symposium on Operating System Design and Implementation, USA (2004)
13. http://hadoop.apache.org/core/
14. http://en.wikipedia.org/wiki/Parallel_computing
15. Elsayed, T., Jimmy, L., Douglas, W.O.: Pairwise Document Similarity in Large Collections with MapReduce. In: Proc. of the 46th Annual Meeting of the Association for Computational Linguistics on Human Language Technologies, pp. 262–268 (2008)

Research on Data Processing of RFID Middleware Based on Cloud Computing

Zheng-Wu Yuan and Qi Li

Sino-Korea Chongqing GIS Research Center,
College of Computer Science and Technology,
Chongqing University of Posts and Telecommunications,
Chongqing 400065, China

Abstract. According to the problem that research on algorithm of RFID middleware mainly focus on removing data redundancies but less on data noises, this paper builds a cube model of RFID middleware data processing, and proposes algorithms of removing data redundancies and data noises, and provides service of produces track to upper system. Cloud Computing that providing powerful computing ability to processes the cube model of data processing. Analyzing to algorithm and model shows that: the cube model describes tags' status in reader areas well, its algorithm enable removing data redundancies and data noises, and provide service of produces track to upper system.

Keywords: Cloud Computing, Internet of Things, RFID Middleware, Cube Model.

1   Introduction

Cloud Computing[1][2][3] is evolved from the Grid Computing. The front application provides storage and computing service to users through the Internet. Cloud back system is a large pool of virtual resources. The pool made of large number of computer cluster which using virtual machines and connecting by high speed Internet. And these virtual resources enable self-management and self-configuration, and ensure high availability of virtual resources by redundancy way, and have the characters of distributed storage, computing, scalability, high availability and user friendly.

IOT (Internet of Things)[4] makes full use of IT technologies among industries, specifically, is to embed sensors and equipment to objects, like the power grids, railways, bridges, tunnels, highways, buildings, water systems, dams, oil and gas pipelines, etc., and then the IOT integrate with the existing Internet to achieve the integration of human society and physical system. In this integrated network, it needs a powerful central computer cluster to process super-massive sensor data, and achieve real-time control and management to personnel, machinery, equipment and infrastructure of integrated network. Cloud Computing provide IOT with powerful computing ability and super-massive storage ability that enable IOT operating normally.

J. Yu et al. (Eds.): RSKT 2010, LNAI 6401, pp. 663–671, 2010.

RFID (Radio Frequency Identification) middleware is the most important part of IOT. It direct faces the massive tag data from the hardware, and filters the tag data, and then submits processed data to upper system. It is called as the nerve center of Internet of Things. Therefore, Cloud Computing has a large development space in IOT, especially in middleware. In the aspect of massive data filtering, it can use the IAAS (Infrastructure as a Service) of Cloud Computing model to provide a powerful computing ability; and in the aspect of massive data storage, it can also use the IAAS of Cloud Computing model to provide massive storage ability[5].

2   RFID Middleware Background

In this paper, it defines two definitions as follows:

Definition 1. Data redundancy[6] is the situation that many times to write in database for one tag by logical reader when the tag in the communication radius of logic reader.

Definition 2. Data Noise is the situation that one tag was read by a logical reader in misread or missed-read. It includes two types: 0 data noise and 1 data noise.

RFID data processing and filtering is recognized as one of the core functions of RFID middleware. Although logical readers collected a large number of initial tag data, meaningful tag data for users only in a small proportion. If the data redundancy was not removed, three loads will appear as follows[6]:

1. the load of network bandwidth;
2. the load of data processing;
3. the load of data storage.

Therefore, RFID data filtering is the most important task of RFID middleware. The module of data processing and filtering is designed to accomplish the task[7][8][9].

Existing filtering methods can be divided into two categories:

1. Event Listing. Detecting each new arrival tag, if it is first appeared, and add it to the list accordingly; if it was appeared, and then updates its arrival time but not add new one; in this way, it can achieve the goal of removing data redundancy.
2. Event Encoding. According to tags status, if tag appeared, and then encode 0; if tag disappeared, and then encode 1. Then add the timer mechanism, ignoring the tags status changes in an effective time of timer mechanism. In this way, it can achieve the goal of removing data redundancy.

All these algorithms enable removing data redundancies well, and reducing loads of upper system. However, in practical, except for removing RFID data redundancies, there are other requirements for data processing, such as removing data noises. In the basis of analyzed and summarized for existing methods, this paper builds a cube model of RFID middleware data processing, and uses the powerful computing ability that provided by Cloud Computing to filter the massive data. Mainly solved problems are removing data redundancies, removing data noises and providing service of produces track to upper system.

3   Cube Model of RFID Middleware Data Processing Based on Cloud Computing

3.1   Structure of RFID Middleware

RFID middleware[10] include edge controller module, user interface module and data storage module (Fig. 1). Where, edge controller module consists of data processing and filtering and reader driver manager, and it is the core of RFID middleware. It is used to connect and control types of readers, and filters the tag data, and then submits uniform data format to upper system. User interface module mainly provides user reader configure and management. Data storage module mainly store collected tag data and configuration information. As this paper research on RFID data processing, research will mainly focus on data processing and filtering of edge controller module.

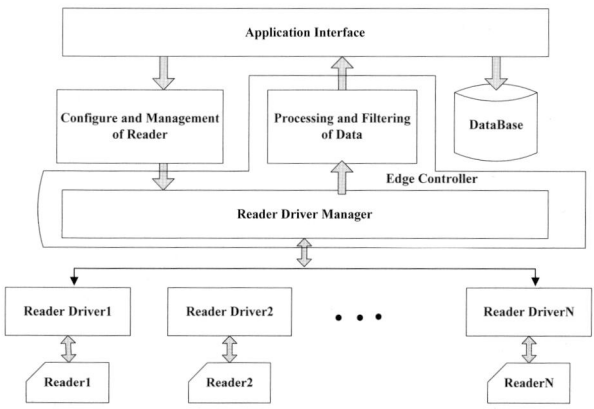

Fig. 1. Structure of RFID middleware

3.2   Building Cube Model of RFID Middleware Data Processing

Where, X-axis denotes read cycles of logical reader. Y-axis denotes RFID tags in the communication radius of logical readers. Z-axis denotes logical readers in the data collection area. This model selects any 10 consecutive read cycles named: Cycle1, Cycle2 ... Cycle10, and 10 tags named: Tag1, Tag2 ... Tag10, and 15 adjacent logical readers named: Reader1, Reader2 ... Reader15.

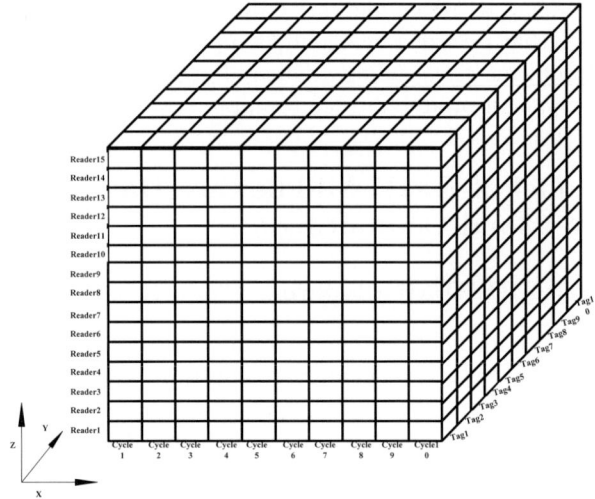

Fig. 2. Cube model of RFID middleware

3.3   Algorithm Description

Fig. 3 describes the situation that 10 tags status in 10 consecutive read cycles of one logical reader. If tag was read by one cycle then set "1", if not then set "0", therefore, a tag event can be denoted by a "01" string. "01" strings of tag event can be divided into follow types:

1. Tag1 event encoding: 1111111111, it denotes Tag1 always in the communication radius of logical reader;
2. Tag2 event encoding: 0000000000, it denotes Tag2 always not in the communication radius of logical reader;

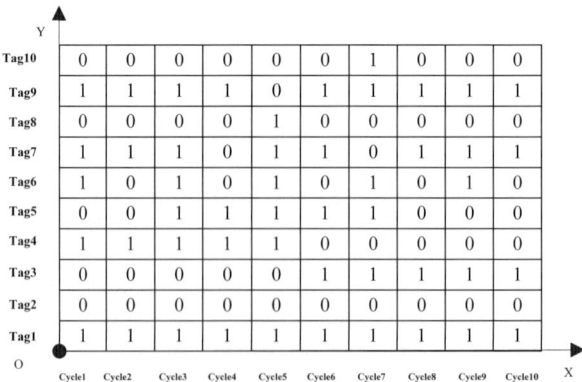

Fig. 3. XOY section of cube model

3. Tag3 event encoding: 0000011111, it denotes Tag3 enter the communication radius of logical reader;
4. Tag4 event encoding: 1111100000, it denotes Tag4 leave the communication radius of logical reader;
5. Tag5 event encoding: 0011111000, it denotes Tag5 through the communication radius of logical reader;
6. Tag6 event encoding: 1010101010, it denotes Tag6 move around on the border of logical reader;

For Tag7, Tag8, Tag9 and Tag10, they appeared data noises. It can be divided two types:

1. 0 data noise, according to traditional filtering algorithm, Tag7 denotes two enter events and two leave events. In fact, it never leave because 0 data noise appeared, and Tag9 in the same situation;
2. 1 data noise, Tag8 denotes a through event. In fact, Tag8 never enter because 1 data noise appeared, and Tag10 in the same situation.

```
1   Filtering algorithm description as follows:
2   Define variant:
3       Tag value: Tag_Value
4       Tag appearance cumulative weight: App_Value
5       Tag disappearance reductive weight: Dis_Value
6       Tag status: Status
7       Tag appearance event trigger threshold: App_Fun
8       Tag disappearance event trigger threshold: Dis_Fun
9       Tag value maximum: Max
10      Tag value minimum: Min
11  Initialization:
12      Tag_Value=0
13      Status=False
14  Algorithms pseudocode expression:
15    Fileter_Fun ()
16    { If (Tag appeared)
17            Tag_Value+=App_Value;
18      Else  If (Tag disappeared)
19            Tag_Value -=Dis_Value;
20      If (Tag_Value >=App_Fun and Status==False)
21          {Triggering tag appearance event;
22           Status=True;}
23      Else  If (Tag_Value <=Dis_Fun and Status==True)
24              {Triggering tag disappearance event;
25               Status=False;}//removing data redundancies
26      If (Status==True and Tag_Value >=Max)
27            Tag_Value=Max;
28      Else  If (Status==False and Tag_Value <=Min)
29            Tag_Value=Min;
30      }//end Filter _Fun
```

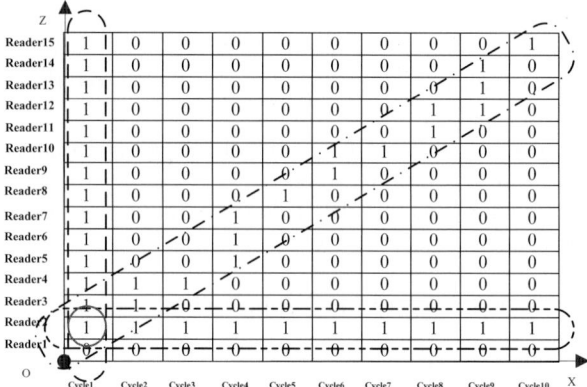

Fig. 4. XOZ section of cube model

In this algorithm, if tag value is larger than tag appearance threshold and tag status is false, then triggering tag disappearance event. When tag move around on the border of logical reader, tags status decide to write tag data or not, like tag6; as data noise appearance times less, tag value hardly larger than tag appearance threshold App_Fun or smaller than tag disappearance Dis_Fun, so not easy to trigger tag event, and prove this algorithm restrain data noise appearance.

When the filtering algorithm finds the tag already exists, if tag value has reached the maximum, it will remain at this value to prevent the tag value too high to judge. When the filtering algorithm finds the tag already leaves, if tag value has reached the minimum, it will remain at this value to prevent the tag value too low to judge.

The cube model is consists of 0 and 1 which was expressing the tag whether in the communication radius of corresponding logical reader or not. From the view of XOZ section, these coordinates that the value is 1 denotes the tag passed through, and the coordinates value denote by the centroid of communication area of logical reader. Tag tracks can be divided into 4 types as Fig. 4.

1. parallel to the X-axis: tag always in the communication radius of Reader2, it denotes the produce in static;
2. parallel to the Z-axis: tag through the communication radius of Reader2 Reader15, it denotes the produce moving in a high speed;
3. ranged in 1 and 2: such as Fig. 4, tag through the communication radius of Reader2 Reader15 in Cycle2 Cycle15, it denotes the produce moving in a middle speed;
4. the red cycle in Fig. 4: it denotes tag was read only once or several times, this type regarded as data noises. This paper proposed another algorithm of moving data noise based on statistics. Such as Tagi.

```
 1  Algorithm of removing data noise based on statistics
      description as follows:
 2  Define variant:
 3      Tag value: Tag_Value
 4      Triggering tag event threshold: Dispose_Value
 5      Number of reader cycle: Cycle_number
 6      Number of logical reader: Reader_number
 7  Initialization:
 8      Tag_Value=0
 9  Algorithm s pseudocode expression:
10  Dispose_Fun()
11  { For (j=1;j<=Cycle_number ;j++)
12    For (k=1;k<=Reader_number ;k++)
13    Statistic read times of tagi in Cycle_number consecutive
      cycles of Reader_number logical reader, that is tag value;
14    If (Tag_Value<=Dispose_Value)
15      Triggering tag event and removing data noise;
16  }//end Dispose_Fun
```

At Cycle1, Tagi in the read range of Reader2, the coordinates of Tagi is the mass center of Reader2 read range, denotes as { Reader2} ; at Cycle2, Tagi in the public read range of Reader3 and Reader4, the coordinates of Tagi is the midpoint of the line that connect the two center of Reader3 and Reader4, denotes as { Reader3, Reader4 }; at Cycle4, Tagi in the public read range of Reader5, Reader6 and Reader7, the coordinates of Tagi is the mass center of the triangle that consist of the 3 center of Reader5, Reader6 and Reader7, denotes as { Reader5, Reader6, Reader7 }. This article lists the three most common situations in Fig. 5, otherwise are like this.

Therefore, the movement trajectory of Tagi in Cycle1$\tilde{}$Cycle10 in Fig. 4 can be expressed as: { Reader2} { Reader3, Reader4} { Reader4} { Reader5, Reader6, Reader7} { Reader8} { Reader9, Reader10} { Reader10} { Reader11, Reader12} { Reader12, Reader13, Reader14} { Reader15}. Generally, the location of the logical reader is fixed coordinates, can be pre-measured into a database, obtaining the mass center coordinates of the public read range by geometric calculation.

4   Model and Algorithm Analysis

4.1   Feature of Model

In practical application, the coverage of data collecting area is certain. The number of logical reader is certain; the number of tag read by a logical reader in one cycle is certain; read cycles of logical reader is able to continue. Therefore, the data process cube model is a dynamic model that Z-axis and Y-axis are certain, X-axis is able to continue. Generally, Y-axis can reach tens or hundreds of tags.

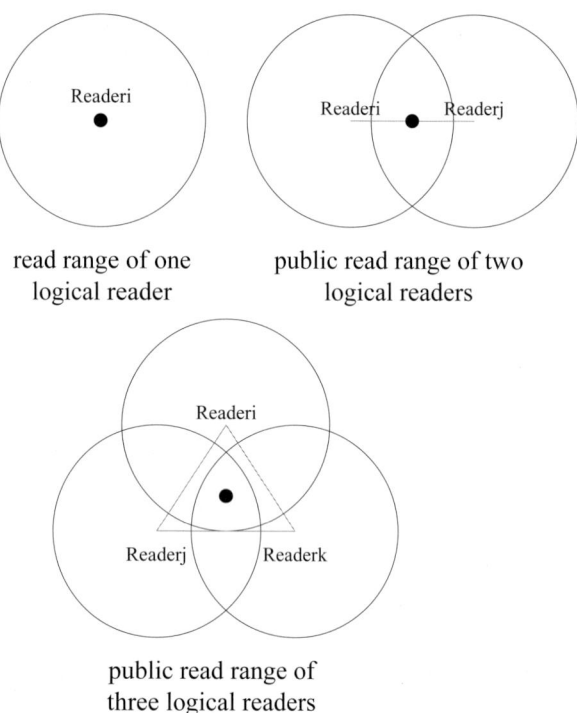

read range of one logical reader

public read range of two logical readers

public read range of three logical readers

Fig. 5. Mass center coordinates of logical reader read range

4.2   Comparison of the Two Algorithms

Fig. 3 shows the case that a set of tags was read in 10 consecutive read cycles of a logical reader, but there are a lot of data redundancies and a small amount of data noises. This paper proposed Filter_Fun algorithm to removing data redundancies and data noises. Fig. 4 shows the case that a tags movement in 10 consecutive read cycles of 15 logical reader read range. In this view, it is easier to find data noise, especially in 1 data noise. This paper proposed Dispose_Fun algorithm of moving data noise based on statistics. Comparison of two algorithms as follows:

1. Filter_Fun enables to remove data redundancies in the collected data set well, and to remove data noises, include 0 data noise and 1 data noise;
2. Dispose_Fun can only remove 1 data noise in the collected data set, but have a good feature of easy to indentify 1 data noise and 0 data noise.

According to the needs of practical applications, these two algorithms can play their own advantages.

5 Conclusion

This paper introduces the enormous potential of Cloud Computing in the application of IOT, and analyzes the development statuses and problems of IOT middleware. In this paper, it builds data processing cube model, and proposed algorithms of removing data redundancies and data noises. Analyzing to processed data, it can find the trajectory of tag, and provides service of produces track to upper system.

References

1. Armbrust, M., Fox, A., Griffith, R., et al.: Above the Clouds: A Berkeley view of Cloud Computing (February 2009),
   http://citeseerx.ist.psu.edu/viewdoc/
   download?doi=10.1.1.150.628&rep=rep1&type=pdf
2. Chen, Q., Deng, Q.-n.: Cloud Computing and its key techniques. Journal of Computer Applications (September 2009)
3. Antoine de Saint-Exupery. Internet of Things (September 2009),
   http://www.sintef.no/upload/IKT/9022/CERP-IoT%20SRA_IoT_v11_pdf.pdf
4. Aitenbichler, E., Behring, A., Bradler, D., et al.: Shaping the Future Internet (December 2009), http://www.ss4sw.org/Publications/2009/IOPTS2009.pdf
5. Jiang, S.-G., Tan, J.: Research of data processing and filtration in RFID middleware. Journal of Computer Application (October 2008)
6. Cheng, X.-z., Li, Y.-c.: Design of RFID middleware architecture. Journal of Computer Application (April 2008)
7. Li, B., Xie, S., Su, X.: Research and Implementation of Embedded RFID Middleware System. Computer Engineering (August 2008)
8. Ding, Z., Li, J., Feng, B., et al.: Survey on RFID Middleware. Computer Engineering (November 2006)
9. EPCglobal. The Application Level Event (ALE) Specification Version 1.1. EPCglobal Standard Specification (February 2008)
10. Chu, W.-J., Tian, Y.-M., Li, W.-p.: Integration Application of RFID Middleware Based on SOA. Computer Engineering (July 2008)

Attribute Reduction for Massive Data Based on Rough Set Theory and MapReduce*

Yong Yang, Zhengrong Chen, Zhu Liang, and Guoyin Wang

Institute of Computer Science & Technology,
Chongqing University of Posts and Telecommunications,
Chongqing, 400065, P.R. China
yangyong@cqupt.edu.cn, {554760686,121495784}@qq.com, wanggy@cqupt.edu.cn

Abstract. Data processing and knowledge discovery for massive data is always a hot topic in data mining, along with the era of cloud computing is coming, data mining for massive data is becoming a highlight research topic. In this paper, attribute reduction for massive data based on rough set theory is studied. The parallel programming mode of MapReduce is introduced and combined with the attribute reduction algorithm of rough set theory, a parallel attribute reduction algorithm based on MapReduce is proposed, experiment results show that the proposed method is more efficiency for massive data mining than traditional method, and it is a effective method effective method effective method for data mining on cloud computing platform.

Keywords: Attribute reduction, rough set, MapReduce.

1 Introduction

Data processing and knowledge discovery for massive data is always a hot topic in data mining. There are many applications associated with massive data, such as gene analysis, remote sense image processing and intrusion detection. Along with the cloud computing era is coming, there are urgent requirements for data mining on the massive data of cloud computing platform, for example, the requirement of business intelligence(BI). Since the data volume of cloud platform would be TB, even to be PB(1000TB), therefore, data mining for cloud platform is a new challenge.

For the purpose of data mining for massive data, there are a lot of research works, and parallel computing mode and parallel algorithm are typical methods in these research works. There are many parallel algorithms, such as divide and conquer method, pipeline method, etc. Feng Hu and Guoyin Wang [1] combined rough set theory and divide and conquer method, and proposed a quick attribute

---

* This paper is partially supported by National Natural Science Foundation of China under Grants No.60773113 and No.60573068, Natural Science Foundation of Chongqing under Grants No.2008BA2017 and No.2008BA2041, CQUPT-ICST Fundation under Grants No.JK-Y-2010002 and CY-CNCL-2009-02.

J. Yu et al. (Eds.): RSKT 2010, LNAI 6401, pp. 672–678, 2010.

reduction method. On the other side, parallel computing modes are alternative methods for massive data mining. MapReduce is a parallel computing mode, and it is a popular computing mode for cloud computing platform. There are many research works related MapReduce combined with the traditional method, for example, John Sharer combined with decision tree and MapReduce mode and proposed a scalable parallel classifier SPRINT [2], Andrew proposed a method combined with MapReduce with PSO [3], and Abhishek proposed a scaling genetic algorithm based on MapReduce [4].

In this paper, rough set theory combined with MapReduce are studied, and a parallel rough set reduction algorithm based on MapReduce is proposed, it is more efficient based on the experiment results. The rest of this paper is organized as follows. In section 2, MapReduce and rough set theory is surveyed. A parallel rough reduction algorithm is proposed in section 3. Simulation experiments and discussion are introduced in section 4. Finally, conclusion is drawn in section 5.

2 Algorithm Basis

In this section, the algorithm basis of MapReduce and rough set theory are surveyed and introduced as follows.

2.1 MapReduce Overview

MapReduce is a parallel programming technique derived from the functional programming concepts and is proposed by Google for large-scale data processing in a distributed computing environment. The authors [5] describe the MapReduce programming model as follows:

The computation task takes a set of *input* key/value pairs, and produces a set of *output* key/value pairs. The user of the MapReduce library expresses the computation task as two functions: *Map* and *Reduce*.

Map, written by the user, takes an input pair and produces a set of *intermediate* key/value pairs. The MapReduce library groups together all intermediate values associated with the same intermediate key and transforms them to the *Reduce* function.

The Reduce function, also written by the user, accepts an intermediate key and a set of values for that key. It merges together these values to form a possibly smaller set of values. Typically, just zero or one output value is produced per Reduce invocation. Fig. 1 shows the data flow and different phases of the MapReduce framework [6].

2.2 Rough Set Theory

Rough Set (RS) is a valid mathematical theory to deal with imprecise, uncertain, and vague information. Rough Set has been applied successfully in such fields as machine learning, data mining, pattern recognition, intelligent data analyzing and control algorithm acquiring, etc, since been developed by professor Z. Pawlak in 1980s [7,8].

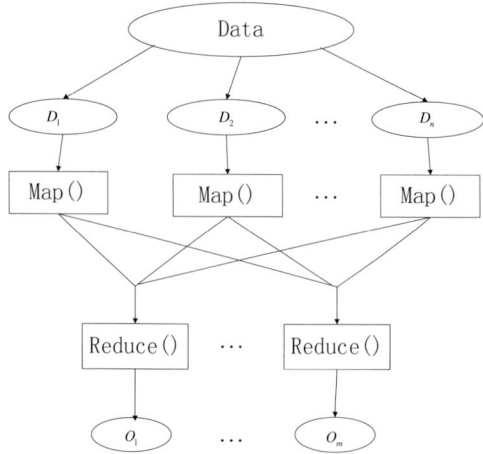

Fig. 1. The MapReduce programming model [6]

The most advantage of RS is its great ability to compute the reductions of information systems. There are a lot of research works on the attribute reduction [9,10,11]. It is very important that do some research on reduction on cloud computing platform since massive data could be shorten and knowledge of the massive data cloud be found accordingly.

3   Proposed Parallel Reduction Algorithm Based on Rough Set and MapReduce

In this section, a parallel reduction algorithm is proposed based on rough set theory and MapReduce programming mode. The proposed algorithm consists of three steps mainly. First of all, a big decision table S is divided into some smaller decision table S_i before a reduction procedure is beginning, that is to say, $S = \cup S_i$. Afterwards, reduction red_i is gotten from sub-decision tale S_i. Secondly, judge whether new attributes $AttrSet$ should be added in $\cup red_i$ and taken as the candidate reduction of S. Thirdly, delete redundant attributes and the reduction Red is gotten accordingly. The MapReduce programming mode is adopt in each step of proposed algorithm. Detailed algorithm is listed as follows.

Algorithm: Proposed parallel reduction algorithm
Input: A decision table $S = \cup S_i$
Output: reduction of S
   1. Compute reduction red_i from sub-decision tale S_i;
   2. Add new attributes $AttrSet$ in $\cup red_i$;
   3. Delete redundant attributes.
For the convenience of introduction, the three steps of the proposed algorithm are called compute, add and delete respectively. Each step of the proposed algorithm is programmed according to MapReduce and is listed as follows.

Compute-Map($key, value$)
Input:$< Null, S_i >$ is taken as $(key, value)$ pairwhere key is null, and value is $S_i =< U_i, C \cup D, V, f >$
Output:$< id, red_i >$ is taken as $(key, value)$ pair, where id can be given a same number to all the map operation for a same reduction iteration, and red_i is the reduction of S_i
   1. Delete the same samples on the sub-decision table S_i;
   2. Adopt traditional reduction algorithm and compute reduction red_i;
   3. MapOutput $< id, red_i >$.

Compute-Reduce($key, value$)
Input:$< id, red_i >$ pair from Compute-Map($key, value$)
Output:$< Null, Red1 >$ pair, where Red1 is the basis of reduction of S
   1. $Red1 = \varnothing$;
   2. for $i=1$ to n do
      $Red1 = Red1 \cup red_i$;
   3. ReduceOutput $< Null, Red1 >$.

Add-Map($key, value$)
Input:$< Null, record_i >$ pair, where $record_i$ is a sample of decision table S
Output:$< Red1\_Attr, record_i >$ pair, where $Red1\_Attr$ is the attribute of $Red1$ and $record_i$ is a sample of decision table S
   1. Extract Red1from conditional set of S and get a subset $Red1 - Attr$;
   2. MapOutput$< Red1\_Attr, record_i >$.

Add-Reduce($key, value$)
Input:$< Red1\_Attr, record\_list >$ pair, where the records forms a list $record\_list$ if they have the same values of $Red1\_Attr$
Output:$< Null, Add\_Attr >$ pair, where Add_Attr is a conditional attribute subset added into $Red1\_Attr$
   1. $Add\_Attr = \varnothing$;
   2. for each $record\_list$, find inconsistent records and their inconsistent $C_k \notin C - Red1\_Attr$;
      $Add\_Attr = Add\_Attr \cup C_k$;
   3. ReduceOutput$< Null, Add\_Attr >$.

Delete-Map($key, value$)
Input:$< Null, S_i >$ pair Output:$< id, S_i\_simplify >$ pair, where id can be given a same number to all the map operation for same reduction iteration, and $S_i\_simplify$ is a sub-table of S
   1. Delete the conditional attribute subset $C - \{Red1\_Attr \cup Add\_Attr\}$ from S;
   2. Delete the same records;
   3. MapOutput$< id, S_i\_simplify >$.

Delete-Reduce($key, value$)
Input:$< id, S_i\_simplify >$ pair, where id can be given a same number to all the map operation for a same reduction iteration
Output:$< Null, Red >$ pair, where Red is the reduction of the decision table S

1. Merge all the $S_i\_simplify$ with the same id and get a simplified decision table $S\_simplify$;

2. Adopt traditional reduction algorithm on $S\_simplify$ and compute reduction Red;

3. ReduceOutput$< Null, Red >$, where Red is taken as the final reduction of the decision table S.

4   Experiments and Discussion

In this section, a series of experiments are taken to test the efficiency of the proposed algorithm.

First of all, the algorithm is running on the Hadoop-Mapreduce platform [12], where Hadoop MapReduce is a programming model and software framework for developing applications that rapidly process massive data in parallel on large clusters of compute nodes. In the experiments, hadoop-0.20.2 version is used, A PC is taken as NameNode, master and jobTracker, and 5 PC and 10 PC are taken as DataNode, slave and taskTracker in compared experiments respectively. All the PCs have 1GB memory and 2.4Ghz CPU. A massive dataset KDDCUP99 [13] is taken as test set, which has 4,898,432 records and 41 attributes. A equal frequency discretization algorithm[14] is used before reduction.

To test the effectiveness and efficiency, the proposed algorithm is compared with three traditional rough set reduction algorithm, the reduction algorithm based conditional entropy proposed by Guoyin Wang [9], the algorithm based on mutual information proposed by Duoqin Miao [10], and the discernibility matrix reduction algorithm proposed by Skowron [11]. The algorithm proposed by Skowron [11] is used in the function of Compute-Map and Delete-Reduce in the proposed algorithm, and the decision table S is divided into S_i with 1000 records. The detailed experiment results are listed as Table 1.

From Table 1, we can find the proposed algorithm is superior to the traditional reduction algorithm for the massive dataset, when 5 or 10 nodes are taken computation nodes and less than half time is used to achieve the same reduction job for the amount of 10,000 records. Moreover, the traditional algorithm

Table 1. Compared experiment results

Data(record)	Proposed algorithm		Alg[9](s)	Alg[10](s)	Alg[11](s)
	5 slave(s)	10 slave(s)			
10,000	72	69	165	205	570
100,000	168	121	Memory overflow	66347	-
Kddcup20% (979,686)	1196	939	-	-	-
Kddcup60% (2,939,059)	3944	3301	-	-	-
Kddcup100% (4,898,432)	6120	5050	-	-	-

seems to be worse capability and can't afford the computation when the data is more than 100,000 records, meanwhile, the proposed algorithm do well for the same job. Since the traditional reduction algorithm is used as reduction steps of Compute-Map and Delete-Reduce in the proposed algorithm, and the proposed algorithm is a traditional rough set reduction algorithm merged in the MapReduce programming mode in substance. We can infer it would be more efficiency if the parallel computation model is used for the above traditional rough set reduction steps. In a words, the proposed algorithm is an effective method for attribute reduction, knowledge discovery based on rough set theory and MapReduce mode for massive dataset, it is more efficiency than traditional rough set method.

From Table 1, it is apparently that 10 nodes are superior to 5 nodes. Unfortunately, we cant afford more nodes for the compared experiments. It is intuitive that it will be more efficiency if more nodes are taken as slaves, but it should be further studied whether there is a bottleneck with the slave nodes added continually.

In order to test the impact of granularity on the algorithm, a compared experiments are taken, granularity varied from 1,000 records to 20,000 records are taken as granularity of the computation respectively, and 20% KDDCUP99 are tested and 10 nodes are taken as slaves. Experiment results are listed as Fig. 2.

From Fig. 2, we can find it is different calculation time when different granularity is adopt, and it is achieve better capability when the granularity is varied from 4,000 records to 9,000 records in the experiments. However, the proposed parallel reduction algorithm is fluctuated in the above region, and there is no clear inflexion in the performance curve. Therefore, the best granularity for the different massive dataset and the method to find the best granularity should be studied continually.

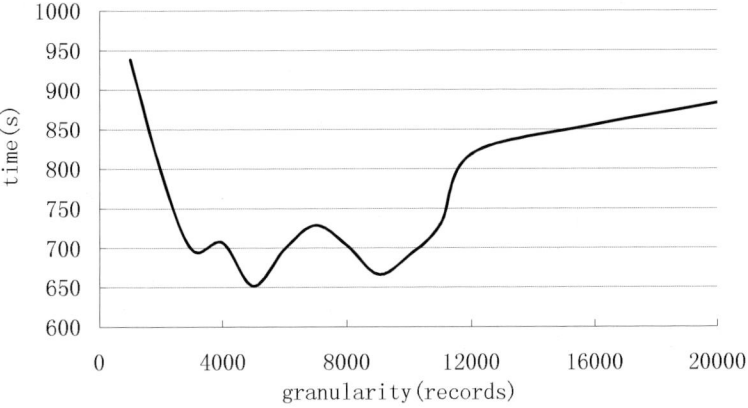

Fig. 2. Compared experiment for granularity

5   Conclusion

In this paper, attribute reduction for massive data based on rough set theory is studied. A parallel attribute reduction algorithm is proposed, which is based on the parallel program mode of MapReduce and the attribute reduction algorithm of rough set theory. Experiment results show that the proposed method is more efficiency for massive data mining than traditional method, and it is a useful method for data mining and knowledge discovery for massive data of cloud computing platform.

References

1. Hu, F., Wang, G.: Quick reduction algorithm based on attribute order. Chinese Journal of Computers 30(8), 1429–1435 (2007)
2. Sharer, J., Agrawal, R., Mehta, M.: SPRINTA Scalable Parallel Classifier for Data Mining. In: Proceedings of the 22th International Conference on Very Large Data Bases, pp. 544–555 (1996)
3. Andrew, W., Christopher, K., Kevin, D.: Parallel PSO Using MapReduce. In: Proceedings of 2007 IEEE Congress on Evolutionary Computation, pp. 7–16 (2007)
4. Abhishek, V., Xavier, L., David, E., Roy, H.: Scaling Genetic Algorithms using MapReduce. In: Proceedings of the 2009 Ninth International Conference on Intelligent Systems Design and Applications, pp. 13–18 (2009)
5. Dean, J., Ghemawat, S.: MapReduce: Simplified Data Processing on Large Clusters. Communications of the ACM 51(1), 107–113 (2008)
6. Jaliya, E., Shrideep, P., Geoffrey, F.: MapReduce for Data Intensive Scientific Analyses. In: Proceedings of Fourth IEEE International Conference on eScience, pp. 277–284 (2008)
7. Pawlak, Z.: On Rough Sets. Bulletin of the EATCS 24, 94–108 (1984)
8. Pawlak, Z.: Rough Classification. International Journal of Man-Machine Studies 20(5), 469–483 (1984)
9. Wang, G.: Rough reduction in algebra view and information view. International Journal of Intelligent System 18(6), 679–688 (2003)
10. Miao, D., Hu, G.: A Heuristic Algorithm for Reduction of Knowledge. Journal of Computer Research and Development 6, 681–684 (1999) (in Chinese)
11. Skowron, A., Rauszer, C.: The Discernibility Functions Matrics and Fanctions in Information Systems. In: Slowinski, R. (ed.) Intelligent Decision Support CHandbook of Applications and Advances of the Rough Sets Theory, pp. 331–362. Kluwer Academic Publisher, Dordrecht (1991)
12. Hadoop MapReduce, http://hadoop.apache.org/mapreduce/
13. KDDCUP99, http://kdd.ics.uci.edu/databases/kddcup99/
14. Wang, G.: Rough Set Theory and Knowledge Acquisition. Xi'an Jiaotong University Press, Xi'an (2001) (in Chinese)

Analysis of Rough and Fuzzy Clustering

Manish Joshi[1], Pawan Lingras[2], and C. Raghavendra Rao[3]

[1] Department of Computer Science, North Maharashtra University
Jalgaon, Maharashtra, India
joshmanish@gmail.com
[2] Department of Mathematics and Computing Science, Saint Mary's University
Halifax, Nova Scotia, B3H 3C3, Canada
pawan@cs.smu.ca
[3] Department of Computer and Information Sciences, University of Hyderabad
Hyderabad, Andhra Pradesh, India
crrcs@uohyd.ernet.in

Abstract. With the gaining popularity of rough clustering, soft computing research community is studying relationships between rough and fuzzy clustering as well as their relative advantages. Both rough and fuzzy clustering are less restrictive than conventional clustering. Fuzzy clustering memberships are more descriptive than rough clustering. In some cases, descriptive fuzzy clustering may be advantageous, while in other cases it may lead to information overload. This paper provides an experimental comparison of both the clustering techniques and describes a procedure for conversion from fuzzy membership clustering to rough clustering. However, such a conversion is not always necessary, especially if one only needs lower and upper approximations. Experiments also show that descriptive fuzzy clustering may not always (particularly for high dimensional objects) produce results that are as accurate as direct application of rough clustering. We present analysis of the results from both the techniques.

Keywords: Rough Clustering, Fuzzy Clustering, Rough K-means, Fuzzy C-means, FuzzyRough Correlation Factors, Cluster Quality.

1 Introduction

The conventional crisp clustering techniques group objects into separate clusters. Each object is assigned to only one cluster. The term crisp clustering refers to the fact that the cluster boundaries are strictly defined and object's cluster membership is unambiguous.

In addition to clearly identifiable groups of objects, it is possible that a data set may consist of several objects that lie on fringes. The conventional clustering techniques mandate that such objects belong to precisely one cluster. In practice, an object may display characteristics of different clusters. In such cases, an object should belong to more than one cluster, and as a result, cluster boundaries necessarily overlap.

J. Yu et al. (Eds.): RSKT 2010, LNAI 6401, pp. 679–686, 2010.

The conventional clustering algorithm like K-means categorizes an object into precisely one cluster. Whereas, fuzzy clustering [2,14] and rough set clustering [6,9,15,16] provide ability to specify the membership of an object to multiple clusters, which can be useful in real world applications.

Fuzzy set representation of clusters, using algorithms such as fuzzy C-means (FCM), make it possible for an object to belongs to multiple clusters with a degree of membership between 0 and 1 [14].

Rough K-means algorithm (RKM) groups objects into lower and upper region of clusters and an object can belongs to an upper region of multiple clusters.

In this paper we present analysis of results obtained using the RKM and the FCM on different data sets including two standard data sets. An important finding is the identification of FuzzyRough correlation factors that determine appropriate conversion of fuzzy clusters into rough clusters. We also compared results for the standard Letter Recognition data set and the Iris data set to analyze the accuracy of both the techniques. While the comparisons are used for deciding the classes of soft clusters, they lay foundations for a more elaborate cost benefit analysis of the decisions [12].

Due to the space limitations the conceptual and algorithmic details of rough and fuzzy clustering can be obtained from [7]. The remaining paper is organized as follows. Section 2 provides information about data sets used for experiments followed by the experimental details. The findings are revealed in section 3 and results of FCM and RKM are compared to analyze accuracy in section 4. Section 5 presents conclusions.

2   Experiments

We used four different data sets for experiments. For each data set we followed a specific procedure in order to identify a correlation among the FCM results and the RKM results. Information of all the four data sets followed by experimental steps applied on these data sets is given in following subsections.

2.1   Synthetic Data Set

We used the synthetic data set developed by Lingras et al.[12] to experiment with the RKM and the FCM algorithms. Sixty objects from a total of 65 objects are distributed over three distinct clusters. However, five objects do not belong to any particular cluster. Since there are three distinct clusters in the data set, the input parameter k for both the algorithms is set to three.

2.2   Library Data Set

We used the data obtained from a public library to analyze the results of both the algorithms. The data consist of members' books borrowed information. The objective is to group members with similar reading habits. The data of 1895 members shows propensity of a member to borrow a book of a certain category. The data is normalized in the range of 0 to 1 to reduce the effect of outliers.

2.3   Letter Recognition Data Set

We used the Letter Recognition [1] data from the University of California Irvine machine learning repository. The data set contains 20,000 events representing character images of 26 capital letters in the English alphabet based on 20 different fonts. Each event is represented using 16 primitive numerical attributes (statistical moments and edge counts) that are scaled to fit into a range of integer values from 0 through 15.

We tailored the data set to consist eight distinctly different characters A, H, L, M, O, P, S, and Z. This data set consists of 6114 events.

2.4   Iris Data Set

This standard data set [1] also obtained from the University of California Irvine machine learning repository. The data set contains three classes of fifty instances each, where each class refers to a type of an Iris plant. One class is linearly separable from the other two; the latter are not linearly separable from each other. Each instance of an Iris plant is represented by four different attributes. So in all there are 150 objects and each object is four-dimensional. A few objects lie on boundary of two classes.

2.5   Experimental Steps

This subsection outlines procedural steps for obtaining an equivalent rough clustering result from a fuzzy clustering output. The FCM and the RKM algorithms are applied to above mentioned four data sets. The RKM algorithm is implemented in Java whereas standard FCM function of the MATLAB software is used to get results for fuzzy clustering. For both the algorithms the termination criterion (δ) and the *iter* values are set to 0.00001 and 100, respectively [7].

The preparatory steps to determine equivalence between rough and fuzzy clustering are as follow:

1. Apply the FCM and the RKM algorithms to selected data set.
2. The fuzzy membership matrix U changes for different values of m (a fuzzification factor).
3. For each data set the FCM results for eight different values of m ($m = 3, m = 2, m = 1.9, ...$) are attained. Hence, for each data set eight different U matrices are available.
4. In order to determine how a fuzzy U matrix can be transformed to an equivalent rough clustering result; the rough clustering result is also represented in a matrix form. We formulated a matrix $Rough$ with n rows and k columns and its values are set as follows:

$$Rough[i][j] = \begin{cases} 1, \text{ if object i belongs to cluster j} \\ 0, \text{ if object i does not belong to cluster j} \end{cases} \tag{1}$$

The $Rough$ matrix we are mentioning is just another way of representation of the rough clustering results we obtained from the RKM. This $Rough$ matrix preserves the rough set property by having only one '1' value in a row, if a corresponding object belongs to a lower region of a cluster and more than one '1' values in a row if that object belongs to an upper region of more than one clusters.

5. The U and the $Rough$ matrices are analyzed to understand the correlation between the FCM results and the RKM results.

6. The observations for different data sets are analyzed to deduce how fuzzy clustering can be converged in to rough clustering.

Results of our comparisons are presented in the next section.

3   Results and Discussions

Fuzzy clustering generates a descriptive fuzzy coefficient matrix U, whereas rough clustering result forms a $Rough$ matrix. We found that fuzzy coefficient values produced by FCM in a matrix U alone could not generate an equivalent rough clustering matrix $Rough$. That is, one can not bifurcate an object into lower or upper region of a cluster just by examining a fuzzy coefficient value u_{ij}.

We identified '$FuzzyRough$ correlation factors' that determine whether an object with a certain fuzzy coefficient value (say 0.44) belongs to a lower region of a cluster or belongs to upper regions of multiple clusters. With the help of these 'FuzzyRough correlation factors' a Fuzzy U matrix can be converted into the $fuzzyRough$ matrix, that is equivalent to the $Rough$ matrix using following steps :

1. $ratio_{ij} = \frac{max(\boldsymbol{u_i})}{u_{ij}}, \forall i = 1$ to n and $\forall j = 1$ to k . The $max(\boldsymbol{u_i})$ is the maximum of u_{ij} over j i.e. the maximum fuzzy coefficient value for an ith object. For an object if the fuzzy coefficients are (0.5, 0.4, 0.1) for 3 clusters then the corresponding ratios are (1(05/0.5), 1.25(0.5/0.4), 5(0.5/0.1)), respectively.

2. If $ratio_{ij} < FuzzyRoughCorrelationFactor$ then $fuzzyRough_{ij} = 1$ else $fuzzyRough_{ij} = 0$.
   This means that if the ratios from the first step are less than a specified 'FuzzyRough correlation factor' then object i belongs to the upper approximation of cluster j. Note that if $fuzzyRough_{ij}$ is equal to '1' for only one cluster j, then object i belongs to the lower approximation of the cluster j as well as the upper approximation.

3. Now the newly formed matrix $fuzzyRough$ contains only 0 or 1 values and is equivalent to $Rough$ matrix that represents rough clustering output.

We obtained the 'FuzzyRough correlation factors' for different values of m and the Table 1 lists all the FuzzyRough correlation factors for Synthetic, Library and Iris data sets.

Though RKM produces sensible results for the Letter Recognition data set, FCM terminates by assigning almost every object same fuzzy coefficients. The

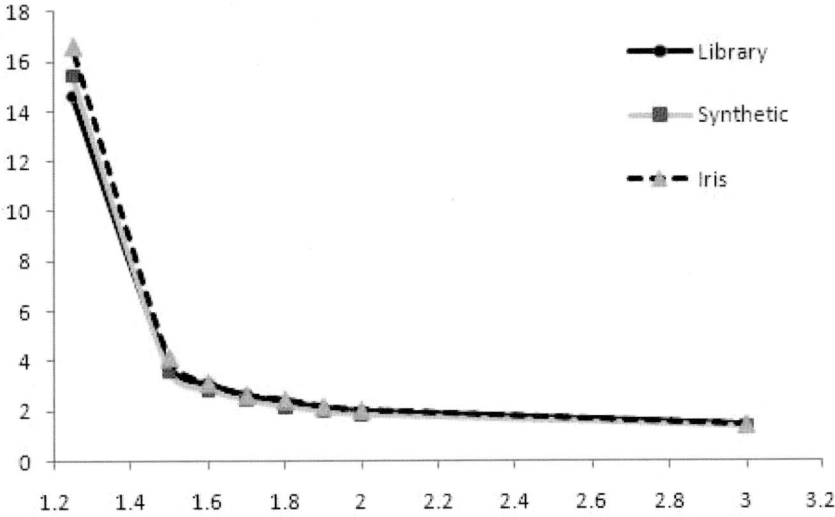

Fig. 1. FuzzyRough Correlation Factors Vs Fuzzification Parameter m

FCM (for m = 2)groups every object of the Letter Recognition data set to each of the eight clusters with the same membership of 0.1250 (1/8). Hence, it was not possible to obtain an appropriate value of FuzzyRough correlation factors for the Letter Recognition data set.

The values of 'FuzzyRough Correlation Factors' that determine whether an object belongs to a lower region of a cluster or upper regions of multiple clusters seem to be more or less the same for different data sets. However, these factors vary for different values of fuzzification parameter m used in the FCM (see Fig. 1).

We can see using a graph in Fig. 1 that the Fuzzy Rough correlation factors are inversely proportional to m. Further study may reveal the effect of number of objects and their distribution in data set on the FuzzyRough correlation factors.

Table 1. FuzzyRough Correlation Factors for different Data Sets

m Value in FCM	Synthetic	Library	Iris
3	1.49	1.401	1.43
2	2.26	1.96	2.04
1.9	2.48	2.112	2.18
1.8	2.78	2.32	2.45
1.7	3.23	2.62	2.63
1.6	3.94	3.07	3.13
1.5	5.21	3.85	4.13
1.25	15.43	15.1	16.58

4   Uncertainty Handling

We apply both the clustering techniques to the standard data sets in order to check how each technique handles uncertainty while grouping objects. These data sets consists of well defined priorly known classes. The Iris data set has three classes whereas the Letter Recognition data set has eight classes.

We calculated the average precision obtained by both the techniques. For the Letter Recognition data set, each cluster is labeled using the character that appears most frequently in the cluster. Whereas for the Iris data set each cluster is labeled with the type of Iris plant that appears most frequently in the cluster.

'Frequency' indicates the number of events in the cluster that match the cluster label. For both the data sets an Average Precision is defined as the sum of the 'Frequency' for all clusters divided by the total number of events in the data set.

Table 2 and Table 3 show the results obtained when both clustering algorithms are applied to the Letter Recognition and the Iris data sets, respectively.

The FCM Frequency specifies the numbers of objects that match with a corresponding cluster label. In case of RKM, the Frequency value represents the total number of objects from lower and upper approximations of a cluster that match with a corresponding cluster label. The matching number of lower approximation objects are explicitly mentioned in the parenthesis.

The average precision for the FCM (Table 2) is around 40% that is more than 60% objects are not grouped correctly. Eventhough, for an object if its cluster membership is uncertain, the FCM's fuzzy coefficient measure is not adequate enough to indicate such uncertainty significantly. This is the main reason for such a poor clustering performance.

Whereas for RKM the uncertainty of an object's cluster membership is discerned (identified) and represented with the help of boundary regions of distinct clusters. Hence we can see that more numbers of objects are matching with the label of clusters in RKM. The Average Precision (Average Precision-2) for RKM is more than 75%. If we would have considered only lower approximation objects of the clusters then the average precision (Average Precision-1) would have

Table 2. Average precision for the Letter Recognition Data Set

FCM			RKM			
Cluster Label	Frequency	Average Precision	Cluster Label	Frequency	Average Precision-1	Average Precision-2
P	21		M	496(174)		
Z	324		H	191(189)		
-	0		L	509(338)		
S	437	39.33	M	526(181)	53.04	75.02
A	667		O	710(594)		
-	0		A	709(642)		
L	390		Z	722(552)		
O	566		P	724(573)		

Table 3. Average precision for Iris Data Set

FCM			RKM			
Cluster Label	Frequency	Average Precision	Cluster Label	Frequency	Average Precision-1	Average Precision-2
Setosa	50		Setosa	50(50)		
Versicolour	48	89.33	Versicolour	49(41)	84	95.33
Virginica	36		Virginica	44(35)		

been only 53% (3243 objects that belong to appropriate clusters with certainty is divided by the total number of objects 6114).

We can also observe that (Table 2) two clusters do not have a single object in the results when the FCM is applied on the Letter Recognition Data set. It means the FCM is not able to group two letters (H and M) prominently in any cluster. The RKM, on the other hand, could not identify the letter 'S' prominently in any cluster.

Table 3 shows results when both the algorithms are applied to the Iris data set. It also shows how proper usage of lower and upper approximations improves RKM precision from 84% to 95%, which is better than the FCM precision of 89%.

We can conclude that for higher dimensional data sets FCM is not able to generate meaningful results whereas RKM not only generated results but showed promise in identifying objects that lie on fringes. Hence the average precision for the RKM is better than the FCM.

5 Conclusions

There is a growing interest in the applications of rough set theory to various areas of machine intelligence. With the help of lower and upper approximations, rough clustering overcomes a limitation of crisp clustering by allowing objects to be members of multiple clusters. Whereas fuzzy clustering does the same by specifying fuzzy coefficients that represent degree of membership to clusters.

Our experiments with rough and fuzzy clustering for four different data sets revealed that for data sets with modest number of dimensions a fuzzy clustering output can be transformed to an equivalent rough clustering output. We proposed an appropriate conversion procedure based on empirically derived 'Fuzzy-Rough correlation factors'. We also presented a comparative analysis of how uncertainty is handled by both the techniques. The comparison included a simple decision theoretic framework for determining classes of soft clusters. The comparison described here can be further extended to include a more elaborate cost benefit analysis of the decisions such as the one described in [12].

References

1. Asuncion, A., Newman, D.J.: UCI Machine Learning Repository. University of California, School of Information and Computer Science, Irvine, CA (2007), http://www.ics.uci.edu/~mlearn/MLRepository.html
2. Bezdek, J.C., Hathaway, R.J.: Optimization of Fuzzy Clustering Criteria using Genetic Algorithms (1994)
3. Bezdek, J.C.: Pattern Recognition with Fuzzy Objective Function Algoritms. Plenum Press, New York (1981)
4. Dunn, J.C.: A Fuzzy Relative of the ISODATA Process and Its Use in Detecting Compact Well-Separated Clusters. Journal of Cybernetics 3, 32–57 (1973)
5. Hartigan, J.A., Wong, M.A.: Algorithm AS136: A K-Means Clustering Algorithm. Applied Statistics 28, 100–108 (1979)
6. Ho, T.B., Nguyen, N.B.: Nonhierarchical Document Clustering by a Tolerance Rough Set Model. International Journal of Intelligent Systems 17, 199–212 (2002)
7. Joshi, M., Lingras, P.: Evolutionary and Iterative Crisp and Rough Clustering I: Theory. In: Chaudhury, S., Mitra, S., Murthy, C.A., Sastry, P.S., Pal, S.K. (eds.) PReMI 2009. LNCS, vol. 5909, pp. 615–620. Springer, Heidelberg (2009)
8. Joshi, M., Lingras, P.: Evolutionary and Iterative Crisp and Rough Clustering II: Experiments. In: Chaudhury, S., Mitra, S., Murthy, C.A., Sastry, P.S., Pal, S.K. (eds.) PReMI 2009. LNCS, vol. 5909, pp. 621–627. Springer, Heidelberg (2009)
9. Lingras, P., West, C.: Interval Set Clustering of Web Users with Rough K-Means. Journal of Intelligent Information Systems 23, 5–16 (2004)
10. Lingras, P., Hogo, M., Snorek, M.: Interval Set Clustering of Web Users using Modified Kohonen Self-Organizing Maps based on the Properties of Rough Sets. Web Intelligence and Agent Systems: An International Journal 2(3), 217–230 (2004)
11. Lingras, P.: Precision of rough set clustering. In: Chan, C.-C., Grzymala-Busse, J.W., Ziarko, W.P. (eds.) RSCTC 2008. LNCS (LNAI), vol. 5306, pp. 369–378. Springer, Heidelberg (2008)
12. Lingras, P., Chen, M., Miao, D.: Rough Multi-category Decision Theoretic Framework. In: Wang, G., Li, T., Grzymala-Busse, J.W., Miao, D., Skowron, A., Yao, Y. (eds.) RSKT 2008. LNCS (LNAI), vol. 5009, pp. 676–683. Springer, Heidelberg (2008)
13. Lingras, P.: Evolutionary rough K-means Algorithm. In: Wen, P., Li, Y., Polkowski, L., Yao, Y., Tsumoto, S., Wang, G. (eds.) RSKT 2009. LNCS, vol. 5589, pp. 68–75. Springer, Heidelberg (2009)
14. Pedrycz, W., Waletzky, J.: Fuzzy Clustering with Partial Supervision. IEEE Trans. on Systems, Man and Cybernetics 27(5), 787–795 (1997)
15. Peters, G.: Some Refinements of Rough k-Means. Pattern Recognition 39, 1481–1491 (2006)
16. Peters, J.F., Skowron, A., Suraj, Z., et al.: Clustering: A Rough Set Approach to Constructing Information Granules. Soft Computing and Distributed Processing, 57–61 (2002)
17. MacQueen, J.: Some Methods for Classification and Analysis of Multivariate Observations. In: Proceedings of Fifth Berkeley Symposium on Mathematical Statistics and Probability, vol. 1, pp. 281–297 (1967)

Autonomous Knowledge-Oriented Clustering Using Decision-Theoretic Rough Set Theory

Hong Yu[1], Shuangshuang Chu[1], and Dachun Yang[2]

[1] Institute of Computer Science and Technology, Chongqing University of Posts and Telecommunications, Chongqing, 400065, P.R. China
[2] Chongqing R&D Institute of ZTE Corp., Chongqing, 400060, P.R. China
yuhong@cqupt.edu.cn

Abstract. This paper processes an autonomous knowledge-oriented clustering method based on the decision-theoretic rough set theory model. In order to get the initial clustering of knowledge-oriented clusterings, the threshold values are produced autonomously in view of physics theory in this paper rather than are subjected by human intervention. Furthermore, this paper proposes a cluster validity index based on the decision-theoretic rough set theory model by considering various loss functions. Experiments with synthetic and standard data show that the novel method is not only helpful to select a termination point of the clustering algorithm, but also is useful to cluster the overlapped boundaries which is common in many data mining applications.

Keywords: clustering, knowledge-oriented clustering,decision-theoretic rough set theory, autonomous, data mining.

1   Introduction

In recent years, clustering has been widely used as a powerful tool to reveal underlying patterns in many areas such as data mining, web mining, geographical data processing, medicine, classification of statistical findings in social studies and so on. Clustering categorizes unlabeled objects into several clusters such that the objects belonging to the same cluster are more similar than those belonging to different clusters.

Some researchers have produced excellent results in the analysis of data with partition clustering techniques and hierarchical techniques [7,9,10,11,14]. Although structurally varied, partition algorithms and hierarchical algorithms share the common property of relying on local data properties to reach an optimal clustering solution, which carries the risk of producing a distorted view of the data structure. Rough set theory introduced by Pawlak [8] moves away from this local dependence and focuses on the idea of using global data properties to establish similarity between the objects in the form of coarse and representative patterns. To combine use of hierarchical clustering and rough set theory, Hirano and Tsumoto [5] present a knowledge-oriented (K-O) clustering method, in which a modification procedure allows clusters to be formed using both local and global properties of the data.

J. Yu et al. (Eds.): RSKT 2010, LNAI 6401, pp. 687–694, 2010.

As we have known, the efficiency and optimality of knowledge-oriented cluster-ing algorithms are dependent on the selection of individual threshold parameters. There are some predetermined thresholds needed to set in the clustering tech-niques discussed above. Therefore, some adaptive algorithms are studied [1,2,3]. For example, Bean and Kambhampati in literature [2,3] use the center of gravity (CoG) to select autonomously the initial threshold values which are on the role of partitioning the objects into two classes. A new method, sort subtractions(SS), will be proposed in this paper in a physical sense, where there are no manual coefficient used any more .

Quality of clustering is an important issue in application of clustering tech-niques to real-world data. A good measure of cluster quality will help in deciding various parameters used in clustering algorithms. For instance, it is common to most clustering algorithms that is the number of clusters. There are many dif-ferent indices of cluster validity [4]. Decision-theoretic rough set (DTRS) model introduced by Yao has been helpful in providing a better understanding of clas-sification [13]. The decision theoretic considers various classes of loss functions. By adjusting loss functions, it is possible to construct a cluster validity index. Lingras, Chen and Miao [6] proposes a decision-theoretic measure of the clusters.

Therefore, this paper describes how to develop a knowledge-oriented cluster validity index from the decision-theoretic rough set model. Based on the DTRS, the proposed cluster validity index is taken as a function of risk for a cluster. Experiments with synthetic and standard data show that the novel method is not only helpful to select a termination point of the clustering algorithm, but also is useful to cluster the overlapped boundaries which is common in many data mining applications.

The structure of the paper is organized as follows. First, we introduce the generic knowledge-oriented clustering framework, and our autonomous initial threshold selecting method, the sort subtractions. Then, a cluster validity index based on DTRS will be proposed, and some experiments with synthetic and standard data will be described in Section 4. Some conclusions and discusses will be given in Section 5.

2   Autonomous Knowledge-Oriented Clustering

2.1   Knowledge-Oriented Clustering Framework

We are interested in an information system, which is a four tuple: $I = (U, A, V, f)$, where $U = \{x_1, x_2, \ldots, x_n\}$ is a finite non-empty universe of objects, $A = \{a_1, a_2, \ldots, a_m\}$ is a finite non-empty set of features, $V = \{V_1, \cdots, V_m\}$ and $f(x_i, a_k) \in V_k$. The indiscernibility relation of an information system is $R = \{(x, y) \in U \times U | \forall a \in A, f(x, a) = f(y, a)\}$, which is an equivalence relation.

In using a simple algorithmic framework, K-O clustering is computationally efficient. Furthermore, the combined use of tools from hierarchical clusterings and rough set theory allows clusters to be formed using both local and global properties of the data.

The K-O clustering algorithm includes the following steps[3,5]:

Step 1. Construct a matrix of similarities $\mathbb{S} = \{sim(x_i, x_j)\}$ between all pairs of objects.

Step 2. Assign an initial indiscernibility relation to each object in the universe. Pool information to obtain an initial clustering U/R.

Step 3. Construct an indiscernibility matrix $\eta = \{\eta(x_i, x_j)\}$ to assess the clustering U/R.

Step 4. Modify clustering to gain a modified clustering U/R_{mod}.

Step 5. Repeat Steps 3 and 4 until a stable clustering is obtained.

An initial equivalence relation R_i for an object $x_i \in U$ is defined by $R_i = \{\{P_i\}, \{U - P_i\}\}$, where $P_i = \{x_j | sim(x_i, x_j) \geq Th_i\}$.

The $sim(x_i, x_j)$ denotes similarity between objects x_i and x_j, and the Th_i denotes a threshold value of similarity for the object x_i. The choice of an appropriate similarity measure is reliant on a number of factors including the size and application of the data as well as statistical properties and it must be chosen accordingly to the conditions of the given clustering problem. Here the Euclidean distance is used simply in the later experiments as following:

$$sim(x_i, x_j) = 1 - \frac{\sqrt{\sum_{p=1}^{m} (f(x_i, a_p) - f(x_j, a_p))^2}}{max_{i,j} \sqrt{\sum_{p=1}^{m} (f(x_i, a_p) - f(x_j, a_p))^2}} \tag{1}$$

where, $0 \leq sim(x_i, x_j) \leq 1$.

2.2 Autonomous Initial Threshold Selecting by Sort Subtractions

The initial clustering of object set is the most important step in a sense for the K-O algorithms. If without the correct initial clustering, the next clustering will not correctly show the true structure of the data, and lead to a meaningless result. Because of the initial clustering is decided by the initial threshold Th_i, so how to select Th_i is the main part of the algorithm.

References [2,3] use the CoG method to select Th_i. The CoG method imitates the means of getting the center of gravity in a physical sense. An initial threshold Th_i corresponding to the object x_i may be obtained by selecting the similarity value $sim(x_i, x_j)$, $j = \{1, 2, \cdots, n\}$, which minimizes the following sum of differences:

$$|\sum_{j=1}^{n} (sim(x_i, x_j) - \omega sim(x_i, x_k))| \quad i, k = 1, 2, \cdots, n \tag{2}$$

where ω is set by the user. This procedure produces a set of initial threshold values corresponding to each object in the universe from which the initial partitions may be obtained. This information is then pooled to obtain the initial clustering of the universe U/R.

If we chose the optimal ω, the number of iteration of the K-O algorithm is small and the result is precise and the time complexity is $O(n^2)$ [2]. Oppositely when chose an improper ω, the number of iteration would grow and the accuracy would reduce. It can be seen that the ω will influence the efficiency and accuracy of the algorithm. Since the ω is decided by the user, the CoG method is also influenced by the user, and the CoG method is not totally impersonal and autonomous.

Therefore, a new method, sort subtractions(SS), will be proposed to select the initial threshold Th_i in this section. Firstly, we sort the element of the every row in the similarity matrix $\mathbb{S} = sim(x_i, x_j)$, and we denote the sorted result as $\mathbb{S}' = sim(x_i, x_j)$. Then we get the Th_i as:

$$Th_i = \{sim(x_i, x_{j+1}) | secondmax(sim(x_i, x_j) - sim(x_i, x_{j+1}))\} \qquad (3)$$

where $j = 1, \cdots, n-1$. That is, the similarities do subtraction gradually. If the threshold is the max subtraction, that means the initial equivalence class of the object x_i are composed of the most similarity object to object x_i according to the physical sense [3]; in other words, the clustering scheme has the smallest granularity.

In order to converge quickly, the second max values is used here; which is proven to be better in most cases by our experiments.

3   Cluster Quality Index Based on DTRS

3.1   Decision-Throretic Rough Set Model

Yao proposes the decision-throetic rough set model [13] which applies the Bayesian decision procedure for the construction of probabilistic approximations. The classification of objects according to approximation operators in rough set theory can be easily fitted into the Bayesian decision-theoretic framework.

Let $\Omega = \{A, A^c\}$ denote the set of states indicating that an object is in A and not in A, respectively. Let $A = \{a_1, a_2, a_3\}$ be the set of actions, where a_1, a_2, and a_3 represent the three actions in classifying an object, deciding $POS(A)$, deciding $NEG(A)$ and deciding $BND(A)$, respectively. Let $\lambda(a_i | x \in A)$ and $\lambda(a_i | x \in A^c)$ denote the loss (cost) for taking the action a_i when the state is A, A^c, respectively. For an object with description x, suppose an action a_i is taken.

The expected loss $R(a_i | [x])$ associated with taking the individual actions can be expressed as: $R(a_1 | [x]) = \lambda_{11} P(A | [x]) + \lambda_{12} P(A^c | [x]), R(a_2 | [x]) = \lambda_{21} P(A | [x]) + \lambda_{22} P(A^c | [x]), R(a_3 | [x]) = \lambda_{31} P(A | [x]) + \lambda_{32} P(A^c | [x])$, where the $\lambda_{i1} = \lambda(a_i | A)$, $\lambda_{i2} = \lambda(a_i | A^c)$, and $i = 1, 2, 3$. And the probabilities $P(A | [x])$ and $P(A^c | [x])$ are the probabilities that an object in the equivalence class $[x]$ belongs to A and A^c, respectively.

3.2   Estimate Risk of Clustering Scheme

To define our framework, we will assume existence of a hypothetical clustering scheme, CS, that partition a set of objects $U = \{x_1, \cdots, x_n\}$ into clusters $CS = \{c_1, \cdots, c_K\}$, where $c_k \subseteq U$. We define the similarity between the object and the cluster as follows.

Definition 1. *Let $x_j \in c_k$, the similarity between the x_i and the c_k:*

$$SimObjClu(x_i, c_k) = \sum_{j=1}^{|c_k|} sim(x_i, x_j)/|c_k| \qquad (4)$$

Consider a clustering scheme CS, let $b_k(CS, x_i)$ be the action that assigns the object x_i to a cluster c_k. The risk associated with the assignment will be denoted by $Risk(c_k, x_i) = Risk(b_k(CS, x_i))$. $Risk(c_k, x_i)$ is obtained assuming that the $P(c_k|x_i)$ is proportional to the similarity between the x_i and the c_k.

According to the DTRS in Section 3.1, when considering the action a_1, classifying an object deciding $POS(c_k)$, we have: $Risk(c_k, x_i) = \lambda_{11} P(c_k|x_i) + \lambda_{12} P(c_k^c|x_i)$. Therefore, the risk of a cluster c_k considering the action a_1 can be given as follows.

Definition 2. *The risk for the cluster c_k is:*

$$Risk(c_k) = \sum_i Risk(c_k, x_i) \qquad (5)$$

Obviously, the smaller the value of the cluster risk, the better the cluster [6]. Consider a special kind of loss functions here. Let $c_k \in CS$, it is reasonable that the loss of classifying an object x belonging to the set c_k into the positive region $POS(c_k)$ is 0; that is to say, there is no loss when x is the positive object. On the contrary, the maximal loss of classifying an object x belonging to c_k into the negative region $NEG(c_k)$ is 1. The loss of classifying an object into boundary region is between 0 and 1. Therefore, we have the following:

$$\lambda_{11} = 0, \lambda_{12} = 1$$
$$\lambda_{21} = 1, \lambda_{22} = 0 \qquad (6)$$
$$0 \le \lambda_{31} < 1, 0 \le \lambda_{32} < 1$$

Based on the hypothesis, each object x_i can be classified into only one region if we know the probability $P(c_k|x_i)$ and the loss function λ_{31} and λ_{32}.

Let $P(c_k^c|x_i)$ denote the probability of assigning an object x_i not being in the c_k into the c_k; in other words, $x_i \notin c_k$. The $P(c_k^c|x_i)$ is computed by the following equation:

$$P(c_k^c|x_i) = \frac{SimObjClu(x_i, c_k)}{\sum_i SimObjClu(x_i, c_k)} \qquad (7)$$

Then, according to above definitions, we have the risk of a cluster as following:

$$Risk(c_k) = \sum_{i=1}^{n} (\lambda_{11} P(c_k|x_i) + \lambda_{12} P(c_k^c|x_i)) = \sum_{i=1}^{n} P(c_k^c|x_i) \qquad (8)$$

Thus, the risk of a clustering scheme is:

$$Risk(CS) = \sum_{k=1}^{K} Risk(c_k) \qquad (9)$$

Table 1. The Risk of Assign the Objects to the Distinct Clusters

Distinct cluster	cluster1	cluster2	cluster3	cluster4
Risk(c_i)	0.532196	0.525859	0.545161	0.525692
Risk($c_i \cup \{x_{101}\}$)	0.525464	0.523025	0.546453	0.531201
Risk($c_i \cup \{x_{102}\}$)	0.525351	0.529737	0.540172	0.531023
Risk($c_i \cup \{x_{103}\}$)	0.532142	0.524472	0.544452	0.525230
Risk($c_i \cup \{x_{104}\}$)	0.534420	0.517509	0.549473	0.524371
Risk($c_i \cup \{x_{105}\}$)	0.535888	0.529969	**0.539448**	**0.520801**

4   Experiments

The first experiment is on a synthetic data set to show that the new method
can cluster the overlapped boundaries which is common in many data mining
applications. Fig.1 pictures the data set. The part (a) describes a clustering
scheme obtained according to the method in Section 2. It is obvious that there
are four distinct clusters. However, five objects do not belong to any particular
cluster, which are numbered from 101 to 105. We will perform the clustering on
the data for different numbers of clusters.

Table 1 shows that the new risk of the clustering after assigning the five objects
to the different distinct clusters respectively, which is calculated according to Eq.
(8). If the risk decreases, we assign the object to the corresponding cluster. For
example, when the object x_{105} is assigned to the cluster3 or cluster4, the risk
decreases, which means the object should belong to both of the clusters. On the
other hand, it is obvious that the x_{105} is not belong to the cluster3 or cluster4
from Fig1., which embodies the increase of risks in Table 1. We can see the
final clustering results in the part (b). Obviously, it looks more reasonable than
assigning an object to one precise cluster.

In order to show that the new method is valuable, we have done another
experiment on some standard data set from UCI repository [12]. The results of
the experiments are shown in Table 2, where the classical FCM algorithm is
performed as well. For the letter data set, which has 26 classes, we choose some
classes of the set denoted as the data set L1, L2 and so on. For example, the set
L4 includes the objects whose class are letter A, I, L, M, W, Y, O or E.

The framework of the new method programmed here is based on the K-O
clustering algorithm. For Step 1 and Step 2, the initial clustering, are computed
by the method introduced in Section 2. For Step 3 and Step 4, the assessment
of a clustering is performed as the above experiment on the synthetic data set.
Actually, after Step 3 and Step 4, the number of clusters are decrease. Here we
simple repeat until the number of clusters is 1. Then we can observe the change
of risks in every iteration, which is calculated according to Eq.(9). The sudden
change can show the appropriate number of clusters, which is very helpful to
decide the termination point of the algorithm.

In Table 2, "DS" is the abbreviation of "data sets". $|U|$, $|A|$ and $|Clusters|$
are the numbers of objects , attributes and clusters in the data set, respec-
tively. $|Itera|$ means the number of iterations of the program execution, and the

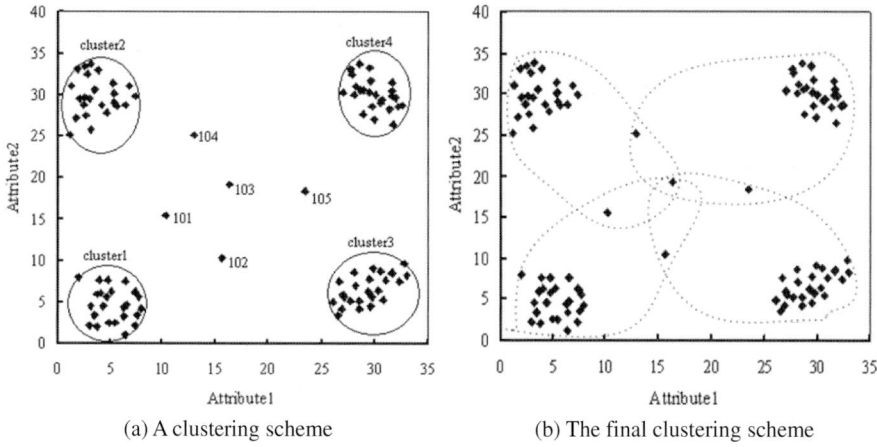

(a) A clustering scheme

(b) The final clustering scheme

Fig. 1. Cluster Results on a Synthetic Data Set

Table 2. Comparison of the CPU time and Results of the Algorithms

DS	$|\mathbf{U}|$	$|A|$	$clusters$	the New Method				Algorithm FCM	
				Accuracy%	$|Itera|,|ConItera|$	CPU(s)	CPU(s)	Accuracy%
iris	150	4	3	89.0	50,48	0.313	0.125	89.3
wine	178	13	3	68.0	18,16	0.202	0.375	68.5
L1	680	16	2	100	277,276	37.86	0.422	97.9
L2	580	16	3	98.2	239,237	23.87	2.156	97.2
L3	101	16	6	99.9	22,17	0.078	0.172	81.2
L4	116	16	7	98.5	29,23	0.141	0.250	81.9
L5	131	16	8	91.1	47,40	0.252	0.562	69.5

$|ConItera|$ means the number of iterations when a good clustering scheme is obtained. The CPU runtime is for all the iterations. Observing the much huger data set, the L1 or L2, because the new method is based on iterations, the time spending is much higher. Thus, how to improve the time efficiency is the further work. From Table 2, we can see that the CPU runtime and the accuracy rate here are reasonable, which shows that the new method is valuable.

5   Conclusion

In some applications, an object should belong to more than one cluster, and as a result, cluster boundaries necessarily to overlap. Besides, many of the clustering methods need to set threshold values by human intervention. Therefore, we focus on to study the autonomous clustering method based on the knowledge-oriented framework, which combines the property of the rough set theory. In order to get the initial clustering, autonomous threshold values are produced in view of physics theory. In addition, this paper proposes a cluster validity index based on the decision-theoretic rough set theory model by considering various loss functions. Experiments show that the novel method is not only helpful to select a termination point of the number of the clusters, but also is useful to cluster the overlapped boundaries. However, to choose the appropriate number of clusters is still based on observing the change of risks, it remains a further work how to formal the change of risks in order to terminate the algorithm autonomously.

Acknowledgments

This work was supported in part by the China NNSF grant(No.60773113), the NSF of Chongqing of China grant (No.2009BB2082 and KJ080510).

References

1. Asharaf, S., Murty, M.N.: An adaptive rough fuzzy single pass algorithm for clustering large data sets. Pattern Recognition (36), 3015–3018 (2003)
2. Bean, C., Kambhampati, C.: Knowledge-Based Clustering: A Semi-Autonomous Algorithm Using Local and Global Data Properties. IEEE International Joint Conference on Neural Networks 11(3), 95–100 (2004)
3. Bean, C., Kambhampati, C.: Autonomous clustering Using Rough Set Theory. International Journal of Automation and Computin 05(1), 90–102 (2008)
4. Halkidi, M., Batistakis, Y., Vazirgianni, M.: Clustering Validity Checking Methods: Part II. ACM SIGMOD Conf. Record. 31(3), 19–27 (2002)
5. Hirano, S., Tsumoto, S.: A Knowledge-oriented Clustering Technique Based on Rough Sets. In: Proceedings of 25th IEEE International Conference on Computer and Software Applications, Chicago, USA, pp. 632–637 (2001)
6. Lingras, P., Chen, M., Miao, D.Q.: Rough Cluster Quality Index Based on Decision Theory. IEEE Transactions on Knowledge and Data Engineering 21(7), 1014–1026 (2009)
7. Ngo, C.L., Nguyen, H.S.: A Method of Web Search Result Clustering Based on Rough Sets. In: IEEE/WIC/ACM International Conference on Web Intelligence (WI 2005), pp. 673–679 (2005)
8. Pawlak, Z.: Rough sets. International Journal of Computer and Information Sciences (11), 341–356 (1982)
9. Peters, G.: Some Refinements of Rough k-Means. Pattern Recognition 39(8), 1481–1491 (2006)
10. Saha, S., Murthy, C.A., Pal, S.K.: Rough Set Based Ensemble Classifier for Web Page Classification. Fundamenta Informaticae 76(1-2), 171–187 (2007)
11. Serban, G., Câmpan, A.: Hierarchical Adaptive Clustering. Informatica 19(1), 101–112 (2008)
12. UCIrvine Machine Learning Repository, http://archive.ics.uci.edu/ml/
13. Yao, Y.Y.: Decision-Theoretic Rough Set Models. In: Yao, J., Lingras, P., Wu, W.-Z., Szczuka, M.S., Cercone, N.J., Ślęzak, D. (eds.) RSKT 2007. LNCS (LNAI), vol. 4481, pp. 1–12. Springer, Heidelberg (2007)
14. Yu, H., Luo, H.: A Novel Possibilistic Fuzzy Leader Clustering Algorithm. In: Sakai, H., Chakraborty, M.K., Hassanien, A.E., Ślęzak, D., Zhu, W. (eds.) RSFDGrC 2009. LNCS, vol. 5908, pp. 423–430. Springer, Heidelberg (2009)

An Attribute Reduction of Rough Set Based on PSO

Hongyuan Shen[1,2], Shuren Yang[1,2], and Jianxun Liu[1]

[1] Key laboratory of knowledge processing and networked manufacturing,
college of hunan province, Xiangtan, 411201, China
[2] Institute of Information and Electrical Engineering,
Hunan University of Science and technology, Xiangtan, 411201, China
sryang86@gmail.com

Abstract. The basic concept of attribute reduction in Rough Sets Theory (RST) and the idea of Particle Swarm Optimization(PSO) are briefly combined. A new reduction algorithm based on PSO is developed. Furthermore, the thought of Cache is introduced into the proposed method, which reduces the algorithm complexity effectively, The experimental results demonstrate that the algorithm is simple and viable.

Keywords: Rough Set Theory, Reduction of Attribute, Particle Swarm Optimization, Information System.

1 Introduction

The Rough Set Theory(RST) was introduced by Z.Pawlak in 1985, which is a strong mathematical tool to deal with fuzzy and uncertain knowledge [1, 2]. Attribute reduction is an important aspect of RST research [2]. many applications of the RST were actively proposed in Knowledge Discovery recently, as an important tool in Data Mining domain, RST has been widely applied in many areas [3, 4, 5, 6, 7], especially in the area of the pattern recognition, knowledge discovery, machine learning, decision analysis. It evaluates the importance of the different attributes, discovers the reduction of the information system, reduces the redundant attributes, and seeks the minimum subset of attributes. When the random data are included in the collected sample dataset of the information system, the existence of redundant attributes is widespread phenomenon, the redundant attributes will be interfere with decision system of Rough Set. Unfortunately, the process of finding reduction has been proven to be the NP-hard problem [8]. At present there have been many kinds of methods to search the optimal attributes subset. The algorithm based on discernibility matrix that through disjunction and conjunction operator between all possible combinations of the attribute columns to achieve the final reduction, this algorithm is simple and easy to understand or to be applied, but the complexity is too high $(O(|U||A|2^{|A|} \lg |U|))$,where the parameter U is the number of the domain, A is the number of all attributes, when the scale of the database is enormous, its feasibility will face a tremendous challenge [9]. The algorithm based on information

J. Yu et al. (Eds.): RSKT 2010, LNAI 6401, pp. 695–702, 2010.
© Springer-Verlag Berlin Heidelberg 2010

quantity or conditional information quantity can be attain the simplest reduction through the introduction of informational theory, Its complexity depends on the polynomial of the importance of attributes, generally, which is $O(|A|\log|A||U|^2)$, but, the algorithm is lack of the completeness [10]. The reduction algorithm with Genetic Algorithm is inefficient and sensitive in parameters selection. Hence, it is hard to guarantee the effectiveness and validity of the results [11]. However, the reduction algorithm based on conditional entropy which is regarded as the heuristic knowledge during the procedure, it have to be calculate the partition of decision table in each iterative, and the complexity is $O(|U|\log|U|)$ [12].The algorithm based on attribute similarity is achieved by calculating importance and similarity between different attributes, yet, the method is not fit to incompatibleincomplete and lager scale dataset [13]. Consequently, in the future, Explore the efficientsimplereasonable and common attributes reduction algorithm in the Rough Set still is a hotspot.

2   Basic Knowledge

2.1   Rough Set

In the general case, for Rough Set Attributes Reduction(RSAS), the core issue is the concepts of indiscernibility. Let $S = (U, A, V, f)$ will be an information system, it is also a two dimension decision table actually, which is consist of discrete-valued datasets, the definitions of these notions are as follows:where U is a finite set of objects,Q is a finite set of the total attributes,$V = \bigcup_{q \in Q} V_q$ and V_q is a domain of the attribute q, and $f : U \times A \rightarrow V$ is an information function,if $\forall a \in A, x \in U$, then $f(x, a) \in V$, Sometimes, information system S can be written as $S = (U, A)$ briefly [1].

Obviously, an information system corresponds to a two-dimensional data table, it is the abstract description of data table. The row of table is the research object, and the rank of table is the attribute, the objects are expressed by the value of different attributes. Consequently, the problem of Rough Set attributes reduction can be translated into numerical problem to solve by existed Optimization Algorithm.

Definition 1. Let $S = (U, A, V, f)$ be an information system, then with any $P \subseteq A$ there is associated an equivalence relation $IND_A(P)$:

$$IND_A(P) = \{(x, y) \in U^2 | \forall a \in P, a(x) = a(y)\} \tag{1}$$

$IND_A(P)$,or denoted $IND(P)$ too, is called the P-indiscernibility relation, its classes are denoted by $[x]_P$. By $X|P$ giving the partition of U defined by the indiscernibility relation $IND(P)$.

Definition 2. Let $P \subset A$ and let $Q \subset A$ be the equivalence relation over U, the positive region of P on Q, defined as $POS_P(Q)$, the form is

$$POS_P(Q) = \cup P_-(X), X \in U|Q \tag{2}$$

Where $P_-(X)$ is the P-lower approximation of X set. In terms of the indiscernibility relation $U|P$, the positive region contains all objects of U that can be classified to blocks of the partition $U|Q$ employing the knowledge in attributes set P.

Definition 3. Let $S = (U, A, V, f)$ be an information system, with $P, Q \subseteq A$, the dependency between the subset P and Q can be written as

$$k = \Upsilon(P, Q) = |POS_P(Q)|/|U| \qquad (3)$$

Where we define that parameter $k(0 \leq k \leq 1)$ is dependency of P on Q, denoted $P \Rightarrow_k Q$ or $\Upsilon(P, Q)$, if $k = 1$ Q depends entirely on P, if $0 < k < 1$ Q depends partially on P , generally, called degree k, and if $k = 0$ Q does not depend on P completely.

2.2   Particle Swarm Optimization

Particle swarm optimization(PSO) is a Swarm Intelligence method, was proposed by Kennedy and Ruseel Eberhart in 1995 [14]. Due to the simple and efficient features of PSO, it has been applied to many function optimization solutions and engineering planning areas successfully [15, 21, 17].

The classical PSO algorithm can be described in form as follows:

$$\nu_{i+1} = w \cdot \nu_i + c_1 \cdot t_1 \cdot (pBest_i - x_i) + c_2 \cdot t_2 \cdot (gBest - x_i) \qquad (4)$$

$$x_{i+1} = x_i + \nu_{i+1} \qquad (5)$$

where ν_i is the previous or inertial rate of the position velocity for the ith particle, x_i is the corresponding position of the ith particle, $pBest_i$ is the particle swarm own previous best position of the ith particle up to the executed iteration, yet, $gBest$ is the particle swarm globally best position of the swarm up to the executed iteration, w is the inertia weight with range [0.1,0.9] [18], generally, which is brought in for balancing the global and local search in generations, c_1 and c_2 are two positive constants numbers, called acceleration coefficients, t_1 and t_2 are two positive random numbers in the range , they introduce the diversify factor into the movements of the particles, ν_{i+1} is the newly computed(the(i+1)th) velocity, and x_{i+1} (the (i+1)th) is the newly computed position of particle swarm [19, 20].

3   Attribute Reduction of Rough Set Based on PSO

In this section, according to the characteristics of Attribute Reduction of Rough Set and PSO algorithm, a new reduction method is presented, which is called Attribute Reduction of Rough Set Based on PSO(ARRSBP).

3.1   Algorithm Thought

Let $S = (U, A, V, f)$ be an information system, U is called domain over S, and A, called attributes set. It is composed of the conditional attribute C and the decision attribute D, denoted $A = [C, D]$.The ARRSBP algorithm is given by using the Eq.(3) to remove redundant conditional attribute from C set, while retaining their information content. the algorithm can be summarized as follows:

First, there is a one-to-one correspondence between each condition attribute C and an exclusive binary string, so, the condition attribute C can be expressed by a unique binary string, and convert these binary strings to the equal positive integers, these integer are corresponding to the particles of PSO, called also the position. Then, a fitness function can be designed, which is used to measure the performance of each 'particle' or positive integer in PSO, the best particle of the swarm can be obtained by applying the defined fitness function. Finally, the reduction results of Rough Set can be obtained by translating the best particle in PSO to the optimal reduction.

3.2   Basic Concepts

3.2.1   The Design of PSO Algorithm

1 The design of fitness function

In practical attributes reduction algorithm of Rough Set, It is necessary that a fitness function be found, which used to evaluate superiority of every potential reduction. Here, the fitness function can be written as

$$f = k + p_0/p \tag{6}$$

The form can be divided into two parts, the first part and the second part is expressed as $k, p_0/p$ respectively. where k is the reliability of condition attributes over decision attributes, its value can be got by the Eq.(3), which checks whether the relative reduction is the reduction or not. But, the first part k is even less able to ensure that the relative reduction is the first best reduction. Besides, the second part of the form is duty cycle, which defines the number of the redundant attributes VS the total number of conditional attributes, where p_0 is the number of the redundant attributes in relative reduction, and p is the number of total conditional attributes. The variable p_0/p optimizes the simplified degree of the potential reduction, which ensures the obtained reduction is simplest finally.

2 The modification of particle value

In the original particle swarm optimizer, the values range of each particle in $[-\infty, \infty]$ is the all real number field, after each iteration, it is necessary to check the velocity and the position of total particle whether the value of them exceed the predefined limit or not, if it is true, the velocity or the position of the exceeded particle will be set to be initial random value or boundary value. Otherwise, when the PSO method is combined with the Rough Set Theory, the new given ARRSBP algorithm has some special properties, unlike the classical PSO algorithms, in the ARRSBP algorithm, the position of each particle is not the continuous. Consequently, in order to modify the position of particle, a operation is introduced, which acquires the integer part of Eq.(5), called Rounding Operation. Yet, the other parameters are the same as those in Eq.(4) and Eq.(5). At last, a modified PSO algorithm can be acquired.

3.2.2 Encoding

Given an information system $S = (U, A, V, f)$, where U is a non-empty set of finite objects; A is a non-empty finite set of attributes such that $a : U \rightarrow V_a$, V_a, being the value set of attribute a. In a decision system, with $A = \{C \cup D\}$, where C is the set of conditional attributes and D is the set of decision attributes, for all subsets of the conditional attributes, them can be converted into the matched binary string, where this work is defined as Encoding, the length of the matched binary codes is equivalent to the number of conditional attributes C. Encoding rules will be described by the following stipulation:

0: For a binary string, it means that there is no conditional attribute in this position, that is to say, the corresponding position of binary string is zero, the corresponding position attributes is not included in reduction or redundancy;

1: Obviously, it indicates that the non-zeros position attributes is included in reduction; that is to say, the non-zeros attributes is irreducible to the decision system.

For instance: there are 8 attributes in the set of conditional attributes C ,that listed as $c_1, c_2, \ldots c_8$, suppose a given binary string was 01001000, makes that it correspond to the form of the set of conditional attributes C_2, C_5. With regard to a subset of conditional attributes, of which we assume that the number is P, then, the total number of condition attributes subset can be got, $2^P - 1$. Therefore, if an information system $S = (U, A, V, f)$ will be given, then the optimal reduction could be obtained among the all subsets or candidacy $[1, 2^P - 1]$, because the empty reduction attribute is meaningless for an information system. As a result, the range of the particles in PSO is limited be $[1, 2^P - 1]$, and the value of the particles is set to be integer, so that the search space is discrete in iteration.

3.2.3 Decoding

For an information system, the Encoding of attributes that makes the attributes of each object will be turned into a positive integer, where another operation is defined, which called Decoding, compared with the Encoding, it is the inverse operation. When Decoding was used, the positive integer can be inverted into some equal binary strings, and these obtained binary strings may express the relative reduction in Rough Set.

3.2.4 The Introduction of Cache

The fitness function $f = k + p_0/p$ is proposed in the above, the second part is very easy to compute,and the compute complexity is low. However, for the first part, after sorting order for the information table, its values can be got by just once more scanning the information table, and the time complexity is $O(p \cdot n \cdot \log n)$, and the space complexity is $O(n)$ [21]; when the discernibility matrix be used to compute the value of the first part, the algorithm has high complexity, respectively, the time complexity and the space complexity is $O(t{\cdot}n_2{\cdot} p{\cdot}n^2), O(p{\cdot}n^2)$,where t is the number of iteration, n^2 is the number of particles, p is the number of conditional attributes, n is the number of objects in domain. In proposed algorithm, since the reduction method based on discernibility matrix is applied, so fa as the complexity of algorithm is concerned, furthermore, in the processing of solving the reduction optimization utilizing PSO, the searching space is discrete, the value of fitness function is few or very finite. Therefore, the double-counting of fitness function is a common phenomenon during the PSO algorithm iteration, especially, in the later stages of the iteration. From what has been discussed in above, it is concluded that exiting of this phenomenon wastes the many CPU resources, and increases the algorithm complexity. Taking into account the analysis in above, in order to improve the performance and reduce the complexity of algorithm, the Cache Thought in Computer Science is introduced into the ARRSBP algorithm. the Key ideas can be found here [22].

3.3   Algorithm for Discovering the Optimal Reduction

As described in above section, first, the problem of Rough Set attribute reduction and the PSO are combined, then, initialize the modified PSO's parameter, the optimal solution will be obtained after meeting to the condition of iteration. At last, for an information system, the optimal solution will be converted into the simplest attributes reduction.

Algorithm Reduce-attributes
   Inputs: An information system or decision table, $S = (U, A, V, f)$, where U is a non-empty set of finite objects, A is a non-empty finite set of attributes, $A = \{C \cup D\}$ where C is the set of conditional attributes and D is the set of decision attributes.
   Outputs: A relative reduction of information system.
   1. Initiation. Set the parameters of PSO, including particle counts n_2, initial velocity V, the position of particles X, iteration times t, and so on.
   2. When the value of some particle is required, first, to check the Cache, if corresponding fitness value is found with a tag, known as a cache hit, direct accessing without computing.
   3. If corresponding fitness values is not found with a tag, known as a cache miss, the Eq.(6)will be used to calculate the function values, and updating the Cache using the least recently used (LRU) replacement policy.
   4. By evaluate fitness of every particle, updating $gBest$ and $pBest_i$.
   5. Calculate the new particle velocity v_{i+1} using Eq.(4), Calculate particle new position x_{i+1}using Eq.(5). Check the new estimated velocity and position, if beyond the range of maximum or minimum value respectively, then, they value will be set to Boundary value or random value.
   6. Check the termination(Iteration times).If iteration times are not over the maximum, then go to step **2** , else continue downward.
   7. Output the simplest reduction.

3.4   Algorithm Analysis

PSO was masterly designed for removing the Redundant attributes. But this proposed method has also disadvantage, such as high time complexity in iteration, the computation of fitness function is one of the most important aspects in the algorithm complexity, respectively, the time complexity and the space complexity is $O(t \cdot n_2 \cdot p \cdot n^2), O(p \cdot n^2)$, all parameters as indicated in above [21]. However, after taking the Cache ideas in Compute Science into account algorithm, the double computing of fitness values is avoided effectively, so that the time complexity is $O(t \cdot l_i \cdot p \cdot n^2)$,, and the space complexity increases by $O(u)$, where l_i is the number of not-found or "cache miss" fitness of the particles in Cache, u is the size of the Cache.

4   Computational Experiment

The RSTR algorithm was implemented by MATLAB programming language, select a decision table showed in table.1 from the paper [9], The selected dataset have 5 conditional attributes, 1 decision attribute, where the discrete search space of the modified PSO is , the algorithm was running 100s for this testing dataset, the fitness value of the global optimal particle was obtained, which is $1.6000(gBest)$, its value is greater than 1. Hence, being the attributes reduction, the corresponding position of the best particle is 12, then convert this Decimal integer into a equal binary string, its value is 01100, which is correspond to the simplest reduction $\{b, c\}$. As a result, the optimal reduction $\{b, c\}$ is obtained, the obtained result is the same to the result in the above paper [9] ,and the results illustrated that the algorithm is valid.

5   Conclusions and Future Research

A novel attribute reduction modeling method is presented which can remove the redundant attributes in rough set, and the proposed algorithm makes the best of the simplicity and efficiency of the PSO, which is simple and easy to be applied. Furthermore, the Cache thought in Compute Science is introduced into the algorithm, which overcome the disadvantage of double-counting of the fitness value, Future research will concentrate on the problem that all optimal attributes reduction will be found in the iteration of the ARRSBP.

Acknowledgements

This work was supported by the grant 90818004(the Natural Science Foundation of China), grant 09k085(the Education Department Foundation of Hunan Provincial.

References

1. Zeng, H.L.: Rough set theory and its application (Revision). Chongqing University Press, Chongqing (1998)
2. Pawlak, Z.: Rough sets: Theoretical aspects of reasoning about data. Springer, Heidelberg (1991)
3. Sadiq, W., Orlowska, M.: Analyzing process models using graph reduction techniques. Information systems 25, 117–134 (2000)
4. Huang, J., Liu, C., Ou, C., Yao, Y., Zhong, N.: Attribute reduction of rough sets in mining market value functions. In: IEEE/WIC International Conference on Web Intelligence, pp. 470–473. IEEE Press, New York (2003)
5. Pawlak, Z.: Rough set theory and its applications. Journal of Telecommunications and information technology 134, 35–42 (2002)
6. Wu, W.: Attribute reduction based on evidence theory in incomplete decision systems. Information Sciences 178, 1355–1371 (2008)
7. Zhang, W., Wei, L., Qi, J., Zhang, W., Wei, L., Qi, J.: Attribute reduction theory and approach to concept lattice. Science in China Series F: Information Sciences 48, 713–726 (2005)
8. Wong, S., Ziarko, W.: On optimal decision rules in decision tables. Bulletin of Polish Academy of Sciences 33, 693–696 (1985)
9. Shao-Hui, L., Qiu-Jian, S., Bin, W., Zhong-Zhi, S., Fei, H.: Research on efficient algorithms for rough set methods. Chinese journal of computers 26, 524–529 (2003)
10. Liang, J.Y., Qu, K.S., Xu, Z.B.: Reduction of Attribute in Information Systems. Systems Engineering-Theory and Practice 21, 76–80 (2001)
11. Tao, Z., Xu, B.D., Wang, D.W.: Rough Set Knowledge Reduction Approach Based on GA. Systems Engineering 21(4), 116–122 (2003)
12. Wang, G.Y., Wang, D.C.: Decision Table Reduction based on Conditional Information Entropy. Chinese Journal of Computers 25(7), 759–766 (2002)
13. Xia, K.W., Liu, M.X., Zhang, Z.W.: An Approach to Attribute Reduction Based on Attribute Similarity. Journal of Hebei Unviersity of Technology 34(4), 20–23 (2005)
14. Kennedy, J., Eberhart, R.: Particle swarm optimization. In: IEEE International Conforence on Neural Networks, Piscataway, NJ, pp. 1942–1948 (1995)
15. Lee, K., Jhang, J.: Application of particle swarm algorithm to the optimization of unequally spaced antenna arrays. Journal of Electromagnetic Waves and Applications 20(14), 2001–2012 (2006)
16. Parsopoulos, K., Vrahatis, M.: Particle swarm optimization method in multiobjective problems, pp. 603–607. ACM, New York (2002)
17. Parsopoulos, K., Vrahatis, M.: On the computation of all global minimizers through particle swarm optimization. IEEE Transactions on Evolutionary Computation 8(3), 211–224 (2004)
18. Cheng, G.C., Yu, J.S.: Particle Swarm Optimization Algorithm. Information and Control 34(3), 318–324 (2005)
19. Shen, H.Y., Peng, X.Q.: A multi-modality function optimization based on PSO algorithm. Journal of Hunan University of Science and Technology(Natural Science Edition) 20(3), 10–14 (2005)
20. Clerc, M.: Particle swarm optimization. ISTE, London (2006)
21. Hoa, S.N., Son, H.N.: Some efficient algorithms for rough set methods. In: The sixth international conference, Information Procesing and Management of Uncertainty in Knowledge-Based Systems, Granada, Spain, pp. 1451–1456 (1996)
22. Cache-Wikipedia, the free encyclopedia, http://en.wikipedia.org/wiki/Cache

Multiple-Category Classification with Decision-Theoretic Rough Sets

Dun Liu[1], Tianrui Li[2], Pei Hu[1], and Huaxiong Li[3]

[1] School of Economics and Management, Southwest Jiaotong University
Chengdu 610031, P.R. China
newton83@163.com, huhupei@126.com
[2] School of Information Science and Technology, Southwest Jiaotong University
Chengdu 610031, P.R. China
trli@swjtu.edu.cn
[3] School of Management and Engineering, Nanjing University
Nanjing 210093, P.R. China
huaxiongli@nju.edu.cn

Abstract. Two stages with bayesian decision procedure are proposed to solve the multiple-category classification problems. The first stage is changing an m-category classification problem into m two-category classification problems, and forming three classes of rules with different actions and decisions by using of decision-theoretic rough sets with bayesian decision procedure. The second stage is choosing the best candidate rules in positive region by using the minimum probability error criterion with bayes decision theory. By considering the levels of tolerance for errors and the costs of actions in real decision procedure, we propose a new approach to deal with the multiple-category classification problems.

Keywords: Decision-theoretic rough set, Probabilistic rough sets, bayesian decision procedure, three-way decisions, multiple-category.

1 Introduction

In many decision making problems, the people simply make a decision in two ways: do it, or not do it. Instead of two-way decisions, three-way decision procedure seems more closer to human being's behavior. Something is considered as correct and it should be done immediately; something is regarded as incorrect and it should not be done; something is hard to judge, so it need further discussion. However, two thresholds are used to distinguish the three parts and there have been many successful applications in different domains e.g., medicinal clinics [5], products inspecting process [11], documents classification [4], model selection criteria [1] and actions for an environmental manager [2]. Therefore, it seems that the three-way decision procedure comes closer to the philosophy of the real decision problems.

Motivated by these above analysis, the three-way decision procedure may be induced into rough set theory (RST) to solve problems. In RST, a pair of certain

J. Yu et al. (Eds.): RSKT 2010, LNAI 6401, pp. 703–710, 2010.

sets named lower approximation and upper approximation are used to describe a vague or imprecise set [6]. Clearly, the two approximations can divide the universe into three pair-wise disjoint regions: positive region, boundary region and negative region. The rules generated by these three regions correspond to the results of a three-way decision that the situation is verified positively, negatively, or undecidedly based on the evidence [12,16].

Unfortunately, in Pawlak rough set model, the set inclusion must be fully correct or certain, and it does not allow any tolerance of errors. In order to overcome the disadvantages, extended probabilistic rough set models are induced by allowing certain acceptable level of errors and a pair of threshold parameters are used to redefine the lower and upper approximations [9,10,13,17]. Specially, Yao induced the three-way decision procedure into probabilistic rough set and proposed the decision-theoretic rough set model (DTRS). In DTRS model, two thresholds can be directly calculated by minimizing the decision cost with bayesian theory. We can use the two parameters to generate three pair-wise disjoint regions and obtain the corresponding decision rules automatically. However, most of the researches about three-way decisions assume that there only exists a two-category classification. The assumption is unreasonable in real decision problems. In this paper, we try to extend the two-category classification into an m-category classification.

In the following, bayes decision theory is induced to solve m-category classification problems. As a fundamental statistical approach to the problem of pattern classification, bayes criterion allows us to choose the best state or action with the minimum error (cost, risk). In two-category classification problems, DTRS focuses on the actions for a certain classification. Bayesian decision procedure is used to choose the best action. However, m-category classification problems focuses the states for all classifications, and the best classification can be selected by using of bayes theory. With these observations, it seems DTRS is not suitable for our research, and a two-stage bayesian decision procedure which combines with action selection and state selection, is proposed to solve multiple-category classification problems. The remainder of the paper is organized as follows: Section 2 provides basic concepts of the probabilistic rough sets model. In Section 3, a new approach with two stages are proposed to solve the multiple-category classification problems, and the detailed modeling process is shown. Then, a case is given to explain our approach in Section 4. The paper ends with conclusions and further research topics in Section 5.

2   Preliminaries

Basic concepts, notations and results of the probabilistic rough sets model are briefly reviewed in this section [9,10,13,17].

Definition 1. [8] *Let* $S = (U, A, V, f)$ *be an information system.* $\forall x \in U, X \subseteq U$, *let:* $Pr(X|[x]) = \frac{|[x] \cap X|}{|[x]|}$, *where,* $|\cdot|$ *stands for the cardinal number of objects in sets,* $Pr(X|[x])$ *is denoted the conditional probability of the classification and we remark it as* α, $\alpha \in (0.5, 1]$.

Definition 2. *Let* $S = (U, A, V, f)$ *be an information system.* $\forall X \subseteq U$ *and* $0 \leq \beta < \alpha \leq 1$, *the* (α, β)-*lower approximation,* (α, β)-*upper approximation are defined as follows:*

$$\underline{apr}_{(\alpha,\beta)}(X) = \{x \in U | Pr(X|[x]) \geq \alpha\};$$
$$\overline{apr}_{(\alpha,\beta)}(X) = \{x \in U | Pr(X|[x]) > \beta\}. \tag{1}$$

From the (α, β)-probabilistic lower and upper approximations, we can obtain the (α, β)-probabilistic positive, boundary and negative regions:

$$\text{POS}_{(\alpha,\beta)}(X) = \{x \in U \mid Pr(X|[x]) \geq \alpha\},$$
$$\text{BND}_{(\alpha,\beta)}(X) = \{x \in U \mid \beta < Pr(X|[x]) < \alpha\},$$
$$\text{NEG}_{(\alpha,\beta)}(X) = \{x \in U \mid Pr(X|[x]) \leq \beta\}. \tag{2}$$

In Pawlak rough set model, the two parameters are defined by using the two extreme values, 0 and 1 as a qualitative nature for probabilities, but the magnitude of the value $Pr(X|[x])$ is not taken into account. By considering the real applications of a probabilistic rough set model, one may directly supply the parameters α and β based on an intuitive understanding of the levels of tolerance for errors [17]. In 0.5 probabilistic rough set model [7], we set $\alpha = \beta = 0.5$, and this model corresponds to the application of the simple majority rule. In addition, two formulations probabilistic rough set models named decision-theoretic rough set model and variable precision rough set model (VPRS) are proposed based on the statistical information of membership function. The difference between DTRS and VPRS is the former can automatically calculate the two thresholds according to bayes theory while the latter cannot do it. Furthermore, VPRS is just a special case of DTRS, and VPRS can be directly derived from DTRS when the decision costs are equal to some certain values. It seems DTRS can be regarded as a representative model for probabilistic rough set and one can choose a suitable rough set approach to satisfy the user's requirements and expectations which fulfill their needs [3].

3   Multiple-Category Classifications Decision Approach

In DTRS model, the two-category classification strategy is induced to solve the decision problem [12], and it use two states $\Omega = \{X, \neg X\}$ to generate the probabilistic rules. However, the states may conclude many choices in a practical decision problem, and the decision maker should select the best choice from the candidates. In order to achieve this goal, Yao believed an alternative method is to change an m-category classification problem into m two-category classification problems [12]. But the problem is that there may have more than one category enter into the position region, and we need more information to choose the best candidate. Moreover, Ślęzak [9] suggested an approach for defining the three probabilistic regions based on pair-wise comparisons of categories. A matrix of threshold values is used, with a pair of different threshold values on each pair of categories. Although the approach is a natural way to solve the problem, one can hardly estimate all threshold values, especially the database is huge. One may simplify the model by using the same pair of threshold values for all pairs of categories, but it is unreasonable in practical decision making.

With the insights gained from the these opinions, we suggest to use the above two approaches to deal with the multiple-category classifications. Firstly, we convert m-category classification problem into m two-category classification problems, and each classification can be divided into one of the three regions by using DTRS. Obviously, the classifications which enter to the positive region may be the candidates for the best choice. Secondly, we choose the best candidate in positive region by using the minimum probability error criterion with Bayesian decision procedure and find the best deicides. The detailed modeling process of the two stage approach is shown as follows.

3.1 m-Category Classification Process

Suppose an m-category classification problem, the set of states is composed by m classifications, which are given by $\mathcal{C} = \{C_1, C_2, \ldots, C_m\}$, where $C_1, C_2, \ldots, C_m$ form a family of pair-wise disjoint subset of U, namely, $C_i \cap C_j = \emptyset$ for $i \neq j$, and $\cup C_i = U$. For each C_i, we can define a two-category classification by $\{C, \neg C\}$, where $C = C_i$ and $\neg C = \neg C_i = \cup_{i \neq j} C_j$ [16]. So, the m-category classification problem can convert to m two-category classification problems, and the results from DTRS model can be immediately applied in our research.

In general, different categories have different losses, and different threshold values should be used for different categories in a model [9]. So, as an m-category classification problem, we need obtain m threshold value pairs (α_i, β_i) according to m different groups of loss or cost functions. In detail, for each $C_i \in \mathcal{C}$, the set of states is given by $\Omega_i = \{C_i, \neg C_i\}$ and the set of actions is given by $\mathcal{A} = \{a_P, a_B, a_N\}$ with respect to the three decision-way. There are 6 loss function regarding the risk or cost: λ_{PC_i}, λ_{BC_i} and λ_{NC_i} denote the losses incurred for taking actions a_P, a_B and a_N, respectively, when an object belongs to C_i. Similarly, $\lambda_{P \neg C_i}$, $\lambda_{B \neg C_i}$ and $\lambda_{N \neg C_i}$ denote the losses incurred for taking the same actions when the object does not belong to C_i. The expected cost $R(a_\bullet|[x])$ associated with taking the individual actions $a_\bullet$ can be expressed as:

$$
\begin{aligned}
R(a_P|[x]) &= \lambda_{PC_i} Pr(C_i|[x]) + \lambda_{P \neg C_i} Pr(\neg C_i|[x]), \\
R(a_B|[x]) &= \lambda_{BC_i} Pr(C_i|[x]) + \lambda_{B \neg C_i} Pr(\neg C_i|[x]), \\
R(a_N|[x]) &= \lambda_{NC_i} Pr(C_i|[x]) + \lambda_{N \neg C_i} Pr(\neg C_i|[x]).
\end{aligned}
\tag{3}
$$

Then, we can easily induce the three-way decision rules when $\lambda_{PC_i} \leq \lambda_{BC_i} < \lambda_{NC_i}$ and $\lambda_{N \neg C_i} \leq \lambda_{B \neg C_i} < \lambda_{P \neg C_i}$ by using the minimum overall risk criterion:

(P') If $Pr(C_i|[x]) \geq \alpha_i$ and $Pr(C_i|[x]) \geq \gamma_i$, decide $x \in \mathrm{POS}(C_i)$;

(B') If $Pr(C_i|[x]) \leq \alpha_i$ and $Pr(C_i|[x]) \geq \beta_i$, decide $x \in \mathrm{BND}(C_i)$;

(N') If $Pr(C_i|[x]) \leq \beta_i$ and $Pr(C_i|[x]) \leq \gamma_i$, decide $x \in \mathrm{NEG}(C_i)$;

Followed by the DTRS model [12,13,16], the three-way decision rules generated by C_i are displayed as follows:

$$\text{If } Pr(C_i|[x]) \geq \alpha_i, \text{ decide } x \in \text{POS}(C_i);$$
$$\text{If } \beta_i < Pr(C_i|[x]) < \alpha_i, \text{ decide } x \in \text{BND}(C_i);$$
$$\text{If } Pr(C_i|[x]) \leq \beta_i, \text{ decide } x \in \text{NEG}(C_i).$$

where, $\alpha_i = \frac{(\lambda_{P\neg C_i} - \lambda_{B\neg C_i})}{(\lambda_{P\neg C_i} - \lambda_{B\neg C_i}) + (\lambda_{BC_i} - \lambda_{PC_i})}$, $\beta_i = \frac{(\lambda_{B\neg C_i} - \lambda_{N\neg C_i})}{(\lambda_{B\neg C_i} - \lambda_{N\neg C_i}) + (\lambda_{NC_i} - \lambda_{BC_i})}$.

As stated above, we can compute the threshold value pairs (α_i, β_i) for $\forall C_i \in \mathcal{C}$, and m pairs of the threshold values can be obtained by repeating the calculational process with DTRS. By considering the fact that the rules generating by positive region make the decision of acceptance, so we choose the classifications which enter to the positive region as the candidates for the best choice. Clearly, only the classifications in positive region are considered to our following discussing, and the process of dimension reduction (the classifications in boundary region and negative region are no longer discussed in Section 3.2) in Section 3.1 makes the problem more easier. Furthermore, in our model, (α_i, β_i) is generated by two classification $\{C_i, \neg C_i\}$. It's easier to obtain the loss function in two classification than the value threshold matrix in m-category classification, and that is the advantage of our approach.

However, depending on the different values of α_i and β_i, an equivalence class in an information system may produce more than one positive rule. Specially for $\alpha_i > 0.5$, each equivalence class produces at most one positive rule. Similarly, an equivalence class may produce several boundary rules and several negative rules. Specially for $\beta_i > 0.5$, each equivalence class produces at most one boundary rule. In general, one has to consider the problem of rule conflict resolution in order to make effective acceptance, rejection, and abstaining decisions. Therefore, it is necessary to further study on the probabilistic regions of a classification, as well as the associated rules [16].

3.2   The Process of Choosing the Best Candidate Classification

Suppose there are m' ($m' < m$) classifications enter into the position region, and we should choose the best one from these candidates. The strategy of our approach is considering all the candidate classifications together and calculating the loss utility by using the bayesian procedure for each candidate and finding the minimum one.

Furthermore, by thinking of a classifier as a device for partitioning feature space into decision regions, we can obtain additional insight into the operation of a Bayes classifier. Considering the m'-category case, and suppose that the classifier C_i has divided the space into two regions, a_P and a_N. There are $(m'-1)$ ways in which a classification error can occur, that is, an observation x falls in C_j and the true state of nature is C_i ($j \neq i$). The probability of error is:

$$P(error) = \sum_{i=1}^{m'} \sum_{j=1, j \neq i}^{m'} P(C_j, C_i) = \sum_{i=1}^{m'} \sum_{j=1, j \neq i}^{m'} P(C_j|C_i)P(C_i) \quad (4)$$

By considering the cost $\lambda(C_i|C_i) = 0$ when doing a right classification, the overall error cost of the misclassification of C_i can be calculated by two parts:

$$\mathbf{Er}(C_i) = \sum_{j=1,j\neq i}^{m'} P(C_j|C_i)P(C_i)\lambda(C_j|C_i) + \sum_{i=1,i\neq j}^{m'} P(C_i|C_j)P(C_j)\lambda(C_i|C_j) \quad (5)$$

where, the first part stands for the costs of rejecting a right classification C_i when the true state is C_i, and the second part stands for the costs of accepting a wrong classification C_i when the true state is not C_i. From the Bayesian point of view, the first part stands for the cost that one looks at information that should not substantially change one's prior estimate of probability, but it actually do it; The second part stands for the cost that one looks at information which should change one's estimate, but it actually not do it. The following work is to find the best classification which has minimum overall conditional risk.

Let $\mathcal{C}' = \{C_1, C_2, \ldots, C_{m'}\}$ ($m' < m$) be the candidate classifications in position region, and the best one C_l satisfies:

$$Decide\ C_l: \ if\ \mathbf{Er}(C_l) \leq \mathbf{Er}(C_i),\ i = 1, 2, \cdots, m';\ i \neq l.$$

The classification C_l which has the minimum overall risk, is the best performance that can be achieved, and the decision maker can choose it as the best choice. In addition, if there are more than one classification satisfy the above condition, a tie-breaking criterion can be used.

4   An Illustration

Let us illustrate the above concepts on a didactic example about a medical diagnose case. The symptoms of the patient are described as {fever, cough, nausea, headache, nose snivel, dysentery} according to a series of carefully diagnoses. Furthermore, the results of the symptom may lead to five possible diseases named H1N1, SARS, Viral influenza, Wind chill and Common cold, respectively. The doctor want to choose the best candidate from the five diseases.

Followed by the analysis process in Section 3, we denote the five possible classifications as $\{C_1, C_2, C_3, C_4, C_5\}$. Firstly, the five-category classification problem can be converted to five two-category classification problems $\{C_i, \neg C_i\}$ ($i = 1, 2, 3, 4, 5$). There are 6 parameters in each model, λ_{PC_i}, λ_{BC_i}, λ_{NC_i} denote the costs incurred for taking actions of accepting C_i, need further observation and refusing C_i when the state is C_i; $\lambda_{P\neg C_i}$, $\lambda_{B\neg C_i}$, $\lambda_{N\neg C_i}$ denote the costs incurred for taking actions of accepting C_i, need further observation and refusing C_i when the state is not C_i. By considering the fact that there is no cost when doing a right decision, we set $\lambda_{PC_i} = \lambda_{N\neg C_i} = 0$. In addition, we have $\lambda_{BC_i} \leq \lambda_{NC_i}$ and $\lambda_{B\neg C_i} \leq \lambda_{P\neg C_i}$ [13]. The detailed cost parameter for the five diseases are presented in Table 1.

By using DTRS, we can directly compute the value of α_i and β_i for the five diseases in Table 1, we have: $\alpha_1 = 0.5556$, $\beta_1 = 0.4$; $\alpha_2 = 0.375$, $\beta_2 = 0.2875$; $\alpha_3 = 0.7143$, $\beta_3 = 0.1667$; $\alpha_4 = 0.4545$, $\beta_4 = 0.3333$; $\alpha_5 = 0.3333$, $\beta_5 = 0.2$.

Table 1. The cost functions for the five diseases

C_i	λ_{PC_i}	λ_{BC_i}	λ_{NC_i}	$\lambda_{P\neg C_i}$	$\lambda_{B\neg C_i}$	$\lambda_{N\neg C_i}$
C_1	0	4	10	9	4	0
C_2	0	5	10	5	4	0
C_3	0	2	7	6	1	0
C_4	0	3	6	4	1.5	0
C_5	0	1	3	1	0.5	0

Table 2. The misclassification probability and cost for correlative parameters

Parameters	C_1	C_3	Parameters	C_1	C_3	Parameters	C_1	C_3	Parameters	C_1	C_3
$P(C_1\|C_i)$	0.3	0.2	$\lambda(C_1\|C_i)$	0	2	$P(C_i\|C_1)$	0.3	0.1	$\lambda(C_i\|C_1)$	0	8
$P(C_2\|C_i)$	0.1	0.2	$\lambda(C_2\|C_i)$	10	4	$P(C_i\|C_2)$	0.05	0.1	$\lambda(C_i\|C_2)$	9	8
$P(C_3\|C_i)$	0.3	0.3	$\lambda(C_3\|C_i)$	8	0	$P(C_i\|C_3)$	0.15	0.3	$\lambda(C_i\|C_3)$	6	0
$P(C_4\|C_i)$	0.1	0.1	$\lambda(C_4\|C_i)$	7	3	$P(C_i\|C_4)$	0.1	0.4	$\lambda(C_i\|C_4)$	5	3
$P(C_5\|C_i)$	0.2	0.2	$\lambda(C_5\|C_i)$	6	1	$P(C_i\|C_5)$	0.05	0.25	$\lambda(C_i\|C_5)$	4	2

Inspired by the simple majority rule, the project which has more than half the votes should be executed. That is, the candidates which $\alpha_i > 0.5$ may enter into the positive region. Hence, the diseases C_1: H1N1 and C_3: Influenza may considered as the positive choices according primary diagnoses.

Then, we should choose the best one and make the final decision from the two candidates. Followed by the idea of Section 3.2, it needs further observation to get some other parameters. After an intensively examination, the symptoms of the patient also include sore throat, body aches, chills and fatigue. So, the correlative parameters are shown in Table 3.

In addition, the prior probability of the five diseases is $P(C_1) = 0.1$, $P(C_2) = 0.05$, $P(C_3) = 0.25$, $P(C_4) = 0.2$, $P(C_5) = 0.4$ according to the historical database. So, we can compute the overall conditional risk of the misclassification of C_1 and C_3 in formula (5). We get $\mathbf{Er}(C_1) = 1.355$, $\mathbf{Er}(C_3) = 0.63$. Due to $\mathbf{Er}(C_3) < \mathbf{Er}(C_1)$. Then we conclude that the patient gets influenza.

5   Conclusions

Three way decision produce is introduced into our research to solve the multiple-category classification problems. Two stages are proposed to design the approach, the first stage is changing an m-category classification problem into m two-category classification problems and the second stage is choosing the best candidate by using the bayesian decision procedure. The process of our approach provides a basic and naive thought for the multiple-category classification problems and the case of medicine diagnosis validates the feasibility of our method. Our future work will focus on how to acquire the parameters automatically and some behavior strategies may be induced to our research.

Acknowledgements

The authors thank Professor Y.Y. Yao for his insightful suggestions. The authors also thank the National Science Foundation of China (No. 60873108), the Doctoral Innovation Fund (200907) and the Scientific Research Foundation of Graduate School of Southwest Jiaotong University (2009LD) for their support.

References

1. Forster, M.: Key concepts in model selection: performance and generalizability. Journal of Mathematical Psychology 44, 205–231 (2000)
2. Goudey, R.: Do statistical inferences allowing three alternative decision give better feedback for environmentally precautionary decision-making. Journal of Environmental Management 85, 338–344 (2007)
3. Joseph, P., Yao, J.: Criteria for choosing a rough set model. Computer and Mathematics with Application 57, 908–918 (2009)
4. Li, Y., Zhang, C., Swan, J.: An information fltering model on the Web and its application in JobAgent. Knowledge-Based Systems 13, 285–296 (2000)
5. Pauker, S., Kassirer, J.: The threshold approach to clinical decision making. The New England Journal of Medicine 302, 1109–1117 (1980)
6. Pawlak, Z.: Rough sets. International Journal of Computer and Information Science 11, 341–356 (1982)
7. Pawlak, Z., Wong, S., Ziarko, W.: Rough sets: probabilistic versus deterministic approach. International Journal of Man-Machine Studies 29, 81–95 (1988)
8. Pawlak, Z., Skowron, A.: Rough membership functions. In: Advances in the D-S Theory of Evidence, pp. 251–271. John Wiley and Sons, New York (1994)
9. Ślęzak, D.: Rough sets and bayes factor. In: Peters, J.F., Skowron, A. (eds.) Transactions on Rough Sets III. LNCS, vol. 3400, pp. 202–229. Springer, Heidelberg (2005)
10. Wong, S., Ziarko, W.: Comparison of the probabilistic approximate classification and the fuzzy set model. Fuzzy Sets and Systems 21, 357–362 (1987)
11. Woodward, P., Naylor, J.: An application of Bayesian methods in SPC. The Statistician 42, 461–469 (1993)
12. Yao, Y.: Three-way decisions with probabilistic rough sets. Information Sciences 180, 341–353 (2010)
13. Yao, Y., Wong, S.: A decision theoretic framework for approximating concepts. International Journal of Man-machine Studies 37(6), 793–809 (1992)
14. Yao, Y.: Probabilistic approaches to rough sets. Expert Systems 20, 287–297 (2003)
15. Yao, Y.: Decision-theoretic rough set models. In: Yao, J., Lingras, P., Wu, W.-Z., Szczuka, M.S., Cercone, N.J., Ślęzak, D. (eds.) RSKT 2007. LNCS (LNAI), vol. 4481, pp. 1–12. Springer, Heidelberg (2007)
16. Yao, Y.Y.: Three-way decisions with probabilistic rough sets. Information Sciences 180, 341–353 (2010)
17. Ziarko, W.: Variable precision rough set model. Journal of Computer and System Sciences 46, 39–59 (1993)

A Multi-agent Decision-Theoretic Rough Set Model

Xiaoping Yang[1,2] and Jingtao Yao[2]

[1] School of Mathematics, Physics & Information Science, Zhejiang Ocean University
Zhoushan, Zhejiang, P.R. China, 316004
[2] Department of Computer Science, University of Regina
Regina, Saskatchewan, Canada S4S 0A2
yxpzyp@sina.com, jtyao@cs.uregina.ca

Abstract. The decision-theoretic rough set (DTRS) model considers cost and risk factors when classifying an equivalence class into a particular region. Using DTRS, informative decisions with rough rules can be made. The current research has a focus on single agent decision makings. We propose a multi-agent DTRS model in this paper. This model seeks synthesized or consensus decision when there are multiple sets of decision preferences and criteria adopted by different agents. A set of rough decision rules that are satisfied by multiple agents can be derived from the multi-agent decision-theoretic rough set model.

Keywords: Decision-theoretic rough sets, probabilistic rough sets, multi-agent decision making.

1 Introduction

Rough set theory was proposed by Pawlak in 1982 [8]. It is a widely used model in machine learning and artificial intelligence [10,11]. In classical rough set, the universe is partitioned by equivalence relationship into positive, negative and boundary regions. The equivalence relationship is restrict. It results in large boundary region which is an uncertain region.

To reduce large boundary region is one of the major concerns for rough set applications. Probabilistic rough set models aim to decrease the boundary region and therefore increase the range of decisions with rough rules. There are many forms of probabilistic rough sets such as the DTRS model [12,14], the variable precision rough set model [18] and the Bayesian rough set model [2].

DTRS model was proposed by Yao et al. [15] in 1990 using Bayesian decision procedures. Within the decision-theoretic framework, the required threshold values can be interpreted and calculated according to concrete notions, such as cost and risk [13]. There are some extensions in recent research regarding to DTRS. Herbert and Yao [3,9] formulated the game theoretic approximation measures and modified conditional risk strategies to provide the user with tolerance levels for their loss functions by using the tolerance values. New thresholds are calculated to provide correct classification regions. Zhou et al. [17] proposed a

J. Yu et al. (Eds.): RSKT 2010, LNAI 6401, pp. 711–718, 2010.

multi-view decision method based on DTRS in which optimistic decision, pessimistic decision and indifferent decision are provided according to the cost of misclassification. Lingras et al. [5] proposed a cluster validity index based on DTRS model by considering various loss functions. It is reported the proposed measures have the ability to incorporate financial considerations in evaluating quality of a clustering scheme. Zhao et al. [16] applied DTRS to emails classification and claimed the proposed model may reduce error ratio by comparing with popular classification methods like Naive Bayes classification.

Multi-agent technology can improve the results of group decision making. It has many intelligent properties such as autonomy, activity and sociality. Agent collaboration is often used to address the uncertainty problem. It can make agents share potential knowledge that has been discovered so that one agent may benefit from the additional opinion of other agents [6]. Keeney and Raiffa [4] provide a comprehensive discussion of multi-attribute utility theory, its axiomatic foundations, and the forms of utility functions that are commonly employed in decision-making problems, both under certainty and uncertainty. Multi-agent technology is widely used[1,7].

Previous research work in DTRS has a focus on single agent decision makings. In this paper, we want to explore the multi-agent DTRS using the concept of multi-agent when many agents have different decisions with DTRS. The main problem is that a consensus is hard to be achieved. We want to explore methods to form a synthesized decision by bringing together the separate decisions of agents. According to decisions made by each single agent, we try to give out a reasonable formulation for overall utilization. We may also take advantage of loss functions to make a decision as one made in DTRS.

2   Formulating DTRS with Single Agent Decision-Making

Suppose U is a finite non-empty set called the universe, and $E \subseteq U \times U$ is an equivalence relation on U. For each element $x \in U$, the equivalence class containing x is denoted by $[x]$, all elements in the equivalence class $[x]$ share the same description. For a given subset $A \subseteq U$, the approximation operators partition U into three disjoint classes $POS(A)$, $NEG(A)$, and $BND(A)$. Furthermore, to enlarge the $POS(A)$, $NEG(A)$, and reduce the $BND(A)$, researchers try to consider how to assign x in $BND(A)$ into the other two regions based on the conditional probability $P(A|[x])$. The Bayesian decision procedure can be immediately applied to solve this problem [2,12,14,15].

Let $P(A|[x])$ be the conditional probability of an object x being in state A given the object description x. For agent $j(j = 1, 2, ..., n)$, the set of actions is given by $S = \{a_P^j, a_N^j, a_B^j\}$ (for consistency and easy understanding, we use super script j to represent jth agent), where a_P^j, a_N^j and a_B^j represent the three actions to classify an object into $POS^j(A)$, $NEG^j(A)$ and $BND^j(A)$, respectively. Let $\lambda^j(a_i|[x])$ denote the loss incurred for taking action a_i^j $(i = P, N, B)$ when an object in fact belongs to A, and let $\lambda^j(a_i|A^c)$ denote the loss incurred for taking the action a_i^j when the object does not belong to A. The expected loss $R^j(a_i|[x])$ associated with taking the individual actions can be expressed as

$$R^j(a_P|[x]) = \lambda_{PP}^j P(A|[x]) + \lambda_{PN}^j (1 - P(A|[x])),$$
$$R^j(a_N|[x]) = \lambda_{NP}^j P(A|[x]) + \lambda_{NN}^j (1 - P(A|[x])), \qquad (1)$$
$$R^j(a_B|[x]) = \lambda_{BP}^j P(A|[x]) + \lambda_{BN}^j (1 - P(A|[x])),$$

where $\lambda_{iP}^j = \lambda^j(a_i|A)$, $\lambda_{iN}^j = \lambda^j(a_i|A^c)$, and $i = P, N, B$. The Bayesian decision procedure leads to the following minimum-risk decision rules:

(P)If $R^j(a_P|[x]) \leq R^j(a_N|[x]), R^j(a_P|[x]) \leq R^j(a_B|[x]),$ decide $POS(A),$
(N)If $R^j(a_N|[x]) \leq R^j(a_P|[x]), R^j(a_N|[x]) \leq R^j(a_B|[x]),$ decide $NEG(A),$ (2)
(B)If $R^j(a_B|[x]) \leq R^j(a_P|[x]), R^j(a_B|[x]) \leq R^j(a_N|[x]),$ decide $BND(A).$

Generally speaking, the loss of classifying an object x belonging to A into the positive region $POS^j(A)$ is less than or equal to the loss of classifying x into the boundary region $BND^j(A)$, and the above two losses are strictly less than the loss of classifying x into the negative region $NEG^j(A)$. Similarly, the loss of classifying an object x not belonging to A into the negative region $NEG^j(A)$ is less than or equal to the loss of classifying x into the boundary region $BND^j(A)$, and the these two losses are strictly less than the loss of classifying x into the positive region $POS^j(A)$. We may express such situation with the following inequalities labelled by (c^j) for the jth agent:

$$(c^j) \qquad \lambda_{PP}^j \leq \lambda_{BP}^j < \lambda_{NP}^j, \quad \lambda_{NN}^j \leq \lambda_{BN}^j < \lambda_{PN}^j.$$

Under condition (c^j), we have $\beta^j < \gamma^j < \alpha^j$, where $\alpha^j = \dfrac{\lambda_{PN}^j - \lambda_{BN}^j}{(\lambda_{BP}^j - \lambda_{BN}^j) - (\lambda_{PP}^j - \lambda_{PN}^j)},$
$\beta^j = \dfrac{\lambda_{PN}^j - \lambda_{NN}^j}{(\lambda_{NP}^j - \lambda_{NN}^j) - (\lambda_{PP}^j - \lambda_{PN}^j)},$ and $\gamma^j = \dfrac{\lambda_{BN}^j - \lambda_{NN}^j}{(\lambda_{NP}^j - \lambda_{NN}^j) - (\lambda_{BP}^j - \lambda_{BN}^j)}.$

So we can formulate the following decision rules based on the set of inequalities (P),(N) and (B) above:

(P^j) If $P(A|[x]) \geq \alpha^j,$ decide $x \in POS^j(A);$
(N^j) If $P(A|[x]) \leq \beta^j,$ decide $x \in NEG^j(A);$ (3)
(B^j) If $\beta^j < P(A|[x]) < \alpha^j,$ decide $x \in BND^j(A).$

These minimum risk decision rules offer us a foundation to classify objects into the three regions.

3   Multi-agent DTRS for Group Decision-Making

Assume that there are n agents to decide which region an object should belong to according to the minimum risk. Similar to one agent in DTRS, for agent j, $(j = 1, 2, 3..., n)$, the decision rules are (P^j), (N^j) and (B^j). If we find two reasonable functions $f(\alpha^1, \alpha^2, ..., \alpha^n)$ and $g(\beta^1, \beta^2, ..., \beta^n)$ such that $\alpha = f(\alpha^1, \alpha^2, ..., \alpha^n),$

$\beta = g(\beta^1, \beta^2, ..., \beta^n)$ and $\alpha > \beta$, we would get the new rules for the multiple agents as follows:

(P') If $P(A|[x]) \geq \alpha$,   decide $x \in POS(A)$;
(N') If $P(A|[x]) \leq \beta$,   decide $x \in NEG(A)$;   (4)
(B') If $\beta < P(A|[x]) < \alpha$, decide $x \in BND(A)$.

We give out some examples for $f(\alpha^1, \alpha^2, ..., \alpha^n)$ and $g(\beta^1, \beta^2, ..., \beta^n)$ in Equations (5)(6)(7)(8)(9).

$$\begin{cases} f(\alpha^1, \alpha^2, ..., \alpha^n) = \max\{\alpha^1, \alpha^2, ..., \alpha^n\}, \\ g(\beta^1, \beta^2, ..., \beta^n) = \min\{\beta^1, \beta^2, ..., \beta^n\}. \end{cases} \quad (5)$$

$$\begin{cases} f(\alpha^1, \alpha^2, ..., \alpha^n) = \min\{\alpha^1, \alpha^2, ..., \alpha^n\}, \\ g(\beta^1, \beta^2, ..., \beta^n) = \min\{\beta^1, \beta^2, ..., \beta^n\}. \end{cases} \quad (6)$$

$$\begin{cases} f(\alpha^1, \alpha^2, ..., \alpha^n) = \max\{\alpha^1, \alpha^2, ..., \alpha^n\}, \\ g(\beta^1, \beta^2, ..., \beta^n) = \max\{\beta^1, \beta^2, ..., \beta^n\}. \end{cases} \quad (7)$$

$$\begin{cases} f(\alpha^1, \alpha^2, ..., \alpha^n) = (\alpha^1 + \alpha^2 + ... + \alpha^n)/n, \\ g(\beta^1, \beta^2, ..., \beta^n) = (\beta^1 + \beta^2 + ... + \beta^n)/n. \end{cases} \quad (8)$$

$$\begin{cases} f(\alpha^1, \alpha^2, ..., \alpha^n) = (k_1\alpha^1 + k_2\alpha^2 + ... + k_n\alpha^n)/(k_1 + k_2 + ... + k_n), \\ g(\beta^1, \beta^2, ..., \beta^n) = (k_1\beta^1 + k_2\beta^2 + ... + k_n\beta^n)/(k_1 + k_2 + ... + k_n). \end{cases} \quad (9)$$

These functions have their own properties which could be used in specific environments. We could select one of these functions according to concrete details. We would discuss these functions further in our later studies.

We could also consider the average risk for all the agents based on expectation of loss $R^j(a_i|[x])$. Denote

$$\begin{aligned} \overline{R}(a_P|[x]) &= \tfrac{1}{n}\sum_{j=1}^n R^j(a_P|[x]) = \overline{\lambda}_{PP}P(A|[x]) + \overline{\lambda}_{PN}(1 - P(A|[x])), \\ \overline{R}(a_N|[x]) &= \tfrac{1}{n}\sum_{j=1}^n R^j(a_N|[x]) = \overline{\lambda}_{NP}P(A|[x]) + \overline{\lambda}_{NN}(1 - P(A|[x])), \quad (10) \\ \overline{R}(a_B|[x]) &= \tfrac{1}{n}\sum_{j=1}^n R^j(a_B|[x]) = \overline{\lambda}_{BP}P(A|[x]) + \overline{\lambda}_{BN}(1 - P(A|[x])), \end{aligned}$$

where

$$\overline{\lambda}_{iP} = \frac{1}{n}\sum_{j=1}^n \lambda_{iP}^j, \quad \overline{\lambda}_{iN} = \frac{1}{n}\sum_{j=1}^n \lambda_{iN}^j, \quad (i = P, N, B).$$

Similar to DTRS, we have

(P'') If $\overline{R}(a_P|[x]) \leq \overline{R}(a_N|[x])$, $\overline{R}(a_P|[x]) \leq \overline{R}(a_B|[x])$, decide $POS(A)$;
(N'') If $\overline{R}(a_N|[x]) \leq \overline{R}(a_P|[x])$, $\overline{R}(a_N|[x]) \leq \overline{R}(a_B|[x])$, decide $NEG(A)$;   (11)
(B'') If $\overline{R}(a_B|[x]) \leq \overline{R}(a_P|[x])$, $\overline{R}(a_B|[x]) \leq \overline{R}(a_N|[x])$, decide $BND(A)$.

Denote

$$\alpha = \frac{\overline{\lambda}_{PN} - \overline{\lambda}_{BN}}{(\overline{\lambda}_{BP} - \overline{\lambda}_{BN}) - (\overline{\lambda}_{PP} - \overline{\lambda}_{PN})},$$

$$\beta = \frac{\overline{\lambda}_{PN} - \overline{\lambda}_{NN}}{(\overline{\lambda}_{NP} - \overline{\lambda}_{NN}) - (\overline{\lambda}_{PP} - \overline{\lambda}_{PN})}, \tag{12}$$

$$\gamma = \frac{\overline{\lambda}_{BN} - \overline{\lambda}_{NN}}{(\overline{\lambda}_{NP} - \overline{\lambda}_{NN}) - (\overline{\lambda}_{BP} - \overline{\lambda}_{BN})}.$$

Under the condition (c^j), we have

$$\overline{\lambda}_{PP} \le \overline{\lambda}_{BP} < \overline{\lambda}_{NP}, \qquad \overline{\lambda}_{NN} \le \overline{\lambda}_{BN} < \overline{\lambda}_{PN}.$$

So we have $\beta < \gamma < \alpha$ and get decision rules based on DTRS as follows:

$$
\begin{aligned}
&(\text{P}''') \text{ If } P(A|[x]) \ge \alpha, && \text{decide } x \in POS(A); \\
&(\text{N}''') \text{ If } P(A|[x]) \le \beta, && \text{decide } x \in NEG(A); \\
&(\text{B}''') \text{ If } \beta < P(A|[x]) < \alpha, && \text{decide } x \in BND(A).
\end{aligned}
\tag{13}
$$

4   An Example of Multi-agent DTRS Model

A wholesaler buys a large amount of products from a factory and sells them to a retailer, earning $\$p_1$ for each product. A retailer buys some of the products from the wholesaler and sells them to customers, earning $\$p_2$ for each product.

For simplicity, we define the product which is good in use during a certain period of time as good product. If a product is not good in use during a certain period of time, the product may be returned to the retailer and the retailer returns the product to the wholesaler, and then the wholesaler returns it to the factory. In this case, the retailer will lose $\$l_2$ for each returned product and the wholesaler will lose $\$l_1$ for it.

Both the wholesaler and the retailer want to make profits. Their target profits are $\$t_1$ and $\$t_2$ for each product, respectively. Only when the product is good can both of them make money. Therefore, we can decide according to the acceptability which kind of products could be sold so that both of them could get the target profits.

Let x be a product of all the products, $[x]$ be a class of products sharing the same attributes as x, A stand for a class of products in good state. $P(A|[x])$ stands for the conditional probabilities of x in good state and it can also be taken as the acceptability of products $[x]$.

Let W and R stand for the profits of the wholesaler and the retailer. W and R are random variables. We list the probability distributions of W and R in Table 1 and Table 2.

Table 1. Distributions table of W

W	p_1	l_1
p	$P(A|[x])$	$1 - P(A|[x])$

Table 2. Distributions table of R

R	p_2	l_2
p	$P(A\|[x])$	$1 - P(A\|[x])$

We get the expectations for W and R:

$$w = E(W) = p_1 P(A|[x]) - l_1(1 - P(A|[x])),$$
$$r = E(R) = p_2 P(A|[x]) - l_2(1 - P(A|[x])). \quad (14)$$

For the wholesaler, if his expectation of profits for selling x can reach or exceed his target profit, $w \geq t_1$, we put x into $POS(A)$; We put x into $NEG(A)$ if $w \leq 0$; We put x in $BND(A)$ if $0 < w < t_1$. We have the rules as follows based on the inequalities above:

$$(\text{P}^1) \text{ If } P(A|[x]) \geq \tfrac{l_1+t_1}{p_1+l_1}, \qquad \text{decide } x \in POS^1(A);$$
$$(\text{N}^1) \text{ If } P(A|[x]) \leq \tfrac{l_1}{p_1+l_1}, \qquad \text{decide } x \in NEG^1(A); \quad (15)$$
$$(\text{B}^1) \text{ If } \tfrac{l_1}{p_1+l_1} < P(A|[x]) < \tfrac{l_1+t_1}{p_1+l_1}, \text{ decide } x \in BND^1(A).$$

Similarly for the retailer, we have the following rules:

$$(\text{P}^2) \text{ If } P(A|[x]) \geq \tfrac{l_2+t_2}{p_2+l_2}, \qquad \text{decide } x \in POS^2(A);$$
$$(\text{N}^2) \text{ If } P(A|[x]) \leq \tfrac{l_2}{p_2+l_2}, \qquad \text{decide } x \in NEG^2(A); \quad (16)$$
$$(\text{B}^2) \text{ If } \tfrac{l_2}{p_2+l_2} < P(A|[x]) < \tfrac{l_2+t_2}{p_2+l_2}, \text{ decide } x \in BND^2(A).$$

We take the functions in the form of equation (1) and get α and β as follows:

$$\alpha = \max\{\frac{l_1+t_1}{p_1+l_1}, \frac{l_2+t_2}{p_2+l_2}\}, \qquad \beta = \min\{\frac{l_1}{p_1+l_1}, \frac{l_2}{p_2+l_2}\}.$$

We have the new rules for both of them

$$(\text{P}''') \text{ If } P(A|[x]) \geq \alpha, \qquad \text{decide } x \in POS(A);$$
$$(\text{N}''') \text{ If } P(A|[x]) \leq \beta, \qquad \text{decide } x \in NEG(A); \quad (17)$$
$$(\text{B}''') \text{ If } \beta < P(A|[x]) < \alpha, \text{ decide } x \in BND(A).$$

We can explain these decision rules in the following way. For a product x in U,

1. If $P(A|[x]) \geq \alpha$, that is, statistically both the wholesaler and the retailer can get the target profits by selling x, we put the x into $POS(A)$;
2. If $P(A|[x]) \leq \beta$, this means neither of them gets any profits, we put the x into $NEG(A)$;
3. If $\beta < P(A|[x]) < \alpha$, that is, only one of them can get his target profit, the other one can not get his profit, we put the x into $BND(A)$.

Thus, U is divided into three regions denoted by $POS(A)$, $NEG(A)$ and $BND(A)$, respectively.

5   Conclusion

Many of the complex problems faced by decision makers involve multiple points of view. Multi-agent decision theory describes how decision makers, who wish to make a reasonable and responsible choice among alternatives, can systematically make a wise decision. Decision-theoretic rough set model is useful in providing rough rules for decision makers. It is based on costs when classifying an equivalence class into a particular region. With the consideration of many decisions made with cooperations amongst experts and managers, we propose a multi-agent decision-theoretic rough set model. We first reformulate DTRS when there is a single agent involved. Based on this formulation, a multi-agent decision-theoretic rough set model is proposed. We suggest some functions to calculate the positive, negative and boundary regions. The decision rules are given out by the form of conditional probabilities which is similar to DTRS. Informative decisions could be made with various functions and region thresholds. The theory is illustrated by a concrete example taken from a deal between the wholesaler and the retailer. This extension to DTRS will broaden application domains of DTRS.

Acknowledgement

Discussing with DTRS research group in computer science department of the University of Regina is greatly appreciated. Special thanks go to Dr. Yiyu Yao for his discussion and suggestions. This work was partially supported by grants from the National Natural Science Foundation of China (No.60673096), the Natural Science Foundation of Zhejiang Province in China (No.Y107262) and the Natural Sciences and Engineering Research Council of Canada.

References

1. Chau, M., Zeng, D., Chen, H., Huang, M., Hendriawan, D.: Design and evaluation of a multi-agent collaborative Web mining system. Decision Support Systems 35(1), 167–183 (2003)
2. Greco, S., Matarazzo, B., Slowinski, R.: Rough membership and Bayesian confirmation measures for parameterized rough sets. In: Ślęzak, D., Wang, G., Szczuka, M.S., Düntsch, I., Yao, Y. (eds.) RSFDGrC 2005. LNCS (LNAI), vol. 3641, pp. 314–324. Springer, Heidelberg (2005)
3. Herbert, J.P., Yao, J.T.: Game-theoretic risk analysis in decision-theoretic rough sets. In: Wang, G., Li, T., Grzymala-Busse, J.W., Miao, D., Skowron, A., Yao, Y. (eds.) RSKT 2008. LNCS (LNAI), vol. 5009, pp. 132–139. Springer, Heidelberg (2008)
4. Keeney, R.L., Raiffa, H.: Decisions with Multiple Objectives: Preferences and Value Trade-Offs. Cambridge University Press, Cambridge (1993)
5. Lingars, P., Chen, M., Miao, D.Q.: Rough cluster quality index based on decision theory. IEEE Transactions on Knowledge and Nowledge and Data Engineering 21(7), 1014–1026 (2009)

6. Liu, Y., Bai, G., Feng, B.: Multi-agent based multi-knowlege acquisition method for rough set. In: Wang, G., Li, T., Grzymala-Busse, J.W., Miao, D., Skowron, A., Yao, Y. (eds.) RSKT 2008. LNCS (LNAI), vol. 5009, pp. 140–147. Springer, Heidelberg (2008)

7. O'Hare, G.M.P., O'Grady, M.J.: Gulliver's Genie: a multi-agent system for ubiquitous and intelligent content delivery. Computer Communications 26(11), 1177–1187 (2003)

8. Pawlak, Z.: Rough sets. International Journal of Computer and Information Sciences 11, 341–356 (1982)

9. Yao, J.T., Herbert, J.P.: A game-theoretic perspective on rough set analysis. Journal of Chongqing University of Posts and Telecommunications (Natural Science Edition) 3, 291–298 (2008)

10. Yao, J.T., Herbert, J.P.: Financial time-series analysis with rough sets. Applied Soft Computing 3, 1000–1007 (2009)

11. Yao, J.T., Herbert, J.P.: Web-based support systems with rough set sanalysis. In: Kryszkiewicz, M., Peters, J.F., Rybiński, H., Skowron, A. (eds.) RSEISP 2007. LNCS (LNAI), vol. 4585, pp. 360–370. Springer, Heidelberg (2007)

12. Yao, Y.Y.: Information granulation and rough set approximation in a decision-theoretical model of rough sets. In: Pal, S.K., Polkowski, L., Skowron, A. (eds.) Rough-neural Computing: Techniques for Computing with Words, pp. 491–518. Springer, Berlin (2003)

13. Yao, Y.Y.: Probabilistic approaches to rough sets. Expert Systems 20, 287–297 (2003)

14. Yao, Y.Y., Wong, S.K.M.: A decision theoretic framework for approximating concepts. International Journal of Man-machine Studies 37, 793–809 (1992)

15. Yao, Y.Y., Wong, S.K.M., Lingras, P.: A decision-theoretic rough set models. In: Ras, Z.W., Zemankova, M., Emrich, M.L. (eds.) Methodologes for Intelligent Systems, vol. 5, pp. 17–24. North-Holland, New York (1990)

16. Zhao, W.Q., Zhu, Y.L.: An email classification scheme based on decision-theoretic rough set theory and analysis of email security. In: Proceedings of IEEE Region 10 Conference (TENCON 2005), Melbourne, Australia, pp. 2237–2242 (2005)

17. Zhou, X.Z., Li, H.X.: A multi-view decision model based on decision-theoretic rough set. In: Wen, P., Li, Y., Polkowski, L., Yao, Y., Tsumoto, S., Wang, G. (eds.) RSKT 2009. LNCS, vol. 5589, pp. 650–657. Springer, Heidelberg (2009)

18. Ziarko, W.: Variable precision rough set model. Journal of Computer and System Science 46, 39–59 (1993)

Naive Bayesian Rough Sets

Yiyu Yao and Bing Zhou

Department of Computer Science, University of Regina
Regina, Saskatchewan, Canada S4S 0A2
{yyao,zhou200b}@cs.uregina.ca

Abstract. A naive Bayesian classifier is a probabilistic classifier based on Bayesian decision theory with naive independence assumptions, which is often used for ranking or constructing a binary classifier. The theory of rough sets provides a ternary classification method by approximating a set into positive, negative and boundary regions based on an equivalence relation on the universe. In this paper, we propose a naive Bayesian decision-theoretic rough set model, or simply a naive Bayesian rough set (NBRS) model, to integrate these two classification techniques. The conditional probability is estimated based on the Bayes' theorem and the naive probabilistic independence assumption. A discriminant function is defined as a monotonically increasing function of the conditional probability, which leads to analytical and computational simplifications.

Keywords: three-way decisions, naive Bayesian classification, Bayesian decision theory, cost-sensitive classification.

1 Introduction

Naive Bayesian classifier and rough set classification are two useful techniques for classification problems. A naive Bayesian classifier is a probabilistic classifier based on Bayesian decision theory with naive independence assumptions [1,2]. As a fundamental statistical approach, Bayesian decision theory is often used for binary classification problems, i.e., each class is associated with a yes/no decision. The Pawlak rough set theory provides a ternary classification method by approximating a set by positive, negative and boundary regions based on an equivalence relation of the universe [7,16].

The qualitative categorization of Pawlak three regions may be too restrictive to be practically useful. This has led to the extension of rough sets by allowing some tolerance of uncertainty. Probabilistic rough set models were proposed [3,5,8,10,11,12,14,17,18,19], in which the degrees of overlap between equivalence classes and a set to be approximated are considered. A conditional probability is used to state the degree of overlapping and a pair of threshold values α and β are used to defined three probabilistic regions. Elements whose probability is above the first threshold α are put into the positive region, between α and the second threshold β in the boundary region, and below β is the negative region. The three regions correspond to a three-way decision of acceptance, deferment, and rejection [16]. The decision-theoretic rough set (DTRS)

J. Yu et al. (Eds.): RSKT 2010, LNAI 6401, pp. 719–726, 2010.

model provides a systematic way to calculate the two threshold values based on the well established Bayesian decision theory, with the aid of more practically operable notions such as cost, risk, benefit etc. [14,17,18].

On the other hand, the estimation of the conditional probability has not received much attention. The rough membership function is perhaps the only commonly discussed way [9]. It is necessary to consider other methods for estimating the probability more accurately. For this purpose, we introduce a naive Bayesian decision-theoretic rough set model, or simply a naive Bayesian rough set (NBRS) model. The conditional probability is estimated based on the Bayes' theorem and the naive probabilistic independence assumption. A discriminant function is defined as a monotonically increasing function of the conditional probability, which leads to analytical and computational simplifications.

2   Contributions of the Naive Bayesian Rough Set Model

The proposed naive Bayesian rough set model is related to several existing studies, but contributes in its unique way. In the Bayesian decision theory, one may identify three important components, namely, the interpretation and computation of the required threshold value when constructing a classifier, the use of Bayes' theorem that connects, based on the likelihood, the *a priori* probability of a class to the *a posteriori* probability of the class after observing a piece of evidence, and the estimation of required probabilities. These three components enable us to show clearly the current status of various probabilistic models of rough sets and the contributions of the naive Bayesian rough set model.

The decision-theoretic rough set model [14,15,17,18] focuses on the first issue, namely, the interpretation and computation of a pair of threshold values on the *a posteriori* probability of class for building a ternary classifier. The later proposed variable precision rough set (VPRS) model [19] uses a pair of threshold values on a measure of set-inclusion to define rough set approximations, which is indeed equivalent to the result of a special case of the DTRS model [14,20]. The more recent parameterized rough set model [3] uses a pair of thresholds on a Bayesian confirmation measure, in addition to a pair thresholds on probability. In contrast to the DTRS model, the last two models suffers from a lack of guidelines and systematic methods on how to determining the required threshold values.

The Bayesian rough set (BRM) model [11,12] is an attempt to resolve the above problem by using the *a priori* probability of the class as a threshold for defining probabilistic regions, i.e., one compares the *a posteriori* probability and the *a priori* probability of the class. Based on the Bayes' theorem, one can show that this is equivalent to comparing two likelihoods [12]. The rough Bayesian (RM) model [10] further explores the second issue of the Bayesian decision theory. A pair of threshold values on a Bayes factor, namely, a likelihood ratio, is used to define probabilistic regions. The Bayesian rough set model, in fact, uses a threshold of 0 on the difference between the *a posteriori* and the *a priori* probabilities, or a threshold of 1 on the likelihood ration; the rough Bayesian model uses a pair of arbitrary threshold values. However, the latter model does not address the problem of how to setting the threshold values. Recently, the Bayes' theorem is introduced into the decision-theoretic rough set model to address this problem [16].

All these probabilistic models do not address the third issue of the Bayesian decision theory, namely, the estimation of the required probabilities. The full implications of Bayesian decision theory and Bayesian inference have not been fully explored, even though the phrases, rough Bayesian model and Bayesian rough sets, have been used. In this paper, we propose a Bayesian decision-theoretic rough set model, or simply a Bayesian rough set model, to cover all three issues of the Bayesian decision theory, and a naive Bayesian rough set model, in particular, to adopt the naive independence assumption in probability estimation. Since the first issue, namely, interpretation and computation of the thresholds, has been extensively discussed in other papers [14,15,16,17], we will concentrate on the contributions of the naive Bayesian rough sets with respect to the other two issues, namely, application of Bayes' theorem and probability estimation.

3   Basic Formulation of Bayesian Rough Sets

We review the basic formulations of probabilistic rough set and Bayesian rough set models in the following subsections.

3.1   Decision-Theoretic Rough Sets

Let $E \subseteq U \times U$ be an equivalence relation on U, i.e., E is reflexive, symmetric, and transitive. Two objects in U satisfy E if and only if they have the same values on all attributes. The pair $apr = (U, E)$ is called an approximation space. The equivalence relation E induces a partition of U, denoted by U/E. The basic building blocks of rough set theory are the equivalence classes of E. For an object $x \in U$, the equivalence class containing x is given by $[x] = \{y \in U \mid xEy\}$. For a subset $C \subseteq U$, one can divide the universe U into three disjoint regions, the positive region $\text{POS}(C)$, the boundary region $\text{BND}(C)$, and the negative region $\text{NEG}(C)$ [6]:

$$\text{POS}(C) = \{x \in U \mid [x] \subseteq C\},$$
$$\text{BND}(C) = \{x \in U \mid [x] \cap C \neq \emptyset \wedge [x] \not\subseteq C\},$$
$$\text{NEG}(C) = \{x \in U \mid [x] \cap C = \emptyset\}. \tag{1}$$

One can say with *certainty* that any object $x \in \text{POS}(C)$ belongs to C, and that any object $x \in \text{NEG}(C)$ does not belong to C. One cannot decide with certainty whether or not an object $x \in \text{BND}(C)$ belongs to C.

The qualitative categorization in the Pawlak rough set model may be too restrictive to be practically useful. Probabilistic rough set model is proposed to enable some tolerance of uncertainty, in which the Pawlak rough set model is generalized by considering degrees of overlap between equivalence classes and a set to be approximated, i.e., $[x]$ and C in equation (1),

$$Pr(C|[x]) = \frac{|C \cap [x]|}{|[x]|}, \tag{2}$$

where $|\cdot|$ denotes the cardinality of a set, and $Pr(C|[x])$ is the conditional probability of an object belongs to C given that the object is in $[x]$, estimated by using the cardinalities of sets. Pawlak and Skowron [9] suggested to call the conditional probability a rough membership function. According to the above definitions, the three regions can be equivalently defined by:

$$\begin{aligned}
\mathrm{POS}(C) &= \{x \in U \mid Pr(C|[x]) = 1\}, \\
\mathrm{BND}(C) &= \{x \in U \mid 0 < Pr(C|[x]) < 1\}, \\
\mathrm{NEG}(C) &= \{x \in U \mid Pr(C|[x]) = 0\}.
\end{aligned} \tag{3}$$

They are defined by using the two extreme values, 0 and 1, of probabilities. They are of a qualitative nature; the magnitude of the value $Pr(C|[x])$ is not taken into account.

A main result of decision-theoretic rough set model is parameterized probabilistic approximations. This can be done by replacing the values 1 and 0 in equation (3) by a pair of threshold values α and β with $\alpha > \beta$. The (α, β)-probabilistic positive, boundary and negative regions are defined by:

$$\begin{aligned}
\mathrm{POS}_{(\alpha,\beta)}(C) &= \{x \in U \mid Pr(C|[x]) \geq \alpha\}, \\
\mathrm{BND}_{(\alpha,\beta)}(C) &= \{x \in U \mid \beta < Pr(C|[x]) < \alpha\}, \\
\mathrm{NEG}_{(\alpha,\beta)}(C) &= \{x \in U \mid Pr(C|[x]) \leq \beta\}.
\end{aligned} \tag{4}$$

The three probabilistic regions lead to three-way decisions [16]. We accept an object x to be a member of C if the probability is greater than α. We reject x to be a member of C if the probability is less than β. We neither accept or reject x to be a member of C if the probability is in between of α and β, instead, we make a decision of deferment.

The threshold values α and β can be interpreted in terms of cost or risk of the three-way classification. They can be systematically computed based on minimizing the overall risk of classification. The details can be found in papers on decision-theoretic rough sets [14,15,17,18].

3.2   Classification Based on Bayes' Theorem

The conditional probabilities are not always directly derivable from data. In such cases, we need to consider alternative ways to calculate their values. A commonly used method is to apply the Bayes' theorem,

$$Pr(C|[x]) = \frac{Pr(C)Pr([x]|C)}{Pr([x])}, \tag{5}$$

where

$$Pr([x]) = Pr([x]|C)Pr(C) + Pr([x]|C^c)Pr(C^c),$$

$Pr(C|[x])$ is the *a posteriori* probability of class C given $[x]$, $Pr(C)$ is the *a priori* probability of class C, and $Pr([x]|C)$ the likelihood of $[x]$ with respect to C. The Bayes' theorem enable us to infer the *a posteriori* probability $Pr(C|[x])$, which is difficulty to estimate, from the *a priori* probability $Pr(C)$ through the likelihood $Pr([x]|C)$, which is easy to estimate.

One may define monotonically increasing functions of the conditional probability to construct an equivalent classifier. This observation can lead to significant analytical and computational simplifications. The probability $Pr([x])$ in equation (5) can be eliminated by taking the odds form of Bayes' theorem, that is,

$$O(Pr(C|[x])) = \frac{Pr(C|[x])}{Pr(C^c|[x])} = \frac{Pr([x]|C)}{Pr([x]|C^c)} \cdot \frac{Pr(C)}{Pr(C^c)} = \frac{Pr([x]|C)}{Pr([x]|C^c)} O(Pr(C)). \tag{6}$$

A threshold value on the probability can indeed be interpreted as another threshold value on the odds. For the positive region, we have:

$$Pr(C|[x]) \geq \alpha \iff \frac{Pr(C|[x])}{Pr(C^c|[x])} \geq \frac{\alpha}{1-\alpha}$$
$$\iff \frac{Pr([x]|C)}{Pr([x]|C^c)} \cdot \frac{Pr(C)}{Pr(C^c)} \geq \frac{\alpha}{1-\alpha}. \tag{7}$$

By applying logarithms to both sides of the equation, we get

$$\log \frac{Pr([x]|C)}{Pr([x]|C^c)} + \log \frac{Pr(C)}{Pr(C^c)} \geq \log \frac{\alpha}{1-\alpha}. \tag{8}$$

Similar expressions can be obtained for the negative and boundary regions. Thus, the three regions can now be written as:

$$\text{POS}^B_{(\alpha',\beta')}(C) = \{x \in U \mid \log \frac{Pr([x]|C)}{Pr([x]|C^c)} \geq \alpha'\},$$

$$\text{BND}^B_{(\alpha',\beta')}(C) = \{x \in U \mid \beta' < \log \frac{Pr([x]|C)}{Pr([x]|C^c)} < \alpha'\},$$

$$\text{NEG}^B_{(\alpha',\beta')}(C) = \{x \in U \mid \log \frac{Pr([x]|C)}{Pr([x]|C^c)} \leq \beta'\}, \tag{9}$$

where

$$\alpha' = \log \frac{Pr(C^c)}{Pr(C)} + \log \frac{\alpha}{1-\alpha},$$
$$\beta' = \log \frac{Pr(C^c)}{Pr(C)} + \log \frac{\beta}{1-\beta}. \tag{10}$$

This interpretation simplifies the calculation by eliminating $Pr([x])$. The detailed estimations of related probabilities need to be further addressed.

3.3   Naive Bayesian Model for Estimating Probabilities

The naive Bayesian rough set model provides a practical way to estimate the conditional probability based on the naive Bayesian classification [1,2]. In the Pawlak rough set model [7], information about a set of objects are represented in an information table with a finite set of attributes [6]. Formally, an information table can be expressed as:

$$S = (U, At, \{V_a \mid a \in At\}, \{I_a \mid a \in At\}),$$

where

U is a finite nonempty set of objects called universe,

At is a finite nonempty set of attributes,

V_a is a nonempty set of values for $a \in At$,

$I_a : U \rightarrow V_a$ is an information function.

The information function I_a maps an object in U to a value of V_a for an attribute $a \in At$, that is, $I_a(x) \in V_a$. Each object x is described by a logic formula $\bigwedge_{a \in At} a = I_a(x)$, where $v_a \in V_a$, and the atomic formula $a = I_a(x)$ indicates that the value of an object on attribute a is $I_a(x)$. For simplicity, we express the description of $[x]$ as a feature vector, namely, $Des([x]) = (v_1, v_2, ..., v_n)$ with respect to the set of attributes $\{a_1, a_2, ..., a_n\}$ where $I_{a_i}(x) = v_i$. For simplicity, we write $Des([x])$ as $[x]$.

Recall that the conditional probability $Pr(C|[x])$ can be reexpressed by the *prior* probability $Pr(C)$, the *likelihood* $Pr([x]|C)$, and the probability $Pr([x])$, where $Pr([x]|C)$ is a joint probabilities of $Pr(v_1, v_2, ..., v_n|C)$, and $Pr([x])$ is a joint probability of $Pr(v_1, v_2, ..., v_n)$. In practice, it is difficult to analyze the interactions between the components of $[x]$, especially when the number n is large. A common solution to this problem is to calculate the likelihood based on the naive conditional independence assumption [2]. That is, we assume each component v_i of $[x]$ to be conditionally independent of every other component v_j for $j \neq i$.

For the Bayesian interpretation of three regions based on equation (8), we can add the following naive conditional independence assumptions:

$$Pr([x]|C) = Pr(v_1, v_2, ..., v_n|C) = \prod_{i=1}^{n} Pr(v_i|C),$$

$$Pr([x]|C^c) = Pr(v_1, v_2, ..., v_n|C^c) = \prod_{i=1}^{n} Pr(v_i|C^c). \qquad (11)$$

Thus, equation (7) can be re-expressed as:

$$\log \frac{Pr([x]|C)}{Pr([x]|C^c)} \geq \log \frac{Pr(C^c)}{Pr(C)} + \log \frac{\alpha}{1 - \alpha}$$

$$\Longleftrightarrow \sum_{i=1}^{n} \log \frac{Pr(v_i|C)}{Pr(v_i|C^c)} \geq \log \frac{Pr(C^c)}{Pr(C)} + \log \frac{\alpha}{1 - \alpha}. \qquad (12)$$

where $Pr(C)$ and $Pr(v_i|C)$ can be easily estimated from the frequencies of the training data by putting:

$$Pr(C) = \frac{|C|}{|U|},$$

$$Pr(v_i|C) = \frac{|m(a_i, v_i) \cap C|}{|C|},$$

where $m(a_i, v_i)$ is called the meaning set. It is defined as $m(a_i, v_i) = \{x \in U | I_{a_i}(x) = v_i\}$, that is, the set of objects whose attribute value equal to v_i with regard to attribute a_i. Similarly, we can estimate $Pr(C^c)$ and $Pr(v_i|C^c)$. We can then rewrite equation (8) as:

$$\text{POS}^B_{(\alpha',\beta')}(C) = \{x \in U \mid \sum_{i=1}^{n} \log \frac{Pr(v_i|C)}{Pr(v_i|C^c)} \geq \alpha'\},$$

$$\text{BND}^B_{(\alpha',\beta')}(C) = \{x \in U \mid \beta' < \sum_{i=1}^{n} \log \frac{Pr(v_i|C)}{Pr(v_i|C^c)} < \alpha'\},$$

$$\text{NEG}^B_{(\alpha',\beta')}(C) = \{x \in U \mid \sum_{i=1}^{n} \log \frac{Pr(v_i|C)}{Pr(v_i|C^c)} \leq \beta'\}. \tag{13}$$

All the related factors in the above equations are easily derivable from data for real applications.

4 Conclusion

This paper proposes a naive Bayesian rough set model to intergrade two classification techniques, namely, naive Bayesian classifier and the theory of rough sets. The conditional probability in the definition three regions in rough sets is interpreted by using the probability terms in naive Bayesian classification. A discriminant function is defined as a monotonically increasing function of the conditional probability, which leads to analytical and computational simplifications. The integration provides a practical solution for applying naive Bayesian classifier to ternary classification problems. Two threshold values instead of one are used, which can be systematically calculated based on loss functions stating how costly each action is.

Acknowledgements

The first author is partially supported by an NSERC Canada Discovery grant. The second author is supported by an NSERC Alexander Graham Bell Canada Graduate Scholarship.

References

1. Duda, R.O., Hart, P.E.: Pattern Classification and Scene Analysis. Wiley, New York (1973)
2. Good, I.J.: The Estimation of Probabilities: An Essay on Modern Bayesian Methods. MIT Press, Cambridge (1965)
3. Greco, S., Matarazzo, B., Słowiński, R.: Parameterized rough set model using rough membership and Bayesian confirmation measures. International Journal of Approximate Reasoning 49, 285–300 (2009)
4. Herbert, J.P., Yao, J.T.: Game-theoretic risk analysis in decision-theoretic rough sets. In: Wang, G., Li, T., Grzymala-Busse, J.W., Miao, D., Skowron, A., Yao, Y. (eds.) RSKT 2008. LNCS (LNAI), vol. 5009, pp. 132–139. Springer, Heidelberg (2008)
5. Herbert, J.P., Yao, J.T.: Game-theoretic rough sets. Fundamenta Informaticae (2009)
6. Pawlak, Z.: Rough sets. International Journal of Computer and Information Sciences 11, 341–356 (1982)
7. Pawlak, Z.: Rough Sets, Theoretical Aspects of Reasoning about Data. Kluwer Academic Publishers, Dordrecht (1991)
8. Pawlak, Z., Wong, S.K.M., Ziarko, W.: Rough sets: probabilistic versus deterministic approach. International Journal of Man-Machine Studies 29, 81–95 (1988)
9. Pawlak, Z., Skowron, A.: Rough membership functions. In: Yager, R.R., Fedrizzi, M., Kacprzyk, J. (eds.) Advances in the Dempster-Shafer Theory of Evidence, pp. 251–271. John Wiley and Sons, New York (1994)
10. Ślęzak, D.: Rough sets and Bayes factor. In: Peters, J.F., Skowron, A. (eds.) Transactions on Rough Sets III. LNCS, vol. 3400, pp. 202–229. Springer, Heidelberg (2005)
11. Ślęzak, D., Ziarko, W.: Bayesian rough set model. In: Procedings of FDM 2002, Maebashi, Japan, pp. 131–135 (December 9, 2002)
12. Ślęzak, D., Ziarko, W.: The investigation of the Bayesian rough set model. International Journal of Approximate Reasoning 40, 81–91 (2005)
13. Yao, Y.Y.: Probabilistic approaches to rough sets. Expert Systems 20, 287–297 (2003)
14. Yao, Y.Y.: Decision-theoretic rough set models. In: Yao, J., Lingras, P., Wu, W.-Z., Szczuka, M.S., Cercone, N.J., Ślęzak, D. (eds.) RSKT 2007. LNCS (LNAI), vol. 4481, pp. 1–12. Springer, Heidelberg (2007)
15. Yao, Y.Y.: Probabilistic rough set approximations. International Journal of Approximation Reasoning 49, 255–271 (2008)
16. Yao, Y.Y.: Three-way decisions with probabilistic rough sets. Information Sciences 180(3), 341–353 (2010)
17. Yao, Y.Y., Wong, S.K.M.: A decision theoretic framework for approximating concepts. International Journal of Man-machine Studies 37, 793–809 (1992)
18. Yao, Y.Y., Wong, S.K.M., Lingras, P.: A decision-theoretic rough set model. In: Ras, Z.W., Zemankova, M., Emrich, M.L. (eds.) Methodologies for Intelligent Systems, vol. 5, pp. 17–24. North-Holland, Amsterdam (1990)
19. Ziarko, W.: Variable precision rough sets model. Journal of Computer and Systems Sciences 46, 39–59 (1993)
20. Ziarko, W.: Probabilistic approach to rough sets. International Journal of Approximate Reasoning 49, 272–284 (2008)

Protein Interface Residues Recognition Using Granular Computing Theory

Jiaxing Cheng[1,*], Xiuquan Du[1], and Jiehua Cheng[1,2]

[1] Key Laboratory of Intelligent Computing and Signal Processing,
Ministry of Education, Anhui University, Hefei, Anhui, China
cjx@ahu.edu.cn, dxqllp@163.com
[2] Department of Computer Science, Anhui Science and Technology University,
Fengyang, Anhui, China
cjh532@163.com

Abstract. Predicting of protein-protein interaction sites (PPIs) merely are researched in a single granular space in which the correlations among different levels are neglected. In this paper, PPIs models are constructed in different granular spaces based on Quotient Space Theory. We mainly use HSSP profile and PSI-Blast profile as two features for granular space, then we use granularity synthesis theory to synthesis PPIs models from different features, finally we also improve the prediction by the number of neighboring residue. With the above method, an accuracy of 59.99% with sensitivity (68.87%), CC (0.2113), F-measure (53.12%) and specificity (47.56%) is achieved after considering different level results. We then develop a post-processing scheme to improve the prediction using the relative location of the predicted residues. Best success is then achieved with sensitivity, specificity, CC, accuracy and F-measure pegged at 74.96%, 47.87%, 0.2458, 59.63% and 54.66%, respectively. Experimental results presented here demonstrate that multi-granular method can be applied to automated identification of protein interface residues.

Keywords: protein-protein interaction, SVM, granular computing, sequence profile.

1 Introduction

Biological functions are performed through the interaction on proteins. Because of identification of interface residues can help the construction of a structural model for a protein complex, many methods are developed for predicting protein interaction sites, such as support vector machines (SVM) [1,2], neural networks [3], conditional random fields [4], Bayesian method [5], random decision forests [6] have been successfully applied in this field. These studies use various protein features.

In our previous report [7], we try to use an SVM to identify interface residues from different granularity using sequence neighbors of a target residue with sequence profile, solvent accessible surface area (ASA) and entropy. In this study,

---

* Corresponding author.

J. Yu et al. (Eds.): RSKT 2010, LNAI 6401, pp. 727–734, 2010.

Instead of inherence of PPIs, only parts of characters of PPIs are emphasized. PPIs model in different granular space are constructed based on Quotient Space Theory [8]. We try to simulate the thought of people and to analyze the attributes of the different PPIs using granularity computing of quotient space theory. Generally, it is a process from granularity coarse to thin step by step. The samples in different granular space will be constructed based on this novel method can mine the essence of information in or among the data. This method not only takes advantage of the high accuracy of categorizations by machine learning method, but also overcomes its shortcomings that the prediction results are not continuous values but discrete values. The experimental results show that using this novel method to predict and analyze PPIs is effective and the prediction accuracy is better than that of commonly used statistical models. Multi-granular method improves the accuracy of PPIs predictions effectively and offers a new way to the application of PPIs prediction, a new model to the research of protein-protein interaction.

2   Materials and Method

2.1   Data Collection

In order to generate a predictor that can capture the general properties of residues locate on a protein interfaces, we extract 594 individual proteins from a set of 814 protein complexes in the study of [4] for our training dataset and testing dataset. These individual proteins are extracted from [4] according to the following criterions. First, in order to get non-redundant hetero-complexes, all those selected chains are compared using BLASTCLUST program. The first chain of each cluster is selected. Second, for each selected chain, we calculate interfacial residues against other chains of this complex. A residue is considered to be an interface residue if the distance between CA atom and any CA atom of its interacting chains is ¡1.2nm. Each chain is selected at most one time. Finally, 594 chains (training dataset (400) and testing dataset (194)) are selected as non-redundant protein chains of hetero-complexes.

2.2   Definition of Protein Interaction Sites

A residue is defined as a surface residue if its solvent accessible surface area (ASA) is at least 16% of its nominal maximum area. The remainder is defined as an inside residues. A surface residues is define as an interaction site residues when its distance between CA atom and any CA atom of its interacting chain is ¡ 1.2nm. According to this definition, we get about 33.9% (35354) of all surface residues (61.37%, 104253) in the dataset.

2.3   Granularity Thinking in Protein-Protein Interaction Sites

The thinking of quotient space granularity computing can be applied well in PPIs field. PPIs based on quotient space granularity theory include two parts:

First, describing protein interface under different granularities to obtain domain partition according to residue attribute features, i.e. form surface residues partition under different granularities; Second, according to granularity combination theory, select a reasonable optimal guide line function to combine domain partition in different levels, then form finally classification could describe well protein interface entirely, and with which to retrieve in order to achieve better effect.

In actual solving process, descriptions under different granularities are needed according to different cognitive phases. Multi-granular description from multi-level could be taken into PPIs process; this paper takes support vector machine (SVM) algorithm as partition method and discusses the PPIs recognition based on multi-granular domain combination according to protein features. Making full use of protein-protein interaction information with its own to extract features is the crucial problem in protein-protein interaction sites recognition (PPIs). In this paper, we extract two level features from protein. These two features include sequence profile from HSSP and sequence profile from PSI-BLAST program.

2.4   Predictor Construction

Here, in order to obtain the sub-quotient space, we need to select the suitable granularity of protein interaction sites. So, we try to construct the granular from protein features. In this paper, we mainly use protein sequence profile and we adopt LibSVM [9] to construct the first quotient space $([X_1],[f_1],[T_1])$ using RBF kernel function with PSI-profile feature input, also the second quotient space $([X_2],[f_2],[T_2])$ with HSSP-profile feature input. The two quotient spaces are obtained according to general classification problem method (i.e. using protein sequence profile as feature and SVM is used to classify the residues). Thus, two quotient spaces are composed of predicted interface residues and non-interface residues. Finally we use granularity synthesis theory to improve the results. In our experiment, we consider only surface residues in the predictor training process; two SVM-based predictors are constructed using PSI-profile and HSSP-profile. The input vector of these predictors is extracted using a window of 11 residues, centered on the target residue and including 5 sequence neighboring residues on each side. Thus, each residue is represented in the predictor by a 220-D vector.

2.5   Prediction of PPIs Based on Multi-granularity Learning

After obtaining domain partition (i.e. two new quotient spaces) under different granularities using SVM algorithms, we can systemize them properly to fully describe protein interface, and to achieve a better classification effect. Quotient space granularity synthesis theory provides stable foundation for this, PPIs recognition method based on quotient space granularity combination can be described as follows.

a) Through HSSP-profile feature level using SVM algorithm, we get the corresponding quotient space is $([X_1],[f_1],[T_1])$, $[f_1]$ is a set of protein features, $[T_1]$ is similarity relation structure among residues, $[X_1]$ is

$$[X_1] = \{ \begin{array}{ll} Y & if \frac{pro(SVM_{r_i}^+ f)}{pro(SVM_{r_i}^- f)} > \theta \\ N & otherwise \end{array} \tag{1}$$

Where Y denotes interface residue, N is non-interface residue, r denotes residues and f is input vector of SVM with PSI-profile and i denotes residue position in the protein sequence. + denotes output probability of interface residue and - denotes opposition, pro denotes probability.

b) Through PSI-profile feature level using SVM algorithm we get the corresponding quotient space is $([X_2],[f_2],[T_2])$, $[f_2]$ is a set of protein features, $[T_2]$ is similarity relation structure among residues and $[X_2]$ is

$$[X_2] = \{ \begin{array}{ll} Y & if \frac{pro(SVM_{r_i}^+ f)}{pro(SVM_{r_i}^- f)} > \theta \\ N & otherwise \end{array} \tag{2}$$

Where Y denotes interface residue, N is non-interface residue, r denotes residues and f is input vector of SVM with HSSP-profile and i denotes residue position in the protein sequence.

c) Domain synthesis $[X_3]$ of different granularity spaces. According domain synthesis principle, for $[X_1]$, $[X_2]$ and $[X_3]$, the synthesis function is obtained as follows:

$$X(r) = judge\_same([X_1](r), [X_2](r), [X_3](r)) \tag{3}$$

Where judge_same() is the optimal guide line function of domain synthesis under the two granularities. If the label of residue r between $[X_1]$ and $[X_2]$ is the same, then label of $[X_3]$ (r) is the same as $[X_1]$ or $[X_2]$, otherwise, go d).

d) According to the synthesis domain function X(r), formula (1) is defined in this paper for PPIs recognition.

$$[X_3](r) = \{ \begin{array}{ll} Y & if \frac{pro_r^{S1}(+)}{pro_r^{S1}(-)} > \frac{pro_r^{S2}(+)}{pro_r^{S2}(-)} \\ N & otherwise \end{array} \tag{4}$$

where S1 denotes quotient space $([X_1],[f_1],[T_1])$. S2 denotes quotient space $([X_2],[f_2],[T_2])$, r denotes residue, + denotes output probability of interface residue and - denotes output probability of non-interface residue.

3   Experimental Results

3.1   The Initial Performance of Our Method versus Other Methods

The performance of our propose method on the testing dataset is shown Table 1. Each experiment is carried out the five times, and adopts average performance finally. From Table 1, it can be seen that our method can get best performance

Table 1. Performances of various methods on the testing dataset

Input feature	Method	Sensitivity	Specificity	CC	Accuracy	F-measure
HSSP-profile	SVM	0.5387	0.4756	0.1964	0.6253	0.4792
(quotient space	NN	0.6044	0.4642	0.1663	0.6006	0.4995
S1)	Bayesian	0.5045	0.4335	0.0960	0.5536	0.4249
PSI-profile	SVM	0.6061	0.4814	0.1979	0.6130	0.5019
(quotient space	NN	0.6596	0.4641	0.1854	0.5940	0.5149
S2)	Bayesian	0.5336	0.4713	0.1666	0.6102	0.4654
Our method		0.6887	0.4756	0.2113	0.5999	0.5312

except specificity, accuracy compared with SVM. In addition, it can further be seen that the difference in performance between our method and NN, Bayesian is very notable. It is superior to them in almost all of the performance measures. In particular, the enhancement by our method is impressive: at least 2.91% in sensitivity, 1.63% in F-measure and 1.34% in correlation coefficient than other all methods. In the aspect of sensitivity, our method can achieve a better performance (higher at least 13%, 8.26% respectively) than quotient space S1 and quotient space S2, although lower (0.58%) than S2 in specificity compared with SVM. These findings indicate that the multi-granularity learning method may be complementary in computational method.

3.2   Specificity versus Sensitivity Tradeoff Curves

It is desirable to predict for interface residues with high sensitivity. This can be met by ROC curves. We plot specificity-sensitivity with different methods when =1. Fig.1.A shows that specificity versus sensitivity curves. Fig.1.B shows that TP rate versus FP rate curves. From the Fig.1.A, we can see our method can obtain better performance between specificity and sensitivity when sensitivity higher than 0.5. Our method gets higher AUC (0.6462) than random classifier (0.5088) in our dataset from the Fig.1.B

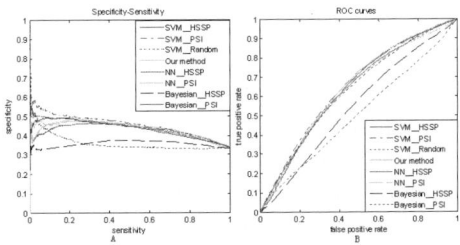

Fig. 1. (A) Specificity-Sensitivity graph. (B) The ROC curves.

3.3   Post-classification Refinement to Improve Prediction Performance

Fig.2.A shows a typical example of a protein with true positives, false negatives, true negatives and false positives predicted by applying the above protocol. We inspect the location of predicted residues on the proteins and observe that some of the FNs (yellow balls) are located in close vicinity of the TPs (red balls). Similarly, many of FPs (blue balls) lies far apart from other positive predictions. So, we reason that distance-dependent refinement of the classification may boost the prediction performance. Chain A, B and D is represented in pink, violet and green, respectively. True positives, false negatives and false positives and true negatives are represented by ball representation in red, yellow, blue and light grey, respectively.

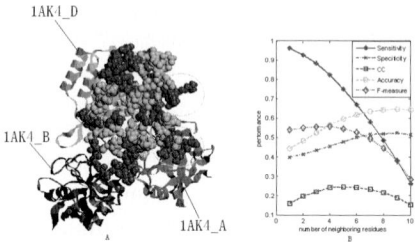

Fig. 2. (A) Prediction of different residues. (B) The overall performance of our method.

Here we make use of this finding to improve prediction by re-classifying a subset of predicted positives as negatives and negatives as positives if less than N interface residues are within 10 of their spatial neighbors. Fig.2.B shows the results of our method when N=1,,10. It can be found that the best classification performance (74.96% sensitivity, 47.87% specificity, 0.2458 CC and 54.66% F-measure) is obtained when N=5 except accuracy (lower 0.36%).

4   Some Examples with Experiment

The first example is complex of active site inhibited human blood coagulation factor viia with human recombi-nant soluble tissue factor with resolution 2.0 Å[10].We use our classifier predict 26 residues to be interface with 78.79% sensitivity and 56.52% specificity (Fig.3.B). SVM with HSSP profile predicts 20 interface residues with 60.61% sensitivity, 55.56% specificity (Fig.3.C) and SVM with PSI-profile predicts 17 interface residues with 51.52% sensitivity, 51.52% specificity (Fig.3.D) while the actual interface residues are 33 (Fig.3.A).

The second example is complex of gamma/epsilon atp synthase from E.coli with 2.1 angstroms x-ray crystal structure [11]. This interface region is accurately identified by our method covering 66.7% of the actual binding site with

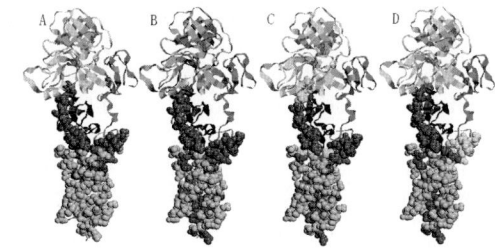

Fig. 3. Predicted interface residues (red color) on protein (PDB: 1DAN_U)

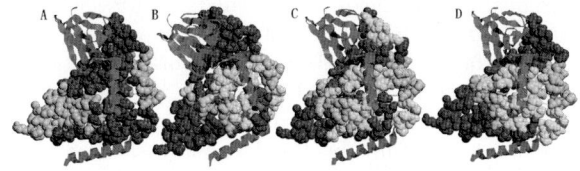

Fig. 4. Predicted interface residues (red color) on protein (PDB: 1FS0_G)

a specificity of 60.67% (Fig.4.B), The prediction result by SVM with HSSP-profile covers only 49.38% of the actual binding site with a specificity of 57.14% (Fig.4.C) and SVM-PSI predictor covers 56.79% sensitivity, 58.23% specificity (Fig.4.D) versus actual interface residues are 81 (Fig.4.A).

5   Discussion and Conclusion

In this paper, some SVM-based predictors have been constructed for inferring interface residues. The results reported here showed the quotient space theory based on different protein sequence profile can successfully predict interface residues. The residue samples in different granular space are constructed based on this novel method can mine the essence of information in or among the data. This method not only takes advantage of the high accuracy of categorizations by machine learning method, but also overcomes its shortcomings that the prediction results are not continuous values but discrete values. Despite the success of our approach, there is still a high false positive ratio. So we will add more features from different level to inferring interface residues and binding protein information will also be considered based on quotient space thinking in the future work.

Acknowledgments. We would like to thank Dr. Chih-Jen Lin from National Taiwan University for providing the original the LIBSVM tool and Christopher Manning and Dan Klein from Stanford University for providing the maximum entropy software package. This work also is supported by the Project of Provincial Natural Scientific Fund from the Bureau of Education of Anhui Province (KJ2007B239) and the Project of Doctoral Foundation of Ministry of Education (200403057002).

References

1. Bradford, J.R., Westhead, D.R.: Improved prediction of protein-protein binding sites using a support vector machines approach. Bioinformatics 21(8), 1487–1494 (2005)
2. Res, I., Mihalek, I., Lichtarge, O.: An evolution based classifier for prediction of protein interfaces without using protein structures. Bioinformatic 21(10), 2496–2501 (2005)
3. Chen, H., Zhou, H.X.: Prediction of interface residues in protein-protein complexes by a consensus neural network method: test against NMR data. Proteins: Structure, Function, and Bioinformatics 61(1) (2005)
4. Li, M.H., Lin, L., Wang, X.L., Liu, T.: Protein-protein interaction site prediction based on conditional random fields. Bioinformatics 23(5), 597 (2007)
5. Bradford, J.R., Needham, C.J., Bulpitt, A.J., Westhead, D.R.: Insights into protein-protein interfaces using a Bayesian network prediction method. Journal of Molecular Biology 362(2), 365–386 (2006)
6. Sikić, M., Tomić, S., Vlahoviček, K.: Prediction of Protein-Protein Interaction Sites in Sequences and 3D Structures by Random Forests. PLoS Comput. Biol. 5(1), e1000278 (2009)
7. Du, X., Cheng, J.: Prediction of protein-protein interaction sites using granularity computing of quotient space theory. In: International conference on computer science and software engineering, vol. 1, pp. 324–328. Inst. of Elec. and Elec. Eng. Computer Society (2008)
8. Zhang, L.: The Quotient Space Theory of Problem Solving. Fundamenta Informaticae 59(2), 287–298 (2004)
9. Chang, C.-C., Lin, C.-J.: LIBSVM: a library for support vector machines (2001)
10. Banner, D.W., D'Arcy, A., Chne, C., Winkler, F.K., Guha, A., Konigsberg, W.H., Nemerson, Y., Kirchhofer, D.: The crystal structure of the complex of blood coagulation factor VIIa with soluble tissue factor (1996)
11. Rodgers, A.J., Wilce, M.C.: Structure of the gamma-epsilon complex of ATP synthase. Nature Structural Biology 7(11), 1051 (2000)

Application of Quotient Space Theory in Input-Output Relationship Based Combinatorial Testing

Longshu Li*, Yingxia Cui, and Sheng Yao

Key Lab of Intelligent Computing and Signal Processing Ministry of Education,
Anhui University, Hefei Anhui, China
cuiyingxia520@163.com
http://www.springer.com/lncs

Abstract. The input-output relationship based combinatorial testing strategy considers the interactions of factors and generates test suite to cover them rather than to cover all t-way interactions. It is a very effective combinatorial testing method. However in some cases, for its characteristic of black-box testing, the identified input-output relationships may actually be inexistent or incomplete and the range of faults diagnosis is wide. To address these issues, we apply the quotient space theory in input-output combinatorial testing and extend the model of the traditional strategy to be used for gray-box testing by identifying the input-output relationships of different granules of program and constructing the test suite for each granule. In the end of the paper, we have evaluated the strategy using a case study, which indicates the effectiveness of it.

Keywords: Quotient space, Input-output analysis, Gray-box testing, Variant strength test suite generation, Fault diagnosis.

1   Introduction

Software faults of a system may be triggered by not only some single factors (input parameters) but also the interactions of these factors, so combinatorial testing as one of such testing approach, which can detect the faults triggered by these factors and their interactions, was proposed by R.mandl in 1985. Combinatorial testing assumes that there is pair-wise or t-way$(t > 2)$interaction among any two (pair-wise combinatorial testing) or t factors (t-way combinatorial testing), but in some practical software systems, the interaction of factors may be very complicated and the strength of these interactions may be different, so that there are some problems in traditional combinatorial testing: (i) it is not always having interactions among any t factors; (ii) there may be some interactions

---

* Supported by Nature Science Foundation of Anhui Province under Grant Nos. 090412054, Department of Education Key Project of Anhui Province under Grant Nos. KJ2009A001Z.

J. Yu et al. (Eds.): RSKT 2010, LNAI 6401, pp. 735–742, 2010.

among factors more than t-way, which means some valid combinations may not be discovered by t-way test suite and some faults may not be detected. For these reasons, the test suite that generated by traditional combinatorial testing still has redundancy and the effectiveness of it can be increased too.

In this situation, Schroeder P J et al proposed a more general case of combinatorial testing, called input-output relationship based combinatorial testing (IO-CT) strategy which focuses on those input combinations that affect a program output parameter, rather than considering all possible input parameter combinations[1][2][3]. Unfortunately, all of these strategies are black-box testing, which leads to the identified input-output (IO) relationships, may actually inexistent or incomplete and further causes the generated test suite redundancy. This nature also makes a wide range of faults diagnosis after having detected faults. Thus, the strategies need to be improved to generate more effective test suite and locate the faults more efficiently.

In this paper, we propose a novel IO-CT strategy by using the granularity model based on quotient space proposed by Professor Zhang Lin [8] [9] [10]. We extend the model of traditional strategy to be used for gray-box testing by identifying the IO relationships of different granules program and generating the tests for each granule, in the sense that the IO relationships are obtained based on the partial information on a system's internals to avoid getting unreal IO relationships. In new strategy, we assume the finer granule test suite as an input parameter when generating the coarser granule tests by using VIOP tool, which we promoted in earlier works [6]. The union of the results from different granules is our ultimate test suite. As part of our research, we study a case using the new strategy on a program and the result has shown that it can significantly reduce the number of combinatorial tests and narrow the range of diagnosis at the same time.

The remainder of the paper is organized as follows. Section 2 briefly reviews the IO-CT strategy and quotient space theory. Section 3 details the new IO-CT strategies. Section 4 gives an example of the strategy. Section 5 concludes the paper.

2   IO-CT and Quotient Space

2.1   IO-CT

Normally, the test suite is referred to as an array, each row represents a test, and each column represents a factor (in the sense that each entry in a column is a value of the parameter represented by the column) [7]. In a t-way combinational of system under test (SUT), for each t-tipple of factors, every combination of valid values of these t factors must be covered by at least one test. IO-CT is a combinatorial testing strategy, which only needs to cover the factors that affect a program output parameter rather than to cover all t-way interactions, and generates variant strength combinatorial test suite according to the interactions of factors.

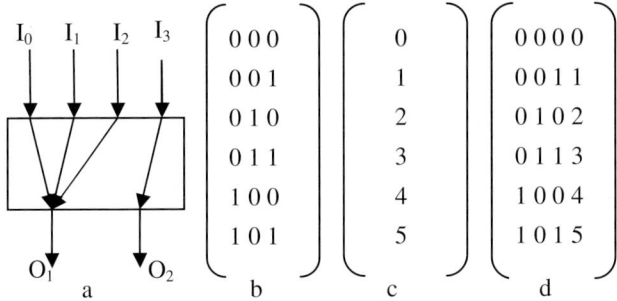

Fig. 1. An example system of IO-CT strategy

Suppose that SUT has n input parameters which influence m output parameters. Let $I = \{I_0, I_1, \ldots, I_{n-1}\}$denote the set of input parameters, and each parameter I_i has $I_{[i]}(0 \leq i \leq n-1)$ discrete valid values,and $V_i = \{0, 1, \ldots, I_{[i-1]}\}(0 \leq i \leq n-1)$ denotes the value set of I_i. Let $O = \{O_0, O_1, \ldots, O_{m-1}\}$ denotes the set of output parameters.

Definition 1. $O_j = \{I_{i1}, I_{i2}, \ldots, I_{ik}\}(I_{i1}\epsilon I, , I_{ik}\epsilon I)(0 \leq j \leq m-1)$ is the subset of input parameters that affect the jth output parameter.

Fig.1(a) shows a simple system of IO-CT strategy, where $I = \{I_0, I_1, I_2, I_3\}$, $O = \{O_0, O_1\}$, $V_0 = \{0, 1\}$, $V_1 = \{0, 1\}$, $V_2 = \{0, 1\}$, $V_3 = \{0, 1, 2, 3, 4, 5\}$, $O_0 = f(I_0, I_1, I_2), O_1 = f(I_3)$. First, generate 3-way test suite for the parameters $I_0, I_1$and I_2which is shown in Fig.1(a), in that $O_0 = f(I_0, I_1, I_2)$, in the sense that they have interactions with each other; then, since $O_1 = f(I_3)$, generate test suite for the parameters I_3which is shown in Fig.1(b); last, merge the two test suite into one test suite that is the ultimate test suite and is shown in Fig.1(c).∗presents "don't care" value, which can be replaced by any value without affecting the coverage of a test suite.

2.2 Quotient Space Theory

The quotient space is a basic concept of set theory, professor Zhang Lin and Zhang Bo promote that one of the basic characteristics in human problem solving is the ability to conceptualize the world at different granules and translate from one abstraction level to the others easily, i.e. deal with then hierarchically [9]. They use a triple(X, F, T) to describe problem space in quotient space, X is the domain of the problem; F is the attribute of domain X; T is the structure of domain X, which are the relationships between elements. When a triple(X, F, T) and the equivalence relation R are given, the definition of quotient space $([X], [F], [T])$ is as follows:

[X]defined by an equivalence relation R on X, that is, an element in [X] is equivalent to a set of elements, namely an equivalence class in X.

$[F] : [X] \to Y$, when $f : X \to Y$;

$[T] : \{z|g^{-1}(z) \in T, z \in X\}$, when $g : X \to [X]$ is a nature projection.

Granule [X] is formed from original domain, then coarser or finer granule from [X], which can form different granules.

3   Application of Quotient Space in IO-CT

3.1   Granules of IO-CT

According to quotient space granularity theory, we can solve the same problem from different granules, which divides original coarser granules into a number of finer granules or combines some finer granules to a few coarser granules to research. The traditional IO-CT is black-box testing, that is, it assumes the whole SUT as a black box, in others words, it only consider the problem from the coarsest granule, which sometimes lead to identified IO relationships that are inexistent or incomplete and the wide range of faults diagnosis.

In order to solve these problems, we identify the IO relationships from coarser granules to finer granules and generate the tests from reverse process. We first use a triple(X, f, T) to represent the IO-CT problem space, where X is the statements set of the program of SUT that is the coarsest granule, define $X = \{s_0 s_1 \ldots s_{n-1}\}$; f is an attribute function for each statement(for example, cycle, function, documents etc.); T is the structure relationships of elements in X, define $T_{ij} = f(s_i) \bigcap f(s_j)$, in which null element stands for that there is no relation between statements s_i and s_j. The finer granule [X] defined on domain X according to relations on different attributes, namely attribute-based method. Then we have a quotient space of(X, F, T) that is ([X],[f],[T]), where [X] -the quotient set of X, [f]-the quotient attribute of F, and [T] -the quotient structure of T. we also build a granule tree for the problem, which is shown in Fig.2.

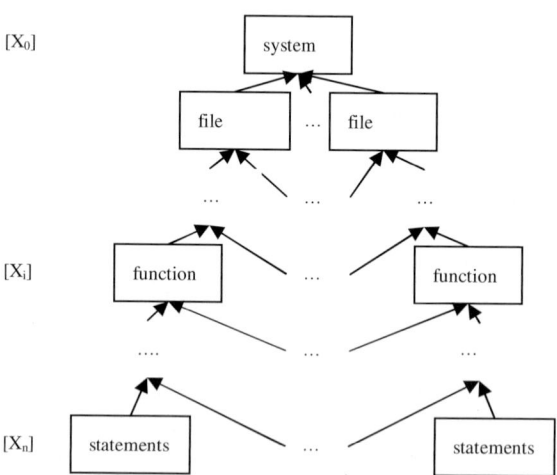

Fig. 2. Granule tree of IO-CT strategy

In Fig.2, we only list 4 different attributes and corresponding 4 different granules, it can have more attributes depending on different research objects.

3.2   IO-CT Test Suite Generation Based on Quotient Space

Definition 2. Suppose A is attribute set of SUT, $a(n)$ is an element of A. statements s_i and s_j have equivalent relationship R about $a(n)$ is defined : $a(n)\varepsilon f(s_i) \bigcap f(s_j)$.

Definition 3. Suppose $([X_0],[F_0],[T_0])$, and then its quotient space $([X_1],[F_1],[T_1])$ on R about $a(n)$ is defined as follows:

$[X_1] = \{s|s\varepsilon[X_0], a(n)\varepsilon f(s)\}$;

$[F_1] = [F_0] - \{a(n-1)\}$, where $a(n-1)$is an attribute $([X_0],[F_0],[T_0])$depends on;

$[T_1] = \{t|t\varepsilon[T_0] \bigcap f(s), s\varepsilon[X_1]\}$.

Definition 4. Suppose B is a program block of SUT, and then its IO relationships is denoted by $IO(B) : \{r|r\varepsilon O\}$, where $O = \{O_0, O_1, \ldots, O_{m-1}\}$.

The IO-CT test suite generation model based on quotient space is established here. In order to make the model generates smaller test suite and has a narrower range of fault diagnosis. It does not build the test suite at once, but builds it through several steps according to the identified IO relationships. Now, we describe the model in details in the following manner:

Step1. Initialize$([X_0],[F_0],[T_0])$, where $[X_0] = \{s_0, s_1, \ldots, s_{n-1}\}$,$[F_0] = \{a(0), a(1), \ldots, a(k)\}$, $[T_0] = \{f(s_0), f(s_1), \ldots, f(s_{n-1})\}$. Initialize $IO(SUT)$: $IO(SUT) = \{O_0 = f(I_0, I_1, \ldots, I_{n-1}), O_1 = f(I_0, I_1, \ldots, I_{n-1}), \ldots, O_{m-1} = f(I_0, I_1, \ldots, I_{n-1})\}$.

Step 2. Compute quotient space $([X_1],[F_1],[T_1])$ of $([X_0],[F_0],[T_0])$ and identify IO[B], where $B\varepsilon[X_1]$.

Step 3. Repeat step 2, where suppose $([X_1],[F_1],[T_1])$ as a new $([X_0],[F_0],[T_0])$, until granule reaches the leaf nodes of the granule tree.

Step 4. Generate variable strength combinatorial test suite for SUT.

We gradually generate test suite and assume the former test suite as a parameter and the elements of it as values in the case of the IO identified relationship included in IO relationship identified from the coarser one.

4   Case Study

For an example of our approach, consider a small software system in Fig.1 and its codes in Fig.3. But suppose the IO relationships of it are unclear, here.

Now we apply our approach to create test suite by following steps:

Step 1.Use a triple$([X_0],[F_0],[T_0])$ to present the problem, where $[X_0] = \{s_0, s_1, s_2, s_3, s_4, s_5\}$, $[F_0] = \{a(0) = system, a(1) = function, a(2) = statement\}$, $[T_0] = Sys$. The identified IO relationships are: $IO(Sys) = \{O_0 = f(I_0, I_1, I_2, I_3), O_1 = f(I_0, I_1, I_2, I_3)\}$.

```
int I₀,I₁,I₂,I₃;              int F2(int I0,int I1
int O₀,O₁;                    int I2)
int F0(int I₀,int I₁)         {…
{…                            s₂ return f2=f0(I0, I1)
s₀ return f0=f2(I₀,I₁)        }
}                             void main()
int F1(int I₀,int I₁)         {…cin>>I₀>>I₁>>I₂>>I₃;
{…                            s₃ O₂=F2(I₀,I₁,I₂);
s₁ return f1=f1(I₁)           s₄ O₀=F0(O₂,I₁);
}                             s₅ O₁=F1(I₂,I₃);
                             cout<<O₀<<O₁;…}
```

Fig. 3. A case for IO-CT based on quotient space

Step 2. Suppose quotient space of$([X_0], [F_0], [T_0])$ on R about a(1) is$([X_1], [F_1],$ $[T_1])$, so $[X_1] = \{\{s_0\}, \{s_1\}, \{s_2\}, \{s_3, s_4, s_5\}\}$, $[F_1] = \{a(1) = function, a(2) = statement\}$, $[T_1] = \{F0, F1, F2, main\}$ according to the definition. At this situation, we identify IO relationships are: $IO(F0) = \{f0 = f(F0(0), F0(1))\}$, $IO(F1) = \{f1 = f(F1(0), F1(1))\}$, $IO(F2) = \{f2 = f(F2(0), F2(1), F2(2))\}$, $IO(main) = \{O_0 = f(I_0, I_1, I_2, I_3), O_1 = f(I_0, I_1, I_2, I_3)\}$, where F(i) presents the ith formal parameter of function F.

Step 3. Suppose $([X_2], [F_2], [T_2])$ is the quotient space of $([X_1], [F_1], [T_1])$ on R about a(2), thus $[X_2] = \{\{s_0\}, \{s_1\}, \{s_2\}, \{s_3\}, \{s_4\}, \{s_5\}\}$, $[F_2] = \{a(2) = statement\}$, $[T_2] = \{s_0, s_1, s_2, s_3, s_4, s_5\}$, and the identified IO relationships are: $IO(\{s_0\}) = \{f0 = f(F0(0), F0(1))\}$, $IO(\{s_1\}) = \{f1 = f(F1(0))\}$, $IO(\{s_2\}) = \{f2 = f(F0(1), F0(1))\}$, $IO(\{s_3\}) = \{O_2 = f(F0(I_0, I_1, I_2))\}$, $IO(\{s_4\}) = \{O_0 = f(F1(O_2, I_2))\}$, $IO(\{s_5\}) = \{O_1 = f(F2(I_2, I_3))\}$.

Step 4. Merge all of the IO relationship sets and get the final IO relationships are $O_0\{O_2\{I_0, I_1\}, I_2\}$,$O_1\{I_3\}$.

Step 5. For IO relationship $O_2\{I_0, I_1\}$, since O_0 is a function of input parameters I_0 and I_1 only, it is these two parameters that will be used to generate combinatorial tests for output parameter O_2, as in Fig.4 (a).

Step 6. For $O_0\{O_2\{I_0, I_1\}, I_2\}$, here we assume the test suite of O_2 as one parameter which has 4 values:$\{O_{00}, O_{01}, O_{02}, O_{03}\}$, so we can get tests like step 5 and the result is shown in Fig.4(b).

Step 7. For $O_1\{I_3\}$ the result is shown in Fig.4(c). Step 8. The union set of these test suites, which is shown in Fig.4(d), is our ultimate goal.

Suppose we use traditional combinatorial testing strategy, since Sys has 4 input parameters, we must first consider how the strength of combinatorial testing to adopt. In order to achieve a high fault-detection probability, the most

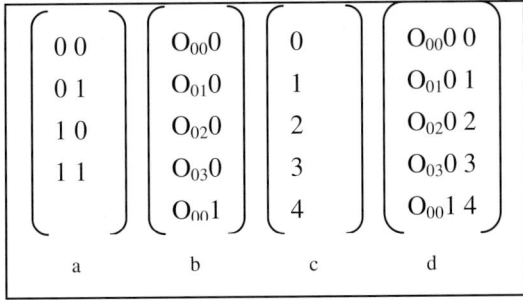

Fig. 4. General test suite generated by the new strategy

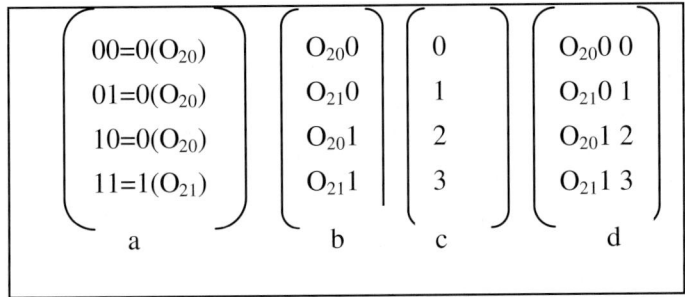

Fig. 5. Special test suite generated by the new strategy

thorough approach is to test every possible combination of the input parameters. In such a case, the size of test suite is $48(2*2*2*6)$, another choice is 3-way combinatorial or pair-wise testing, which respectively requires 24 tests or 12 tests. Further more , since the IO relationships are $O_0\{O_2\{I_0, I_1\}, I_2\}$ and $O_1\{I_3\}$, so there is no need to cover the interactions between I_3 and $\{I_0, I_1, I_2\}$. Thus, regardless of what strength adopted that the tests still redundant.

In Fig.4, suppose that O_2, f2 computes $\wedge$, then the tests of Fig.4(d) is actually 6, the details are shown in Fig.5. While the test suite shown in Fig.1(d) does achieve coverage of *Sys*, it has redundant tests. This reduces the size of test suite also narrows the range of fault diagnosis. Suppose that f2 is incorrectly implemented as $\vee$ instead of $\wedge$.The test suite in Fig.4(a) reveals this fault, so we only need to obtain the factors that cause the faults in F2 while test suite in Fig.1(b) does, we need to obtain the factors in F0 and F2. Thus, the new strategy not only optimizes the test suite but also narrows the range of fault diagnosis.

5   Conclusions

In this paper, we propose a novel IO-CT strategy by using the granularity model based on quotient space, which extends the traditional strategy to be used for gray-box testing. We identify the IO relationships of different granules from a granule tree which we build for the problem and generate the tests for each granule. We assume the finer granule test suite as an input parameter when generating the coarser granule tests and our ultimate target is the union of the results. As part of our research, we conduct an experiment using the new strategy on a program and the experimental data shows the advantage of the strategy.

Our future work will focus in exploring new applications of quotient space on testing and developing a test tool supporting these strategies.

References

1. Schroeder, P.J.: Black-box Test Reduction Using Input-output Analysis. In: Harold, M.J. (ed.) Proc. of the Int'l. Symp. on Software Testing and Analysis (ISSTA), pp. 173–177. ACM Press, New York (2000)
2. Schroeder, P.J., Korel, B., Faherty, P.: Generating Expected Results for Automated Blackbox Testing. In: Proceedings of the International Conference on Automated Software Engineering, pp. 139–148. IEEE Press, Washington (2002)
3. Cheng, C.T., Dumitrescu, A., Schroeder, P.J.: Generating Small Combinatorial Test Suites to Cover Input-output Relationships. In: Proceedings of 3rd International Conference on Quality Software, pp. 76–82. IEEE Press, Washington (2003)
4. Wang, Z., Nie, C., Xu, B.: Generating Combinatorial Test Suite for Interaction relationship. In: Proceeding of 4th International Workshop on Software Quality Assurance, pp. 55–61. ACM Press, New York (2007)
5. Cohen, M., Colbourn, C., Collofello, J., Gibbons, P., Mugridge, W.: Variable Strength Interaction Testing of Components. In: Proceedings of the 27th Annual International Computer Software and Applications Conference, pp. 413–418. IEEE Press, Washington (2003)
6. Cui, Y., Li, L., Yao, S.: A New Strategy for Pairwise Test Case Generation. In: Proceedings of the 3rd International Conference on Intelligent Information Technology Application, pp. 303–306. IEEE Press, Piscataway (2009)
7. Lei, Y., Kacker, R., Kuhn, D.R., Okun, V., Lawrence, J.: IPOG: A General Strategy for t-way Software Testing. In: Proc. of the 14th Annual IEEE International Conference and Workshops on the Engineering of Computer-Based Systems, pp. 549–556. IEEE Press, Los Alamitos (2007)
8. Zhang, B., Zhang, L.: Theory and application of problem solving. Tsinghua University Press, Beijing (1990)
9. Zhang, L., Zhang, B.: Granule Computing Based on Quotient Space Model. Journal of Software, Beijing, 770–776 (2003)
10. Zhang, L., Zhang, B.: The Quotient Space Theory of Problem Solving. In: Fundamental Informaticae, pp. 287–298. IOS Press, Amsterdam (2004)
11. Li, H., Jin, H., Zhang, D.: Application of Quotient Space Theory in Cases Organization. In: International Symposium on Intelligent Information Technology Application Workshops, pp. 975–978. IEEE Computer Society, Washington (2008)
12. Zhao, S., Zhang, Y.-p., Zhang, L., Chen, J., Wan, Z., Zhang, Y.-c., Zhang, C.-x.: The Quotient Structure in Granule Computing. In: 2005 IEEE International Conference on Granular Computing, pp. 359–362. IEEE Press, Los Alamitos (2005)

Granular Analysis in Clustering Based on the Theory of Fuzzy Tolerance Quotient Space

Lunwen Wang[1,2], Lunwu Wang[2], and Zuguo Wu[2]

[1] National Lab of Information Control Technology for Communication System,
Jiaxing 314033, China
[2] 309 Research Room of Electronic Engineering Institute, Hefei, 230037
wanglunwen@163.com

Abstract. Clustering is defining an equivalence relation between the samples in nature, and two samples are equivalent if they belong to one class. The rough and fine of the granularity reflect the similarity threshold in clustering. In this paper, the disadvantage of granular analysis in clustering based on the theory quotient space, which can't solve the problem when there are intersections between classes, is pointed out, and the theory of fuzzy tolerance quotient space is introduced, and granular analysis in clustering based on the theory of fuzzy tolerance quotient space is presented. The results of the experiment about clustering radio communication signals show the efficiency of the algorithm.

1   Introduction

Clustering is an important research subject in data mining. At present, there are many clustering algorithms, such as partitioning method (k-means, CLARANS), hierarchical method (BIRCH, CURE), density-based method (DBSCAN, OPTICS), grid-based method (STING, CLIQUE), model method (COBWEB, SOM). Whatever the methods are, there are two problems need to be disposed, which are the option of similarity function in clustering and the decision of similarity threshold. Similarity function is related with the measure standard, the distance and the fuzzy equivalence relation are the common standards. The clustering algorithms are different because of the difference of similarity measure standards. For example, the equivalence relation of distance is used in the partitioning method while the fuzzy equivalence relation is used in the fuzzy clustering method [1].

The decision of the similarity threshold is another key problem in clustering, so granularity analysis in clustering is generated. Since "granularity computation" is proposed by Zadeh, the research of granularity computation becomes popular in the artificial intelligence [2]. Because the concept of class is explained by granularity visually, and the similarity threshold in clustering can be decided by the selection of roughness and fineness of granularity, granularity is used in clustering by more scholars. In [3], the license plate binary analyzing algorithm based on information granularity is achieved. In [4], according to initial equivalence relation and the equivalence class which are formed by evaluating relative similarity to each object, a method of rough clustering based on granularity is proposed. In [5], the significance of granular analysis in clustering is illuminated. The result of clustering is adjusted according to the selection of

J. Yu et al. (Eds.): RSKT 2010, LNAI 6401, pp. 743–750, 2010.
© Springer-Verlag Berlin Heidelberg 2010

granularity size based on the quotient space theory. When the granularity is rough, the number of class is small. When the granularity is fine, the number of class is large. The fit clustering number is got through selecting the best granularity. Though these clustering methods have many advantages, the class is determined by selecting an equivalence relation and there is a definite gap between classes. But there are intersections between classes in some case, there are some disadvantages when the boundary point in clustering is disposed by granular analysis based on the quotient space. So the theory of fuzzy tolerance quotient space is presented to solve this problem.

In section 2, granular analysis in clustering based on quotient space is introduced briefly and its disadvantage is pointed out. In section 3, granular analysis in clustering based on fuzzy quotient space is discussed in detail. In section 4, the procedures of the algorithm are given. Finally, the method is applied to clustering of radio monitoring data and the results of the experiment show this algorithm's efficiency.

2   Quotient Space and Granular Analysis in Clustering

The theory of quotient space is a method that describes the different granularity world. We can use a triple form(X, f, T) to depict a problem. X, the universe of discourse, is the set of objects. f is a set of attributes of basic elements. T, defined as the relation between each element in the universe, denotes the structure of universe. The problem (X, f, T) is investigated in different granularity (angle, hierarchy), which gives an equivalence relation R in X, and quotient set [X] is generated by R. Then the relative problems ([X], [f], [T]) are researched. [f] represents the relative quotient attribute function in the quotient set [X], while [T] represents the relative quotient structure. ([X], [f], [T]) is called quotient space of (X, f, T). The different granularity world of the problem (X, f, T) is composed of the different quotient set and the relative quotient space in X. More details can be referenced to [6].

Clustering is similar to the quotient space partition. The partitioning of the quotient space is that the equivalent elements of the assigned assemble are partitioned together based on the equivalence relation. However, clustering is to gather similar samples by the certain criterion. The samples in one class are equivalent while the samples between classes are different. Granular analysis in clustering can be analyzed from different granularity partitions in the quotient space.

The clustering is adjusted according to the granularity of the equivalence relation, so there is no intersection between classes. From Fig. 1, ▲ belong to class A while ■ belong to class B. There is definite division between classes, so the clustering can be completed through defining the equivalence relation RA and RB. In Fig.2, if there are

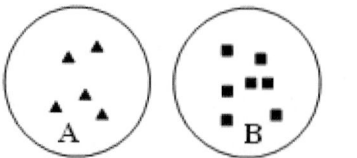
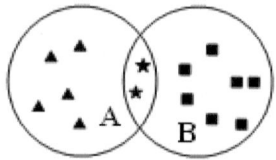
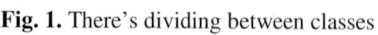

Fig. 1. There's dividing between classes   **Fig. 2.** No dividing between classes

intersections between classes, ★maybe belong to class A or class B to some extent. The equivalence relation R_A or R_B. should be rough under this circumstance. If ★ are partitioned to class A or class B simply, it is ill-considered in fact.

3 Granular Analysis Based on Fuzzy Tolerance Quotient Space

The theory of fuzzy tolerance quotient space is presented by Professor Zhang Ling to solve the problem above. The disadvantage of the granularity based on the equivalence class is that there is no intersection between classes. But in fact, there are intersections between classes. From the view of application or the continuation of theory, the model of granularity computation based on the equivalence class should be generalized, for example, tolerance, similarity, fuzzy tolerance. It is necessary to illustrate the relation between tolerance and equivalence. In [7], it is approved that the necessary and sufficient condition of fuzzy equivalence relation is fuzzy tolerance relation. In order to assure completeness of the thesis, some concepts are introduced. The depiction of the problem is similar to a triple form(X, f, T).

Definition 1. Suppose $C_i, i = 1, ... n$ is the subset of X, if $\bigcup_{i=1}^{n} C_i = X$, then $\{C_i, i = 1, ..., n\}$ is defined as a cover of X.

Definition 2. Suppose the function $R : X \times X \to \{0,1\}$ satisfies: ① reflexivity, $\forall x \in X$, $R(x,x) = 1$, ② symmetry, $\forall x, y \in X, R(x, y) = R(y, x)$. Then R is called as a tolerance relation of X.

Definition 3. Suppose $\{C_i, i = 1, ..., n\}$ is a cover of X, define the function $R : X \times X \to \{0,1\}$: $R(x, y) = 1$, if $\exists C_i, x, y \in C_i$, else define $R(x, y) = 0$.

According to definition 3, there are many covers corresponding to a tolerance relation. But the maximum cover and the tolerance relation are one-to-one correspondence. The cover is the maximum cover usually, except under special illustration.

Definition 4. Suppose the function $R : X \times X \to [0,1]$ satisfies: ① reflexivity, $\forall x \in X$, $R(x,x) = 1$, ② symmetry, $\forall x \in X, R(x, y) = R(y, x)$. Then R is called as a fuzzy tolerance relation of X.

The difference between fuzzy tolerance relation and general tolerance relation is that the ranges are changed from $\{0,1\}$ to $[0,1]$.

Definition 5. Suppose R_1 , R_2 are two fuzzy tolerance relations of X. If $\forall (x, y) \in (X \times X)$, $R_2(x, y) \le R_1(x, y)$, then R_2 is finer than R_1 , denoted by $R_1 < R_2$.

Under the relation "<"defined above, all of the fuzzy tolerance relations in the universe of discourse forms a complete partial order case.

Definition 6. Suppose there are a series of tolerance relations $\{R_1, R_2, ..., R_n\}$ in the universe X, which satisfy $R_1 \leq R_2 \leq ... \leq R_n$, then $\{R_1, R_2, ..., R_n\}$ is called as a hierarchical cover chain.

After introducing the theory of fuzzy tolerance quotient space, let's review the problem above. When there are intersections between classes, the clustering can be achieved by defining fuzzy tolerance relation. Take ★in Fig.2 for example, define fuzzy tolerance relation R_A and R_B , R_A (★, ▲) and R_B (★, ■) are decimals between 0 and 1, which denote the relationship between ★and A, B. Then, the class of data point can be determined by comparing fuzzy tolerance relation, and it also can be determined by the expert knowledge.

    In a sense, the difference between the theory of fuzzy tolerance quotient space and the theory of quotient space is that the partition of the different granularity is not a definite line, but a region.

4   Procedures of the Algorithm

In order to adjust granularity better, the transformation of the granularity is discussed firstly. In this thesis, two definitions are introduced.

Definition 7. Let that R_1 and R_2 denote two tolerance relations on universe X. If R is also a tolerance relation in X, where $R_1 < R$ and $R_2 < R$, and there is another R' which satisfies $R_1 < R'$, $R_2 < R'$ and $R < R'$, R is regarded as the arithmetic product of R_1 and R_2 , denoted by $R = R_1 \otimes R_2$.

Definition 8. Let that R_1 and R_2 denote two tolerance relations on universe X. If R is also a tolerance relation in X, where $R < R_1$ and $R < R_2$, and there is another R' which satisfies $R' < R_1$, $R' < R_2$ and $R' < R$, R is regarded as the sum of R_1 and R_2 , denoted by $R = R_1 \oplus R_2$.

From the above definition we know $R_1 \otimes R_2$ is the roughest division that can subdivide R_1 and R_2 , $R_1 \oplus R_2$ is the finest division that can be subdivided by R_1 and R_2. That is, $R_1 \otimes R_2$ is the roughest upper bound divided R_1 and R_2, and $R_1 \oplus R_2$ is the finest lower bound.

    When it comes to a specific clustering problem, the granularity of clustering is adjusted by definition 7 and definition 8 in order to cluster appropriately. The selection of clustering threshold can be solved by this method that analyzes clustering based on the granularity selection. Fig. 3 is the flow chart of algorithm, the procedures are as follows:

①According to the needs of the problem, an assemble that is corresponding to the partition of fuzzy tolerance relation R_0 is preset (Δ_0 is the corresponding granularity) in order to get fuzzy tolerance quotient space S_0, then the initial conclusion A_0 is obtained.

②According to this conclusion A_0 and granularity Δ_0, determine whether the cluster is rough or fine. If it is rough, go to③. If it is fine, go to④. Otherwise, go to⑤.

③ According to present conclusion, choose an tolerance relation R_0' (finer)and $R_1 = R_0 \otimes R_0'$. Analyze again in R_1, and obtain clustering granularity Δ_1 and conclusion A_1. If the granularity is appropriate, go to⑤. If it is still rough, repeat③. If it is fine, go to④.

④ According to present conclusion, choose an tolerance relation R_0' (rougher) and $R_1 = R_0 \oplus R_0'$. Analyze again in R_1, and obtain clustering granularity Δ_1 and conclusion A_1. If the granularity is appropriate, go to⑤. If it is still fine, repeat④. If it is rough, go to③.

⑤ Analyze the samples in the intersection region, determine the class according to the fuzzy tolerance relation and knowledge of expert.

⑥ End.

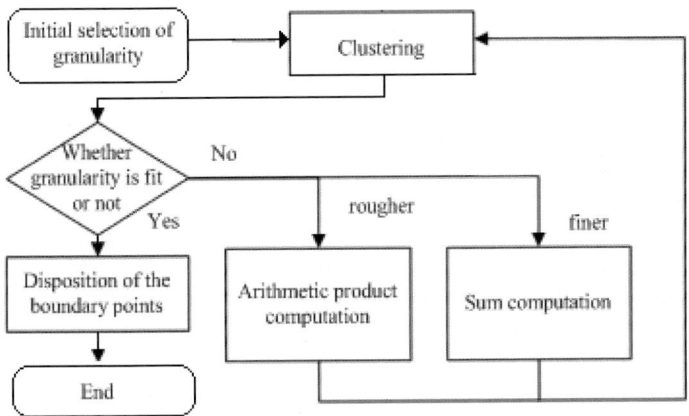

Fig. 3. Flow chart

5   The Experiments and the Results

The platform of experiments is PC with PentiumIV3.0GHz, EMS memory is 1G, and the operation system is Window XP. The data is the same as [5]. We take the radio monitoring data for clustering. The short wave receiver, remote controlled by the

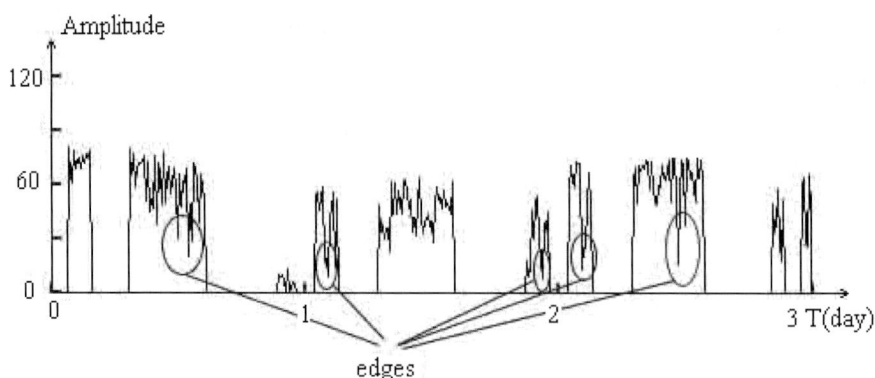

Fig. 4. The time-amplitude figure of 15.06MHz

computer, searches from 15MHz to 16MHz. The step is 3 kHz and there are 333 frequency points. Then sampling signals from the intermediate frequency output port of the receiver are transformed to frequency-domain data by the Fast Fourier Transform.

We judge whether there is a signal on any frequency point through clustering. If signals do exist, it should be judged how many signals there are. Because the powers of the stations are different, the sampling values of the different receiver are varied. The appropriate clustering granularity should be selected by the method based on granular analysis in clustering. The "large-signals" are clustered by rough granular analysis, while the "small-signals" are clustered by fine granular analysis.

The results of the experiments based on two granularities in clustering are shown in Table 1. From table 1, the results of clustering based on fuzzy tolerance quotient space are better, compared with quotient space. The detailed reasons are discussed as follows.

Fig.4 is the time-amplitude figure of 15.06MHz. The data are divided into two classes by whether consisting of signals. The boundary points that should be disposed in clustering are marked by ellipses. These data points are affected by the noise and the propagation path and disposed difficult in clustering. There is a definite boundary between classes when granularity is selected by quotient space, and these boundary points belong to the class of no signals. The results of the experiments are shown in Fig.5, and the results in circle are incorrect.

The boundary samples marked in Fig.4 are in the boundary region when granular analysis in clustering based on fuzzy tolerance quotient space is used. These points belong to two classes according to a given degree of membership. Their belongingness should be judged according to specific situation. We can judge the class of the

Table 1. Results of the experiments based on different methods

Clustering method	Total signals	Incorrect signals of clustering	Incorrect radio of clustering
Quotient space	333	8	2.4%
Fuzzy tolerance quotient space	333	4	1.2%

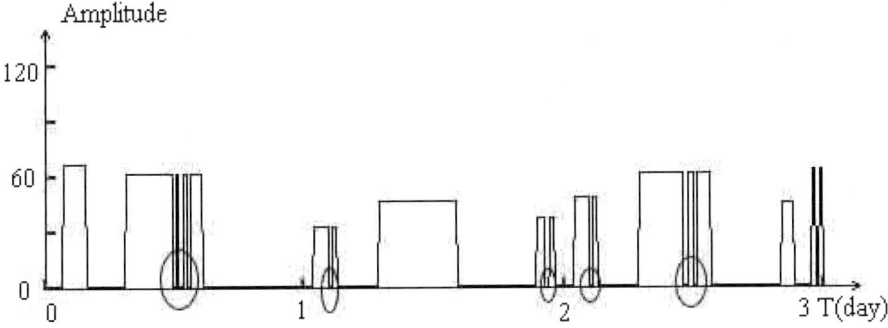

Fig. 5. The time-amplitude after clustering with QS

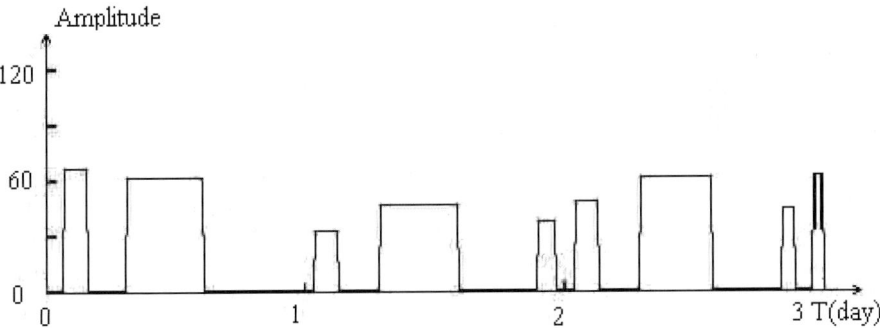

Fig. 6. The time-amplitude after clustering with TQS

boundary points according to time association (continuity) and the amplitude range of the signals. In Fig.6, the results of the experiments based on fuzzy tolerance quotient space are better than results of Fig.5.

In the paper, the disadvantages of granular analysis in clustering based on quotient space are discussed, and the reasons of them are analyzed. The theory of fuzzy tolerance quotient space is used in granular analysis in clustering. The problems of the traditional granular analysis in clustering are settled. The experimental results show its validity.

References

[1] Ji-Gui, S., Jie, L., Lian-Yu, Z.: Clustering Algorithms Research. Journal of Software 19(1), 48–61 (2008)
[2] Guo-Yin, W., Qing-Hua, Z., Jun, H.: An Overview of Granular Computing. CAAI Transactions on Intelligent Systems 2(6), 8–26 (2007)
[3] Jie, C., Ying-Chun, Z., Yan-Ping, Z., Ling, Z.: Analysis and Application of Clustering Based on Information Granularity. Journal of Image and Graphics 12(1), 87–91 (2007)

[4] Ming, H.: Analytical Method of Rough Clustering Based on Granulation. Computer Engineering 34(8), 203–204 (2008)

[5] Lun-Wen, W.: Study of Granular Analysis in Clustering. Computer Engineering and Applications, 29–31 (2006)(5)

[6] Ling, Z., Bo, Z.: Theory of Problem Solving and Its Application—Theory and Application of Granular Computation Based on Quotient Space, pp. 1–188. Tsinghua University Press, Beijing (2007)(3)

[7] Dasa, M., Chakraborty, M.K., Ghoshal, T.K.: Fuzzy tolerance relation, fuzzy tolerance space and basis. Fuzzy Sets and Systems 97, 361–369 (1998)

Computing the Point-to-Point Shortest Path: Quotient Space Theory's Application in Complex Network

Fugui He*, Yanping Zhang, Shu Zhao, and Ling Zhang

School of Computer Science and Technology, Anhui University, Hefei 230039, China
{fuguihe}@163.com

Abstract. The quotient space theory can represent the world at different granularity sizes and deal with complicated problems hierarchically. We present significant improvement to point-to-point shortest path based on quotient space theory in complex large-scale network. We propose the shortest path algorithm that is a heuristic method, in which evaluation function is based on community and hierarchical granularity decomposition of quotient space theory. In preprocessing, we decompose large-scale network into some communities using hierarchical granularity decomposition of quotient space theory, compute and store the minimum spanning trees in the communities and the shortest distance among communities. The implementation works on the large-scale road network. From experimental results, we know the proposed algorithm is effective and efficient in the road network of US.

Keywords: quotient space theory, point-to-point shortest path, complex network.

1 Introduction

Granular computing[1,2,3,4,5,6] is one of foci in AI fields. The quotient space theory[6] imitates the human's thought of solving complex problems and proposes some approaches for the granular world model and some applies them to heuristic search, path planning, etc.

For the shortest path finding in network, Dijkstra and Floyd algorithm are used widely and are all global searching algorithm basically. Due to large space demanding and time-consuming, Dijkstra and Floyd algorithm aren't suitable for solving the path problem in the large-scale complex network. A* search algorithm[7], a heuristic optimization technique in the field of AI, is a goal-directed approach to solve point-to-point shortest-path problem. It adds an evaluation function as the sense of direction to the search process. Landmark-based A* search (ALT) algorithm[8,9,10] uses a small number of vertices (landmarks)

---

* Supported by the National Basic Research Program of China under Grant Nos.2007CB311003 (973 Project), by the Innovative Research Team of 211 Project in Anhui University.

J. Yu et al. (Eds.): RSKT 2010, LNAI 6401, pp. 751–758, 2010.

and the triangle inequality to compute evaluation function. A crucial point in ALT algorithm is how to find good landmarks. In large-scale complex network, some speedup techniques[10,11,12] divide the procedure of path finding into two parts: preprocessing and query step.

The goal of the work presented in this paper is to test the point-to-point shortest path finding based on quotient space theory in complex large-scale network. In preprocessing, the idea of hierarchical granularity decomposition is a good approach to simplify complex problem. We analyze network global layout information: topological centrality and community. According to topological centrality of network, the communities' centers are selected. Communities are identified according to modularity as community structure evaluation criteria. Then we compute and store the minimum spanning trees in the communities and the shortest distance among communities. In query, according to the minimum spanning trees in the communities and the shortest distance among communities, it changes computing evaluation function of the traditional heuristic algorithm such as A* search algorithm and avoids the difficult selecting of landmarks of ALT. In the experiment, we concentrate on point-to-point shortest path in transportation road network. We test our algorithm comparing with A* and ALT in five US road networks. Experimental results presented in Section 4 are effective and efficient.

2   Hierarchical Granularity Decomposition of Quotient Space Theory

The granular world in the quotient space theory[6] is different from the information granule. Information granules are partitioned according to equivalence relation, this only changes the granule of universe of problems and attributed relationship, and the space structure does not been discussed and changed. The representation of different granular worlds-quotient space in this paper changes both the universe, attributed relationship and the space structure of a problem.

In this paper, for the shortest path finding in large-scale network, our aim is to decompose it into some small sub-graphs. It reduces the scale of the problem solving. In [13], according to different weights of edges of weighted network, Zhang et al construct a series of different quotient space networks which are hierarchical. The edge information is considered to construct the quotient space to avoid losing path information in network topology. It leads to little change of network scale among adjacent granular worlds in the hierarchical space chain. In this paper, according to network global layout information we plan to construct quotient space to speed up decomposing network into some sub-graphs.

In the field of complex network, most networks show community structure. Sub-graphs of vertices that have a higher density of edges within them while a lower density of edges between sub-graphs. A community is thought as a group, and compared to other communities, it is a whole. During decomposing network, using hierarchical granularity decomposition of quotient space theory, we think that a community is a granule, and an equivalence class. With quotient space theory to represent network decomposition, a community is an element of quotient set. In the quotient topology, we consider link relation among communities, and note edges information among communities. So the large-scale complex network is decomposed into some communities and a quotient space. If scale of the quotient space is too large, we then decompose it in the same way. If communities have moderate scale, it is suitable for the classical algorithms to solve the shortest path in large-scale network. Community structure discovery is a technique of network global layout information mining. Our goal is to extract moderate scale communities.

In recent years, there are some researches on community structure discovery: spectral bisection method[14], GN[15], fast divisive algorithm[16] and information centrality algorithm [17]. Due to high time complexity, the above algorithms which require over $O(n^2)$ time, aren't suitable for community structure discovery in the large-scale complex network. In [18], Newman and Girvan propose modularity as community structure evaluation criteria. But Brandes et al[19] prove that modularity maximization is NP-hard. In optimize modularity methods, scale of identified communities is similar[20]. We decompose network into some communities taking modularity as community structure evaluation criteria.

Hierarchical granularity decomposition to construct quotient space: Computing network structure centrality property of each vertex, and selecting some local core vertices from the local maximum property value; For each local core vertex as community's center, the community absorbs core vertex's surrounding vertices. For controversial vertices among communities, determine the community which the vertices belongs to according to modularity; According to modularity, adjust a part of communities: Merge some small-scale communities.

3   Point-to-Point Shortest Path Algorithm

3.1   Preprocessing

Our goal is to introduce a novel evaluation function which is different from the evaluation function of traditional heuristic search algorithm. The preprocessing entails decomposing network into some communities by hierarchical granularity decomposition, labeling margin vertices, computing and storing the minimum spanning trees in communities and the shortest distances among communities.

Given a graph $G(V, E)$ with non-negative edge weights. We decompose $G(V, E)$ into some communities by hierarchical granularity decomposition. It is the process of constructing quotient space $G_1(V_1, E_1)$ of $G(V, E)$. A community is one element of quotient space $G_1(V_1, E_1)$. Edges among communities reflect the quotient topology. In $G(V, E)$, compute network structure centrality property of each vertex, and select some local core vertices from the local maximum property value as community's centers. On complex network research, except for degree centrality and topological centrality, computing other network structure centrality properties are concerned with shortest path in the whole network and require over $O(n^2)$ time. Degree centrality only considers the out-degree characteristic while ignores in-degree characteristic of each vertex[21]. To reduce network decomposition time complexity, we select topological centrality as network structure centrality property. Taking each local core vertex as community's center, communities absorb local core vertex's surrounding vertices.

Let a community as an equivalence class, with quotient space theory to represent network decomposition, a community is an element of quotient set. In the quotient topology, edges information among communities is regarded as structure of quotient set. We note multiple vertices of a community contact with other communities as attribute of quotient set. So the quotient space $G_1(V_1, E_1)$ is a hyper graph which each vertex is one community of initial graph $G(V, E)$.

For one community as basic unit, we compute the minimum spanning trees in the communities. In the hyper graph of quotient space, we compute the shortest distance among communities by the classical shortest path algorithm. After preprocessing, we store the minimum spanning trees of communities and the shortest distances among communities in $G_1(V_1, E_1)$. It helps to compute evaluation function in query step for computing the shortest path from the source vertex to the destination vertex.

If scale of quotient space $G_1(V_1, E_1)$ isn't suitable for the classical path finding algorithm, we will decompose quotient space to a moderate scale by repeating the above steps. According to the scale of different quotient spaces, a quotient space chain is constructed by granularity level. We compute the shortest distance between two elements of the coarsest quotient space and the minimum spanning trees in other quotient spaces of the quotient space chain. This decomposition is bottom-up while the point-to-point shortest path problem solving is top-down according to the quotient space chain.

3.2 Query

In $G(V, E)$, given two vertices, the source s and the destination t. A* search method [7] uses evaluation function of vertex v: $e(v) = d_s(v) + \pi_t(v)$ to solve point-to-point shortest-path problem. $d_s(v)$ denotes the shortest path's distance from s to v. $\pi_t(v)$ denotes an estimate on the distance from v to t. In our method's query step, we use the minimum spanning trees of communities and the shortest distance between two communities in quotient space $G_1(V_1, E_1)$ to compute evaluation function.

For the scanning vertex v, if estimated value $\pi_t(v)$ close to the true shortest distance from v to t, it reduces the searched area and visited vertices during query step. We refer to A* search method that use an optimal function $\pi_t(.)$. Let us discuss how to select an optimal function $\pi_t(.)$: First, define lower bound, $\pi_t(t) = 0$. For the scanning vertex v, according to the shortest distance between two communities in initial network $G(V, E)$, we firstly identify the community $C(v)$ which vertex v belongs to. If vertex v and vertex t belong to the same community $(C(v) = C(t))$, according to the minimum spanning trees of the community, $\pi_t(v)$ don't need to compute and the shortest path from vertex v to vertex t can be searched. Otherwise, we compute the shortest distance $d_t(i)$ from vertex t to vertex i which is one of all vertices of the community $C(v)$

contact with other communities. Let $d_{C(v)}(C(t))$ is the shortest distance between community $C(v)$ and $C(t)$ in $G_1(V_1, E_1)$. $\pi_t(v) = d_{C(v)}(C(t)) + d_t(i)$.

Queries for our method work as follows, its pseudo-code is given in Algorithm.

```
Algorithm: (G(V,E), source s and destination t)
Q.add(s,0);//priority queue
//array for distance of vertices from source s
distance[];
while (!Q.empty() or Q.minElement()!=t )
     u=Q.deleteMin();
     for all outgoing edges e from u
          v= e.head;
          If distance[u]+ e.weight()< distance[v]
               distance[v]=distance[u]+ e.weight();
               for all vertices i of the community C(v)
                    e(v)=min(distance[v]+ d_{C(v)}(C(t)) + d_{t}(i));
               End for
               Enqueue(v, e(v));
          End if
     End for
End while
```

4   Experimental Results

To illustrate practical implications of the above techniques, we here concentrate on road network. The road networks are parts of states and a district in America which are taken from the DIMACS Challenge homepage. Each vertex has a latitude and a longitude. Table 1 gives more details of the networks used, as well as the shorthand names we use to report data. For each graph, distances of edges are calculated according to the actual Euclidean length of the road segments.

Our implementation is written in Visual C++. Our tests execute on one core of an Intel. Pentium 4 running Windows XP SP2. The machine is clocked at 3.00GHz, has 2GB of RAM. In preprocessing, we decompose network into some communities by hierarchical granularity decomposition. According to topological centrality, the communities' centers are selected. Communities are identified taking modularity as community structure evaluation criteria. Kruskal algorithms

Table 1. Transportation road network of parts of states and district in America

Name	Description	No. of vertex	No. of edges	Latitude and longitude range
DC	District of Columbia	9559	14909	[38.7, 38.9]/[-76.9,-77.1]
HI	Hawaii	64892	76809	[18.9, 22.2]/[-15.4,-16.0]
AL	Alabama	566843	661487	[30.2, 35.0]/[-84.9,-88.4]
CA	California	1613325	1989149	[32.5,42.0]/[-114.1,-123.9]
TX	Texas	2073870	2584159	[25.8,36.5]/[-93.5,-106.6]

Table 2. Topological centrality of five road network

Role of vertex		0	0.25	0.33	0.4	0.5	0.6	0.67	0.75	0.8	0.83	1.0
	DC	1950	127	2344	1	1840	15	1300	1511	46	3	422
	HI	23124	40	8545	0	9527	7	16089	3127	21	0	4412
No. of vertex	AL	209034	239	59857	11	94840	279	132622	17042	94	1	52825
	CA	593134	576	146665	27	248203	586	379605	83056	124	17	161332
	TX	797642	829	188533	32	296267	428	460860	131569	186	13	197511

are used to compute the minimum spanning trees in the community and Floyd algorithm is used to compute the shortest distances among communities in the quotient space.

In order to determine the communities' centers, according to topological centrality, vertices can play different roles: core vertex, margin vertex, bridge vertex and mediated vertex [21]. The role of vertex i is distinguished by the ratio of $\sharp l_i$ and $\sharp h_i$. $\sharp l_i$ denotes the number of neighbor vertices of i with topological centrality degrees lower than vertex i. $\sharp h_i$ denotes the number of neighbor vertices of i with topological centrality degrees higher than vertex i. Table 2 gives distribution of role of vertices in five road networks.

For topological centrality of network, we focus on vertices' distribution of the role of vertex as 1 which is core vertices. In Table 2, the number of core vertices is listed. In order to avoid occurring small-scale communities, we merge small-scale communities. From Fig. 1, the scale of communities is suitable to use the classical minimum spanning tree algorithm. For very large-scale network, the number of communities will be large if there are many small-scale of communities. It brings to some trouble during computing the shortest distance among communities. Such as in CA and TX network, we carry out decomposition of the hyper graph which vertex is one community and four-levels quotient space chain is constructed by granularity level. We compute the shortest distance between

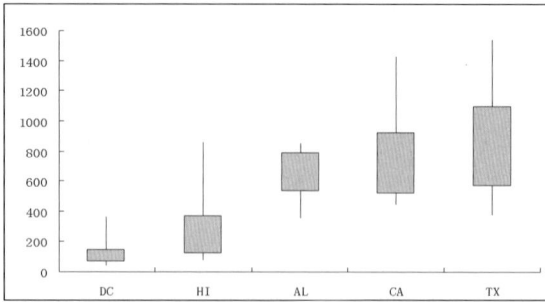

Fig. 1. The scale distribution of communities in the five road networks: minimum, maximum, average minimum and average maximum of scale

Table 3. Preprocessing and query results of our algorithm in five road networks: preprocessing time(PT), total space in disk required by the preprocessed data(TS), No. of community(NC), average ratio of the number of scanned vertices and the number of vertices on the shortest path(ARatio) and average query time(AQT).

Input	PT[s]	TS[MB]	NC	ARatio	AQT[ms]
DC	3.819	0.18	132	2.71	8.76
HI	43.56	1.32	386	3.36	27.37
AL	766.04	5.21	766	3.73	70.78
CA	1696.4	20.06	2016	3.80	129.83
TX	2310.7	28.69	2475	3.56	164.29

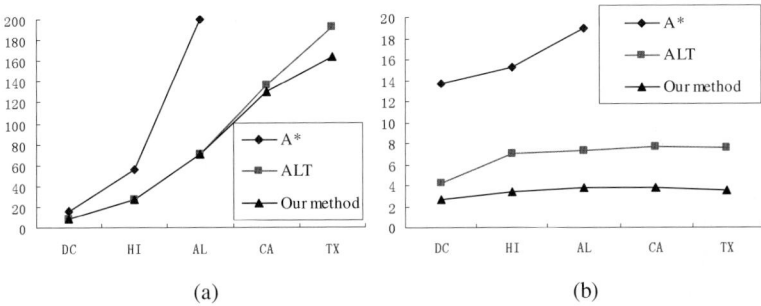

(a)   (b)

Fig. 2. (a)comparing with average running time,(b)comparing with average ratio of the number of scanned vertices and the number of vertices on the shortest path of A*, ALT and our algorithm in 100 source-sink pairs

two elements of the coarsest quotient space and the minimum spanning trees in other quotient spaces of the quotient space chain. In experiment, we test our algorithm comparing with A*, ALT. For five road networks, we pick a random set of 100 source-sink pairs and run the point-to-point shortest path problem. Table 3 gives preprocessing and query results of our proposed algorithm in five road networks. Fig. 2 reports query results comparing with A*, ALT. In ALT algorithm, due to memory requirements we use *avoid* algorithm[9] to select 32 landmarks. Due to the time complexity, we only select three road network (DC, HI, AL) to test A* algorithm.

References

1. Lin, T.Y.: Granular Computing: Fuzzy Logic and Rough Sets. In: Zadeh, L.A., Kacprzyk, J. (eds.) Computing with words in information- intelligent systems, pp. 183–200. Physica-Verlag, Heidelberg (1999)
2. Zadeh, L.A.: Towards a theory of fuzzy information granulation and its centrality in human reasoning and fuzzy logic. Fuzzy Sets and Systems 19, 111–127 (1997)

3. Bargiela, A., Pedrycz, W.: Human centric information processing through granular modelling. Springer, Berlin (2009)
4. Yao, Y.Y.: Triarchic theory of granular computing. In: Zhang, Y.P., et al. (eds.), pp. 115–143. Science press, Beijing (2010) (in Chinese)
5. Pawlak, Z.: Rough Sets Theoretical Aspects of Reasoning about Data. Kluwer Academic Publishers, Dordrecht (1991)
6. Zhang, L., Zhang, B.: Theory and Applications of Problem Solving– the Quotient Space granular computation theory and Applications (the second version). Tsinghua University Press, Beijing (2007) (in Chinese)
7. Hart, P.E., Nilsson, N.J., Raphael, B.: A formal basis for the heuristic determination of minimum cost paths. IEEE Trans. on Sys. Sci. and Cyber. 4(2), 100–107 (1968)
8. Goldberg, A.V., Harrelson, C.: Computing the shortest path: A* search meets graph theory. In: 16th ACM-SIAM symposium on discrete algorithms, pp. 156–165 (2005)
9. Goldberg, A.V., Werneck, R.F.: Computing point-to-point shortest paths from external memory. In: Workshop on Algorithm Engineering and Experiments, pp. 26–40 (2005)
10. Goldberg, A.V., Kaplan, H., Werneck, R.F.: Better landmarks within reach. In: Demetrescu, C. (ed.) WEA 2007. LNCS, vol. 4525, pp. 38–51. Springer, Heidelberg (2007)
11. Sanders, P., Schultes, D.: Engineering Highway Hierarchies. In: Azar, Y., Erlebach, T. (eds.) ESA 2006. LNCS, vol. 4168, pp. 804–816. Springer, Heidelberg (2006)
12. Bauer, R., Delling, D.: SHARC:Fast and robust unidirectional routing. In: Workshop on Algorithm Engineering and Experiments (2008)
13. Zhang, L., He, F.G., Zhang, Y.P., et al.: A new algorithm for optimal path finding in complex networks based on the quotient space. Fundamenta Informaticae 93(4), 459–469 (2009)
14. Pothen, A., Simon, H., Liou, K.P.: Partitioning sparse matrices with eigenvectors of Graphs. SIAMJ. Matrix Anal. Appl. 11(3), 430–452 (1990)
15. Girvan, M., Newman, M.E.J.: Community structure in social and biological network. Proc. Natl. Acad. Sci. 99, 7821–7826 (2001)
16. Radicchi, F., Castellano, C., Cecconi, F., et al.: Defining and identifying communities in networks. Eur. Phys. J. B. 38, 373–380 (2004)
17. Fortunato, S., Latora, V., Marchiori, M.: A method for finding community structure based on information centrality. Phys. Rev. E 70, 056104 (2004)
18. Newman, M.E.J., Girvan, M.: Finding and evaluating community structure in networks. Phys. Rev. E 69, 026113 (2004)
19. Ulrik, B., Daniel, D., Marco, G., et al.: On Modularity Clustering. IEEE Transactions on In Knowlegde Data Engineering 20(2), 172–188 (2007)
20. Fortunato, S., Barthelemy, M.: Resolution limit in community detection. Proc. of the National Academy of Sciences of the United States of America 104(1), 36–41 (2007)
21. Zhuge, H., Zhang, J.S.: Topological centrality and its applications (2009), http://arxiv.org/abs/0902.1911v1

Fuzzy Measures and Granular Computing

Ling Zhang* and Bo Zhang

School of Computer Science and Technology, Anhui University, Hefei, China
Tsinghua National Lab for Information Science and Technology, Tsinghua University,
Beijing, China
zling@ahu.edu.cn
http://www.springer.com/lncs

Abstract. From the view point of granular computing, by analyzing
the structure of fuzzy measures based on the quotient space analytical
method, we have the following results: (1) the necessary and sufficient
condition of the isomorphism of fuzzy measure functions in fuzzy mathematics; (2) the necessary and sufficient condition of the fuzzy and granular monotony of fuzzy measures. The results open out the essence of
fuzzy measures and provide a simple way to constructing fuzzy measures. The results also show that the quotient space analytical method
is ubiquitous and effective in many fields.

Keywords: fuzzy set, fuzzy measure, granular computing, quotient
space.

1   Introduction

After Zadeh[1] presented the concept of the probabilistic measure of fuzzy events,
people tried to explore the fuzzy measures of fuzzy sets from different angles.
On the one hand, the concepts, properties and propositions of the measures in
function theory are extended to fuzzy sets mathematically. On the other hand,
due to the impact of the concept of the information entropy the fuzzy measure is
regarded as a measurement of the fuzziness of fuzzy sets. In mathematical side,
Sugeno presented the concepts of fuzzy integrals[2] and max-measures[3] as fuzzy
measures of fuzzy sets. Zhang[4] discussed the convergence of sequences of fuzzy
measures and generalized convergence theorems of fuzzy integrals. Wang[5] discussed the fuzzy integrals and their absolute continuity. Mesiar[6] investigated
the probabilistic measure of fuzzy events and its integral. Klir[7] studied the
construction of fuzzy measures in expert systems. Ha[8] introduced the concept
of the fundamental convergence of sequences of fuzzy measures and their properties. Li[9], Liu[10], Jiang[11], Mesiar[12,17], Ban[13,14], Krasdhmer[15] and
Guo[16] discussed the definition and mathematical properties of different fuzzy

* Supported by the National Natural Science Foundation of China, Grant
Nos.90820305, the National Basic Research Program of China, Grant Nos.
2007CB311003.

J. Yu et al. (Eds.): RSKT 2010, LNAI 6401, pp. 759–765, 2010.
© Springer-Verlag Berlin Heidelberg 2010

measures, respectively. In a word, the works above intended to extend the measures of common sets to the family of fuzzy sets from different angles. On the other hand, Chakrabarty[18] presented the measure of fuzziness of fuzzy sets by using the relationship between fuzzy sets and their nearby crisp sets. He[19] defined the concept of fuzzy entropy by the linear fuzzy measure. Wang[20] presented a new fuzzy entropy measure based on the concept of information entropy. Liang[21,22] introduced the formal definition of fuzzy entropy measure of rough sets. Zhao[23] presented the fuzzy and rough entropy measures of fuzzy rough sets. Qi[24] defined a fuzzy measure of fuzzy rough sets by using a kind of information entropy and discussed its property. Miao[25] defined a fuzzy measure based on the concept of entropy. Delgado[26] introduced the concepts of upper and lower fuzzy measures. Banerj[27] defined the roughness of fuzzy sets as $\rho_A^{\alpha,\beta} \equiv 1 - \frac{|A_\alpha|}{|\overline{A_\alpha}|}$.Chakrabarty [18] defined the fuzzy measure of rough sets as $F_X^R(u) = \frac{|[u]_R \cap X|}{|[u]_R|}$. Liang[28]introduced the concept of the fuzzy measures in information systems. The concepts of fuzzy measures were appeared in fuzzy mathematics as well. In [29,30], the fuzzy measure is defined by formula $d(F_X^B) = g(\sum_{i=1}^n c_i f(\mu_X^B(x_i)))$,and Mao[31] showed that all fuzzy measures (roughness, rough entropy, or fuzzy entropy) defined in [18,19,20,21,22,23,24,25] can be represented by the above formula as long as choosing proper functions f and g.

In conclusion, although many fuzzy measures were defined and their properties were discussed carefully, there still is lack of their essential comprehension. As we known, an object can be represented at a sequence of quotient spaces with different grain-sizes. And a fuzzy space consists of a sequence of common quotient spaces based on the fuzzy quotient space theory presented in [32]. Thus, a fuzzy measure can be regarded as a coarse view of the object. Simply speaking, the fuzzy measure is a measurement of fuzziness of a fuzzy set. Compared to a membership function, a fuzzy measure is a coarser description of fuzzy sets. It depicts the order relation among fuzzy sets according to their fuzziness. Certainly, a membership function is the finest description of a fuzzy set but sometime it's too detailed to be used. In the paper, from the well-known formula of the fuzzy measure in fuzzy mathematics, we discuss the necessary and sufficient condition for constructing the fuzzy measures with some characteristics based on granular computing view. The result will provide a simple way to constructing fuzzy measures.

2   Basic Concepts

Definition 1. *Give a finite set X, All fuzzy sets on X are denoted by $F(X) = [0,1]^X$. A point on $F(X)$ is regarded as a function on X. For two functions $h(x), g(x) \in F(X)$, if $\forall x \in X, h(x) \leq g(x)$, then we define $h \prec g$. Obviously, under the definition all fuzzy sets on X compose a semi-order set. In fuzzy mathematics [29][30], in order to define the fuzzy measures, the following axiom is introduced first.*

Definition 2. *Assume that $d(A) : F(X) \to [0,1]$ satisfies: (1) $d(A) = 0, A \in P(X)$, where $P(X)$ is the power set of X, (2) $\forall A \in F(X), d(A) = d(A^c)$, (3) $d((0.5, \cdots, 0.5)) = 1$, (4) $A(x_i), B(x_i) \leq 0.5, i = 1, \cdots, n, A \prec B \Rightarrow d(A) \leq d(B)$ (Monotony). Then, d is called a fuzzy measure on $F(X)$.*

A fuzzy measure satisfying the above axiom is not necessarily consistent with the human's intuition of fuzziness. For example, given two fuzzy sets with $f = (0.4, 0.5)$ and g=(0.6,0.8) as their membership functions respectively, Intuitionally the fuzzy set with membership function f should be fuzzier than the later with g as its membership function. But from the semi-order relation of functions, we have $f \prec g$. So it's needed to properly transform the membership functions so that the order of the functions is in proportion to the fuzziness of the corresponding fuzzy sets. That is, when the value of the membership function f is 0.5, the corresponding element of a fuzzy set is the fuzziest one, when its value is greater or smaller than 0.5, the corresponding element will be crisper than the fuzziest one. One of the possible transformations is the following.

$$h_1 : [0,1] \to [0,1], h_1(x) = 2min\{x, 1-x\} \tag{1}$$

By the transformation, the four conditions of the axiom can be replaced by one condition as follows. $g(x) \prec f(x) \to d(g) \leq d(f), f(0) = 0, f(1) = 1$. Thus, in [17] the fuzzy measure d is defined as the isomorphic transformation from $(F(X), \prec)$ to $([0,1], \leq)$. In order to have more useful information, we adopt the following definition of the fuzzy measures.

Definition 3. *Assume $d : F(X) \to [0,1]$. If $d(0) = 0$, $d(1) = 1$ and d is strictly monotonic, i.e., $\forall g \prec f \in F(X), d(g(x)) \leq d(f(x)), \forall g \prec f \in F(X), g \neq f, d(g(x)) < d(f(x))$, then d is a fuzzy measure of $F(X)$.*

Now, we investigate the fuzzy measure from the structural point of view.

Definition 4. *Two fuzzy measures d_1 and d_2 are isomorphic $\iff \forall x, y \in X, d_1(x) < d_1(y) \leftrightarrow d_2(x) < d_2(y)$.*

Proposition 1. *Assume that two fuzzy measures d_1 and d_2 are isomorphic. Then $\forall x, y \in X, d_1(x) < d_1(y) \leftrightarrow d_2(x) < d_2(y)$ and $\forall x, y \in X, d_1(x) = d_1(y) \leftrightarrow d_2(x) = d_2(y)$.*

Definition 5. *Given fuzzy measure d, define an equivalence relation R: $xRy \leftrightarrow d(x) = d(y)$. Then R is called an equivalence relation corresponding to d.*

Proposition 2. *If two fuzzy measures d_1 and d_2 are isomorphic, then their corresponding equivalence relations $R_1 = R_2$.*

Definition 6. *Assume that (X, T) is a complete semi-order set. R is an equivalence relation on X and its corresponding quotient space is $[X]$. Let $p : X \to [X]$ be a natural projection. If $\forall x \prec y, p(x) \prec p(y)$ and $\forall x \prec y, x \neq y, p(x) \prec p(y), p(x) \neq p(y)$, then R is strictly order-preserving.*

Proposition 3. *Assume that d is a fuzzy measure on X. Let R be its corresponding equivalence relation. Then, R is strictly order-preserving in semi-order set X. Contrarily, if [X] is a strictly order-preserving (totally order) quotient space on X, then the fuzzy measure on [X] will define a fuzzy measure on X.*

From the concept of quotient spaces, the fuzzy measure is a coarse view of fuzzy sets.

3   Fuzzy Measure of Rough Sets

Chakrabarty[18] introduce the concept of fuzzy measures of rough sets. His definition is the following.

Definition 7. *Given a quotient space [U] generated from the attribute set B of [U] and a set [X] on U. Define a membership function:*

$$\forall x \in X, \mu_X^B(x) = \frac{|X \cap [x]_B|}{|[x]_B|} \tag{2}$$

In fuzzy mathematics, the fuzzy measure is generally defined as follows

$$d : F(U) \to [0, 1], d(F_X^B) = g\left(\sum_{i=1}^{n} c_i f_i(\mu_X^B(x_i))\right) \tag{3}$$

Where c_i is a positive real number; $g : [0, a] \to [0, 1]$ satisfies (1) $g(a) = 1, g(0) = 0, a = \sum_{i=1}^{n} c_i f_i(0.5)$, (2) $g(t)$ is strictly increasing on $[0, a]$; $f_i : [0, 1] \to [0, 1]$ satisfies (1) $\forall f_i(0) = 0$, (2) $f_i(t)$ is strictly increasing on $[0, 0.5]$, (3) $\forall x \in [0, 1], f(x) = f(1 - x)$.

In fact, after transformation h_1, in formula (3) the three conditions that function f satisfied can be come down to one condition, i.e., $f[0, 1] \to [0, 1], f(0) = 0, f(1) = 1$ is a strictly increasing function.

In the following discussion, the membership functions of fuzzy sets are always assumed to be transformed by formula (1). The fuzzy axiom cab be summed up as

(1) (Fuzzy) monotonic assumption

By the assumption, the fuzzy measures of elements in the comparable direction are totally ordered. But how to define the fuzzy measures of elements in different directions, we introduce the second assumption as follows.

(2) Isotropic assumption

$\forall x_i, x_j \in X, a_i = a_j$, let the kth component of f_k be a_k, otherwise 0. Then, $d(f_i) = d(f_j)$. We introduce the third assumption again from granular computing view.

(3) (Granulation) monotonic assumption

In formula (3), the coarser the quotient space the higher its fuzziness.

Under the three assumptions, we discuss the property of the fuzzy measure defined by formula (3). Substituting form (2) into (1), we have

$h_1(\mu_X^B(x)) = 2min\{\mu_X^B(x), 1 - \mu_X^B(x)\} = 2min\frac{|X \cap [x]_B|}{|[x]_B|}, 1 - \frac{|X \cap [x]_B|}{|[x]_B|} =$

$2min\frac{|X \cap [x]_B|}{|[x]_B|}, \frac{|X^C \cap [x]_B|}{|[x]_B|}$. Simply,$\mu_X^B(x) = \frac{2min(|X \cap [x]_B|, |X^C \cap [x]_B|)}{|[x]_B|}$.

Proposition 4. *Assume that g_1 and g_2 are two functions satisfying formula (3). d_1 and d_2 are their corresponding fuzzy measures. Then, d_1 and d_2 are isomorphic.*

From proposition 4, it's known that all fuzzy measures defined by function g satisfying (3) are isomorphic, so we can choose a simplest one $g(x) = cx$. Thus, formula (4) is reduced to

$$d(F_X^B) = \sum_i c_i f_i(\mu_X^B(x_i)) \tag{4}$$

Proposition 5. *If the fuzzy measure defined by formula (5) is isotropic, then*

$$d(F_X^B) = \sum_i f(\mu_X^B(x_i)) \tag{5}$$

Theorem 1. *f_1 and f_2 are strictly increasing and continuous functions on [0,1]. d_1 and d_2 are fuzzy measures defined by formula (6). We have (1) $n = 1$, for any f_1 and f_2, d_1 and d_2 are isomorphic. (2)$n > 1$, d_1 and d_2 are isomorphic $\Leftrightarrow \forall x \in [0,1], f_1(x) = f_2(x)$.*

Theorem 2. *(**Basic Theorem**). The necessary and sufficient condition that fuzzy measures defined by form (6) have fuzzy and granular monotony is $f; [0,1] \to [0,1]$ is a strictly increasing function and its derivative f' is a monotonically decreasing and continuous function.*

Corollary 1. *If $f(x) = x^\alpha, 0 < \alpha \leq 1$, then the fuzzy measures defined by form (6) satisfy the fuzzy and granular monotony.*

The proof of theorems 1 and 2 is omitted.

4 Conclusion

We present three assumptions on fuzzy measures of fuzzy sets. Based on the assumptions, from quotient space theory, we have the following result: the requirements for the fuzzy and granular monotony of fuzzy measures are equivalent to the requirements for analytical property of functions. This provides a simple way to constructing fuzzy measures.

References

1. Zadeh, L.A.: Probability measures of fuzzy events, J. Math. Anal. Appl. 23, 421–427 (1968)
2. Sugeno, M.: Theory of fuzzy integrals and its application, Ph.D. Dissertation, Tokyo Institute of Technology (1974)
3. Sugeno, M., Murofushi, T.: Pseudo-additive measures and integrals. J. Math. Anal. Appl. 122, 197–222 (1987)

4. Zhang, D.L., Guo, C.M.: On the convergence of sequence of fuzzy measures and generalized convergence theorems of fuzzy integrals. Fuzzy Sets and Systems 72, 349–356 (1995)
5. Wang, Z.Y., Klir, G.J., Wang, W.: Fuzzy measures defined by fuzzy Integral and their absolute continuity. Journal of Mathematical Analysis and Applications 203, 150–165 (1996)
6. Mesiar, R.: Possibility measures, integration and fuzzy possibility measures. Fuzzy Sets and Systems 92, 191–196 (1997)
7. Klir, G.J., Wang, Z.Y., Harmanc, D.: Constructing fuzzy measures in expert systems. Fuzzy Sets and Systems 92, 231–264 (1997)
8. Ha, M.H., Wang, X.Z., Wu, C.X.: Fundamental convergence of sequences of measurable functions on fuzzy measure space. Fuzzy Sets and Systems 95, 77–81 (1998)
9. Li, X.Q., Wang, H., Li, R.X., et al.: Extensions of a class of semi-continuous fuzzy measures. Fuzzy Sets and Systems 94, 397–401 (1998)
10. Liu, Y.K., Zhang, G.Q.: On the completeness of fuzzy measure space. Fuzzy Sets and Systems 102, 345–351 (1999)
11. Jiang, Q.S., Suzuki, H.: Fuzzy measures on metric space. Fuzzy Sets and Systems 102, 345–351 (1999)
12. Mesiar, R.: Generalizations of k-order additive discrete fuzzy measures. Fuzzy Sets and Systems 102, 423–428 (1999)
13. Ban, A.I., Gal, S.G.: On the defect of complementarity of fuzzy measures. Fuzzy Sets and Systems 127, 353–362 (2002)
14. Ban, A.I., Gal, S.G.: On the defect of additivity of fuzzy measures. Fuzzy Sets and Systems 131, 365–380 (2002)
15. Krasdhmer, V.: When fuzzy measures are upper envelopes of probability measures. Fuzzy Sets and Systems 138, 455–468 (2003)
16. Guo, C.M., Zhang, D.L.: On set-valued fuzzy measures. Information Sciences 160, 13–25 (2004)
17. Mesiar, R.: Fuzzy measures and integrations. Fuzzy Sets and Systems 156, 365–370 (2005)
18. Chakrabarty, K., Biswas, R., Nanda, S.: Fuzziness in rough sets. Fuzzy Sets and Systems 110, 247–251 (2000)
19. He, Y.Q.: New method for measuring fuzziness in rough sets. Transaction of Nanjing University of Aeronautics and Astronautics 21, 31–34 (2004)
20. Wang, G.Y., et al.: On the uncertainty of fuzzy sets under different knowledge granularities. Journal of Computer Science and Technology 31(9), 1588–1596 (2008) (in Chinese)
21. Liang, J.Y., Chin, K.S., Dang, C.Y.: A new method for measuring uncertainty and fuzziness in rough set theory. International Journal of General Systems 31(4), 331–342 (2002)
22. Liang, J.Y., Shi, Z.Z.: The information entropyrough entropy and knowledge granulation in rough set theory. International Journal of Uncertainty, Fuzziness and Knowledge Based Systems 12(1), 37–46 (2004)
23. Zhao, X.F., et al.: The uncertainty measures of fuzzy rough sets. Transaction of Ninxia University 28(3), 206–209 (2007) (in Chinese)
24. Qi, X.D., et al.: The uncertainty measures of general fuzzy rough sets. Fuzzy Systems and Mathematics 21(2), 136–139 (2007)
25. Miao, D.Q.: Information Entropy and Granular Computing. In: Miao, D.Q. (ed.) Granular Computing: Past, Present and Future, pp. 142–178. Scientific Publishing Company (2007) (in Chinese)

26. Delgado, M., Moral, S.: Upper and lower fuzzy measures. Fuzzy Sets and Systems 33, 191–200 (1989)
27. Mohua, Banerje, Sankarkpai: Roughness of a fuzzy set. Informatics and Computer Science 93, 235–246 (1996)
28. Liang, J.Y., Li, D.Y.: The Uncertainty and Knowledge Acquisition of Information Systems. Scientific Publishing Company, Beijing (2005) (in Chinese)
29. Hu, B.Q.: The Foundation of Fuzzy Theory (second version). Transaction of Wuhan University, Wuhan (2005) (in Chinese)
30. Yang, L.B., Gao, Y.Y.: The Principle and Application of Fuzzy Mathematics. South China University of Technology Publishing Company (2004) (in Chinese)
31. Mao, J.J., Wu, T., Zhang, L.: The interpretation of knowledge fuzzy measures of rough sets. Submit to Journal of Computer Science and Technology (in Chinese)
32. Zhang, L., Zhang, B.: The fuzzy quotient space theory. Journal of Software 14(4), 770–776 (2003) (in Chinese)

Identifying Protein-Protein Interaction Sites Using Granularity Computing of Quotient Space Theory

Yanping Zhang, Yongcheng Wang, Jun Ma, and Xiaoyan Chen

Key Lab of Ministry of Education for CI & SP, Anhui University,
Hefei, Anhui, China
zhongguokd@163.com

Abstract. The function of protein-protein interaction is very important to cell activity. Studying protein-protein interaction can help us understand life activities and pharmaceutical design. In this study, a kernel covering algorithm combined with the theory of granular computing of quotient space for predicting protein-protein interaction sites is proposed, (i.e. KCA-GS Model). This method achieves good performances, and the Sensitivity, Specificity, Accuracy and Correlation coefficient are 52.97%, 53.92%, 70.27%, 24.61%, respectively. It is indicated that our method is effective, potential and promising to identify protein-protein interaction sites.

Keywords: Protein-protein interaction, Kernel covering algorithms, Granularity synthesis, Quotient space, Sequence profile, Entropy, Residue accessible area.

1 Introduction

Organism's function is carried out through protein-protein interaction between the biological molecules. Especially, the so-called interaction sites play a vital role in those interactions, so the identification of those sites is extremely essential to interpret protein function mechanism. In recent years, some experimental methods have been adopted to predict protein-protein interaction sites, such as protein affinity chromatography[1], immunoprecipiation[2]. However, those experiments generate high numbers of false positives and false negatives[3], and they are also time-consuming and monotonous. So many people try their best to look for computational methods to deal with prediction task, such as neural networks[4], support vector machine (SVM) method[5],[6],[7] and so on. Those methods have achieved good performances to some extent and provided other researchers useful information about interaction sites.

In this study, we refine the covering algorithm[8],[9] and obtain the kernel covering algorithm which is employed as the classifier for prediction. Recently, Zhang Ling and Zhang Bo have proposed the model and theory of granularity computing of quotient space[10],[11]. In this Study, by selecting proper protein features, we combine the kernel covering algorithm and the theory of granularity computing of quotient space to identify protein-protein interaction sites.

J. Yu et al. (Eds.): RSKT 2010, LNAI 6401, pp. 766–771, 2010.

2   Materials and Methods

2.1   Dataset

In the research, the dataset is composed of 113 protein-protein interaction pairs which were used by Fariselli. et al[12] before. However, the dataset have lots of redundancies, so we remove redundant chains by using BLASTCLUST program[13]. As a result, the obtained non-redundant dataset contains 74 chains which are used in the experiments.

2.2   Definition of Surface Residues and Interface Residues

Following most researchers, in the study, we mainly focus on surface residues. A residue is viewed as a surface residue if the accessible surface area (ASA) is at least 16% of the nominal maximum area[14], and a surface residue is defined as an interface residue if its CASA of surface residue is less than its MASA by at least $1\mathring{A}^2$[15], where CASA represents the ASA in the complex, MASA means the ASA in unbound chain and they can be computed by using DSSP program[16]. As a result, the experimental data contains 10879 surface residues and 3375 interface residues.

2.3   Evaluation Measures of Classifier Performance

In the experiment, four widely used measures of Sensitivity, Specificity, Accuracy, CC (Correlation Coefficient) are adopted to evaluate the performances of the results. They can be represented as follows:

$$Sensitivity = \frac{TP}{TP + FN} \tag{1}$$

$$Specificity = \frac{TP}{TP + FP} \tag{2}$$

$$Accuracy = \frac{TP + FN}{TP + FN + TN + FP} \tag{3}$$

$$CC = \frac{TP * TN - FP * FN}{\sqrt{(TP + FN) * (TP + FP) * (TN + FP) * (TN + FN)}} \tag{4}$$

Where TP, FP, TN and FN denote the numbers of true positives, false positives, true negatives and false negatives, respectively. Sensitivity measures the fraction of interface residues that are predicted as such, Specificity represents the fraction of the predicted interface residues that are actually interface residues, and CC (Correlation Coefficient) measures how well the predicted class labels correlate with the actual class labels.

2.4   The Theory of Quotient Space

In the model of Quotient Space, a problem is represented by a triplet (X, f, T), where X denotes its domain, f is the attributes of domain X and T is its structure. (X, f, T) is called problem space or space for short. Assume R is an equivalence relation on X, [X] is a quotient set under R. Considering [X] as new domain, then the new problem space is ([X], [f], [T]), where [f] and [T] are the corresponding quotient attributes and quotient structure of X, respectively. Thus, the worlds with different grain size are represented by a set of quotient spaces. In a word, the theory of quotient is discussing representations and properties of domains, attributes and structures under different grain size and it also discusses the relation between those representations and properties. In this paper, we mainly use the granularity synthesis model to predict protein-protein interaction sites according to the information obtained from prediction results under different granularities. Assume that the research object is A, i.e. (X, f, T) and there is some hierarchical information about A. (X_1, f_1, T_1) and (X_2, f_2, T_2) is two levels of A where X_1 and X_2 is quotient set of X.

Definition 1 *(The synthesis of domain)*
Assume (X_1, f_1, T_1) and (X_2, f_2, T_2) is quotient space of (X, f, T), R_1 and R_2 is the correspondingly equivalence relation of X_1 and X_2. Define X_3 is synthesis space of X_1 and X_2, R_3 is equivalence relation of X_3 and $xR_3y \Longleftrightarrow xR_1y$ and xR_2y. Then

$$X_3 = \{a_i \cap b_j \mid a_i \in X_1, b_j \in X_2\}$$

Definition 2 *(The principle of attribute function synthesis)*
Assume knowing (X_1, f_1, T_1) and (X_2, f_2, T_2), the synthesis space (X_3, f_3, T_3) should conform to the following conditions:
(1)$p_i f_3 = f_i, i = 1, 2$
Where p_i: $(X_3, f_3, T_3) \rightarrow (X_i, f_i, T_i)$ is natural projection, i=1,2
(2)Assume D(f, f_1, f_2) is a known judging criterion, then
D (f_3, f_1, f_2) = min D(f, f_1, f_2) or max D(f, f_1, f_2).
Where min (max) can be got from all attribute functions f on X_3 which satisfy from (1).

2.5   KCA-GS Model Algorithm

The covering algorithm is proposed by Zhang Ling and Zhang Bo[8],[9]. We refine it and obtain the kernel covering algorithm according to gaussian kernel function. In the theory of quotient space, a problem can be described by different quotient spaces, for example, (X_1, f_1, T_1) , (X_2, f_2, T_2) , ..., (X_n, f_n, T_n) and those quotient spaces can be synthesized by using the principles of synthesis mentioned above. In our research, we choose different protein features to construct different quotient spaces, obtain different classification results and synthesize those results using principles of synthesis. The model predicting protein-protein interaction sites based on granularity synthesis of quotient space is represented as follows:

Step 1. Choosing some features from ASA, entropy and profile to construct the input samples for classifier by using 11 windows(the target residue is in the middle and there are five sequentially neighboring residues on each side.)

Step 2. Thick classification. Firstly, using ASA and entropy to build feature vectors for the kernel covering algorithm, then obtaining classification labels of all testing set. $[X_1]$ represents partition after classification and constructing the granularity space $([X_1], [f_1], [T_1])$. Secondly, using profile to construct feature vectors for the kernel covering algorithm, then obtaining classification labels of all testing set. $[X_2]$ represents partition after classification and constructing the granularity space $([X_2], [f_2], [T_2])$.

Step 3. Synthesizing different granularity spaces. Comparing each residue of X_1 and X_2, the label keeps up if they have the same label. Otherwise, C_k is used to represented composition of this residue.

Step 4. According to Definition 2, partition C_k again. Constructing new feature vectors for each residue of C_k for partitioning. In this new vector, compute distance d_1/d_2 between new vector and $[X_1]/ [X_2]$. The residue of C_k belongs to the class of the min distance. Thus, $[X_3]$ is synthesized.

Step 5. Statistic $[X_3]$, compute results.

3   Results and Discussion

In this study, the kernel covering algorithm is used as classifier to predict the class of each residue using fivefold cross-validation method on the selected dataset. Some judging criterions, for example, Sensitivity, Specificity, Accuracy and CC (Correlation Coefficient) are adopted as evaluation measures of classifier performance. The results are shown in Table 1. It can be seen that KCA-GS (The classifier using kernel covering algorithm based on granularity synthesis of quotient space) achieves almost all performances better than that of ASA + entropy classifier and profile classifier except Sensitivity. However, as long as Sensitivity, KCA-GS outperforms ASA + entropy classifier (by 5.58%) and is less than profile predictor (by 0.55%). Furthermore, for KCA-GS, Table 1 shows that of the residues predicted to be interface, 53.92% is actually interface residues, and 52.97% of interface residues are identified as such and the Accuracy of it is

Table 1. Four performances of the classifier

	Sensitivity	Specificity	Accuracy	Correlation Coefficient
ASA + entropy	0.4739	0.4261	0.6174	0.1945
profile	0.5352	0.4534	0.6528	0.2198
KCA-GS	0.5297	0.5392	0.7027	0.2461

70.27% with a Correlation Coefficient of 24.61%. If only judged by Accuracy, KCA-GS performs best (70.27%), profile classifier is second (65.28%).

Furthermore, the numbers of chains for each evaluation measure of KCA-GS are obtained. The distributions of evaluation measure values are depicted in Figure 1. It can be seen that, for Sensitivity, its values of 41(55.4%) chains are greater than 50%. When Specificity values share the same threshold, its number is 40. In addition, it also can be seen that Accuracy values are higher than 30% for all chains and for 71.6% of proteins the Correlation Coefficient is greater than 0.0. From the analysis above, it suggests that our KCA-GS is an effective model of predicting protein-protein interaction sites.

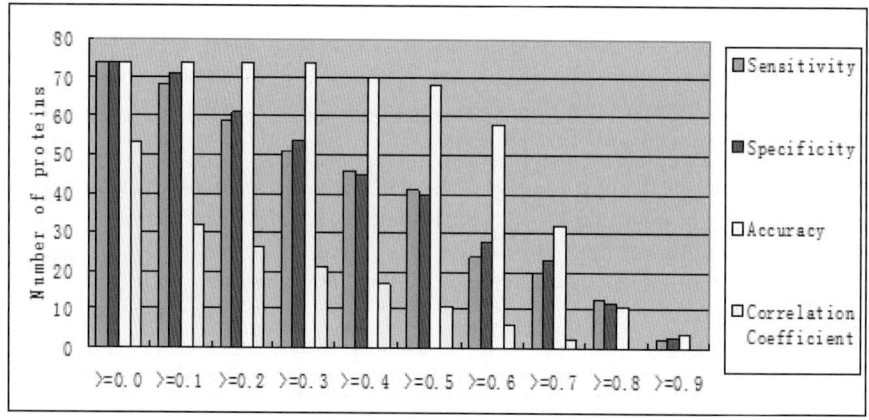

Fig. 1. The distributions of prediction performance measure values of KCA-GS for 74 proteins

4   Conclusion and Future Work

As far as prediction of protein-protein interaction sites, it's an extremely difficult work because of lack of adequate useful information about protein partners. However, some research on this field has achieved great performances in recent years, which provides instrumental methods for other researchers. In this paper, after selecting proper features including profile, entropy and ASA (residue accessible area), we adopt the kernel covering algorithm as the predictor to get different classification results which are used to construct different quotient spaces by using the principles of granularity synthesis of quotient space. The experimental results show that our proposed KCA-GS model algorithm is an effective, promising method to identify protein-protein sites. In this paper, during the procedure of the kernel covering algorithm, the covering center of a cover is selected randomly each time, which may influences the classification result, so we will look for some other methods to obtain proper covering center of a cover in the future. In addition, we will also seek useful attributes to help predict protein-protein interaction sites.

Acknowledgments. This research is supported by the National Natural Science Foundation of China (NO.60675031), the National Grand Fundamental Research 973 Program of China under Grant No.2004CB318108, No.2007CB311003, High School Surpass Talent Foundation of Anhui No.2009SQRZ020ZD, the Innovative Research Team of 211 Project in Anhui University of China, and the Program of superior Teacher Team in Anhui University of China. The authors would like to thank Fariselli. et al for experimental data.

References

1. Mitchell, D.A., Marshall, T.K., Deschenes, R.J.: Vectors for the inducible overexpression of glutathione S-transferase fusion proteins in yeast. Yeast 9(7), 715–722 (1993)
2. Harlow, E., Whyte, P., Franza, B.R., et al.: Association of adenovirus early-region 1A proteins with cellular polypeptides. MolCellBiol. 6(5), 157–1589 (1986)
3. Mrowka, R., Patzak, A., Herzel, H.: Is there a bias in proteome research?, vol. 11, pp. 1971–1973. Cold Spring Harbor Laboratory Press, New York (2001)
4. Huang, X.Z., Shan, Y.B.: Prediction of Protein Interaction Sites From Sequence Profile and Residue Neighbor List. Proteins: Structure, Function, and Genetics 44, 336–343 (2001)
5. James, R., Bradford, D.R.: Westhead. Improved Prediction of Protein-protein Binding Sites Using a Support Vector Machines Approach. Bioinformatics 21(8), 1487–1494 (2005)
6. Wang, B., San Wong, H., Huang, D.S.: Inferring protein-protein interacting sites using residue conservation and evolutionary information. Protein Pept. Lett. 13, 999–1005 (2006)
7. Zhu, H., Domingues, F.S., Sommer, I., Lengauer, T.: NOXclass: prediction of protein-protein interaction types. BMC Bioinformatics 7, 27 (2006)
8. Zhang, L., Zhang, B.: A geometrical representation of McCulloch-Pitts neural model andits applications. IEEE Trans. Neural Netw. 10, 925–929 (1999)
9. Zhang, L., Zhang, B., Yin, H.F.: An alternative covering design algorithm of multilayer neural networks. J. Soft. 10, 737–742 (1999)
10. Zhang, B., Zhang, L.: Theory and Applications of problem solving. Tsinghua University Press, Beijing (1990)
11. Zhang, L., Zhang, B.: The quotient space theory of problem solving. In: Wang, G., Liu, Q., Yao, Y., Skowron, A. (eds.) RSFDGrC 2003. LNCS (LNAI), vol. 2639, pp. 11–15. Springer, Heidelberg (2003)
12. Fariselli, P., Pazos, F., Valencia, A., et al.: Prediction of protein-protien interaction sites in heterocompleses with neural networks. Eur. J. Bichem. 269(5), 1356–1361 (2002)
13. Altschul, S.F., Madden, T.L., Schaffer, A.A., et al.: Gapped BLAST and PSI-BLAST: a new generation of protein database search programs. Nucleic Acids Res. 25(17), 3389–3402 (1997)
14. Rost, B., Sander, C.: Conservation and prediction of solvent accessibility in protein families. Proteins 20, 216–226 (1994)
15. Young, L., Jernigan, R.L., Covell, D.G.: A role for surface hydrophobicity in protein-protein recognition. Protein Sci. 3, 717–729 (1994)
16. Young, L., Jernigan, R.L., Covell, D.G.: A role for surface hydrophobicity in protein-protein recognition. Protein Sci. 3, 717–729 (1994)

Moving Object Detection Based on Gaussian Mixture Model within the Quotient Space Hierarchical Theory

Yanping Zhang, Yunqiu Bai, and Shu Zhao

Key Lab of Ministry of Education for CI & SP,
Anhui University, Hefei Anhui, China
baiyunqiu120@163.com

Abstract. Based on the deficiencies of the Gaussian mixture model (GMM), the improvement is proposed in this paper. The image of video is partitioned into coarse Granularities by equivalence relation R, and the Quotient space can be obtained. Then the moving object is detected within it. The experiments show that the algorithm can improve the detection rate of the moving object without influencing to identify the object.

1 Introduction

Moving-object detection is a very important part in the machine vision. It covers image processing, pattern recognition and signal processing, control theory as well as other disciplines. And it is used widely in the intelligent monitoring [1], the field of human-computer interaction system and so on.

The most commonly moving target detection methods are divided into three categories [2]: optical flow [1], temporal difference [3], and background subtraction [4]. Among them, optical flow needs larger amount of calculation, higher hardware requirements, and timeliness is also poor. Temporal difference can quickly extract the object, but it is difficult to detect the less obvious changes in two adjacent frames. While background subtraction is one of the popular methods, and the GMM which was proposed by Stauffer and Grimson[5],[6] has been extensively studied. Because it can overcome the impact of slight changes. However, it takes much more time to detect moving-object, and it may produce the situation that the object converts to background when it stays the position more than a certain time.

Quotient space theory [7],[8] has also been used in route planning, data mining, remote sensing image analysis, and is causing widespread concern [9],[10]. In this paper, the improvements is made combining with the advantages of the Quotient Space theory, and it is illustrated by the experiments.

J. Yu et al. (Eds.): RSKT 2010, LNAI 6401, pp. 772–777, 2010.

2   Quotient Space Theory

2.1   The Quotient Space Hierarchical Theory

In quotient space theory, a problem is represented by a triplet(X,f,T), where X is the domain of the problem, f is the attribute of elements and T is the structure of X. When X is complicated, people can get a coarser granularity world [X] by regulating granularity, then the original problem (X,f,T) turns out to be a new problem([X],[f],[T]). So the problem can be sovled in the new and coarser granularity world.

In [8], Zhang and Zhang discussed the relationship between the original problem(X,f,T) and new problem([X],[f],[T]) in detail, and gave us the two important propositions as follows:

Proposition 1. p: $(X,T) \rightarrow ([X],[T])$ *is the natural projection, so p is continuous. If $A \subseteq X$ and A is the connected set in X, then* p(A) *is the connected set in* [X].

Proposition 1 shows that the new problem also has a solution in the appropriate domain [X],if the problem has a solution in the original domain X. On the contrary, if the problem has no solution in the coarse-grained universe, then the original problem will be no solution.

Proposition 2. *Set* (X,T) *is a semi-ordered space (or a pseudo-semi-order space),* R *is compatible, if* $x, y \in (X, T)$ *and* $x < y,$ *then* $[x] < [y],$ *where* $[x], [y] \in ([X], [T]).$

Proposition 2 shows that [X] can been gained by introducing a category R if the original domain X is complex. If the structure T of X and R are compatible, then quotient semi-ordered [T] is induced by R. So the problem which seeks x to y is transformed into the new problem that seeks $[x]$ to $[y]$ in [X]. Because R is compatible ,thus p:$(X,T) \rightarrow ([X],[T])$ has order preserving. That is, the problem is discussed in the coarse-grained universe by appropriate classification technology, if the problem has no solution, then the original problem also has no solution in fine-grained universe, since the coarse-grained universe is usually simpler than original space, thus it narrows the scope of the solution and accelerates the progress of the solution.

2.2   Object Detection Problem Description Based on Quotient Space Hierarchical Theory

In the object detection, the domain X of the original problem (X,f,T) is composed of the miage's pixels. f is the pixel value, and T is the relationship between pixels. Then the pixel of X is modeled independently by using the GMM. In the present high-resolution video capture device, it will take much more processing time and won't reach the purpose of real-time detection. Yet the original problem is converted to a new problem ([x],[f],[T]) by a proper equivalence relation R which is given in the paper. Then the object is detected in the new problem.

According to Proposition 1, if moving object can be detected in the domain X, it should be detected in the appropriate coarse grain universe. Conversely, if it can not be detected in [X](ie all background), then it must not be detected in the fine grain universe.

In accordance with proposition 2, the algorithm can improve the target detection rate by introducing a proper classification R in the X.

3   Moving-Object Detection

3.1   Moving Object Detection Based on Quotient Space Hierarchical Theory

Supposing the resolution of video is M*N, the domain of original problem is the set $X = \{(x,y), 1 \leq x \leq M, 1 \leq y \leq N, x, y \in N^+\}$, the attribution f=I(x,y,i), where I is the pixel value of the ith frame on (x,y). The equivalence relation R can be acquired by giving a partition $block(X,L)$ of X. The $block(X,L)$ means that X is divided into many disjoint rectangles from top to bottom, left to right, where L is the size of rectangle. The rectangle is defined as granule in the two-dimensional space[11]. So a new and coarser granularity world is constructed. The new domain is [X], attribute function$[f] = \frac{1}{L*L} \sum_{x=1}^{L} \sum_{y=1}^{L} I(x,y,i)$, denoted, X_t.

According to the GMM[6],[7], at any time, t, given a particular pixel (x_0, y_0), $\{X_1, \cdots, X_t\} = \{I(x_0, y_0, i) | 1 \leq i \leq t\}$ is its history, where I is the image sequence. The recent history of each granule $\{X_i, \cdots, X_t\}$ are modeled respectively. The probability of observing the current attribute value of X_t is

$$P(X_t) = \sum_{i=1}^{K} w_{i,t} \eta(X_t, \mu_{i,t}, \Sigma_{i,t}) \tag{3.1}$$

where K is the number of distributions, $w_{i,t}$ is the weight of the ith Gaussian model at time t with mean $\mu_{i,t}$ and standard deviation $\Sigma_{i,t} = \delta_i I$, and η is a Gaussian probability density function

$$\eta(X_t, \mu_{i,t}, \Sigma_{i,t}) = \frac{1}{(2\pi)^{\frac{1}{2}} |\Sigma_{i,t}|^{\frac{1}{2}}} e^{\frac{-1}{2}(X_t - \mu_{i,t})^T \Sigma_{i,t}^{-1}(X_t - \mu_{i,t})}, i = 1, 2, \cdots, K \tag{3.2}$$

The parameters are updated as follows:

$$w_{i,t} = (1 - \alpha)w_{i,t-1} + \alpha M_{i,t} \tag{3.3}$$

$$\mu_{i,t} = (1 - \rho)\mu_{i,t-1} + \rho X_t \tag{3.4}$$

$$\delta_{i,t}^2 = (1 - \rho)\delta_{i,t-1}^2 + \rho(X_t - \mu_{i,t-1})^T(X_t - \mu_{i,t-1}) \tag{3.5}$$

Where $\rho = \alpha \eta(X_t, \mu_{i,t-1}, \Sigma_{i,t})$, α is the learning rate and $M_{i,t}$ is 1 for the model which matched, 0 for the remaining models.

If none of the K distributions match that granular attribute value, the least probable component is replaced by a distribution with the current value as its mean, an initially high variance, and a low weight parameter.Then the K distributions are ordered based on the fitness value ω_i/δ_i and the first B distributions are used as a model of the background of the scene where B is estimated as

$$B = \arg\min_b(\Sigma_{i=1}^b \omega_{i,t} > T) \tag{3.6}$$

where T is a measure of the minimum portion of the data that should be accounted for by the background. In other words, it is the minimum prior probability that the background is in the scene. Background subtraction is performed by marking a foreground pixel any pixel that is more than 2.5 standard deviations away from any of the B distributions.

3.2   The Algorithm Based on Quotient Space Hierarchical Theory

The algorithm based on quotient space hierarchical theory is given as follows:

1. Converting color image into gray image.
2. Initializing the Gaussian Mixture Model.
3. Transforming the original problem into the new problem([X],[f],[T]) by R of section 3.1.
4. Training the model, and determining whether X_t obeys the Gaussian distribution.
   (a) if no, the value of current granular attribute is assigned the mean value of the model, other parameters as same as initialization, and get a new model.
   (b) else, according to formula (3.3)-(3.5), updating parameters of the model.
5. Detecting moving-object.
6. Return to 3, and continue to the next frame.

4   Experimental Results and Analyses

In this paper, the algorithm is verified by the human motion video person.avi and public test video pets2000.avi. Their resolution is respectively 320*240 and 768*576. Experimental environment: PC, cpu is pentium (R) 4 2.8G, memory 1G; matlab R2008a version. Experimental result as fig.1 and fig.2. Table 1 lists the processing time to detect each frame, and unit is second.

From (b), (e) and Table 1 can be seen, there is a lot of noise in the figure, and detection rate is very slow. The moving-object can be extracted using our method in the (c) and (f). The detection rate is much better than the former with a lot of noise reduced. Foremost, the detection results doesn't affect the identification of object. Here the granular size can take 3, 5, 7 and so on. But the compromise will be made between detection rate and effectiveness. If the granularity is too coarse, it may not extract the moving object completely, and then it affects to identify the target. But if it takes too small, the detection rate will not be improved. So the paper takes 5 as the granular size.

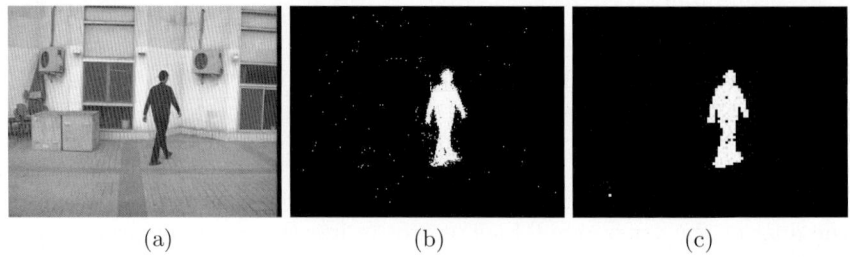

(a)   (b)   (c)

Fig. 1. The detection results of video person.(a) is the 102nd frame of video person. (b) means the detection result of GMM, and (c) shows the detection result of the Gaussian Mixture Model based on Quotient Space Hierarchical theory.

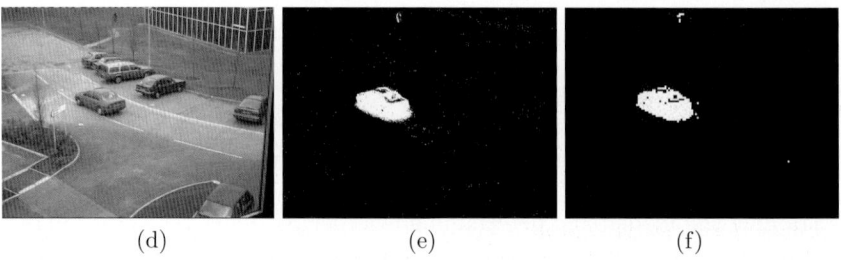

(d)   (e)   (f)

Fig. 2. The detection results of video pets2000.(d) is the 56th frame of video Pets2000.(e) shows the detection result of GMM,and (f) means the detection result of the Gaussian Mixture Model based on Quotient Space Hierarchical theory.

Table 1. The Runtime of Detecting by Two Methods

Video name	the runtime of GMM	the runtime of our method
person	3.57	0.14
pets2000	22.18	0.82

Additional Moving-object can be extracted in (c) and (f), it shows that the Proposition 1 of the quotient space hierarchical theory is validated from experimental angle. That is, moving object can be detected in the original space. If given a appropriate equivalence relation R, it can be detected in the new problem([X],[f],[T]). The detection rate is much better than the original universe, and it also validates the correctness of Proposition 2. The boundary is not clear, but it can be explained from proposition 2. Since the original problem(X, f, T) transforms into the new problem([X],[f],[T]). Then some information may be lost in the original problem space, so there is not a clear boundary. But the distortion can be accepted from the experiment, because the main purpose is to improve the target detection rate without affecting the identification.

5   Conclusions

In this paper, the Gaussian Mixture Model has been improved by combining with the advantage of Quotient Space hierarchical theory, and the algorithm has been validated by the experiments. It shows that the algorithm can improve the rate of detection object. The next work, we will pay attention to the hierarchical model which pixels combine with blocks by the quotient space theory in the moving-object detection.

Acknowledgments. This work is supported by the National Grand Fundamental Research 973 Program of China under Grant No.2004CB318108, No.2007CB3 11003; the National Natural Science Foundation of China (NO.60675031); High School Surpass Talent Foundation of Anhui No.2009SQRZ020ZD; the Innovative Research Team of 211 Project in Anhui University of China; the Program of superior Teacher Team in Anhui University of China.

References

1. Dai, K.X., Li, G.H.: Prospects and Current Studies on Background Subtraction Techniques for Moving Objects Detection from Surveillance Video. Journal of Image and Graphics 11, 920–927 (2006)
2. Chen, Z.J., Chen, X.J.: Moving Object Detection Based on Improved Mixture Gaussian Models. Journal of Image and Graphics 12, 1585–1589 (2007)
3. Liu, X., Liu, H.: Adaptive Background Modeling Based on Mixture Gaussian Model and Frame Subtraction. Journal of Image and Graphics 13, 729–734 (2008)
4. Yu, C.Z., Zhu, J.: Video object detection based on background subtraction. Journal of Southeast University (Natural Science Edition) 35, 159–161 (2005)
5. Stauffer, C., Grimson, W.: Adaptive background mixture models for real-time tracking. In: Proceedings of IEEE Conference on Computer Vision and Pattern Recognition, Fort Collis Colorado, USA, vol. 2, pp. 246–252 (1999)
6. Stauffer, C., Grimson, W.: Learning patterns of activity using real-time tracking. IEEE Trans. on Pattern Analysis and Machine Intelligence 22, 747–757 (2000)
7. Zhang, L., Zhang, B.: Theory and Applications of Problem Solving. Tsinghua University Press, Beijing (2007)
8. Zhang, L., Zhang, B.: A Quotient Space Approximation Model of Multiresolution Signal Analysis. J. Comput. Sci. & Technol. 20, 90–94 (2005)
9. Zhang, Y.P., Zhang, L., Wu, T.: The representation of different granular worlds: A quotient space. Chinese Journal of Computers 27, 328–333 (2004)
10. Zhang, Y.P., Zhang, L.: A Constructive Kernel Covering Algorithm and Applying It to Image Recognition. Journal of Image and Graphics 9, 1304–1308 (2004)
11. Yao, Y.Y., Zhong, N.: Granular computing. In: Wah, B.W. (ed.) Wiley Encyclopedia of Computer Science and Engineering, vol. 3, pp. 1446–1453 (2009)

Author Index

Printing: Mercedes-Druck, Berlin
Binding: Stein+Lehmann, Berlin